Fundamentals of Computer Architecture and Design

Ahmet Bindal

Fundamentals of Computer Architecture and Design

Second Edition

Springer

Ahmet Bindal
Computer Engineering Department
San Jose State University
San Jose, CA, USA

ISBN 978-3-030-00222-0 ISBN 978-3-030-00223-7 (eBook)
https://doi.org/10.1007/978-3-030-00223-7

Library of Congress Control Number: 2018953318

This Springer imprint is published by the registered company Springer Nature Switzerland AG
The registered company address is: Gewerbestrasse 11, 6330 Cham, Switzerland

For my mother who always showed me the right path.

Preface

This book is written for young professionals and graduate students who have prior logic design background and want to learn how to use logic blocks to build a complete system from design specifications. My two-decade-long industry experience has taught me that engineers are "shape-oriented" people, and that they tend to learn from charts and diagrams. Therefore, the teaching method you will see in this textbook caters this mid-set: a lot of circuit schematics, block diagrams, timing diagrams, and examples supported by minimal text.

The book has eight chapters. The first three chapters give a complete review of the logic design principles since rest of the chapters significantly depend on this review. Chapter 1 concentrates on the combinational logic design. It describes basic logic gates, De Morgan's theorem, truth tables, and logic minimization. The chapter uses these key concepts in order to design megacells, namely various types of adders and multipliers. Chapter 2 introduces sequential logic components, namely latches, flip-flops, registers and counters. It introduces the concept of timing diagrams to explain the functionality of each logic block. Moore and Mealy-type state machines, counter–decoder-type controllers, and the construction of simple memories are also explained in this chapter. Chapter 2 is the first chapter that illustrates the design process: how to develop architectural logic blocks using timing diagrams and how to build a controller from a timing diagram to govern data flow. Chapter 3 focuses on the review of asynchronous logic design, which includes state definitions, primitive flow tables, and state minimization. Racing conditions in asynchronous designs, how to detect and correct them are also explained in this chapter. The chapter ends with designing an important asynchronous timing block: the C (or the Mueller) element and describes an asynchronous timing methodology that leads to a complete design using timing diagrams.

From Chaps. 4 to 8, computer architecture-related topics are covered. Chapter 4 examines a very essential system element: system bus and communication protocols between system modules. This chapter studies parallel and serial bus architectures, defines bus master and bus slave concepts, and examines their bus interfaces. Read and write bus protocols, bus handover, and bus arbitration are also examined in this chapter. System memories, namely static random-access memory (SRAM), synchronous dynamic-random access memory (SDRAM), electrically erasable programmable read-only memory (E^2PROM), and flash memory are examined in Chap. 5. This chapter also shows how to design bus interface for each memory type using timing diagrams and state machines.

Chapter 6 is about the design of a simple reduced instruction set computer (RISC) central processing unit (CPU). This chapter has been expanded in the second edition to cover a variety of subjects in the floating-point unit and cache memory. In the first part of this chapter, fixed-point instructions are introduced. This section first develops a dedicated hardware (data-path) to execute each RISC instruction and then groups several instructions together in a set and designs a common data-path to execute user programs that use this instruction set. In this section, fixed-point-related structural, data and program control hazards are described, and the methods of how to prevent each type are explained. The second part of this chapter is dedicated to the IEEE single and double-precision floating-point formats, leading to the simplified designs of floating-point adder and multiplier. These designs are then integrated with the fixed-point hardware to obtain a RISC CPU capable of executing both fixed-point and floating-point arithmetic instructions. In the same section, floating-point-related data hazards are described. A new floating-point architecture is proposed based on a simplified version of the Tomasulo algorithm in order to reduce and eliminate these hazards. In the third part, various techniques to increase the program execution efficiency are discussed. The trade-offs between static and dynamic pipelines, single-issue versus dual-issue and triple-issue pipelines are explained with examples. Compiler enhancement techniques, such as loop unrolling and dynamic branch prediction methods, are illustrated to reduce overall CPU execution time. The last section of this chapter explains different types of cache memory architectures, including direct-mapped, set-associative and fully associative caches, their operation and the trade-off between each cache structure. The write-through and write-back mechanisms are discussed and compared with each other, using design examples.

Furthermore, Chap. 6 now contains static and dynamic, single and multiple issue CPUs, the advantages of out-of-order execution and register renaming. The final phase of this chapter is dedicated to multi-core CPUs with a central memory and distributed memories. The data update and replacement policy are described for each CPU architecture to maintain cache coherency.

The design of system peripherals, namely direct memory access (DMA), interrupt controller, system timers, serial interface, display adapter, and data controllers are covered in Chap. 7. The interrupt controller has been expanded in this new edition to cover context switching. The design methodology for constructing the data-paths using timing diagrams shown in Chap. 2 is closely followed to design the bus interface for each peripheral in this chapter. Chapter 8 describes the field-programmable gate array (FPGA), and the fundamentals of data-driven processors as special topics.

At the end of the book, there is a small appendix that introduces the Verilog language. Verilog is a widely used Hardware Design Language (HDL) to build and verify logic blocks, mega cells and systems. Interested readers are encouraged to go one step beyond and learn system Verilog to be able to verify large logic blocks.

San Jose, USA Prof. Ahmet Bindal

Contents

About the Author

Ahmet Bindal received his M.S. and Ph.D. degrees in Electrical Engineering from the University of California, Los Angeles, CA. His doctoral research was the material characterization of HEMT GaAs transistors. During his graduate studies, he was a research associate and a technical consultant for Hughes Aircraft Company. In 1988, he joined the technical staff of IBM Research and Development Center in Fishkill, NY, where he worked as a device design and characterization engineer. He developed asymmetrical MOS transistors and ultrathin Silicon-On-Insulator (SOI) technologies for IBM. In 1993, he transferred to IBM at Rochester, MN, as a senior circuit design engineer to work on the floating-point unit of AS-400 mainframe processor. He continued his circuit design career at Intel Corporation in Santa Clara, CA, where he designed 16-bit packed multipliers and adders for the MMX unit in Pentium II processors. In 1996, he joined Philips Semiconductors in Sunnyvale, CA, where he was involved in the designs of instruction/data caches and various SRAM modules for the TriMedia processor. His involvement with VLSI architecture also started in Philips Semiconductors and led to the design of the Video-Out unit for the same processor. In 1998, he joined Cadence Design Systems as a VLSI architect and directed a team of engineers to design a self-timed asynchronous processor. After approximately 20 years of industry work, he joined the Computer

Engineering faculty at San Jose State University in 2002. His current research interests range from nanoscale electron devices to robotics. He has over 30 scientific journal and conference publications and 10 invention disclosures with IBM. He currently holds three US patents with IBM and one with Intel Corporation. On the light side of things, he is a model aircraft builder and an avid windsurfer for more than 30 years.

Review of Combinational Logic Circuits

1

Logic gates are the essential elements of digital design, and ultimately constitute the building blocks for digital systems. A good understanding in designing complex logic blocks from primitive logic gates and mastering the design tools and techniques that need to be incorporated in the design process is a requirement for the reader before moving on to the details of computer architecture and design.

This chapter starts with defining logic gates and the concept of truth table which then leads to the implementation of basic logic circuits. Later in the chapter, the concept of Karnaugh maps is introduced in order to minimize gate count, thereby completing the basic requirements of combinational logic design. Following the minimization techniques, various fundamental logic blocks such as multiplexers, encoders, decoders and one-bit adders are introduced so that they can be used to construct larger scale combinational logic circuits. The last section of this chapter is dedicated to the design of mega cells. These include different types of adders such as ripple carry adder, carry-look-ahead adder, carry-select adder, and the combination of all three types depending on the goals of the design: gate count, circuit speed and power consumption. Subtractors, linear and barrel shifters, array and Booth multipliers constitute the remaining sections of this chapter.

It is vital for the reader to also invest time to learn a hardware design language such as Verilog while studying this chapter and the rest of the chapters in this book. A simulation platform incorporating Verilog and a set of tools that work with Verilog such as design synthesis, static timing analysis, and verification is an effective way to check if the intended design is correct or not. There is nothing more valuable than trying various design ideas on a professional design environment, understanding what works and what does not while working with different tool sets, and most importantly learning from mistakes. An appendix introducing the basic principles of Verilog is included at the end of this book for reference.

1.1 Logic Gates

AND gate

Assume that the output, OUT, in Fig. 1.1 is at logic 0 when both switches, A and B, are open. Unless both A and B close, the output stays at logic 0.

A. Bindal, *Fundamentals of Computer Architecture and Design*,
https://doi.org/10.1007/978-3-030-00223-7_1

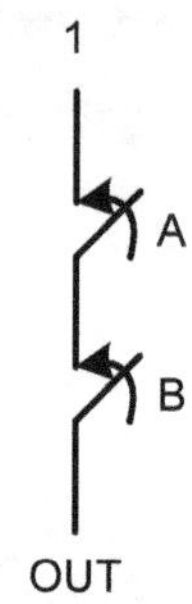

Fig. 1.1 Switch representation of a two-input AND gate

A two-input AND gate functions similarly to the circuit in Fig. 1.1. If any two inputs, A and B, of the AND gate in Fig. 1.2 are at logic 0, the gate produces an output, OUT, at logic 0. Both inputs of the gate must be equal to logic 1 in order to produce an output at logic 1. This behavior is tabulated in Table 1.1, which is called a "truth table".

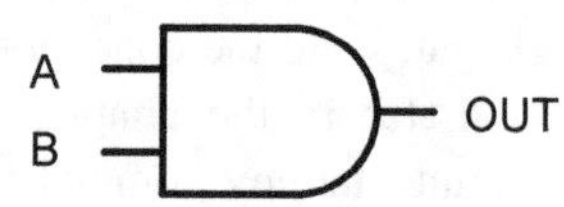

Fig. 1.2 Two-input AND gate symbol

Table 1.1 Two-input AND gate truth table

A	B	OUT
0	0	0
0	1	0
1	0	0
1	1	1

The functional representation of the two-input AND gate is:

$$\mathrm{OUT} = \mathrm{A}\,.\,\mathrm{B}$$

Here, the symbol "." between inputs A and B represents the AND-function.

OR gate

Now, assume a parallel connectivity between switches A and B as shown in Fig. 1.3. OUT becomes to logic 1 if any of the switches close; otherwise the output will stay at logic 0.

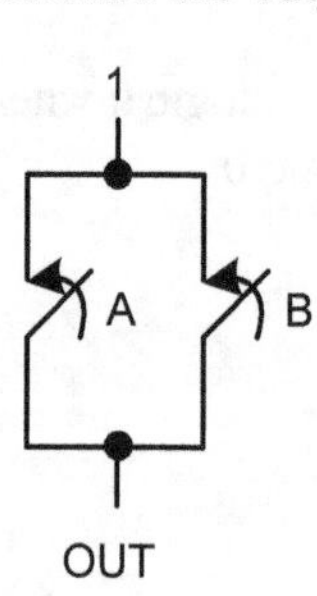

Fig. 1.3 Switch representation of two-input OR gate

A two-input OR gate shown in Fig. 1.4 also functions similarly to the circuit in Fig. 1.3. If any two inputs are at logic 1, the gate produces an output, OUT, at logic 1. Both inputs of the gate must be equal to logic 0 in order to produce an output at logic 0. This behavior is tabulated in the truth table, Table 1.2.

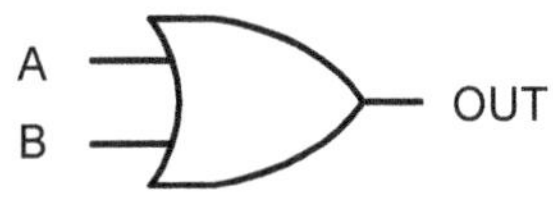

Fig. 1.4 Two-input OR gate symbol

Table 1.2 Two-input OR gate truth table

A	B	OUT
0	0	0
0	1	1
1	0	1
1	1	1

The functional representation of the two-input OR gate is:

$$\text{OUT} = \text{A} + \text{B}$$

Here, the symbol "+" between inputs A and B signifies the OR-function.

Exclusive OR gate

A two-input Exclusive OR gate, XOR gate, is shown in Fig. 1.5. The XOR gate produces a logic 0 output if both inputs are equal. Therefore, in many logic applications this gate is used to compare the input logic levels to see if they are equal. The functional behavior of the gate is tabulated in Table 1.3.

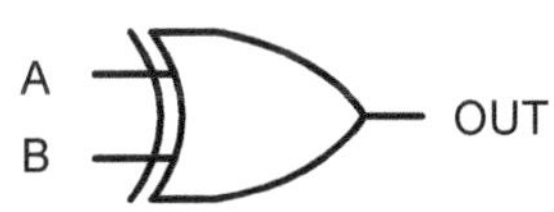

Fig. 1.5 Two-input XOR gate symbol

Table 1.3 Two-input XOR gate truth table

A	B	OUT
0	0	0
0	1	1
1	0	1
1	1	0

The functional representation of the two-input XOR gate is:

$$OUT = A \oplus B$$

Here, the symbol "⊕" between inputs A and B signifies the XOR-function.

Buffer

A buffer is a single input device whose output is logically equal to its input. The only use of this gate is to be able to supply enough current to logic gate inputs connected to its output. The logical representation of this gate is shown in Fig. 1.6.

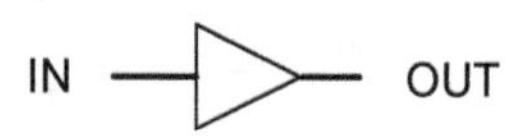

Fig. 1.6 Buffer symbol

Complementary Logic Gates

All basic logic gates need to have complemented forms. If a single input needs to be complemented, an inverter shown in Fig. 1.7 is used. The inverter truth table is shown in Table 1.4.

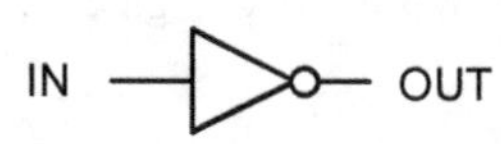

Fig. 1.7 Inverter symbol

Table 1.4 Inverter truth table

IN	OUT
0	1
1	0

The functional representation of the inverter is:

$$OUT = \overline{IN}$$

Here, the symbol "-" on top of the input, IN, represents the complement-function.

The complemented form of two-input AND gate is called two-input NAND gate, where "N" signifies negation. The logic representation is shown in Fig. 1.8, where a circle at the output of the gate means complemented output. The truth table of this gate is shown in Table 1.5. Note that all output values in this table are exact opposites of the values given in Table 1.1.

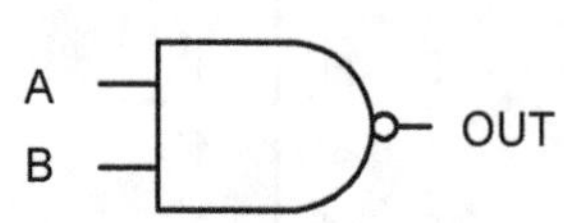

Fig. 1.8 Two-input NAND gate symbol

Table 1.5 Two-input NAND gate truth table

A	B	OUT
0	0	1
0	1	1
1	0	1
1	1	0

The functional representation of the two-input NAND gate is:

$$OUT = \overline{A.B}$$

Similar to the NAND gate, two-input OR and XOR gates have complemented configurations, called two-input NOR and XNOR gates, respectively.

The symbolic representation and truth table of a two-input NOR gate is shown in Fig. 1.9 and Table 1.6, respectively. Again, all the outputs in Table 1.6 are exact complements of Table 1.2.

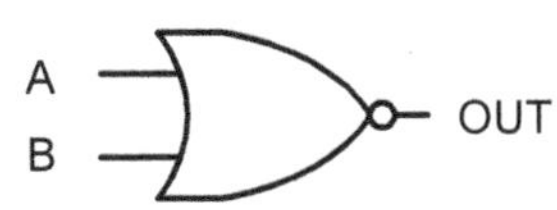

Fig. 1.9 Two-input NOR gate symbol

Table 1.6 Two-input NOR gate truth table

A	B	OUT
0	0	1
0	1	0
1	0	0
1	1	0

The functional representation of the two-input NOR gate is:

$$OUT = \overline{A + B}$$

The symbolic representation and truth table of a two-input XNOR gate is shown in Fig. 1.10 and Table 1.7, respectively. This gate, like its counterpart two-input XOR gate, is often used to detect if input logic levels are equal.

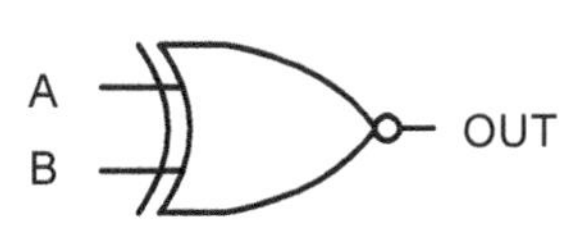

Fig. 1.10 Two-input XNOR gate symbol

Table 1.7 Two-input XNOR gate truth table

A	B	OUT
0	0	1
0	1	0
1	0	0
1	1	1

The functional representation of the two-input XNOR gate is:

$$OUT = \overline{A \oplus B}$$

Tri-State Buffer and Inverter

It is often necessary to create an open circuit between the input and the output of a logic gate if the gate is not enabled. This need creates two more basic logic gates, the tri-state buffer and tri-state inverter.

The tri-state buffer is shown in Fig. 1.11. Its truth table in Table 1.8 indicates continuity between the input and the output terminals if the control input, EN, is at logic 1; when EN is lowered to logic 0, an open circuit exists between IN and OUT, which is defined as a high impedance, HiZ, condition at the output terminal.

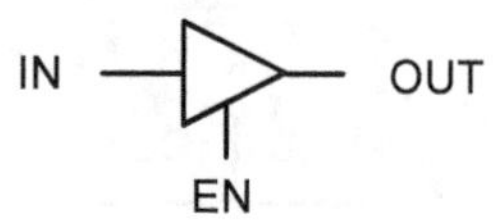

Fig. 1.11 Tri-state buffer symbol

Table 1.8 Tri-state buffer truth table

EN	IN	OUT
0	0	HiZ
0	1	HiZ
1	0	0
1	1	1

The tri-state inverter is shown in Fig. 1.12 along with its truth table in Table 1.9. This gate behaves like an inverter when EN input is at logic 1; however, if EN is lowered to logic 0, its output disconnects from its input.

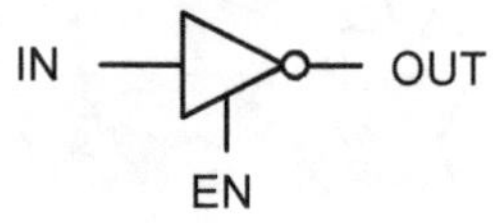

Fig. 1.12 Tri-state inverter symbol

Table 1.9 Tri-state inverter truth table

EN	IN	OUT
0	0	HiZ
0	1	HiZ
1	0	1
1	1	0

The control input, EN, to tri-state buffer and inverter can also be complemented in order to produce an active-low enabling scheme.

The tri-state buffer with the active-low enable input in Fig. 1.13 creates continuity when EN = 0.

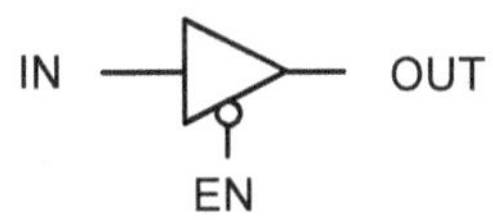

Fig. 1.13 Tri-state buffer symbol with complemented enable input

The tri-state inverter with the active-low input in Fig. 1.14 also functions like an inverter when EN is at logic 0, but its output becomes HiZ when EN is changed to logic 1.

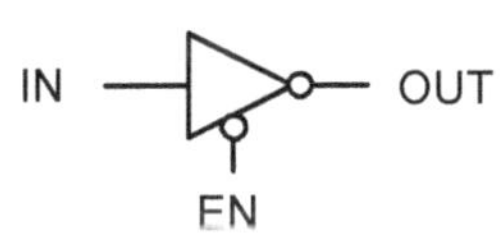

Fig. 1.14 Tri-state inverter symbol with complemented enable input

1.2 Boolean Algebra

It is essential to be able to reconfigure logic gates to suit our design goals. Logical reconfigurations may be as simple as re-grouping the inputs of a single gate or complementing the inputs of several gates to reach a design objective.

Identity, commutative, associative, distributive laws and DeMorgan's negation rules are used to perform logical manipulations. Table 1.10 tabulates these laws.

Table 1.10 Identity, commutative, associative, distributive and DeMorgan's rules

$A . 1 = A$	Identity
$A . 0 = 0$	
$A . A = A$	
$A . \overline{A} = 0$	
$A + 1 = 1$	
$A + 0 = A$	
$A + A = A$	
$A + \overline{A} = 1$	
$\overline{\overline{A}} = A$	
$A . B = B . A$	Commutative
$A + B = B + A$	
$A . (B . C) = (A . B) . C$	Associative
$A + (B + C) = (A + B) + C$	
$A . (B + C) = A . B + A . C$	Distributive
$A + B . C = (A + B) . (A + C)$	
$\overline{A . B} = \overline{A} + \overline{B}$	DeMorgan's
$\overline{A + B} = \overline{A} . \overline{B}$	

Example 1.1 Reduce $OUT = A.\overline{B}.C + A.B.C + A.\overline{B}$ using algebraic rules.

$$
\begin{aligned}
OUT &= A.\overline{B}.C + A.B.C + A.\overline{B} \\
&= A.C.(\overline{B} + B) + A.\overline{B} \\
&= A.(C + \overline{B})
\end{aligned}
$$

Example 1.2 Reduce $OUT = A + \overline{A}.B$ using algebraic rules.

$$
\begin{aligned}
OUT &= A + \overline{A}.B \\
&= (A + \overline{A}).(A + B) \\
&= A + B
\end{aligned}
$$

Example 1.3 Reduce $\mathrm{OUT} = \mathrm{A}\,.\,\mathrm{B} + \overline{\mathrm{A}}\,.\,\mathrm{C} + \mathrm{B}\,.\,\mathrm{C}$ using algebraic rules.

$$\begin{aligned}
\mathrm{OUT} &= \mathrm{A}\,.\,\mathrm{B} + \overline{\mathrm{A}}\,.\,\mathrm{C} + \mathrm{B}\,.\,\mathrm{C} \\
&= \mathrm{A}\,.\,\mathrm{B} + \overline{\mathrm{A}}\,.\,\mathrm{C} + \mathrm{B}\,.\,\mathrm{C}\,.\,(\mathrm{A} + \overline{\mathrm{A}}) \\
&= \mathrm{A}\,.\,\mathrm{B} + \overline{\mathrm{A}}\,.\,\mathrm{C} + \mathrm{A}\,.\,\mathrm{B}\,.\,\mathrm{C} + \overline{\mathrm{A}}\,.\,\mathrm{B}\,.\,\mathrm{C} \\
&= \mathrm{A}\,.\,\mathrm{B}\,.\,(1 + \mathrm{C}) + \overline{\mathrm{A}}\,.\,\mathrm{C}\,.\,(1 + \mathrm{B}) \\
&= \mathrm{A}\,.\,\mathrm{B} + \overline{\mathrm{A}}\,.\,\mathrm{C}
\end{aligned}$$

Example 1.4 Reduce $\mathrm{OUT} = (\mathrm{A} + \mathrm{B})\,.\,(\overline{\mathrm{A}} + \mathrm{C})$ using algebraic rules.

$$\begin{aligned}
\mathrm{OUT} &= (\mathrm{A} + \mathrm{B})\,.\,(\overline{\mathrm{A}} + \mathrm{C}) \\
&= \mathrm{A}\,.\,\overline{\mathrm{A}} + \mathrm{A}\,.\,\mathrm{C} + \overline{\mathrm{A}}\,.\,\mathrm{B} + \mathrm{B}\,.\,\mathrm{C} \\
&= \mathrm{A}\,.\,\mathrm{C} + \overline{\mathrm{A}}\,.\,\mathrm{B} + \mathrm{B}\,.\,\mathrm{C} \\
&= \mathrm{A}\,.\,\mathrm{C} + \overline{\mathrm{A}}\,.\,\mathrm{B} + \mathrm{B}\,.\,\mathrm{C}\,.\,(\mathrm{A} + \overline{\mathrm{A}}) \\
&= \mathrm{A}\,.\,\mathrm{C} + \overline{\mathrm{A}}\,.\,\mathrm{B} + \mathrm{A}\,.\,\mathrm{B}\,.\,\mathrm{C} + \overline{\mathrm{A}}\,.\,\mathrm{B}\,.\,\mathrm{C} \\
&= \mathrm{A}\,.\,\mathrm{C}\,.\,(1 + \mathrm{B}) + \overline{\mathrm{A}}\,.\,\mathrm{B}\,.\,(1 + \mathrm{C}) \\
&= \mathrm{A}\,.\,\mathrm{C} + \overline{\mathrm{A}}\,.\,\mathrm{B}
\end{aligned}$$

Example 1.5 Convert $\mathrm{OUT} = (\mathrm{A} + \mathrm{B})\,.\,\overline{\mathrm{C}\,.\,\mathrm{D}}$ into an OR-combination of two-input AND gates using algebraic laws and DeMorgan's theorem.

$$\begin{aligned}
\mathrm{OUT} &= (\mathrm{A} + \mathrm{B})\,.\,\overline{\mathrm{C}\,.\,\mathrm{D}} \\
&= (\mathrm{A} + \mathrm{B})\,.\,(\overline{\mathrm{C}} + \overline{\mathrm{D}}) \\
&= \mathrm{A}\,.\,\overline{\mathrm{C}} + \mathrm{A}\,.\,\overline{\mathrm{D}} + \mathrm{B}\,.\,\overline{\mathrm{C}} + \mathrm{B}\,.\,\overline{\mathrm{D}}
\end{aligned}$$

Example 1.6 Convert $\mathrm{OUT} = \mathrm{A}\,.\,\mathrm{B} + \mathrm{C}\,.\,\mathrm{D}$ into an AND-combination of two-input OR gates using algebraic laws and DeMorgan's theorem.

$$\begin{aligned}
\mathrm{OUT} &= \mathrm{A}\,.\,\mathrm{B} + \mathrm{C}\,.\,\mathrm{D} \\
&= \overline{\overline{\mathrm{A}\,.\,\mathrm{B} + \mathrm{C}\,.\,\mathrm{D}}} \\
&= \overline{(\overline{\mathrm{A}} + \overline{\mathrm{B}})\,.\,(\overline{\mathrm{C}} + \overline{\mathrm{D}})}
\end{aligned}$$

1.3 Designing Combinational Logic Circuits Using Truth Tables

A combinational circuit is cascaded form of basic logic gates without any feedback from the output to any input. The logic function is obtained from a truth table that specifies the complete functionality of the digital circuit.

Example 1.7 Using the truth table given in Table 1.11 determine the output function of the digital circuit.

Table 1.11 An arbitrary truth table with four inputs

A	B	C	D	OUT
0	0	0	0	1
0	0	0	1	1
0	0	1	0	1
0	0	1	1	0
0	1	0	0	0
0	1	0	1	1
0	1	1	0	0
0	1	1	1	0
1	0	0	0	1
1	0	0	1	1
1	0	1	0	1
1	0	1	1	0
1	1	0	0	0
1	1	0	1	0
1	1	1	0	0
1	1	1	1	0

The output function can be expressed either as the OR combination of AND gates or the AND combination of OR gates.

If the output is expressed in terms of AND gates, all output entries that are equal to one in the truth table must be grouped together as a single OR gate.

$$\begin{aligned}\mathrm{OUT} &= \overline{\mathrm{A}}.\overline{\mathrm{B}}.\overline{\mathrm{C}}.\overline{\mathrm{D}} + \overline{\mathrm{A}}.\overline{\mathrm{B}}.\overline{\mathrm{C}}.\mathrm{D} + \overline{\mathrm{A}}.\overline{\mathrm{B}}.\mathrm{C}.\overline{\mathrm{D}} + \overline{\mathrm{A}}.\mathrm{B}.\overline{\mathrm{C}}.\mathrm{D} \\ &\quad + \mathrm{A}.\overline{\mathrm{B}}.\overline{\mathrm{C}}.\overline{\mathrm{D}} + \mathrm{A}.\overline{\mathrm{B}}.\overline{\mathrm{C}}.\mathrm{D} + \mathrm{A}.\overline{\mathrm{B}}.\mathrm{C}.\overline{\mathrm{D}}\end{aligned}$$

This expression is called the Sum Of Products (SOP), and it contains seven terms each of which is called a "minterm". In the first minterm, $\overline{\mathrm{A}}.\overline{\mathrm{B}}.\overline{\mathrm{C}}.\overline{\mathrm{D}}$, each A, B, C and D input is complemented to produce OUT = 1 for the A = B = C = D = 0 entry of the truth table. Each of the remaining six minterms also complies with producing OUT = 1 for their respective input entries.

The resulting combinational circuit is shown in Fig. 1.15.

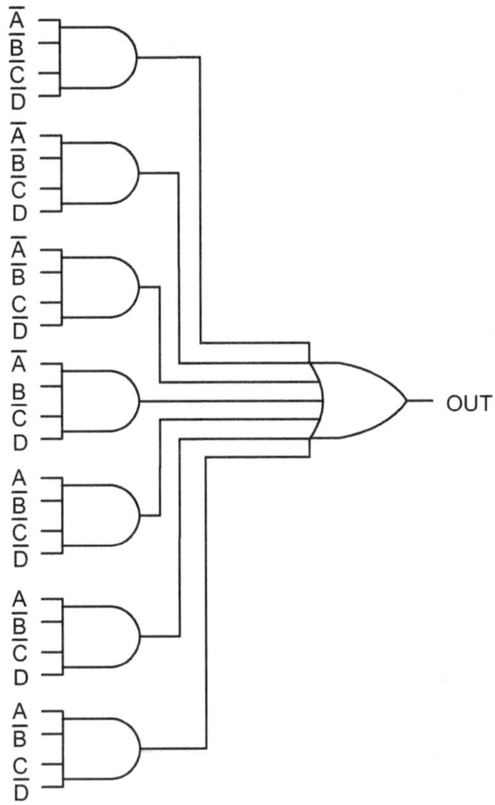

Fig. 1.15 AND-OR logic representation of the truth table in Table 1.11

If the output function needs to be expressed in terms of OR gates, all the output entries that are equal to zero in the truth table must be grouped as a single AND gate.

$$\begin{aligned} OUT = & \left(A+B+\overline{C}+\overline{D}\right).\left(A+\overline{B}+C+D\right).\left(A+\overline{B}+\overline{C}+D\right) \\ & .\left(A+\overline{B}+\overline{C}+\overline{D}\right).\left(\overline{A}+B+\overline{C}+\overline{D}\right).\left(\overline{A}+\overline{B}+C+D\right) \\ & .\left(\overline{A}+\overline{B}+C+\overline{D}\right).\left(\overline{A}+\overline{B}+\overline{C}+D\right).\left(\overline{A}+\overline{B}+\overline{C}+\overline{D}\right) \end{aligned}$$

This expression is called the Product Of Sums (POS), and it contains nine terms each of which is called a "maxterm". The first maxterm, $A+B+\overline{C}+\overline{D}$, produces OUT = 0 for the ABCD = 0011 entry of the truth table. Since the output is formed with a nine-input AND gate, the values of the other maxterms do not matter to produce OUT = 0. Each of the remaining eight maxterms generates OUT = 0 for their corresponding truth table input entries.

The resulting combinational circuit is shown in Fig. 1.16.

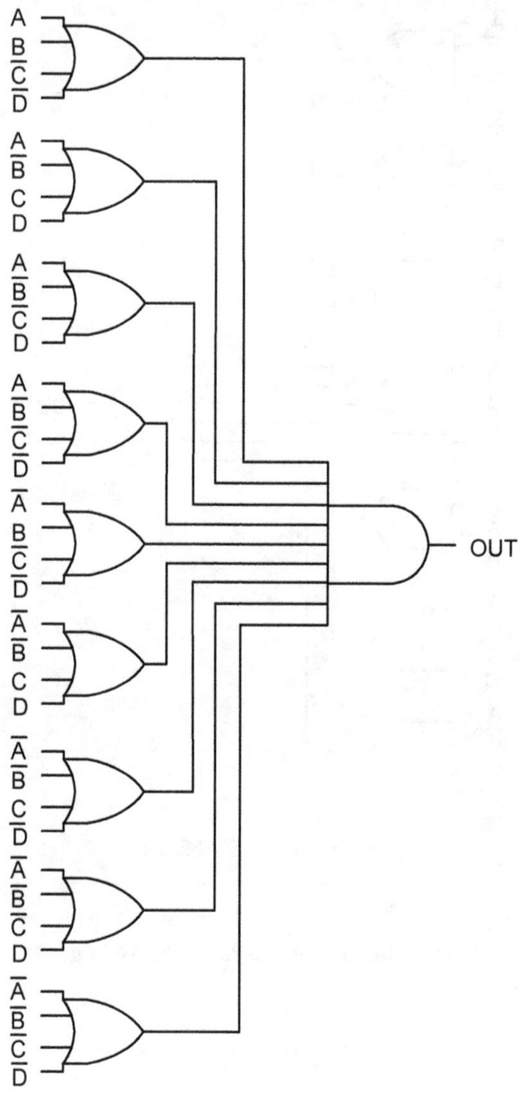

Fig. 1.16 OR-AND logic representation of the truth table in Table 1.11

1.4 Combinational Logic Minimization—Karnaugh Maps

One of the most useful tools in logic design is the use of Karnaugh maps (K-map) to minimize combinational logic functions.

Minimization can be performed in two ways. To obtain SOP form of a minimized logic function, the entries with logic 1 in the truth table must be grouped together in the K-map. To obtain POS form of a minimized logic function, the entries with logic 0 must be grouped together in the K-map.

Example 1.8 Using the truth table in Table 1.12, determine the minimized SOP and POS output functions. Prove them to be identical.

Table 1.12 An arbitrary truth table with three inputs

A	B	C	OUT
0	0	0	1
0	0	1	1
0	1	0	1
0	1	1	0
1	0	0	1
1	0	1	1
1	1	0	0
1	1	1	0

The K-map formed according to the truth table groups 1s to obtain the minimized output function, OUT, in SOP form in Fig. 1.17.

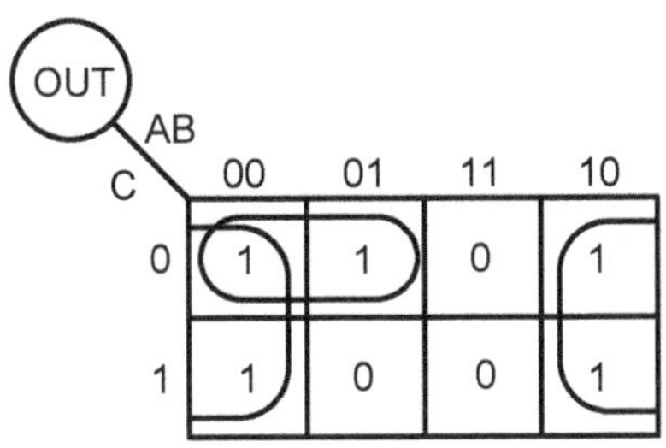

Fig. 1.17 K-map of the truth table in Table 1.12 to determine SOP

Grouping 1s takes place among neighboring boxes in the K-map where only one variable is allowed to change at a time. For instance, the first grouping of 1s combines ABC = 000 and ABC = 010 as they are in neighboring boxes. Only B changes from logic 0 to logic 1 while A and C stay constant at logic 0. To obtain OUT = 1, both A and C need to be complemented; this produces the first term, $\overline{A}.\overline{C}$, for the output function. Similarly, the second grouping of 1s combines the neighboring boxes, ABC = 000, 001, 100 and 101, where both A and C change while B stays constant at logic 0. To obtain OUT = 1, B needs to be complemented; this generates the second term, $\overline{B}$, for the output function.

This means that either the term $\overline{A}.\overline{C}$ or $\overline{B}$ makes OUT equal to logic 1. Therefore, the minimized output function, OUT, in the SOP form is:

$$OUT = \overline{B} + \overline{A}.\overline{C}$$

Grouping 0s produces the minimized POS output function as shown in Fig. 1.18.

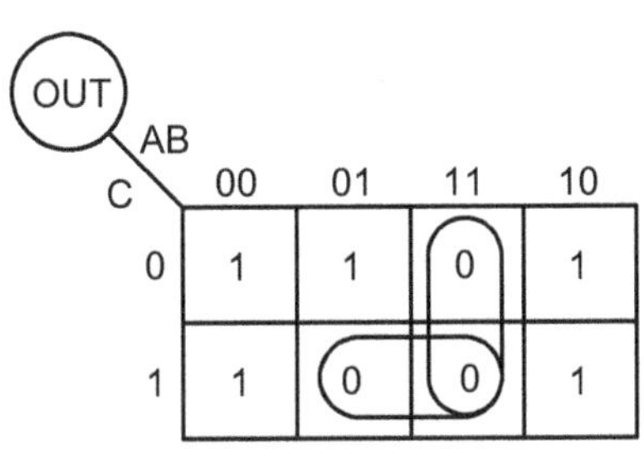

Fig. 1.18 K-map of the truth table in Table 1.12 to determine POS

This time, the first grouping of 0s combines the boxes, ABC = 011 and 111, where A changes from logic 0 to logic 1 while B and C stay constant at logic 1. This grouping targets OUT = 0, which requires both B and C to be complemented. As a result, the first term of the output function, $\overline{B}+\overline{C}$, is generated. The second grouping combines ABC = 110 and 111 where C changes value while A and B are equal to logic 1. To obtain OUT = 0, both A and B need to be complemented. Consequently, the second term, $\overline{A}+\overline{B}$, is generated.

Either of the terms $\overline{B}+\overline{C}$ or $\overline{A}+\overline{B}$ makes OUT = 0. Therefore, the minimized output function in the POS form becomes:

$$\mathrm{OUT} = \left(\overline{B}+\overline{C}\right).\left(\overline{A}+\overline{B}\right)$$

To find out if the SOP and POS forms are identical, one can manipulate the POS using the algebraic rules given earlier.

$$\begin{aligned}\mathrm{OUT} &= \left(\overline{B}+\overline{C}\right).\left(\overline{A}+\overline{B}\right)\\ &= \overline{A}.\overline{B}+\overline{B}.\overline{B}+\overline{A}.\overline{C}+\overline{B}.\overline{C}\\ &= \overline{A}.\overline{B}+\overline{B}+\overline{A}.\overline{C}+\overline{B}.\overline{C}\\ &= \overline{B}.\left(\overline{A}+1+\overline{C}\right)+\overline{A}.\overline{C}\\ &= \overline{B}+\overline{A}.C\end{aligned}$$

However, this is the SOP form of the output function derived above.

Example 1.9 Using the truth table in Example 1.7 determine the minimized SOP and POS output functions.

To obtain an output function in SOP form, 1s in the K-map in Fig. 1.19 is grouped together as shown below.

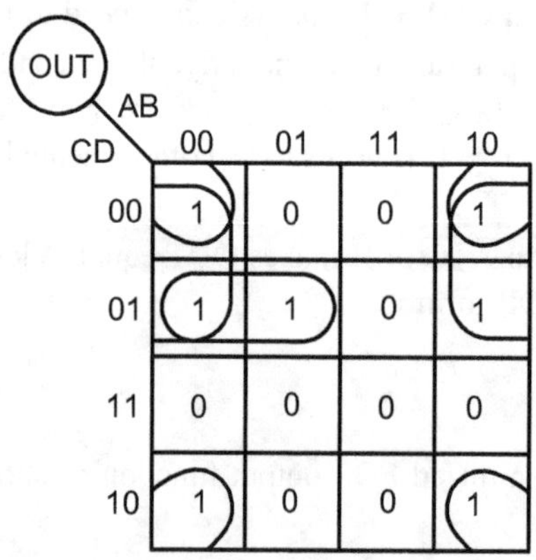

Fig. 1.19 K-map of the truth table in Table 1.11 to determine SOP

The minimized output function contains only three minterms compared to seven minterms in Example 1.7. Also, the minterms are reduced to groups of two or three inputs instead of four.

$$\mathrm{OUT} = \overline{\mathrm{B}}.\overline{\mathrm{C}} + \overline{\mathrm{A}}.\overline{\mathrm{C}}.\mathrm{D} + \overline{\mathrm{B}}.\overline{\mathrm{D}}$$

The resultant combinational circuit is shown in Fig. 1.20.

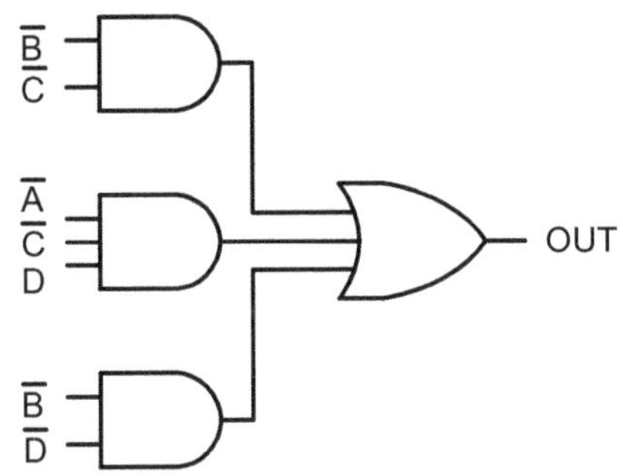

Fig. 1.20 Minimized logic circuit in SOP form from the K-map in Fig. 1.19

Further minimization can be achieved algebraically, which then reduces the number of terms from three to two.

$$\mathrm{OUT} = \overline{\mathrm{B}}.(\overline{\mathrm{C}} + \overline{\mathrm{D}}) + \overline{\mathrm{A}}.\overline{\mathrm{C}}.\mathrm{D}$$

The corresponding combinational circuit is shown in Fig. 1.21.

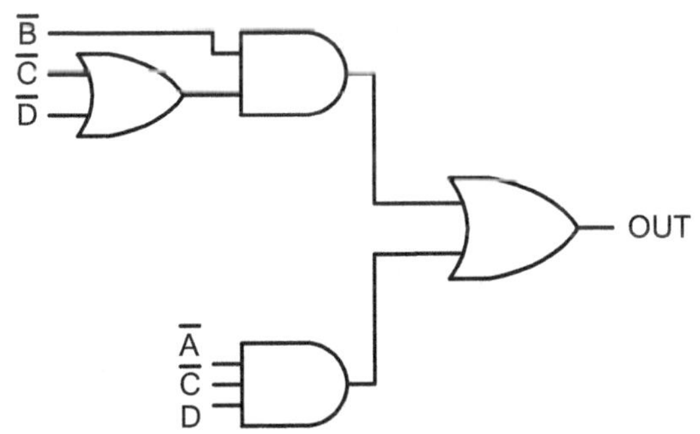

Fig. 1.21 Logic circuit in Fig. 1.20 after algebraic minimizations are applied

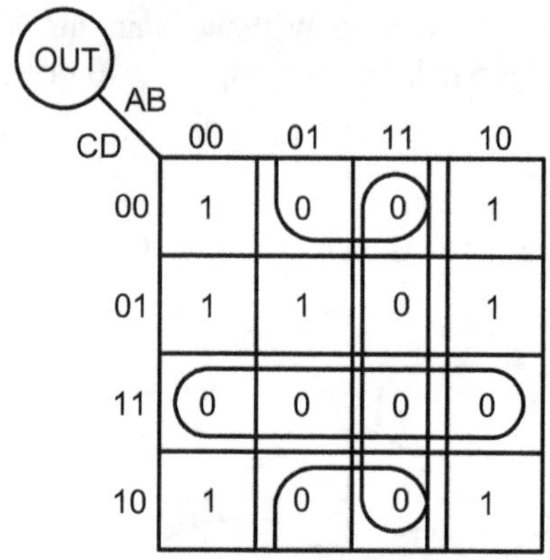

Fig. 1.22 K-map of the truth table in Table 1.11 to determine POS

To obtain a POS output function, 0s are grouped together as shown in Fig. 1.22.

The minimized output function contains only three maxterms compared to nine in Example 1.7. Also, the maxterms are reduced to groups of two inputs instead of four.

$$OUT = (\overline{C}+\overline{D}).(\overline{A}+\overline{B}).(\overline{B}+D)$$

The resultant combinational circuit is shown in Fig. 1.23.

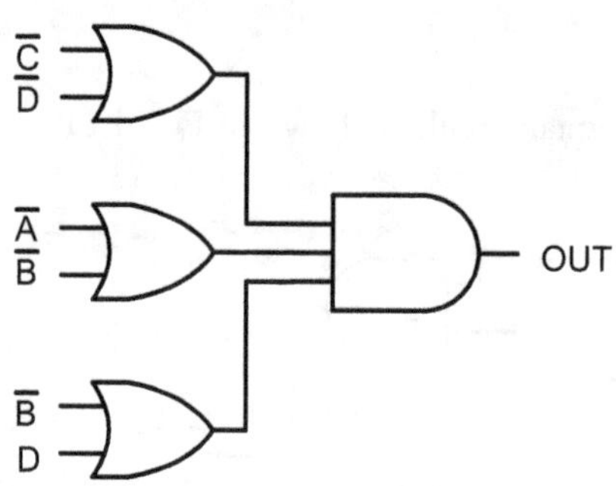

Fig. 1.23 Minimized logic circuit in POS form from the K-map in Fig. 1.22

Example 1.10 Determine if the minimized SOP and POS output functions in Example 1.9 are identical to each other.

Rewriting the POS form of OUT from Example 1.9 and using the algebraic laws shown earlier, this expression can be re-written as:

$$\begin{aligned}OUT &= (\overline{C}+\overline{D}).(\overline{A}+\overline{B}).(\overline{B}+D)\\ &= (\overline{A}.\overline{C}+\overline{A}.\overline{D}+\overline{B}.\overline{C}+\overline{B}.\overline{D}).(\overline{B}+D)\\ &= \overline{A}.\overline{B}.\overline{C}+\overline{A}.\overline{B}.\overline{D}+\overline{B}.\overline{C}+\overline{B}.\overline{D}\\ &\quad +\overline{A}.\overline{C}.D+\overline{B}.\overline{C}.D\\ &= \overline{B}.\overline{C}+\overline{B}.\overline{D}+\overline{A}.\overline{C}.D\end{aligned}$$

The result is identical to the SOP expression given in Example 1.9.

Example 1.11 Determine the minimal SOP and POS forms of the output function, OUT, from the K-map in Fig. 1.24. Note that the "X" sign corresponds to a "don't care" condition that represents either logic 0 or logic 1.

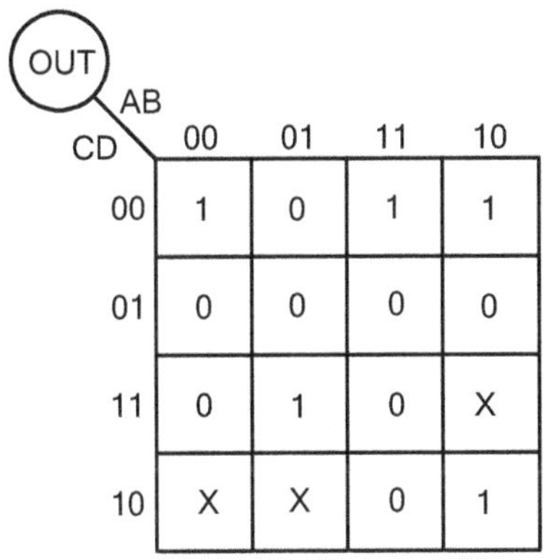

Fig. 1.24 An arbitrary K-map with "don't care" entries

For SOP, we group 1s in the K-map in Fig. 1.25. Boxes with "don't care" are used as 1s to achieve a minimal SOP expression.

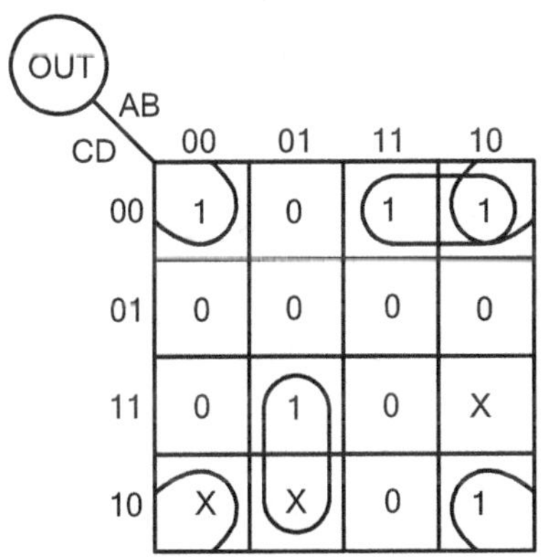

Fig. 1.25 Grouping to determine SOP form for the K-map in Fig. 1.24

The SOP functional expression for OUT is:

$OUT = A.\overline{C}.\overline{D} + \overline{A}.B.C + \overline{B}.\overline{D}$

For POS, we group 0s in the K-map in Fig. 1.26. Boxes with "don't care" symbols are used as 0s to achieve a minimal POS expression.

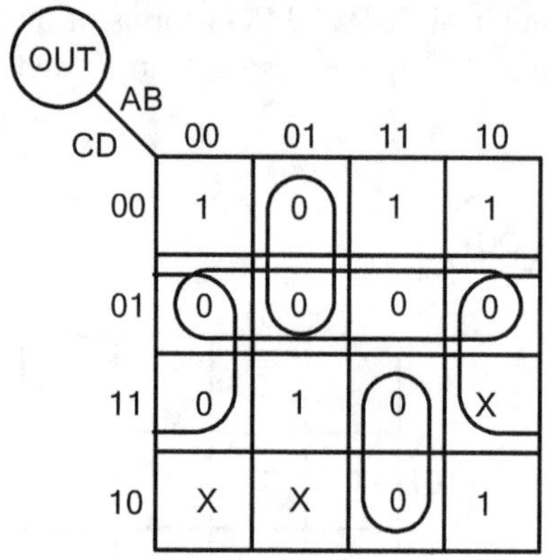

Fig. 1.26 Grouping to determine POS form for the K-map in Fig. 1.24

The POS functional expression for OUT is:

$$OUT = (C+\overline{D}).(B+\overline{D}).(A+\overline{B}+C).(\overline{A}+\overline{B}+\overline{C})$$

To show that the SOP and POS expressions are identical, we start with the POS expression using the algebraic manipulations described earlier in Table 1.10.

$$\begin{aligned}
OUT &= (C+\overline{D}).(B+\overline{D}).(A+\overline{B}+C).(\overline{A}+\overline{B}+\overline{C})\\
&= (B.C+C.\overline{D}+B.\overline{D}+\overline{D}).(A.\overline{B}+A.\overline{C}+\overline{A}.\overline{B}+\overline{B}+\overline{B}.\overline{C}+\overline{A}.C+\overline{B}.C)\\
&= (B.C+\overline{D}).(A.\overline{C}+\overline{A}.C+\overline{B})\\
&= \overline{A}.B.C+A.\overline{C}.\overline{D}+\overline{A}.C.\overline{D}+\overline{B}.\overline{D}\\
&= \overline{A}.B.C+A.\overline{C}.\overline{D}+\overline{B}.\overline{D}+\overline{A}.C.\overline{D}.(B+\overline{B})\\
&= \overline{A}.B.C+A.\overline{C}.\overline{D}+\overline{B}.\overline{D}+\overline{A}.B.C.\overline{D}+\overline{A}.\overline{B}.C.\overline{D}\\
&= \overline{A}.B.C.(1+\overline{D})+A.\overline{C}.\overline{D}+\overline{B}.\overline{D}.(1+\overline{A}.C)\\
&= \overline{A}.B.C+A.\overline{C}.\overline{D}+\overline{B}.\overline{D}
\end{aligned}$$

The result is identical to the minimal SOP expression for OUT shown above.

1.5 Basic Logic Blocks

2-1 Multiplexer

A 2-1 multiplexer (MUX) is one of the most versatile logic elements in logic design. It is defined as follows:

$$OUT = \begin{bmatrix} A & \text{if} & \text{sel} = 1 \\ B & \text{else} & \end{bmatrix}$$

A functional diagram of the 2-1 MUX is given in Fig. 1.27. According to the functional description of this device, when sel = 1 input A is passed through the device to become its output. When sel = 0 input B is passed through the device to become its output.

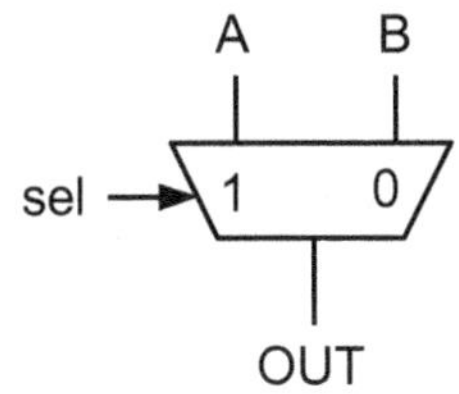

Fig. 1.27 2-1 MUX symbol

According to this definition, the truth table in Table 1.13 can be formed:

Table 1.13 2-1 MUX truth table

sel	A	B	OUT
0	0	0	0
0	0	1	1
0	1	0	0
0	1	1	1
1	0	0	0
1	0	1	0
1	1	0	1
1	1	1	1

Now, let us transfer the output values from the truth table to the K-map in Fig. 1.28.

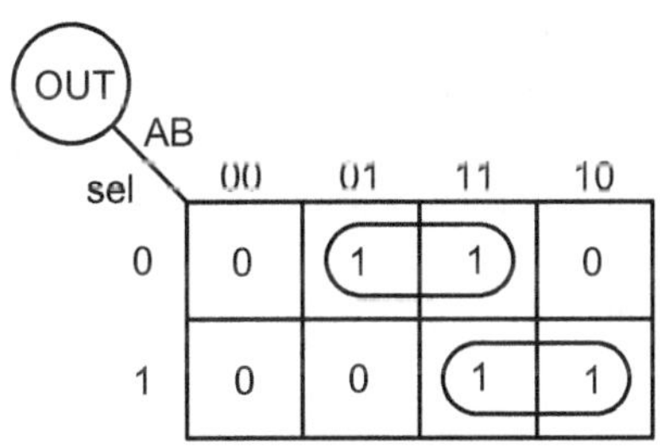

Fig. 1.28 2-1 MUX K-map

Grouping 1s in the K-map reveals the minimal output function of the 2-1 MUX in SOP form:

$$OUT = sel \,.\, A + \overline{sel} \,.\, B$$

The corresponding combinational circuit is shown in Fig. 1.29.

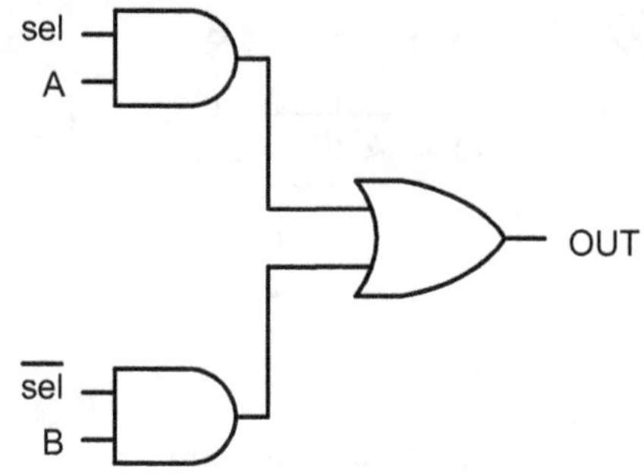

Fig. 1.29 2-1 MUX logic circuit

4-1 Multiplexer

Considering A, B, C and D are the inputs, the functional description of a 4-1 MUX becomes as follows:

$$\text{OUT} = \begin{bmatrix} \text{A} & \text{if sel1} = 0 \text{ and sel2} = 0 \\ \text{B} & \text{if sel1} = 0 \text{ and sel2} = 1 \\ \text{C} & \text{if sel1} = 1 \text{ and sel2} = 0 \\ \text{D} & \text{else} \end{bmatrix}$$

According to this description, we can form a truth table and obtain the minimal SOP or POS expression for OUT. However, it is quite easy to decipher the SOP expression for OUT from the description above. The AND-combination of A, complemented sel1 and complemented sel2 inputs constitute the first minterm of our SOP. The second minterm should contain B, complemented sel1 and uncomplemented sel2 according to the description above. Similarly, the third minterm contains C, uncomplemented sel1 and complemented sel2. Finally, the last minterm contains D, uncomplemented sel1 and uncomplemented sel2 control inputs. Therefore, the SOP expression for the 4-1 MUX becomes equal to the logic expression below and is implemented in Fig. 1.30.

$$\text{OUT} = \overline{\text{sel1}}\,.\,\overline{\text{sel2}}\,.\,\text{A} + \overline{\text{sel1}}\,.\,\text{sel2}\,.\,\text{B} + \text{sel1}\,.\,\overline{\text{sel2}}\,.\,\text{C} + \text{sel1}\,.\,\text{sel2}\,.\,\text{D}$$

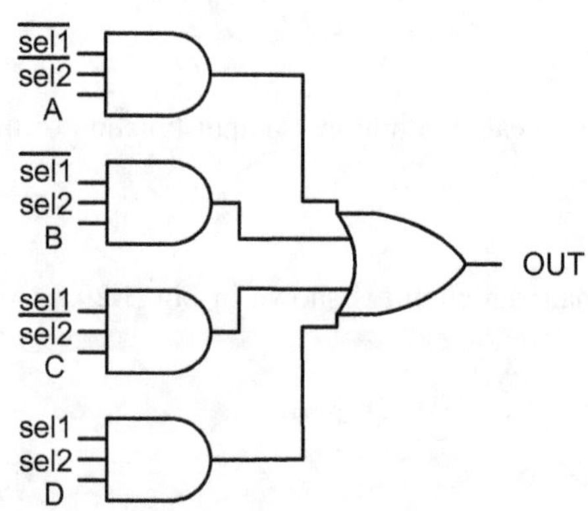

Fig. 1.30 4-1 MUX logic circuit in SOP form

However, implementing 4-1 MUX this way is not advantageous due to the amount of gate delays; a three-input AND gate is a serial combination of a three-input NAND and an inverter, and similarly a four-input OR connects a four-input NOR to an inverter. Therefore, we obtain a minimum of four gate delays instead of two according to this circuit.

Logic translations are possible to reduce the gate delay. The first stage of this process is to complement the outputs of all four three-input AND gates. This necessitates complementing the inputs of the four-input OR gate, and results in a circuit in Fig. 1.31.

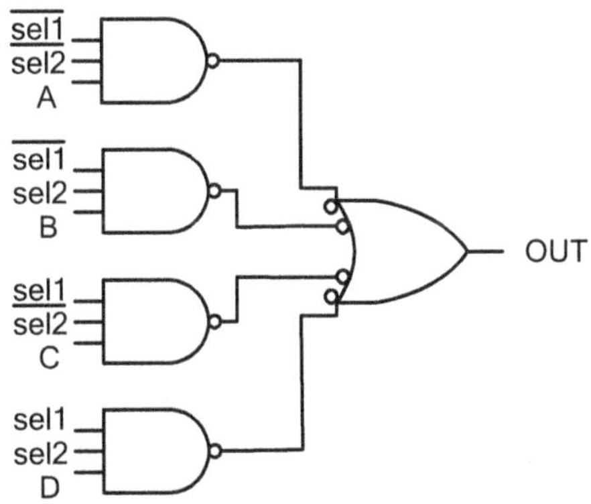

Fig. 1.31 Logic conversion of 4-1 MUX in Fig. 1.30

However, an OR gate with complemented inputs is equivalent to a NAND gate. Therefore, the circuit in Fig. 1.32 becomes optimal for implementation purposes because the total MUX delay is only the sum of a three-input NAND and a four-input NAND gate delays instead of the earlier four gate delays.

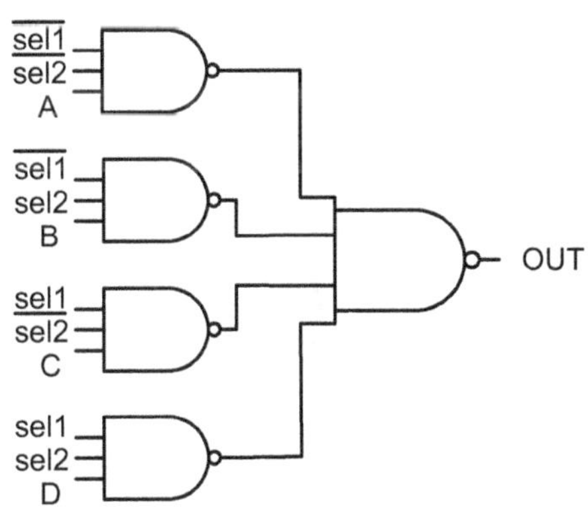

Fig. 1.32 4-1 MUX logic circuit in NAND-NAND form

The symbolic diagram of the 4-1 MUX is shown in Fig. 1.33.

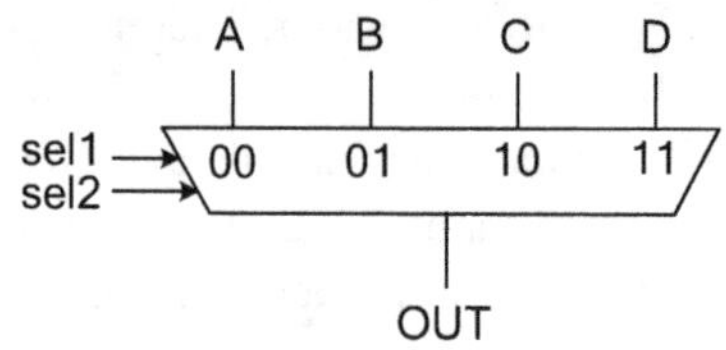

Fig. 1.33 4-1 MUX symbol

Encoders

Encoders are combinational logic blocks that receive 2^N number of inputs and produce N number of encoded outputs.

Example 1.12 Generate an encoding logic from the truth table given in Table 1.14.

Table 1.14 An arbitrary encoder truth table with four inputs

IN1	IN2	IN3	IN4	OUT1	OUT2
0	0	0	0	0	1
0	0	0	1	1	1
0	0	1	0	1	1
0	0	1	1	0	1
0	1	0	0	1	0
0	1	0	1	0	0
0	1	1	0	0	0
0	1	1	1	1	0
1	0	0	0	0	1
1	0	0	1	0	0
1	0	1	0	0	0
1	0	1	1	0	1
1	1	0	0	1	0
1	1	0	1	1	1
1	1	1	0	1	1
1	1	1	1	1	0

The K-maps in Fig. 1.34 groups 1s and produces the minimized SOP expressions for OUT1 and OUT2.

$$\begin{aligned}\mathrm{OUT1} &= \mathrm{IN1}\,.\,\mathrm{IN2} + \overline{\mathrm{IN1}}\,.\,\overline{\mathrm{IN2}}\,.\,\mathrm{IN3}\,.\,\overline{\mathrm{IN4}} + \overline{\mathrm{IN1}}\,.\,\overline{\mathrm{IN2}}\,.\,\overline{\mathrm{IN3}}\,.\,\mathrm{IN4} + \mathrm{IN2}\,.\,\overline{\mathrm{IN3}}\,.\,\overline{\mathrm{IN4}} + \mathrm{IN2}\,.\,\mathrm{IN3}\,.\,\mathrm{IN4}\\ &= \mathrm{IN1}\,.\,\mathrm{IN2} + \overline{\mathrm{IN1}}\,.\,\overline{\mathrm{IN2}}\,.\,(\mathrm{IN3} \oplus \mathrm{IN4}) + \mathrm{IN2}\,.\,(\overline{\mathrm{IN3} \oplus \mathrm{IN4}})\end{aligned}$$

$$\begin{aligned}\mathrm{OUT2} &= \overline{\mathrm{IN1}}\,.\,\overline{\mathrm{IN2}} + \overline{\mathrm{IN2}}\,.\,\overline{\mathrm{IN3}}\,.\,\overline{\mathrm{IN4}} + \overline{\mathrm{IN2}}\,.\,\mathrm{IN3}\,.\,\mathrm{IN4} + \mathrm{IN1}\,.\,\mathrm{IN2}\,.\,\overline{\mathrm{IN3}}\,.\,\mathrm{IN4} + \mathrm{IN1}\,.\,\mathrm{IN2}\,.\,\mathrm{IN3}\,.\,\overline{\mathrm{IN4}}\\ &= \overline{\mathrm{IN1}}\,.\,\overline{\mathrm{IN2}} + \overline{\mathrm{IN2}}\,.\,(\overline{\mathrm{IN3} \oplus \mathrm{IN4}}) + \mathrm{IN1}\,.\,\mathrm{IN2}\,.\,(\mathrm{IN3} \oplus \mathrm{IN4})\end{aligned}$$

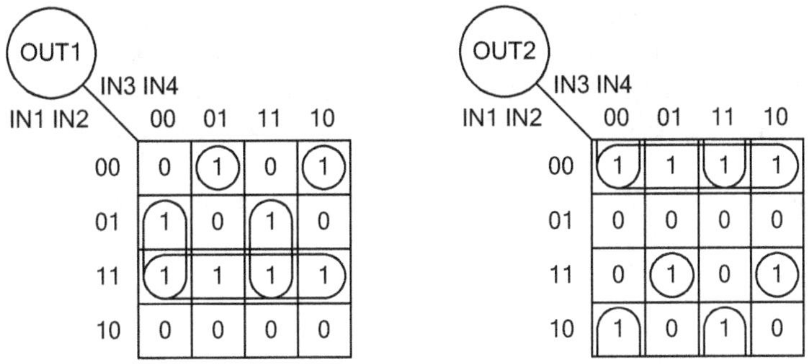

Fig. 1.34 K-map of the truth table in Table 1.14

Decoders

Decoders are combinational logic blocks used to decode encoded inputs. An ordinary decoder takes N inputs and produces 2^N outputs.

Example 1.13 Design a line decoder in which an active high enable signal activates only one of eight independent outputs according to truth table in Table 1.15. When the enable signal is lowered to logic 0, all eight outputs are disabled and stay at logic 0.

Table 1.15 Truth table of a line decoder with three inputs with enable

EN	IN[2]	IN[1]	IN[0]	OUT[7]	OUT[6]	OUT[5]	OUT[4]	OUT[3]	OUT[2]	OUT[1]	OUT[0]
0	0	0	0	0	0	0	0	0	0	0	0
0	0	0	1	0	0	0	0	0	0	0	0
0	0	1	0	0	0	0	0	0	0	0	0
0	0	1	1	0	0	0	0	0	0	0	0
0	1	0	0	0	0	0	0	0	0	0	0
0	1	0	1	0	0	0	0	0	0	0	0
0	1	1	0	0	0	0	0	0	0	0	0
0	1	1	1	0	0	0	0	0	0	0	0
1	0	0	0	0	0	0	0	0	0	0	1
1	0	0	1	0	0	0	0	0	0	1	0
1	0	1	0	0	0	0	0	0	1	0	0
1	0	1	1	0	0	0	0	1	0	0	0
1	1	0	0	0	0	0	1	0	0	0	0
1	1	0	1	0	0	1	0	0	0	0	0
1	1	1	0	0	1	0	0	0	0	0	0
1	1	1	1	1	0	0	0	0	0	0	0

In this table, all outputs become logic 0 when the enable signal, EN, is at logic 0. However, when EN = 1, activation of the output starts. For each three-bit input entry, there is always one output at logic 1. For example, when IN[2] = IN[1] = IN[0] = 0, OUT[0] becomes active and equals to logic 1 while all other outputs stay at logic 0. IN[2] = IN[1] = IN[0] = 1 activates OUT[7] and disables all other outputs.

We can produce each output expression from OUT[7] to OUT[0] simply by reading the input values from the truth table. The accompanying circuit is composed of eight AND gates, each with four inputs as shown in Fig. 1.35.

$$\mathrm{OUT}[7] = \mathrm{EN}\,.\,\mathrm{IN}[2]\,.\,\mathrm{IN}[1]\,.\,\mathrm{IN}[0]$$
$$\mathrm{OUT}[6] = \mathrm{EN}\,.\,\mathrm{IN}[2]\,.\,\mathrm{IN}[1]\,.\,\overline{\mathrm{IN}[0]}$$
$$\mathrm{OUT}[5] = \mathrm{EN}\,.\,\mathrm{IN}[2]\,.\,\overline{\mathrm{IN}[1]}\,.\,\mathrm{IN}[0]$$
$$\mathrm{OUT}[4] = \mathrm{EN}\,.\,\mathrm{IN}[2]\,.\,\overline{\mathrm{IN}[1]}\,.\,\overline{\mathrm{IN}[0]}$$
$$\mathrm{OUT}[3] = \mathrm{EN}\,.\,\overline{\mathrm{IN}[2]}\,.\,\mathrm{IN}[1]\,.\,\mathrm{IN}[0]$$
$$\mathrm{OUT}[2] = \mathrm{EN}\,.\,\overline{\mathrm{IN}[2]}\,.\,\mathrm{IN}[1]\,.\,\overline{\mathrm{IN}[0]}$$
$$\mathrm{OUT}[1] = \mathrm{EN}\,.\,\overline{\mathrm{IN}[2]}\,.\,\overline{\mathrm{IN}[1]}\,.\,\mathrm{IN}[0]$$
$$\mathrm{OUT}[0] = \mathrm{EN}\,.\,\overline{\mathrm{IN}[2]}\,.\,\overline{\mathrm{IN}[1]}\,.\,\overline{\mathrm{IN}[0]}$$

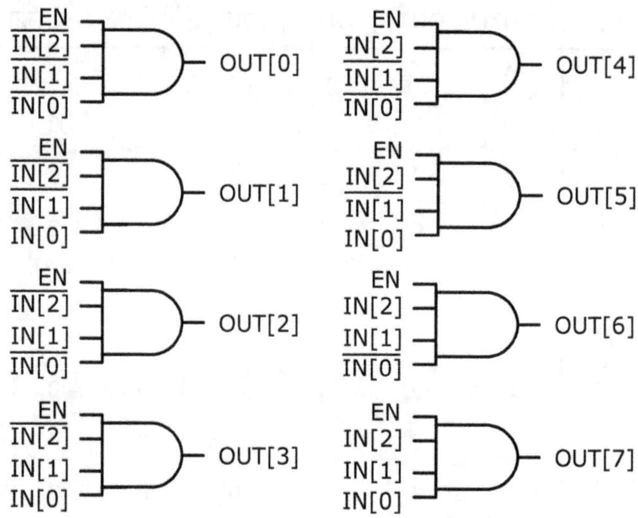

Fig. 1.35 Logic circuit of a line decoder in Table 1.15

One-Bit Full Adder

A one-bit full adder has three inputs: A, B, and carry-in (CIN), and two outputs: sum (SUM) and carry-out (COUT). The symbolic representation of a full adder is shown in Fig. 1.36.

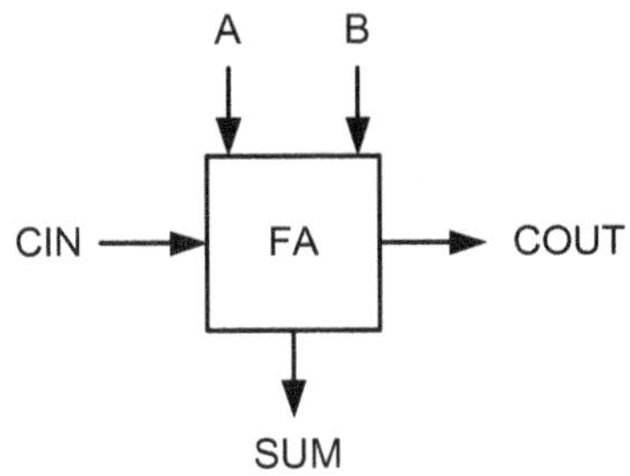

Fig. 1.36 One-bit full adder symbol

A one-bit full adder simply adds the contents of its two inputs, A and B, to the contents of CIN, and forms the truth table given in Table 1.16.

Table 1.16 One-bit full adder truth table

CIN	A	B	SUM	COUT
0	0	0	0	0
0	0	1	1	0
0	1	0	1	0
0	1	1	0	1
1	0	0	1	0
1	0	1	0	1
1	1	0	0	1
1	1	1	1	1

We can obtain the minimized SOP expressions for SUM and COUT from the K-maps in Figs. 1.37 and 1.38.

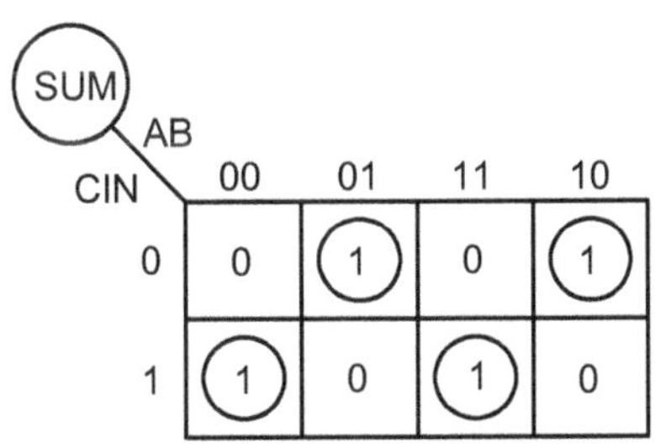

Fig. 1.37 SUM output of a one-bit full adder

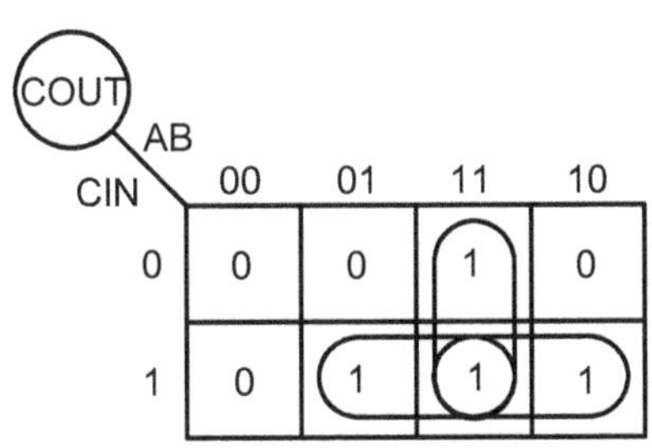

Fig. 1.38 COUT output of a one-bit full adder

Consequently,

$$\begin{aligned}
\mathrm{SUM} &= \overline{\mathrm{A}}\,.\,\overline{\mathrm{B}}\,.\,\mathrm{CIN} + \overline{\mathrm{A}}\,.\,\mathrm{B}\,.\,\overline{\mathrm{CIN}} + \mathrm{A}\,.\,\mathrm{B}\,.\,\mathrm{CIN} + \mathrm{A}\,.\,\overline{\mathrm{B}}\,.\,\overline{\mathrm{CIN}} \\
&= \mathrm{CIN}\left(\overline{\mathrm{A}}\,.\,\overline{\mathrm{B}} + \mathrm{A}\,.\,\mathrm{B}\right) + \overline{\mathrm{CIN}}\,.\,\left(\overline{\mathrm{A}}\,.\,\mathrm{B} + \mathrm{A}\,.\,\overline{\mathrm{B}}\right) \\
&= \mathrm{CIN}\left(\overline{\mathrm{A} \oplus \mathrm{B}}\right) + \overline{\mathrm{CIN}}\,.\,(\mathrm{A} \oplus \mathrm{B}) \\
&= \mathrm{A} \oplus \mathrm{B} \oplus \mathrm{CIN}
\end{aligned}$$

Thus,

$$\begin{aligned}
\mathrm{COUT} &= \mathrm{CIN}\,.\,\mathrm{B} + \mathrm{A}\,.\,\mathrm{B} + \mathrm{A}\,.\,\mathrm{CIN} \\
&= \mathrm{CIN}\,.\,(\mathrm{A} + \mathrm{B}) + \mathrm{A}\,.\,\mathrm{B}
\end{aligned}$$

The resultant logic circuits for SUM and COUT are shown in Fig. 1.39.

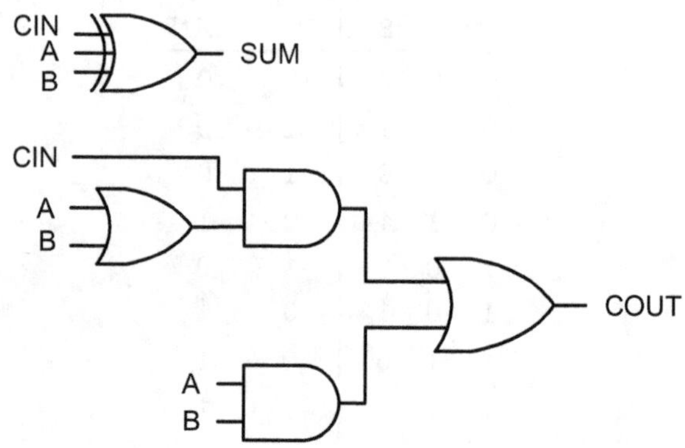

Fig. 1.39 One-bit full adder logic circuit

One-Bit Half Adder

A one-bit half adder has only two inputs, A and B with no CIN. A and B inputs are added to generate SUM and COUT outputs. The symbolic representation of a half-adder is shown in Fig. 1.40.

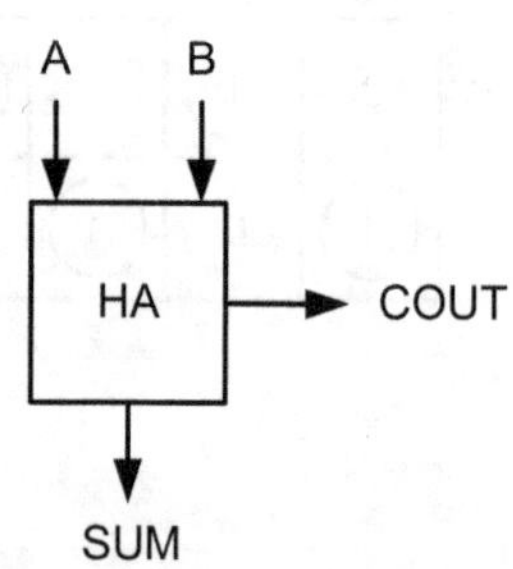

Fig. 1.40 One-bit half-adder symbol

The truth table given in Table 1.17 describes the functionality of the half adder.

Table 1.17 One-bit half-adder truth table

A	B	SUM	COUT
0	0	0	0
0	1	1	0
1	0	1	0
1	1	0	1

From the truth table, the POS expressions for SUM and COUT can be written as:

$$\mathrm{SUM} = \mathrm{A} \oplus \mathrm{B}$$

$$\mathrm{COUT} = \mathrm{A}\,.\,\mathrm{B}$$

Therefore, we can produce SUM and COUT circuits as shown in Fig. 1.41.

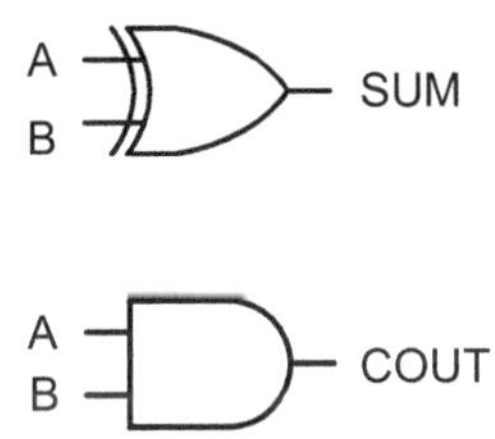

Fig. 1.41 One-bit half adder logic circuit

1.6 Combinational Mega Cells

Adders

One-bit full-adders can be cascaded serially to produce multiple-bit adder configurations. There are three basic adder types:

Ripple-Carry Adder
Carry-Look-Ahead (CLA) Adder
Carry-Select Adder

However, different hybrid adder topologies can be designed by combining these configurations. For the sake of simplicity, we will limit the number of bits to four and explain each topology in detail.

Ripple-Carry Adder

The ripple-carry adder is a cascaded configuration of multiple one-bit full adders. The circuit topology of a four-bit ripple carry adder is shown in Fig. 1.42. In this figure, the carry-out output of a one-bit full adder is connected to the carry-in input of the next full adder to propagate carry from CIN[0] to higher bits.

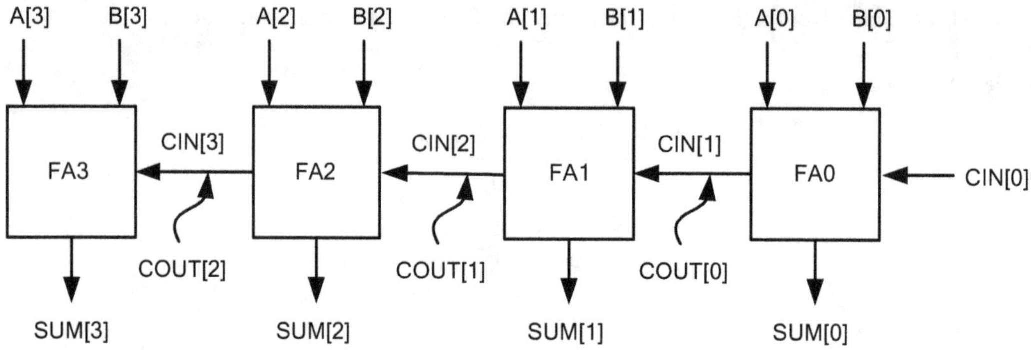

Fig. 1.42 Four-bit ripple-carry adder

For the 0th bit of this adder, we have:

$$\mathrm{SUM}[0] = \mathrm{A}[0] \oplus \mathrm{B}[0] \oplus \mathrm{CIN}[0]$$
$$\mathrm{COUT}[0] = \mathrm{CIN}[1] = \mathrm{A}[0]\,.\,\mathrm{B}[0] + \mathrm{CIN}[0]\,.\,(\mathrm{A}[0] + \mathrm{B}[0]) = \mathrm{G}[0] + \mathrm{P}[0]\,.\,\mathrm{CIN}[0]$$

where,

$\mathrm{G}[0] = \mathrm{A}[0]\,.\,\mathrm{B}[0]$ as the zeroth order generation term
$\mathrm{P}[0] = \mathrm{A}[0] + \mathrm{B}[0]$ as the zeroth order propagation term

For the first bit:

$$\mathrm{SUM}[1] = \mathrm{A}[1] \oplus \mathrm{B}[1] \oplus \mathrm{CIN}[1] = \mathrm{A}[1] \oplus \mathrm{B}[1] \oplus (\mathrm{G}[0] + \mathrm{P}[0]\,.\,\mathrm{CIN}[0])$$
$$\mathrm{COUT}[1] = \mathrm{CIN}[2] = \mathrm{G}[1] + \mathrm{P}[1]\,.\,\mathrm{CIN}[1]$$
$$= \mathrm{G}[1] + \mathrm{P}[1]\,.\,(\mathrm{G}[0] + \mathrm{P}[0]\,.\,\mathrm{CIN}[0]) = \mathrm{G}[1] + \mathrm{P}[1]\,.\,\mathrm{G}[0] + \mathrm{P}[1]\,.\,\mathrm{P}[0]\,.\,\mathrm{CIN}[0]$$

where,

$\mathrm{G}[1] = \mathrm{A}[1]\,.\,\mathrm{B}[1]$ as the first order generation term
$\mathrm{P}[1] = \mathrm{A}[1] + \mathrm{B}[1]$ as the first order propagation term

For the second bit:

$$\mathrm{SUM}[2] = \mathrm{A}[2] \oplus \mathrm{B}[2] \oplus \mathrm{CIN}[2] = \mathrm{A}[2] \oplus \mathrm{B}[2] \oplus \{\mathrm{G}[1] + \mathrm{P}[1]\,.\,(\mathrm{G}[0] + \mathrm{P}[0]\,.\,\mathrm{CIN}[0])\}$$
$$= \mathrm{A}[2] \oplus \mathrm{B}[2] \oplus (\mathrm{G}[1] + \mathrm{P}[1]\,.\,\mathrm{G}[0] + \mathrm{P}[1]\,.\,\mathrm{P}[0]\,.\,\mathrm{CIN}[0])$$
$$\mathrm{COUT}[2] = \mathrm{CIN}[3] = \mathrm{G}[2] + \mathrm{P}[2]\,.\,\mathrm{CIN}[2]$$
$$= \mathrm{G}[2] + \mathrm{P}[2]\,.\,\{\mathrm{G}[1] + \mathrm{P}[1]\,.\,(\mathrm{G}[0] + \mathrm{P}[0]\,.\,\mathrm{CIN}[0])\}$$
$$= \mathrm{G}[2] + \mathrm{P}[2]\,.\,\mathrm{G}[1] + \mathrm{P}[2]\,.\,\mathrm{P}[1]\,.\,\mathrm{G}[0] + \mathrm{P}[2]\,.\,\mathrm{P}[1]\,.\,\mathrm{P}[0]\,.\,\mathrm{CIN}[0]$$

where,

$G[2] = A[2] . B[2]$ as the second order generation term
$P[2] = A[2] + B[2]$ as the second order propagation term

And for the third bit:

$$\begin{aligned} SUM[3] &= A[3] \oplus B[3] \oplus CIN[3] = A[3] \oplus B[3] \oplus \{G[2] + P[2] . \{G[1] + P[1] . (G[0] + P[0] . CIN[0])\}\} \\ &= A[3] \oplus B[3] \oplus (G[2] + P[2] . G[1] + P[2] . P[1] . G[0] + P[2] . P[1] . P[0] . CIN[0]) \end{aligned}$$

These functional expressions of SUM and COUT also serve to estimate the maximum gate delays for each bit of the adder.

The circuit diagram in Fig. 1.43 explains the maximum delay path through each bit.

The maximum gate delay from A[0] or B[0] inputs to SUM[0] is $2T_{XOR2}$, where T_{XOR2} is a single two-input XOR gate delay.

The maximum gate delay from A[0] or B[0] to COUT[0] is $2T_{OR2} + T_{AND2}$ where T_{OR2} and T_{AND2} are two-input OR gate and two-input AND gate delays, respectively.

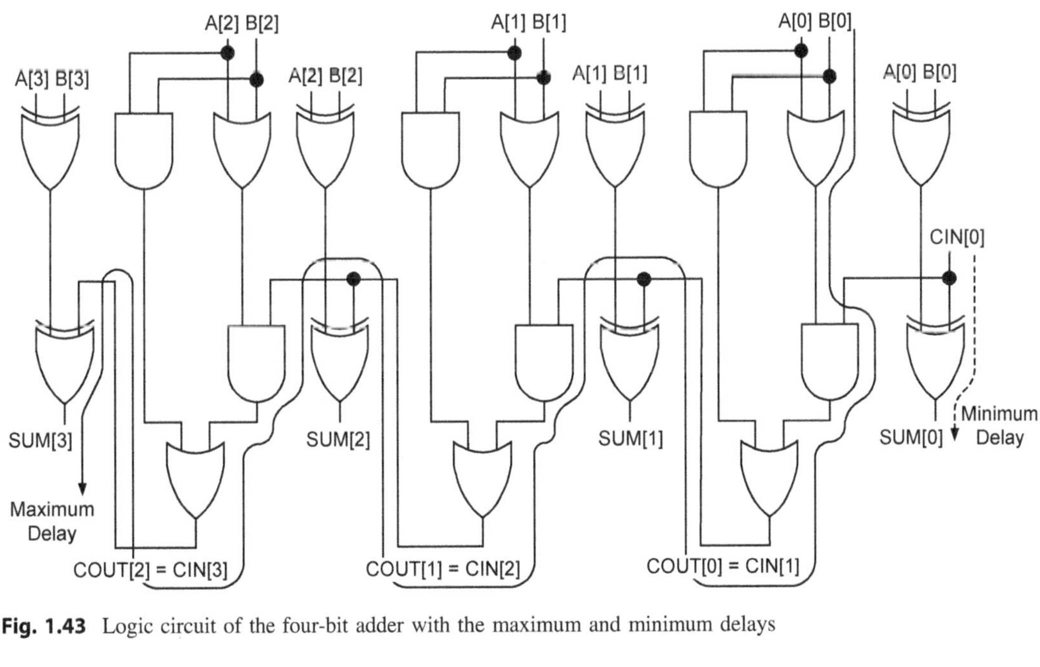

Fig. 1.43 Logic circuit of the four-bit adder with the maximum and minimum delays

The gate delay from A[1] or B[1] to SUM[1] is still $2T_{XOR2}$; however, the delay from A[0] or B[0] to SUM[1] is $2T_{OR2} + T_{AND2} + T_{XOR2}$, which is more than $2T_{XOR2}$ and must be considered the maximum gate delay for this particular bit position and for more significant bits.

The maximum gate delay from A[0] or B[0] to COUT[1] is $3T_{OR2} + 2T_{AND2}$. It may make more sense to expand the expression for COUT[1] as COUT[1] = G[1] + P[1] . G[0] + P[1] . P[0] . CIN [0], and figure out if the overall gate delay, $T_{OR2} + T_{AND3} + T_{OR3}$, is smaller compared to

$3T_{OR2} + 2T_{AND2}$. Here, T_{AND3} and T_{OR3} are single three-input AND and OR gate delays, respectively.

The maximum gate delay from A[0] or B[0] to SUM[2] is $3T_{OR2} + 2T_{AND2} + T_{XOR2}$. When the expression for SUM[2] is expanded as SUM[2] = A[2] ⊕ B[2] ⊕ (G[1] + P[1] . G[0] + P[1] . P[0] . CIN[0]), we see that this delay becomes $T_{OR2} + T_{AND3} + T_{OR3} + T_{XOR2}$, and may be smaller than the original delay if $T_{AND3} < 2T_{AND2}$ and $T_{OR3} < 2T_{OR2}$.

The maximum gate delay from A[0] or B[0] to COUT[2] is $4T_{OR2} + 3T_{AND2}$. When COUT[2] is expanded as COUT[2] = G[2] + P[2] . G[1] + P[2] . P[1] . G[0] + P[2] . P[1] . P[0] . CIN[0], the maximum delay becomes $T_{OR2} + T_{AND4} + T_{OR4}$, and may be smaller than the original delay if $T_{AND4} < 3T_{AND2}$ and $T_{OR4} < 3T_{OR2}$. Here, T_{AND4} and T_{OR4} are single four-input AND and OR gate delays, respectively.

Finally, the maximum delay from A [0] or B [0] to SUM [3] is $4T_{OR2} + 3T_{AND2} + T_{XOR2}$, which is also the maximum propagation delay for this adder. When the functional expression for SUM[3] is expanded as SUM[3] = A[3] ⊕ B[3] ⊕ (G[2] + P[2] . G[1] + P[2] . P[1] . G[0] + P[2] . P[1] . P[0] . CIN[0]), the total propagation delay becomes $T_{OR2} + T_{AND4} + T_{OR4} + T_{XOR2}$, and again it may be smaller compared to the original delay if $T_{AND4} < 3T_{AND2}$ and $T_{OR4} < 3T_{OR2}$.

Carry-Look-Ahead Adder

The idea behind carry-look-ahead (CLA) adders is to create a topology where carry-in bits to all one-bit full adders are available simultaneously. A four-bit CLA circuit topology is shown in Fig. 1.44. In this figure, the SUM output of a more significant bit does not have to wait until the carry bit ripples from the least significant bit position, but it gets computed after some logic delay. In reality, all carry-in signals are generated by complex combinational logic blocks called Carry-Look-Ahead (CLA) hook-ups, as shown in Fig. 1.44. Each CLA block adds a certain propagation delay on top of the two-input XOR gate delay to produce a SUM output.

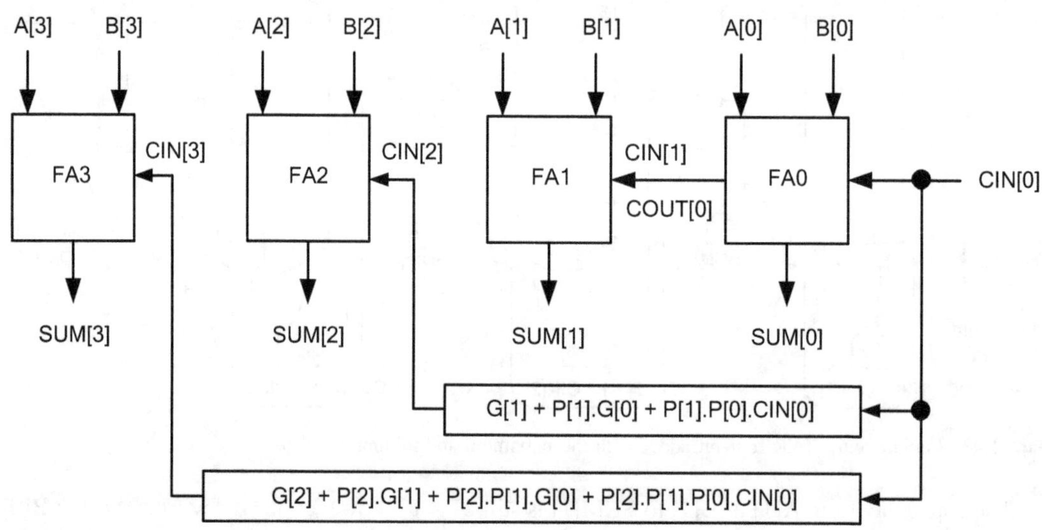

Fig. 1.44 A four-bit carry-look-ahead adder

The earlier SUM and CIN expressions derived for the ripple carry adder can be applied to the CLA adder to generate its functional equations.

Therefore,

$$
\begin{aligned}
SUM[0] &= A[0] \oplus B[0] \oplus CIN[0] \\
SUM[1] &= A[1] \oplus B[1] \oplus CIN[1] \\
SUM[2] &= A[2] \oplus B[2] \oplus CIN[2] \\
SUM[3] &= A[3] \oplus B[3] \oplus CIN[3]
\end{aligned}
$$

where,

$$
\begin{aligned}
CIN[1] &= G[0] + P[0] \,.\, CIN[0] \\
CIN[2] &= G[1] + P[1] \,.\, G[0] + P[1] \,.\, P[0] \,.\, CIN[0] \\
CIN[3] &= G[2] + P[2] \,.\, G[1] + P[2] \,.\, P[1] \,.\, G[0] + P[2] \,.\, P[1] \,.\, P[0] \,.\, CIN[0]
\end{aligned}
$$

Therefore, CIN[1] is generated by the COUT[0] function within FA0. However, CIN[2] and CIN [3] have to be produced by separate logic blocks in order to provide CIN signals for FA2 and FA3.

According to Fig. 1.44, once a valid CIN[0] becomes available, it takes successively longer times to generate valid signals for higher order SUM outputs due to the increasing complexity in CLA hook-ups.

Assume that T_{SUM0}, T_{SUM1}, T_{SUM2} and T_{SUM3} are the propagation delays corresponding to the bits 0, 1, 2 and 3 with respect to the CIN[0] signal. We can approximate $T_{SUM0} = T_{XOR2}$. To compute T_{SUM1}, we need to examine the expression for CIN[1]. In this expression, P[0] . CIN[0] produces a two-input AND gate delay, and G[0] + (P[0] . CIN[0]) produces a two-input OR gate delay to be added on top of T_{XOR2}. Therefore, T_{SUM1} becomes $T_{SUM1} = T_{AND2} + T_{OR2} + T_{XOR2}$. Similarly, the expressions for CIN[2] and CIN[3] produce $T_{SUM2} = T_{AND3} + T_{OR3} + T_{XOR2}$ and $T_{SUM3} = T_{AND4} + T_{OR4} + T_{XOR2}$, respectively.

The maximum propagation delay for this adder is, therefore, $T_{SUM3} = T_{AND4} + T_{OR4} + T_{XOR2}$.

Despite the CLA adder's advantage of being faster than the ripple-carry adder, in most cases the extra CLA logic blocks make this adder topology occupy a larger chip area if the number of adder bits is above eight.

Carry-Select Adder

Carry-Select Adders require two rows of identical adders. The identical adders can be as simple as two rows of identical ripple-carry adders or CLA adders depending on the design requirements. Figure 1.45 shows the circuit topology of a four-bit carry-select adder composed of two rows of ripple carry adders.

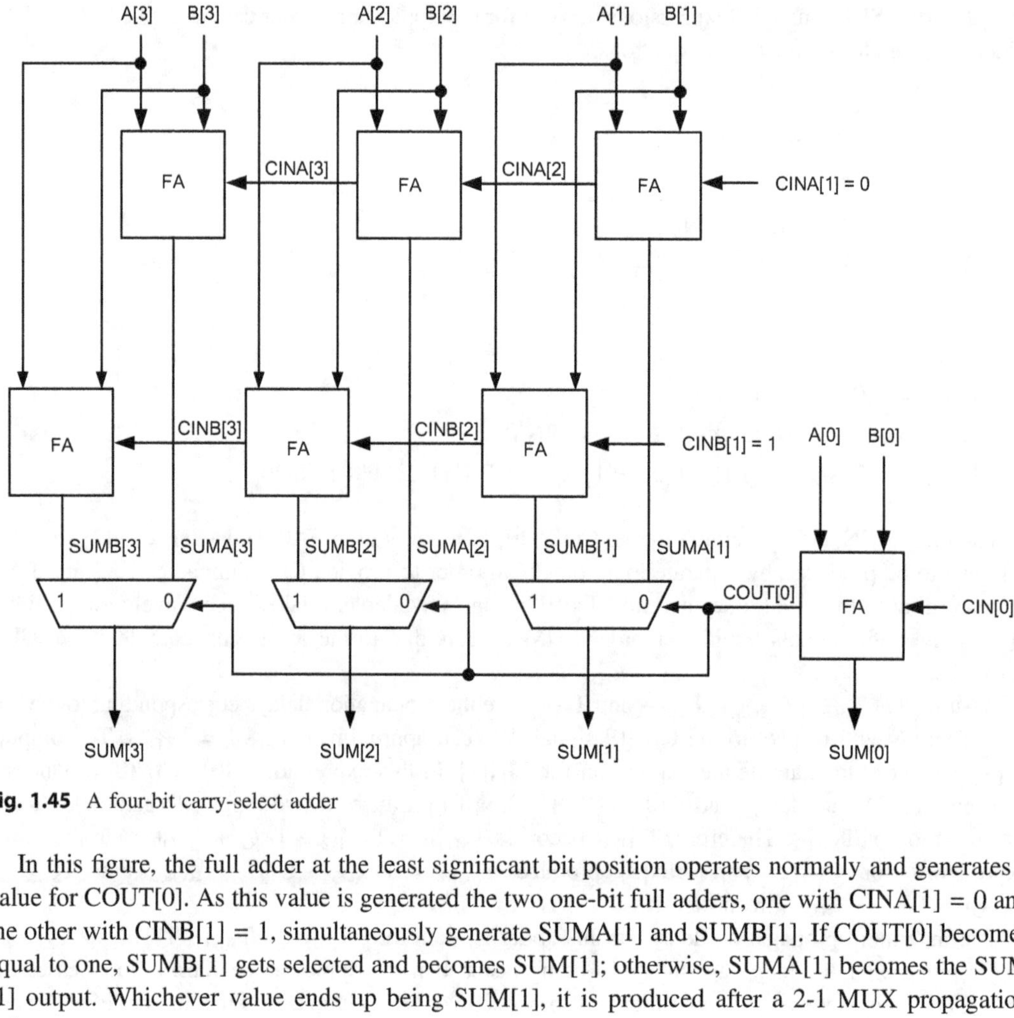

Fig. 1.45 A four-bit carry-select adder

In this figure, the full adder at the least significant bit position operates normally and generates a value for COUT[0]. As this value is generated the two one-bit full adders, one with CINA[1] = 0 and the other with CINB[1] = 1, simultaneously generate SUMA[1] and SUMB[1]. If COUT[0] becomes equal to one, SUMB[1] gets selected and becomes SUM[1]; otherwise, SUMA[1] becomes the SUM [1] output. Whichever value ends up being SUM[1], it is produced after a 2-1 MUX propagation delay.

However, we cannot say the same in generating SUM[2] and SUM[3] outputs in this figure. After producing SUM[1], carry ripples through both adders normally to generate SUM[2] and SUM[3]; hence, the speed advantage of having two rows of adders becomes negligible. Therefore, we must be careful when employing a carry-select scheme before designing an adder, as this method practically doubles the chip area.

Even though carry-select topology is ineffective in speeding up this particular four-bit adder, it may be advantageous if employed to an adder with greater number of bits in conjunction with another adder topology such as the CLA.

Example 1.14 Design a 32-bit carry-look-ahead adder. Compute the worst-case propagation delay in the circuit.

We need to be careful in dealing with the CLA hook-ups when generating higher order terms because the complexity of these logic blocks can "grow" exponentially in size while they may only provide marginal speed gain when compared to ripple-carry scheme.

Therefore, the first step of the design process is to separate the adder into eight-bit segments with full CLA hook-ups. The proposed topology is shown in Fig. 1.46.

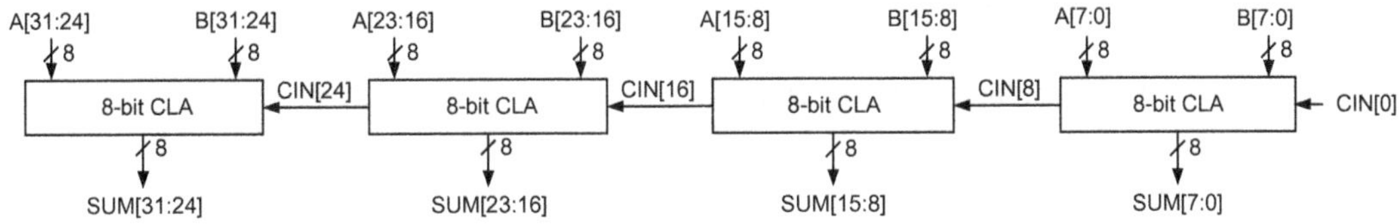

Fig. 1.46 A 32-bit carry-look-ahead topology

Each eight-bit CLA segment contains six CLA hook-ups from CLA0 to CLA5 as shown in Fig. 1.47.

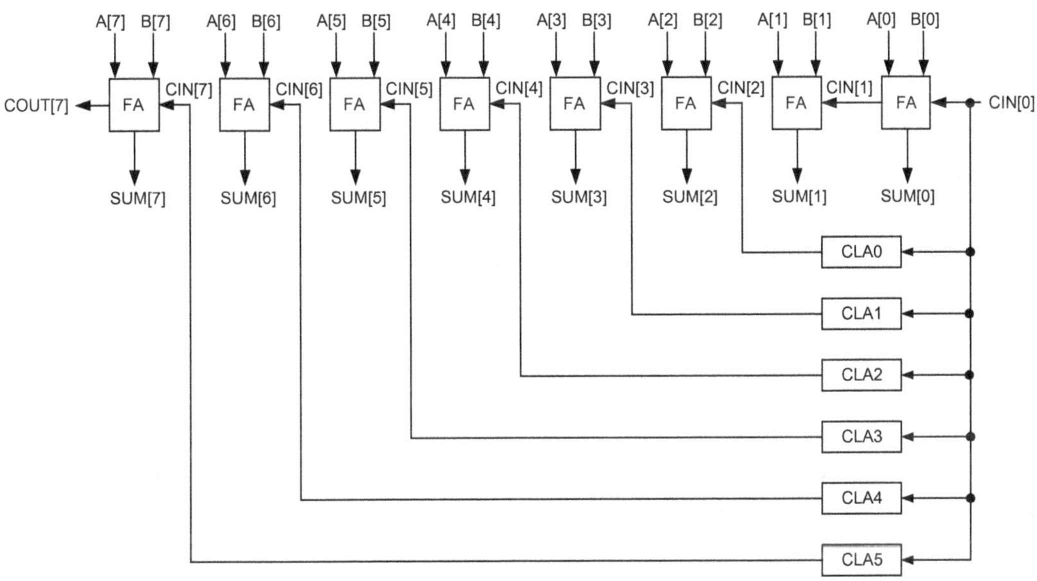

Fig. 1.47 An eight-bit segment of the carry-look ahead adder in Fig. 1.46

CIN and SUM expressions from bit 0 through bit 7 are given below.

$$SUM[0] = A[0] \oplus B[0] \oplus CIN[0]$$

$$SUM[1] = A[1] \oplus B[1] \oplus CIN[1]$$

where,

$$CIN[1] = G[0] + P[0] \,.\, CIN[0]$$

$$SUM[2] = A[2] \oplus B[2] \oplus CIN[2]$$

where,

$$\mathrm{CIN}[2] = \mathrm{G}[1] + \mathrm{P}[1]\,.\,\mathrm{G}[0] + \mathrm{P}[1]\,.\,\mathrm{P}[0]\,.\,\mathrm{CIN}[0]$$

$$\mathrm{SUM}[3] = \mathrm{A}[3] \oplus \mathrm{B}[3] \oplus \mathrm{CIN}[3]$$

where,

$$\mathrm{CIN}[3] = \mathrm{G}[2] + \mathrm{P}[2]\,.\,\mathrm{G}[1] + \mathrm{P}[2]\,.\,\mathrm{P}[1]\,.\,\mathrm{G}[0] + \mathrm{P}[2]\,.\,\mathrm{P}[1]\,.\,\mathrm{P}[0]\,.\,\mathrm{CIN}[0]$$

$$\mathrm{SUM}[4] = \mathrm{A}[4] \oplus \mathrm{B}[4] \oplus \mathrm{CIN}[4]$$

where,

$$\mathrm{CIN}[4] = \mathrm{G}[3] + \mathrm{P}[3]\,.\,(\mathrm{G}[2] + \mathrm{P}[2]\,.\,\mathrm{G}[1] + \mathrm{P}[2]\,.\,\mathrm{P}[1]\,.\,\mathrm{G}[0] + \mathrm{P}[2]\,.\,\mathrm{P}[1]\,.\,\mathrm{P}[0]\,.\,\mathrm{CIN}[0])$$

$$\mathrm{SUM}[5] = \mathrm{A}[5] \oplus \mathrm{B}[5] \oplus \mathrm{CIN}[5]$$

where,

$$\begin{aligned}\mathrm{CIN}[5] &= \mathrm{G}[4] + \mathrm{P}[4]\,.\,\{\mathrm{G}[3] + \mathrm{P}[3]\,.\,(\mathrm{G}[2] + \mathrm{P}[2]\,.\,\mathrm{G}[1] + \mathrm{P}[2]\,.\,\mathrm{P}[1]\,.\,\mathrm{G}[0] \\ &\quad + \mathrm{P}[2]\,.\,\mathrm{P}[1]\,.\,\mathrm{P}[0]\,.\,\mathrm{CIN}[0])\} \\ &= \mathrm{G}[4] + \mathrm{P}[4]\,.\,\mathrm{G}[3] + \mathrm{P}[4]\,.\,\mathrm{P}[3]\,.\,(\mathrm{G}[2] + \mathrm{P}[2]\,.\,\mathrm{G}[1] + \mathrm{P}[2]\,.\,\mathrm{P}[1]\,.\,\mathrm{G}[0] \\ &\quad + \mathrm{P}[2]\,.\,\mathrm{P}[1]\,.\,\mathrm{P}[0]\,.\,\mathrm{CIN}[0])\end{aligned}$$

$$\mathrm{SUM}[6] = \mathrm{A}[6] \oplus \mathrm{B}[6] \oplus \mathrm{CIN}[6]$$

where,

$$\begin{aligned}\mathrm{CIN}[6] &= \mathrm{G}[5] + \mathrm{P}[5]\,.\,\{\mathrm{G}[4] + \mathrm{P}[4]\,.\,\mathrm{G}[3] + \mathrm{P}[4]\,.\,\mathrm{P}[3]\,.\,(\mathrm{G}[2] + \mathrm{P}[2]\,.\,\mathrm{G}[1] + \mathrm{P}[2]\,.\,\mathrm{P}[1]\,.\,\mathrm{G}[0] \\ &\quad + \mathrm{P}[2]\,.\,\mathrm{P}[1]\,.\,\mathrm{P}[0]\,.\,\mathrm{CIN}[0])\} \\ &= \mathrm{G}[5] + \mathrm{P}[5]\,.\,\mathrm{G}[4] + \mathrm{P}[5]\,.\,\mathrm{P}[4]\,.\,\mathrm{G}[3] + \mathrm{P}[5]\,.\,\mathrm{P}[4]\,.\,\mathrm{P}[3]\,.\,(\mathrm{G}[2] + \mathrm{P}[2]\,.\,\mathrm{G}[1] \\ &\quad + \mathrm{P}[2]\,.\,\mathrm{P}[1]\,.\,\mathrm{G}[0] + \mathrm{P}[2]\,.\,\mathrm{P}[1]\,.\,\mathrm{P}[0]\,.\,\mathrm{CIN}[0])\end{aligned}$$

$$\mathrm{SUM}[7] = \mathrm{A}[7] \oplus \mathrm{B}[7] \oplus \mathrm{CIN}[7]$$

where,

$$\begin{aligned}\mathrm{CIN}[7] &= \mathrm{G}[6] + \mathrm{P}[6]\,.\,\{\mathrm{G}[5] + \mathrm{P}[5]\,.\,\mathrm{G}[4] + \mathrm{P}[5]\,.\,\mathrm{P}[4]\,.\,\mathrm{G}[3] + \mathrm{P}[5]\,.\,\mathrm{P}[4]\,.\,\mathrm{P}[3]\,.\,(\mathrm{G}[2] \\ &\quad + \mathrm{P}[2]\,.\,\mathrm{G}[1] + \mathrm{P}[2]\,.\,\mathrm{P}[1]\,.\,\mathrm{G}[0] + \mathrm{P}[2]\,.\,\mathrm{P}[1]\,.\,\mathrm{P}[0]\,.\,\mathrm{CIN}[0])\}\end{aligned}$$

And finally,

$$\begin{aligned}\mathrm{COUT}[7] = \mathrm{CIN}[8] &= \mathrm{G}[7] + \mathrm{P}[7]\,.\,\{\mathrm{G}[6] + \mathrm{P}[6]\,.\,\{\mathrm{G}[5] + \mathrm{P}[5]\,.\,\mathrm{G}[4] + \mathrm{P}[5]\,.\,\mathrm{P}[4]\,.\,\mathrm{G}[3] \\ &\quad + \mathrm{P}[5]\,.\,\mathrm{P}[4]\,.\,\mathrm{P}[3]\,.\,(\mathrm{G}[2] + \mathrm{P}[2]\,.\,\mathrm{G}[1] + \mathrm{P}[2]\,.\,\mathrm{P}[1]\,.\,\mathrm{G}[0] + \mathrm{P}[2]\,.\,\mathrm{P}[1]\,.\,\mathrm{P}[0]\,.\,\mathrm{CIN}[0])\}\} \\ &= \mathrm{G}[7] + \mathrm{P}[7]\,.\,\mathrm{G}[6] + \mathrm{P}[7]\,.\,\mathrm{P}[6]\,.\,\{\mathrm{G}[5] + \mathrm{P}[5]\,.\,\mathrm{G}[4] + \mathrm{P}[5]\,.\,\mathrm{P}[4]\,.\,\mathrm{G}[3] \\ &\quad + \mathrm{P}[5]\,.\,\mathrm{P}[4]\,.\,\mathrm{P}[3]\,.\,(\mathrm{G}[2] + \mathrm{P}[2]\,.\,\mathrm{G}[1] + \mathrm{P}[2]\,.\,\mathrm{P}[1]\,.\,\mathrm{G}[0] + \mathrm{P}[2]\,.\,\mathrm{P}[1]\,.\,\mathrm{P}[0]\,.\,\mathrm{CIN}[0])\}\end{aligned}$$

In these derivations, particular attention was paid to limit the number of inputs to four in all AND and OR gates since larger gate inputs are counterproductive in reducing the overall propagation delay.

From these functional expressions, maximum propagation delays for SUM[7] and COUT[7] are estimated using the longest logic strings in Fig. 1.48.

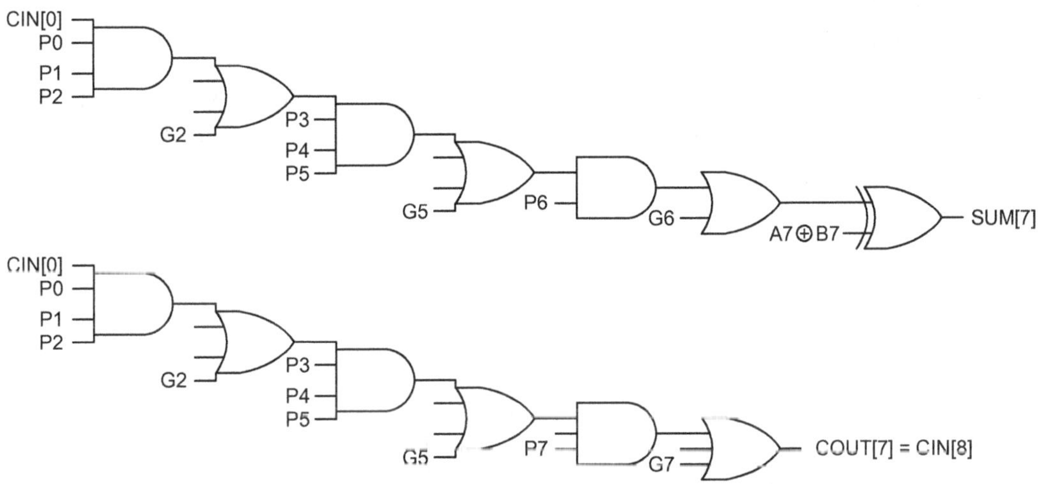

Fig. 1.48 Propagation delay estimation of the eight-bit carry-look-ahead adder in Fig. 1.47

For SUM[7], the minterm, P[2] . P[1] . P[0] . CIN[0], generates the first four-input AND gate. This is followed by a four-input OR-gate whose minterms are G[2], P[2] . G[1], P[2] . P[1] . G[0], and P[2] . P[1] . P[0] . CIN[0]. This gate is followed by a four-input AND, four-input OR, two-input AND, two-input OR and two-input XOR-gates in successive order. The entire string creates a propagation delay of $T_{SUM7} = 2(T_{AND4} + T_{OR4}) + T_{AND2} + T_{OR2} + T_{XOR2}$ from CIN[0] to SUM[7].

For COUT[7], the longest propagation delay is between CIN[0] to COUT[7] as shown in Fig. 1.48, and it is equal to $T_{COUT7} = 2(T_{AND4} + T_{OR4}) + T_{AND3} + T_{OR3}$. The delays for the rest of the circuit in Fig. 1.46 become easy to determine since the longest propagation delays have already been evaluated.

The delay from CIN[0] to SUM[15], T_{SUM15}, simply becomes equal to the sum of T_{COUT7} and T_{SUM7}. In other words, $T_{SUM15} = 4(T_{AND4} + T_{OR4}) + T_{AND3} + T_{OR3} + T_{AND2} + T_{OR2} + T_{XOR2}$. Similarly, the delay from CIN[0] to COUT[15], T_{COUT15}, is equal to $T_{COUT15} = 4(T_{AND4} + T_{OR4}) + 2(T_{AND3} + T_{OR3})$.

The remaining delays are evaluated in the same way and lead to the longest propagation delay in this circuit, which is from CIN[0] to SUM[31], T_{SUM31}.

Thus,

$$T_{SUM31} = 3[2(T_{AND4} + T_{OR4}) + T_{AND3} + T_{OR3}] + 2(T_{AND4} + T_{OR4}) + T_{AND2} + T_{OR2} + T_{XOR2}$$

or

$$T_{SUM31} = 8(T_{AND4} + T_{OR4}) + 3(T_{AND3} + T_{OR3}) + T_{AND2} + T_{OR2} + T_{XOR2}$$

Example 1.15 Design a 32-bit hybrid carry-select/carry-look-ahead adder. Compute the worst-case propagation delay in the circuit.

Large adders are where the carry-select scheme shines! This is a classical example in which the maximum propagation delay is reduced considerably compared to the CLA scheme examined in Example 1.14.

As mentioned earlier, a twin set of an adder configuration is required by the carry-select scheme. The adders can be ripple-carry, carry-look-ahead or the combination of the two.

In this example, the 32-bit adder is again divided in eight-bit segments where each segment consists of a full CLA adder as shown in Fig. 1.49. The first segment, CLA-0, is a single unit which produces COUT[7] with full CLA hook-ups. The rest of the eight-bit segments are mirror images of each other and they are either named A-segments (A1, A2 and A3) or B-segments (B1, B2 and B3).

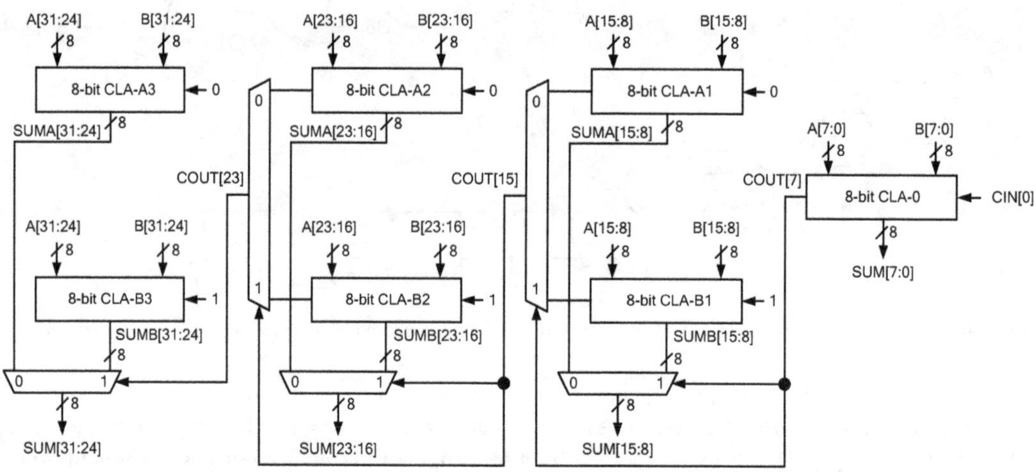

Fig. 1.49 A 32-bit carry-look-ahead/carry-select adder

As the CLA-0 generates a valid COUT[7], the CLA-A1 and CLA-B1 simultaneously generate COUTA[15] and COUTB[15]. When COUT[7] finally forms, it selects either COUTA[15] or COUTB[15] depending on the value of CIN[0]. This segment produces COUT[15] and SUM[15:8].

COUT[15], on the other hand, is used to select between COUTA[23] and COUTB[23], both of which have already been formed when COUT[15] arrives at the 2-1 MUX as a control input. COUT [15] also selects between SUMA[23:16] and SUMB[23:16] to determine the correct SUM[23:16].

Similarly, COUT[23] is used as a control input to select between SUMA[31:24] and SUMB [31:24]. If there is a need for COUT[31], COUT[23] can serve to determine the value of COUT[31].

The maximum propagation delay for the 32-bit carry-select/CLA adder can be found using the logic string in Fig. 1.50. The first section of this string from CIN[0] to COUT[7] is identical to the eight-bit CLA carry delay in Fig. 1.48. There are three cascaded MUX stages which correspond to the selection of COUT[15], COUT[23], and SUM[31].

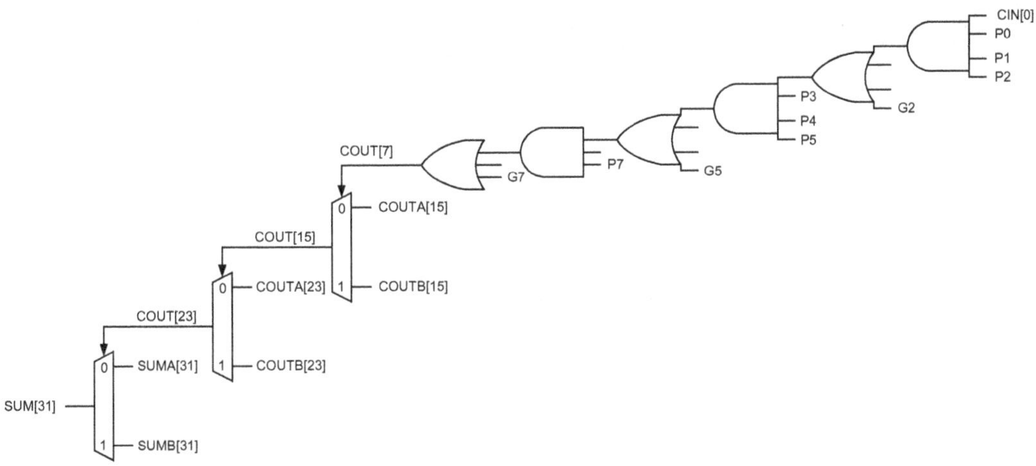

Fig. 1.50 Maximum delay propagation of the 32-bit adder in Fig. 1.49

Considering that a 2-1 MUX propagation delay consists of a two-input AND gate followed by a two-input OR gate, we obtain the following maximum delay for this 32-bit adder:

$$T_{SUM31} = 2(T_{AND4} + T_{OR4}) + T_{AND3} + T_{OR3} + 3(T_{AND2} + T_{OR2})$$

Considering the maximum propagation delay in Example 1.14, this delay is shorter by at least 6 $(T_{AND4} + T_{OR4})$. Larger carry-select/carry-look-ahead adder schemes provide greater speed benefits at the cost of approximately doubling the adder area.

Subtractors

Subtraction is performed by a technique called twos (2s) complement addition. Twos complement addition first requires complementing one of the adder inputs (1s complement) and then adding 1 to the least significant bit.

Example 1.16 Form −4 using 2s complement addition using four bits.

A negative number is created by first inverting every bit of +4 (1s complement representation) and then adding 1 to it. +4 is equal to 0100 in four-bit binary form.

Its 1s complement is 1011. Its 2s complement is 1011 + 0001 = 1100.

Therefore, logic 0 signifies a positive sign, and logic 1 signifies a negative sign at the most significant bit position.

Example 1.17 Add +4 to −4 using 2s complement addition

+4 = 0100 in binary form
−4 = 1100 in 2s complement form of +4.
Perform +4–4 = 0100 + 1100 = 1 0000

Where, logic 1 at the overflow bit position is neglected in a four-bit binary format.

Therefore, we obtain 0000 = 0 as expected.

Subtractors function according to 2s complement addition. We need to form 1s complement of the adder input to be subtracted and use CIN[0] = 1 at the adder's least significant bit position to perform subtraction.

Figure 1.51 illustrates the topology of a 32-bit subtractor where input B is complemented and CIN [0] is tied to logic 1 to satisfy 2s complement addition and produce A − B.

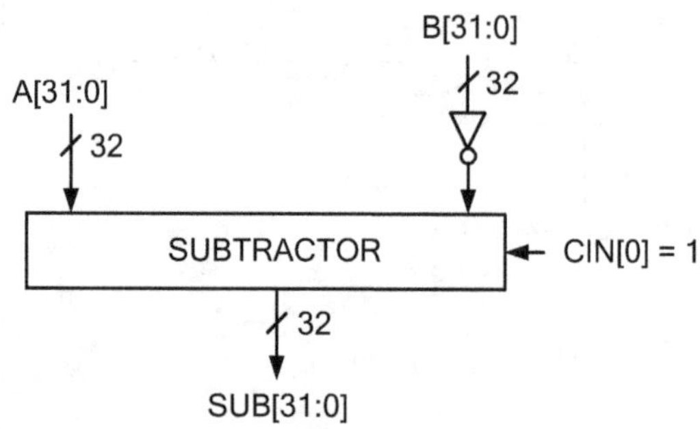

Fig. 1.51 A symbolic representation of a 32-bit subtractor

Shifters

There are two commonly used shifters in logic design:

Linear shifters
Barrel shifters

Linear Shifters

A linear shifter shifts its inputs by a number of bits to the right or to the left, and routes the result to its output.

Example 1.18 Design a four-bit linear shifter that shifts its inputs to the left by one bit and produces logic 0 at the least significant output bit when SHIFT = 1. When SHIFT = 0, the shifter routes each input directly to the corresponding output.

The logic diagram for this shifter is given in Fig. 1.52. In this figure, each input is connected to the port 0 terminal of the 2-1 MUX as well as the port 1 terminal of the next MUX at the higher bit position. Therefore, when SHIFT = 1, logic 0, IN[0], IN[1], and IN[2] are routed through port 1 terminal of each 2-1 MUX and become OUT[0], OUT[1], OUT[2], and OUT[3], respectively. When SHIFT = 0, each input goes through port 0 terminal of the corresponding 2-1 MUX and becomes the shifter output.

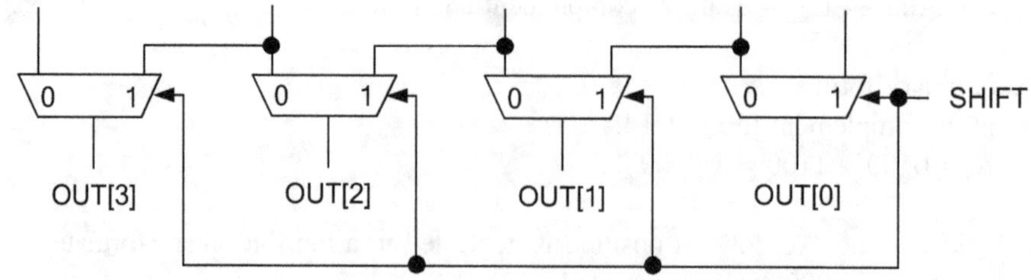

Fig. 1.52 Four-bit linear shifter

Barrel Shifters

Barrel shifters rotate their inputs either in clockwise or counterclockwise direction by a number of bits but preserve all their inputs when generating an output.

Example 1.19 Design a four-bit barrel shifter that rotates its inputs in a clockwise direction by one bit when SHIFT = 1. When SHIFT = 0, the shifter routes each one of its four inputs to its corresponding output.

The logic diagram for this shifter is given in Fig. 1.53. The only difference between this circuit and the linear shifter in Fig. 1.52 is the removal of logic 0 from the least significant bit, and connecting this input to the IN[3] pin instead. Consequently, this leads to OUT[0] = IN[3], OUT[1] = IN[0], OUT[2] = IN[1] and OUT[3] = IN[2] when SHIFT = 1, and OUT[0] = IN[0], OUT[1] = IN[1], OUT[2] = IN[2] and OUT[3] = IN[3] when SHIFT = 0.

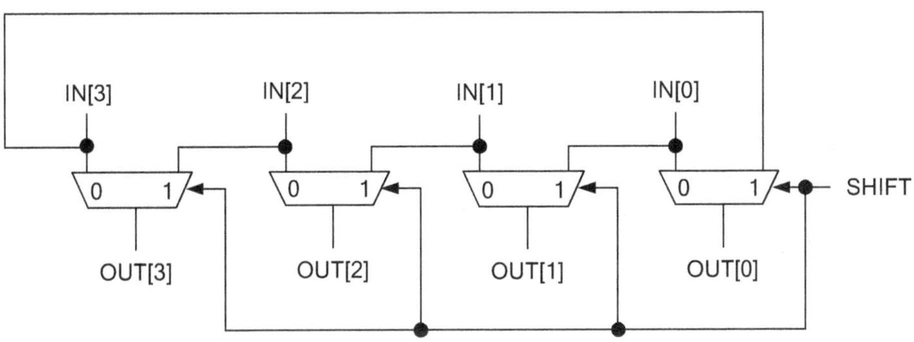

Fig. 1.53 Four-bit barrel shifter

Example 1.20 Design a four-bit barrel shifter that rotates its inputs clockwise by one or two bits.

First, there must be three control inputs specifying "no shift", "shift 1 bit" and "shift 2 bits". This requires a two control-bit input, SHIFT[1:0], as shown in Table 1.18. All assignments in this table are done arbitrarily. However, it makes sense to assign a "No shift" to SHIFT[1:0] = 0, "Shift 1 bit" to SHIFT[1:0] = 1 and "Shift 2 bits" to SHIFT[1:0] = 2 for actual rotation amount.

Table 1.18 A four-bit barrel shifter truth table

SHIFT[1]	SHIFT[0]	OPERATION	OUT[3]	OUT[2]	OUT[1]	OUT[0]
0	0	No shift	IN[3]	IN[2]	IN[1]	IN[0]
0	1	Shift 1 bit	IN[2]	IN[1]	IN[0]	IN[3]
1	0	Shift 2 bits	IN[1]	IN[0]	IN[3]	IN[2]
1	1	No shift	IN[3]	IN[2]	IN[1]	IN[0]

According to this table, if there is no shift, each input bit is simply routed to its own output. If "shift 1 bit" input is active, then each input is routed to the neighboring output at the next significant bit position. In other words, IN[3] rotates clockwise and becomes OUT[0]; IN[0], IN[1] and IN[2] shift 1 bit to the left and become OUT[1], OUT[2] and OUT[3], respectively. If "shift 2 bits" becomes active, then each input is routed to the output of the neighboring bit which is two significant bits higher. This gives the impression that all input bits have been rotated twice before being routed to the output; Thus, OUT[0] = IN[2], OUT[1] = IN[3], OUT[2] = IN[0] and OUT[3] = IN[1].

Therefore, using Table 1.18, we can conclude the logic diagram in Fig. 1.54.

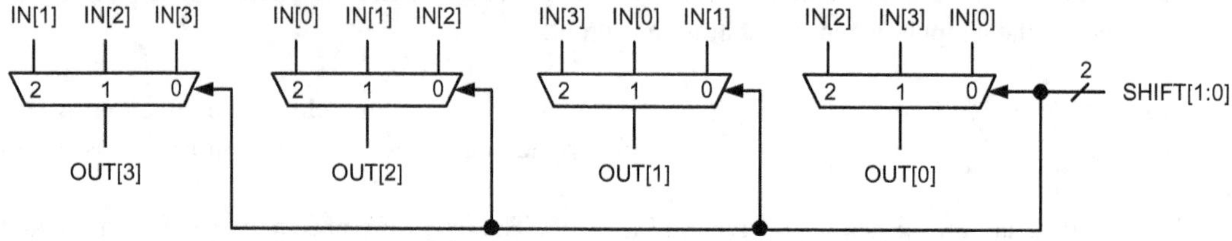

Fig. 1.54 Logic diagram of the barrel shifter in Table 1.18

A more detailed view of Fig. 1.54 is given in Fig. 1.55.

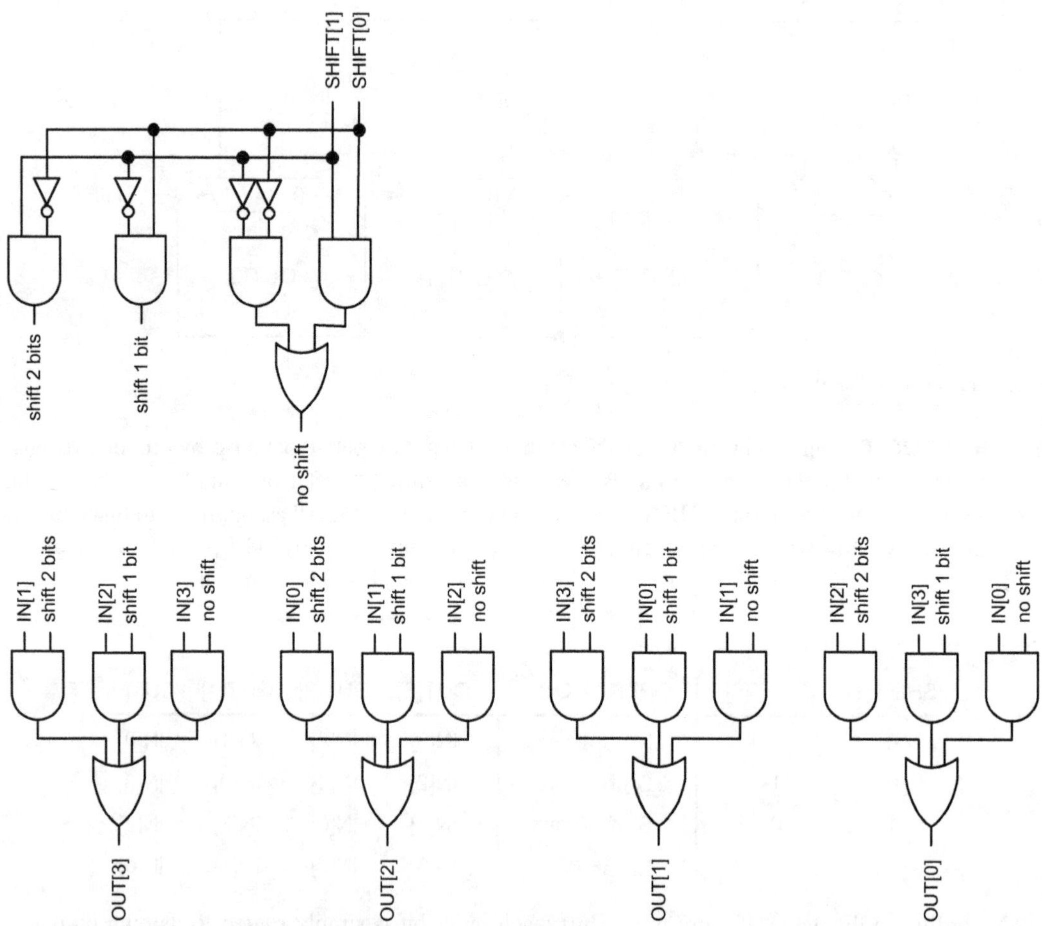

Fig. 1.55 Logic circuit of the barrel shifter in Fig. 1.54

Multipliers

There are two types of multipliers:

Array multiplier
Booth multiplier

An array multiplier is relatively simple to design, but it requires a large number of gates. A Booth multiplier, on the other hand, requires fewer gates but its implementation follows a rather lengthy algorithm.

Array Multiplier

Similar to our everyday hand multiplication method, an array multiplier generates all partial products before summing each column in the partial product tree to obtain a result. This scheme is explained in Fig. 1.56 for a four-bit array multiplier.

				A[3]	A[2]	A[1]	A[0]	MULTIPLICAND
				× B[3]	B[2]	B[1]	B[0]	MULTIPLIER
				B[0].A[3]	B[0].A[2]	B[0].A[1]	B[0].A[0]	0th PARTIAL PRODUCT
			B[1].A[3]	B[1].A[2]	B[1].A[1]	B[1].A[0]		1st PARTIAL PRODUCT
		B[2].A[3]	B[2].A[2]	B[2].A[1]	B[2].A[0]			2nd PARTIAL PRODUCT
+	B[3].A[3]	B[3].A[2]	B[3].A[1]	B[3].A[0]				3rd PARTIAL PRODUCT
SUM[7]	SUM[6]	SUM[5]	SUM[4]	SUM[3]	SUM[2]	SUM[1]	SUM[0]	SUM OUTPUT

Fig. 1.56 4 × 4 array multiplier algorithm

The rules of partial product generation are as follows:

1. The zeroth partial product aligns with multiplicand and multiplier bit columns.
2. Each partial product is shifted one bit to the left with respect to the previous one once it is created.
3. Each partial product is the exact replica of the multiplicand if the multiplier bit is one. Otherwise, it is deleted.

Example 1.21 Multiply 1101 and 1001 according to the rules of array multiplication.

Suppose 1101 is a multiplicand and 1001 is a multiplier. Then, for a four-bit multiplier four partial products are formed. The bits in each column of the partial product are added successively. Carry bits are propagated to more significant bit positions. This process is illustrated in Fig. 1.57.

				1	1	0	1	MULTIPLICAND
			×	1	0	0	1	MULTIPLIER
				1	1	0	1	0th PARTIAL PRODUCT
			0	0	0	0		1st PARTIAL PRODUCT
		0	0	0	0			2nd PARTIAL PRODUCT
+	1	1	0	1				3rd PARTIAL PRODUCT
	1	1	1	0	1	0	1	SUM OUTPUT

Fig. 1.57 4 × 4 array multiplier algorithm example

Example 1.22 Design the partial product tree for a four-bit array multiplier.

Following the convention in Fig. 1.56 and the rules of partial product generation for an array multiplier, we can implement the partial product tree as shown in Fig. 1.58.

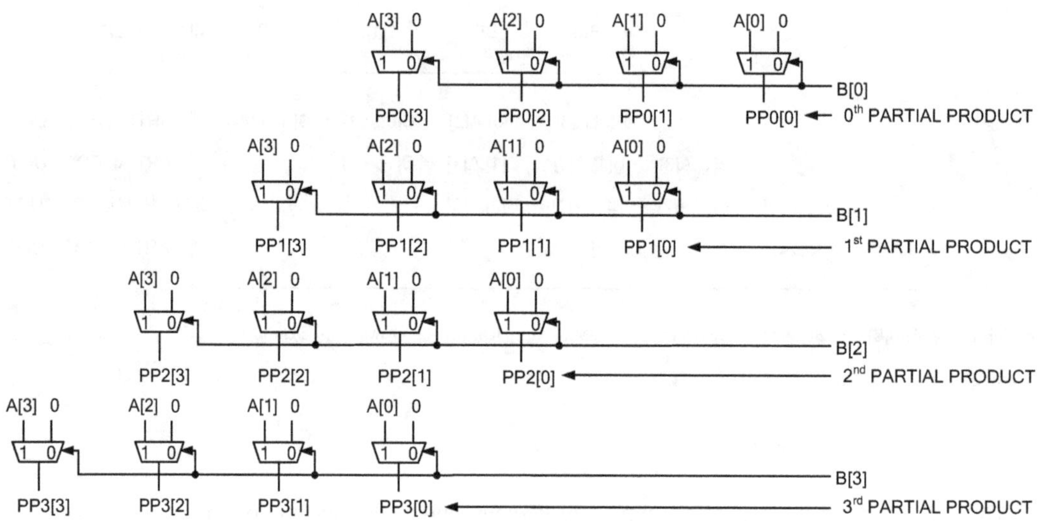

Fig. 1.58 4 × 4 array multiplier bit selector tree

In this figure, partial product elements of the zeroth partial product, B[0] . A[3], B[0] . A[2], B[0] . A[1] and B[0] . A[0], are replaced by PP0[3:0] for purposes of better illustration. Similarly, PP1[3:0], PP2[3:0] and PP3[3:0] are the new partial product outputs corresponding to the rows one, two and three.

Example 1.23 Design a full adder tree responsible for adding every partial product in the partial product tree for a four-bit array multiplier.

After generating partial products, the next design step is to add partial product elements column by column to generate SUM outputs, SUM[7:0], while propagating carry bits for higher order columns. Following the naming convention in Fig. 1.58, all 16 partial product elements are fed to the full adder tree in Fig. 1.59. The box outlined by dashed lines shows how the carry propagation takes place from one column to the next.

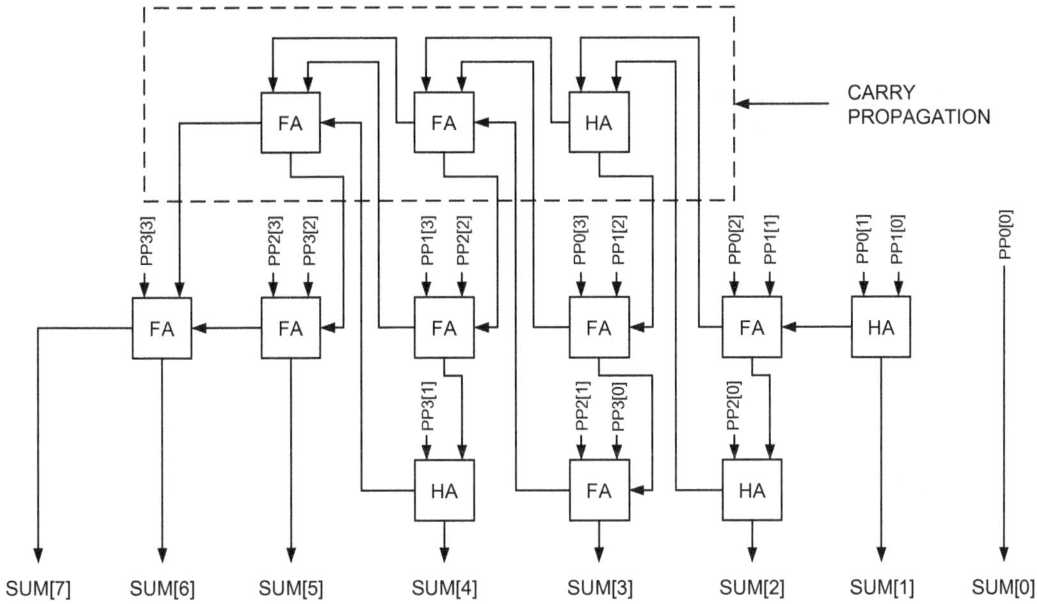

Fig. 1.59 An eight-bit propagate adder for the bit selector tree in Fig. 1.58

Booth Multiplier

The Booth multiplier halves the number of partial products using a lengthy algorithm given below.

Assume that the product of two binary integers, X and Y, forms P = X . Y where X is a multiplicand and Y is a multiplier.

In binary form, Y is expressed in powers of two:

$$Y = \sum_{k=0}^{n-1} y_k 2^k$$

where, the most significant bit, y_{n-1}, corresponds to the sign bit. When $y_{n-1} = 0$, Y is considered a positive number, otherwise it is a negative number as mentioned earlier in the 2s complement representation of integers.

In this section, we examine the Booth multiplication algorithm when Y is both positive and negative numbers.

CASE 1: $Y > 0$, thus $y_{n-1} = 0$.

We can express a *k*th term of Y as:

$$y_k 2^k = (2y_k - y_k)2^k = \left(2y_k - \frac{1}{2}2y_k\right)2^k$$
$$= y_k 2^{k+1} - 2y_k 2^{k-1}$$

Thus:

$$
\begin{aligned}
Y &= \sum_{k=0}^{n-1} y_k 2^k = \sum_{k=0}^{n-1} \left(y_k 2^{k+1} - 2y_k 2^{k-1}\right) \\
&= y_{n-1}2^n - 2y_{n-1}2^{n-2} \\
&\quad + y_{n-2}2^{n-2} \\
&\quad + y_{n-3}2^{n-2} - 2y_{n-3}2^{n-4} \\
&\quad + y_{n-4}2^{n-4} \\
&\quad \cdot \\
&\quad \cdot \\
&\quad \cdot \\
&\quad + y_3 2^4 - 2y_3 2^2 \\
&\quad + y_2 2^2 \\
&\quad + y_1 2^2 - 2y_1 2^0 \\
&\quad + y_0 2^0
\end{aligned}
$$

Regrouping the terms of the same power yields:

$$
\begin{aligned}
Y &= y_{n-1}2^n + 2^{n-2}(-2y_{n-1} + y_{n-2} + y_{n-3}) \\
&\quad + 2^{n-4}(-2y_{n-3} + y_{n-4} + y_{n-5}) \\
&\quad \cdot \\
&\quad \cdot \\
&\quad \cdot \\
&\quad + 2^2(-2y_3 + y_2 + y_1) \\
&\quad + 2^0(-2y_1 + y_0 + y_{-1})
\end{aligned}
$$

But, $y_{n-1} = 0$ and $y_{-1} = 0$ since $Y > 0$

$$
\begin{aligned}
Y &= 2^{n-2}(-2y_{n-1} + y_{n-2} + y_{n-3}) \\
&\quad + 2^{n-4}(-2y_{n-3} + y_{n-4} + y_{n-5}) \\
&\quad \cdot \\
&\quad \cdot \\
&\quad \cdot \\
&\quad + 2^2(-2y_3 + y_2 + y_1) \\
&\quad + 2^0(-2y_1 + y_0 + y_{-1})
\end{aligned}
$$

Now, let's define a new set of coefficients:

$$
z_k = -2y_{k+1} + y_k + y_{k-1}
$$

Then:
$Y = \sum_{k=0}^{n-2} z_k 2^k$ where, k = 0, 2, …, (n − 4), (n − 2).
When X is multiplied by Y, one obtains:

$$P = X \,.\, Y = \sum_{k=0}^{n-2} (z_k.X) \,.\, 2^k$$

where, the number of partial products in the product term, P, is reduced by half.

Each $z_k = -2y_{k+1} + y_k + y_{k-1}$ depends on the value of three adjacent bits, y_{k+1}, y_k and y_{k-1}. This is tabulated in Table 1.19.

Table 1.19 Booth encoder truth table

y_{k+1}	y_k	y_{k-1}	z_k
0	0	0	0
0	0	1	1
0	1	0	1
0	1	1	2
1	0	0	-2
1	0	1	-1
1	1	0	-1
1	1	1	0

Therefore, each partial product, $(z_k.X)$, in P is:

$$(z_k.X) = 0, +X, -X, +2X, -2X$$

These partial products can easily be obtained by the following methods:

For $z_k.X = 0$, all multiplicand bits are replaced by 0.
For $z_k.X = +X$, all multiplicand bits are multiplied by one.
For, $z_k.X = +2X$, all multiplicand bits are shifted left by one bit
For $z_k.X = -X$, the multiplicand is 2s complemented.
For $z_k.X = -2X$, all multiplicand bits are shifted left by one bit to form +2X, and then 2s complemented to form −2X.

Now, the time has come to investigate when Y is negative.

CASE 2: $Y < 0$, thus $y_{n-1} = -1$

The first step is to sign-extend Y by one bit. Sign extension does not change the actual value of Y but increases its terms from n to (n + 1). Thus:

$$Y = -2^n + \sum_{k=0}^{n-1} y_k 2^k$$

Let us consider

$$\sum_{k=0}^{n-1} 2^k = 1 + 2 + 2^2 + \cdots + 2^{n-1} = f$$

Then,

$$2f = 2 + 2^2 + 2^3 + \cdots + 2^n$$

Thus:

$$f = 2f - f = 2^n - 1$$

or

$$2^n = 1 + f = 1 + \sum_{k=0}^{n-1} 2^k$$

Substituting -2^n into Y yields:

$$Y = -2^n + \sum_{k=0}^{n-1} y_k \,.\, 2^k = -1 - \sum_{k=0}^{n-1} 2^k + \sum_{k=0}^{n-1} y_k \,.\, 2^k = -1 + \sum_{k=0}^{n-1} (y_k - 1) 2^k$$

or

$$-Y = 1 + \sum_{k=0}^{n-1} (1 - y_k) 2^k$$

However, $(1 - y_k) = 1$ when $y_k = 0$, and $(1 - y_k) = 0$ when $y_k = 1$. That means:

$$(1 - y_k) = \overline{y_k}$$

Thus:

$$-Y = 1 + \sum_{k=0}^{n-1} \overline{y_k} \,.\, 2^k$$

The same mathematical manipulation applied to $y_k \,.\, 2^k$ in CASE 1 can also be applied to $\overline{y_k} \,.\, 2^k$.

$$\overline{y_k} \,.\, 2^k = (2.\overline{y_k} - \overline{y_k}) 2^k = \overline{y_k} \,.\, 2 \,.\, 2^k - \overline{y_k} \frac{1}{2} \,.\, 2 \,.\, 2^k$$

Therefore,

$$
\begin{aligned}
\sum_{k=0}^{n-1} \overline{y_k} &= \sum_{k=0}^{n-1} \left(\overline{y_k}.\,2\,.\,2^k - \overline{y^k}.\frac{1}{2}.\,2\,.\,2^k \right) \\
&= \sum_{k=0}^{n-1} \overline{y_k}.\,2^{k+1} - \overline{y_k}.\,2\,.\,2^{k-1} \\
&= \overline{y_{n-1}}.\,2^n - \overline{y_{n-1}}.\,2\,.\,2^{n-2} \\
&\quad + \overline{y_{n-2}}.\,2^{n-2} \\
&\quad + \overline{y_{n-3}}.\,2^{n-2} - \overline{y_{n-3}}.\,2\,.\,2^{n-4} \\
&\quad + \overline{y_{n-4}}.\,2^{n-4} \\
&\quad . \\
&\quad . \\
&\quad . \\
&\quad + \overline{y_3}.2^4 - \overline{y_3}.\,2.\,2^2 \\
&\quad + \overline{y_2}.\,2^2 \\
&\quad + \overline{y_1}.\,2^2 - \overline{y_1}.\,2\,.\,2^0 \\
&\quad + 2^0.\overline{y_0} \\
&= \overline{y_{n-1}}.\,2^n \\
&\quad + (-2.\overline{y_{n-1}} + \overline{y_{n-2}} + \overline{y_{n-3}})2^{n-2} \\
&\quad + (-2.\overline{y_{n-3}} + \overline{y_{n-4}} + \overline{y_{n-5}})2^{n-4} \\
&\quad . \\
&\quad . \\
&\quad . \\
&\quad + (-2.\overline{y_3} + \overline{y_2} + \overline{y_1})2^2 \\
&\quad + (-2.\overline{y_1} + \overline{y_0} + \overline{y_{-1}})2^0
\end{aligned}
$$

Here, $\overline{y_{-1}} = 0$.
Then,

$$
\begin{aligned}
-Y &= 1 + \sum_{k=0}^{n-1} \overline{y_k}.\,2^k \\
&= \overline{y_{n-1}}.\,2^n \\
&\quad + (-2.\overline{y_{n-1}} + \overline{y_{n-2}} + \overline{y_{n-3}})2^{n-2} \\
&\quad + (-2.\overline{y_{n-3}} + \overline{y_{n-4}} + \overline{y_{n-5}})2^{n-4} \\
&\quad . \\
&\quad . \\
&\quad . \\
&\quad + (-2.\overline{y_3} + \overline{y_2} + \overline{y_1})2^2 \\
&\quad + (-2.\overline{y_1} + \overline{y_0} + \overline{y_{-1}})2^0 + 1
\end{aligned}
$$

However, the sign-extended term, $\overline{y_{n-1}}\cdot 2^n = -1$
Thus,

$$
\begin{aligned}
-Y &= 1 + \sum_{k=0}^{n-1} \overline{y_k}\cdot 2^k \\
&= (-2.\overline{y_{n-1}} + \overline{y_{n-2}} + \overline{y_{n-3}})2^{n-2} \\
&\quad + (-2.\overline{y_{n-3}} + \overline{y_{n-4}} + \overline{y_{n-5}})2^{n-4} \\
&\quad . \\
&\quad . \\
&\quad . \\
&\quad + (-2.\overline{y_3} + \overline{y_2} + \overline{y_1})2^2 \\
&\quad + (-2.\overline{y_1} + \overline{y_0} + \overline{y_{-1}})2^0
\end{aligned}
$$

Let

$$w_k = -2.\overline{y_{k+1}} + \overline{y_k} + \overline{y_{k-1}}$$

Then,
$-Y = \sum_{k=0}^{n-2} w_k 2^k$ where, $k = 0, 2, \ldots, (n-4), (n-2)$.
But,

$$
\begin{aligned}
w_k &= -2.\overline{y_{k+1}} + \overline{y_k} + \overline{y_{k-1}} = -2(1 - y_{k+1}) + (1 - y_k) + (1 - y_{k-1}) \\
&= 2y_{k+1} - y_k - y_{k-1} = -z_k
\end{aligned}
$$

Then,

$$-Y = \sum_{k=0}^{n-2} w_k 2^k = -\sum_{k=0}^{n-2} z_k 2^k$$

or
$Y = \sum_{k=0}^{n-2} z_k 2^k$ which is the same equation for $Y > 0$.
Therefore, for both positive and negative values of Y, we have:
$Y = \sum_{k=0}^{n-2} z_k 2^k$ where, $k = 0, 2, \ldots, (n-4), (n-2)$.
where, $z_k = -2y_{k+1} + y_k + y_{k-1}$

Example 1.24 Starting from the generation of its partial products, design an eight-bit Booth multiplier

The multiplier term, Y, for the eight-bit Booth multiplier follows the encoded expression:

$$Y = \sum_{k=0}^{6} z_k 2^k = z_0 2^0 + z_2 2^2 + z_4 2^4 + z_6 2^6$$

Where, the encoded multiplier coefficients are:

$$z_0 = -2y_1 + y_0 + y_{-1} = -2y_1 + y_0$$
$$z_2 = -2y_3 + y_2 + y_1$$
$$z_4 = -2y_5 + y_4 + y_3$$
$$z_6 = -2y_7 + y_6 + y_5$$

Thus, P = X . Y yields:

$$P = 2^0(X . z_0) + 2^2(X . z_2) + 2^4(X . z_4) + 2^6(X . z_6)$$

This reduces the number of partial products from eight to four as shown in Fig. 1.60. In this figure, u0, u1, u2, and u3 are added to the least significant bit position of each partial product to handle cases where the partial product becomes −X, −2X.

	sign extension							9-bit partial product due to a possible left shift								
z_0X →	a_8	a_8	a_8	a_8	a_8	a_8	a_8	a_8	a_7	a_6	a_5	a_4	a_3	a_2	a_1	a_0
																u_0
z_2X →	b_8	b_8	b_8	b_8	b_8	b_8	b_7	b_6	b_5	b_4	b_3	b_2	b_1	b_0	0	0
														u_1		
z_4X →	c_8	c_8	c_8	c_8	c_7	c_6	c_5	c_4	c_3	c_2	c_1	c_0	0	0	0	0
												u_2				
z_6X →	d_8	d_8	d_7	d_6	d_5	d_4	d_3	d_2	d_1	d_0	0	0	0	0	0	0
+										u_3						
SUM →	s_{15}	s_{14}	s_{13}	s_{12}	s_{11}	s_{10}	s_9	s_8	s_7	s_6	s_5	s_4	s_3	s_2	s_1	s_0

Fig. 1.60 Partial product tree of an eight-bit Booth multiplier

For cases +X, +2X and 0, all u-terms become equal to zero. All partial products are sign-extended and nine bits in length to be able to handle ±2X.

The calculation of the final product can be further simplified if the sign extension terms are eliminated. Let's sum all the sign extension terms and form the term, SE, below:

$$SE = a_8 . (2^{15} + \cdots + 2^8) + b_8 . (2^{15} + \ldots 2^{10}) + c_8 . (2^{15} + \ldots 2^{12}) + d_8 . (2^{15} + 2^{14})$$

But,

$$2^{15} + \cdots + 2^8 = 2^8 . (2^7 + \cdots + 1) = 2^8 . (2^8 - 1) = 2^{16} - 2^8$$
$$2^{15} + \cdots + 2^{10} = 2^{10} . (2^5 + \cdots + 1) = 2^{10} . (2^6 - 1) = 2^{16} - 2^{10}$$
$$2^{15} + \cdots + 2^{12} = 2^{12} . (2^3 + \cdots + 1) = 2^{12} . (2^4 - 1) = 2^{16} - 2^{12}$$
$$2^{15} + 2^{14} = 2^{14} . (2 + 1) = 2^{14} . (2^2 - 1) = 2^{16} - 2^{14}$$

$$\begin{aligned} SE &= a_8 . 2^{16} - a_8 . 2^8 + b_8 . 2^{16} - b_8 . 2^{10} + c_8 . 2^{16} - c_8 . 2^{12} + d_8 . 2^{16} - d_8 . 2^{14} \\ &= 2^{16} . (a_8 + b_8 + c_8 + d_8) - a_8 . 2^8 - b_8 . 2^{10} - c_8 . 2^{12} - d_8 . 2^{14} \end{aligned}$$

But,

$$-a_8 = \overline{a_8} - 1$$

$$-b_8 = \overline{b_8} - 1$$

$$-c_8 = \overline{c_8} - 1$$

$$-d_8 = \overline{d_8} - 1$$

Thus,

$$\begin{aligned} SE &= 2^{16} . (a_8 + b_8 + c_8 + d_8) + 2^8 . (\overline{a_8} - 1) + 2^{10} . (\overline{b_8} - 1) + 2^{12} . (\overline{c_8} - 1) + 2^{14} . (\overline{d_8} - 1) \\ &= 2^{16} . (a_8 + b_8 + c_8 + d_8) + 2^8 . \overline{a_8} + 2^{10} . \overline{b_8} + 2^{12} . \overline{c_8} + 2^{14} . \overline{d_8} - 2^8 - 2^{10} - 2^{12} - 2^{14} \end{aligned}$$

$$\begin{aligned} SE &= 2^{16} . (a_8 + b_8 + c_8 + d_8) + 2^8 . \overline{a_8} + 2^{10} . \overline{b_8} + 2^{12} . \overline{c_8} + 2^{14} . \overline{d_8} \\ &\quad - 2^8 - 2^{10} - 2^{12} - 2^{14} - 2^{16} + 2^{16} + (1 - 1) \\ &= 2^{16} . (a_8 + b_8 + c_8 + d_8 - 1) + 2^8 . \overline{a_8} + 2^{10} . \overline{b_8} + 2^{12} . \overline{c_8} + 2^{14} . \overline{d_8} \\ &\quad - 2^8 - 2^{10} - 2^{12} - 2^{14} + (2^{16} - 1) + 1 \end{aligned}$$

But,

$$\begin{aligned} 2^{16} - 1 &= (1 + 2 + 2^2 + \cdots + 2^7) + (2^8 + 2^9 + \cdots + 2^{15}) \\ &= (2^8 - 1) + (2^8 + 2^9 + \cdots + 2^{15}) \end{aligned}$$

Then,

$$\begin{aligned} SE &= 2^{16} . (a_8 + b_8 + c_8 + d_8 - 1) + 2^8 . \overline{a_8} + 2^{10} . \overline{b_8} + 2^{12} . \overline{c_8} + 2^{14} . \overline{d_8} \\ &\quad - 2^8 - 2^{10} - 2^{12} - 2^{14} + 1 + (2^8 - 1) + (2^8 + 2^9 + 2^{10} + 2^{11} + 2^{12} + 2^{13} + 2^{14} + 2^{15}) \end{aligned}$$

Regrouping SE with the same power yields:

$$\begin{aligned} SE &= 2^{16} . (a_8 + b_8 + c_8 + d_8 - 1) + 2^8 . (\overline{a_8} + 1) + 2^9 + 2^{10} . \overline{b_8} + 2^{11} \\ &\quad + 2^{12} . \overline{c_8} + 2^{13} + 2^{14} . \overline{d_8} + 2^{15} \end{aligned}$$

But, we don't care what happens to the term 2^{16} since this is the overflow bit in the multiplier sum. Thus,

$$SE = 2^8 . (\overline{a_8} + 1) + 2^9 + 2^{10} . \overline{b_8} + 2^{11} + 2^{12} . \overline{c_8} + 2^{13} + 2^{14} . \overline{d_8} + 2^{15}$$

Then the partial product tree in Fig. 1.60 simplifies a lot more as shown in Fig. 1.61.

In Fig. 1.60, $a_i = z_0 . x_i$, $b_i = z_2 . x_i$, $c_i = z_4 . x_i$, and $d_i = z_6 . x_i$ where i = 0, 1, 2, …, 7 and x_i represents each term in the multiplicand, X.

The complemented a_8, b_8, c_8 and d_8 in Fig. 1.61 are the "reserved bits" in case of a one-bit left shift of the partial product.

	sign extension							MSB	Mid bits							LSB
z_0X →	0	0	0	0	0	0	0	$\overline{a_8}$	a_7	a_6	a_5	a_4	a_3	a_2	a_1	a_0
	0	0	0	0	0	0	0	1	0	0	0	0	0	0	0	u_0
z_2X →	0	0	0	0	0	$\overline{b_8}$	b_7	b_6	b_5	b_4	b_3	b_2	b_1	b_0	0	0
	0	0	0	0	0	0	1	0	0	0	0	0	0	u_1	0	0
z_4X →	0	0	0	$\overline{c_8}$	c_7	c_6	c_5	c_4	c_3	c_2	c_1	c_0	0	0	0	0
	0	0	0	0	1	0	0	0	0	0	0	u_2	0	0	0	0
z_6X →	0	$\overline{d_8}$	d_7	d_6	d_5	d_4	d_3	d_2	d_1	d_0	0	0	0	0	0	0
+	1	0	1	0	0	0	0	0	0	u_3	0	0	0	0	0	0
SUM →	s_{15}	s_{14}	s_{13}	s_{12}	s_{11}	s_{10}	s_9	s_8	s_7	s_6	s_5	s_4	s_3	s_2	s_1	s_0

Fig. 1.61 Partial product tree of an eight-bit Booth multiplier after minimization

Now, the time has come to implement the components of the eight-bit Booth multiplier: the encoder, the partial product tree and the full-adder tree.

The Booth encoder is a logic block that forms each partial product. Earlier, we obtained the Booth coefficient, $z_k = -2y_{k+1} + y_k + y_{k-1}$, to aid the generation of each partial product in Fig. 1.61. One can use the neighboring multiplier bits, y_{k+1}, y_k and y_{k-1}, as inputs to z_k to obtain the encoder outputs. Table 1.19 is slightly modified to form the truth table for a Booth encoder as shown in Table 1.20.

Following Table 1.20, the Booth encoder is implemented in Fig. 1.62.

Table 1.20 Modified Booth encoder truth table

Encoder inputs				
y_{k+1}	y_k	y_{k-1}	z_k	Encoder outputs
0	0	0	0	$ZERO_k = 1$
0	0	1	1	$P1_k = 1$
0	1	0	1	$P1_k = 1$
0	1	1	2	$P2_k = 1$
1	0	0	-2	$M2_k = 1$
1	0	1	-1	$M1_k = 1$
1	1	0	-1	$M1_k = 1$
1	1	1	0	$ZERO_k = 1$

In this figure, $ZERO_k$, $P1_k$, $M1_k$, $P2_k$ and $M2_k$ correspond to the Booth coefficients, 0, +1, −1, +2 and −2, to be multiplied with the multiplicand, respectively.

Each partial product in Fig. 1.61 contains a u-term, namely u0, u1, u2 or u3 in case of a 2s complement conversion of the partial product.

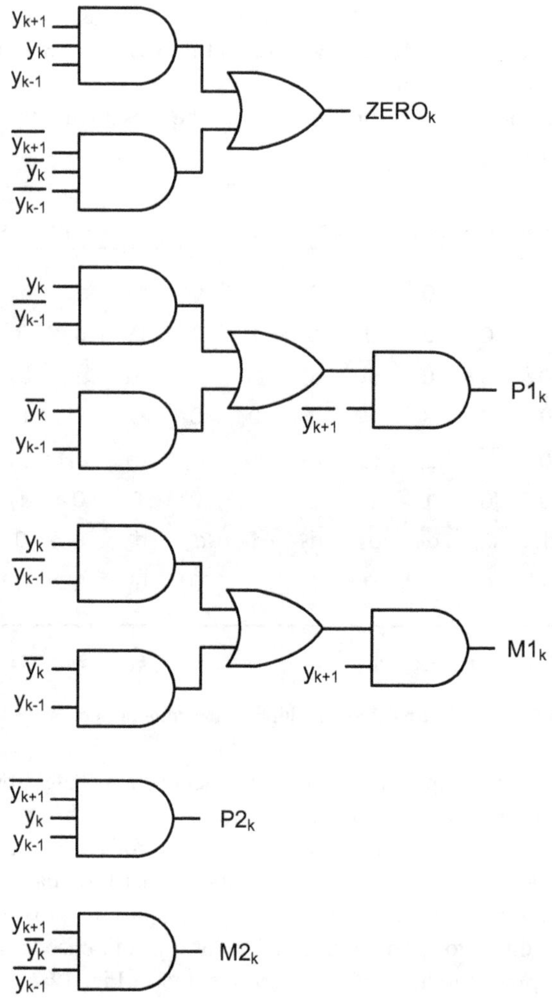

Fig. 1.62 Booth encoder logic circuit

Therefore, for k = 0, 1, 2 or 3, u_k becomes equal to one if $M1_k$ or $M2_k = 1$ (if the multiplicand is multiplied by −1), else it is equal to zero (if the multiplicand is multiplied by zero or +1). The u_k–terms are implemented as shown in Fig. 1.63.

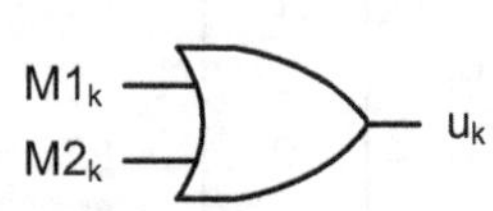

Fig. 1.63 Implementation of u_k

The LSBs, namely the terms a_0, b_0, c_0 and d_0, in Fig. 1.61 form according to Table 1.21.

Table 1.21 Truth table of the LSB for Fig. 1.61

Encoder output	LSB
$ZERO_k = 1$	$a_0 = b_0 = c_0 = d_0 = 0$
$P1_k = 1$	$a_0 = b_0 = c_0 = d_0 = x_0$
$M1_k = 1$	$a_0 = b_0 = c_0 = d_0 = \overline{x_0}$
$P2_k = 1$	$a_0 = b_0 = c_0 = d_0 = 0$
$M2_k = 1$	$a_0 = b_0 = c_0 = d_0 = 1$

According to this table, when the multiplicand is multiplied by zero, the LSB of the partial product becomes equal to zero. When the multiplicand is multiplied by +1 or −1, the LSB is simply equal to the LSB of the multiplicand, x_0, or its complement, respectively. Finally, when the multiplicand is multiplied by +2 or −2, the partial product is shifted one bit to the left; the LSB becomes either zero or one depending on the sign. Therefore, the LSB of the partial product is generated using a 5-1 MUX as in Fig. 1.64.

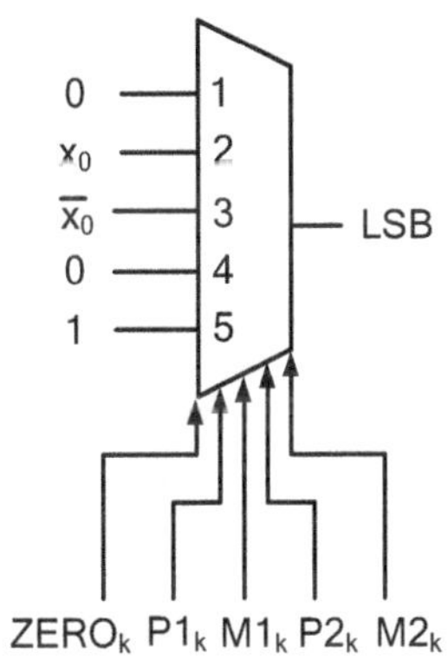

Fig. 1.64 Logic diagram of the LSB for Fig. 1.61

The mid bits are a_1 through a_7, b_1 through b_7, c_1 through c_7, and d_1 through d_7 according to Fig. 1.61. Table 1.22 defines the value of each mid bit as a function of Booth encoder outputs.

Table 1.22 Truth table of the mid bits for Fig. 1.61

Encoder output	Mid bits	
$ZERO_k = 1$	$a_i = b_i = c_i = d_i = 0$	where i = 1, 2,...7
$P1_k = 1$	$a_i = b_i = c_i = d_i = x_i$	where i = 1, 2,...7
$M1_k = 1$	$a_i = b_i = c_i = d_i = \overline{x_i}$	where i = 1, 2,...7
$P2_k = 1$	$a_i = b_i = c_i = d_i = x_{i-1}$	where i = 1, 2,...7
$M2_k = 1$	$a_i = b_i = c_i = d_i = \overline{x_{i-1}}$	where i = 1, 2,...7

According to this table, all mid bits of a partial product are equal to zero if the multiplicand is multiplied by zero. Mid bits become equal to multiplicand bits or their complements depending on the multiplicand is multiplied by +1 or −1, respectively. When the multiplicand is multiplied by +2 or −2, each term in the partial product is shifted one bit to the left. Therefore, each partial product bit becomes equal to the lesser significant multiplicand bit or its complement. Thus, each mid bit can be implemented as shown in Fig. 1.65.

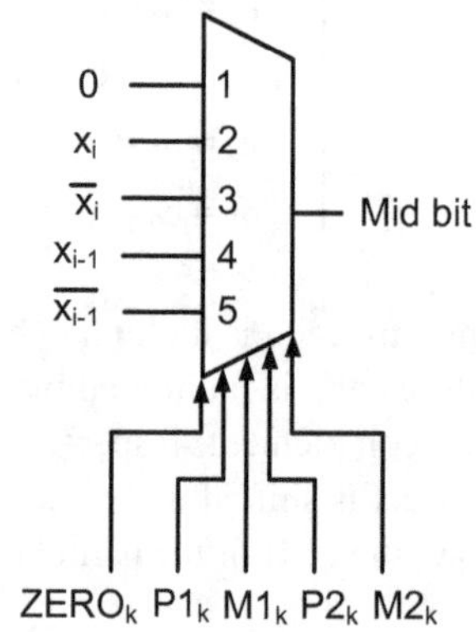

Fig. 1.65 Logic diagram of the mid bits for Fig. 1.61

The MSBs, namely the terms $\overline{a_8}, \overline{b_8}, \overline{c_8}$ or $\overline{d_8}$, in Fig. 1.61 form according to Table 1.23.

Table 1.23 Truth table of the MSB for Fig. 1.61

Encoder output	MSB
$ZERO_k = 1$	$\overline{a_8} = \overline{b_8} = \overline{c_8} = \overline{d_8} = 0$
$P1_k = 1$	$\overline{a_8} = \overline{b_8} = \overline{c_8} = \overline{d_8} = x_7$
$M1_k = 1$	$\overline{a_8} = \overline{b_8} = \overline{c_8} = \overline{d_8} = \overline{x_7}$
$P2_k = 1$	$\overline{a_8} = \overline{b_8} = \overline{c_8} = \overline{d_8} = x_7$
$M2_k = 1$	$\overline{a_8} = \overline{b_8} = \overline{c_8} = \overline{d_8} = \overline{x_7}$

In this table, when the multiplicand is multiplied by zero, the MSB of the partial product becomes equal to zero. When the multiplicand is multiplied by +1 or −1, the MSB is simply equal to the sign-extended value of the most significant multiplicand bit, x_7, or its complement, respectively. When the multiplicand is multiplied by +2 or −2, the partial product is shifted to the left by one bit. Consequently, the LSB becomes equal to the most significant multiplicand bit or its complement, respectively. The MSB of a partial product can therefore be implemented as in Fig. 1.66.

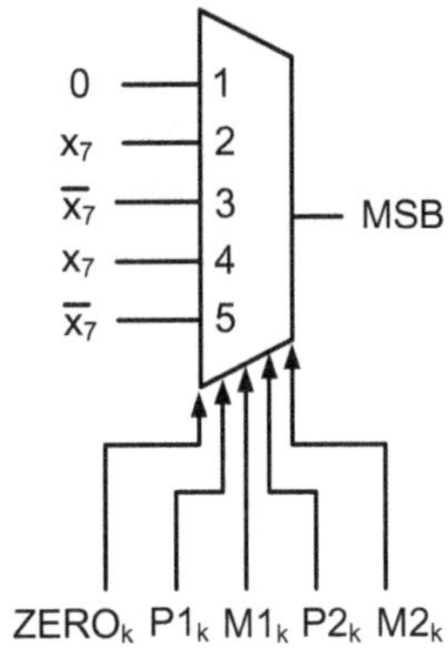

Fig. 1.66 Logic diagram of the MSB for Fig. 1.61

When all of the components are integrated, we obtain the circuit topology in Fig. 1.67 for an eight-bit Booth multiplier.

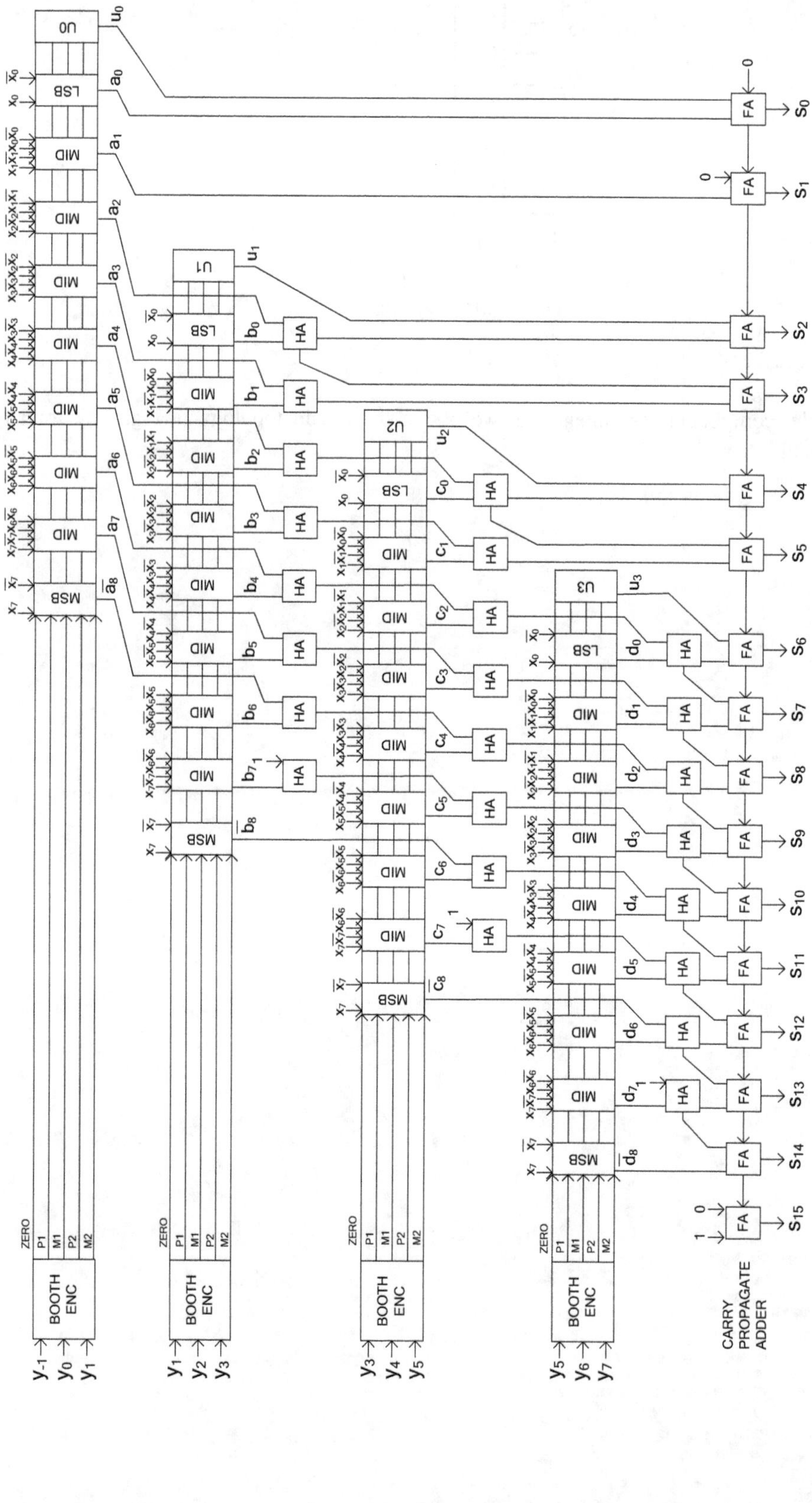

Fig. 1.67 8 × 8 Booth multiplier

Example 1.25 Multiply A = 10110101 (multiplicand) by B = 01110010 (multiplier) using the Booth algorithm. The multiplier bits are as follows:

$$y_0 = 0, y_1 = 1, y_2 = 0, y_3 = 0, y_4 = 1, y_5 = 1, y_6 = 1, y_7 = 0$$

Therefore, the Booth coefficients become:

$$z_0 = -2y_1 + y_0 + y_{-1} = -2\,.\,(1) + 0 + 0 = -2$$
$$z_2 = -2y_3 + y_2 + y_1 = -2\,.\,(0) + 0 + 1 = +1$$
$$z_4 = -2y_5 + y_4 + y_3 = -2\,.\,(1) + 1 + 0 = -1$$
$$z_6 = -2y_7 + y_6 + y_5 = -2\,.\,(0) + 1 + 1 = +2$$

According to these coefficients, the partial product tree forms as in Fig. 1.68.

					Multiplicand =				1	0	1	1	0	1	0	1	
×					Multiplier =				0	1	1	1	0	0	1	0	
	0	0	0	0	0	0	0	0	1	0	0	1	0	1	1	0	-2 = Take 2s complement of the multiplicand and shift left by 1 bit
	1	1	1	1	1	1	1	0	1	1	0	1	0	1			+1 = Replicate the multiplicand
	0	0	0	0	0	1	0	0	1	0	1	1					-1 = Take 2s complement of the multiplicand
+	1	1	0	1	1	0	1	0	1	0							+2 = Shift the multiplicand left by 1 bit
	1	1	0	1	1	1	1	0	1	0	0	1	1	0	1	0	

Fig. 1.68 A numerical example of an 8 × 8 Booth multiplier

In this figure, the zeroth partial product (top row) is generated by multiplying the multiplicand by −2. This requires first taking the 2s complement of the multiplicand, and then shifting it left by one bit. In other words, if the multiplicand is equal to 10110101, then its 2s complement becomes 01001011. Shifting this value to the left by one bit reveals a nine-bit value of 010010110. Since the sign bit of 01001011 is zero before any shifting takes place, sign extending 010010110 after the shift in a 16-bit field yields 0000000010010110. The first partial product (second row from the top) is produced by multiplying the multiplicand by +1. This simply replicates the multiplicand bits, 10110101, in the partial product. Since the sign bit of 10110101 is one, sign extending this value in a 14-bit field yields 11111110110101 as the partial product. The second partial product (third row from the top) is formed by multiplying the multiplicand by −1. This simply requires taking the 2s complement of the multiplicand, which is 01001011. Sign extending this value in a 12-bit field yields 000001001011 as the partial product. The third partial product (last row) is obtained by multiplying the multiplicand by +2, which shifts the multiplicand one bit to the left. Since the multiplicand is 10110101, a nine-bit value of 101101010 is obtained after the shift. Sign extending this value in ten bits yields 1101101010.

Review Questions

1. Implement the following gates:

(a) Implement a two-input XOR gate using two-input NAND gates and inverters.
(b) Implement a two-input AND gate using two-input XNOR gates and inverters.

2. Simplify the equation below:

$$\text{out} = (\overline{\text{A}+\text{B}}).(\overline{\overline{\text{A}}+\overline{\text{B}}})$$

3. Simplify the equation below:

$$\text{out} = (\overline{\text{A}}+\text{C}).(\overline{\text{A}}+\overline{\text{C}}).(\overline{\text{A}+\text{B}+\overline{\text{C}}})$$

4. Obtain the SOP and POS expressions for the following function:

$$\text{out} = (\text{A}.\text{B}+\text{C}).(\text{B}+\text{A}.\text{C})$$

5. Implement the following function using NAND gates and inverters:

$$\text{out} = \text{A}.\overline{\text{C}}+\text{B}.\text{C}+\overline{\text{A}}.\overline{\text{B}}.\text{D}$$

6. Implement the following function using NOR gates and inverters:

$$\text{out} = (\text{A}\oplus\text{B}).(\overline{\text{C}\oplus\text{D}})$$

7. Implement the following 2-1 multiplexer using AND and OR gates:
Note that the function of this complex gate must produce the following:
If En = 1 then out = A else (when En = 0) out = B.

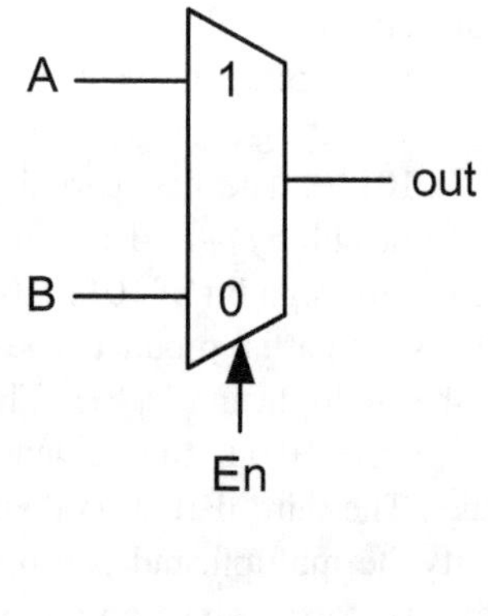

8. Implement the following 3-1 multiplexer using AND and OR gates:
Note that the function of this complex gate must produce the following:
If En = 2 then out = A; if En = 1 then out = B else (when En = 0 or En = 3) out = C.

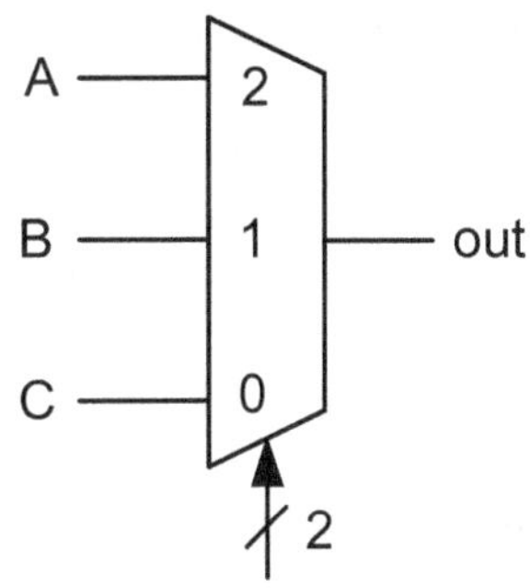

9. Implement a two-bit ripple-carry adder with inputs A[1:0] and B[1:0], and an output C[1:0] using one-bit half and full-adders. Preserve the overflow bit at the output as C[2].

10. Implement a two-bit ripple-carry subtractor with inputs A[1:0] and B[1:0], and an output C[1:0] using one-bit half and full-adders. Preserve the overflow bit at the output as C[2].

11. Implement a two-bit multiplier with inputs A[1:0] and B[1:0], and an output C[3:0] using one-bit half and full-adders.

12. Construct a four-bit comparator with inputs A[3:0] and B[3:0] using a subtractor. The comparator circuit should identify the following cases with active-high outputs:

$$A[3:0] = B[3:0]$$
$$A[3:0] > B[3:0]$$
$$A[3:0] < B[3:0]$$

13. Implement a two-bit decoder that produces four outputs.
When enabled the decoder generates the following outputs:
in[1:0] = 0 then out[3:0] = 1
in[1:0] = 1 then out[3:0] = 2
in[1:0] = 2 then out[3:0] = 4
in[1:0] = 3 then out[3:0] = 8
When disabled the out[3:0] always equals to zero regardless of the input value.

14. Design a 64-bit adder in ripple-carry form and compare it against the carry-look-ahead, carry-select, and carry-look-ahead/carry-select hybrid forms in terms of speed and the number of gates, the latter defining the circuit area. Divide the 64-bit carry-look-ahead/carry-select hybrid into four-bit, eight-bit, 16-bit and 32-bit carry-look-ahead segments. Indicate which carry-look-ahead/carry-select hybrid produces the optimum design.

15. Implement a 4 × 4 Booth multiplier. Design the Booth encoders for partial products and draw the entire schematic of the multiplier. Compare this implementation with the 4 × 4 array multiplier explained in this chapter. List the advantages of both designs in terms of speed and circuit area.

Projects

1. Implement the 4-1 multiplexer and verify its functionality using Verilog.
2. Implement the encoder circuit and verify its functionality using Verilog.
3. Implement the decoder circuit and verify its functionality using Verilog.
4. Implement the four-bit Ripple-Carry Adder and verify its functionality using Verilog.
5. Implement the four-bit Carry-Look-Ahead adder and verify its functionality using Verilog.
6. Implement the four-bit Carry-Select adder and verify its functionality using Verilog.
7. Implement the four-bit Carry-Select/Carry-Look-Ahead adder and verify its functionality using Verilog.
8. Implement the four-bit barrel shifter and verify its functionality using Verilog.
9. Implement the four-bit array multiplier and verify its functionality using Verilog.
10. Implement the eight-bit Booth multiplier and verify its functionality using Verilog.

Review of Sequential Logic Circuits 2

The definition of clock and system timing are integral parts of a sequential digital circuit. Data in a digital system moves from one storage device to the next by the virtue of a system clock. During its travel, data is routed in and out of different combinational logic blocks, and becomes modified to satisfy a specific functionality.

This chapter is dedicated to reviewing the memory devices that store data momentarily or permanently. The process of designing sequential circuits that require clock input will also be explained in detail. The chapter begins with the introduction of two basic memory elements, the latch and the flip-flop. It then explains how data travels between different memory elements using timing diagrams, and analyzes timing violations as a result of unexpected combinational logic delays on the data path or in the clock line. Later in the chapter, the basic sequential building blocks such as registers, shift registers and counters are examined. Moore-type and Mealy-type state machines that control data movement are also studied and compared against counter-decoder type controllers for various design tasks. The concept of memory block and how it is used in a digital system is introduced at the end of this chapter. The chapter concludes with a comprehensive example which demonstrates the transfer of data from one memory block to the next, the use of timing diagrams in the development of the design, and how to incrementally build a data-path and controller using timing diagrams.

2.1 D Latch

The D Latch is the most basic memory element in logic design. It has a data input, D, a clock input and a data output, Q, as shown at the top portion of Fig. 2.1. It contains a tri-state inverter at its input stage followed by two back-to-back inverters connected in a loop configuration, which serves to store data.

The clock signal connected to the enable input of the tri-state inverter can be set either to be active-high or active-low. In Fig. 2.1, the changes at the input transmit though the memory element and become the output during the low phase of the clock. In contrast, the changes at the input are blocked during the high phase of the clock, and no data transmits to the output. Once the data is stored in the back-to-back inverter loop, it does not change until different data is introduced at the input. The buffer at the output stage of the latch is used to drive multiple logic gate inputs.

A. Bindal, *Fundamentals of Computer Architecture and Design*,
https://doi.org/10.1007/978-3-030-00223-7_2

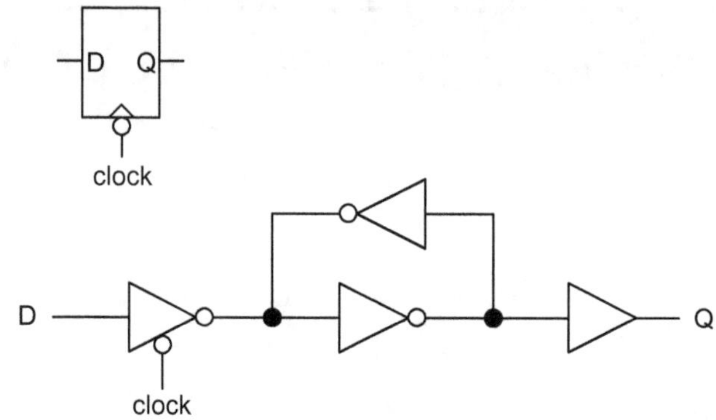

Fig. 2.1 Logic and circuit diagrams of a D latch

The operation of the D latch is shown in Fig. 2.2. During the low phase of the clock, the tri-state inverter is enabled. The new data transmits through the tri-state inverter, overwrites the old data in the back-to-back inverter stage and reaches the output. When the clock switches to its high phase, the input-output data transmission stops because the tri-state buffer is disabled and blocks any new data transfer. Therefore, if certain data needs to be retained in the latch, it needs to be stored in the latch some time before the rising edge of the clock. This time interval is called the set-up time, t_S, and it is approximately equal to the sum of delays through the tri-state inverter and the inverter in the memory element. At the high phase of the clock, the data stored in the latch can no longer change as shown in Fig. 2.2.

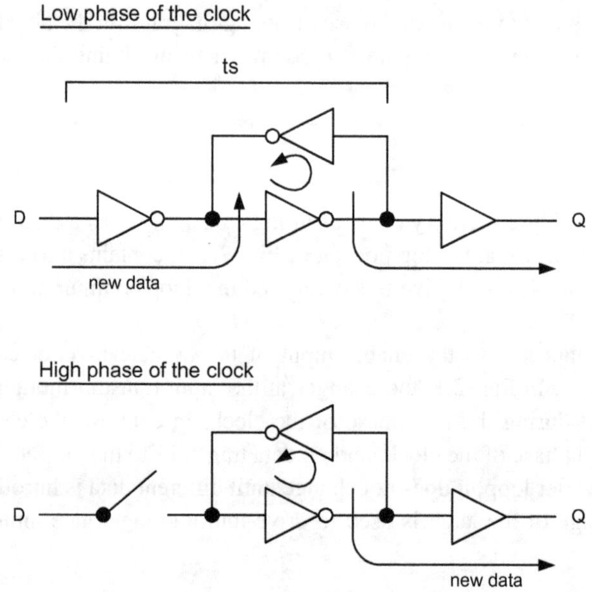

Fig. 2.2 Operation of D latch

2.2 Timing Methodology Using D Latches

Timing in logic systems is maintained by pipeline structures. A pipeline that consists of combinational logic blocks bounded by D latches is shown in the top portion of Fig. 2.3. The main purpose of pipelines is to process several data packets within the same clock cycle and maximize the data throughput.

To illustrate the concept of pipeline, data is stored at each latch boundary for half a clock cycle, but allowed to propagate from one combinational logic stage to the next at the high and low phases of clock.

The bottom part of Fig. 2.3 shows the timing diagram of a data transfer for a set of data packets ranging from D1 to D3 at the IN terminal. The first data packet, $D1^0$, retains its original value during the high phase of the clock (Cycle 1H) at the node A. $D1^0$ then propagates through the T1 stage and settles at the node B in a modified form, $D1^1$, sometime before the falling edge of the clock. Similarly, $D1^1$ at the node C retains its value during the low phase of the clock while its processed form, $D1^2$, propagates through the T2 stage and arrives at the node D before the rising edge of the clock. This data is processed further in the T3 stage and transforms into $D1^3$ before it becomes available at the OUT terminal at the falling edge of the clock in Cycle 2L.

Similarly, the next two data packets, $D2^0$ and $D3^0$, are also fed into the pipeline at the subsequent positive clock edges. Both of these data propagate through the combinational logic stages, T1, T2 and T3, and become available at the OUT terminal at the falling edge of Cycle 3L and Cycle 4L, respectively.

The total execution time for all three data packets takes four clock cycles according to the timing diagram in Fig. 2.3. If we were to remove all the latch boundaries between nodes A and F and wait until all three data packets, D1, D2 and D3, were processed through the sum of the three combinational

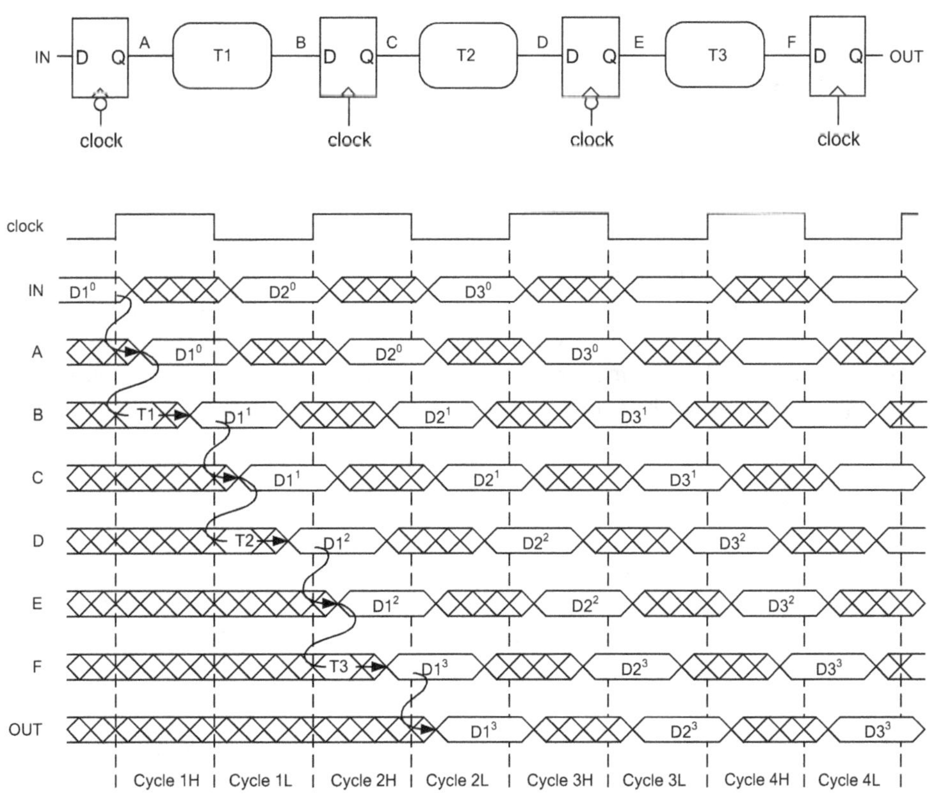

Fig. 2.3 Timing methodology using D latches

logic stages, T1, T2 and T3, the total execution time would have been $3 \times 1.5 = 4.5$ clock cycles as each combinational logic stage requires half a clock cycle to process data. Therefore, pipelining technique can be used both to process data in a shorter amount of time and to increase data throughput.

2.3 D Flip-Flop

D flip-flop is another important timing element in logic design to maintain timely propagation of data from one combinational logic block to the next.

Similar to a latch, a flip-flop also has a data input, D, a clock input and a data output, Q, as shown at the top portion of Fig. 2.4.

The bottom portion of Fig. 2.4 shows the circuit schematic of a typical flip-flop which contains two latches in series. The first latch has an active-low clock input, and it is called the master. The second latch has an active-high clock input, and it is called the slave. The master accepts new data during the low phase of the clock, and transfers this data to the slave during the high phase of the clock.

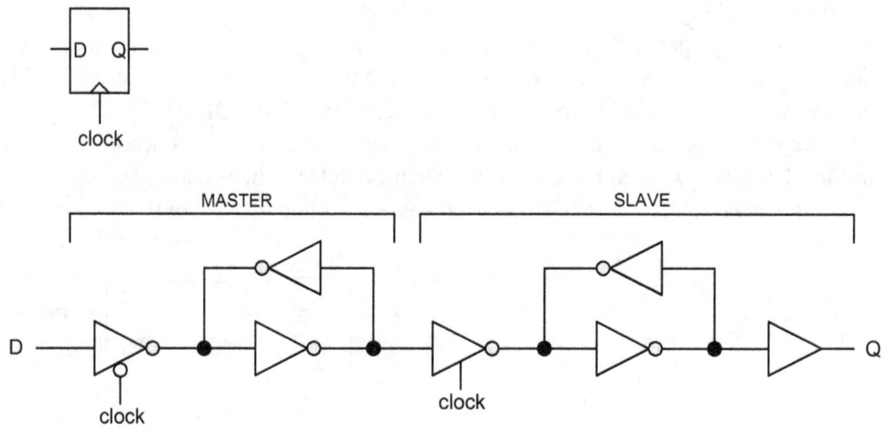

Fig. 2.4 Logic and circuit diagrams of a D flip-flop

Figure 2.5 shows the timing attributes of a flip-flop. The set-up time, t_S, is the time interval for valid data to arrive and settle in the master latch before the rising edge of the clock. Hold time, t_H, is the time interval after the positive edge of the clock when valid data needs to be kept steady and unchanged. The data stored in the master latch propagates through the slave latch and becomes the flip-flop output some time after the rising edge of the clock, and it is called clock-to-q delay or t_{CLKQ}.

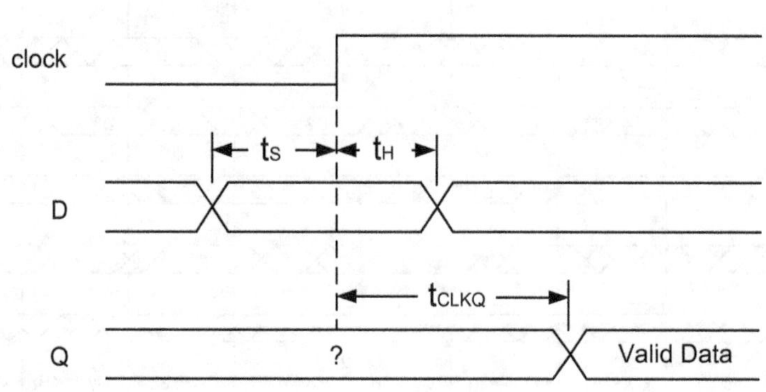

Fig. 2.5 Timing attributes of a D flip-flop

The operation of the flip-flop in two different phases of clock is shown in Fig. 2.6. During the low phase of the clock, new data enters the master latch, and it is stored. This data cannot propagate beyond the master latch because the tri-state inverter in the slave latch acts as an open circuit during the low phase of the clock. The flip-flop output reveals only the old data stored in the slave latch. When the clock goes high, the new data stored in the master latch transmits through the slave and arrives at the output. One can approximate values of t_S and t_{CLKQ} using the existing gate delays in the flip-flop.

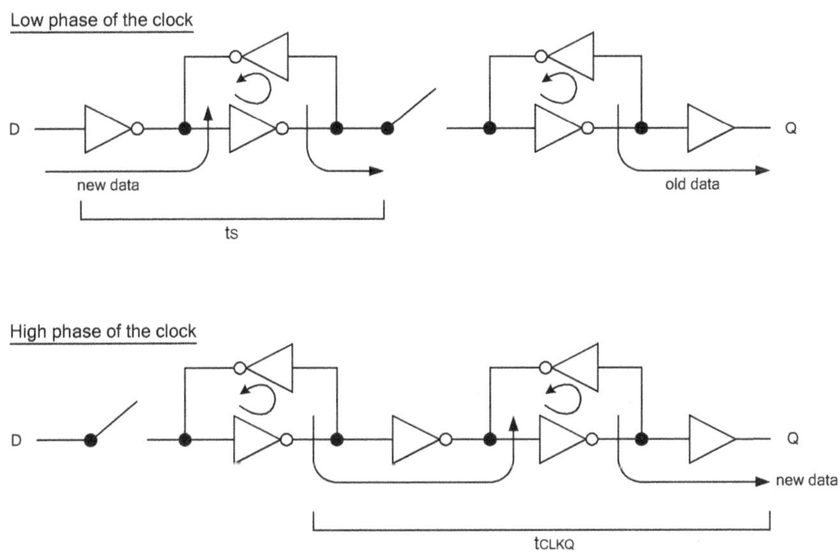

Fig. 2.6 Operation of D flip-flop

2.4 Timing Methodology Using D Flip-Flops

Data propagation through a pipeline with D flip-flops is shown in Fig. 2.7. The bottom part of Fig. 2.7 shows the timing diagram of data transfer through combinational logic blocks, T1, T2 and T3, bounded by D flip-flops for a set of data packets ranging from D1 to D3.

The first data packet, $D1^0$, at the IN terminal has to be steady and valid during the set-up and hold periods of the flip-flop, but it is free to change during the remaining part of the clock period as shown by oscillatory lines. Once the clock goes high, the valid $D1^0$ starts to propagate through the combinational logic block of T1 and reaches the second flip-flop boundary. The processed data, $D1^1$, has to arrive at the second flip-flop input, B, no later than the set-up time of the flip-flop. Otherwise, the correct data cannot be latched. $D1^1$ propagates through the second (T2) and third (T3) combinational logic stages, and becomes $D1^2$ and $D1^3$, respectively, before exiting at the OUT terminal as shown in the timing diagram in Fig. 2.7.

The subsequent data packets, $D2^0$ and $D3^0$ are similarly fed into the pipeline stage from the IN terminal following $D1^0$. They are processed and modified by the T1, T2 and T3 combinational logic stages as they propagate through the pipeline, and emerge at the OUT terminal.

The total execution time for three input data packets, $D1^0$, $D2^0$ and $D3^0$, takes six clock cycles, including the initial three clock cycle build-up period before $D1^3$ emerges at the OUT terminal. If we were to remove all the flip-flop boundaries between the nodes A and F, and wait for these three data packets to be processed without any pipeline structure, the total execution time would have been $3 \times 3 = 9$ clock cycles, assuming each T1, T2 or T3 logic stage imposes one clock cycle delay.

Once again, the pipelining technique considerably reduces the overall processing time and data throughput whether the timing methodology is latch-based or flip-flop-based.

The advantage of using latches as opposed to flip-flops is to be able to borrow time from neighboring stages. For example, the propagation delay in T1 stage can be extended at the expense of shortening the propagation delay in T2. This flexibility does not exist in a flip-flop based design in Fig. 2.7.

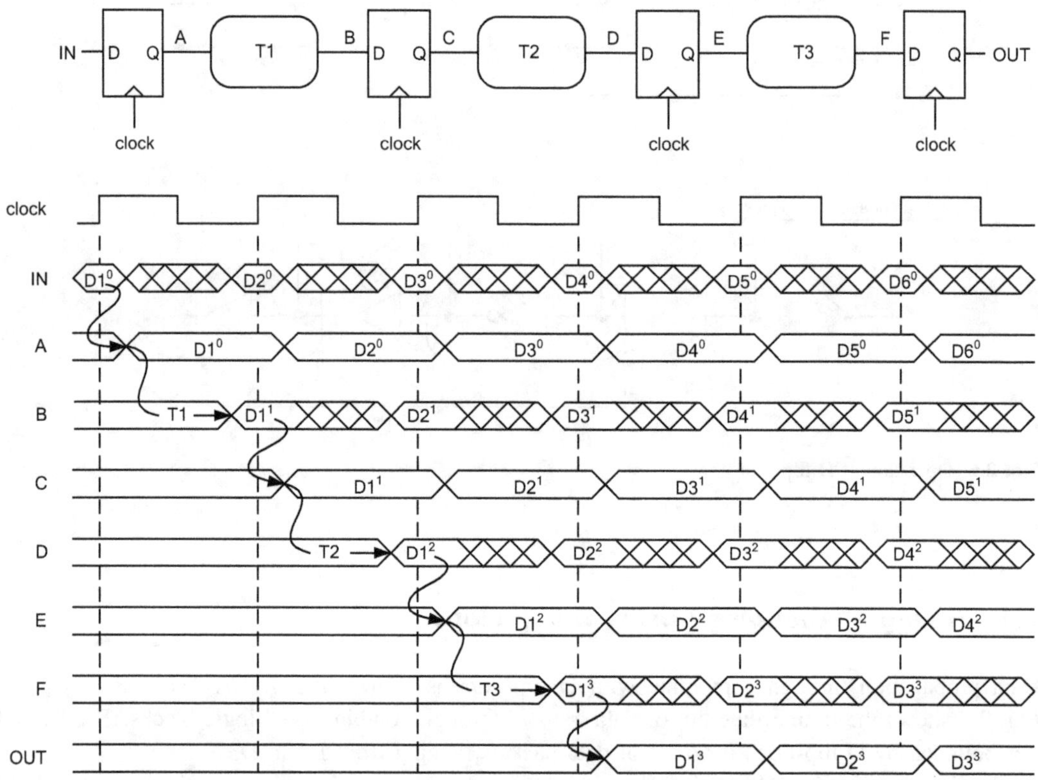

Fig. 2.7 Timing methodology using D flip-flops

2.5 Timing Violations

Although pipelining scheme helps reducing the overall data processing time, we still need to watch out possible timing violations because of the unexpected delays in the data-path and the clock network. Therefore, this section examines the set-up and hold timing violations in a pipeline controlled by flip-flops.

Figure 2.8 shows a section of a pipeline where a combinational logic block with a propagation delay of T_{COMB} is sandwiched between two flip-flop boundaries. At the rising edge of the clock, the valid data that meets the set-up and hold time requirements is introduced at the IN terminal. After t_{CLKQ} delay, the data emerges at the node A and propagates through the combinational logic block as shown in the timing diagram. However, the data arrives at the node B too late and violates the allocated set-up time of the flip-flop. This is called set-up violation. The amount of violation is dependent on the clock period and is calculated as follows:

Set-up violation = $t_S - [T_C - (t_{CLKQ} + T_{COMB})]$

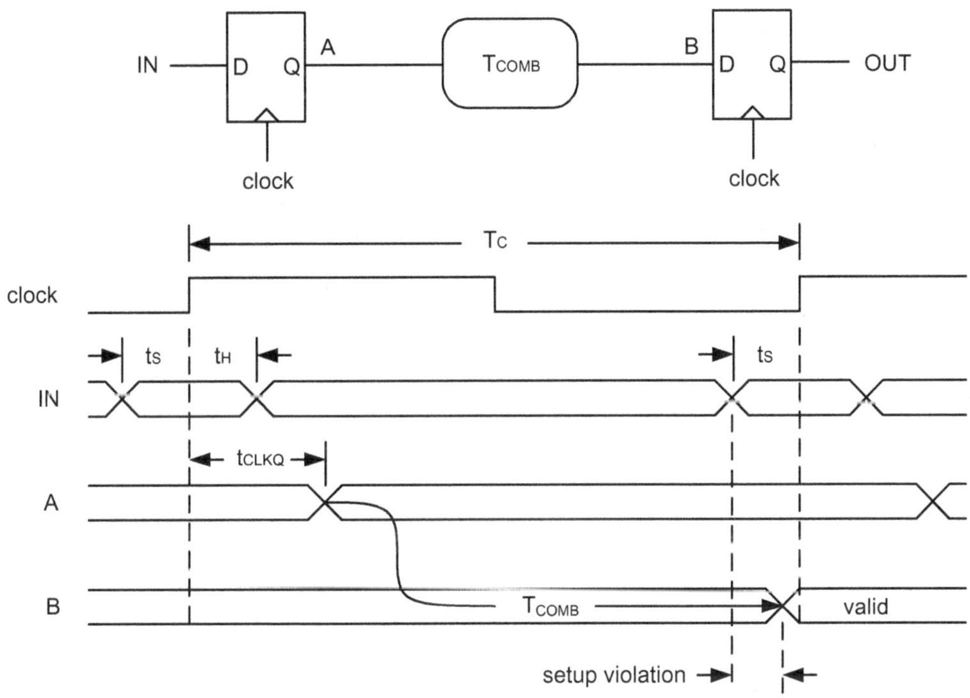

Fig. 2.8 Setup violation

Figure 2.9 describes the hold time violation where the clock shifts by T_{CLK} due to an unexpected delay in the clock line. In the timing diagram, the valid data is introduced to the pipeline from the IN terminal, and it arrives at the node B after t_{CLKQ} and T_{COMB} delays. Due to the shifted clock, the data arrives at the node B early. This creates a substantial set-up time slack equal to ($T_C + T_{CLK} - t_S - t_{CLKQ} - T_{COMB}$) but produces a hold time violation at the delayed clock edge. The amount of violation is dependent on the clock delay and is calculated as follows:

Hold violation = $(T_{CLK} + t_H) - (t_{CLKQ} + T_{COMB})$

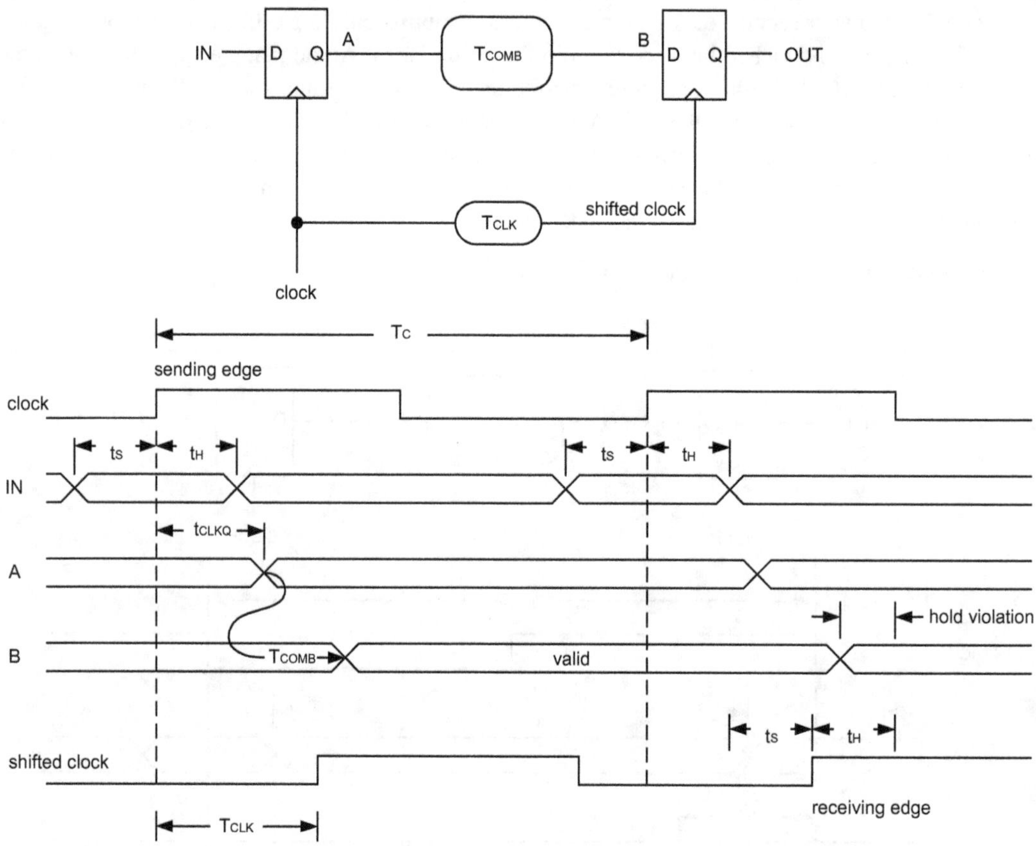

Fig. 2.9 Hold violation

Set-up violations can be recovered simply by increasing the clock period, T_C. However, there is no easy way to fix hold violations as they need to be searched at each flip-flop input. When they are found, buffer delays are added to the combinational logic block, T_{COMB}, in order to avoid the violation.

The schematic in Fig. 2.10 examines the timing ramifications of two combinational logic blocks with different propagation delays merging into one block in a pipeline stage. The data arrives at the node C much earlier than the node D as shown in the timing diagram. The data at the nodes C and D propagate through the last combinational block and arrive at the node E. This scenario creates minimum and maximum delay paths at the node E. We need to focus on the maximum path, (T2 + T3), when examining the possibility of a set-up violation and the minimum path, (T1 + T3), when examining the possibility of a hold violation at the next flip-flop boundary.

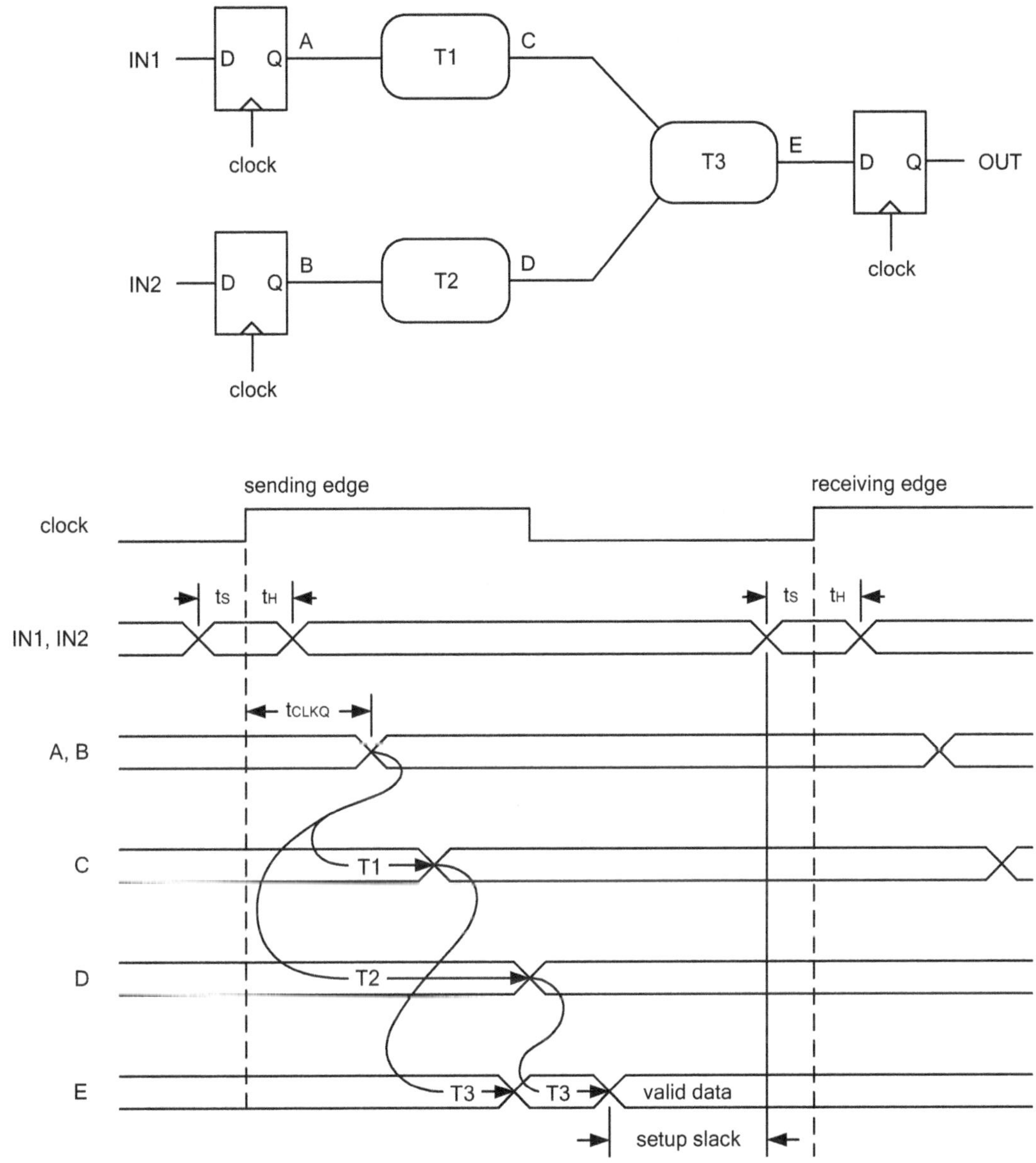

Fig. 2.10 A timing example combining two independent data-paths

To further illustrate the timing issues through multiple combinational logic blocks, an example with logic gates is given in Fig. 2.11. In this example, the inputs of a one-bit adder are connected to the nodes A and B. The adder is bypassed with the inclusion of a 2-1 MUX which selects either the output of the adder or the bypass path.

The propagation delays of the inverter, T_{INV}, and the two-input NAND gate, T_{NAND2}, are given as 100 ps and 200 ps, respectively. The set-up, hold and clock-to-q delays are also given as 100 ps, 0 ps and 300 ps, respectively.

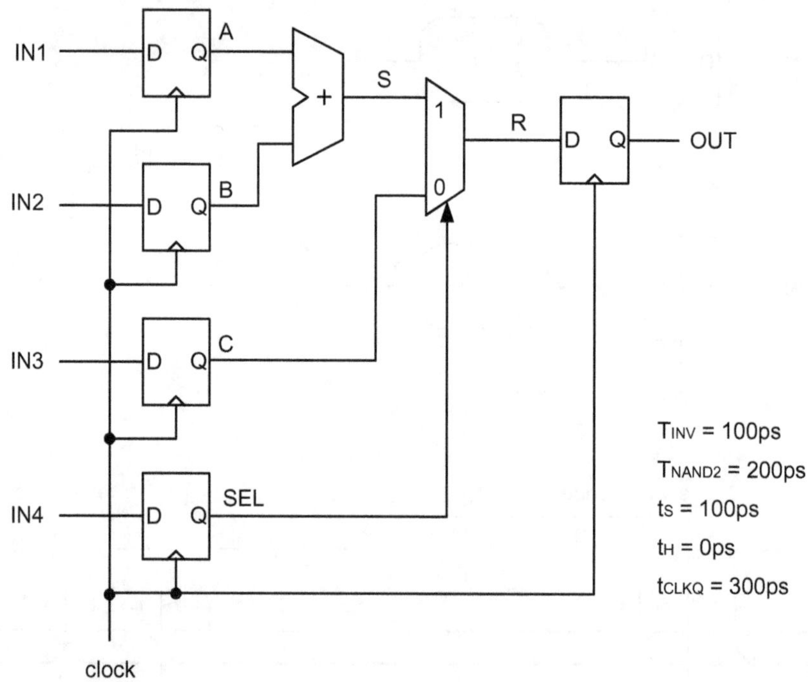

Fig. 2.11 An example with multiple propagation paths

One-bit adder and the 2-1 MUX are shown in terms of inverters and two-input NAND gates in Fig. 2.12. We obtain a total of seven propagation paths all merging at the node R. However, we only need to search for the maximum and the minimum delay paths to locate possible set-up and hold violations.

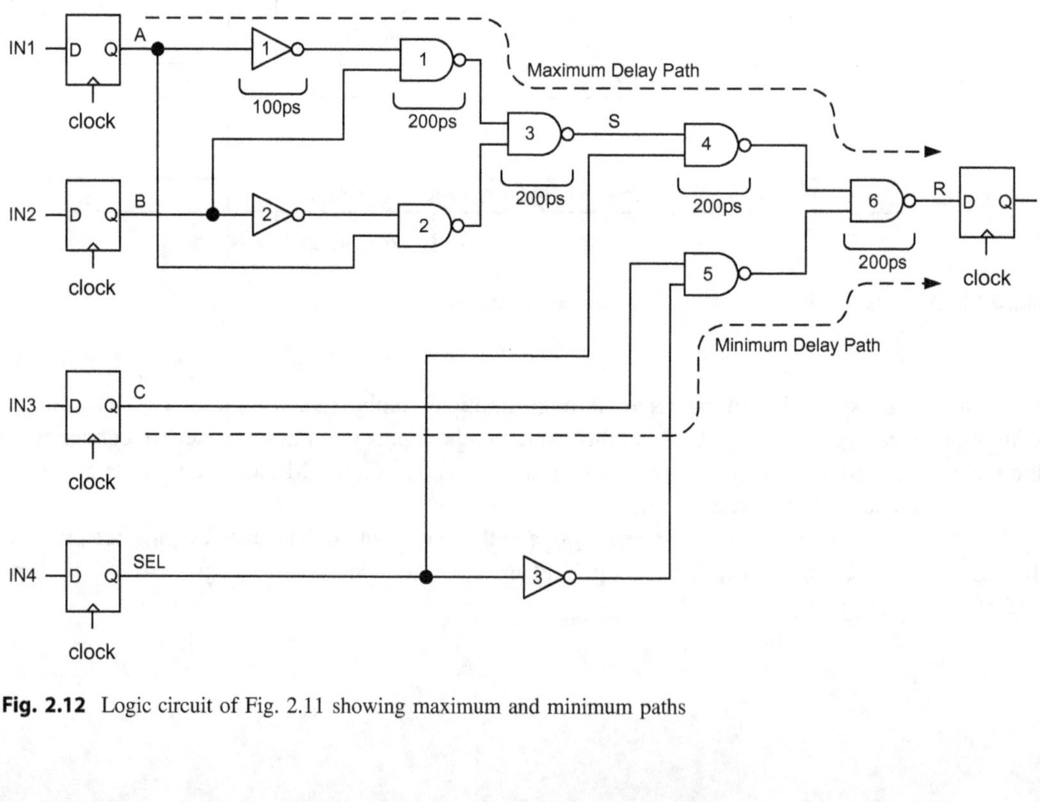

Fig. 2.12 Logic circuit of Fig. 2.11 showing maximum and minimum paths

The maximum delay path consists of the inverter 1 and a series of four two-input NAND gates numbered as 1, 3, 4 and 6 shown in Fig. 2.12. This path results in a total delay of 900 ps. The minimum delay path, on the other hand, contains the two two-input NAND gates numbered as 5 and 6, and it produces a delay of 400 ps. Placing these delays in the timing diagram in Fig. 2.13 yields a set-up slack of 100 ps at the node R when a clock period of 1400 ps is used. There is no need to investigate for hold violations because there is no shift in the clock edge.

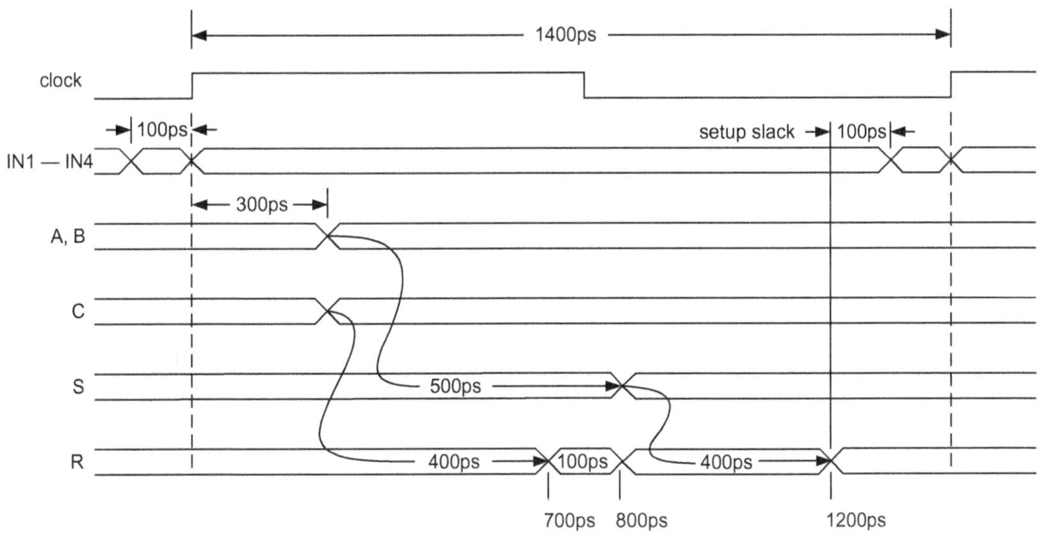

Fig. 2.13 Timing diagram of the circuit in Fig. 2.12

2.6 Register

While a flip-flop holds data only for one clock cycle until new data arrives at the next clock edge, the register can hold data perpetually until the power is turned off.

Figure 2.14 shows the circuit diagram of a one-bit register composed of a flip-flop and a 2-1 MUX. The Write Enable pin, WE, is a selector input to the 2-1 MUX, and transfers new data from the IN terminal to the flip-flop input if WE = 1. When WE = 0, any attempt to write new data to the register is blocked; the old data stored in the register simply circulates around the feedback loop from one clock cycle to the next.

The timing diagram at the bottom of Fig. 2.14 describes the operation of the one-bit register. The data at the IN terminal is blocked until the WE input becomes logic 1 in the middle of the second clock cycle. At this point, the new data is allowed to pass through the 2-1 MUX and renews the contents of the register at the beginning of the third clock cycle. The WE input transitions to logic 0 before the end of the third clock cycle, and causes the register output, OUT, to stay at logic 1 during the fourth clock cycle.

A 32-bit register shown in Fig. 2.15 is composed of 32 one-bit registers. All 32 registers have a common clock and WE input. Therefore, any new 32-bit data introduced at the register input changes the contents of the register at the rising edge of the clock if the WE input is at logic 1.

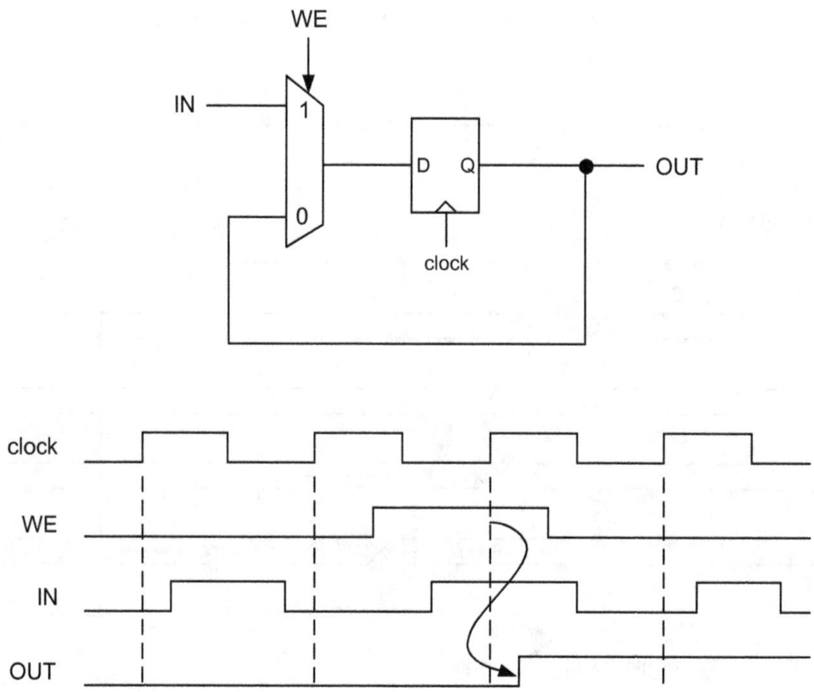

Fig. 2.14 One-bit register and a sample timing diagram

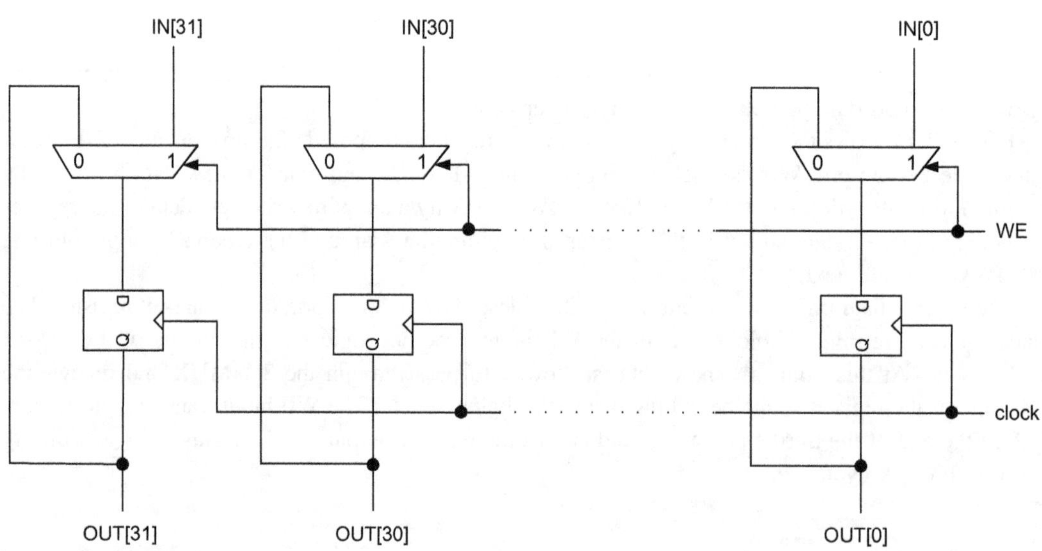

Fig. 2.15 32-bit register

2.7 Shift Register

The shift register is a particular version of an ordinary register and specializes in shifting data to the right or left according to design needs.

Figure 2.16 shows the circuit schematic of a four-bit shift register that shifts serial data at the IN terminal to the left at every positive clock edge.

The operation of this shift register is explained in the timing diagram in Fig. 2.17. In cycle 1, SHIFT = 0. Therefore, the change at the IN terminal during this cycle does not affect the register outputs. However, when the SHIFT input transitions to logic 1 in the middle of cycle 2, it allows IN = 1 to pass to the least significant output bit, OUT[0], at the beginning of the third clock cycle. From the middle of cycle 2 to cycle 13, SHIFT is kept at logic 1. Therefore, any change at the IN node directly transmits to the OUT[0] node at the positive edge of every clock cycle. The other outputs, OUT[1], OUT[2] and OUT[3], produce the delayed versions of the data at OUT[0] by one clock cycle because the output of a lesser significant bit in the shift register is connected to the input of a greater significant bit.

When the SHIFT input becomes logic 0 from the middle of cycle 13 to cycle 17, the shift register becomes impervious to any new data entry at the IN terminal, and retains the old values from the beginning of cycle 13 to cycle 18 as seen in Fig. 2.17. From the middle of cycle 17 onwards, the SHIFT input becomes logic 1 again, and the shift register distributes all new data entries at the IN terminal to its outputs.

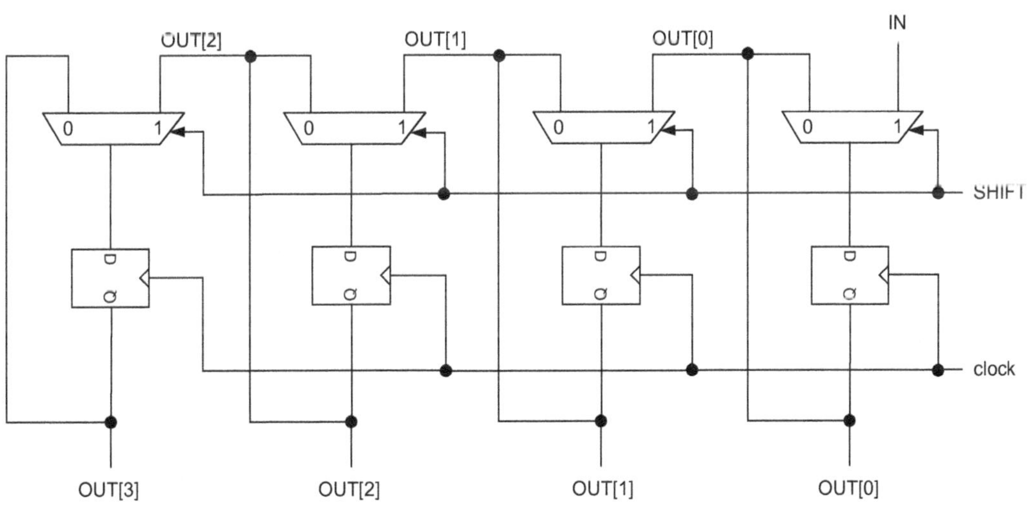

Fig. 2.16 Four-bit shift register

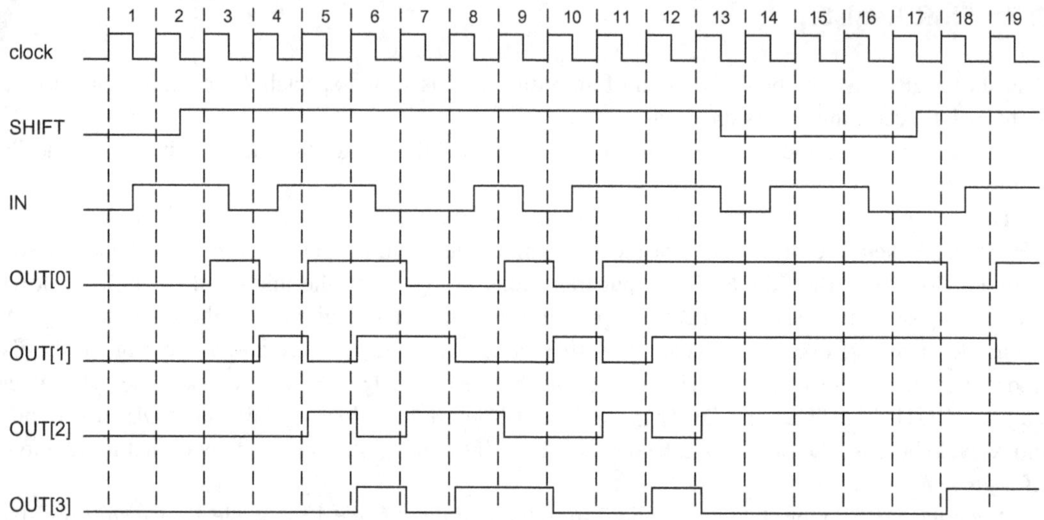

Fig. 2.17 A sample timing diagram of the four-bit shift register in Fig. 2.16

2.8 Counter

The counter is a special form of a register which is designed to count up (or down) at each rising edge of the clock.

The counter in Fig. 2.18 shows a typical 32-bit up-counter with two control inputs, COUNT and LOAD. The COUNT = 1, LOAD = 0 combination selects the C-port of the 3-1 MUX, and enables the counter to count upwards at the rising edge of the clock. The COUNT = 0, LOAD = 1 combination, on the other hand, selects the L-port, and loads new data at the IN[31:0] terminal. Once loaded, the counter output, OUT[31:0], increments by one at every positive clock edge until all the output bits become logic 1. The next increment automatically resets the counter output to logic 0. When LOAD = COUNT = 0, the counter selects the I-port of the 3-1 MUX. At this combination, it neither loads new data nor counts upwards, but stalls, repeating the old output value.

The sample timing diagram at the bottom of Fig. 2.18 illustrates its operation. Prior to the first clock edge, the LOAD input is at logic 1 which allows an input value, IN = 3, to be stored in the counter. This results in OUT[31:0] = 3 at the positive edge of the first clock cycle. The LOAD = 0 and COUNT = 1 combination in the same cycle prompts the counter for the up-count process, and the contents of the output subsequently increments by one in the next cycle. The result, 3 + 1 = 4, passes through the C-port of the 3-1 MUX and arrives at the flip-flop inputs. At the positive edge of the second clock cycle, this new value overwrites the old registered value, making OUT[31:0] equal to 4. In the next cycle, the counter goes through the same process and increments by one. However, in the same cycle, the COUNT input also transitions to logic 0, activating the I-port of the 3-1 MUX and preventing any new data from entering the up-counter. As a result, the counter stops incrementing and stalls at OUT[31:0] = 5 in the following clock cycles.

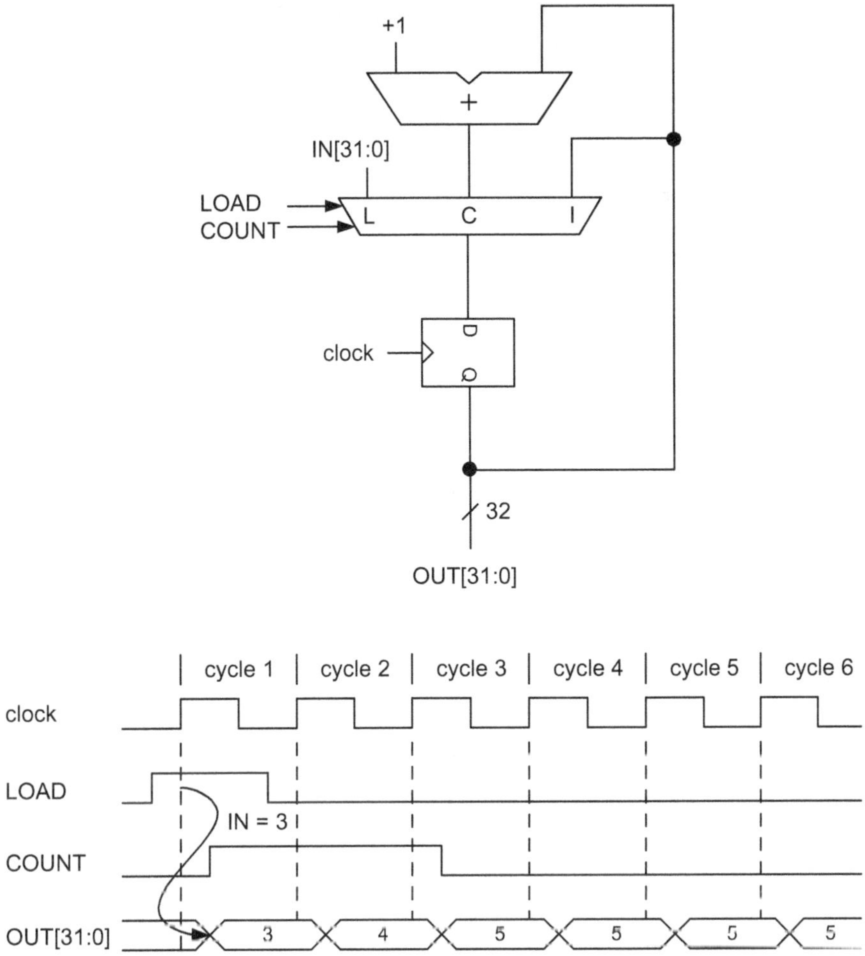

Fig. 2.18 A 32-bit counter and a sample timing diagram

2.9 Moore Machine

State machines are mainly used in digital designs to control the proper data-flow. Their topology is mainly composed of one or multiple flip-flops and feedback loops that connect flip-flop outputs to flip-flop inputs. There are two types of state machines: Moore-type and Mealy-type.

Figure 2.19 shows the Moore-type state machine topology consisting of a flip-flop and a feedback loop. In this configuration, the feedback loop includes a combinational logic block that accepts both the flip-flop output and external inputs. If there are multiple flip-flops, the combination of all flip-flop outputs constitutes the "present" state of the machine. The combination of all flip-flop inputs is called the "next" state because at the positive edge of the clock these inputs become the flip-flop outputs, and form the present state. Flip-flop outputs are processed further by an additional combinational logic block to form present state outputs.

The basic state diagram of a Moore machine, therefore, includes the present state (PS) and the next state (NS) as shown on the right side of Fig. 2.19. The machine can transition from the PS to the NS if the required present state inputs are supplied. The outputs of the Moore machine are solely generated by the present state. Therefore, machine outputs emerge only from each state as shown in the basic state diagram.

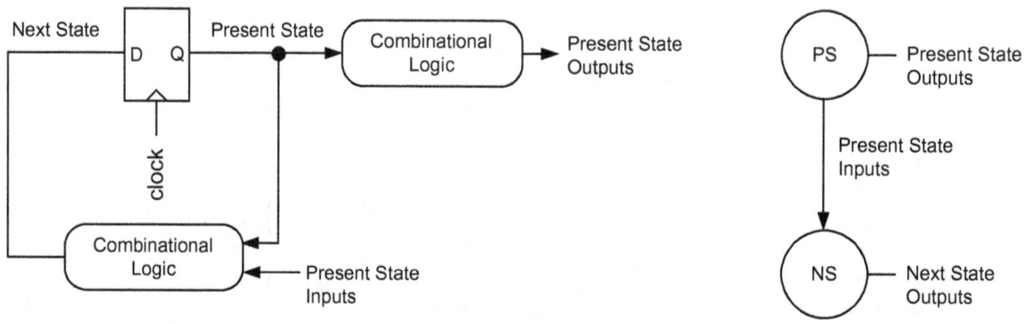

Fig. 2.19 Block diagram and state representation of Moore machine

The state diagram in Fig. 2.20 shows an example of a Moore-type machine with four states. Note that every state-to-state transition in the state diagram requires a valid present state input entry, and every node generates one present state output.

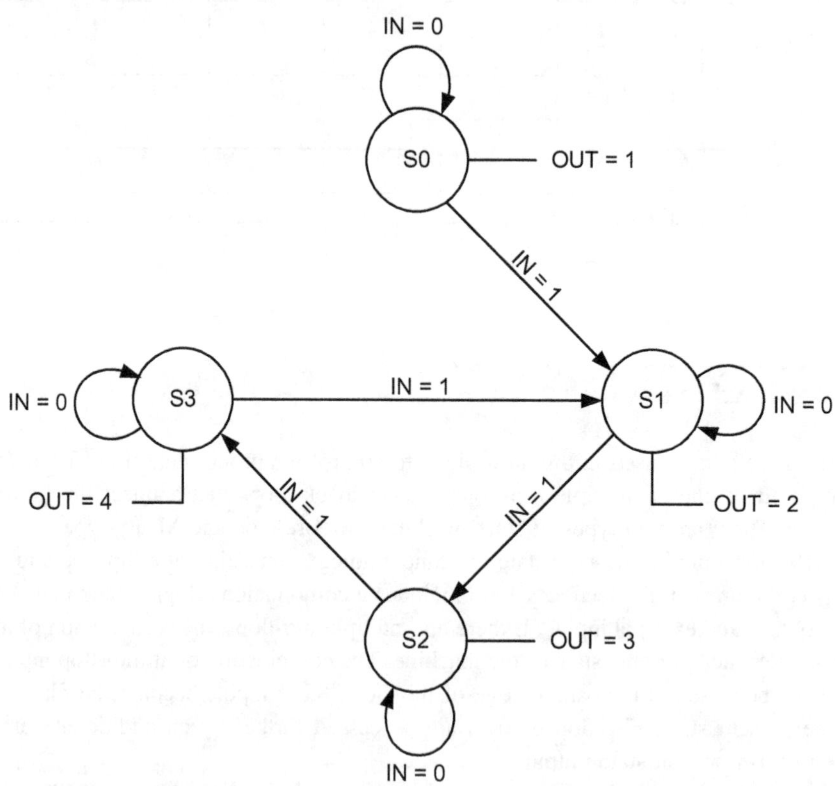

Fig. 2.20 State diagram of a Moore machine with four states

The state 0, S0, produces a present state output, OUT = 1, regardless of the value of the present state input, IN. When IN = 1, the state S0 transitions to the next state S1. Otherwise, it circulates back to itself. The state S1 produces OUT = 2. Its next state becomes the state S1 if IN = 0; otherwise, it transitions to a new state S2. The state S2 also produces a present state output, OUT = 3, and transitions to the state S3 if IN = 1. The state S2 remains unchanged if IN = 0. In the fourth and final state S3, the present state output, OUT = 4, is produced. The machine stays in this state if IN stays at 0; otherwise, it goes back to the state S1.

The present state inputs and outputs of this Moore machine and its states can be tabulated in a table called the "state table" given in Fig. 2.21. In this table, the first column under PS lists all the possible present states of the state diagram in Fig. 2.20. The middle two columns contain the next state entries for IN = 0 and IN = 1. The last column lists the present state outputs, one for each present state.

	NS		
PS	IN = 0	IN = 1	OUT
S0	S0	S1	1
S1	S1	S2	2
S2	S2	S3	3
S3	S3	S1	4

Fig. 2.21 State table of the Moore machine in Fig. 2.20

The binary state assignment is performed according to Fig. 2.22 where only one bit is changed between adjacent states.

States	NS1	NS0
S0	0	0
S1	0	1
S2	1	1
S3	1	0

Fig. 2.22 Bit representations of states S0, S1, S2 and S3

The binary form of the state table in Fig. 2.21 is reconstructed in Fig. 2.23 according to the state assignment in Fig. 2.22. This table, called the "transition table", includes the binary representation of the next state and the present state outputs.

		IN = 0		IN = 1				
PS1	PS0	NS1	NS0	NS1	NS0	OUT2	OUT1	OUT0
0	0	0	0	0	1	0	0	1
0	1	0	1	1	1	0	1	0
1	1	1	1	1	0	0	1	1
1	0	1	0	0	1	1	0	0

Fig. 2.23 Transition table of the Moore machine in Fig. 2.20

Forming this machine's K-maps for the NS0, NS1, OUT0, OUT1 and OUT2 requires grouping all the input terms, PS1, PS0 and IN, according to the table in Fig. 2.23. The K-maps and their corresponding SOP representations are shown in Fig. 2.24.

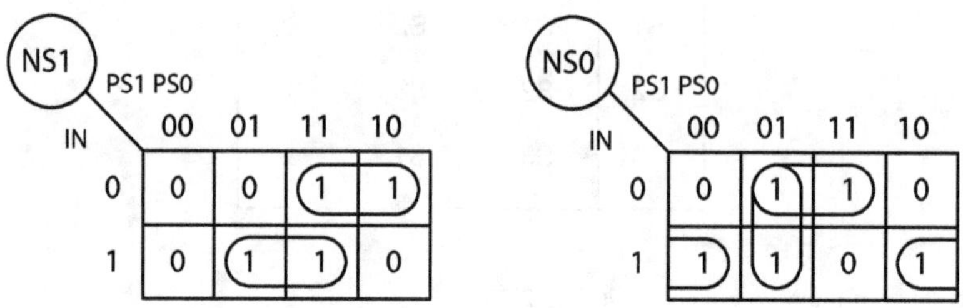

$$NS1 = PS0.IN + PS1.\overline{IN}$$

$$NS0 = PS0.\overline{IN} + \overline{PS1}.PS0 + \overline{PS0}.IN$$

$$= (PS0 \oplus IN) + \overline{PS1}.PS0$$

$$OUT2 = PS1.\overline{PS0}$$

$$OUT1 = \overline{PS1}.PS0 + PS1.PS0 = PS0$$

$$OUT0 = \overline{PS1}.\overline{PS0} + PS1.PS0 = \overline{PS0 \oplus PS1}$$

Fig. 2.24 K-maps and SOP expressions for the Moore machine in Fig. 2.20

The next step is to generate the circuit diagram that produces all five outputs of the Moore machine in Fig. 2.24. This circuit diagram is given in Fig. 2.25.

In order to generate this circuit, the individual combinational logic blocks for the NS0 and NS1 must be constructed first in terms of PS0, PS1 and IN. Then each NS0 and NS1 is connected to the corresponding flip-flop input to form the feedback loops of the state machine. The logic blocks for the OUT0, OUT1 and OUT2 are generated directly from PS0 and PS1.

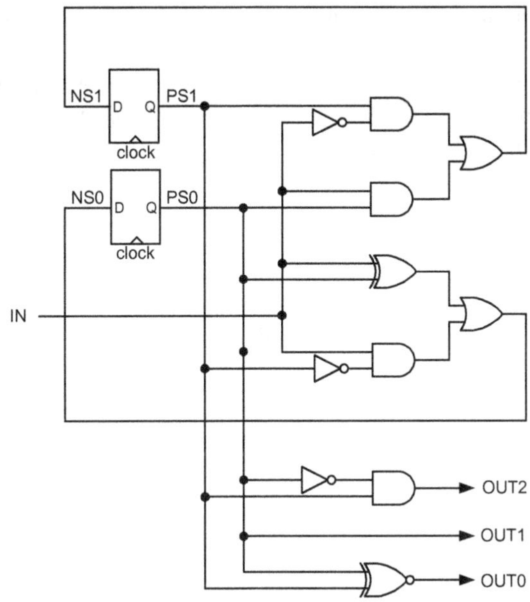

Fig. 2.25 Logic circuit of the Moore machine in Fig. 2.20

2.10 Mealy Machine

The Mealy machine shares the same circuit topology with the Moore machine. The machine also contains one or more flip-flops and feedback loops as shown in Fig. 2.26. However, the present state outputs are generated from the combinational logic block in the feedback loop rather than from the present states in the Moore-type machines.

As a result of this topology, the basic state diagram of a Mealy machine includes the present state, the next state and the input condition that makes the state-to-state transition possible as shown on the right side of Fig. 2.26. Present state outputs do not emerge from each present state; instead, they are functions of present state inputs and present state.

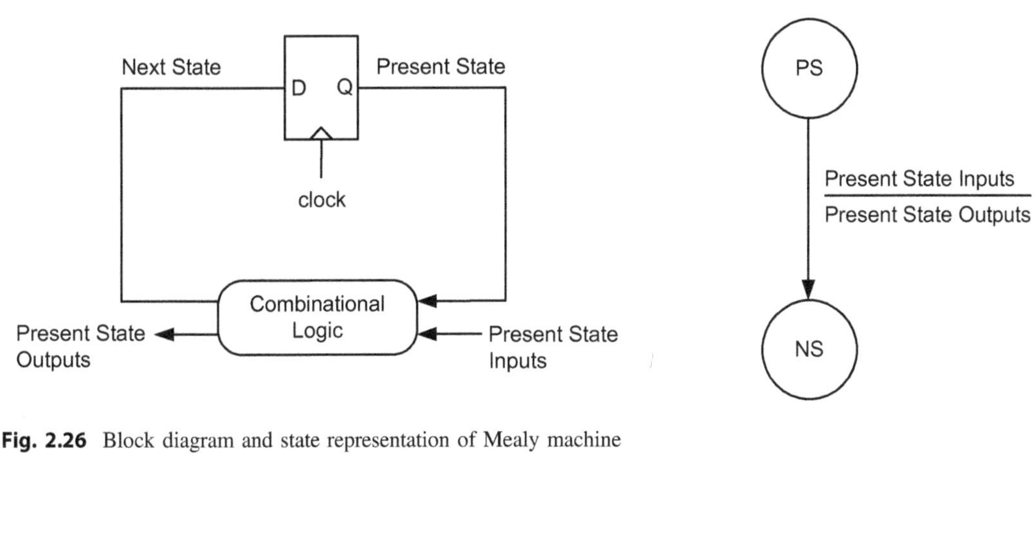

Fig. 2.26 Block diagram and state representation of Mealy machine

The Mealy state diagram in Fig. 2.27 exhibits similar characteristics compared to the Moore state diagram in Fig. 2.20, and all the state names and the state-to-state transitions in this diagram are kept the same for comparison purposes. However, each arrow connecting one state to the next carries the value of present state output as a function of present state input as indicated in Fig. 2.26. As a result, the Mealy state table in Fig. 2.28 contains two separate columns that tabulate the values of NS and OUT for IN = 0 and IN = 1. The binary state assignment is the same as in Fig. 2.22, which results in a transition table in Fig. 2.29.

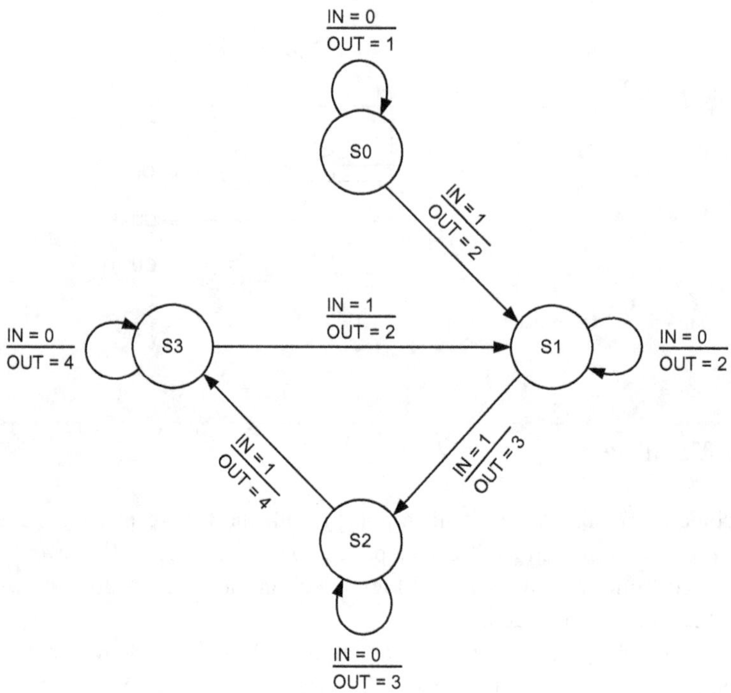

Fig. 2.27 State diagram of a Mealy machine with four states

PS	NS		OUT	
	IN = 0	IN = 1	IN = 0	IN = 1
S0	S0	S1	1	2
S1	S1	S2	2	3
S2	S2	S3	3	4
S3	S3	S1	4	2

Fig. 2.28 State table of the Mealy machine in Fig. 2.27

		IN = 0		IN = 1		IN = 0			IN = 1		
PS1	PS0	NS1	NS0	NS1	NS0	OUT2	OUT1	OUT0	OUT2	OUT1	OUT0
0	0	0	0	0	1	0	0	1	0	1	0
0	1	0	1	1	1	0	1	0	0	1	1
1	1	1	1	1	0	0	1	1	1	0	0
1	0	1	0	0	1	1	0	0	0	1	0

Fig. 2.29 Transition table of the Mealy machine in Fig. 2.27

The K-maps for the NS0, NS1, OUT0, OUT1 and OUT2 are formed according to the table in Fig. 2.29 and shown in Fig. 2.30 with the corresponding SOP expressions. Figure 2.31 shows the circuit diagram of this machine according to the expressions in Fig. 2.30. The methodology used to construct this circuit diagram is identical to the methodology used in the circuit diagram for the Moore machine in Fig. 2.25.

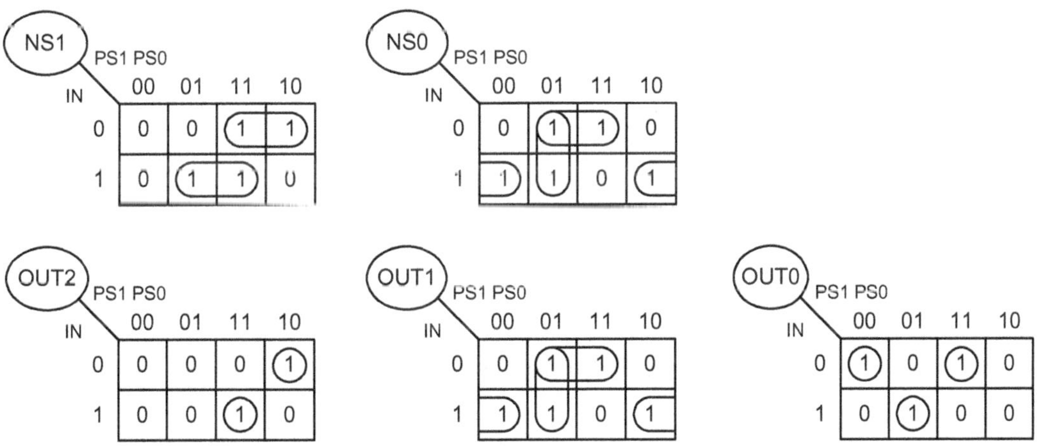

$$NS1 = PS0.IN + PS1.\overline{IN}$$

$$NS0 = PS0.\overline{IN} + \overline{PS1}.PS0 + \overline{PS0}.IN$$

$$= (PS0 \oplus IN) + \overline{PS1}.PS0$$

$$OUT2 = PS1.\overline{PS0}.\overline{IN} + PS1.PS0.IN = PS1.\overline{(PS0 \oplus IN)}$$

$$OUT1 = (PS0 \oplus IN) + \overline{PS1}.PS0 = NS0$$

$$OUT0 = \overline{PS1}.\overline{PS0}.\overline{IN} + \overline{PS1}.PS0.IN + PS1.PS0.\overline{IN} = \overline{PS1}.\overline{(PS0 \oplus IN)} + PS1.PS0.\overline{IN}$$

Fig. 2.30 K-maps and SOP expressions for the Mealy machine in Fig. 2.27

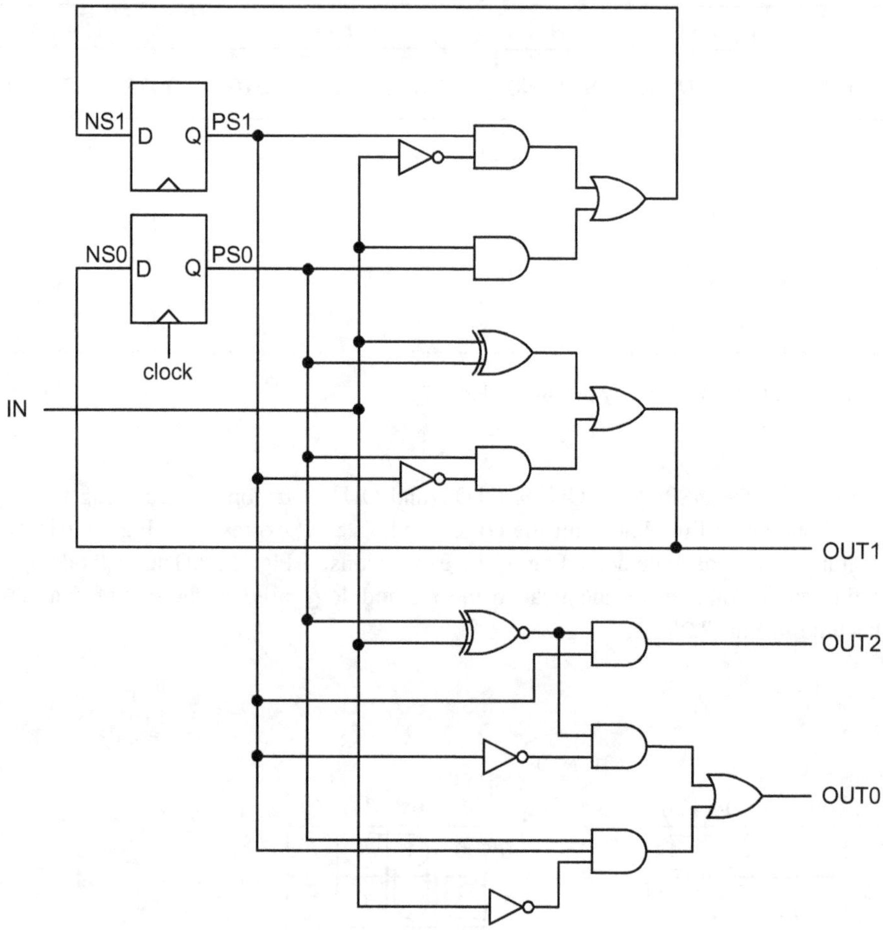

Fig. 2.31 Logic circuit of the Mealy machine in Fig. 2.27

2.11 Controller Design: Moore Versus Counter-Decoder Scheme

Both Mealy and Moore-type state machines have practical implementation limits when it comes to design. A large ring-style state machine composed of N states such as in Fig. 2.32 may have multiple outputs attached to each state, making its implementation nearly impossible with conventional state machine implementation techniques. However, these types of designs are excellent candidates for the counter-decoder type of designs where each state in the state diagram is associated with a counter output value. Therefore, as the counter increments the desired PS outputs of the state machine in Fig. 2.32 can easily be replicated using a set of decoders at the counter output.

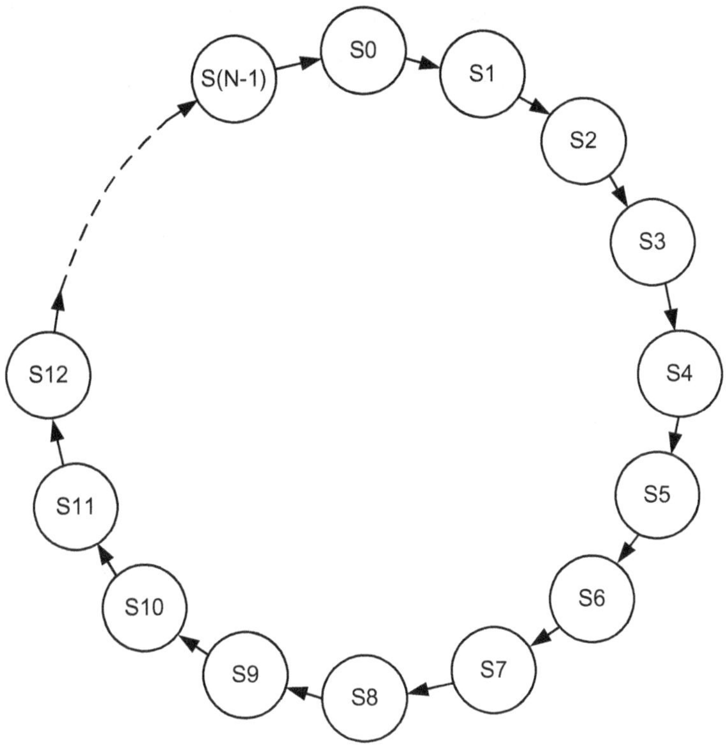

Fig. 2.32 State diagram of a counter with N states

To illustrate this theory, a controller that generates the timing diagram in Fig. 2.33 will be implemented using both the Moore-type state machine and the counter-decoder approach.

From the timing diagram below, this state machine generates a single active-high output, Out = 1, once in every 8 cycles as long as Stop = 0. When Stop = 1, however, the machine stalls and retains its current state.

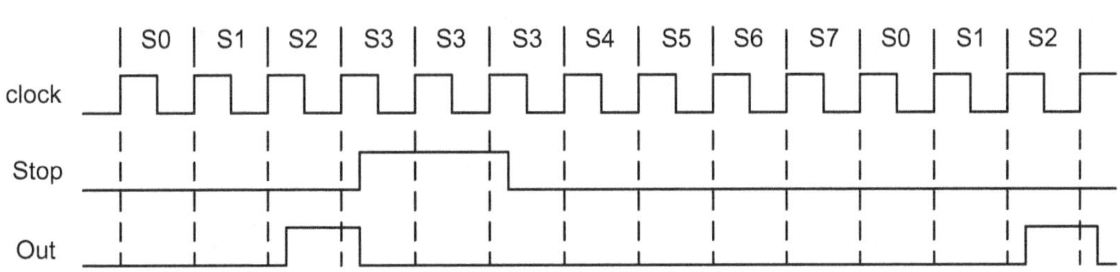

Fig. 2.33 Timing diagram of a state machine with a single input, Stop, and a single output

Once the state assignments are made for each clock cycle in the timing diagram Fig. 2.33, the state diagram for a Moore-type state machine emerges in Fig. 2.34.

The states S0 and S1 in the timing diagram are assigned to the first and second clock cycles, respectively. The third clock cycle is assigned to the S2 state where Out = 1. The fourth clock cycle corresponds to the S3 state. The machine stays in the S3 state as long as Stop = 1. This ranges from the fourth to the sixth clock cycle in the timing diagram. The state assignments from the seventh to the tenth clock cycles become the states S4, S5, S6 and S7. The eleventh clock cycle returns to the S0 state.

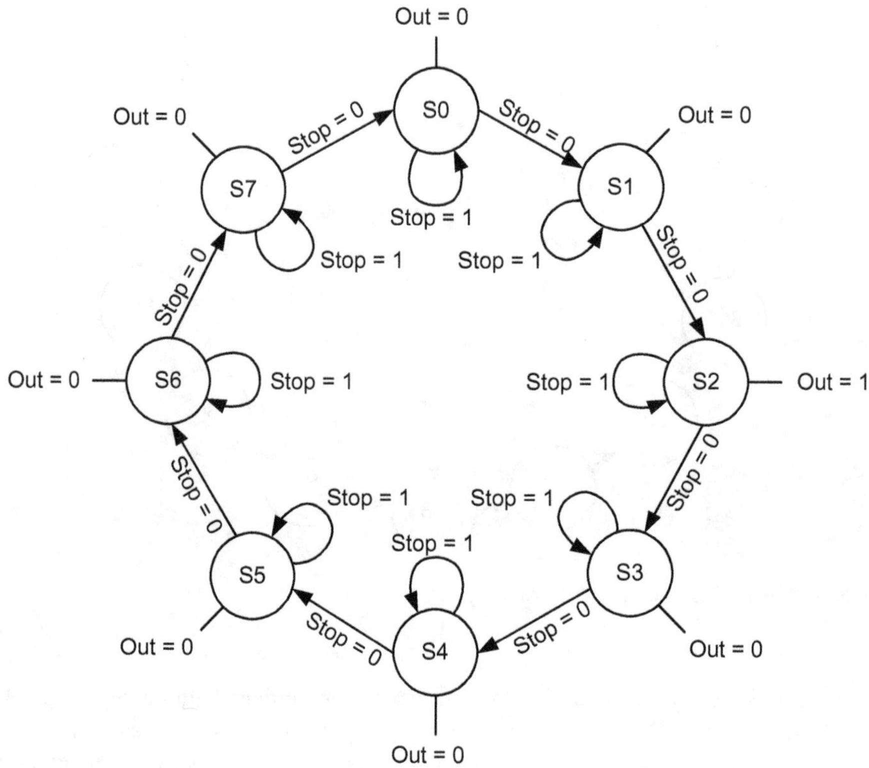

Fig. 2.34 Moore representation of the timing diagram in Fig. 2.33

Implementing the state diagram in Fig. 2.34 follows a lengthy process of producing the state tables, transition tables, and K-maps that results in a total of four outputs (three flip-flop outputs due to eight states and one output for Out). However, using a counter-decoder approach minimizes this design task considerably and reveals a rather explicit circuit implementation.

When the timing diagram in Fig. 2.33 is redrawn to implement the counter-decoder design approach, it yields a simple three-bit counter which counts from zero to seven as shown in Fig. 2.35. The counter output, CountOut, is included in this figure to show the relationships between the state assignments, S0 through S7, the input, Stop, and the output, Out. The figure also shows the clock cycle where the counter resets itself when its output reaches seven.

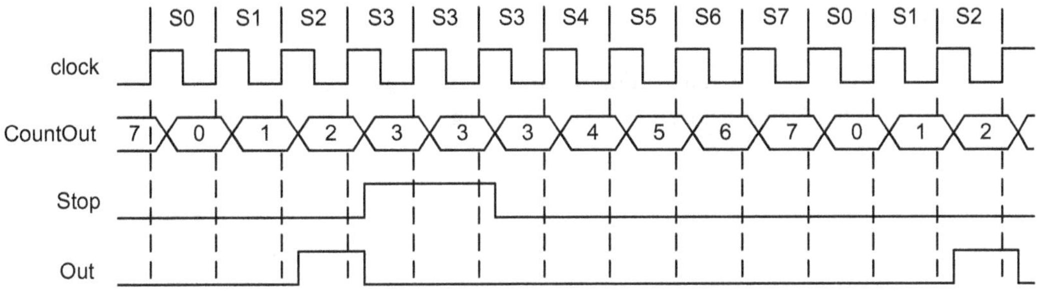

Fig. 2.35 Timing diagram of a three-bit counter with a single input, Stop, and a single output

The first task for the design is to construct a three-bit up-counter as shown in Fig. 2.36. The counter in this figure is derived from a general counter topology, and it consists of a three-bit adder, three 2-1 MUXes and three flip-flops. A three-input AND-gate is used as a decoder at the counter output to implement Out = 1 when the CountOut node reaches 2. Therefore, this method follows a simple, step-by-step design approach in producing the final circuit that does not require implicit logic design techniques.

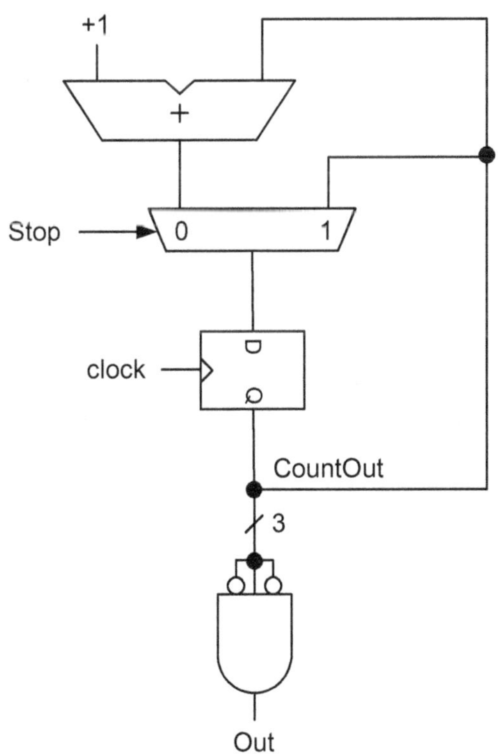

Fig. 2.36 Counter-decoder representation of the timing diagram in Fig. 2.35

2.12 Memory

Small memory blocks can be assembled from one-bit registers in a variety of topologies. For example, a 32-bit wide, 16-bit deep memory block shown in Fig. 2.37 can be built by stacking 16 rows of 32-bit registers on top of each other. Each 32-bit register contains tri-state buffers at its output to be used during read as shown in Fig. 2.38.

All inputs to each column in Fig. 2.37 are connected together to write data. For example, the input terminal, IN[0], in Fig. 2.37 is connected to all the input pins, In[0], from row 0 to row 15 in Fig. 2.38 to be able to write a single bit of data to a selected row. The same is true for the remaining inputs, IN[1] through IN[31].

Similarly, all outputs of each column in Fig. 2.37 are connected together to read data from the memory block. For example, the output pin, OUT[0], is connected to all output pins, Out[0], from row 0 to row 15 in Fig. 2.38 to be able to read one bit of data from a selected row. The same is true for the remaining output pins, OUT[1] through OUT[31].

Every row of the memory block in Fig. 2.37 is accessed by individual Write Enable (WE) and Read Enable (RE) inputs for writing or reading data, respectively.

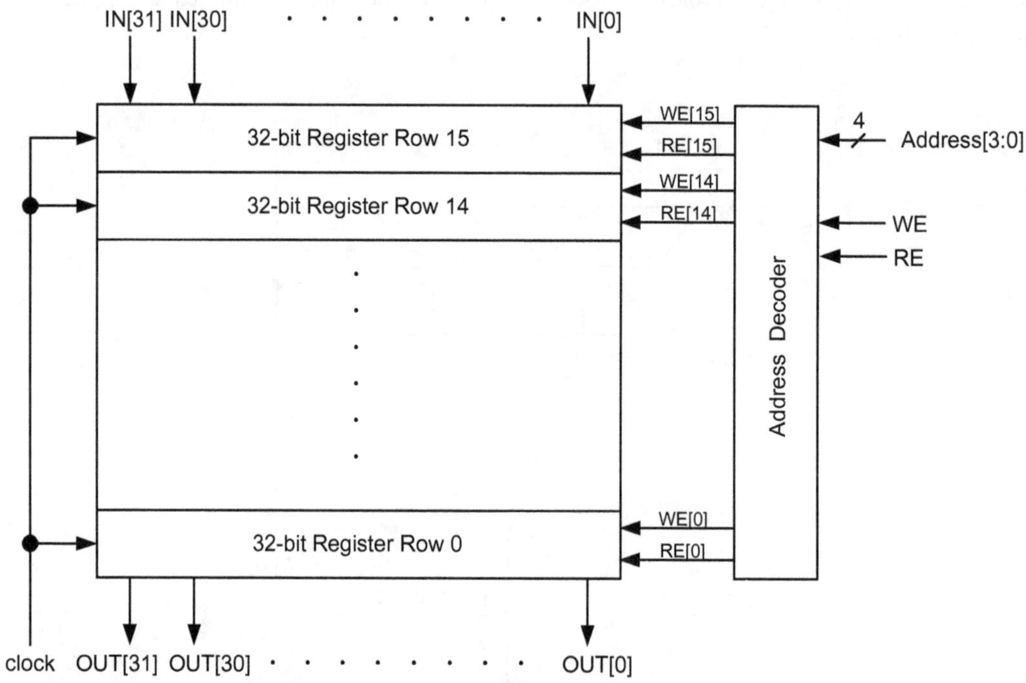

Fig. 2.37 A 32 × 16 memory and the truth table of its address decoder

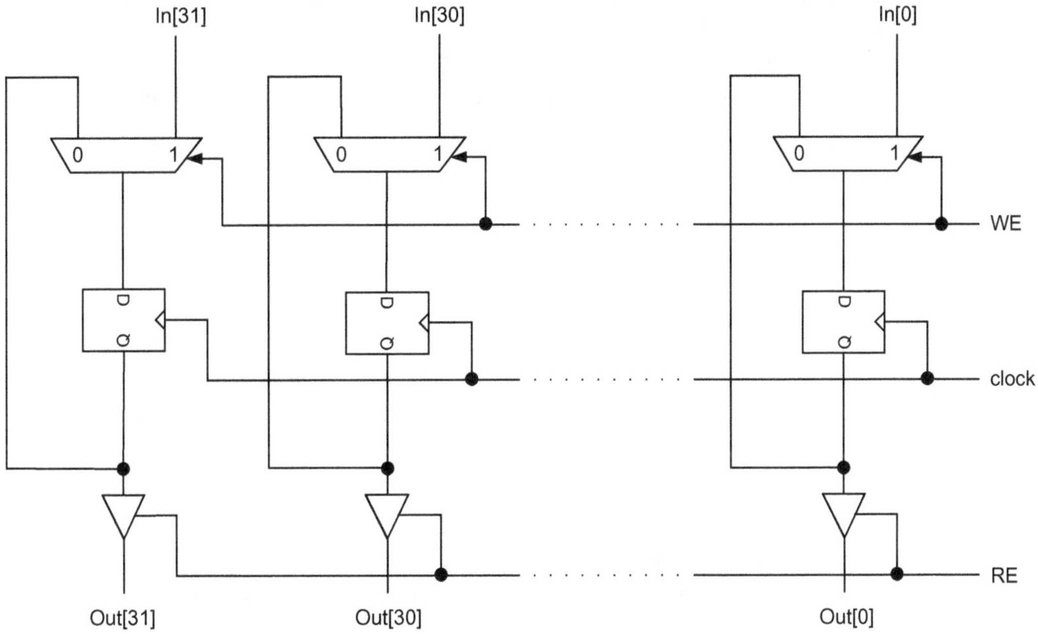

Fig. 2.38 A 32-bit register slice at every row of Fig. 2.37

In order to generate the WE inputs, WE[0] to WE[15], an address decoder is used. This decoder enables only one row while deactivating all the other rows using a four-bit address, Address[3:0], and a single WE input according to the truth table in Fig. 2.39. For example, a 32-bit data is written to row 0 if WE = 1 and Address[3:0] = 0000 at the decoder input. However, WE = 0 blocks writing data to all rows of the memory block regardless of the input address as shown in the truth table in Fig. 2.40.

Address[3:0]	WE[15]	WE[14]	WE[13]		WE[2]	WE[1]	WE[0]
0 0 0 0	0	0	0	. . . .	0	0	1
0 0 0 1	0	0	0	. . . .	0	1	0
0 0 1 0	0	0	0	. . . .	1	0	0
⋮		⋮				⋮	
1 1 1 0	0	1	0	. . . .	0	0	0
1 1 1 1	1	0	0	. . . .	0	0	0

Fig. 2.39 The address decoder for the 32 × 16 memory in Fig. 2.37 when WE = 1

Address[3:0]	WE[15]	WE[14]	WE[13]		WE[2]	WE[1]	WE[0]
0 0 0 0	0	0	0	· · · · · · · · · · · · ·	0	0	0
0 0 0 1	0	0	0	· · · · · · · · · · · · ·	0	0	0
0 0 1 0	0	0	0	· · · · · · · · · · · · ·	0	0	0
⋮		⋮				⋮	
1 1 1 0	0	0	0	· · · · · · · · · · · · ·	0	0	0
1 1 1 1	0	0	0	· · · · · · · · · · · · ·	0	0	0

Fig. 2.40 The address decoder for the 32 × 16 memory in Fig. 2.37 when WE = 0

The RE inputs, RE[0] through RE[15], use an address decoder similar to the one in Figs. 2.39 and 2.40 to read a block of data from a selected row. The read operation is achieved with a valid input address and RE = 1 according to the truth table in Fig. 2.41. An RE = 0 entry disables reading data from any row regardless of the value of the input address as shown in Fig. 2.42.

Address[3:0]	RE[15]	RE[14]	RE[13]		RE[2]	RE[1]	RE[0]
0 0 0 0	0	0	0	· · · · · · · · · · · · ·	0	0	1
0 0 0 1	0	0	0	· · · · · · · · · · · · ·	0	1	0
0 0 1 0	0	0	0	· · · · · · · · · · · · ·	1	0	0
⋮		⋮				⋮	
1 1 1 0	0	1	0	· · · · · · · · · · · · ·	0	0	0
1 1 1 1	1	0	0	· · · · · · · · · · · · ·	0	0	0

Fig. 2.41 The address decoder for the 32 × 16 memory in Fig. 2.37 when RE = 1

Address[3:0]	RE[15]	RE[14]	RE[13]		RE[2]	RE[1]	RE[0]
0 0 0 0	0	0	0	. . . .	0	0	0
0 0 0 1	0	0	0	. . . .	0	0	0
0 0 1 0	0	0	0	. . . .	0	0	0
⋮		⋮				⋮	
1 1 1 0	0	0	0	. . . .	0	0	0
1 1 1 1	0	0	0	. . . .	0	0	0

Fig. 2.42 The address decoder for the 32 × 16 memory in Fig. 2.37 when RE = 0

Therefore, one must provide a valid input address and the control signals, RE and WE, to perform a read or a write operation, respectively. The WE = 0, RE = 1 combination reads data from the selected row. Similarly, the WE = 1 and RE = 0 combination writes data to a selected row. The WE = 0 and RE = 0 combination disables both reading and writing to the memory block. The control input entry, WE = 1 and RE = 1, is not allowed, and it should be interpreted as memory read.

2.13 A Design Example Using Sequential Logic and Memory

This design example combines the data-path and controller design concepts described in this chapter and in Chap. 1. It also introduces the use of important sequential logic blocks such as flip-flop, register, counter and memory in the same design.

Every design starts with gathering the small or large logic blocks to construct a data-path for a proper data-flow according to design specifications. Once the data-path is set, then the precise data movements from one logic block to the next is shown in a timing diagram. Any architectural change in the data-path should follow a corresponding change in the timing diagram or vice versa.

When the data-path design and its timing diagram are complete, and fully associate with each other, the next step in the design process is to build the controller circuit that governs the data-flow. To define the states of the controller, clock periods that generate different sets of controller outputs are separated from each other and named as distinct states. Similarly, clock periods with identical controller outputs can be combined under the same state. The controller design can be Moore-type or Mealy-type state machine according to the design needs. The design methodology of building the data-path, timing diagram and controller shown here will be repeated in every design throughout this book, especially when designing peripherals for a computer system in Chap. 7.

The example design in this section reads two eight-bit data packets from an 8 × 8 source memory (memory A), processes them and stores the result in an 8 × 4 target memory (memory B). The processing part depends on the relative contents of each data packet: if the contents of the first data packet are larger than the second, the contents of the data packets are subtracted from each other before the result is stored. Otherwise, they are added, and the result is stored.

The block diagram in Fig. 2.43 demonstrates the data-path required for this memory-to-memory data transfer as described above. The timing diagram in Fig. 2.44 needs to accompany the data-flow in Fig. 2.43 since it depicts precise data movements at each clock cycle.

To be able to write data to a memory address in Fig. 2.37, a valid data and address must be available within the same clock cycle. In a similar fashion, data is read from the memory core a cycle after a valid address is introduced.

Therefore, counter A generates the addresses 0 to 7 for the memory A and writes the data packets A0 to A7 through DataInA[7:0] port. This is shown in the timing diagram in Fig. 2.44 from clock cycles 1 through 8. When this task is complete, counter A resets and reads the first data packet A0 from AddrA[2:0] = 0 in clock cycle 9. In the next clock cycle, A0 becomes available at DOut1, and the counter A increments by one. In cycle 11, AddrA[2:0] becomes 2, the data packet A1 is read from DOut1[7:0], and the data packet A0 transfers to DOut2[7:0]. In this cycle, the contents of the data packets A0 and A1 are compared with each other by subtracting A1 (at DOut1) from A0 (at DOut2). If the contents of A0 are less than A1, then the sign bit, Sign, of (A0-A1) becomes negative. Sign = 1 selects (A0 + A1) at ADDOut[7:0] and routes this value to DataInB[7:0]. However, if the contents of A0 are greater than A1, (A0 – A1) becomes positive. Sign = 0 selects (A0 – A1) and routes this value from SUBOut[7:0] to DataInB[7:0]. The result at DataInB[7:0] is written at AddrB[1:0] = 0 of memory B at the positive edge of clock cycle 12. In the same cycle, A1 is transferred to DOut2[7:0],

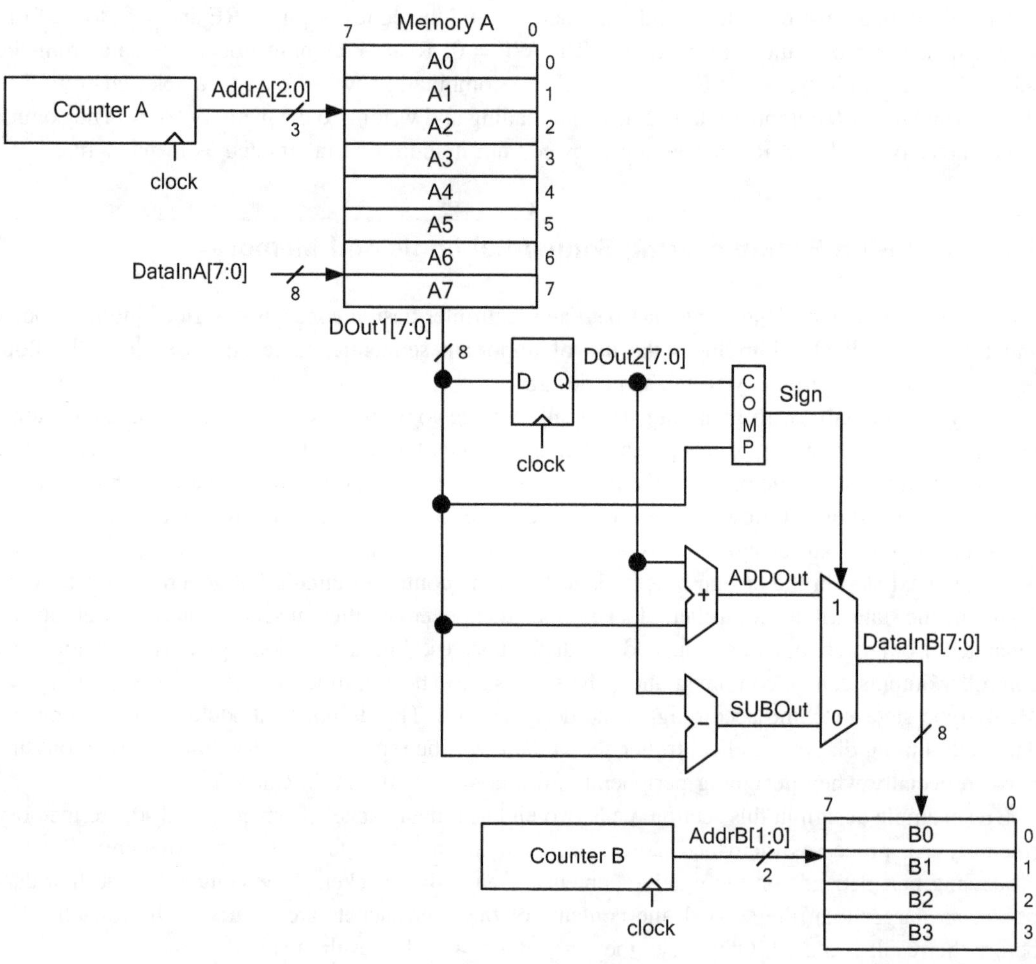

Fig. 2.43 Data-path of a memory transfer example

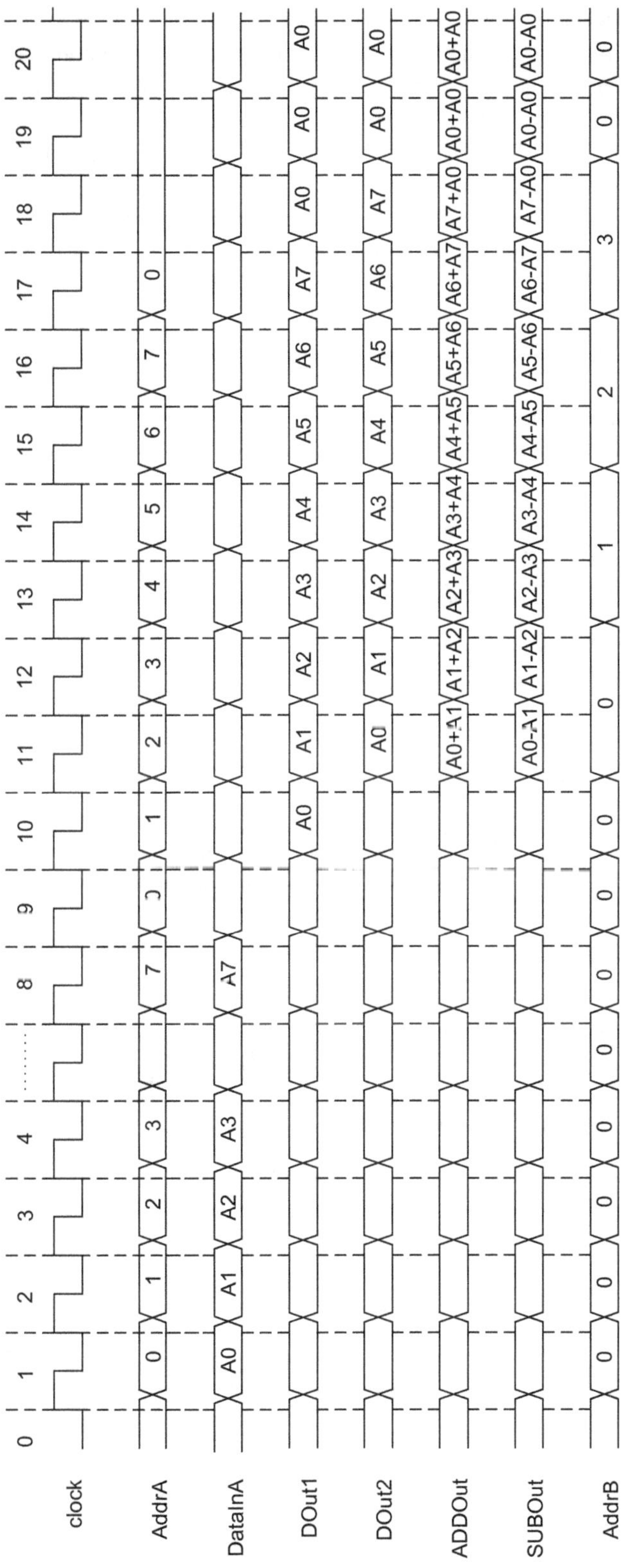

Fig. 2.44 Timing diagram for the memory transfer data-path in Fig. 2.43

and A2 becomes available at DOut1[7:0]. A comparison between A1 and A2 takes place, and either (A1 + A2) or (A1 – A2) is written to memory B depending on the value of the Sign node. However, this is an unwarranted step in the data transfer process because the design requirement states that the comparison has to be done only once between data packets from memory A. Since A1 is used in an earlier comparison with A0, A1 cannot be used in a subsequent comparison with A2, and neither (A1 + A2) nor (A1 − A2) should be written to memory B. The remaining clock cycles from 13 through 18 compare the values of A2 with A3, A4 with A5, and A6 with A7, and write the added or subtracted results to memory B. After clock cycle 19, all operations on this data-path suspend, the counters are reset and all writes to the memory core are disabled.

To govern the data-flow in Fig. 2.44, a Moore-type state machine (or a counter-decoder-type controller) is used. A Mealy-type state machine for a controller design is usually avoided because the present state inputs of this type of state machine may change during the clock period and result in jittery outputs.

The inclusion of the controller identifies the control signals for the data-flow in Fig. 2.45. These signals increment the counters A and B, and enable writes to memory A or B when necessary. Thus,

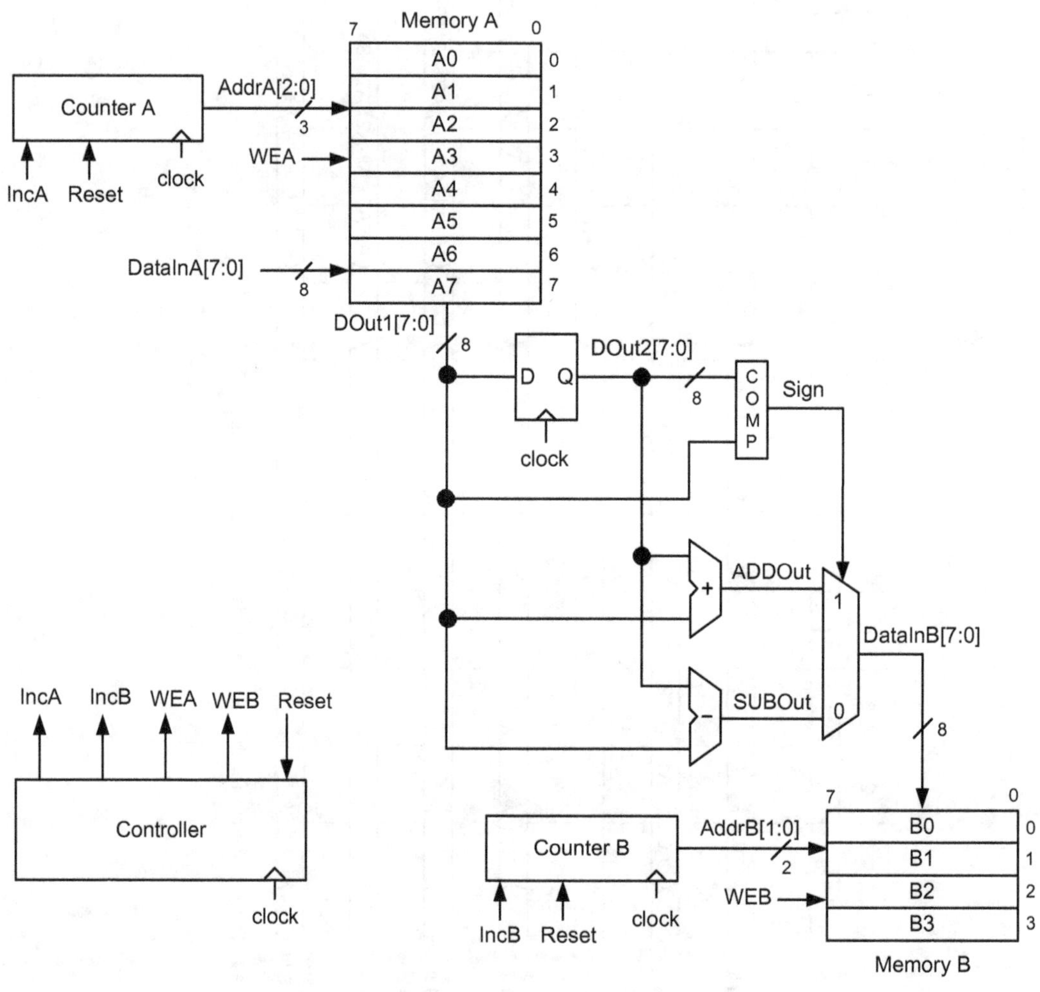

Fig. 2.45 Complete block diagram of the memory transfer example with controller

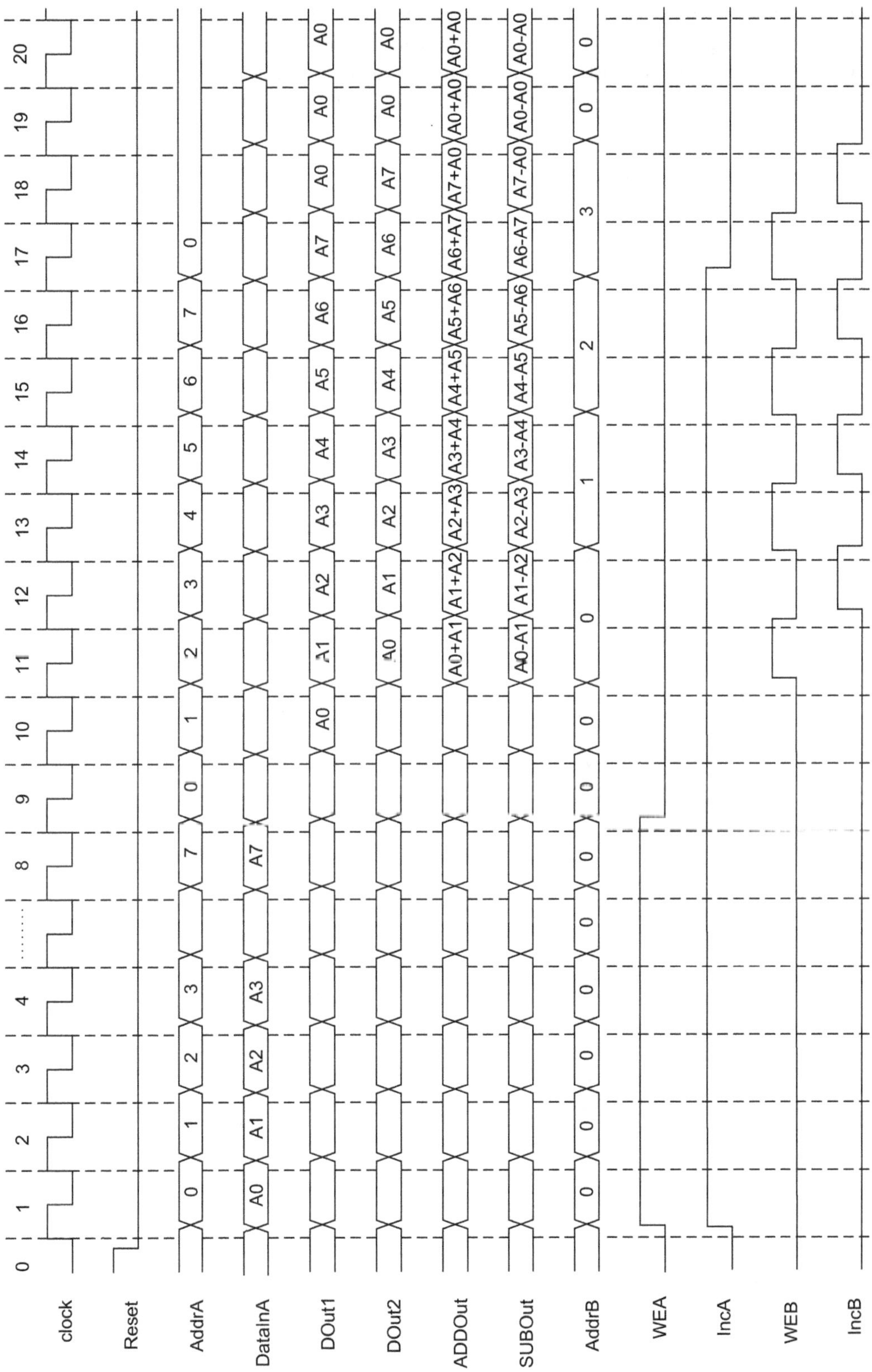

Fig. 2.46 The complete timing diagram for the memory transfer in Fig. 2.45

the timing diagram in Fig. 2.44 is expanded to include the control signals, IncA, IncB, WEA and WEB, as shown in Fig. 2.46, and it provides a complete picture of the data transfer process from memory A to memory B in contrast to the earlier timing diagram in Fig. 2.44.

The controller in Fig. 2.45 is implemented either by a Moore-type state machine in Fig. 2.47 or counter-decoder-type design in Fig. 2.48.

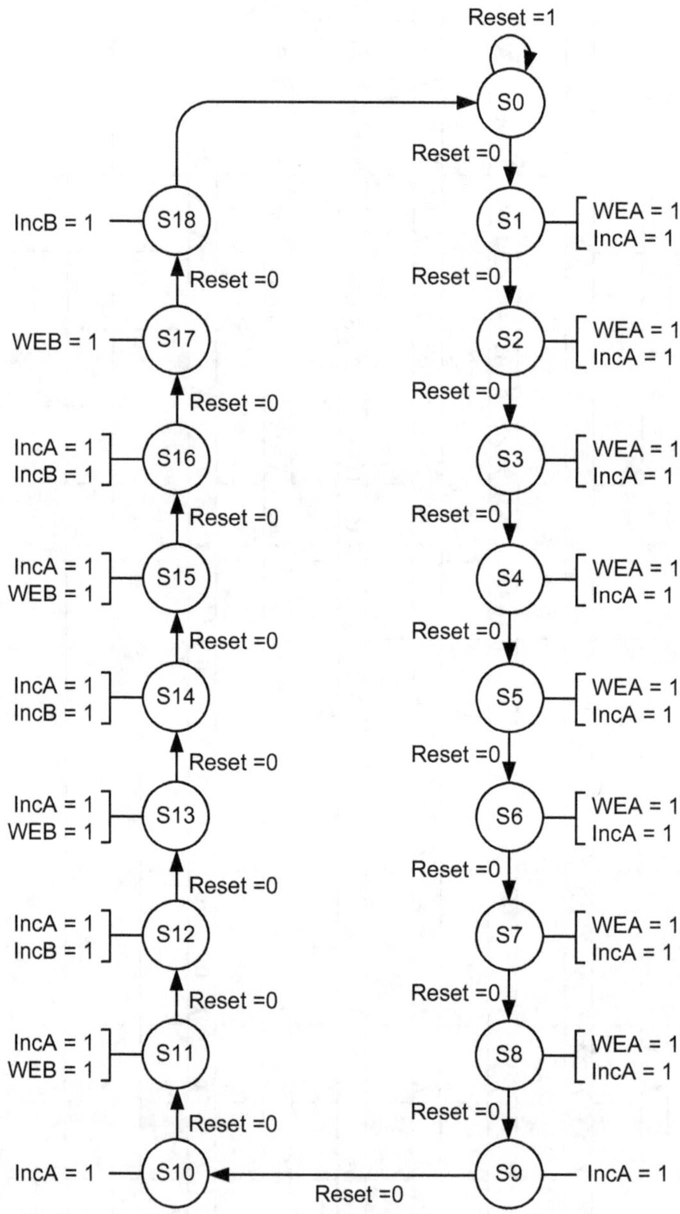

Fig. 2.47 Moore representation of the controller unit in Fig. 2.45 (only control signals equal to logic 1 are included to avoid complexity)

In the Moore type design, the states from S1 through S18 are assigned to each clock cycle of the timing diagram in Fig. 2.46. The values of the present state outputs, WEA, IncA, WEB and IncB, in each clock cycle are read from the timing diagram and attached to each state in Fig. 2.47. The reset state, S0, is included in the Moore machine in case the data-path receives an external reset signal to interrupt an ongoing data transfer process. Whichever state the machine may be in, Reset = 1 always makes the state machine transition to the S0 state. These transitions are not included in Fig. 2.47 for simplicity.

The counter-decoder style design in Fig. 2.48 consists of a five-bit counter and four decoders to generate WEA, IncA, WEB and IncB control signals. To show the operation of this design to generate WEA, for example, this particular decoder includes eight five-input AND gates, one for each clock cycle from cycle 1 to cycle 8 in order to keep WEA = 1 in Fig. 2.46. The five-bit counter implicitly receives a reset signal from its output when it reaches clock cycle 18, and resets counter A, counter B and the rest of the system in Fig. 2.45.

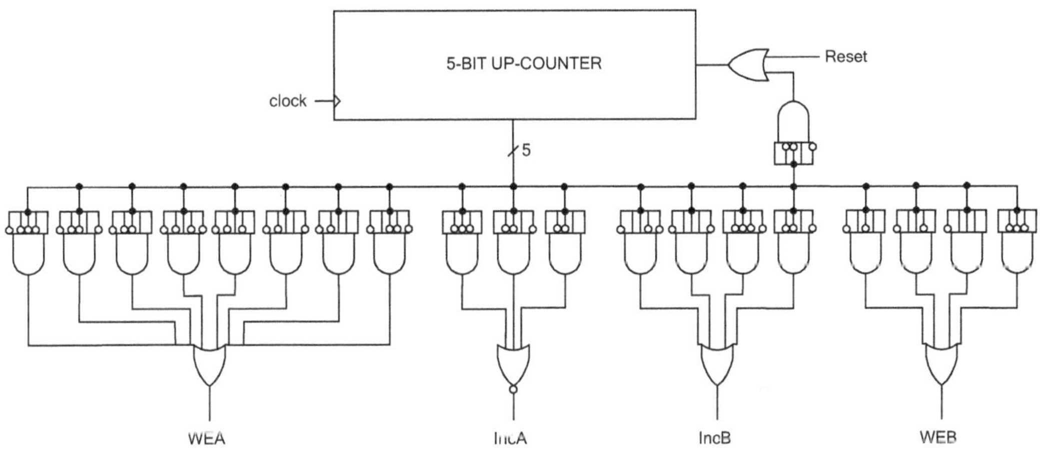

Fig. 2.48 Counter-decoder representation of the controller unit in Fig. 2.45

Review Questions

1. Implement the following Moore machine:

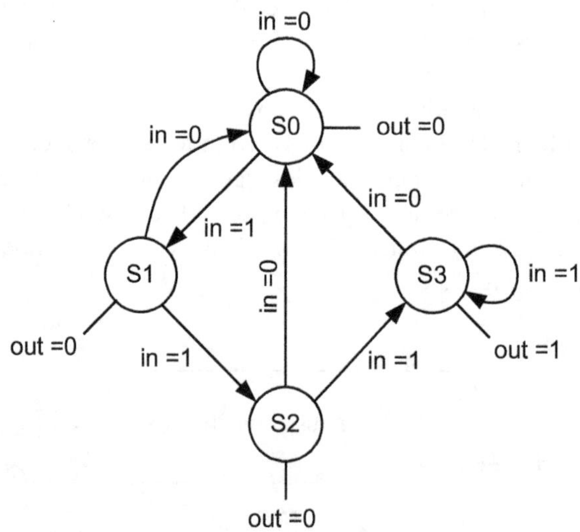

2. Implement the following Moore machine using a timer. The timer is initiated when In = 1. The state machine goes to the A state and stays there for 10 cycles. In the tenth cycle, the state machine transitions to the B state and stays in this state for only one cycle before switching to the IDLE state. One implementation scheme is to construct a four-bit up-counter to generate the timer. When the counter output reaches 9, the decoder at the output of the counter informs the state machine to switch from the A state to the B state.

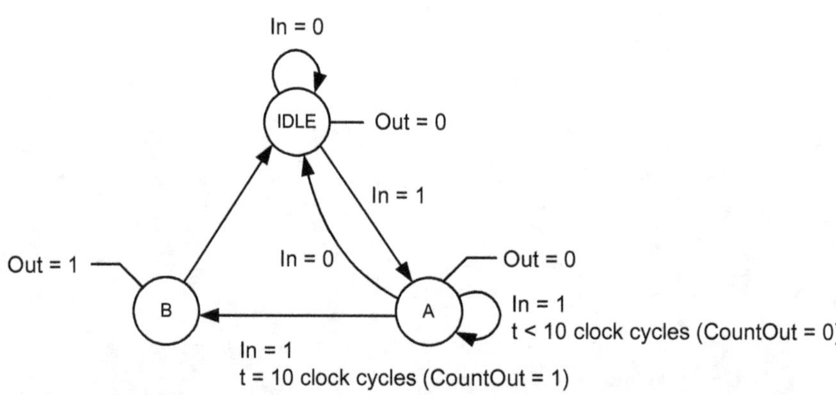

3. The following truth table needs to be implemented using two-input NAND gates and inverters.

A	B	C	Out
0	0	0	0
0	0	1	1
0	1	0	2
0	1	1	3
1	0	0	3
1	0	1	2
1	1	0	1
1	1	1	0

T_{NAND} (two-input NAND gate delay) = 500 ps
T_{INV} (inverter delay) = 100 ps
Tclk-q (clock-to-q delay) = 200 ps
tsu (setup time) = 200 ps
th (hold time) = 300 ps

(a) Implement this truth table between two flip-flop boundaries.
(b) Find the maximum clock frequency using a timing diagram.
(c) Shift the clock by 500 ps at the receiving flip-flop boundary. Show whether or not there is a hold violation using a timing diagram.

4. A block diagram is given below:

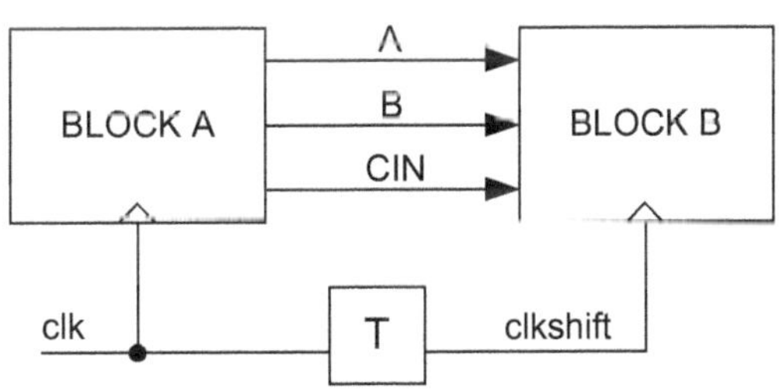

Block B contains a one-bit adder with SUM = A ⊕ B ⊕ CIN, COUT = A . B + CIN . (A + B), and two flip-flops as shown below:

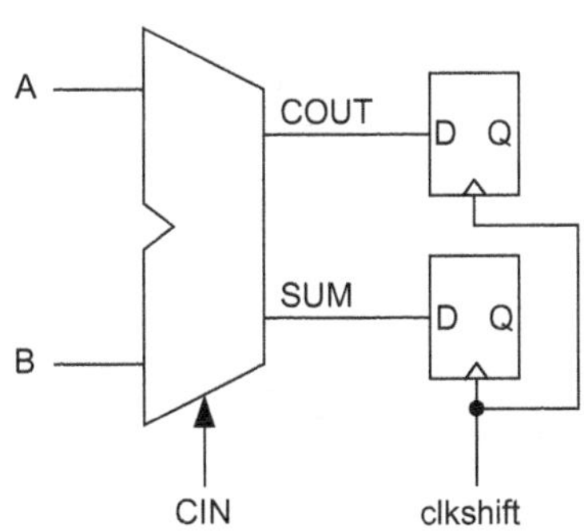

(a) Using the gates with propagation delays above, determine the setup time of A, B, and CIN with respect to clkshift.

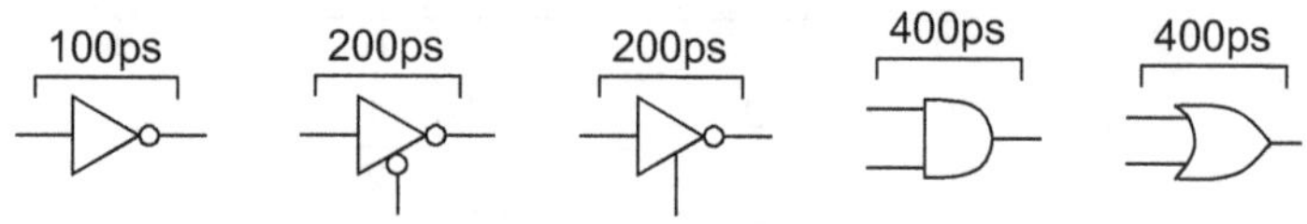

(b) Assuming T = 0 ns and T_{CLK} (clock period) = 5 ns, if data at A, B and CIN become valid and stable 4 ns after the positive edge of clkshift, will there be any timing violations? Assume t_H (hold time) = 3 ns for the flip-flop.

(c) How can you eliminate the timing violations? Show your calculations and draw a timing diagram with no timing violations.

5. A schematic is given below:

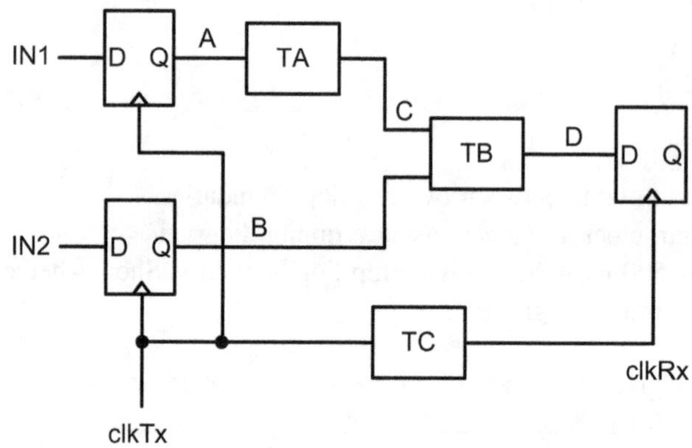

(a) If t_{SU} (setup time) = 200 ps, t_H (hold time) = 200 ps and $t_{CLK\text{-}Q}$ (clock-to-q delay) = 300 ps for the flip-flop, and TA = 1000 ps, TB = 100 ps for the internal logic blocks on the schematic, show if there is any timing violation or timing slack in a detailed timing diagram if TC = 0 ps.

(b) What happens if TC = 400 ps? Show it in a separate timing diagram.

6. The state diagram of a Moore machine is given below:

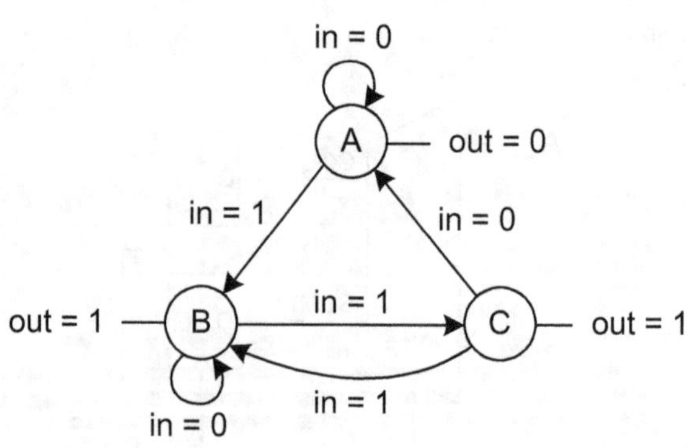

The assignment of the states A, B and C are indicated as follows:

states	PS[1]	PS[0]
A	0	0
B	0	1
C	1	1

(a) Implement this state machine using inverters, two-input and three-input AND gates and two-input OR gates.

(b) Find the maximum operating frequency of the implementation in part (a) if the following timing assignments are applied:
t_{SU} (setup time) = 100 ps, t_H (hold time) = 100 ps, $t_{CLK\text{-}Q}$ (clock-to-q delay) = 200 ps, T_{INV} = 200 ps, T_{AND2} = 300 ps, T_{AND3} = 400 ps, T_{OR2} = 400 ps.

7. Data is transferred from Memory Tx to Memory Rx starting from the address 0x00 and ending at the address 0x0F as shown below. Once a valid address is produced for Memory Tx, the data is read from this address at the next positive clock edge. On the other hand, data is written to the Memory Rx at the positive edge of the clock when a valid address is produced. The operating clock frequency of Memory Tx is twice the clock frequency of Memory Rx.

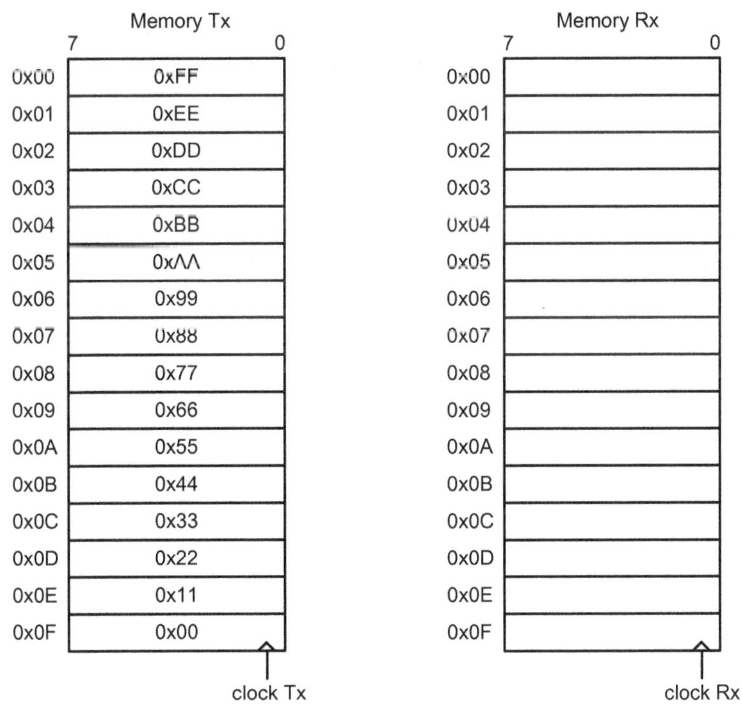

(a) Assuming address generators for Memory Tx and Memory Rx start generating valid addresses at the same positive clock edge, show which data is actually stored in Memory Rx using a timing diagram. Indicate all the address and data values for Memory Tx and Memory Rx in the timing diagram.

(b) Now, assume that the operating clock frequency of Memory Tx is four times higher than the clock frequency of Memory Rx, and the write to Memory Rx takes place at the negative edge of the clock when a valid address is present. Redraw the timing diagram indicating all address and data values transferred from Memory Tx to Memory Rx.

8. Serial data is transferred to program four eight-bit registers. The start of the transfer is indicated by a seven-bit sequence = {1010101} immediately followed by the address of the register (two bits) and the data (eight bits). The transfer stops after programming the last register. After this point, all other incoming bits at the serial input are ignored. Design this interface by developing a data-path and a timing diagram simultaneously. Implement the state diagram. Can this controller be implemented by a counter-decoder scheme?

Projects

1. Implement the one-bit register and verify its functionality using Verilog. Use timing attributes in the flip-flop and the multiplexer to create gate propagation delays. Change the clock frequency until set-up time violation is produced.
2. Implement the four-bit shift register and verify its functionality using Verilog. Use timing attributes for flip-flops and the multiplexers to create gate propagation delays. Examine the resulting timing diagram.
3. Implement the 32-bit counter and verify its functionality using Verilog.
4. Implement the four-state Moore-type state machine and verify its functionality using Verilog.
5. Implement the four-state Mealy-type state machine and verify its functionality using Verilog.
6. Implement the three-bit counter-decoder and verify its functionality using Verilog and examining the resulting timing diagram.
7. Implement the 32 × 16 memory block using Verilog. How can this memory be verified functionally?
8. Implement the memory-to-memory transfer circuit at the end of this chapter and verify its functionality using Verilog.

3 Review of Asynchronous Logic Circuits

A digital system is often comprised of different time domains. Some domains work with clock, and data is sequentially transferred from one flip-flop (or latch) boundary to the next. In other domains, data is asynchronously processed and handled without the aid of a clock.

This chapter introduces asynchronous circuits that require no clock input. The complete design methodology is given in terms of state assignments from timing diagrams, construction of flow tables and gate minimization, which then leads to the implementation of fundamental mode circuits. The chapter concludes with an asynchronous timing methodology with C (Mueller) elements that allows data propagation between logic blocks without any clock input.

3.1 S-R Latch

A common storage element in asynchronous circuits is a Set-Reset (S-R) latch. This circuit is composed of two NAND gates whose outputs are connected to their inputs as shown in Fig. 3.1.

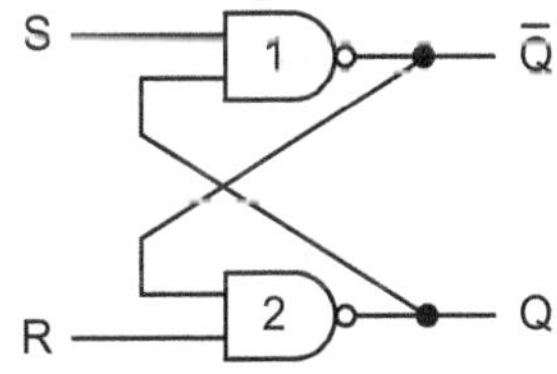

Fig. 3.1 S-R latch

Initially, both S and R inputs may be at logic 0, producing $Q = \overline{Q} = 1$. If the S input transitions to logic 1 while R = 0, Q stays at logic 1 and $\overline{Q}$ transitions to logic 0. This state is called the set state of the S-R latch. If, on the other hand, the R input goes to logic 1 while the S input stays at 0, Q transitions to logic 0 and $\overline{Q}$ stays at logic 1. This state is called the reset state. Simultaneously changing both the S and R inputs from logic 0 to logic 1 causes a racing condition. If NAND gate number 1 has a shorter gate delay than the NAND gate number 2, $\overline{Q}$ switches to logic 0 first, and forces Q to stay at logic 1. If NAND gate number 2 has a shorter gate delay, Q switches to logic 0 first. Therefore, simultaneously switching more than one input in asynchronous circuits creates unexpected outputs due to multiple racing paths in a circuit. The fundamental-mode design methodology corrects this problem by permitting only one input to change, and eliminates all unwanted transitions in the circuit.

A. Bindal, *Fundamentals of Computer Architecture and Design*,
https://doi.org/10.1007/978-3-030-00223-7_3

3.2 Fundamental-Mode Circuit Topology

An asynchronous circuit requires no clock input to operate. The circuit is composed of a combinational logic block and a delay block. The delay block is another combinational logic circuit and its inputs constitute the present state for the circuit. The outputs of the delay block are fed back to the inputs of the combinational logic to form the next state as shown in Fig. 3.2.

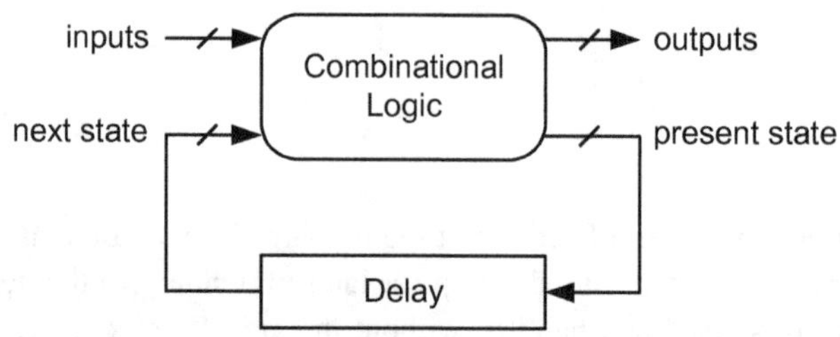

Fig. 3.2 Fundamental mode asynchronous circuit topology

Designing an asynchronous circuit requires to follow a certain procedure. The first task is to form a primitive state flow table tabulating all possible states and transitions that the asynchronous circuit can produce. This table must contain only one stable state per row to maintain the fundamental mode of operation. Similarly, another table, containing only the outputs of the circuit, is formed. This table should include every output change as the circuit makes a transition from one state to another. Implication tables aim to minimize the primitive state and output tables prior to producing a final circuit. In asynchronous circuit design, it is common to create race conditions where circuit delays produce multiple simultaneous state transitions resulting in unwanted outputs. An effective method to eliminate racing conditions is to work on the minimized state table and remove such state transitions that cause the circuit to depart from its fundamental-mode of operation.

3.3 Fundamental-Mode Asynchronous Logic Circuits

In this section, an example will be used to present the entire process designing fundamental-mode asynchronous circuits from the creation of primitive flow tables to the removal of racing conditions.

The circuit in this example consists of two inputs, in1 and in2, and a single output. The output is at logic 0 whenever in1 = 0. The first change in in2 produces out = 1 while in1 = 1. The output transitions back to logic 0 when in1 switches to logic 0.

The timing diagram in Fig. 3.3 summarizes the behavior of this circuit. State assignments in the timing diagram follow the basic rule that requires the change in only one input per state. All stable states are numbered and circled.

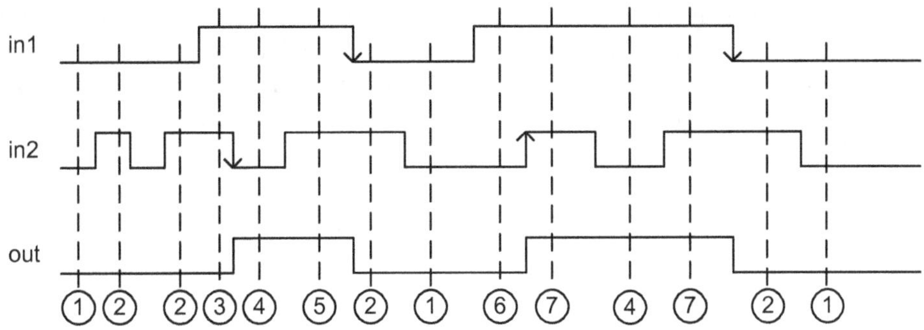

Fig. 3.3 Timing diagram and state assignments

The state ① is when in1 = in2 = 0 and out = 0. Any change in in2 transitions the circuit to the state ② with out = 0. Additional ripples at in2 while in1 = 0 change the state of the circuit between the states ① and ②. Switching in1 from logic 0 to logic 1 while in2 = 1 forms a new state, state ③, and produces out = 0. The first change in in2 while in1 = 1 creates another new state, state ④, with out = 1. As in2 transitions back from logic 0 to logic 1 while in1 = 1, the state of the circuit changes to the state ⑤, but the output stays at out = 1. Further ripples at in2 while in1 = 1 create no change at the output, and the circuit ends up transitioning between the states ④ and ⑤. When in1 transitions to logic 0 while in2 = 1, the state of the circuit switches back to the state ②. Finally, when in2 also switches back to logic 0, the state of the circuit becomes the state ①.

Now, let us assume that in1 transitions to logic 1 first while in2 is steady at logic 0. This condition creates a new state, state ⑥, with an output value of out = 0. When in2 also transitions to logic 1 while in1 = 1, the state of the circuit changes from the state ⑥ to a new state, state ⑦, and out becomes logic 1. As soon as in2 goes back to logic 0 while in1 = 1, the circuit also switches back to the state ④ but out stays at logic 1. Further ripples in in2 while in1 = 1 simply change the state of the circuit between the states ④ and ⑦. As soon as in1 transitions back to logic 0, the circuit goes to the state ②. When in2 also changes to logic 0, the circuit goes back to the state ①.

Constructing primitive state table and output flow table is the direct extension of the timing diagram in Fig. 3.3. When transferring stable, circled states from the timing diagram to the primitive state table in Fig. 3.4, the fundamental-mode rule that enforces one stable state per row is strictly observed. The non-circled states in this table are considered "transitionary" states: the circuit momentarily stays in these states until it makes a permanent move to a stable state. For example, as in2 transitions from logic 0 to logic 1 while in1 stays at logic 0 in the first row of Fig. 3.4, the state of the circuit changes from the stable state ① to the transitionary state 2. The circuit stays in this transitionary state only for a brief moment until it finally transitions to the stable state ② in row 2. Similarly, as in1 switches from logic 0 to logic 1 while in2 = 1, the circuit transitions from the state ② to the state ③ through a transitionary state 3 in the second row. All simultaneous dual input transitions are forbidden because of the primary rule in the fundamental mode of operation. Therefore, a transition from in1 = in2 = 0 to in1 = in2 = 1 in the first row is not allowed. In this figure, the boxes marked with "-" indicate the forbidden transitions.

Primitive state and output flow tables are usually integrated to produce a compact table as shown in Fig. 3.5. Furthermore, labeling the present and next states clarifies the state transitions and the output values during each transition. Figure 3.5 also separates the states that produce out = 0 from the states that produce out = 1. This design practice comes in handy during the minimization step when equivalence classes are formed.

in1 in2: 00	01	11	10
①	2	-	6
1	②	3	-
-	2	③	4
1	-	5	④
-	2	⑤	4
1	-	7	⑥
-	2	⑦	4

in1 in2: 00	01	11	10
0			
	0		
		0	
			1
		1	
			0
		1	

Fig. 3.4 Primitive state and output flow tables for Fig. 3.3

ps	in1 in2				in1 in2			
	00	01	11	10	00	01	11	10
①	①	2	-	6	0			
②	1	②	3	-		0		
③	-	2	③	4			0	
⑥	1	-	7	⑥				0
④	1	-	5	④				1
⑤	-	2	⑤	4			1	
⑦	-	2	⑦	4			1	
	ns				out			

Fig. 3.5 Integrated primitive state and output flow tables

The first step towards state minimization is to form an implication table as shown in Fig. 3.6. This table includes all permitted, forbidden and "implied" states that can either go into the permitted or forbidden categories.

2	✓					
3	4-6	✓				
6	✓	3-7	3-7 4-6			
4	4-6	3-5	3-5	X		
5	4-6	3-5	X	5-7 4-6	✓	
7	4-6	3-7	X	4-6	5-7	✓
	1	2	3	6	4	5

Fig. 3.6 The implication table for Fig. 3.5

Figure 3.6 must display all the present states in Fig. 3.5 except the state 7 (the last one on the list) in the horizontal axis and the same states except the state 1 (the first in the list) in its vertical axis.

The box located at (2, 1) in Fig. 3.6 is checked because there are no other states involved in a transition from the state ① to the state ② in Fig. 3.5. The box at (3, 1) contains 4–6, because the "implied" states 6 and 4 in the column, in1 in2 = 10, of Fig. 3.5 must be traversed in order to go from the state ① to the state ③. The columns, in1 in2 = 00 and 11, contain forbidden transitions, and they cannot be used as implied states to allow a transition from the state ① to the state ③. The only other column for a transition from the state ① to the state ③ is at the column, in1 in2 = 01, and it requires a transition from the state 2 to the state 2, which is not possible. The box at (4, 6) in Fig. 3.6 contains an "X" mark because the outputs produced at the states ④ and ⑥ are different. The same applies to boxes at (5, 3) and (7, 3). The box at (6, 3) contains two implied transitions, 3–7 and 4–6, which correspond to the columns, in1 in2 = 11 and 10, respectively.

We can further eliminate some of the implied state entries in the implication table of Fig. 3.6 if these states are "related" to the boxes that have already been crossed out. Figure 3.7 shows the new implication table after eliminations. In this table, the boxes that contain the transitionary states 4–6, 3–5 and 3–7 are safely crossed out since they are related to the boxes at (4, 6), (5, 3) and (7, 3).

2	✓					
3	X	✓				
6	✓	X	X			
4	X	X	X	X		
5	X	X	X	X	✓	
7	X	X	X	X	5-7	✓
	1	2	3	6	4	5

Fig. 3.7 The implication table after eliminations

Forming the equivalence class table is the next step for minimization. First, all states in the horizontal axis of the implication table in Fig. 3.7 are repeated backwards under the "States" column in the equivalence class table in Fig. 3.8.

Column 5 in Fig. 3.7 contains a checked box. This box corresponds to a joint state, (5–7), listed as an equivalence class in the first row of Fig. 3.8. Column 4 in Fig. 3.7 consists of a checked box at (4, 5) and a second box containing an implied state 5–7 at (4, 7). Therefore, the second row of Fig. 3.8 contains two new joint states, (4, 5) and (4, 7), as well as the earlier joint state, (5, 7), from the first row. Since in this row the joint states, (5, 7), (4, 7) and (4, 5), overlap, they can be combined in a compact joint state (4, 5, 7). Columns 6 and 3 in Fig. 3.7 have crossed-out boxes that contain no implied states. Therefore, the third and the fourth row of Fig. 3.8 do not have new state entries, but only a single combined state carried over from the second row. Column 2 in Fig. 3.7 has also crossed-out boxes except at (2, 3), and this produces a new joint state, (2, 3), in the fifth row of Fig. 3.8. Finally, column 1 in Fig. 3.7 contains two checked boxes at (1, 2) and (1, 6), which form two additional joint states, (1, 2) and (1, 6), in the last row of Fig. 3.8. Therefore, the final equivalence class list includes the joint states, (4, 5, 7), (2, 3) and (1, 6). The joint state (1, 2) is a shared state between (2, 3) and (1, 6), and it is absorbed in the final list.

States	Equivalence Classes
5	(5, 7)
4	(5, 7) (4, 7) (4, 5) = (4, 5, 7)
6	(4, 5, 7)
3	(4, 5, 7)
2	(4, 5, 7) (2, 3)
1	(4, 5, 7) (2, 3) (1, 2) (1, 6)
Final List	(4, 5, 7) (2, 3) (1, 6)

Fig. 3.8 Equivalent class table

The final equivalence class list indicates the presence of only three states. Therefore, the same number of states must be present in the minimal state flow table in Fig. 3.9, where the joint states, (1, 6), (2, 3) and (4, 5, 7) are assigned to the states A, B and C, respectively.

ps	in1 in2				in1 in2			
	00	01	11	10	00	01	11	10
(1, 6) = A	Ⓐ	B	C	Ⓐ	0	-	-	0
(2, 3) = B	A	Ⓑ	Ⓑ	C	-	0	0	-
(4, 5, 7) = C	A	B	Ⓒ	Ⓒ	-	-	1	1
	ns				out			

Fig. 3.9 Minimized integrated primitive state and output flow tables

In Fig. 3.5, the present states ① and ⑥ correspond to the states ① and 1 when in1 in2 = 00. However, the states ① and 1 now belong to the new assigned state A in Fig. 3.9. Therefore, the present state A transitions to a stable state A with out = 0 when in1 in2 = 00. Similarly, the present states ① and ⑥ in Fig. 3.5 correspond to the states 2 and "-" when in1 in2 = 01, and produce no output. In the new table, this translates to a transition from the present state A to the stable state B when in1 in2 = 01. The other entries generating the next state, ns, and out in Fig. 3.9 are formed in a similar manner.

The three states, A, B and C, in Fig. 3.9 require two next state bits, ns1 and ns2. The state assignments are shown in Fig. 3.10.

	ns1	ns2
A	0	0
B	0	1
C	1	1

Fig. 3.10 State assignments

Employing the state assignment table in Fig. 3.10 in the combined minimal state and output flow table in Fig. 3.9 leads to the state and output K-maps in Fig. 3.11. The crossed-out entries in this figure correspond to "don't care" conditions, and they are treated as either logic 0 or logic 1 to achieve the minimal sum of products (SOP) expression for each K-map.

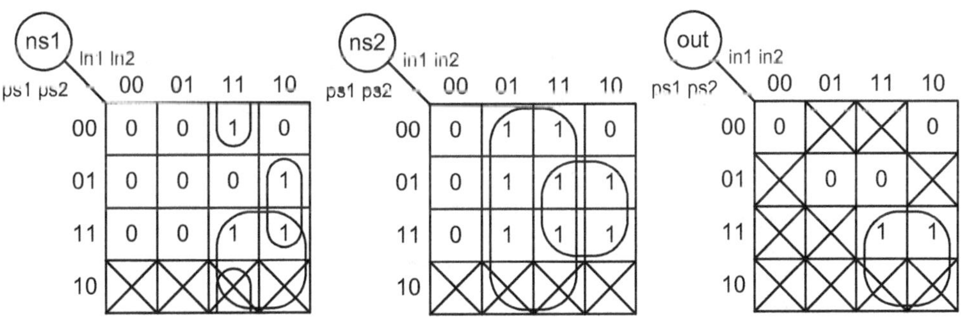

$$ns1 = in1.ps1 + in1.\overline{in2}.ps2 + in1.in2.\overline{ps2}$$
$$ns2 = in2 + in1.ps2$$
$$out = in1.ps1$$

Fig. 3.11 K-maps for the next states, ns1 and ns2, and the output, out

Finally, the SOP expressions for ns1, ns2 and out in Fig. 3.11 generate the circuit diagram in Fig. 3.12. To draw the complete circuit, first combinational logic blocks for ns1, ns2 and out are formed using their SOP expressions. Then, each next state, ns1 and ns2, is connected to its corresponding present state, ps1 and ps2, to complete the circuit diagram in Fig. 3.12.

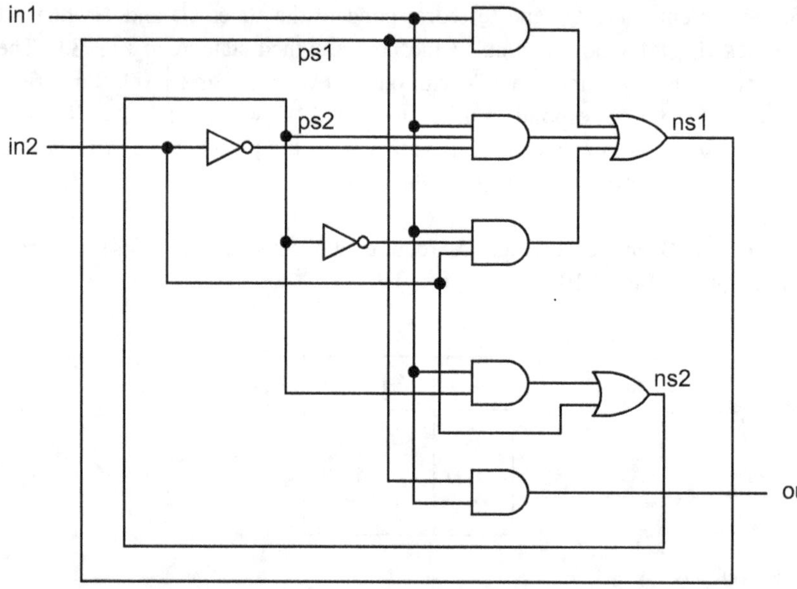

Fig. 3.12 Fundamental mode asynchronous circuit for Fig. 3.3

Even though the circuit in Fig. 3.12 represents the required state behavior, it may not eliminate all possible racing conditions. When the state table in Fig. 3.9 is transformed into a state diagram in Fig. 3.13, the forward and backward transitions between the states A and C may induce racing conditions because two inputs change simultaneously.

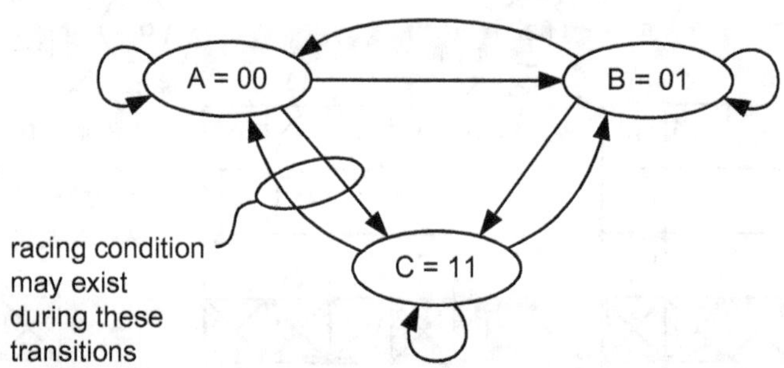

Fig. 3.13 State diagram showing possible racing conditions

To prevent racing conditions, a fictitious fourth state is introduced between the states A and C in Fig. 3.13. This fourth state, α, forms a bridge when going from the state A to the state C (or vice versa) and allows only one state input to change to prevent possible hazards as shown in Fig. 3.14.

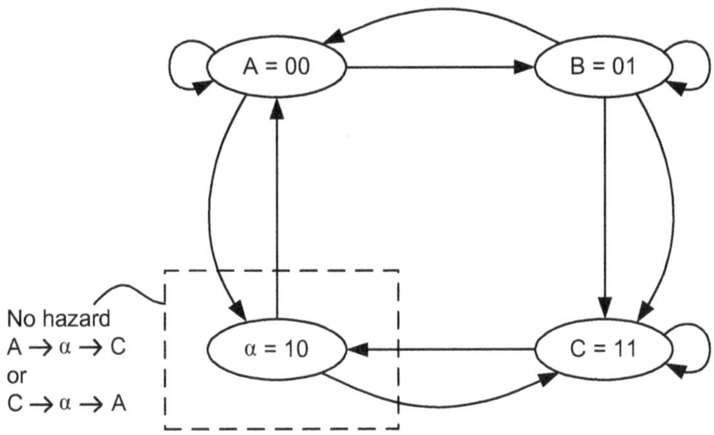

Fig. 3.14 Reconfiguration of the state diagram to eliminate racing conditions

However, the inclusion of the new state, α, necessitates reconfiguring the original state table in Fig. 3.9. The new state table contains the state α as a transitionary state in Fig. 3.15, and the transitions into this state do not produce any output.

ps	ns: in1 in2 = 00	01	11	10	out: in1 in2 = 00	01	11	10
a	(a)	b	α	(a)	0	-	-	0
b	a	(b)	(b)	c	-	0	0	-
c	α	b	(c)	(c)	-	-	1	1
α	a	-	c	-	-	-	-	-

Fig. 3.15 Integrated state and output flow tables without racing conditions

When the hazard-free state and output K-maps are formed based on the flow table in Fig. 3.15, the resultant SOP expressions for ns1 and ns2 in Fig. 3.16 contain only an additional term with respect to the ones in Fig. 3.11. This is a small price to pay in the total gate count for the benefit of eliminating all racing conditions.

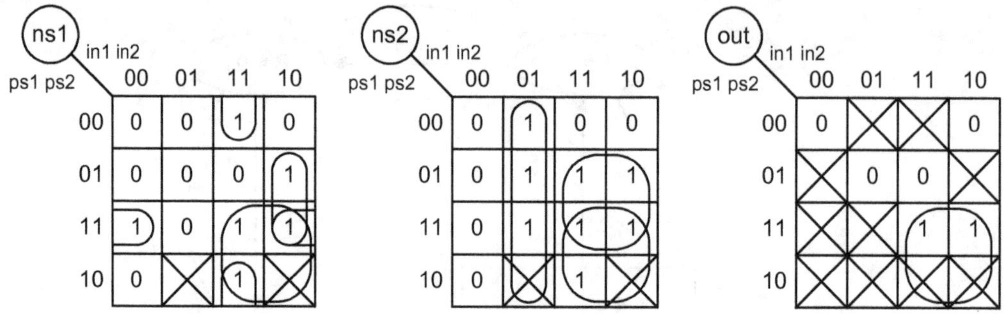

$$ns1 = in1.ps1 + in1.\overline{in2}.ps2 + \overline{in2}.ps1.ps2 + in1.in2.\overline{ps2}$$

$$ns2 = \overline{in1}.in2 + in1.ps2 + in1.ps1$$

$$out = in1.ps1$$

Fig. 3.16 Next state and output K-maps producing no racing conditions

3.4 Asynchronous Timing Methodology

Asynchronous data propagation through combinational logic blocks can be achieved using C-elements (Mueller elements) as shown in Fig. 3.17.

In this figure, combinational logic blocks, CL1, CL2 and CL3, are connected through flip-flops to propagate data. However, data propagation through the combinational logic does not have to be complete within a fixed clock period as in conventional sequential circuits. The C-elements in conjunction with inverting delay blocks, D1 through D3, allow variable data propagation to take place for each stage.

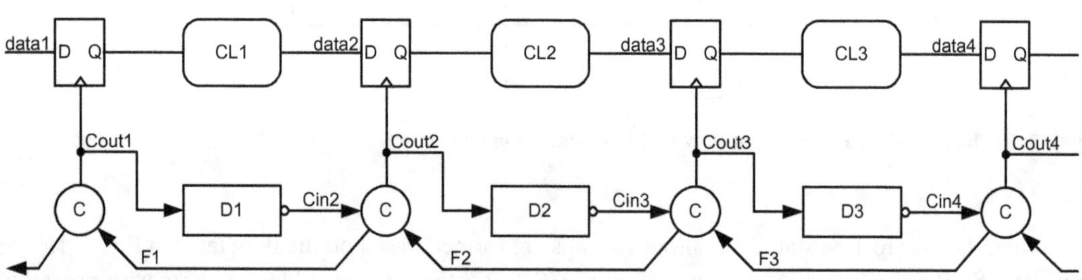

Fig. 3.17 Asynchronous timing methodology using C-elements

As data propagates through a particular combinational logic block, the positive edge produced at the Cout terminal of a C-element also propagates through the corresponding delay circuit. When data reaches the next flip-flop boundary, the C-element in this stage also produces a positive Cout edge for the flip-flop to fetch the incoming data.

The details of variable data propagation in Fig. 3.17 are illustrated in the timing diagram in Fig. 3.18. In this figure, the C-element produces a positive edge at Cout1 and enables the flip-flop to dispatch data1 from its output after a clock-to-q delay. As this data propagates through the combinational logic block, CL1, the positive edge of Cout1 also travels through the delay block, D1, and reaches the next C-element to form a positive edge at Cout2 to fetch the incoming data.

Data propagation in the second stage and the positive edge formation of Cout2 is identical to the first stage with the exception of longer propagation delays, CL2 and D2. The third stage presents a much smaller propagation delay, CL3, and requires a smaller delay element, D3.

Even though each combinational data-path delay in Fig. 3.18 is approximately twice as large as the propagation delay of its corresponding delay element, the flip-flop set-up time, tsu, must be taken into account to fine-tune the length of delay for each delay element.

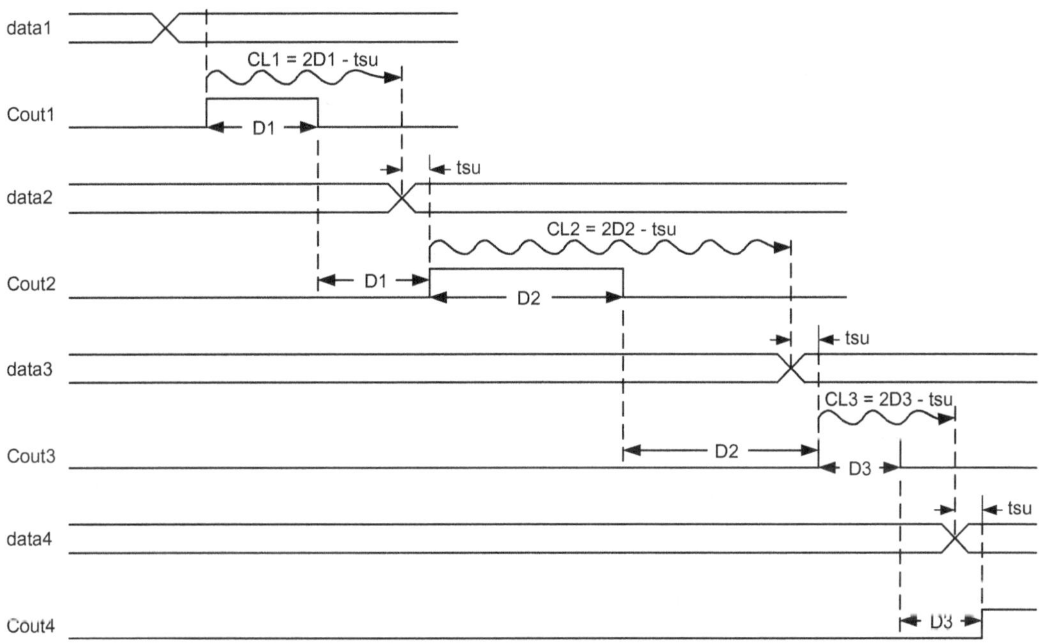

Fig. 3.18 A timing diagram with variable clock lengths and stage delays

Figure 3.19 shows the detailed Input/Output timing diagram of the C-element. In the first stage of the data-path (CL1 in Fig. 3.17), a positive edge at Cout1 travels though the inverting delay element, D1, and produces a negative edge at Cin2 for the next C-element. The C-element is designed such that the negative edge at its Cin input creates a negative edge from its F output. Therefore, the negative edge at the F1 node is fed back to the first C-element as an input and forces the first C-element to lower its Cout output, resulting in Cout1 = 0 and creating a pulse with duration of D1. The negative edge at Cout1, on the other hand, travels through the inverting delay element, D1, the second time, and produces a positive edge at Cin2. This positive edge, in turn, enables the second C-element to generate positive edges at F1 and Cout2 to latch the valid data at the data2 port.

As the data propagates through the second stage, the sequence of timing events that took place between the first and second C-elements repeat once again between the second and the third

C-elements that define the boundaries of CL2. This results in generating a positive pulse at the Cout2, two negative pulses at Cin3 and F2, and a positive Cout3 edge to be able to receive a new data at data3.

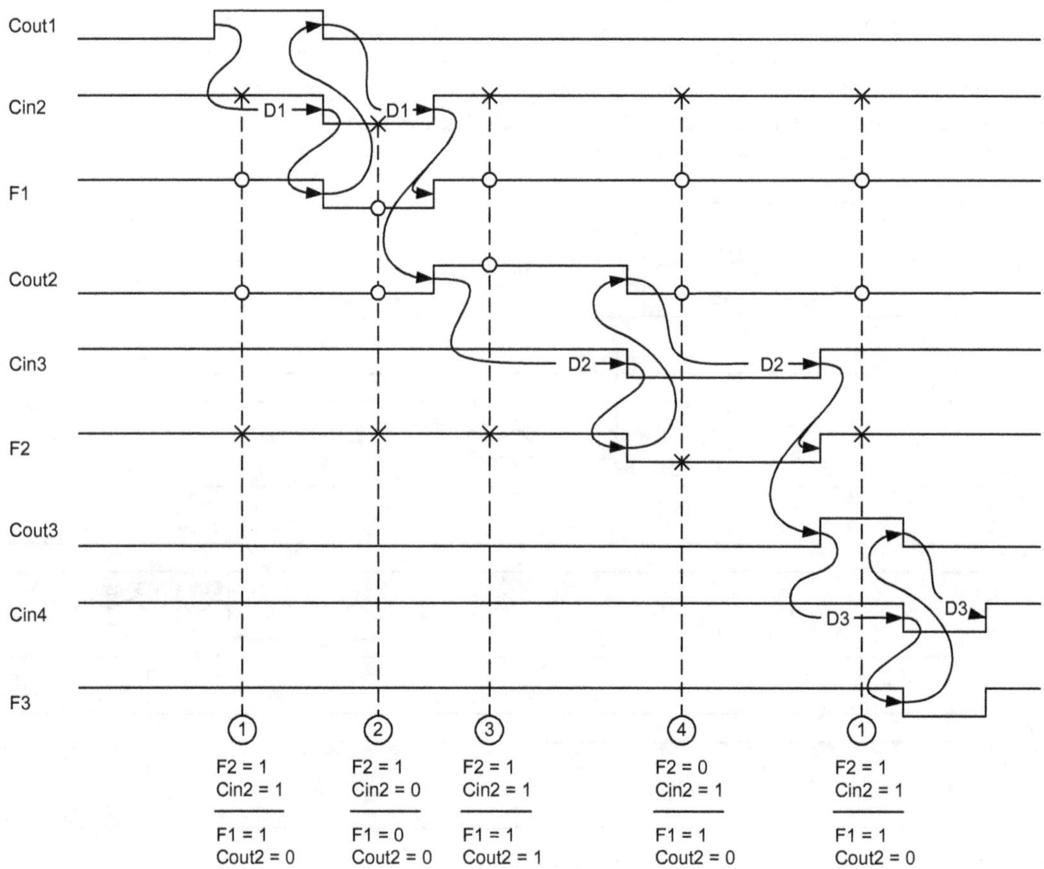

Fig. 3.19 C-element and delay-element input/output activity in Fig. 3.17

The vertical slicing in Fig. 3.19 helps to define all possible stable states in designing the C-element. Even though Fig. 3.19 only samples the inputs and the outputs of the second C-element, all C-elements in Fig. 3.17 yield identical results. Every stable state from the state ① to the state ④ allows only one input change as the fundamental-mode rule in asynchronous design methodology. The state ① is entered when Cin2 = F2 = 1, and produces F1 = 1 and Cout2 = 0. As Cin2 transitions to logic 0, the circuit goes into the state ② where it yields F1 = 0 and Cout2 = 0. The start of the pulse at Cout2 defines the state ③ where Cin2 transitions back to logic 1, and both outputs, F1 and Cout2, change to logic 1. The last state, state ④, emerges when F2 switches to logic 0. This state also causes Cout2 to change to logic 0, but retains F1 at logic 1. The transition of F2 to logic 1 prompts the C-element to go back to the state ①.

All possible state and output changes of the C-element in Fig. 3.19 are condensed in an integrated state and output flow table in Fig. 3.20. In this figure, the "?" mark indicates that the C-element never reaches these transitionary cases; the "-" mark again defines the forbidden states where the fundamental-mode of design is violated.

ps1 ps2	F2 Cin2				F2 Cin2			
	00	01	11	10	00	01	11	10
①	-	?	①	2			10	
②	?	-	3	②				00
③	-	4	③	?			11	
④	?	④	1	-		10		
	ns				F1 Cout2			

Fig. 3.20 Integrated primitive state and output flow tables for C-element

The state assignments of the four stable states in Fig. 3.20 are shown in Fig. 3.21. This is a vital step since the C-element states in Fig. 3.20 are still symbolic and have not yet been converted into binary values.

	ns1	ns2
①	0	0
②	0	1
③	1	1
④	1	0

Fig. 3.21 State assignments for Fig. 3.20

Since the fundamental-mode design rule of changing only one input between state transitions is fully observed, the state table in Fig. 3.22 shows no potential racing hazards in the current C-element design.

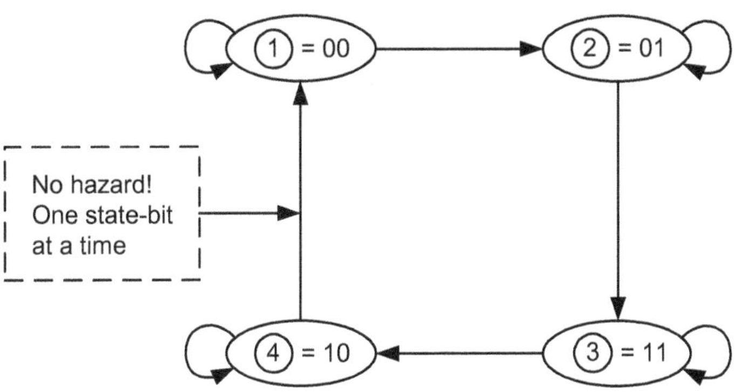

Fig. 3.22 Hazard-free state diagram of the C-element

Once the primitive flow table and state assignments are complete, the next state and output K-maps of the C-element can be constructed as shown in Fig. 3.23.

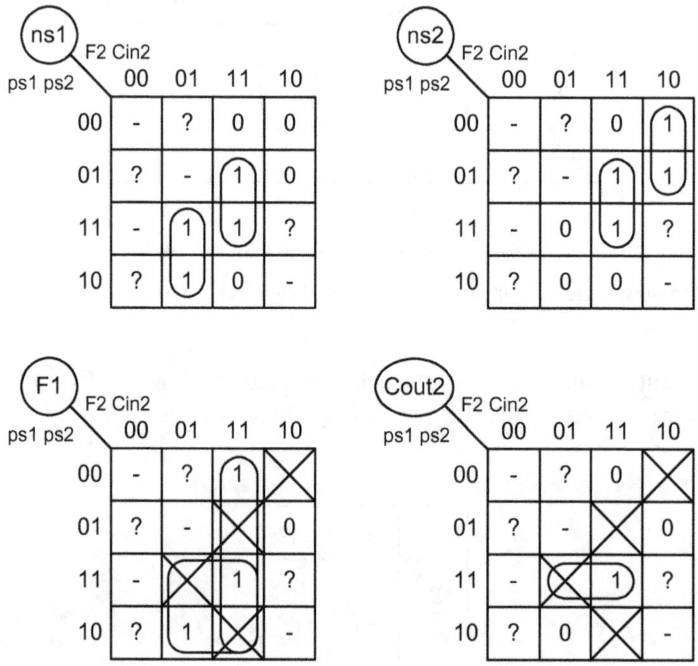

$$ns1 = F2.Cin2.ps2 + \overline{F2}.Cin2.ps1 = Cin2.(F2.ps2 + \overline{F2}.ps1)$$

$$ns2 = F2.Cin2.ps2 + F2.\overline{Cin2}.\overline{ps1} = F2.(Cin2.ps2 + \overline{Cin2}.\overline{ps1})$$

$$F1 = Cin2.ps1 + F2.Cin2 = Cin2.(ps1 + F2)$$

$$Cout2 = Cin2.ps1.ps2$$

Fig. 3.23 Next state and output K-maps of the C-element

In this figure, the cases marked by "?" and "-" are directly transferred from the primitive flow table in Fig. 3.20. When generating the SOP expressions for the ns1, ns2, F1 and Cout2, these cases are deliberately excluded from the terms in Fig. 3.23. This ensures that unwanted state transitions and outputs do not take place in the final circuit in Fig. 3.24.

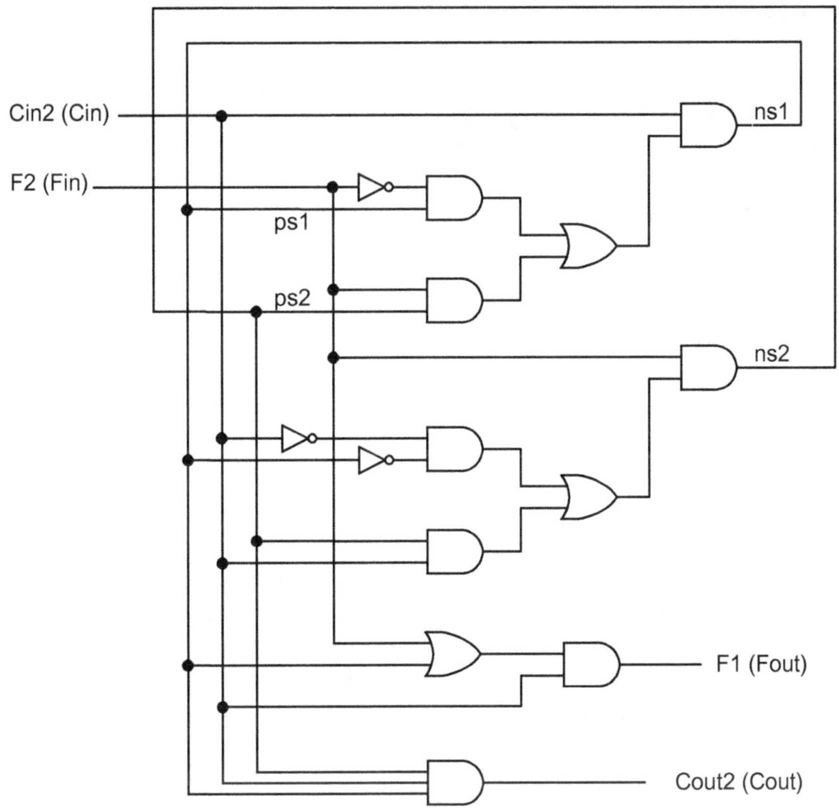

Fig. 3.24 C-element circuit according to the fundamental mode design rules

The input and output names of the second C-element in Fig. 3.17 are also changed for a generic C-element. According to Fig. 3.24, the inputs, Cin2 and F2, have become Cin and Fin, and the outputs, F1 and Cout2, have become Fout and Cout of a generic C-element, respectively.

Review Questions

1. An asynchronous circuit has two inputs, in1 and in2, and an output, out. When in1 = 1, the first transition at in2 from logic 0 to logic 1 generates out = 1 in the waveform below. Output stays at logic 1 unless in1 goes back to logic 0. The first transition at in2 from logic 1 to logic 0 while in1 = 0 switches out back to logic 0. If out is initially at logic 0, transitions at in2 do not affect the value of out as long as in1 = 0.
   A sample waveform is given below.

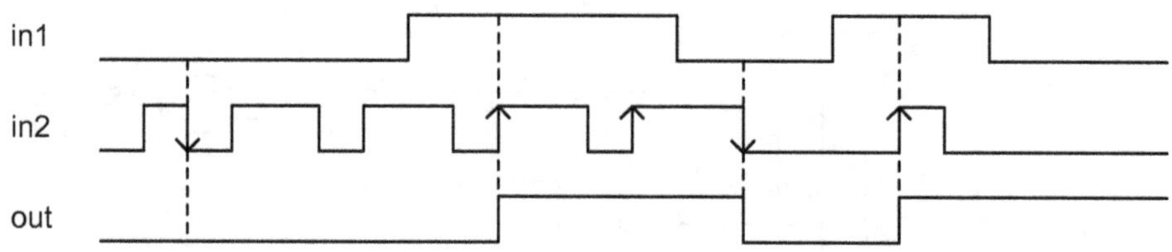

   Define all possible states using the waveform above and form an integrated primitive state and output flow table. Form an associated implication table leading to the minimization of states and outputs. Design the resultant fundamental mode asynchronous circuit.
2. An asynchronous circuit has two inputs, in1 and in2, and an output, out. When in1 = 0, the first transition at in2 produces out = 0 as shown in the waveform below. When in1 = 1, a transition from logic 0 to logic 1 at in2 increments the value of out by one. When out = 3 and in1 = 1, an additional logic 0 to logic 1 transition at in2, produces out = 0.

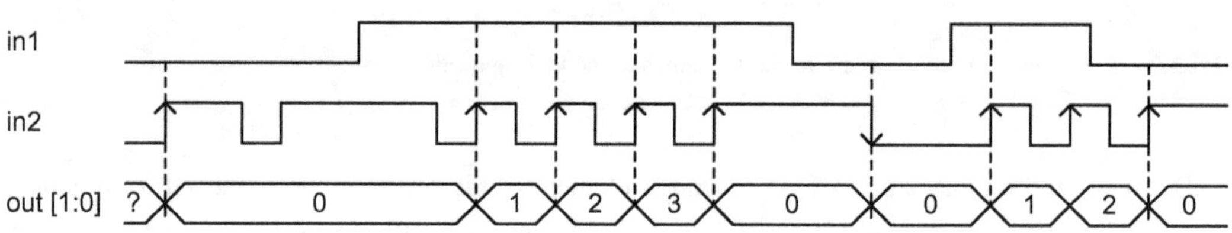

   Define the possible states from the waveform above, and form the primitive state and output flow tables. Define the resultant implication table to minimize the initial states and outputs. Design the resultant fundamental mode asynchronous circuit.
3. An asynchronous circuit has three inputs, in1, in2 and in3, and an output, out. When all inputs are at logic 0, the first logic 0 to logic 1 transition at any input causes the output to display the input ID. For example, a logic 0 to logic 1 transition at in1 while in2 = in3 = 0 produces out = 1 because this value is the ID number of in1. Similarly, the first logic 0 to logic 1 transition at in2 while in1 = in3 = 0 produces out = 2. Logic 0 to logic 1 transition at any input while one or more inputs are at logic 1 does not change the output value. Similarly, logic 1 to logic 0 transition does not affect neither state of the circuit nor the output value.

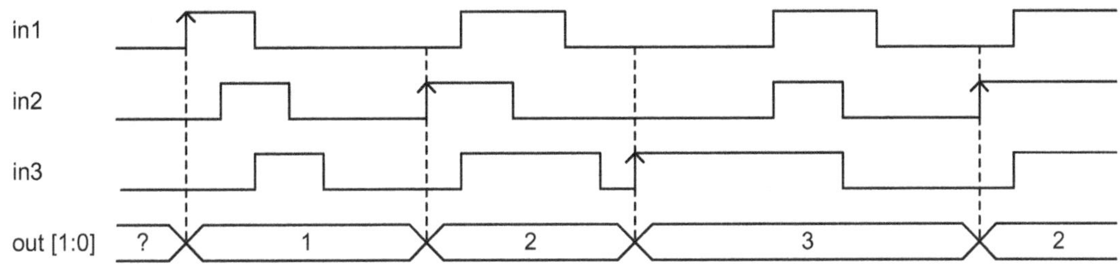

Form the primitive state and output flow tables and the implication table to minimize the initial state and output assignments on the waveform above. Design the fundamental mode asynchronous circuit.

4. The schematic below is a data-path of an asynchronous system controlled by C-elements. The combinational delays are shown by T1 through T4 blocks, each of which has a single input and output. There are also junction delays, J1 through J4, which accept two or more inputs and generate a single output.

   (a) Compute D1 and D2 in terms of combinational and junction delays, T1, T2, T3, T4, J1, J2, J3 and J4.
   (b) Show the data-flow that includes the signals from Cin1 to Cin3, and from Cout1 to Cout3 in a timing diagram. Assume the clock-to-q delay is equal to Tc in the timing diagram.

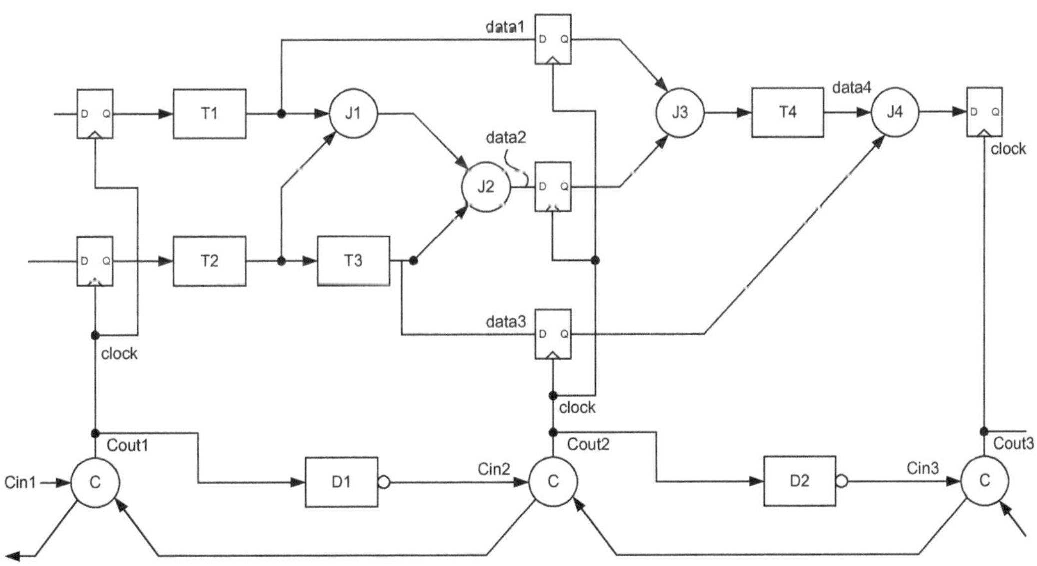

5. Data is transferred from a 32 × 8 source memory to a 32 × 8 destination memory as shown in the schematic below. When a 32-bit data is fetched from the source memory, the high (HI) and the low (LO) 16 bits are multiplied by an integer multiplier, and the product is delivered at a destination address. The initial values of the source and destination addresses are zero and seven, respectively. During the data transfer the source address increments by one while the destination address decrements by one until all eight data packets in the source memory are processed. Assuming that source and destination memories are asynchronous in nature and neither needs a

clock input to read or write data, include C elements in the circuit schematic to make this data transfer possible. Assume Tacc is the access time to fetch data from the source memory, Twrite is the time to write data to the destination memory, and Tclkq is the time to produce address from the address pointer 1 or address pointer 2.

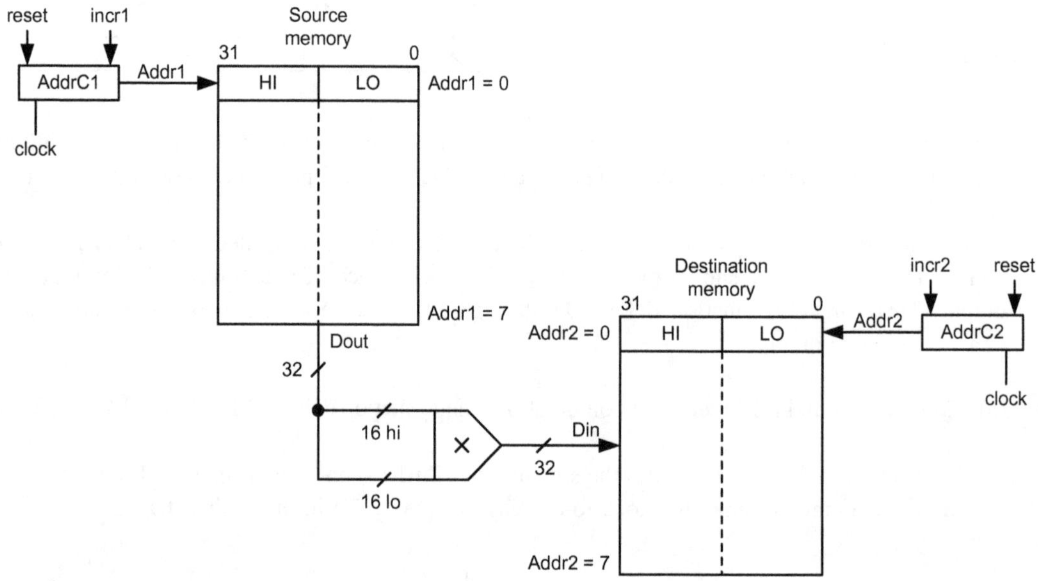

Projects

1. Implement the S-R latch and verify its functionality using Verilog.
2. Implement the fundamental mode asynchronous circuit and verify its functionality using Verilog.
3. Implement the C-element and verify its functionality using Verilog.

System Bus 4

A system bus is responsible for maintaining all communication between the Central Processing Unit (CPU), system peripherals and memories. The system bus operates with a certain bus protocol to exchange data between a bus master and a bus slave. The bus protocol ensures to isolate all other system devices from interfering the bus while the bus master exchanges data with a bus slave. Bus master initiates the data transfer, and sends or receives data from a slave device or a system memory. Bus slave, on the other hand, does not have the capability to start data transfer, but only to respond to bus master to exchange data.

There are two types of bus architectures. Serial bus architecture is essentially composed of a single data wire between a master and a slave where data bits are exchanged one bit at a time. A parallel bus, on the other hand, is comprised of many wires, and multitude of bits are sent or received all at once. In this chapter, we will describe several serial and parallel bus protocols and priority schemes.

4.1 Parallel Bus Architectures

There are two types of parallel bus architectures in a typical system: unidirectional bus and bidirectional bus. A unidirectional bus contains two separate paths for data: one that originates from a bus master and ends at a slave, and the other that starts from a slave and ends at the master. A bidirectional bus, on the other hand, shares one physical data path which allows data to flow in both directions. However, this type of bus requires logic overhead and control complexity.

Figure 4.1 describes a 32-bit unidirectional bus architecture containing two bus masters and three slaves. In this figure, the two unidirectional data-paths are highlighted with thicker lines. The first path is the "write" path, which a bus master uses to write data to a slave. This path requires a Write Data (WData) port from every master and slave. The second path is the "read" path for reading data from a slave. This also requires a Read Data (RData) port from each master and slave. Both the bus master and the slave device have address and control ports that define the destination address, the direction of data transfer, the data width and the length of data transfer.

All bus masters have to negotiate with a bus arbiter to gain the ownership of the bus before starting a data transfer. When there are pending requests from multiple bus masters, the arbiter decides which bus master should start the data transfer first according to a certain priority scheme, and issues an acknowledgement signal to the bus master with the highest priority. Therefore, every bus master has Request (Req) and Acknowledge (Ack) ports to communicate with the arbiter. Once the permission is granted, the master sends out the address and control signals to the selected slave in the first bus cycle, and writes or reads data in the next cycle. The decoder (DEC) connected to the address bus generates "Enable" (EN) signal to activate the selected slave. Every master and slave device has a Ready port that indicates if the selected slave is ready to transmit or receive data.

A. Bindal, *Fundamentals of Computer Architecture and Design*,
https://doi.org/10.1007/978-3-030-00223-7_4

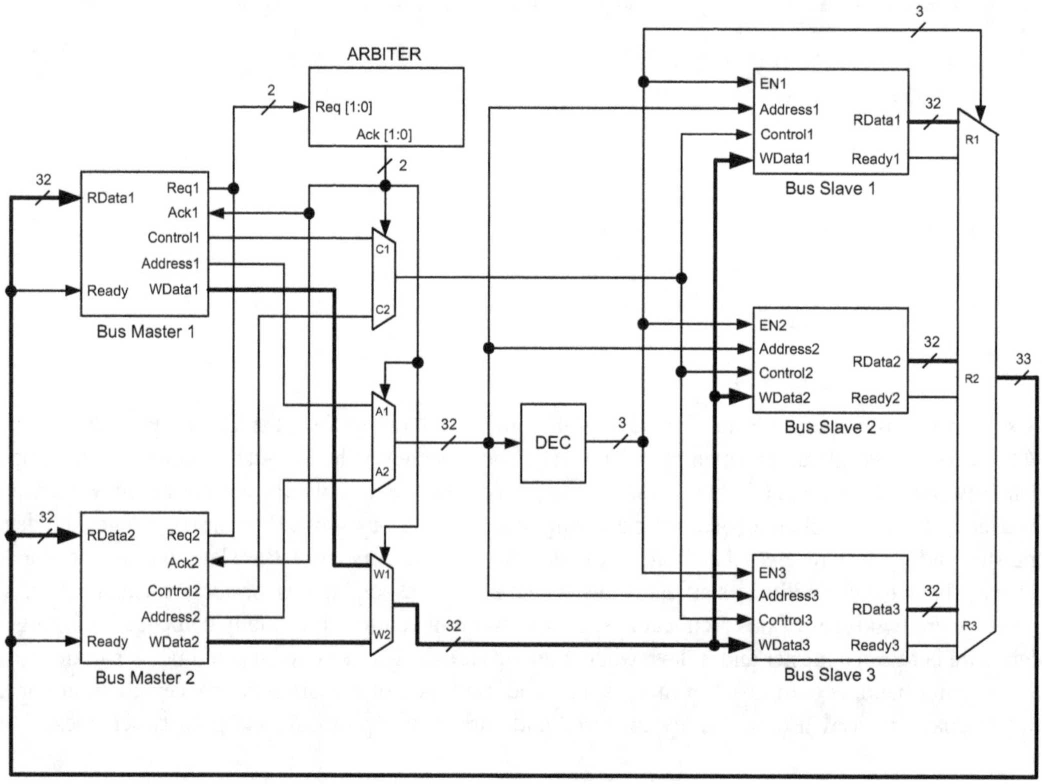

Fig. 4.1 A unidirectional bus structure with two bus masters and three slaves

A 32-bit bidirectional bus architecture is shown in Fig. 4.2. The number of masters and slaves are kept the same as in Fig. 4.1 for comparison purposes. The only difference between the two figures is the replacement of the unidirectional data bus in the earlier architecture with a bidirectional bus for reading and writing data. Tri-state buffers on data lines are essential for bidirectional bus architectures to isolate nonessential system devices from the bus when data transfer takes place exclusively between a master and a slave. The address bus in Fig. 4.2 can be also integrated with the data bus to allow both address and data to be exchanged on the same bus. However, this scheme is much slower and requires extra control logic overhead to maintain proper data-flow and management.

Figure 4.3 shows all the Input/Output (I/O) ports of a typical bus master. The Req and Ack ports are used to communicate with the arbiter as mentioned earlier. Bus master uses the "Ready" port to determine if the slave is ready to transmit or receive data. The WData, RData and Address ports are used for writing and reading data, and specifying the slave address, respectively. The control signals, Status, Write, Size and Burst, describe the nature of the data transfer.

The Status port is a two-bit bus that describes the status of the bus master as shown in Table 4.1. According to this table, the bus master may initiate a new data transfer by issuing the START signal. If the master is in the midst of exchanging data with a slave, it issues the Continue (CONT) signal.

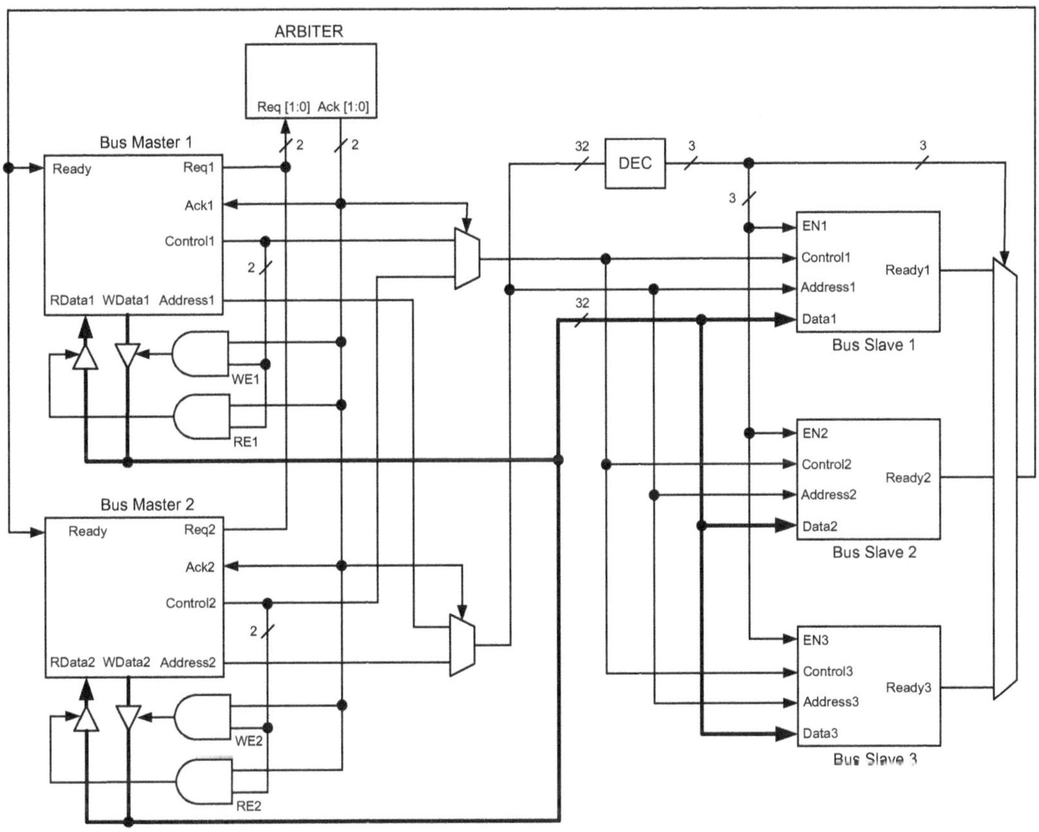

Fig. 4.2 A bidirectional bus structure with two bus masters and three slaves

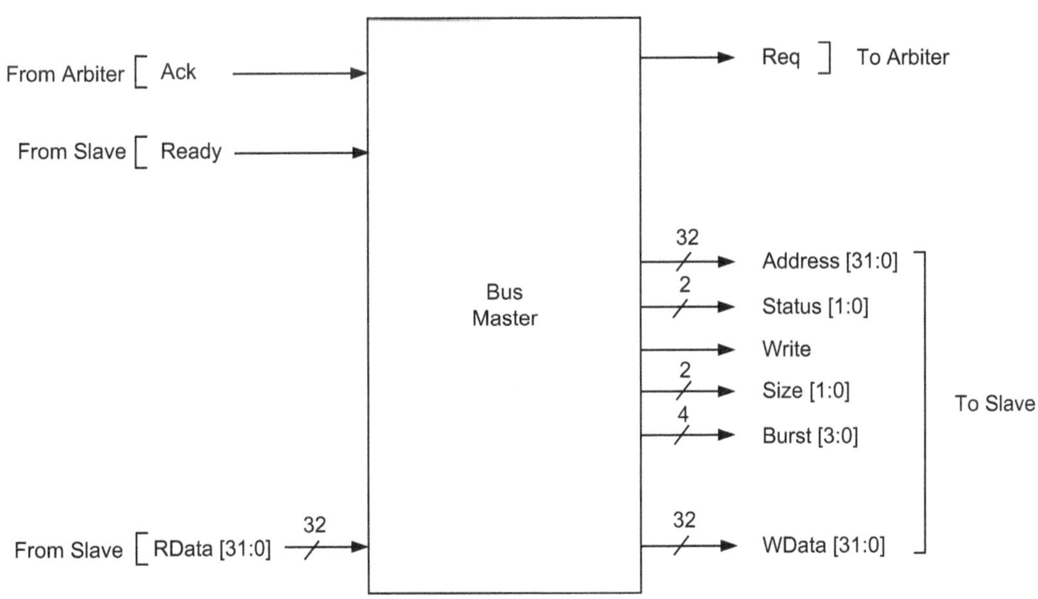

Fig. 4.3 Bus master interface

The IDLE signal indicates that the bus master has finished the data transfer. Once this signal is produced, the master keeps issuing the last address and control signals of the current data transfer until a new data transfer materializes on the bus. The bus master may also be in the midst of an internal operation while exchanging data with a slave. For this particular event, the master may momentarily stall the data transfer by issuing the BUSY signal.

Table 4.1 Bus master status control

Status[1:0]	Bus master status
0 0	Start Transfer (START)
0 1	Continue Transfer (CONT)
1 0	Finish Transfer (IDLE)
1 1	Pause Transfer (BUSY)

The Write port, as its name implies, describes if the master is in the process of writing or reading data as shown in Table 4.2.

Table 4.2 Bus master write control

Write	Bus master write
0	Read
1	Write

The Size port describes the data bit width during a transfer and is shown in Table 4.3. A bus master is allowed to transmit or receive data in eight bits (byte), 16 bits (half-word), 32 bits (word) or 64 bits (double-word), which cannot be changed during the transfer.

Table 4.3 Bus master size control

Size[1:0]	Number of bits
0 0	8
0 1	16
1 0	32
1 1	64

The Burst port describes the number of data packets to be sent or received by the bus master according to Table 4.4. In this table, a bus master can transfer from one packet to over 32,000 packets of data in a single burst.

Table 4.4 Bus master burst control

Burst[3:0]	Number of data packets
0 0 0 0	1
0 0 0 1	2
0 0 1 0	4
0 0 1 1	8
0 1 0 0	16
0 1 0 1	32
0 1 1 0	64
0 1 1 1	128
1 0 0 0	256
1 0 0 1	512
1 0 1 0	1024
1 0 1 1	2048
1 1 0 0	4096
1 1 0 1	8192
1 1 1 0	16384
1 1 1 1	32768

Figure 4.4 shows the I/O ports of a typical bus slave. The Req and Ack ports are omitted since the slave is not authorized to initiate a data transfer. The Ready signal indicates if the slave is ready to transmit or receive data once the transfer is initiated by the bus master. The WData, RData and Address ports are used to write data, read data, and specify a destination address, respectively. The control inputs, Status, Write, Size and Burst, describe the nature of the transfer as mentioned above.

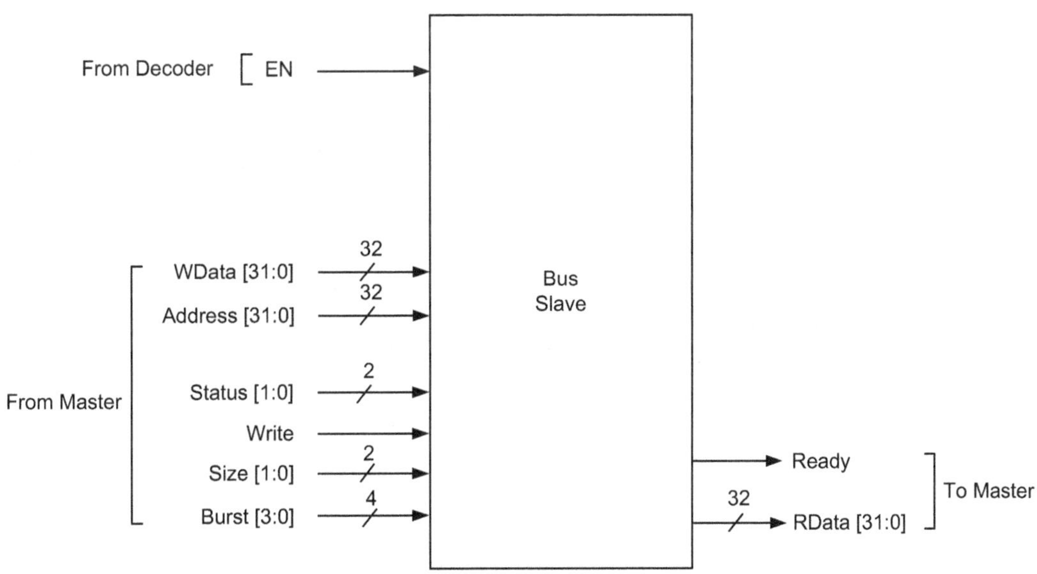

Fig. 4.4 Bus slave interface

The Enable (EN) input is produced by the address decoder, and based on the address generated by the bus master to activate a particular slave.

4.2 Basic Write Transfer

From this point forward, we will be using timing diagrams as a standard tool to show the bus activity between a master and a slave.

The bus protocol for write describes how a bus master writes data to a slave using a unidirectional bus shown by the timing diagram in Fig. 4.5.

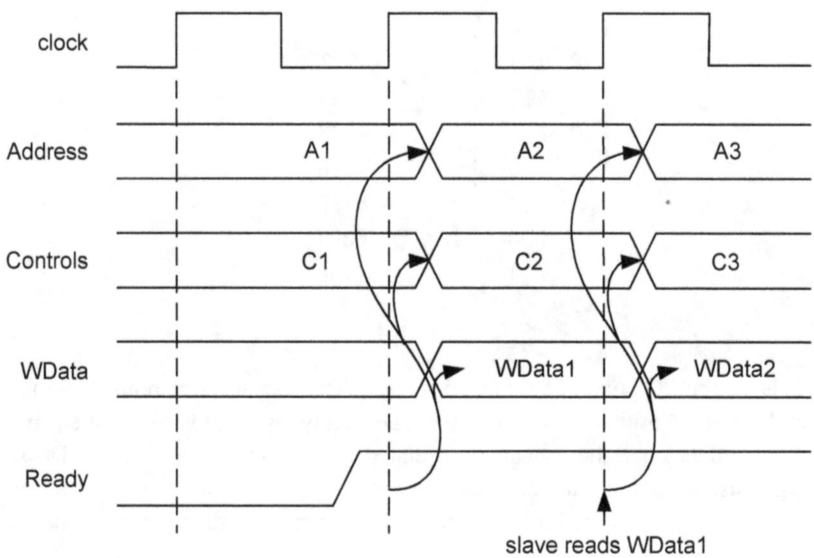

Fig. 4.5 Basic write transfer

In the first clock cycle, the bus master sends out the destination address and control signals, A1 and C1, to the slave regardless of the slave status. If the slave status is "Ready", the actual data packet, WData1, is sent in the second cycle along with the address and the control signals, A2 and C2, of the next data packet. The slave should be able to read WData1 at the positive edge of the third clock cycle if it is ready. However, there are instances where the slave may not be ready to receive or send data. As an example, the slave changes its status to "Not Ready" in the second cycle of Fig. 4.6. As soon as the slave's status is detected at the positive edge of the third clock cycle, the master stalls the write transfer. This means that the current data packet, WData2, and the next address and control signals, A3 and C3, are repeated as long as the slave keeps its Not Ready status. The normal data transfer resumes when the slave becomes Ready to receive the remaining data.

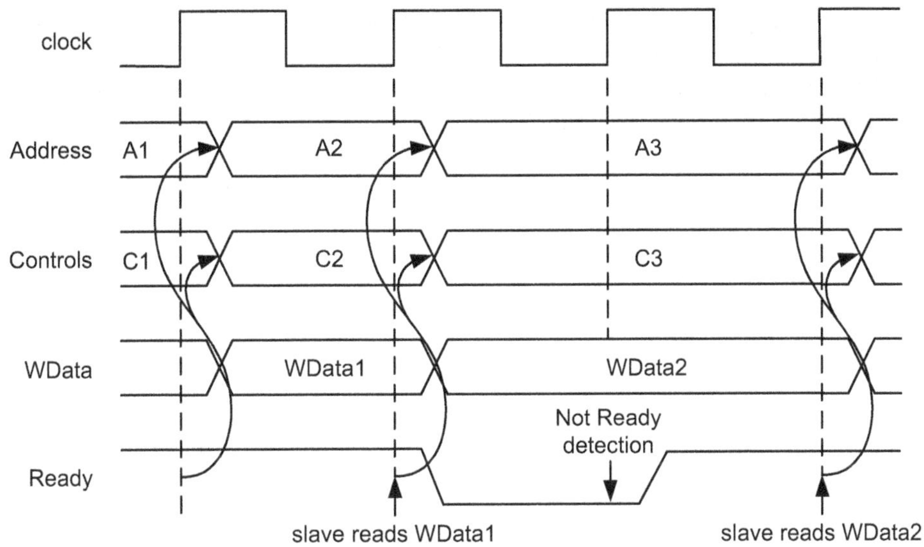

Fig. 4.6 Basic write transfer including the case Ready = 0

Example 4.1 What happens to the write sequence when the slave changes its status frequently?

When the slave changes its status to Not Ready during a clock cycle, the master detects this change at the next positive clock edge and holds the current data, next address and control signals until the slave becomes Ready again.

An example where the slave changes its status frequently is shown in Fig. 4.7. In this figure, the slave is Not Ready in the first cycle. Therefore, the first address and control packets, A1 and C1, are prolonged, and no data is sent to the slave. When the slave produces a Ready signal during the second

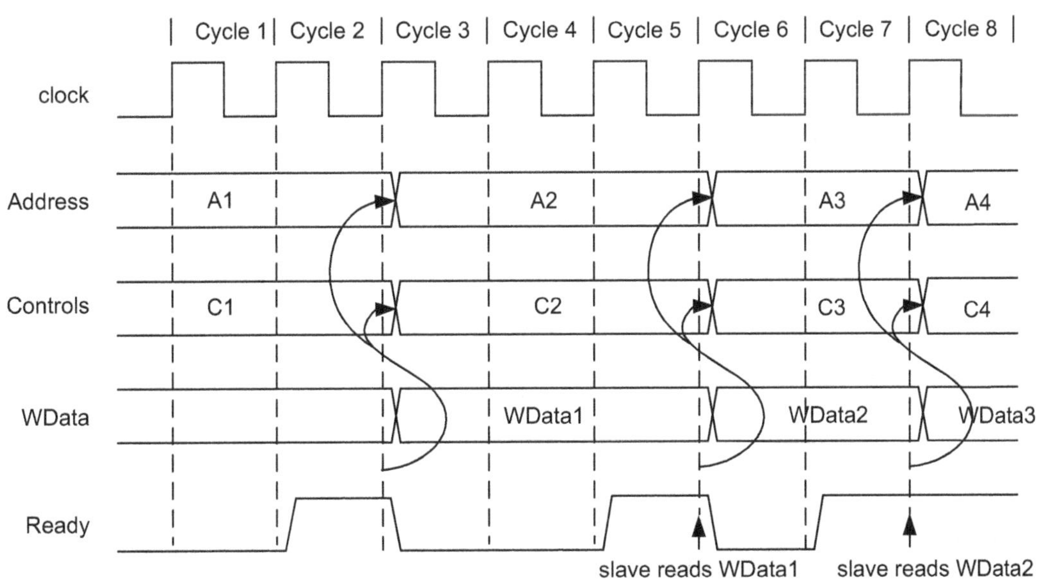

Fig. 4.7 A basic write transfer with varying Ready signal

cycle, the bus master produces the first data packet, WData1, and changes the address and control signals to A2 and C2 at the positive edge of the third cycle. However, the slave decides to change its status to Not Ready again during the third and fourth cycles. The master detects the status change at the positive edge of the fourth and fifth clock cycles and responds by not changing A2, C2 and WData1. The Ready signal in the fifth cycle prompts the master to produce A3, C3 and WData2 at the beginning of the sixth cycle. The master holds these values until the beginning of the eighth cycle when the slave changes its status to Ready again. At this point, the master sends the new A4, C4 and WData3.

4.3 Basic Read Transfer

A basic read transfer is shown in Fig. 4.8. According to this figure, the slave produces data for the master anytime after it issues the Ready signal. Once the master detects the Ready signal at a positive edge, it reads the slave's response at the next positive clock edge, and produces the next address and control signals for the slave.

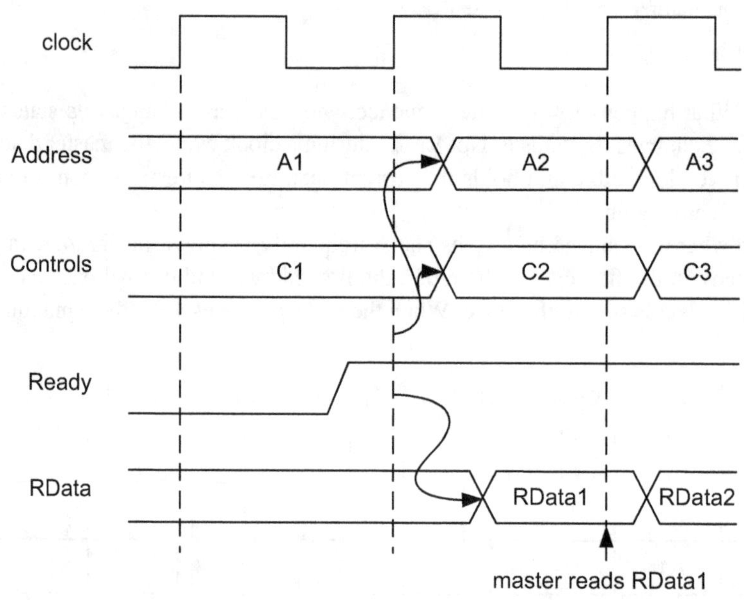

Fig. 4.8 Basic read transfer

Example 4.2 What happens to the read sequence when the slave changes its status frequently?

Figure 4.9 shows an example where the slave changes its status frequently. In this figure, the slave is Not Ready prior to the first cycle. Therefore, the master holds the first address and control packets, A1 and C1, until the slave becomes Ready. When the master detects a Ready signal at the positive edge of the third cycle, it responds by issuing a new set of address and control signals, A2 and C2, for the slave. Within the third cycle, the slave also issues RData1 for the master. The master reads the data at the positive edge of the sixth cycle after detecting a Ready signal from the slave.

The rest of the read transactions in Fig. 4.9 follow the same protocol described above. In other words, the master produces a new set of address and control signals for the slave every time it detects a Ready signal from the slave; the slave issues a new data packet for the master after it becomes Ready; and the master reads slave's data when the slave is Ready.

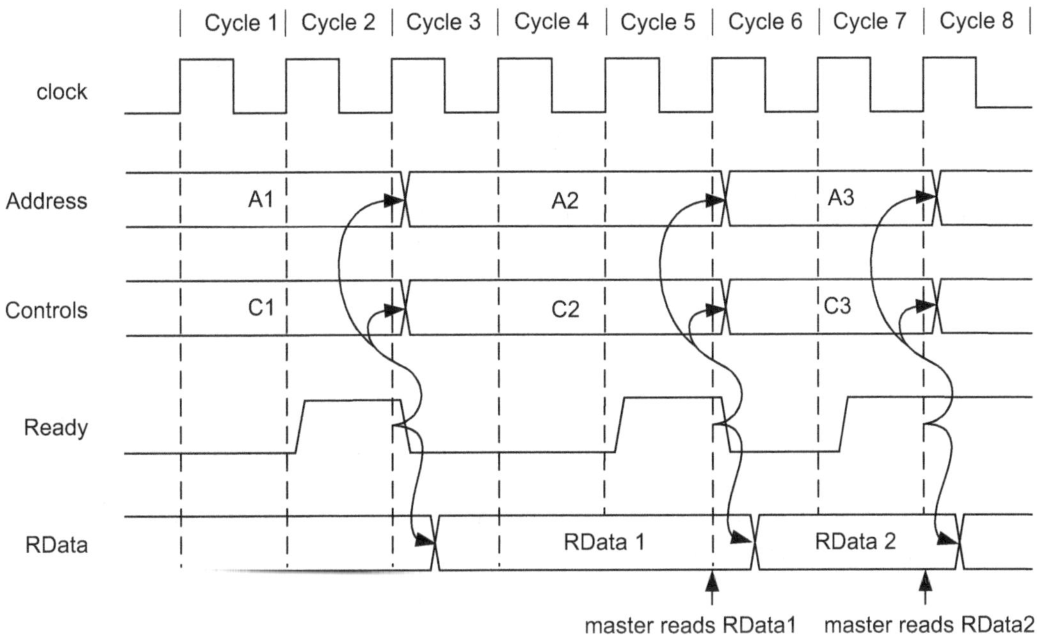

Fig. 4.9 A basic read transfer with varying Ready signal

4.4 Bus Master Status Change

It is possible that the bus master may be intermittently busy during a data transfer. In the event the bus master is busy to carry out its own internal tasks, the bus protocol requires the bus master to hold the address, control and data values as long as it is busy.

Figure 4.10 illustrates an example where the bus master becomes Busy in clock cycles 2, 9 and 10 while writing data to a slave. The master starts the data transfer by issuing Status = 00, which corresponds to the Start condition according to Table 4.1, and promptly sends out the first address, A1. In the second cycle, the bus master becomes busy and issues a Busy signal (Status = 11). As a result, it repeats the previous address, A1, but is unable to dispatch any data even though Ready = 1 is produced by the slave during this period. In the third cycle, all internal operations cease, and the bus master continues the normal data transfer by generating the Cont signal. The master also detects that the slave is Ready at the positive edge of the third cycle and issues the second address, A2 along with the first write data, WD1. In the next cycle, the master repeats A2 and WD1 because the slave was not Ready at the positive edge of the fourth cycle. Despite the slave showing the Not Ready condition in cycle 7, the normal data transfer sequence continues until cycle 9 where the bus master changes its status to Busy again. This change, in turn, causes the bus master to extend the address, A5, and the data, WD4, during cycles 9 and 10 irrespective of the slave status.

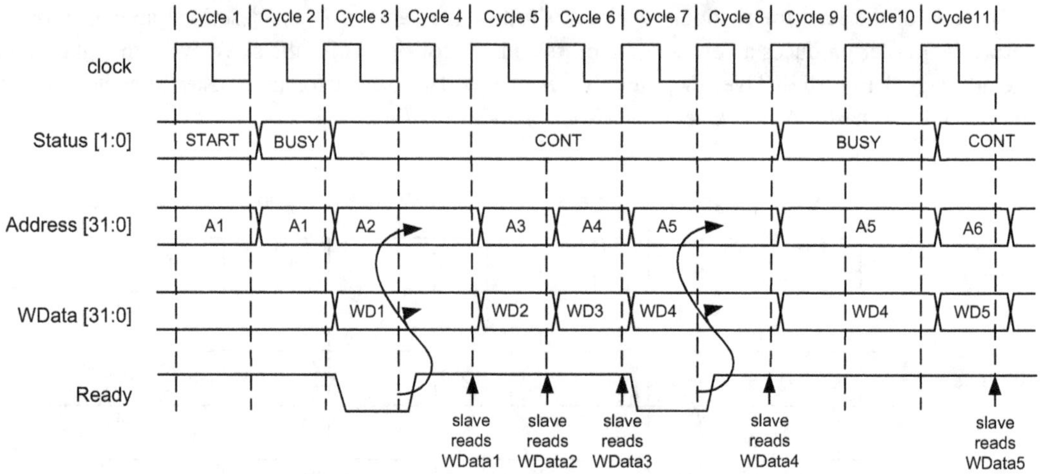

Fig. 4.10 Bus master Status control change

Example 4.3 What happens to the data transfer when the master changes its status frequently?

Assume that the bus master transfers two half words (16-bit wide data packets) to the addresses 0x20 and 0x22 of a memory block followed by 4 words (32-bit wide data packets) to the addresses 0x5c, 0x60, 0x64 and 0x68. The address map of this byte addressable memory is shown in Fig. 4.11 where the numbers in each box indicate an individual address.

20	21	22	23
⋮	⋮	⋮	⋮
5c	5d	5e	5f
60	61	62	63
64	65	66	67
68	69	6a	6b

Fig. 4.11 A byte-addressable memory

During the data transfer, the master issues frequent Busy signals during cycles 2, 3, 4, 7 and 8 as shown in the timing diagram in Fig. 4.12. Note that the slave is continuously Ready from cycle 2 until the end of the data transfer.

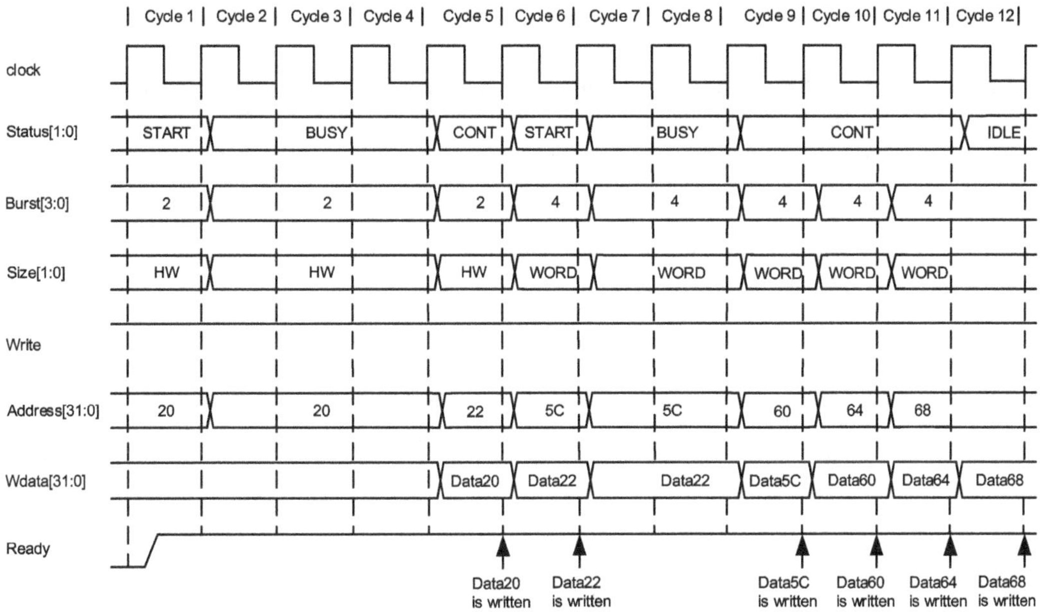

Fig. 4.12 A write transfer example to the byte addressable memory in Fig. 4.11

The bus master starts transferring the first data packet by issuing Status = 00 (START), Burst = 0001 (two total data packets), Size = 01 (half word), Write = 1 and Address = 0x20 in the first cycle according to Tables 4.1, 4.2, 4.3 and 4.4. Since the slave is Ready at the beginning of the second cycle, the master prepares to dispatch the next address, 0x22, and the first write data, Data20. However, in this cycle the master also becomes busy with internal operations until the beginning of the fifth cycle and issues a Busy signal as shown in Fig. 4.12. The Busy condition requires the bus master to repeat its control, address and data signals during this period. Therefore, when the master finally changes its status to Cont in the fifth cycle, it is able to send Data20 in the same cycle, and Data22 in the following cycle.

As soon as the first data transfer finishes, the master starts another write transfer in the sixth cycle by issuing Status = 00 (START), Burst = 0010 (four total data packets), Size = 10 (word), Write = 1 and Address = 0x5C. Since the slave's status is Ready, the master prepares to issue the next address and data packets at the beginning of the seventh cycle. However, its internal operations interfere with this process once again until the beginning of the ninth cycle. The master issues a Busy signal, and repeats its control, address and data outputs. When the master finally changes its status to Cont in the ninth cycle, it delivers the second address, 0x60, and the first data, Data5C. In the tenth cycle, the next address, 0x64, and data, Data60, are issued, respectively. The master writes Data64 in the eleventh cycle, and finishes the transfer by writing the last data, Data68, in the twelfth cycle. In this cycle, the bus master changes its status to Idle, indicating the end of the data transfer.

4.5 Bus Master Handshake

Each bus master communicates with the arbiter using request-acknowledge signals, which form the basic "handshake" protocol. The master starting a bus transfer issues a request signal, Req, requesting the ownership of the bus from the arbiter. The arbiter grants this request by an acknowledgement signal, Ack. If there is no ongoing data transfer, the acknowledgement is usually issued in the following cycle after the master generates a request. However, the Ack signal may not be generated many cycles after the Req signal is issued due to an existing data transfer.

Figure 4.13 shows the timing diagram of a handshake mechanism between a bus master and the arbiter before the bus ownership is granted to the master. The double "~" sign on the Ack signal signifies that this signal is generated many cycles after the arbiter has received a request from the particular bus master. As soon as the master receives the Ack signal in the nth cycle, it changes its status to Start in the (n + 1)th cycle, and sends out the first address, A1, regardless of the slave's status. If the slave is Ready, the master subsequently sends out the second address, A2, and the first data, WData1, in the following cycle.

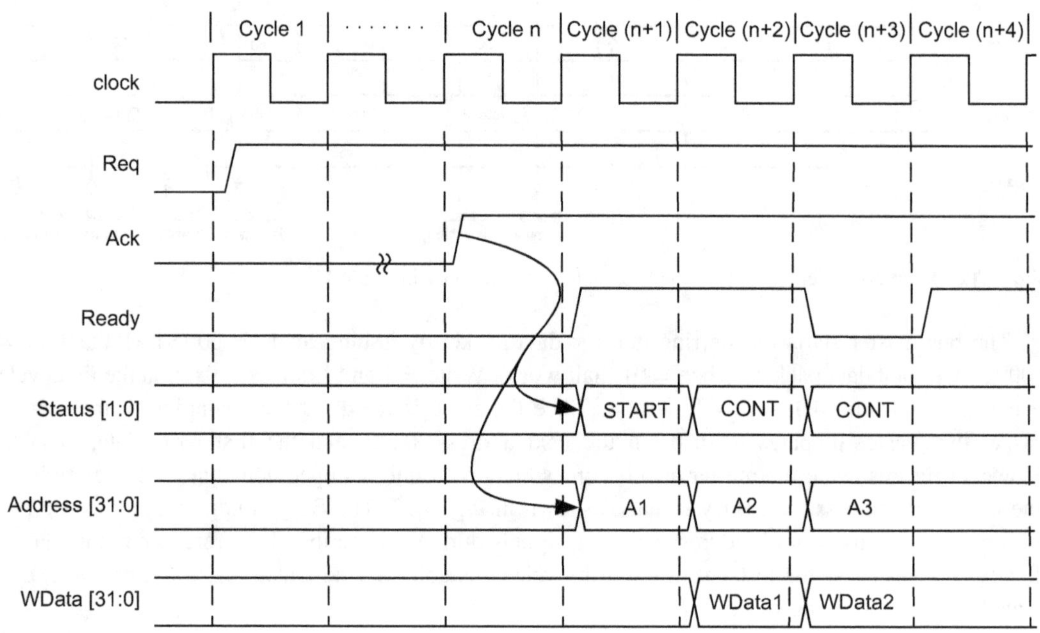

Fig. 4.13 Bus master-arbiter handshake protocol

4.6 Arbiter

Bus arbitration is an essential part of bus management if there are more than one bus master requesting the ownership of the bus. The arbitration is either hardware-coded and implemented as a state machine or programmable and register-based.

Table 4.5 explains a hardware-coded bus arbitration mechanism between two bus masters. When there are no requests to the arbiter, no Ack is generated to either bus master. However, if two requests are issued at the same time, the acknowledge is issued to bus master 1 according to this table since bus master 1 is assumed to have higher priority than bus master 2 as shown in the last row.

Table 4.5 Bus arbitration table for two bus masters

From IDLE State → Req1	Req2	Ack1	Ack2
0	0	0	0
0	1	0	1
1	0	1	0
1	1	1	0

The Table 4.5 is implemented as a state machine in Fig. 4.14. In this figure, the shorthand representation of Req = (Req1 Req2) corresponds to bus master request inputs 1 and 2. Similarly, Ack = (Ack1 Ack2) corresponds to the acknowledge signals generated by the arbiter for bus masters 1 and 2, respectively.

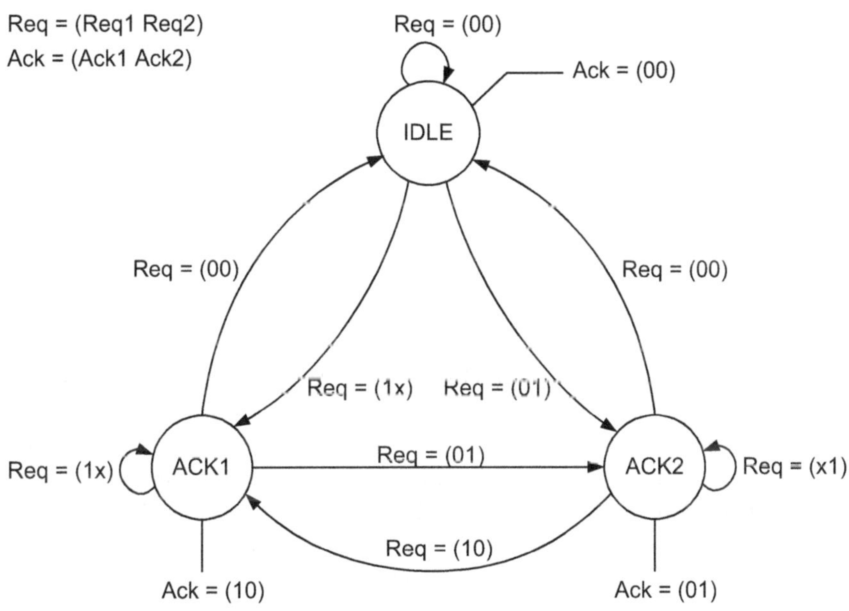

Fig. 4.14 Bus arbiter with two bus masters

The arbiter is normally in the IDLE state when there are no pending requests. If there are simultaneous requests from bus masters 1 and 2, the arbiter moves from the IDLE state to the ACK1 state, generates Ack1 = 1 for bus master 1, and ignores the request from bus master 2 by Ack2 = 0 according to Table 4.5. The inputs for this state-to-state transition are shown by Req = (1 x), where Req1 = 1 and Req2 = x (don't care). When bus master 1 terminates the data transfer by issuing Req1 = 0, the arbiter either stays in the ACK1 state if there is another pending request from bus master 1 or moves back to the IDLE state if there are no requests. However, if the arbiter receives Req1 = 0 and Req2 = 1 while in the ACK1 state, it transitions to the ACK2 state, and issues Ack2 = 1 to bus master 2.

In a similar fashion, the transition from the IDLE state to the ACK2 state requires Req2 = 1 and Req1 = 0. Once in the ACK2 state, the arbiter grants the usage of the bus to bus master 2 by issuing Ack2 = 1 and Ack1 = 0. When bus master 2 finishes the transfer by issuing Req2 = 0, the arbiter either goes back to the IDLE state or transitions to the ACK1 state if Req1 = 1. In case the higher priority bus master, bus master 1, requests the ownership of the bus by Req1 = 1 while the lower priority bus master is in the middle of a transfer, the arbiter stays in the ACK2 state as long as Req2 = 1 from bus master 2, ensuring the data transfer is complete.

Example 4.4 Design a hardware-coded arbiter with three bus masters where bus master 1 has the highest priority followed by bus masters 2 and 3.

According to this definition, the bus master priorities can be tabulated in Table 4.6.

Table 4.6 Bus arbitration table for three bus masters

From IDLE State →

Req1	Req2	Req3	Ack1	Ack2	Ack3
0	0	0	0	0	0
0	0	1	0	0	1
0	1	0	0	1	0
0	1	1	0	1	0
1	0	0	1	0	0
1	0	1	1	0	0
1	1	0	1	0	0
1	1	1	1	0	0

This table generates no acknowledge signal if there are no requests from any of the bus masters (the top row). If only bus master 3 requests the bus, Ack3 = 1 is generated for bus master 3 (the second row from the top). If there are two pending requests from bus masters 2 and 3, Ack2 = 1 is issued for bus master 2 because it has higher priority than bus master 3 (fourth row from the top). If all three bus masters request the ownership of the bus, the arbiter grants the bus to bus master 1 by Ack1 = 1 because it has the highest priority with respect to the remaining bus masters (the last row).

The implementation of this priority table is shown in Fig. 4.15 as a state machine. The naming convention in representing request and acknowledge signals in Fig. 4.15 is the same as in Fig. 4.14. Therefore, Req = (Req1 Req2 Req3) corresponds to bus master requests 1, 2 and 3, and Ack = (Ack1 Ack2 Ack3) corresponds to the arbiter acknowledge signals for bus masters 1, 2 and 3, respectively. Normally, the arbiter is in the IDLE state when there are no requests. If there are three simultaneous requests from bus masters 1, 2 and 3, the arbiter transitions to the ACK1 state where it generates Ack1 = 1 since bus master 1 has the highest priority. When bus master 1 finishes the data transfer, the arbiter can either stay in the ACK1 state or transition to the ACK2 state or the ACK3 state depending on the requests from all three bus masters. If there are no pending requests, the arbiter goes back to the IDLE state.

The arbiter does not issue acknowledge signals to higher priority bus masters until an ongoing data transfer of a lower priority bus master is complete. For example, in the ACK3 state the requests from bus masters 1 and 2 are ignored as long as bus master 3 keeps its request high to continue transferring data.

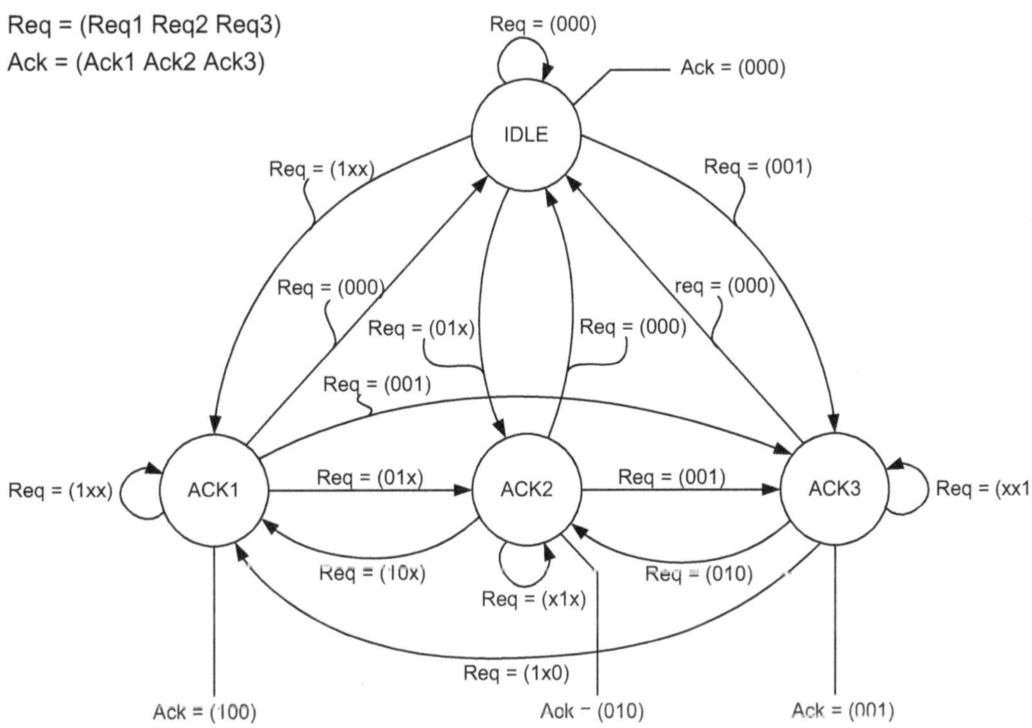

Fig. 4.15 Bus arbiter with three bus masters

4.7 Bus Master Handover

The bus may be handed over to a different bus master if the current bus master lowers its request.

Figure 4.16 describes how this bus ownership takes place in a unidirectional bus. In this timing diagram, the current bus master, bus master 1, starts a new transfer by generating a Start signal a cycle after it receives Ack1 = 1 from the arbiter. The write transfer continues until the eighth cycle when the bus master delivers its last address, A4. In cycle nine, the bus master delivers its last data, WA4, lowers its request, and changes its status to Idle, thus terminating the data transfer. At the positive edge of the tenth cycle, the arbiter detects Req1 = 0 and Req2 = 1, and switches the bus ownership by issuing Ack1 = 0 and Ack2 = 1. The new master, bus master 2, starts a new transfer in the eleventh cycle and generates its first address, B1. The write transfer continues until the fourteenth cycle when bus master 2 delivers its last data, WB2.

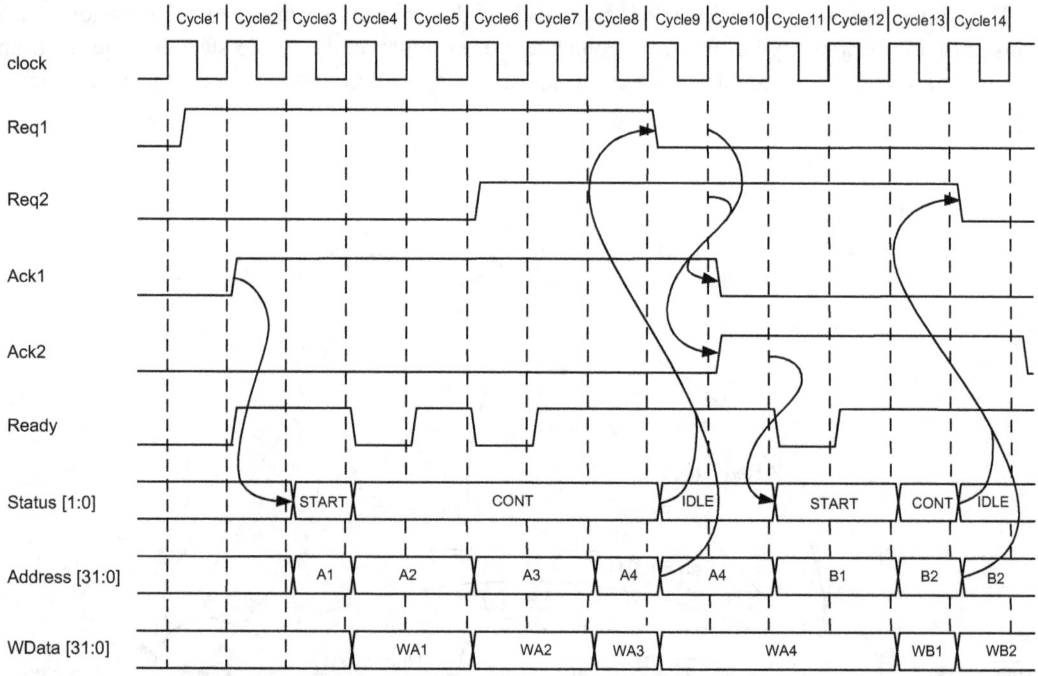

Fig. 4.16 Bus master handover protocol

4.8 Serial Buses

Peripheral devices and external buffer memories that operate at low frequencies communicate with the processor using a serial bus.

There are currently two popular serial buses used in low-speed communication. The Serial Peripheral Interface (SPI) was introduced in 1979 by Motorola as an external microprocessor bus for the well-known Motorola 68000 microprocessor. The SPI bus normally requires four wires; however, wire count increments by one every time a peripheral device is added to the system. The second bus, Inter-Integrated Circuit (I^2C), was developed by Philips in 1982 to connect Philips CPUs to peripheral chips in a TV set. This bus requires only two wires, but it is considerably slower compared to the SPI bus.

Serial Peripheral Interface (SPI)

SPI is designed as a very straightforward serial bus. Four signals establish all the serial communication between a CPU and a peripheral device. The SPI clock signal, SCK, is distributed to all the slaves in a system, and forces each peripheral to be synchronous with a single master. The Slave Select signal (SS) is an active-low signal, and is used to enable a particular slave prior to data transfer. Serial-Data-Out (SDO) or Master-Out-Slave-In (MOSI) port is what the master uses to send serial data to a slave. Serial-Data-In (SDI) or Master-In-Slave-Out (MISO) port is what the master uses to read serial data from a slave.

Figure 4.17 shows the serial bus configuration between the bus master and a single slave. All SPI signals with the exception of SDI must be initiated by the bus master.

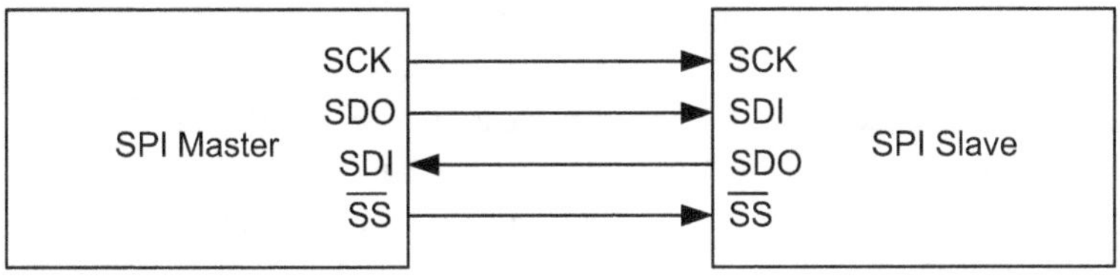

Fig. 4.17 SPI bus between a master and a single slave

When the bus master is connected to a multitude of slaves, it needs to generate an active-low Slave Select signal for each slave as shown in Fig. 4.18.

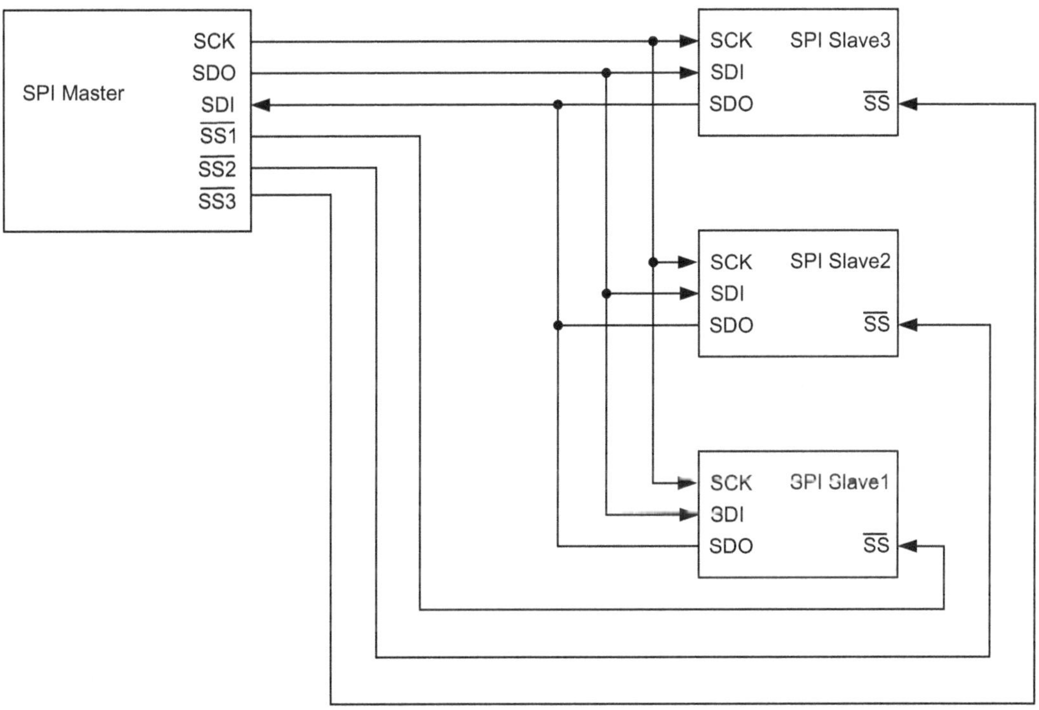

Fig. 4.18 SPI bus between a master and three slaves

SPI is considered a single-master serial communication protocol. This means that only one master is assigned to initiate and carry out all serial communications with slaves. When the SPI master wishes to send or request data from a slave, first it selects a particular slave by lowering the corresponding $\overline{SS}$ signal to logic 0, and then producing a clock signal for the slave as shown in Fig. 4.19. Once the select and clock signals are established, the master can send out serial data to the selected slave at the negative edge of each SCK from its SDO port while simultaneously sample slave data at the positive edge of each SCK at the SDI port. According to the SPI protocol, the slave is capable of sending and receiving data except it cannot generate SCK.

In the example in Fig. 4.19, the master sends out serial data, DataM1 (most significant bit) to DataM4 (least significant bit), from its SDO port at the negative edge of SCK, and samples slave data, DataS1 (most significant bit) to DataS4 (least significant bit), at its SDI port at the positive edge of SCK. The slave, on the other hand, can also dispatch serial data packets, DataS1 (most significant bit) to DataS4 (least significant bit), from its SDO port at the negative edge of SCK, and sample master data, DataM1 (most significant bit) to DataM4 (least significant bit), at its SDI port at the positive edge of SCK.

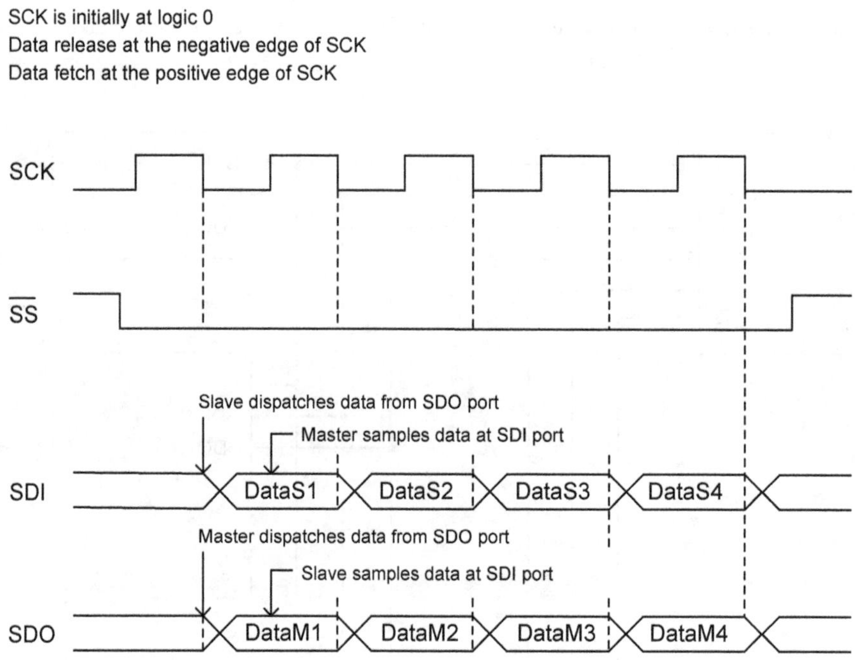

Fig. 4.19 SPI bus protocol between a master and a single slave

There are four communication modes available for the SPI bus protocol. Each protocol is categorized according to the initial SCK level (the logic level at which SCK resides at steady state) and the data generation edge of the SCK. Each communication mode is shown in Fig. 4.20.

MODE 0 communication protocol assumes that the steady state level for SCK is at logic 0. Each data bit is generated at the negative edge of SCK by the master (or the slave), and is sampled at the positive edge. A good example of the MODE 0 protocol is shown in Fig. 4.19.

MODE 1 still assumes the steady state level of SCK to be at logic 0, but the data generation takes place at the positive edge of SCK. Both the master and the slave read data at the negative edge in this mode.

MODE 2 switches the steady state level of SCK to logic 1. Data is released at the positive edge of SCK and is sampled at the negative edge as shown in Fig. 4.20.

MODE 3 also assumes the steady state level of SCK at logic 1. However, data is released at the negative edge of SCK and is sampled at the positive edge.

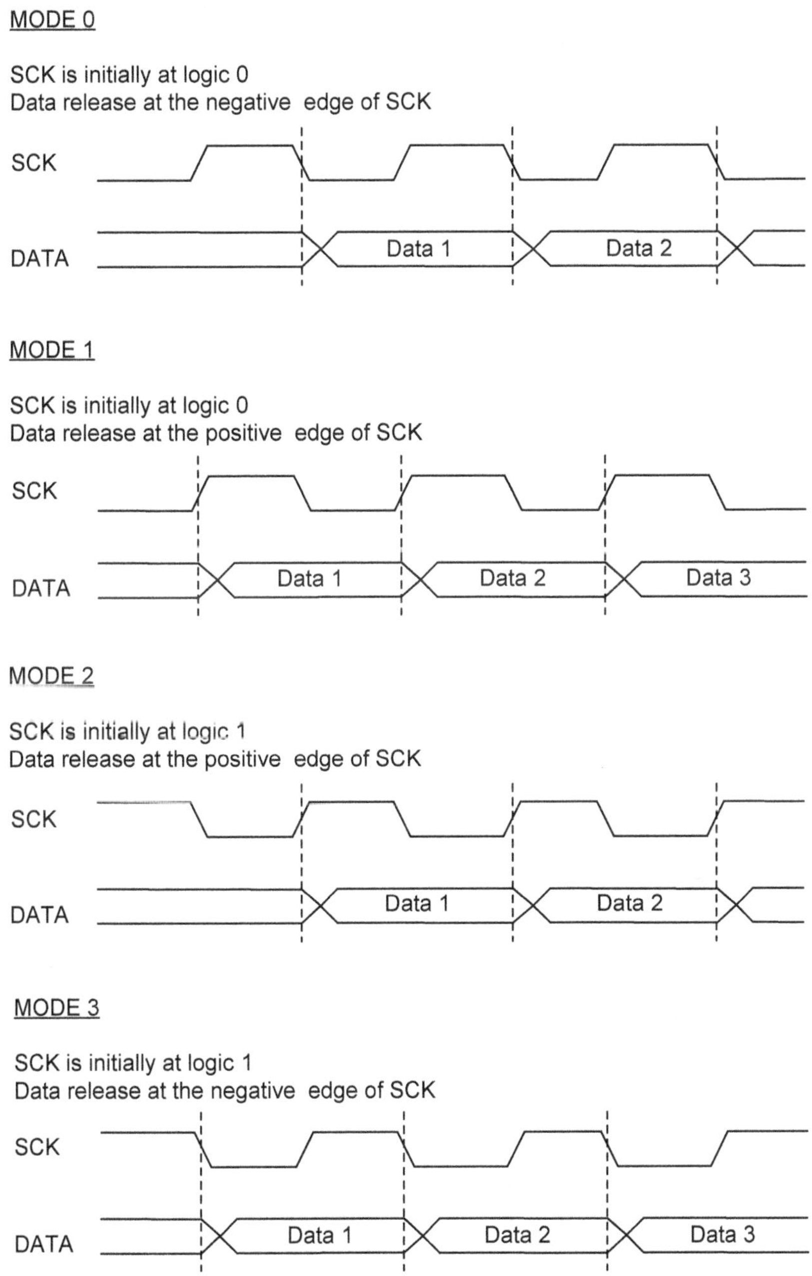

Fig. 4.20 SPI bus protocol modes

A master-slave pair must use the same mode during a data exchange. If multiple slaves are used, and each slave uses a different communication mode, the master has to reconfigure itself each time it communicates with a different slave.

SPI bus has neither an acknowledgement mechanism to confirm receipt of data nor it offers any other data-flow control. In reality, an SPI bus master has no knowledge if a physical slave exists on the receiving end, or the data it sends is properly received by the slave. Most SPI implementations pack a byte of data in a clock burst which is eight clock periods long. However, many variants of SPI today use 16 or even 32 clock cycles to send more data bits in a burst in order to gain speed.

Inter Integrated Circuit (I^2C)

Inter Integrated Circuit (I^2C) is a multi-master bus protocol that exchanges data between bus devices (masters and slaves) using only two lines, Serial Clock (SCL) and Serial Data (SDA).

Slave selection with slave select signals, address decoders or arbitrations is not necessary for this particular bus protocol. Any number of slaves and masters can be employed in an I^2C bus using only two lines.

The data rate is commonly at 100 Kbps which is the standard mode. However, the bus can operate as fast as 400 Kbps or even at 3.4 Mbps at high speed mode.

Physically, the I^2C bus consists of the two active wires, SDA and SCL, between a master and a slave device as shown in Fig. 4.21. The clock generation and data-flow are both bidirectional. This protocol assumes the device initiating the data transfer to be the bus master; all the other devices on the I^2C bus are regarded as bus slaves.

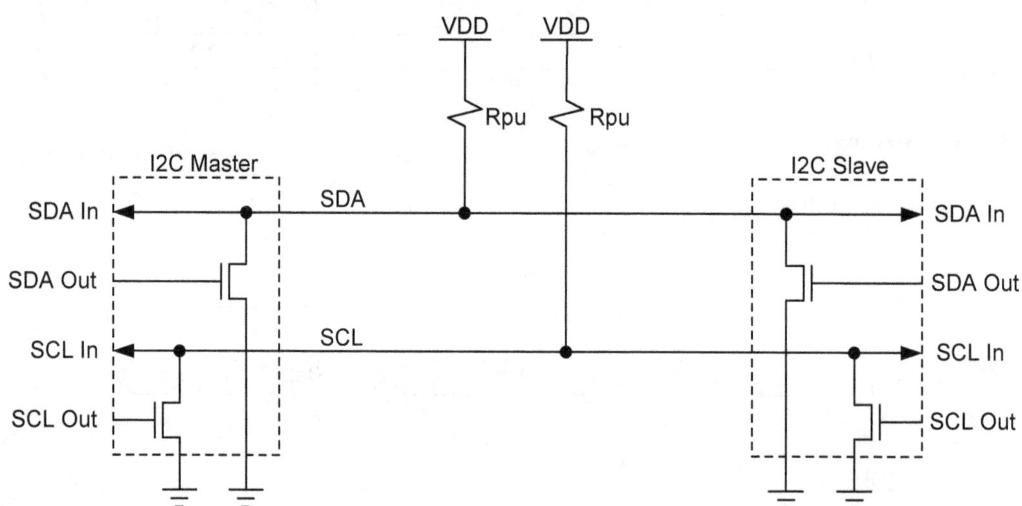

Fig. 4.21 I^2C architecture

In a typical I^2C bus, both the bus master and the slave have two input ports, SCL In and SDA In, and two output ports, SCL Out and SDA Out, as shown in Fig. 4.21. When a master issues SCL Out = 1 (or SDA Out = 1), the corresponding n-channel MOSFET turns on, and pulls the SCL line (or SDA line) to ground. When the master issues SCL Out = 0 (or SDA Out = 0), it causes the corresponding n-channel transistor to turn off, resulting SCL (or SDA) afloat. However, neither SCL nor SDA is truly left floating in an undetermined voltage level. The pull-up resistor, Rpu, immediately lifts the floating line to the power supply voltage level, VDD. On the other side of the bus, the I^2C slave detects the change at the SCL In (or SDA In) port, and determines the current bus value.

Each slave on the I^2C bus is defined by an address field of either seven bits or ten bits as shown in Fig. 4.22. Each data packet following the address is eight bits long. There are only four control signals that regulate the data flow: Start, Stop, Write/Read and Acknowledge.

The seven-bit and ten-bit versions of read and write data transfers are shown in Fig. 4.22. The top sequence in this figure explains how a bus master writes multiple bytes of data to a slave that uses a seven-bit address. The master begins the sequence by generating a Start bit. This acts as a "wake-up" call to all the slave devices and enables them to watch for the incoming address. This step is followed by a seven-bit long slave address. The bus master sends the most significant address bit first. The remaining address bits are released one bit at a time until the least significant bit. At this point, all slave devices compare the bus address just sent out with their own addresses. If the address does not match, the slave simply ignores the rest of the incoming bits at the SDA bus, and waits for the beginning of the next bus transfer. If the addresses match, however, the addressed slave waits for the next bit that indicates the type of the transfer from the master. When the master sends out a Write bit, the slave responds with an acknowledge signal, SAck, by pulling the SDA line to ground. The master detects the acknowledge signal, and sends out the first eight-bit long data packet. The format for transmitting data bits is the same as the address: the most significant bit of the data packet is sent out first followed by the intermediate bits and the least significant bit. The slave produces another acknowledgement when all eight data bits are successfully received. The data delivery continues until the master completes sending all of its data packets. The transfer ends when the master generates a Stop signal.

The second entry in Fig. 4.22 shows the write transfer to a bus slave whose address is ten bits long. Following the Start bit, the bus master sends out a five-bit preamble, 11110, indicating that it is about to send out a ten-bit slave address. Next, the master sends out the two most significant address bits

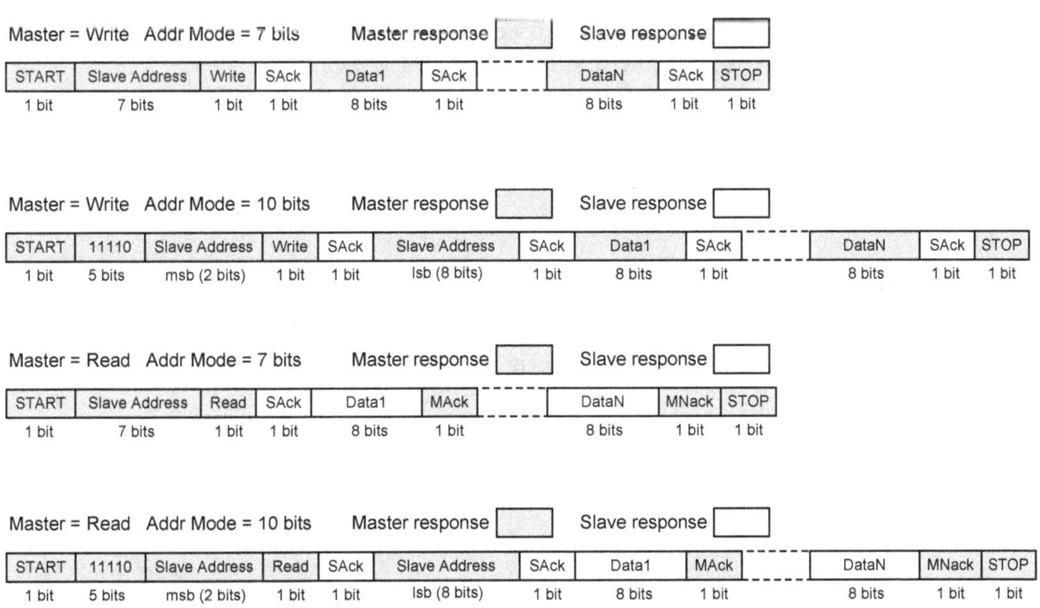

Fig. 4.22 I^2C modes of operation

followed by the Write bit. When the delivery of all these entries is acknowledged by the slave, the master sends out the remaining eight address bits. This is followed by another slave acknowledgement, and the master transmitting all of its data bytes to the designated slave. The data transfer completes when the bus master generating a Stop bit.

The third and the fourth entries in Fig. 4.22 show the seven-bit and ten-bit read sequences initiated by the bus master. In each sequence, after receiving the Start bit and the address, the designated slave starts sending out data packets to the master. After successfully receiving the first data byte, the master responds to the slave with an acknowledge signal, MAck, after which the slave transmits the next byte. The transfer continues until the slave delivers all of its data bytes to the master. However, right before the master issues a Stop bit, it generates a no-acknowledgement signal, MNack, signaling the end of the transfer as shown in Fig. 4.22.

The Start and Stop signals are generated by the combination of SCL and SDA values as shown in Fig. 4.23. According to this figure, a Start signal is produced when the SDA line is pulled to ground by the bus master while SCL = 1. Similarly, a Stop signal is created when the bus master releases the SDA line while SCL = 1.

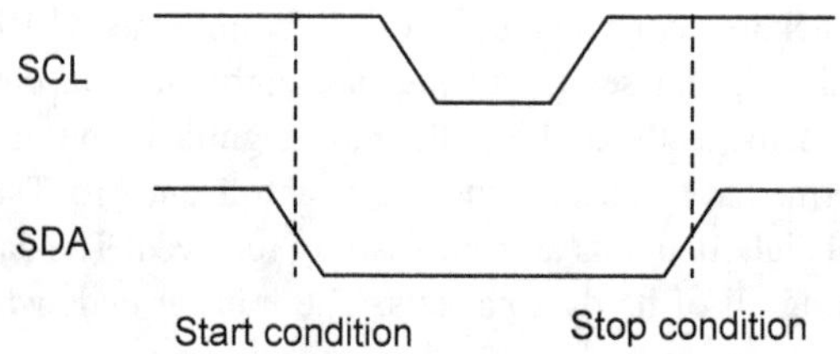

Fig. 4.23 I²C data stream start and stop conditions

Figure 4.24 shows when data is permitted to change, and when it needs to be steady. The I²C protocol only allows data changes when SCL is at logic 0. If the data on SDA changes while SCL is at logic 1, this may be interpreted as a Start or a Stop condition depending on the data transition. Therefore, the data on SDA is not allowed to change as long as SCL = 1.

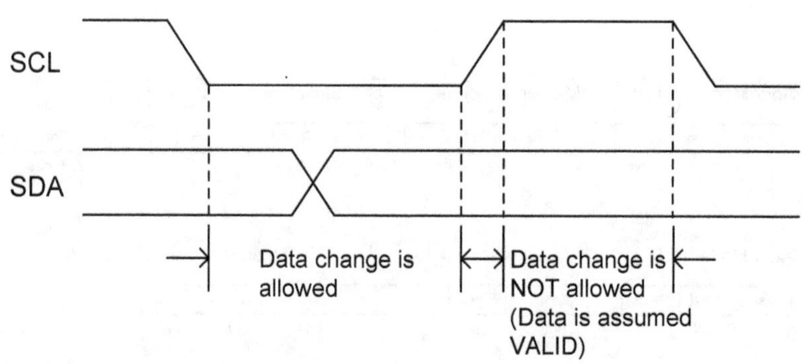

Fig. 4.24 I²C data change conditions

Figure 4.25 explains the timing diagram in which the bus master writes two bytes of data to a slave with a seven-bit address. According to this figure, the write process starts with transitioning the value at the SDA to logic 0 while SCL = 1. Following the Start bit, the slave address bits are delivered sequentially from the most significant bit, SA[6], to the least significant bit, SA[0]. Each address bit is introduced to the SDA only when SCL is at logic 0 according to the I^{2}C bus protocol shown in Fig. 4.24. The Write command and the subsequent slave acknowledgement are generated next. The data bits in Byte 0 and Byte 1 are also delivered to the SDA starting from the most significant data bit, D[7]. The write sequence finishes with the SDA transitioning to logic 1 while SCL = 1.

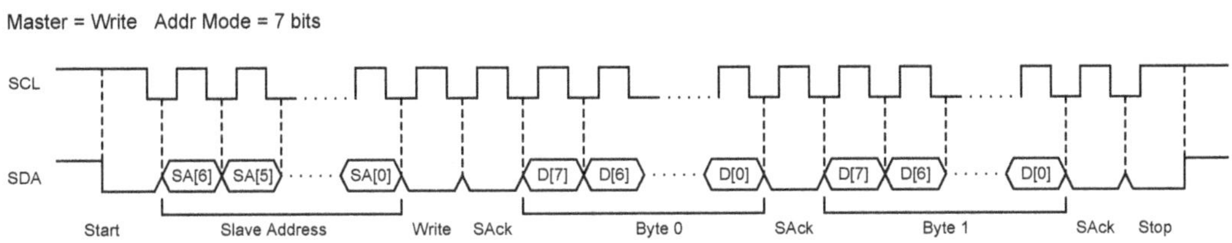

Fig. 4.25 I^{2}C write timing diagram

Figure 4.26 shows the timing diagram of reading two bytes of data from a slave. Following the Start bit and the slave address, the master issues the Read command by SDA = 1. Subsequently, bytes of data are transferred from the slave to the master with the master acknowledging the delivery of each data byte. The transfer ends with the master not acknowledging the last byte of data, MNack, and generating the Stop bit.

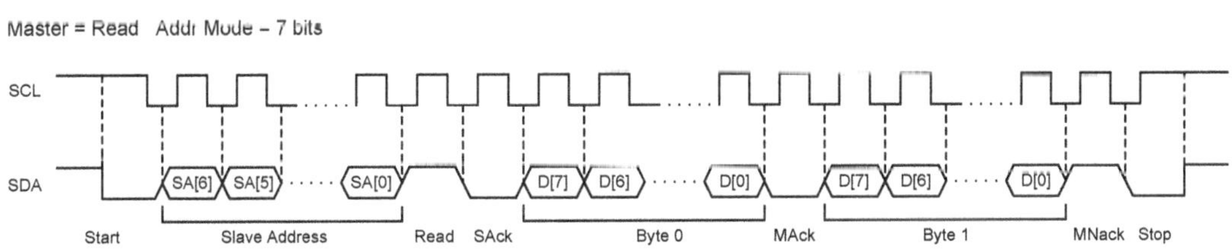

Fig. 4.26 I^{2}C read timing diagram

Here comes the reason why the I^{2}C bus protocol excels in maintaining flawless communication between any number of masters and slaves using only two physical wires. For example, what happens if two or more devices are simultaneously trying to write data on the SDA? At the electrical level, there is actually no contention between multiple devices trying to simultaneously enter a logic level on the bus. If a particular device tries to write logic 0 to the bus while the other issues logic 1, then the physical bus structure with pull-up resistors in Fig. 4.21 ensures that there will be no contention between these two devices, and the bus transitions to logic 0. In other words, in any conflict, logic 0 always wins!

This physical structure of the I^2C bus also allows bus masters to be able to read values from the bus or write values onto the bus freely without any danger of collision. In case of a conflict between two masters (suppose one is trying to write logic 0 and the other logic 1), the master that tries to write logic 0 gains the use of the bus without even being aware of the conflict. Only the master that tries to write logic 1 will know that it has lost the bus access because it reads logic 0 from the bus while trying to write logic 1. In most cases, this device will just delay its access and try it later.

Moreover, this bus protocol also helps to deal with communication problems. Any device present on the bus listens to the bus activity, particularly the presence of Start and Stop bits. Potential bus masters on the I^2C bus detecting a Start signal will wait until they detect the Stop signal before attempting to access the bus. Similarly, unaddressed slaves go back to hibernation mode until the Stop bit is issued.

Similarly, the master-slave pair is aware of each other's presence by the active-low acknowledge bit after delivering each byte. If anything goes wrong and the device sending the data does not detect acknowledgment from the recipient, the device sending data simply issues a Stop bit to stop the data transfer and releases the bus.

An important element of the I^2C communication is that the master device determines the clock speed in order to synchronize with the slave. If there are situations where an I^2C slave device is not able to keep up with the master because the clock speed is too high, the master can lower the frequency by a mechanism referred to as "clock stretching". According to this mechanism, an I^2C slave device is allowed to hold the SCL at logic 0 if it needs the bus master to reduce the bus speed. The master is required to observe the SCL signal level at all times and proceeds with the data transfer until the line is no longer pulled to logic 0 by the slave.

All in all, both SPI and I^2C offer good support for low speed peripheral communication. SPI is faster and better suited for single bus master applications where devices stream data to each other, while I^2C is slower and better suited for multi-master applications. The two protocols offer the same level of robustness and have been equally successful among vendors producing Flash memories, Analog-to-Digital and Digital-to-Analog converters, real-time clocks, sensors, liquid crystal display controllers etc.

Review Questions

1. A CPU reads three bursts of data from a 32-bit wide byte-addressable memory in the following manner:

- It reads four bytes with the starting address 0xF0,
- Immediately after the first transaction, the CPU reads two half-words from the starting address 0xF4,
- Immediately after the second transaction, it reads one word from the starting address 0xF8.

The contents of the memory are as follows:

31			0	
0xCC	0xDD	0xEE	0xFF	0xF0
0x88	0x99	0xAA	0xBB	0xF4
0x44	0x55	0x66	0x77	0xF8
0x00	0x11	0x22	0x33	0xFC

The unidirectional bus protocol states that the data communication between a bus master and a slave requires generating the address and control signals in the first cycle, and the data in the second cycle. The bus master always issues a Start signal to indicate the start of a data transmission. After an initial address, the master changes its status to Cont to indicate the continuation of the data transfer. The bus master issues Idle to indicate the end of the data transfer or Busy to indicate its incapability to produce address and control signals (and data if applicable). Any time the bus master is Busy, it repeats its address and control signals (and data if applicable) from the previous clock cycle. Simi larly, if a Ready signal is not generated by the slave, the bus master also repeats its address and control signals (and data if applicable) in the next clock cycle. At the end of a data transfer when the bus master finishes issuing new addresses, it transitions to the Idle state even though there may be a residual data read or a write still taking place in the subsequent clock cycle(s).

Fill in the blanks of the following timing diagram to complete all three data read bursts in the order specified above.

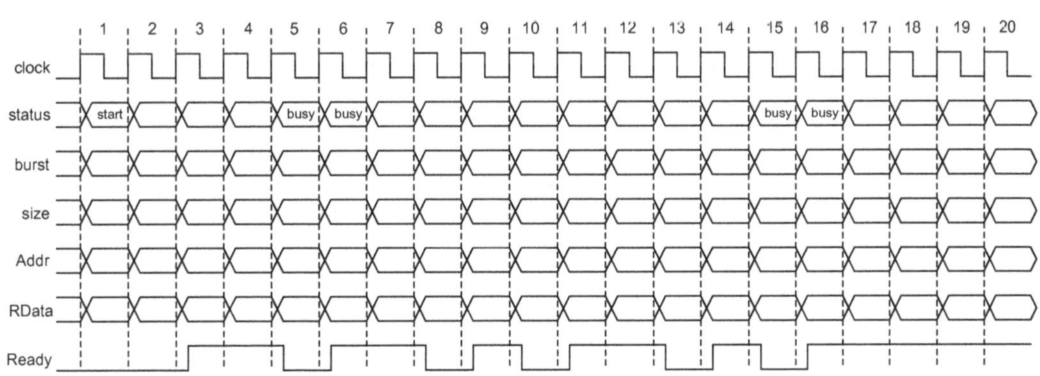

2. The following bidirectional bus maintains the communication between a CPU and a memory.

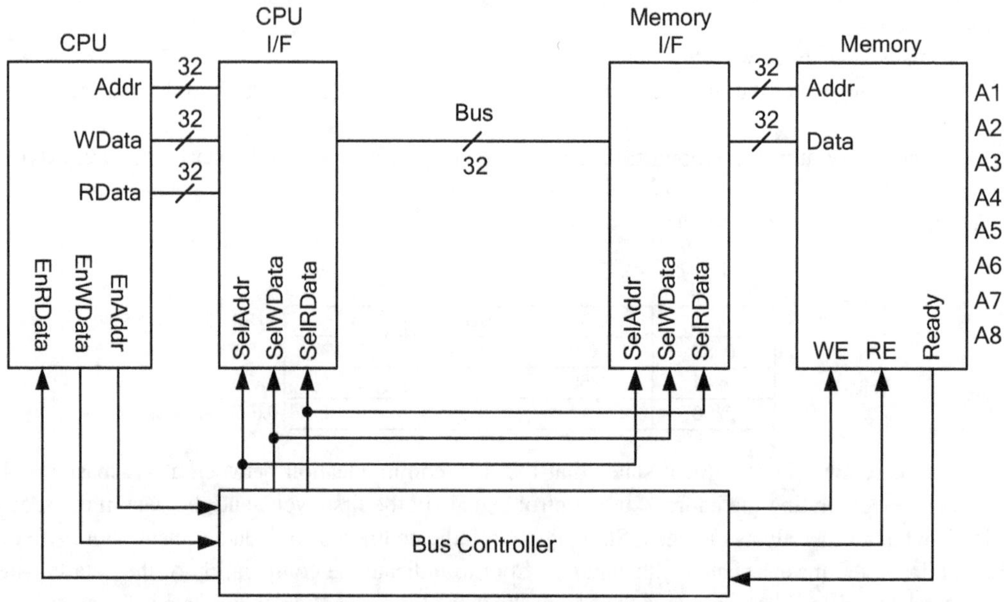

The CPU has the Addr, WData outputs to dispatch address and write-data, respectively. It also uses the RData output to receive data from the memory.

The memory, on the other hand, has the Addr input to receive address from the CPU, and a bidirectional Data port to receive and send data.

To validate the address and write-data, the EnAddr and EnWData signals are issued to the bus controller, respectively. To validate read-data, the memory dispatches the Ready signal to the controller. With all these inputs from the CPU and the memory, the controller generates the WE and RE signals for the memory to write and read data, and the EnRData signal for the CPU to validate the read data. The signals, SelAddr, SelWData and SelRData are also generated by the bus controller to manage the timely distribution of address, write-data and read-data using a single 32-bit wide bidirectional bus as shown in the figure above.

The write process to the memory requires a valid address with data as shown below:

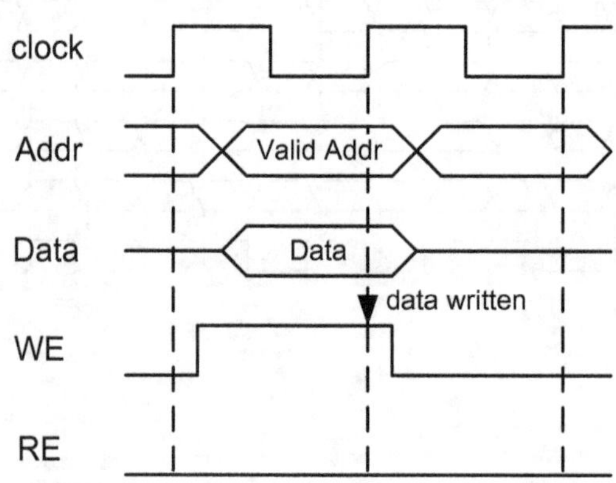

The read process from the memory requires a cycle delay to produce valid data once an address is issued. This is shown below:

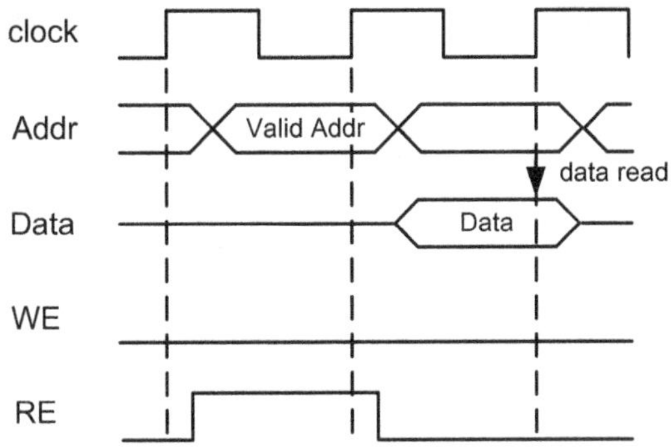

(a) Since the bus protocol requires data following a valid address, construct a timing diagram to write W1 into A1, W2 into A2, W3 into A3 and W4 into A4. Without any delay, perform a read sequence to fetch data packets, R1, R2, R3 and R4, from the memory addresses A5, A6, A7 and A8, respectively. Plot the 32-bit bus, Bus[31:0], and the control signals WE, RE, SelAddr, SelWData and SelRData for the write and read sequences.

(b) Design the CPU and the memory interfaces with the bidirectional bus such that these two data transfers are possible (note that these interfaces are not state machines).

3. A bus master writes four bytes of data to the following address locations of a 32-bit wide byte-addressable memory (slave) organized in a Little Endian format:

Address	Data
0x0D	0x11
0x11	0x22
0x15	0x33
0x19	0x44

Following the write cycle, the same bus master reads data (words) from the following slave addresses:

Address	Data
0x3C	0xAABBCCDD
0x40	0x55667788

(a) Draw the memory contents after writing and reading take place.

(b) If the bus master generates the first address during the first clock cycle and keeps generating new addresses every time the slave responds with a Ready signal, what will be the values of the address, control and data entries in the timing diagram below? Assume that the bus master does not produce any Status signal comprised of Start, Cont, Busy and Idle.

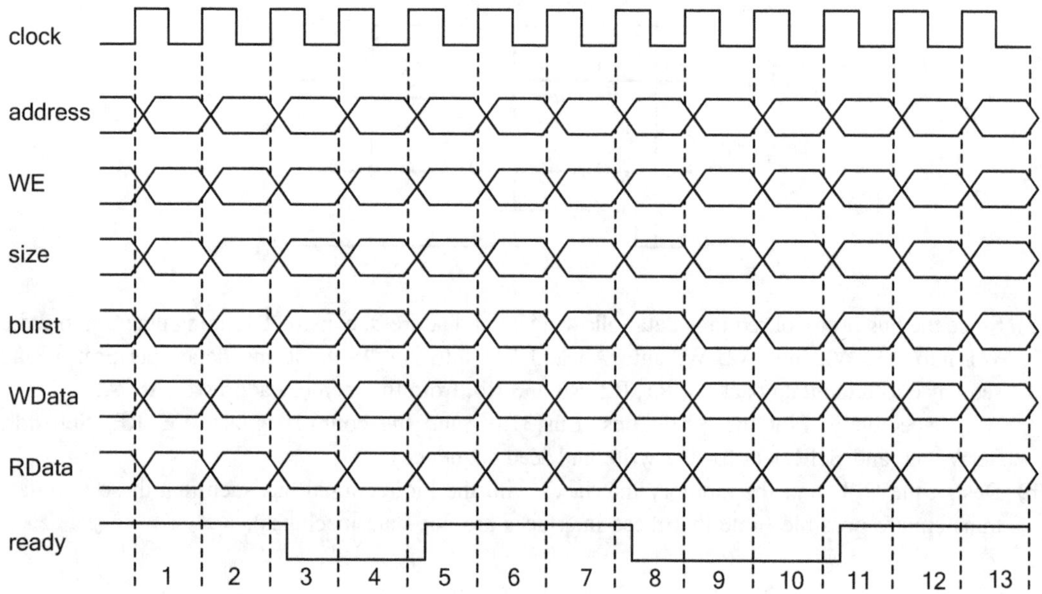

4. A bus master reads four data packets starting from the address, 0x00, and ending at the address, 0x03, from an eight-bit wide memory. Immediately after this transaction, the bus master writes 0x00, 0x11, 0x22 and 0x33 into the addresses 0x04, 0x05, 0x06 and 0x07 respectively. This memory contains the following data after this operation:

	7 0
00	0xAA
01	0xBB
02	0xCC
03	0xDD
04	0x00
05	0x11
06	0x22
07	0x33

Assuming the unidirectional bus protocol is the same as described in question 1, fill in the blanks of the timing diagram below to accommodate each read and write transfer.

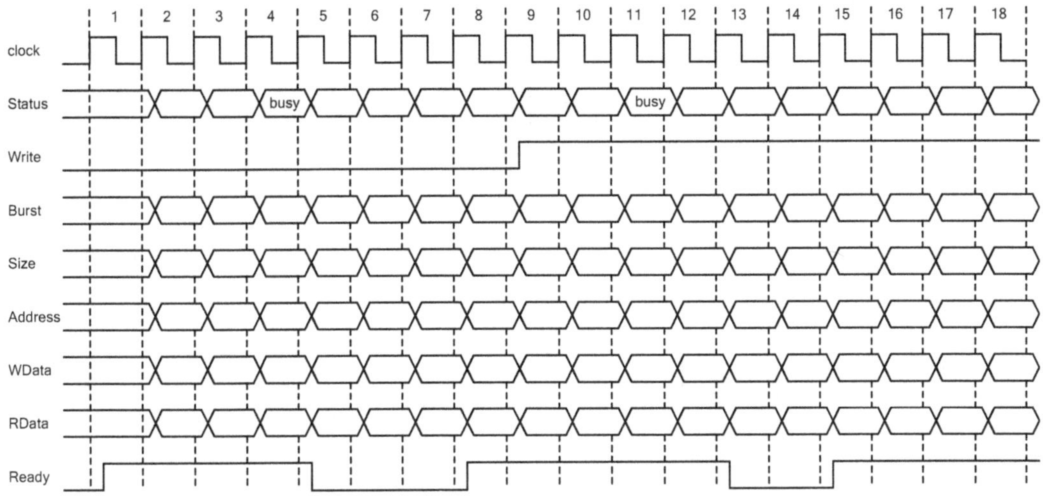

5. A bus master is connected to four memory blocks acting as bus slaves in a bidirectional bus where 32-bit address, write data and read data are sent or received on the same bus. The I/O ports of the bus master and the slaves are shown below:

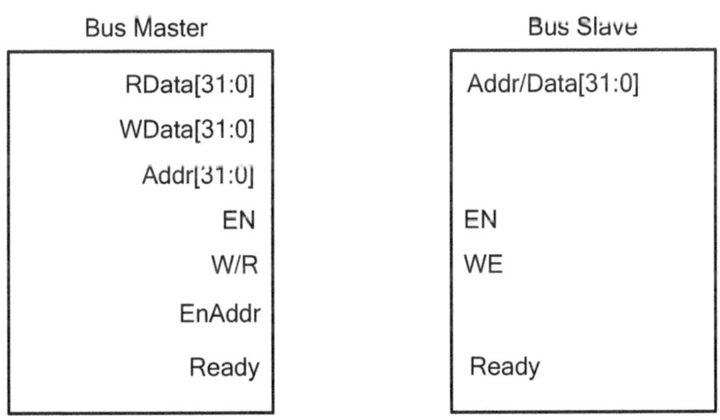

The bus master has separate read and write data ports to receive and transmit data, respectively. The Enable, and W/R ports enable the bus master to write data, i.e. EN = 1 and W/R = 1. Similarly, EN = 1 and W/R = 0 enable the bus master to read. Since address and data entries share the same bus, the bus master provides a third control signal, EnAddr, to enable the address. The bus master determines the slave's readiness through its Ready signal.

The slave, in contrast, has only one port for receiving address or data. EN = 1 and WE = 1 writes data to the slave. If data needs to be read from the slave, then EN = 1 and WE = 0 are used.

Draw the architectural diagram of such a system. Make sure to use the most significant address bits, Addr[31:30], to select one of the slaves for the bus master to read or write data.

6. A 16-bit digital system with unidirectional data and address buses is given below.

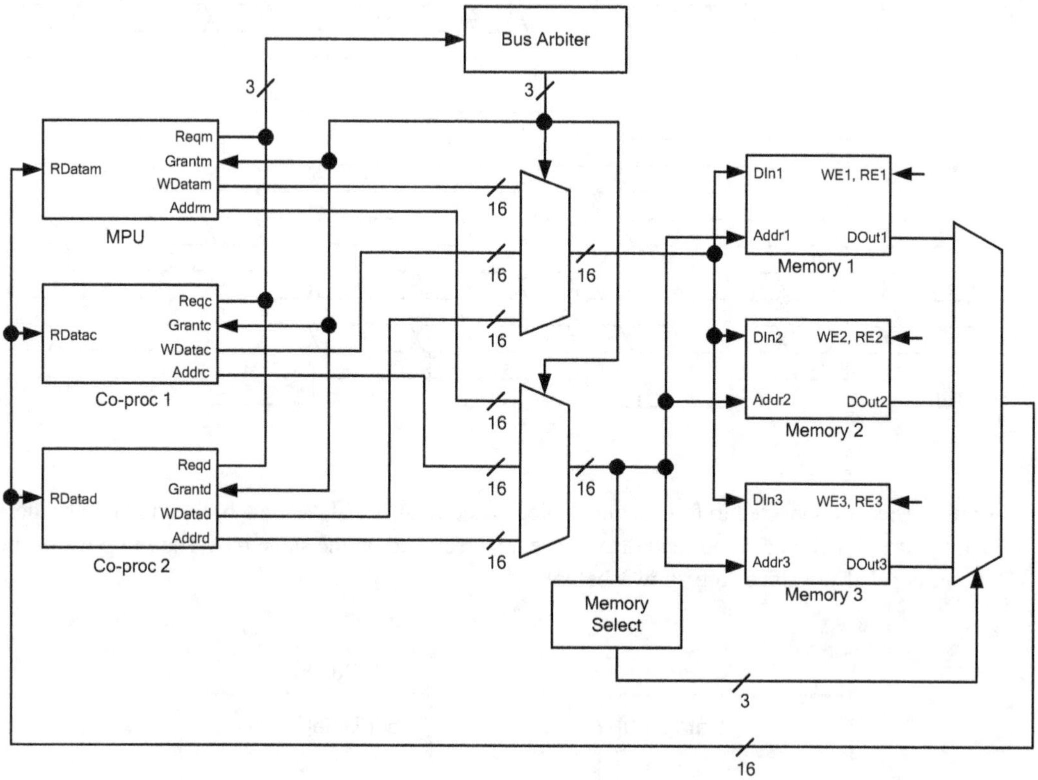

This system contains three bus masters, a Microprocessor Unit (MPU), Co-processor 1 and Co-processor 2. It also contains three slaves, Memory 1, Memory 2 and Memory 3.

A bus arbiter is responsible for prioritizing the ownership of the bus among the three bus masters. The MPU has the highest and Co-processor 2 has the lowest priority to use the bus. When the arbiter gives the ownership of the bus to a bus master, the bus master is free to exchange data with a slave as long as it keeps its request signal at logic 1. When the bus master lowers its request signal after finishing a data transfer, the arbiter also lowers its grant to assign the bus to another bus master according to the priority list.

As long as a bus master owns the bus, it can send a 16-bit wide write data (WData) to the selected slave memory at a specific address (Addr). Similarly, the bus master can read data from the slave using the 16-bit wide read data (RData) bus.

For the sake of simplicity, the control signals on the schematic are not shown; however, each memory has the Read Enable (RE) and Write Enable (WE) ports to control data storage.

(a) With the description above, draw the state diagram of the bus arbiter.

(b) While the MPU and Co-processor 1 remain at idle, Co-processor 2 requests two data transfers.

Note that Co-processor 2 and Memory 1 operate at a clock frequency twice as high as the clock frequency for Memory 2. Both memories have a read latency (access time) of one clock cycle, i.e. data becomes available in the next clock cycle after issuing a valid address with RE = 1. Write happens within the same clock cycle when the address is valid and WE = 1.

Including Co-processor 2's request (reqd), grant (grantd), address (Addrd) and the control signals (RE1, WE1, RE2, WE2) for each memory, create a timing diagram that shows data transfers between Memory 1 and Memory 2 using the timing diagram template below.

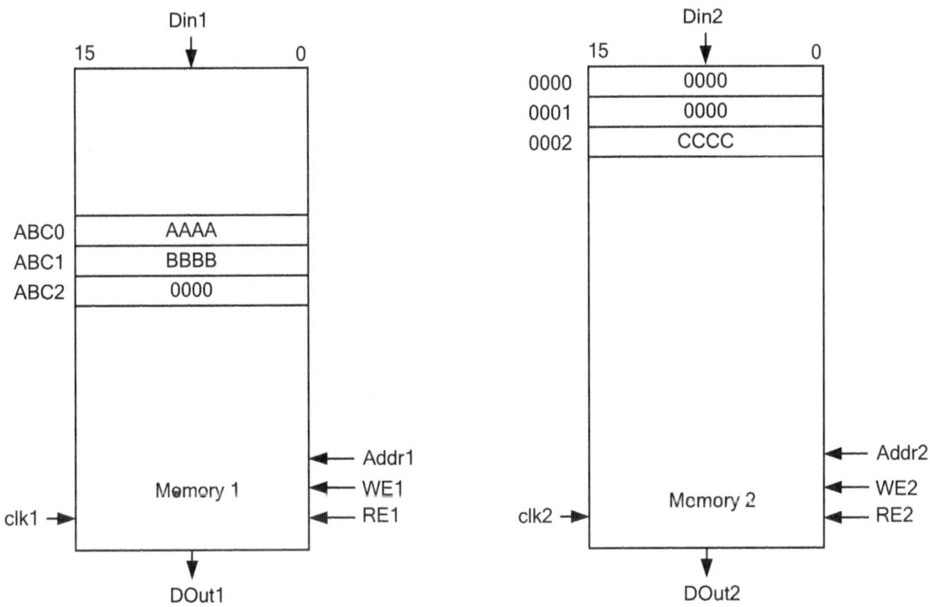

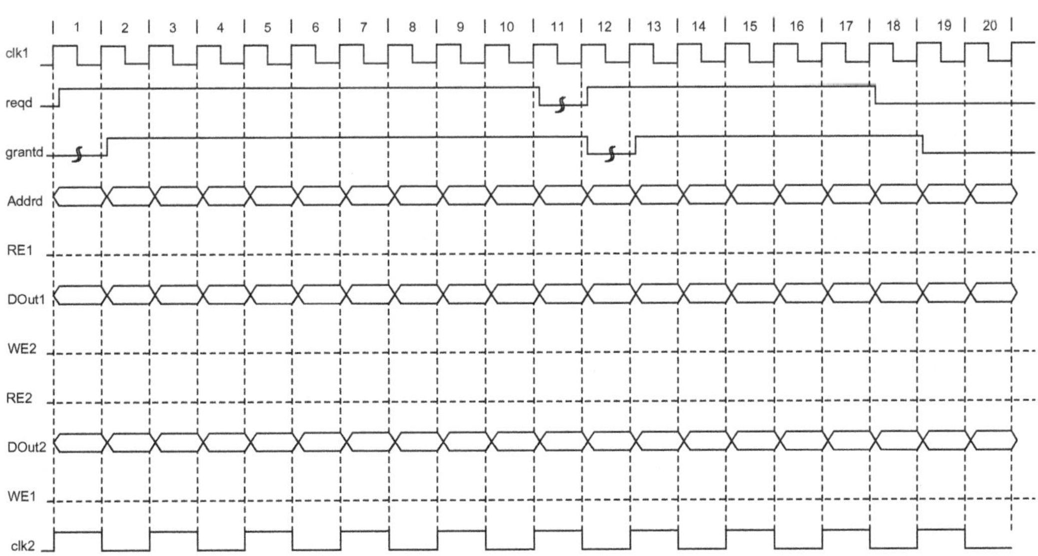

7. The waveforms below describe serial transmission of data using two known bus protocols, I^2C and SPI.

(a) A bus master is writing data to an I^2C compliant slave according to the timing diagram below. Assuming that the bus master uses the negative edge of SCLK to produce data on the SDA bus, determine the slave address and the data packets in binary format.

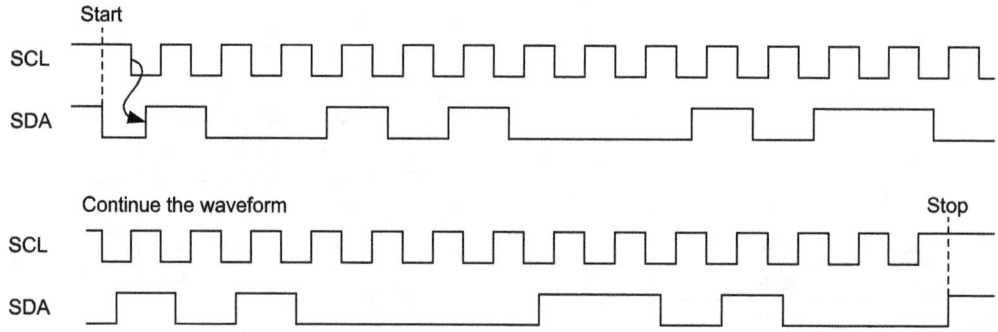

(b) Now, the bus master transmits the same data packets on the SPI bus. Using the timing diagram below, show the value of each data bit at the master's Serial Data Out (SDO) terminal. The bus master uses Mode3 convention and produces data at the negative edge of SCK. Note that, the master may pause in between sending data packets, but the SDO bus should retain the value of the least significant bit of each data packet during a data transfer.

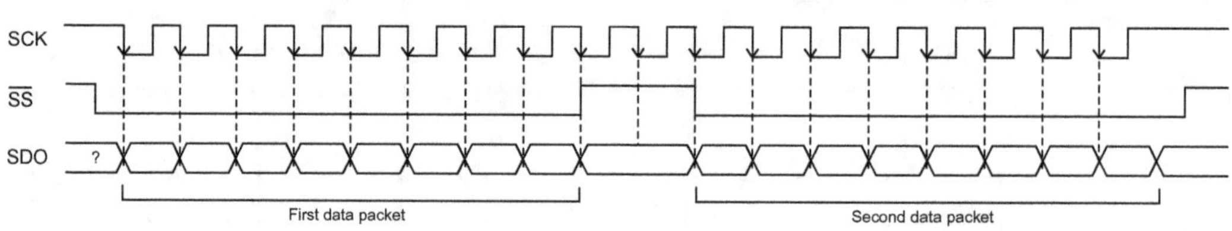

Projects

1. Implement the unidirectional bus with two bus masters and three slaves using Verilog. Make sure both of the bus masters are able to produce status signals, START, CONT, BUSY and IDLE to transfer data packets from one byte to two words (64 bits) on a 32-bit wide bus. Similarly, ensure that all the slaves are able to generate Ready signals compliant to the parallel bus protocol given in this chapter. Design the bus arbiter accordingly. Verify each individual block, i.e. the bus master, the slave and the arbiter, and the overall system functionality.
2. Implement the bidirectional bus with two bus masters and three slaves using Verilog. Make sure that the bus masters and slaves are fully compliant to the parallel bus protocol given in this chapter. Design this bus arbiter. Again, verify each individual block, i.e. the bus master, the slave and the arbiter, and the entire system functionality.
3. Implement the SPI bus with one bus master and three slaves using Verilog. Verify the system functionality with timing diagrams.
4. Implement the I^2C bus with seven-bit addressing mode using Verilog. Verify the system functionality with timing diagrams.

5 Memory Circuits and Systems

Basic serial and parallel bus structures and different forms of data transfer between a bus master and a slave were explained in Chap. 4. Regardless of the bus architecture, the bus master is defined as the logic block that initiates the data transfer while the slave is defined as the device that can only listen and exchange data with the master on demand. Both devices, however, include some sort of a memory, and in slave's case this can be a system memory or a buffer memory that belongs to a peripheral device.

Depending on the read and write speed, capacity and permanence of data, system memories and peripheral buffers can be categorized into three different forms. If fast read and write times are desired, Static Random Access Memory (SRAM) is used despite its relatively large cell size compared to other types of memory. SRAM is commonly used to store small temporary data, and it is typically connected to a high speed parallel bus in a system. If large amounts of storage are required, but slow read and write speed can be tolerated, then Dynamic Random Access Memory (DRAM) should be the main memory type to use. DRAMs are still connected to the high speed parallel bus and typically operate with receiving or delivering bursts of data. A typical DRAM cell is much smaller than an SRAM cell with significantly lower power consumption. The main drawbacks of DRAM are the high data read and write latencies, the complexity of memory control and management of data.

The permanence of data yet calls for a third memory type whose cell type consists of a double-gated Metal-Oxide-Semiconductor (MOS) transistor. Data is permanently stored in the floating gate of the device until it is overwritten with new data. Electrically-Erasable-Programmable-Read-Only-Memory (E^2PROM) or Flash memory fit into this type of device category. The advantage of this memory type is that it keeps the stored data even after the system power is turned off. However, this memory is the slowest compared to all other memory types, and it is subject to a limited number of read and write cycles. Its optimal usage is, therefore, to store permanent data for Built-In-Operating-Systems (BIOS), especially in hand-held devices where power consumption is critical. A typical computing system can contain one or all three types of memories depending on the usage and application software.

The basic functionality of SDRAM, E^2PROM and Flash memories in this chapter is inspired from Toshiba memory datasheets [1–6]. The more recent serial Flash memory with SPI interface is based on the datasheet of an Atmel Flash memory [7]. In each case, the functionality of the memory block has been substantially simplified (and modified) compared to the original datasheet in order to increase reader's comprehension for the subject matter. The purpose here is to show how each memory type operates in a system, covering only the basic modes of operation to train the reader rather than going into the details of the actual datasheets. The address, data and control timing constraints for each memory have also been simplified compared to the datasheets. This allows us to

A. Bindal, *Fundamentals of Computer Architecture and Design*,
https://doi.org/10.1007/978-3-030-00223-7_5

design the bus interface for each memory type with ease. For the sake of simplicity, we avoided duplicating the port names, exact timing requirements and functionality details that can be found in the actual datasheets. After reading this chapter, interested readers are encouraged to study the referenced datasheets prior to carrying out their design tasks.

5.1 Static Random Access Memory

Static Random Access Memory (SRAM) is one of the most fundamental memory blocks in digital design. Among all different types of memory, SRAM ranks the fastest; however, its large memory cell size limits its usage for a variety of applications.

A typical SRAM architecture shown in Fig. 5.1 is composed of four different blocks: the SRAM core, the address decoder, the sense amplifier and the internal SRAM controller. The memory core retains immediate data. The sense amplifier amplifies the cell voltage to full logic levels during read. The address decoder generates 2^N Word Lines (WL) from an N-bit address. Finally, the controller generates self-timed pulses required during a read or write cycle.

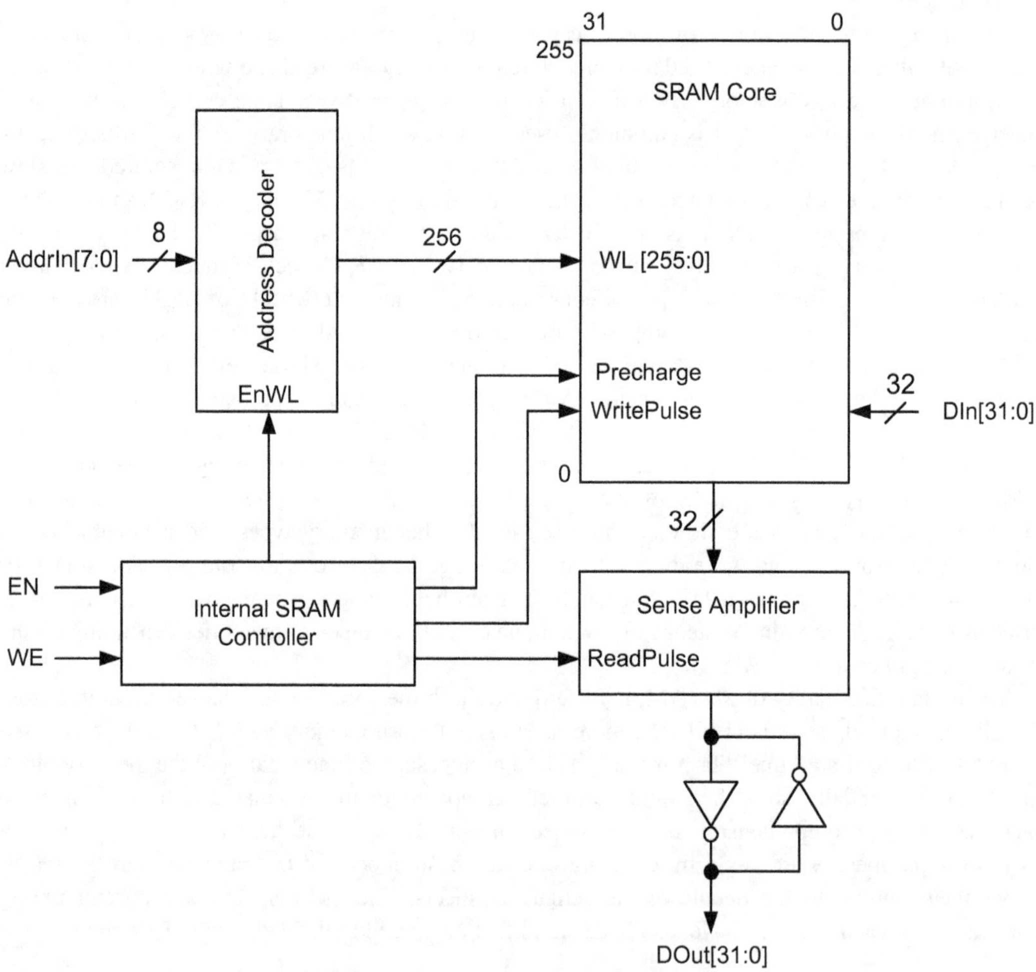

Fig. 5.1 A typical SRAM architecture with eight-bit address and 32-bit data

Each SRAM cell is composed of two back-to-back inverters like ones used in a latch, and two N-channel Metal Oxide Semiconductor (NMOS) pass-gate transistors to isolate the existing data in the cell or allow new data into the cell as shown in Fig. 5.2. When data needs to be written into a cell, WL = 1 turns on both NMOS transistors, allowing the true and complementary data to be simultaneously written into the cell from the Bit and Bitbar inputs. If we assume node A is initially at logic 0, node B at logic 1 and WL = 0, the logic level at WL turns off both NMOS transistors, and the latch becomes completely isolated from its surroundings. As a result, logic 0 level is contained in the cell. But, if WL = 1, Bit node = 1 and Bitbar = 0, the logic level at WL turn on both of the NMOS transistors, allowing the values at the Bit and Bitbar overwrite the existing logic levels at the nodes A and B, changing the stored bit in the cell from logic 0 to logic 1.

Similarly, if the data needs to be read from the cell, both NMOS transistors are turned on by WL = 1, and the small differential potential developed between the Bit and Bitbar outputs are amplified by the sense amplifier to reach to a full logic level at the SRAM output.

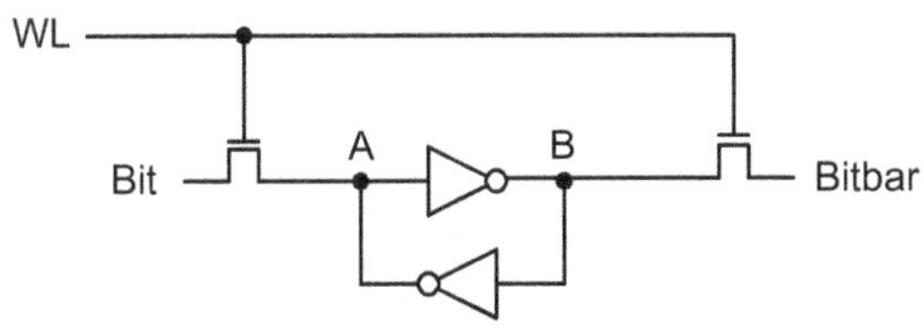

Fig. 5.2 SRAM memory cell

The data write sequence starts with EN (Enable) = 1 and Write Enable (WE) = 1. This combination precharges the Bit and Bitbar nodes in the SRAM core to a preset voltage value and prepares the memory for a write. When the precharge cycle is complete, the controller enables the address decoder by EnWL = 1 as shown in Fig. 5.1. The decoder activates a single WL input out of 256 WLs according to the value provided at AddrIn[7:0]. Within the same time period, the controller also produces WritePulse = 1, which allows the valid data at DIn[31:0] to be written to the specified address.

Reading data from the SRAM core is performed by EN = 1 and WE = 0. Similar to the write operation, the controller first precharges the SRAM core prior to reading data, and then turns on the address decoder. According to the address value at AddrIn port, the WL input to a specific row is activated, and the data is read from each cell to the corresponding Bit and Bitbar nodes from the designated row. The sense amplifier amplifies the cell voltage to full logic levels and delivers the data to DOut port.

The SRAM I/O timing can be synchronized with clock as shown in Figs. 5.3 and 5.4. In Fig. 5.3, when EN and WE inputs are raised to logic 1, SRAM goes into the write mode, and the valid data is written to a specified address at the next positive clock edge. In Fig. 5.4, when EN = 1 and WE = 0, SRAM is enabled and operates in the read mode. The core delivers valid data sometime after the next positive edge of the clock.

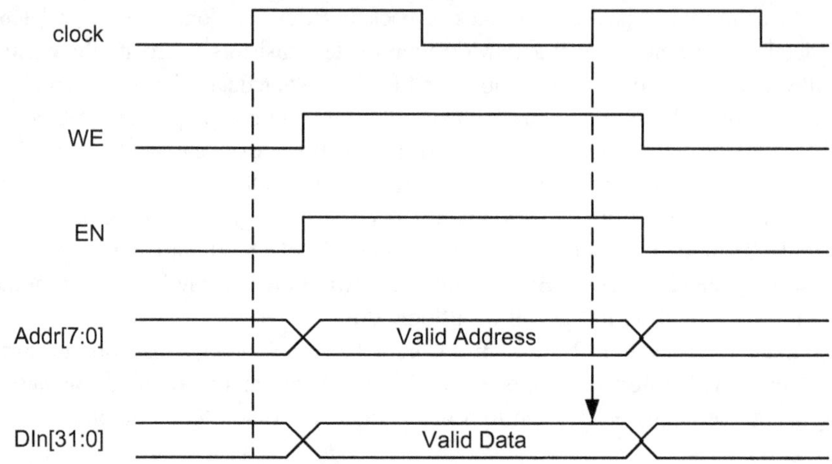

Fig. 5.3 SRAM I/O timing for write

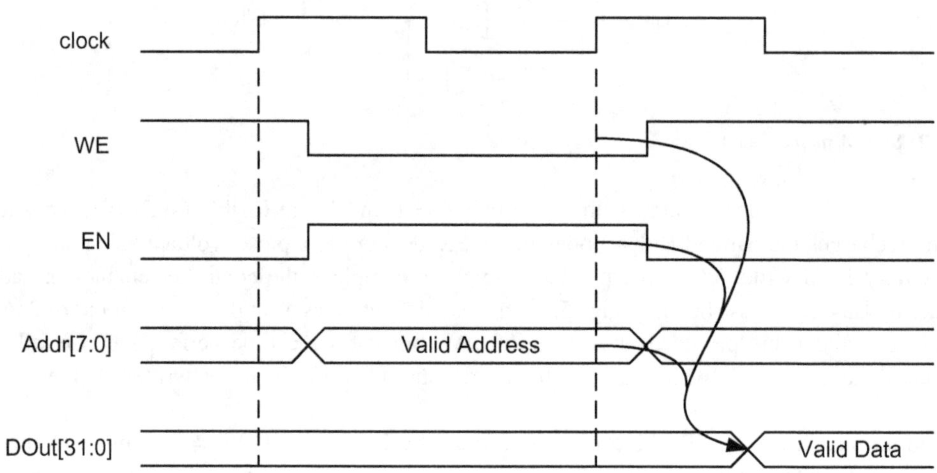

Fig. 5.4 SRAM I/O timing for read

One of the important tasks to integrate an SRAM module to an existing system is to design its bus interface. Figure 5.5 shows the block diagram of such an implementation. The bus interface basically translates all bus control signals to SRAM control signals (and vice versa), but seldom makes any modifications on address or data. In the unidirectional bus protocol described in Chap. 4, SRAM is considered to be a bus slave that exchanges data with the bus master on the basis of a Ready signal. Also as mentioned in Chap. 4, a bus master has four control signals to configure the data transfer. The Status signal indicates if the bus master is sending the first data packet (START) or is in the process of sending remaining data packets (CONT). The bus master may also send IDLE or BUSY signals to indicate if it has finished the current data transfer or is busy with an internal task, respectively. The Write signal specifies if the bus master intends to write data to a slave or read from a slave. The Burst signal designates the number of data packets in the transaction, and the Size signal defines the width of the data.

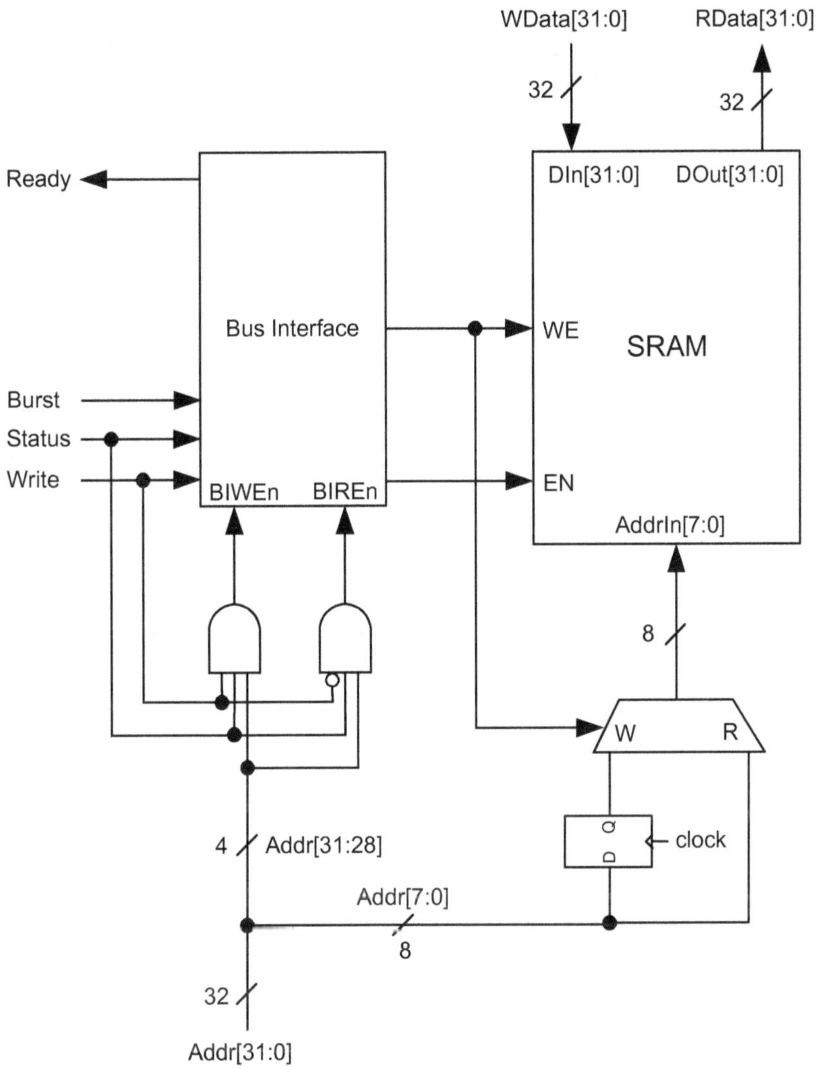

Fig. 5.5 SRAM bus interface block diagram (the counter keeping track of the number of issued SRAM addresses is omitted for simplicity)

The timing diagram in Fig. 5.6 shows how to write four data packets, W1 to W4, to four consecutive SRAM addresses, A1 to A4, as an example for the bus interface in Fig. 5.5. To initiate a write sequence, the bus master issues a valid address, Status = START and Write = 1 in the first clock cycle, and enables the bus interface for a write by producing an active-high Bus Interface Write Enable (BIWEn) signal. Upon receiving BIWEn = 1, the bus interface produces Ready = 1 in the next cycle, and prompts the bus master to change the address and control signals in the third cycle. As the bus master changes its address from A1 to A2, it also sends its first data packet, W1, according to the unidirectional bus protocol explained in Chap. 4. However, in order to write to an SRAM address, a valid data must be available within the same cycle as the valid address as shown in Fig. 5.3. Therefore, a set of eight flip-flops are added at Addr port of the SRAM in Fig. 5.5 so that the address, A1, is delayed for one clock cycle, and aligned with the current data, W1. The bus interface also produces EN = WE = 1 in the third cycle so that W1 is written to A1 at the positive edge of the fourth clock cycle. The next write is accomplished in the same way: the SRAM address is delayed for one cycle in order to write W2 to the address A2 at the positive edge of the fifth cycle. In the sixth cycle, the bus interface lowers the Ready signal so that the bus master stops incrementing the slave address. However, it keeps EN = WE = 1 to be able to write W4 to A4.

The bus interface state diagram for write in Fig. 5.7 is developed as a result of the timing diagram in Fig. 5.6. The first state, Idle state, is the result of the bus interface waiting to receive BIWEn = 1

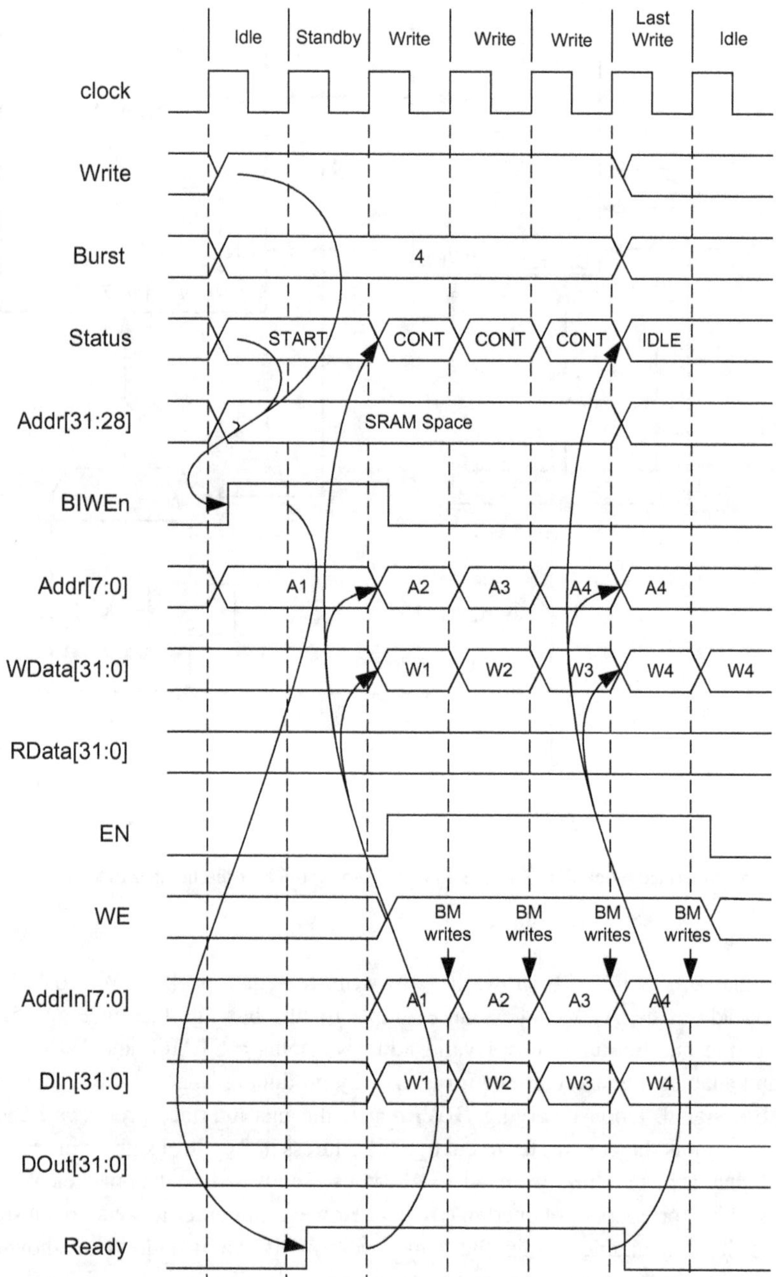

Fig. 5.6 SRAM bus interface timing diagram for write

from the bus master, which corresponds to the first clock cycle of the timing diagram in Fig. 5.6. The next state that follows the Idle state is the Standby state where the bus interface generates Ready = 1. This state is one clock cycle long and represents the second clock cycle in the timing diagram. The Write state is the state during which the actual write sequence takes place: EN and WE are kept at logic 1 as long as the number of write addresses issued by the bus master is less than Burst length. This state corresponds to the third, fourth and fifth clock periods in the timing diagram. When the number of write addresses reaches the value of the Burst length, the bus interface goes to the Last Write stage and Ready signal becomes logic 0. The bus master writes the final data packet to the last SRAM address in the sixth clock cycle.

In order to initiate a read sequence, the bus master issues a valid SRAM address, Status = START and Write = 0 signals in the first clock cycle of Fig. 5.8. This combination produces an active-high Bus Interface Read Enable, BIREn = 1, which is interpreted as the bus master intending to read data from an SRAM address. Consequently, the bus interface generates EN = 1, WE = 0, Ready = 1 in

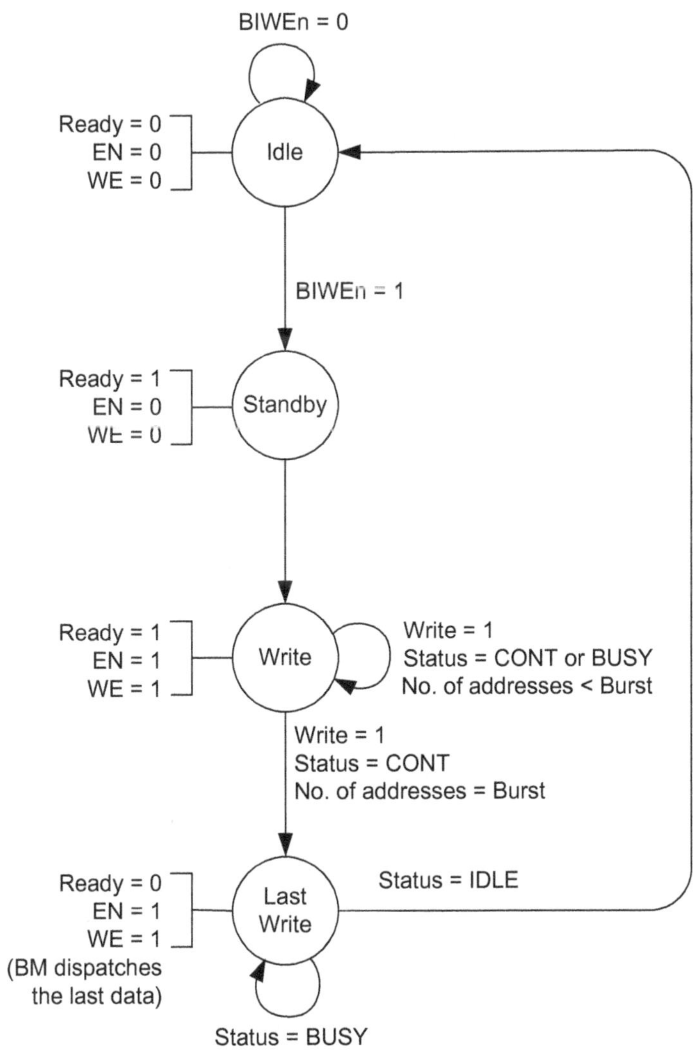

Fig. 5.7 SRAM bus interface for write

the second cycle. This fetches the first data, R1, from the SRAM address, B1, in the third cycle. The read transactions in the fourth and fifth cycles are identical to the third, as the the bus master reads R2 and R3 from the addresses, B2 and B3, respectively. In the sixth cycle, the bus interface retains Ready = 1 so that the bus master is still able to read the last data, R4, from the address, B4.

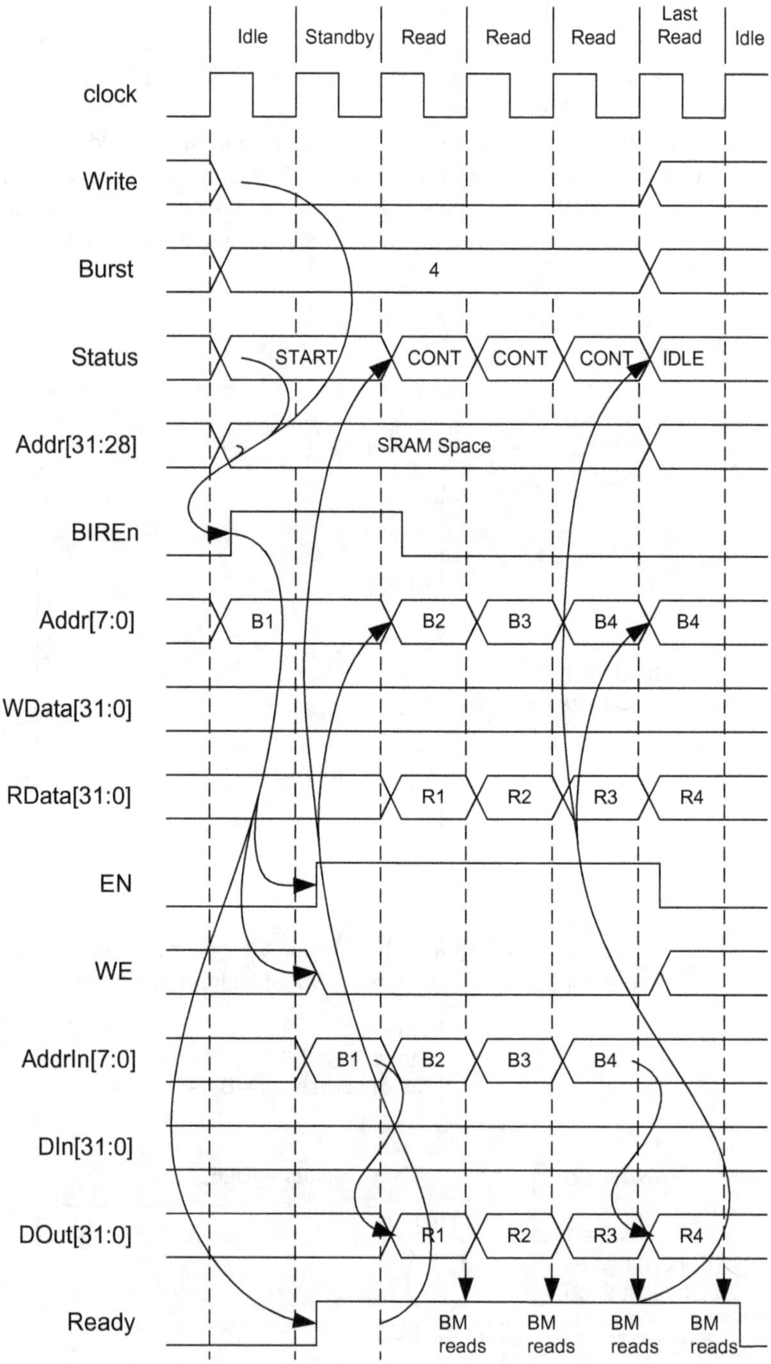

Fig. 5.8 SRAM bus interface timing diagram for read

As in the write case, the read bus interface in Fig. 5.9 is also a direct consequence of the timing diagram in Fig. 5.8. The Idle state corresponds to the first clock cycle of the timing diagram in Fig. 5.8. As soon as BIREn = 1 is generated, the bus interface transitions to the Standby state where it produces EN = 1, WE = 0 and Ready = 1. The interface enters the Read state in the third cycle and produces the same outputs as before in order for the bus master to read its first data, R1, and to send a new address in the next cycle. The interface stays in the Read state until the number of read addresses issued by the bus master is less than the Burst length. The Read state covers from the third to the fifth cycle in the timing diagram in Fig. 5.8. When the number of read addresses reaches the Burst length, the bus interface transitions to the Last Read state in cycle six where it continues to generate Ready = 1. This is done so that the bus master is still able to read the last data as mentioned earlier. The interface unconditionally goes back to the Idle state in the following cycle.

Increasing SRAM capacity necessitates employing extra address bits. In the example shown in Fig. 5.10, the SRAM capacity is increased from 32 × 16 bits to 32 × 64 bits by appending two extra

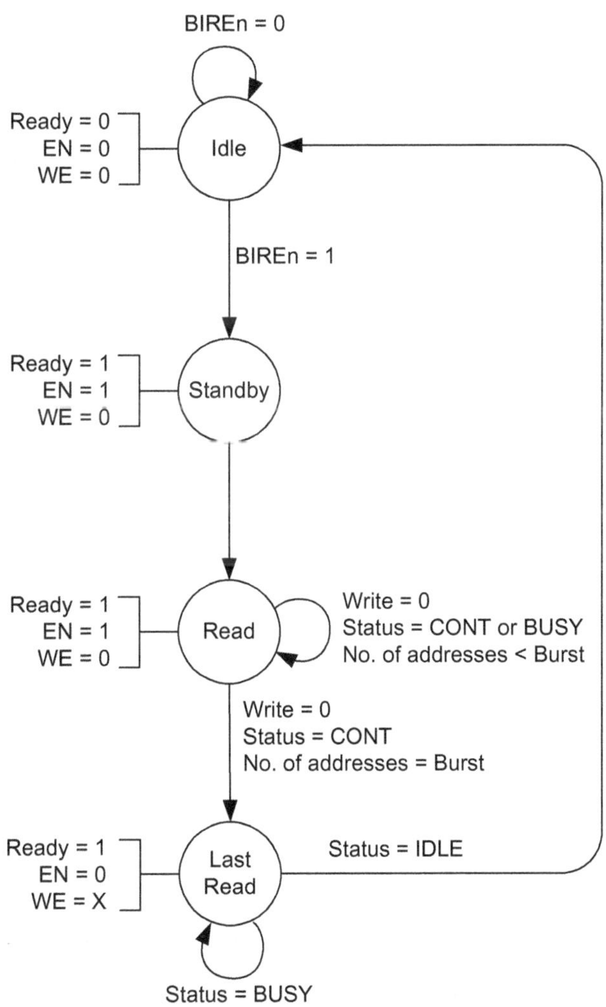

Fig. 5.9 SRAM bus interface for read

address bits, Addr[5:4], which serves to access one of the four SRAM blocks. In this figure, even though Addr[3:0] points to the same address location for all four 32 × 16 SRAM blocks, Addr[5:4] in conjunction with EN enables only one of the four blocks. Furthermore, the data read from the selected block is routed through the 4-1 MUX using Addr[5:4] inputs. Addr[5:4] = 00 selects the contents of DOut0 port and routes the data through port 0 of the 4-1 MUX to Out[31:0]. Similarly, Addr[5:4] = 01, 10 and 11 select ports 1, 2 and 3 of the 4-1 MUX, and route data from DOut1, DOut2 and DOut3 ports to Out[31:0], respectively.

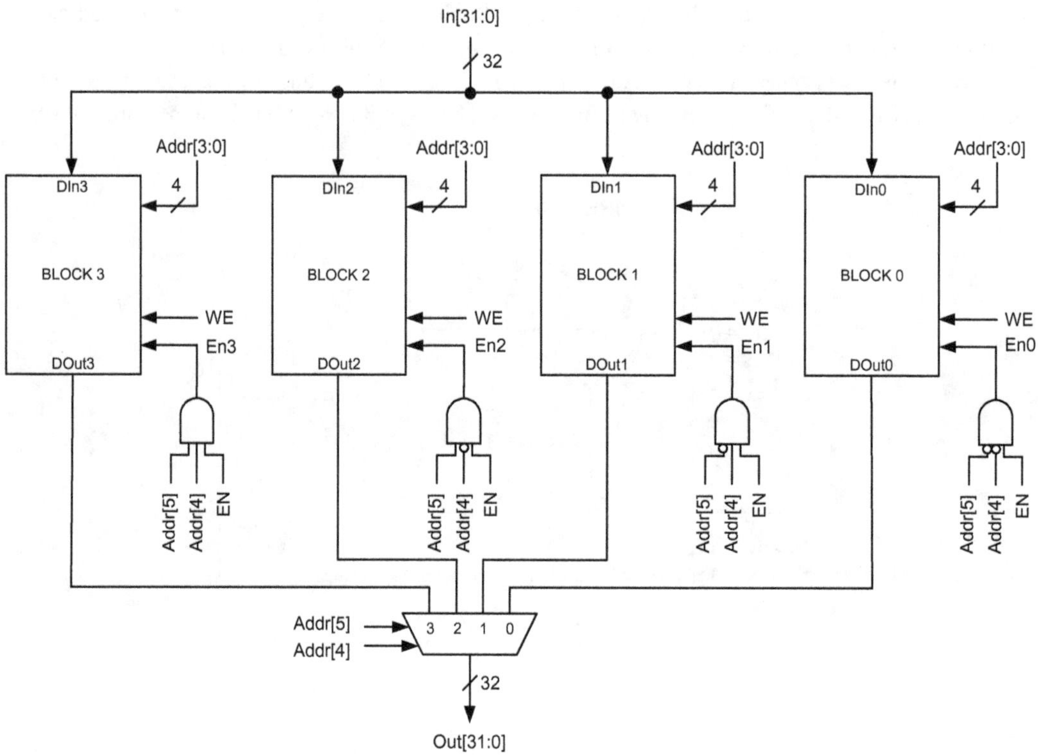

Fig. 5.10 Increasing SRAM address space

5.2 Synchronous Dynamic Random Access Memory

Synchronous Dynamic Random Access Memory (SDRAM) is a variation of the older DRAM, and it constitutes the main memory of almost any computing system. Even though its capacity can be many orders of magnitude higher than SRAM, it lacks speed. Therefore, its usage is limited to storing large blocks of data when speed is not important.

An SDRAM module is composed of four blocks. The memory core is where data is stored. The row and column decoders locate the data. The sense amplifier amplifies the cell voltage during read. The controller manages all the read and write sequences.

The block diagram in Fig. 5.11 shows a typical 32-bit SDRAM architecture composed of four memory cores, called banks, accessible by a single bidirectional input/output port. Prior to operating the memory, the main internal functions, such as addressing modes, data latency and burst length, must be stored in the Address Mode Register. Once programmed, the active-low Row Address Strobe, $\overline{RAS}$, Column Address Strobe, $\overline{CAS}$, and Write Enable, $\overline{WE}$, signals determine the functionality of the memory as shown in Table 5.1. The data at the input/output port of a selected bank can be masked at the Read/Write Logic block before it reaches the data Input/Output port, DInOut.

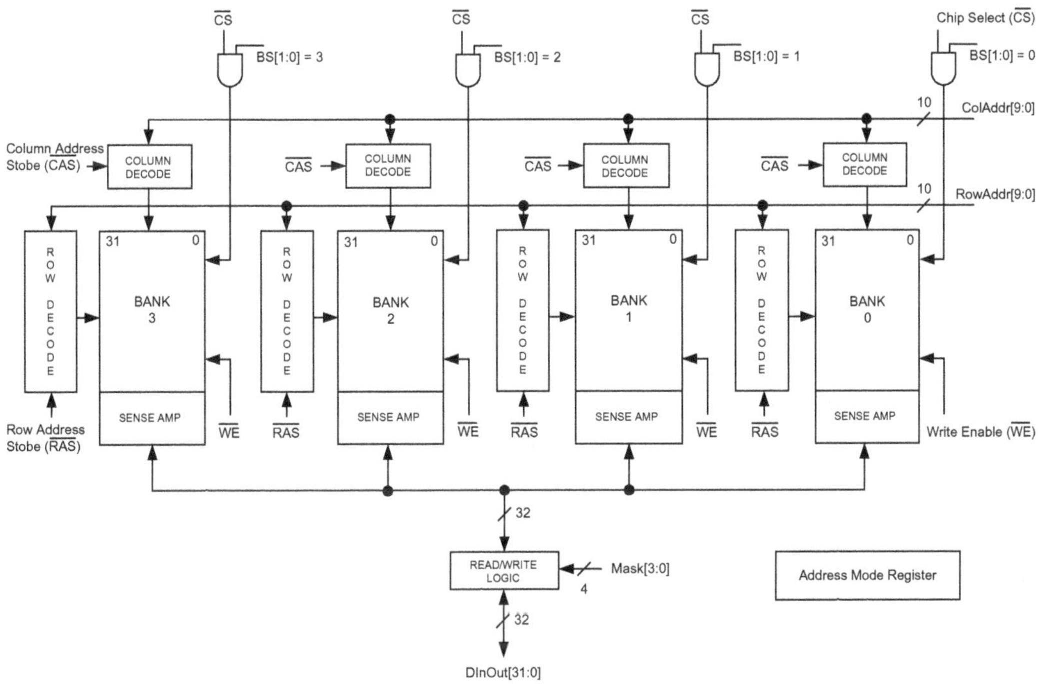

Fig. 5.11 A typical SDRAM architecture

Table 5.1 SDRAM modes of operation

$\overline{CS}$	$\overline{RAS}$	$\overline{CAS}$	$\overline{WE}$	OPERATION
0	0	0	0	Program Addr. Mode Register
0	0	0	1	Self Refresh
0	0	1	0	Precharge a Bank with BS[1:0]
0	0	1	1	Activate a Bank with BS[1:0]
0	1	0	0	Write into a Bank with BS[1:0]
0	1	0	1	Read from a Bank with BS[1:0]
0	1	1	0	Burst Stop
0	1	1	1	Reserved
1	X	X	X	SDRAM Deselect

The SDRAM cell is a simple device composed of an NMOS pass-gate transistor to control the data-flow in and out of the cell and a capacitor to store data as shown in Fig. 5.12. When new data needs to be written into the cell, the NMOS transistor is turned on by Control = 1, and the data at the DIn/Out terminal is allowed to overwrite the old data at the Cell node. Reading data from the cell, on the other hand, requires activation of the sense amplifier prior to turning on the pass-gate transistor. When data needs to be preserved, the NMOS transistor is simply turned off by Control = 0. However, the charge on the cell capacitor slowly leaks through its insulator, resulting in a reduced cell voltage. Thus, an automatic or manual cell refresh cycle becomes necessary during SDRAM operation to preserve the bit value in the cell.

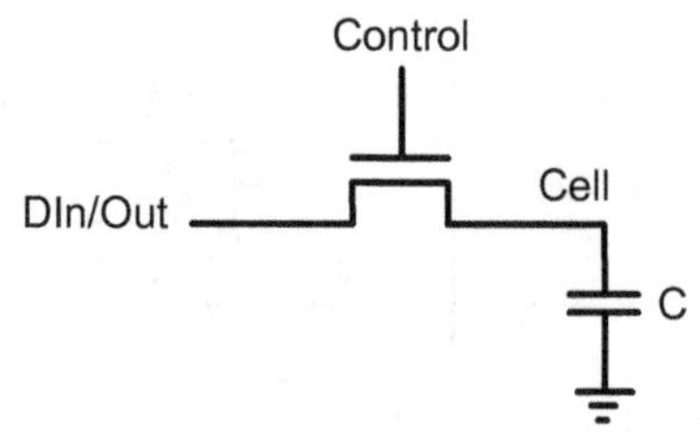

Fig. 5.12 SDRAM memory cell

The first row of the truth table in Table 5.1 indicates how to program the internal Address Mode Register. At the positive edge of the clock, $\overline{CS}$, $\overline{RAS}$, $\overline{CAS}$ and $\overline{WE}$ signals are pulled low to logic 0 to program the Address Mode Register as shown in Fig. 5.13. In the program mode, the address bits, A[2:0], define the data burst length as shown in Table 5.2. Burst length can range from one word to full page, which is equal to the contents of the entire bank. The address bit, A [3], defines how the SDRAM address increments for each data packet. Sequential addressing is achieved simply by incrementing the starting address by one, and eliminating the carry bit according to the size of the burst length.

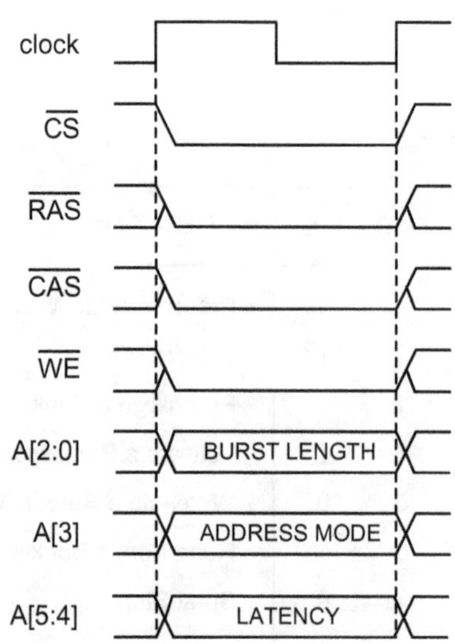

Fig. 5.13 Timing diagram for programming the Address Mode Register

Table 5.2 Truth tables for programming the Address Mode Register

A[2]	A[1]	A[0]	Burst Length
0	0	0	1 Word
0	0	1	2 Words
0	1	0	4 Words
0	1	1	8 Words
1	0	0	16 Words
1	0	1	32 Words
1	1	0	64 Words
1	1	1	Full Page

A[3]	Addressing Mode
0	Sequential
1	Linear

A[5]	A[4]	Latency
0	0	2
0	1	3
1	0	4
1	1	5

For example, if the starting address is 13 and the burst length is two words, the carry bit from the column A[0] is eliminated and the next address becomes 12 as shown in Table 5.3. In the same table, if the burst length is increased to four, the carry bit from the column A [1] is eliminated, and the address values after the starting address 13 become 14, 15 and 12. If the burst length becomes eight, the carry bit from the column A [2] is eliminated, and the address values of 13, 14, 15, 8, 9, 10, 11 and 12 are produced successively. Sequential addressing confines reading or writing of data within a predefined, circulatory memory space, convenient for specific software applications.

The linear addressing mode is a simplified version of the actual interleave addressing mode in various SDRAMs, and increments the SDRAM address linearly as shown in Table 5.4. In this table, if the starting address is 13 and the burst length is two, the next address will be 14. If the burst length is increased to four, the next three addresses following 13 will be 14, 15 and 16. In contrast to the sequential addressing mode, the linear addressing mode increments SDRAM address one bit at a time, not confining the data in a circulatory address space.

The second row of the truth table in Table 5.1 shows how to initiate a manual refresh cycle. In manual refresh mode, SDRAM replenishes node voltage values at each cell because the charge across the cell capacitor leaks through its dielectric layer over time. The time duration between refresh cycles depends on the technology used, the quality of the oxide growth and the thickness of the dielectric used between capacitor plates as shown in Fig. 5.14.

Table 5.3 SDRAM sequential mode addressing for burst lengths of 2, 4 and 8

Starting Address = 13, Burst Length = 2, Mode = Sequential

A[9]	A[8]	A[7]	A[6]	A[5]	A[4]	A[3]	A[2]	A[1]	A[0]	
0	0	0	0	0	0	1	1	0	1	= 13
									+ 1	
0	0	0	0	0	0	1	1	0	0	= 12
						delete the carry bit				

Starting Address = 13, Burst Length = 4, Mode = Sequential

A[9]	A[8]	A[7]	A[6]	A[5]	A[4]	A[3]	A[2]	A[1]	A[0]	
0	0	0	0	0	0	1	1	0	1	= 13
									+ 1	
0	0	0	0	0	0	1	1	1	0	= 14
									+ 1	
0	0	0	0	0	0	1	1	1	1	= 15
									+ 1	
0	0	0	0	0	0	1	1	0	0	= 12
					delete the carry bit					

Starting Address = 13, Burst Length = 8, Mode = Sequential

A[9]	A[8]	A[7]	A[6]	A[5]	A[4]	A[3]	A[2]	A[1]	A[0]	
0	0	0	0	0	0	1	1	0	1	= 13
									+ 1	
0	0	0	0	0	0	1	1	1	0	= 14
									+ 1	
0	0	0	0	0	0	1	1	1	1	= 15
									+ 1	
0	0	0	0	0	0	1	0	0	0	= 8
			delete the carry bit						+ 1	
0	0	0	0	0	0	1	0	0	1	= 9
									+ 1	
0	0	0	0	0	0	1	0	1	0	= 10
									+ 1	
0	0	0	0	0	0	1	0	1	1	= 11
									+ 1	
0	0	0	0	0	0	1	1	0	0	= 12

Table 5.4 SDRAM linear addressing mode for burst lengths of 2 and 4

Starting Address = 13, Burst Length = 2, Mode = Linear

A[9]	A[8]	A[7]	A[6]	A[5]	A[4]	A[3]	A[2]	A[1]	A[0]		
0	0	0	0	0	0	1	1	0	1	=	13
0	0	0	0	0	0	1	1	1	0	=	14

Starting Address = 13, Burst Length = 4, Mode = Linear

A[9]	A[8]	A[7]	A[6]	A[5]	A[4]	A[3]	A[2]	A[1]	A[0]		
0	0	0	0	0	0	1	1	0	1	=	13
0	0	0	0	0	0	1	1	1	0	=	14
0	0	0	0	0	0	1	1	1	1	=	15
0	0	0	0	0	1	0	0	0	0	=	16

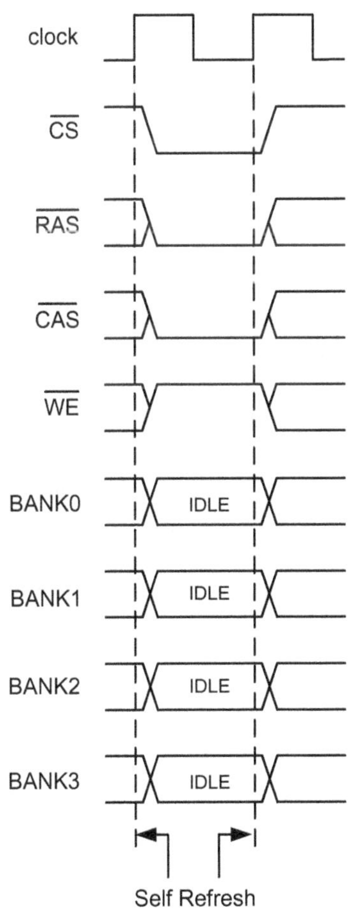

Fig. 5.14 Timing diagram for self-refresh

Rows three through six in Table 5.1 constitute the read and the write sequences as shown in Fig. 5.15. In this figure, a read or a write sequence always starts with precharging all the rows and columns of the SDRAM core. This is followed by an activation cycle where the row address is generated. In the last cycle, the column address is generated, and the data is either written or read from the memory according to the control signals, $\overline{CS}$, $\overline{RAS}$, $\overline{CAS}$ and $\overline{WE}$.

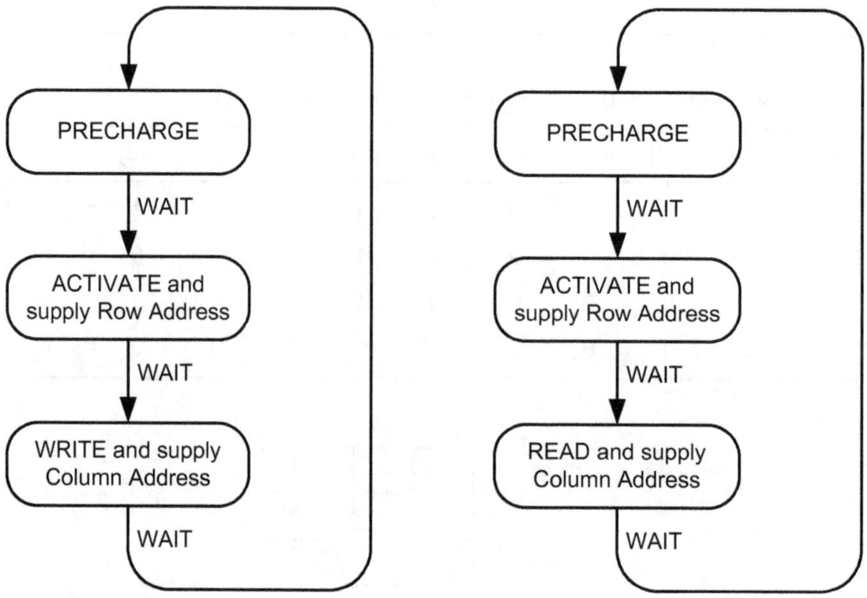

Fig. 5.15 Write and read operation cycles

Prior to a read or a write, all the rows and columns of a bank must be precharged to a certain voltage level for a period of one clock cycle as shown in Fig. 5.16. During precharge, $\overline{CS}$, $\overline{RAS}$ and $\overline{WE}$, must be lowered to logic 0, and $\overline{CAS}$ must be kept at logic 1 as shown in the third row of Table 5.1. The value of the precharge voltage can be anywhere between 0 V and the full supply voltage depending on the technology and the requirements of the circuit design. The activation cycle starts a certain time after precharging the bank. The time interval between the precharge and activation cycles is called the precharge time period, t_{PRE}, as shown in Fig. 5.16. The activation cycle is enabled by lowering $\overline{CS}$ and $\overline{RAS}$ to logic 0, but keeping $\overline{CAS}$ and $\overline{WE}$ at logic 1 as shown in the fourth row of Table 5.1. Following the activation cycle, the next precharge period must not start until after a certain time period has elapsed for the same bank. This time interval is called the RAS time period, t_{RAS}, as shown in Fig. 5.16.

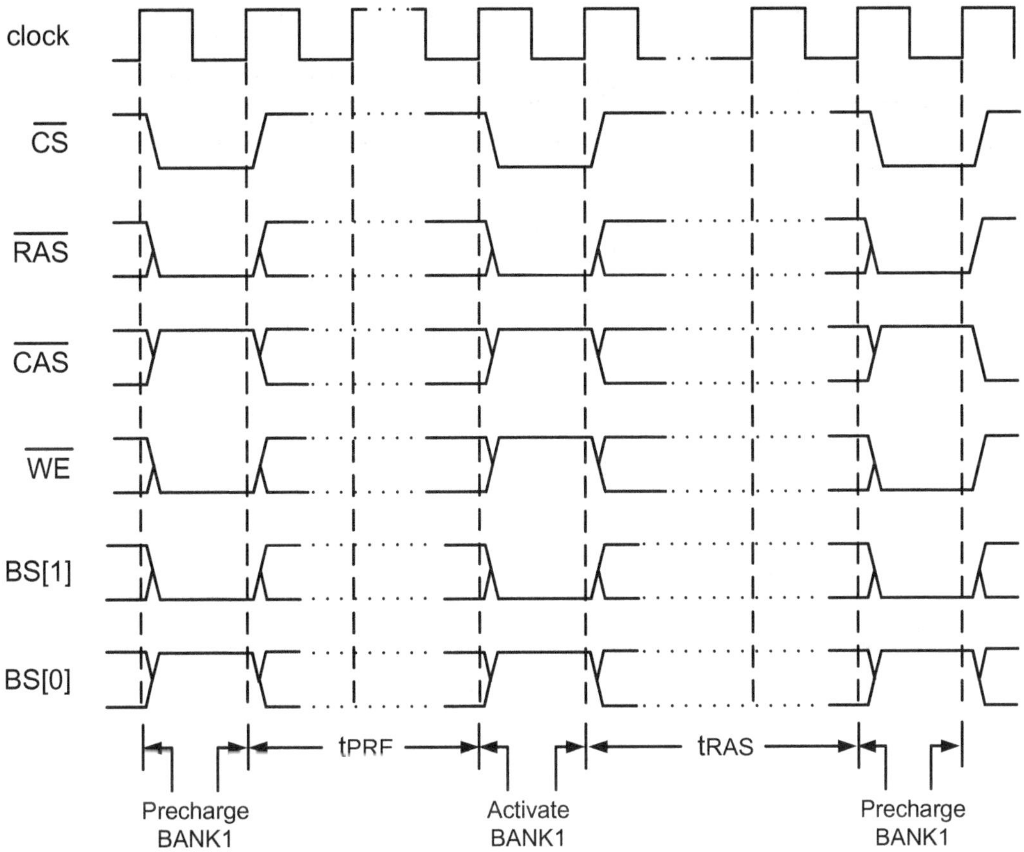

Fig. 5.16 Bank precharge and activation cycles

The fifth row of Table 5.1 shows how to write into a selected bank when $\overline{CS} = \overline{CAS} = \overline{WE} = 0$ and $\overline{RAS} = 1$. The actual write takes place in the last phase of the write sequence in Fig. 5.15 following the precharge and activation cycles. To illustrate the write sequence in detail, a single write example is given in Fig. 5.17. In this figure, the write cycle starts with precharging Bank 1. After $t = t_{PRE}$,, the activation period starts and the row address is supplied to the SDRAM. When the column address supplied after a time period of t_{CAS}, four data packets, D(0) through D(3), are written to SDRAM in four consecutive clock cycles. Note that in this figure if the same bank is used for another write, a new time period, t_{RAS}, needs to be placed between the bank activation cycle and the next precharge period.

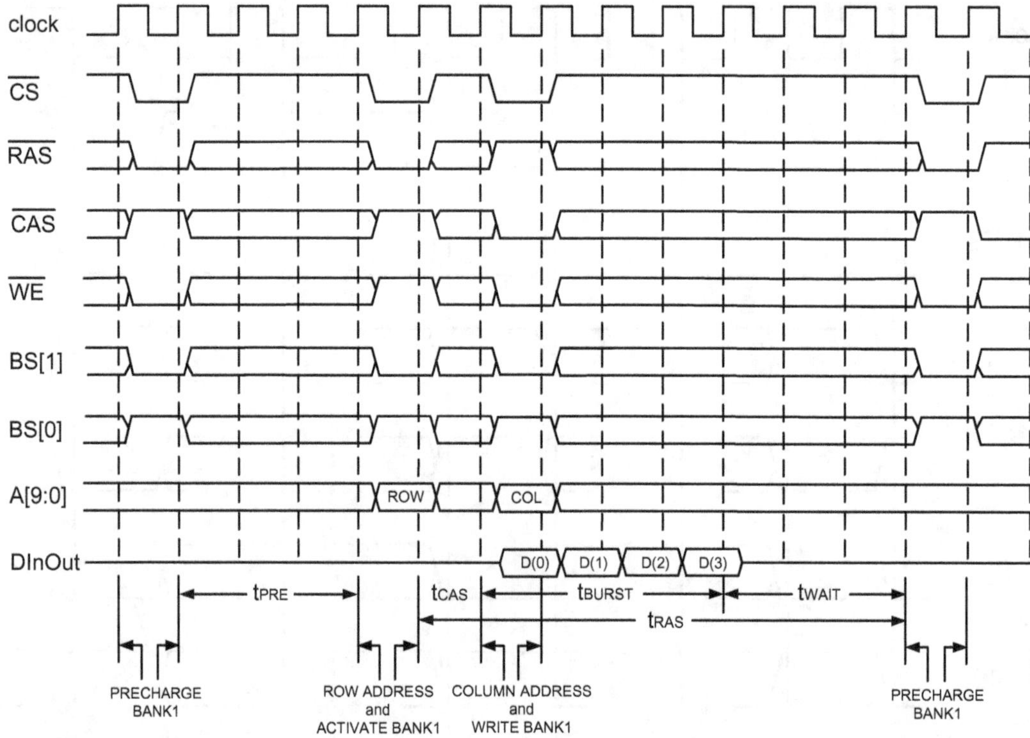

Fig. 5.17 A single write cycle

The example in Fig. 5.18 shows two separate write sequences into two different banks. When writing takes place to more than one bank, interleaving each bank's precharge and activation time periods with each other produces a time saving scenario where one write burst takes place immediately after the other, resulting in a shorter overall write cycle. In this figure, this technique allows to write four words to bank 1 immediately after writing four words to bank 0 without any cycle loss. Therefore, writing to two (or more) banks is a preferred method over writing to a single bank because this process eliminates all unnecessary waiting periods between precharge cycles. However, as the burst length involves a lot more than four words, the placement of bank precharge periods in the timing diagram becomes less and less important.

The sixth row of Table 5.1 shows how to initiate a read cycle from a selected bank. Reading words from SDRAM involves a latency period, and it needs to be programmed in the Address Mode Register. The example in Fig. 5.19 shows the start of a read burst after a latency period of three clock cycles once the read command and the address are given. A latency of three clock cycles means that the data becomes available at the output of SDRAM in the third clock cycle after the read command and the address are issued.

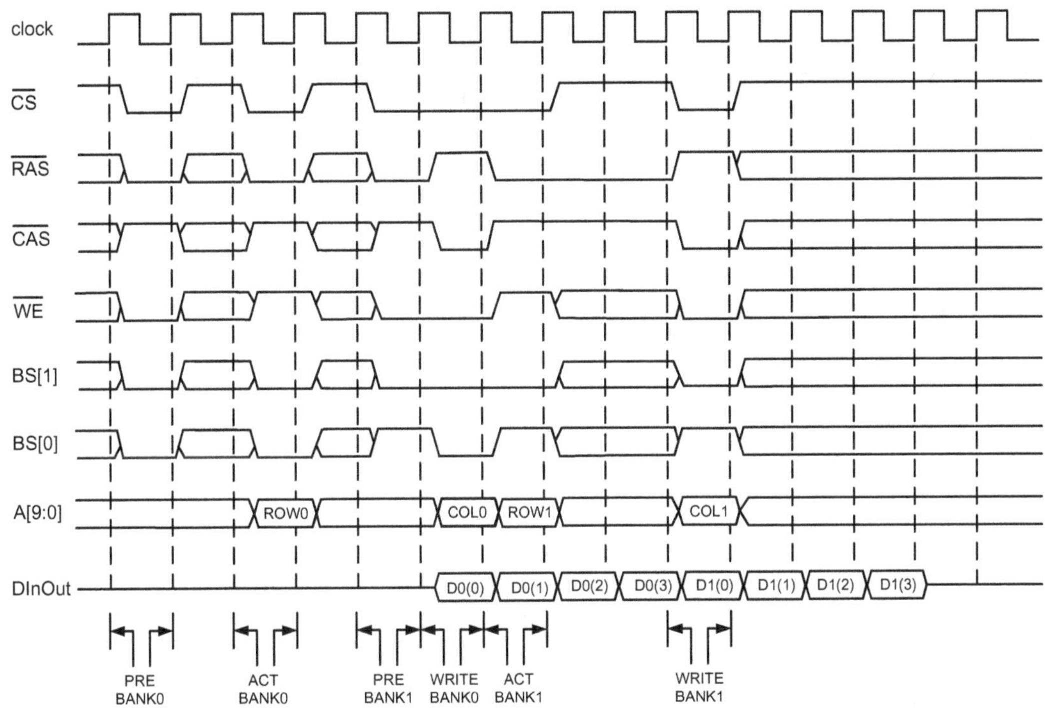

Fig. 5.18 Multiple write cycles to different banks (t_{PRE} = 1 cycle, t_{CAS} = 2 cycles)

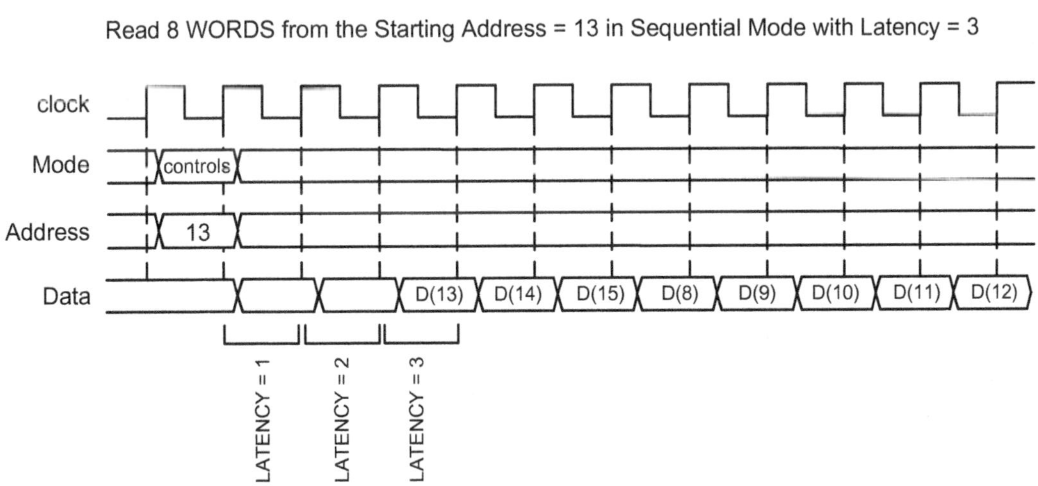

Fig. 5.19 Definition of latency during a read cycle

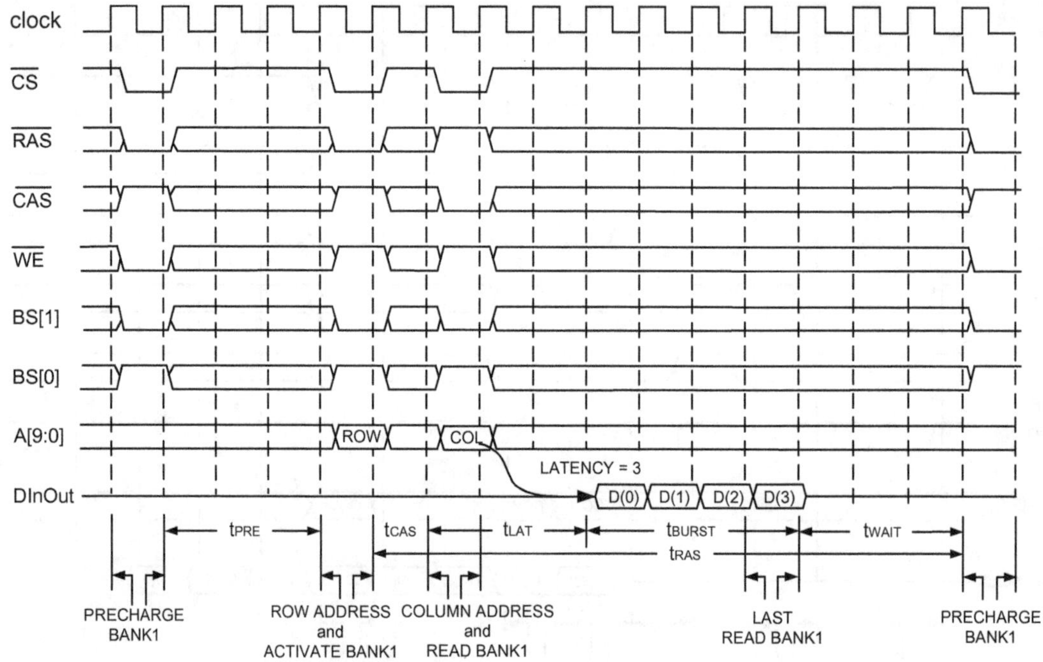

Fig. 5.20 Single read cycle

The example in Fig. 5.20 shows a single read sequence from bank 1. Until the read command and the column address are issued, the read and write sequences follow identical paths. However after this point, the read burst takes a different route and waits for the end of the programmed latency period. Similar to the write process, a certain t_{RAS} period must elapse for the read process before additional streams of data can be read from the same bank. In this figure, t_{WAIT} corresponds the waiting period between the last data packet and the start of the next precharge period.

The example in Fig. 5.21 describes multiple reads from the same bank and assumes t_{WAIT} is equal to zero. This scenario produces a burst read of four words, D(0) through D(3), from bank 1, and precharges the same bank during the last data packet delivery. The second burst read from bank 1 follows the same pattern as the first one, and delivers D(4) through D(7) after a programmed latency of two clock periods. If t_{WAIT} is different from zero, then the second precharge period in this figure follows the pattern in Fig. 5.20 and starts after the t_{WAIT} period expires.

The interleaving technique of reading data from two different banks in Fig. 5.22 is not any different from the one in Fig. 5.18 when writing takes place to two different banks. As with the write case, the placement of the second precharge cycle in the timing diagram is important to achieve two consecutive read bursts, D0(0) to D0(3) from bank 0 and D1(0) to D1(3) from bank 1, without any cycle loss and to accomplish the shortest possible time to fetch data from SDRAM.

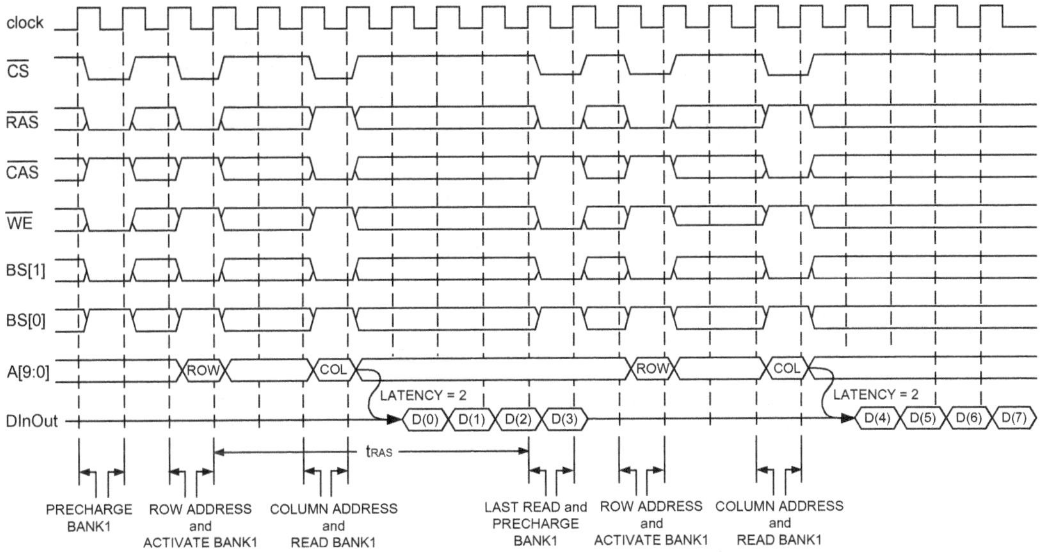

Fig. 5.21 Multiple read cycles from the same bank (t_{PRE} = 1 cycle, t_{CAS} = 2 cycles, t_{WAIT} = 0 cycle)

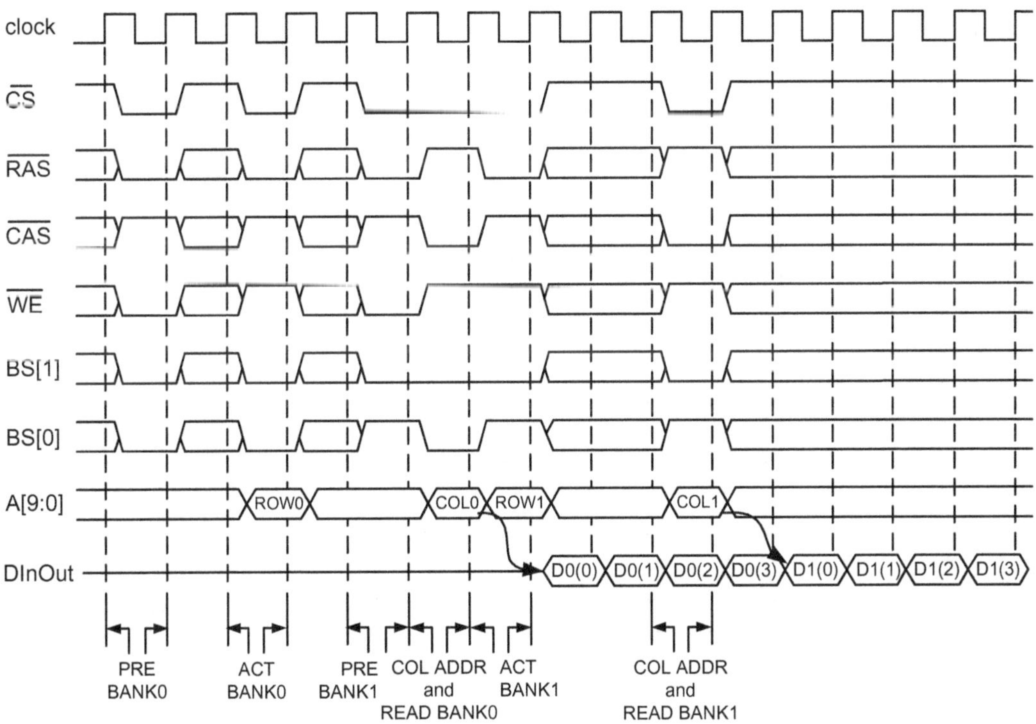

Fig. 5.22 Multiple read cycles from different banks (t_{PRE} = 1 cycle, t_{CAS} = 2 cycles)

The seventh row of Table 5.1 shows how to stop a read or a write burst. Figure 5.23 shows a single write sequence when the burst stop command is issued in the middle of a data burst. Upon receiving this command, the selected bank goes into the standby mode and waits for the next precharge command.

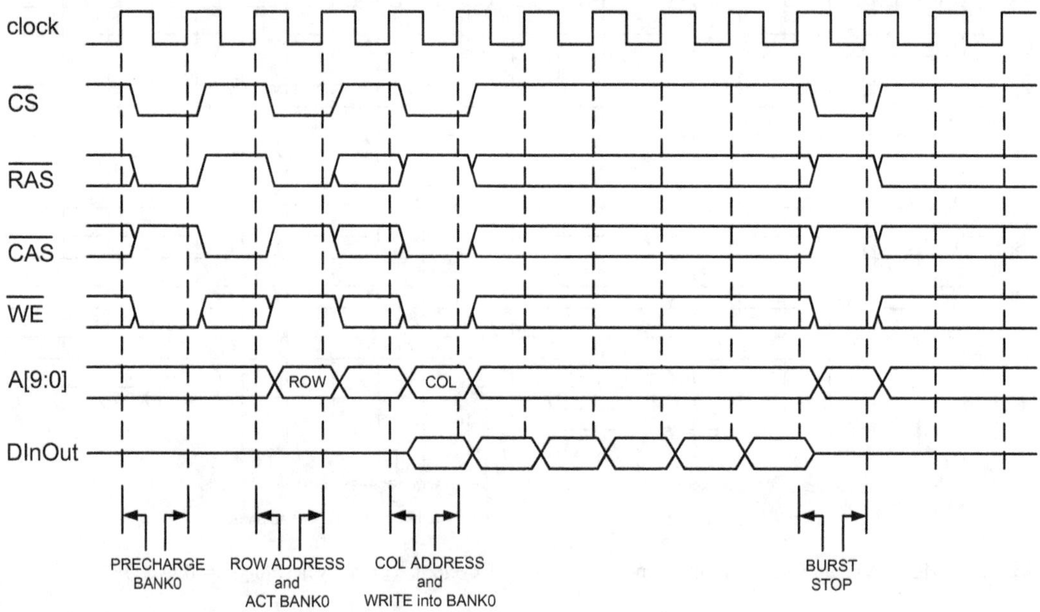

Fig. 5.23 Burst stop during write

When the burst stop command is given in the middle of a read, the last data packet is still delivered at the clock edge following the burst stop command as shown in Fig. 5.24.

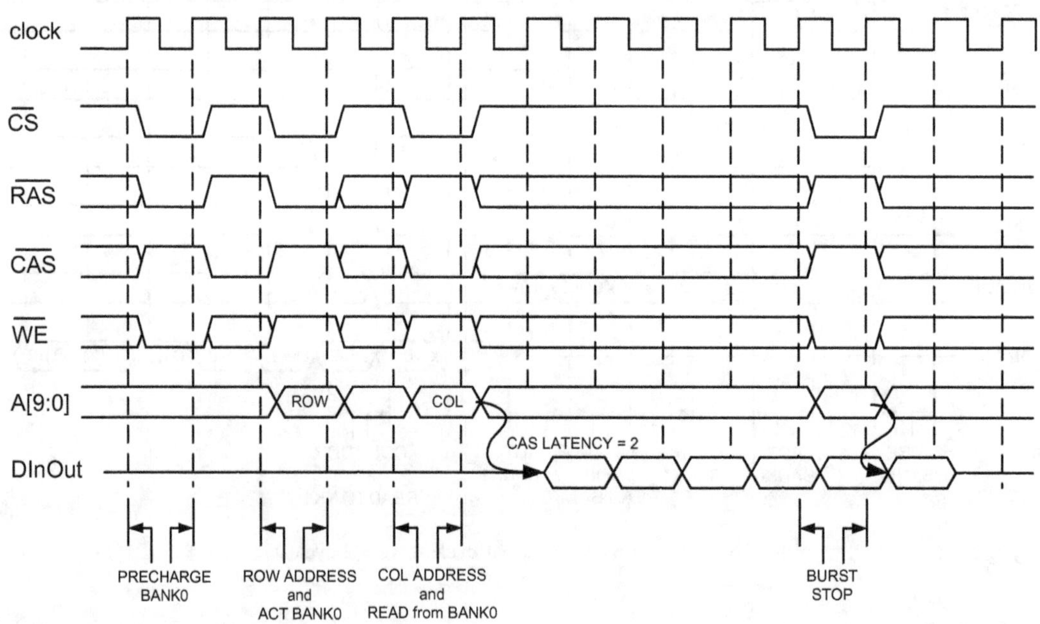

Fig. 5.24 Burst stop during read

The Input/Output data can be masked with the Read/Write Logic block in Fig. 5.11. The truth table in Table 5.5 lists all the possible cases of blocking and transmitting the incoming data from the SDRAM core. When Mask[3:0] = 0000, for example, no mask is applied to the output data; all 32 bits of data are allowed to be written into the selected bank or read from it. The case, Mask [3:0] = 1111, on the other hand, blocks all four bytes of data, and allows no byte to be written or read from the selected address.

Table 5.5 Truth table for data output mask

Mask [3]	Mask [2]	Mask [1]	Mask [0]	MASKED BITS
0	0	0	0	None
0	0	0	1	DInOut[7:0]
0	0	1	0	DInOut[15:8]
0	0	1	1	DInOut[15:0]
0	1	0	0	DInOut[23:16]
0	1	0	1	DInOut[23:16] and DInOut[7:0]
0	1	1	0	DInOut[23: 8]
0	1	1	1	DInOut[23:0]
1	0	0	0	DInOut[31:24]
1	0	0	1	DInOut[31:24] and DInOut[7:0]
1	0	1	0	DInOut[31:24] and DInOut[15:8]
1	0	1	1	DInOut[31:24] and DInOut[15:0]
1	1	0	0	DInOut[31:16]
1	1	0	1	DInOut[31:16] and DInOut[7:0]
1	1	1	0	DInOut[31:8]
1	1	1	1	DInOut[31:0]

Figure 5.25 shows an example of the data-path and the controller of the SDRAM bus interface. In this example, each ten-bit wide bus interface register containing the precharge (t_{PRE}), CAS (t_{CAS}), burst (t_{BURST}), latency (t_{LAT}), and wait (t_{WAIT}) periods must be programmed through a 10-bit program bus prior to operating SDRAM. The precharge, CAS and wait registers contain the number of clock cycles to achieve the required waiting period. The burst register should store the number of data packets of the data transfer. Therefore, the value in this register must be identical to the value programmed in the Address Mode Register. The latency register specifies the number of clock cycles prior to reading the first data from an SDRAM address. The details of how the programming takes place prior to the normal SDRAM operation and the required hardware are omitted from Fig. 5.25 to avoid complexity.

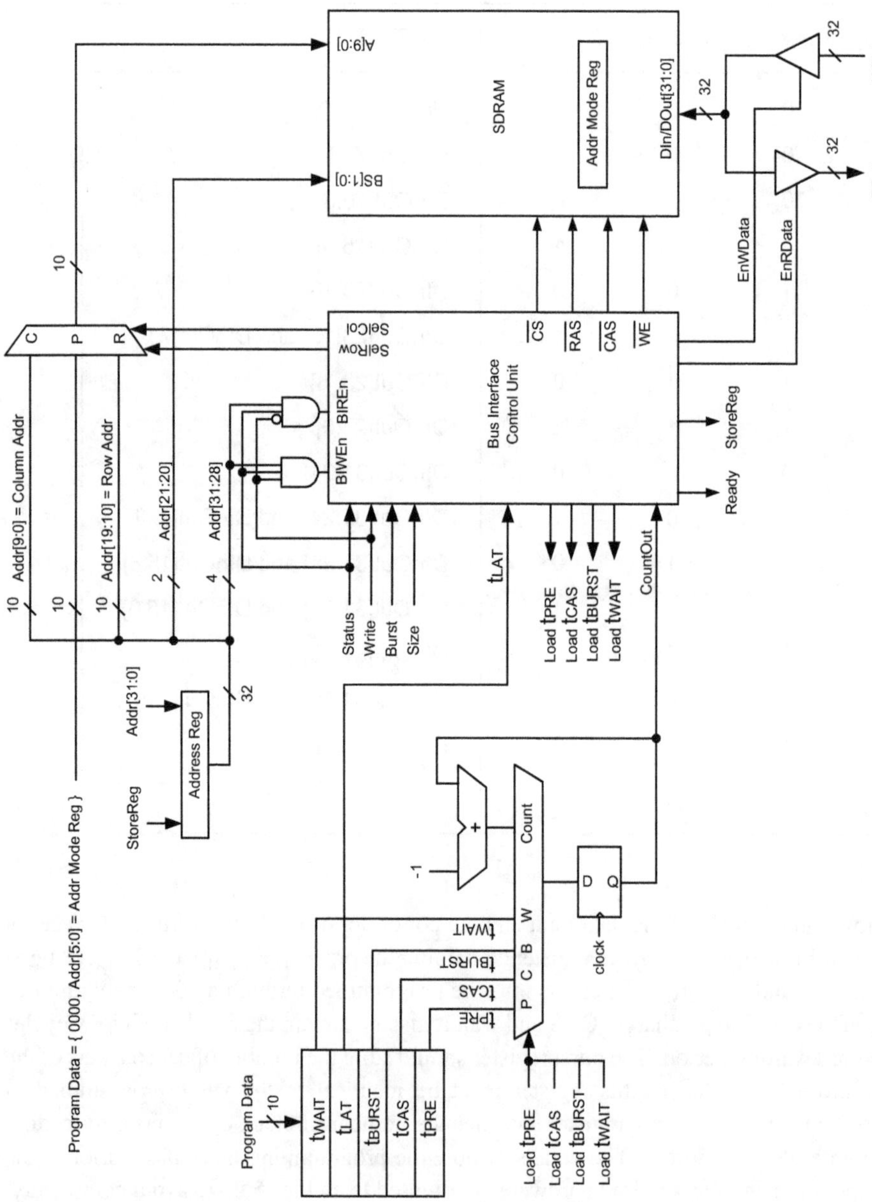

Fig. 5.25 SDRAM bus interface block diagram

The Address Mode Register (or a set of registers defining basic SDRAM functionality) exists in many older SDRAMs. However, recent SDRAM modules omit the mode register completely and rely on the bus interface unit to store such information. To manage the precharge, CAS, burst and wait periods the memory controller continuously interacts with a down-counter in Fig. 5.25 since these periods are often many clock cycles long.

For normal SDRAM operation, the bus address, Addr[31:0], has to be divided into several segments. In the example in Fig. 5.26, the four most significant bits of the SDRAM address, Addr [31:28], indicate the SDRAM chip identification, and it is used to activate the corresponding bus interface. Addr[21:20] is used to select the SDRAM bank, BS[1:0]. Addr[19:10] and Addr[9:0] specify the row and column addresses, respectively.

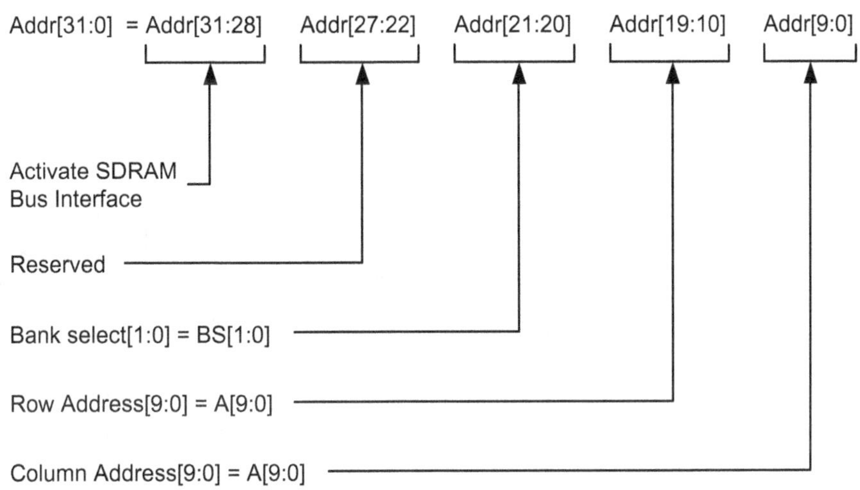

Fig. 5.26 SDRAM bus interface address mapping

Figure 5.27 shows a typical SDRAM write sequence. In this timing diagram, all five SDRAM interface registers must be programmed prior to the IDLE/PROG clock cycle as mentioned earlier. The SDRAM write sequence starts with the system bus sending Status = START, Write = 1 and the starting SDRAM address. These three signals cause the Bus Interface Write Enable signal, BIWEn, to transition to logic 1, which in turn, enables the bus interface for write in the first cycle of Fig. 5.27. Once enabled, the bus interface stores the starting SDRAM address in the Address Reg, and issues the precharge command by $\overline{CS} = 0$, $\overline{RAS} = 0$, $\overline{CAS} = 1$ and $\overline{WE} = 0$ for the selected bank. Within the same cycle, the counter is loaded with the precharge wait period, t_{PRE}, by $Loadt_{PRE} = 1$ as shown in the timing diagram.

The Precharge wait period is calculated by multiplying the number of clock cycles by the clock period. The counter in this design is a down-counter. When its output value, CountOut, becomes one, the controller initiates the activation cycle for the selected SDRAM bank and dispatches the row address. The activation period starts with loading the value of t_{CAS} to the down-counter by $Loadt_{CAS} = 1$. Within the same clock cycle, the row address, Addr[19:10], is transferred from the

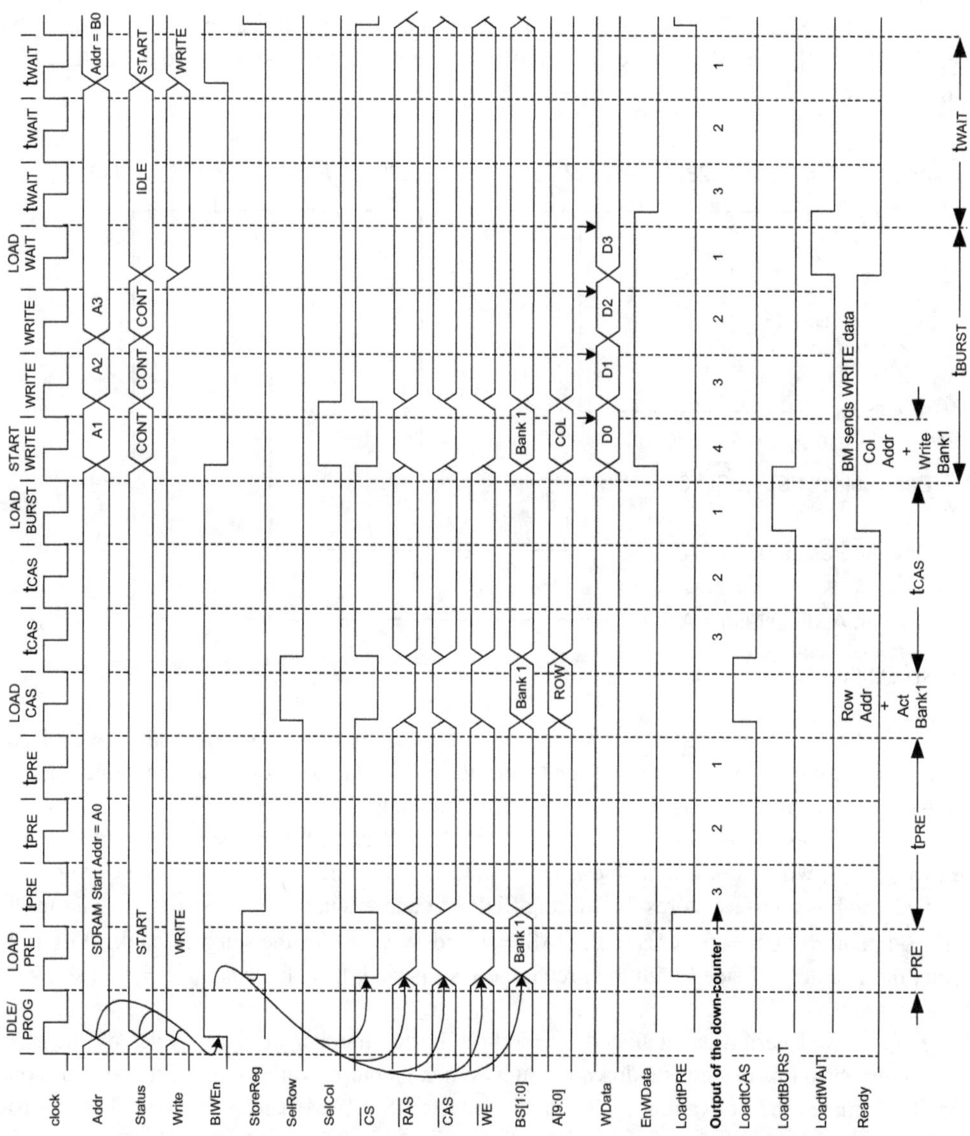

Fig. 5.27 Write cycle via SDRAM bus interface

Address Reg to the SDRAM through the R-port of the 3-1 MUX by SelRow = 1. When the activation wait period expires, the controller uses the Loadt_{BURST} input to load the length of the write burst (the number of data packets) to the down-counter and subsequently initiates the write sequence in the next cycle.

During the START WRITE period in the timing diagram in Fig. 5.27, the controller transfers the column address, Addr[9:0], from the Address Reg to the A[9:0] input of the SDRAM through the C-port of the 3-1 MUX in Fig. 5.25 by generating SelCol = 1. In the same cycle, the controller also generates $\overline{CS} = 0$, $\overline{RAS} = 0$, $\overline{CAS} = 1$ and $\overline{WE} = 1$, and enables the tri-state buffer by EnWData = 1 in order to write the first write data packet, D0, to the SDRAM. To be able to write the remaining data packets, the controller issues Ready = 1 from this point forward. When the sequence comes to the LOAD WAIT period in Fig. 5.27 (where the last write takes place), the controller lowers the Ready signal, but keeps the EnWData signal at logic 1 in order to write the last data packet, D3. This clock cycle also signifies the start of the wait period, t_{WAIT}. The controller issues Loadt_{WAIT} = 1 to load t_{WAIT} into the down-counter if another write sequence needs to take place for the same bank.

The remaining control signals, Burst and Size, are omitted from the timing diagram for simplicity. During the entire data transfer process, Burst is set to four and Size is set to 32 in this example. For byte and half-word transfers, Size needs to be defined with masking in place as described in Table 5.5.

The state diagram of the controller for write is shown in Fig. 5.28. In this diagram, when the interface receives BIWEn = 1, the controller transitions from the IDLE/PROG state, which corresponds to the first cycle of the timing diagram in Fig. 5.27, to the LOAD PRE state, which corresponds to the second clock cycle in the same timing diagram. In the LOAD PRE state, the controller resets $\overline{CS}$, $\overline{RAS}$ and $\overline{WE}$, but sets $\overline{CAS}$ for the selected bank to start the precharge process. In this state, two additional signals are generated: StoreReg = 1 to store the bus address in the Address Reg, and Loadt_{PRE} = 1 to start the precharge wait period. The controller remains in the precharge wait state, t_{PRE}, until CountOut = 1. The controller then transitions to the LOAD CAS state where it activates the selected bank, issues SelRow = 1 to transfer the row address to the SDRAM, and produces Loadt_{CAS} = 1 to initiate the activation wait period. The next state, t_{CAS}, is another wait state where the controller waits until the activation period expires. Once this period is over, the controller first goes into the LOAD BURST state, and then to the START WRITE state to initiate writing data to the SDRAM. The latter corresponds to the state where the first data packet is written to the SDRAM core as mentioned earlier. The subsequent writes take place when the controller transitions to the WRITE state. The controller stays in this state until CountOut = 2, which signifies one more data packet to be written to the SDRAM. The last data packet is finally written when the controller moves to the LOAD WAIT state. Before attempting another write process, the controller waits in the t_{WAIT} state until CountOut = 1.

Note that all the state names in Fig. 5.28 and the cycle names on top of Fig. 5.27 are kept the same to make one-to-one correspondence between the timing diagram and the state diagram.

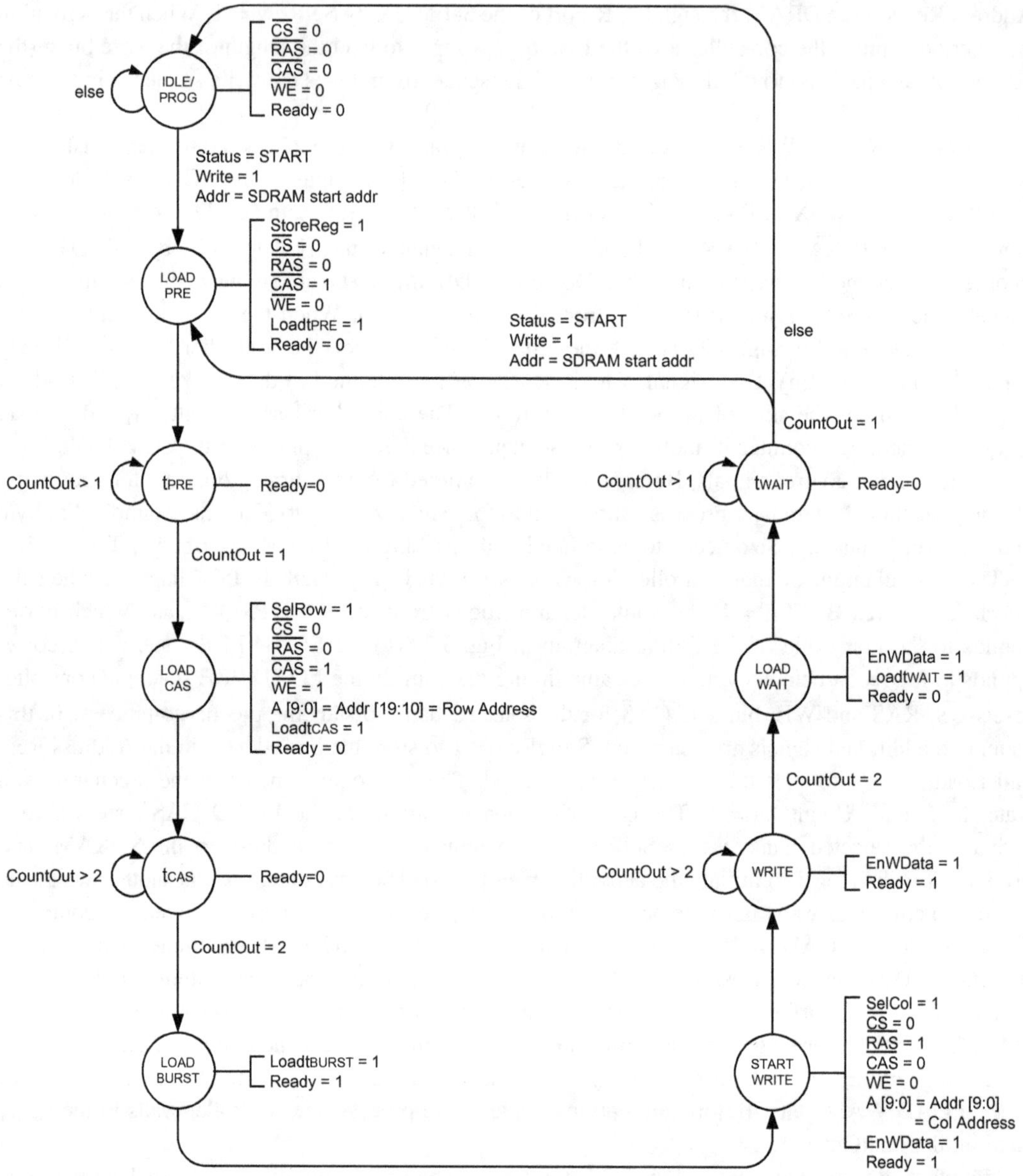

Fig. 5.28 SDRAM bus interface for write (control signals equal to logic 0 is omitted for simplicity)

The SDRAM read sequence also starts with the system bus sending Status = START, Write = 0 and an initial SDRAM address. This combination sets the Bus Interface Read Enable signal, BIREn = 1, to enable the bus interface to read data from the SDRAM core in the first cycle of the timing diagram in Fig. 5.29. The remainder of the read process is identical to the write process until the controller issues the read command during the START READ cycle in Fig. 5.29, and sends the column address of the

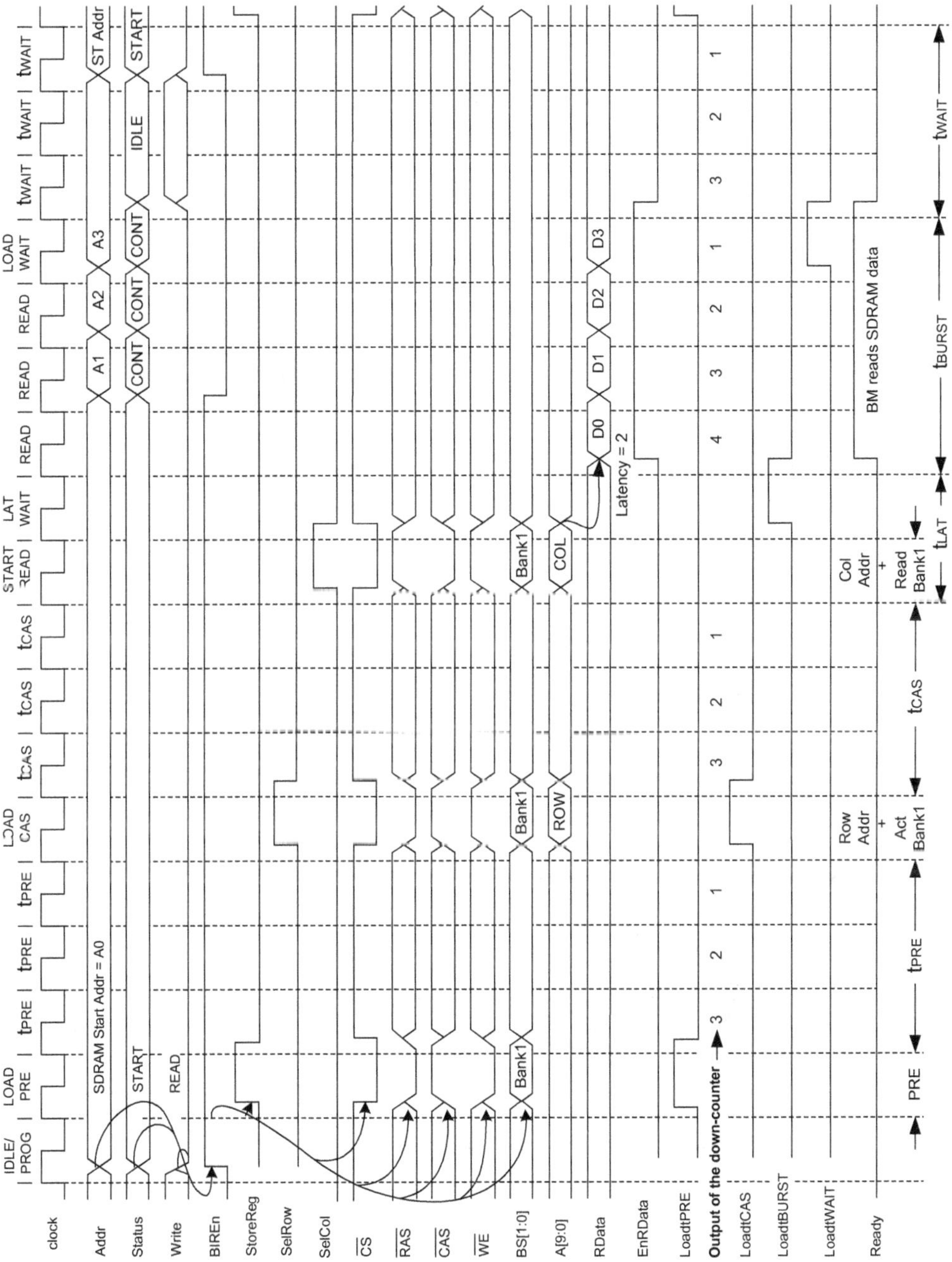

Fig. 5.29 Read cycle via SDRAM bus interface

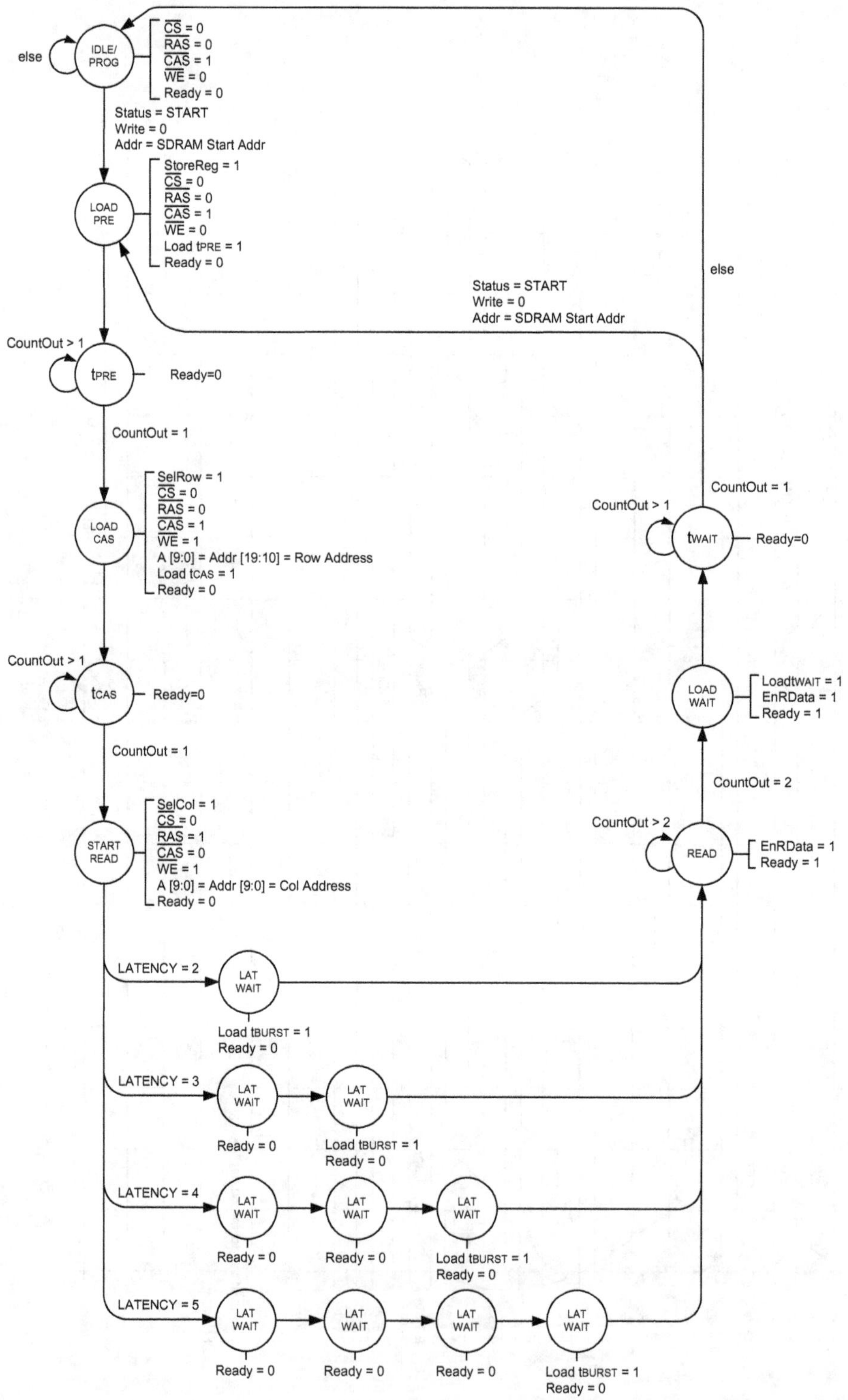

Fig. 5.30 SDRAM bus interface for read (control signals equal to logic 0 is omitted for simplicity)

selected SDRAM bank. Since SDRAM data becomes available after a latency period, the controller must replicate this exact delay prior to a read burst and produce the control signals during and after the burst. For example, a cycle before the latency period expires, the controller needs to generate $\text{Loadt}_{\text{BURST}} = 1$ to load the burst duration to the down-counter to be able to detect the beginning and the end of burst data. As a result, the controller can determine when to generate EnRData = 1 for the tri-state buffer to read data packets, D0 to D3, from the SDRAM RData output. During the last data delivery, the controller issues $\text{Loadt}_{\text{WAIT}} = 1$ to load the value of t_{WAIT} to the down-counter in the event the same bank is selected for another read.

The state diagram in Fig. 5.30 for the read sequence is a direct result of the timing diagram in Fig. 5.29. In this diagram, when the interface receives BIREn = 1, the controller transitions from the IDLE state which corresponds to the first cycle in Fig. 5.29, to the LOAD PRE state which corresponds to the second clock cycle in the same timing diagram. In the LOAD PRE state, the precharge process is initiated by $\overline{\text{CS}} = 0$, $\overline{\text{RAS}} = 0$, $\overline{\text{WE}} = 0$ and $\overline{\text{CAS}} = 1$ for the selected bank. In this state, the controller stores the valid bus address in the Address Reg by StoreReg = 1, and loads the precharge wait period into the down-counter by $\text{Loadt}_{\text{PRE}} = 1$. The controller stays in the t_{PRE} state until the precharge value in the down-counter expires. Subsequently, the controller transitions to the LOAD CAS state where it activates the selected bank by $\overline{\text{CS}} = 0$, $\overline{\text{RAS}} = 0$, $\overline{\text{CAS}} = 1$ and $\overline{\text{WE}} = 1$, issues SelRow = 1 to transfer the row address from the Address Reg to the SDRAM, and generates $\text{Loadt}_{\text{CAS}} = 1$ to load the activation wait period to the down-counter. The CAS wait period corresponds to the t_{CAS} state in Fig. 5.30. When this period is over at CountOut = 1, the controller transitions to the START READ state where it issues $\overline{\text{CS}} = 0$, $\overline{\text{RAS}} = 1$, $\overline{\text{CAS}} = 0$ and $\overline{\text{WE}} = 1$ to initiate the data read and produces SelCol = 1 to transfer the column address from Address Reg to the SDRAM address port. This state is followed by four individual latency states to select the programmed read latency period. Since the read latency in Fig. 5.29 is equal to two, the state machine traces through the single LAT WAIT state. In the LAT WAIT state, the controller issues $\text{Loadt}_{\text{BURST}} = 1$ and loads the value of the data burst, t_{BURST}, to the down-counter. Following the latency states, the state machine transitions to the READ state where it stays until CountOut = 2, signifying the end of the read burst. In this state, it produces EnRData = 1 to enable the data output buffer and Ready = 1 to validate the read data. At the end of the burst period, the state machine moves to the LOAD WAIT state and issues $\text{Loadt}_{\text{WAIT}} = 1$ to load the required wait period to the down-counter until the next precharge takes place. Subsequently, the state machine transitions to the t_{WAIT} state and stays there until the wait period is over.

5.3 Electrically-Erasable-Programmable-Read-Only-Memory

Electrically-Erasable-Programmable-Read-Only-Memory (E^2PROM) is historically considered the predecessor of Flash memory and also the slowest memory in a computing system. Its greatest advantage over the other types of memories is its ability to retain data after the system power is turned off due to the floating-gate MOS transistor in its memory core. Its relatively small size compared to electromechanical hard disks makes this device an ideal candidate to store Built-In-Operating-Systems (BIOS) especially for hand-held computing platforms.

A typical E^2PROM memory is composed of multiple sectors, each of which contains multiple pages as shown in the example in Fig. 5.31. A single word in E^2PROM can be located by specifying its sector address, page address and row address. Sector address indicates which sector a particular word resides. Page address locates the specific page inside a sector. Finally, the row address points to the location of data byte inside a page. There are five control signals in E^2PROM to perform read, write or erase operations. The active-low Enable signal, $\overline{EN}$, places a particular page in standby mode and prepares it for an upcoming operation. The active-low Command Enable signal, $\overline{CE}$, is issued with a command code, such as read, write (program) or erase. The active-low Address Enable signal, $\overline{AE}$, is issued when an address is provided. Finally, the active-low Write Enable signal, $\overline{WE}$, and the Read Enable signal, $\overline{RE}$, are issued for writing and reading data, respectively.

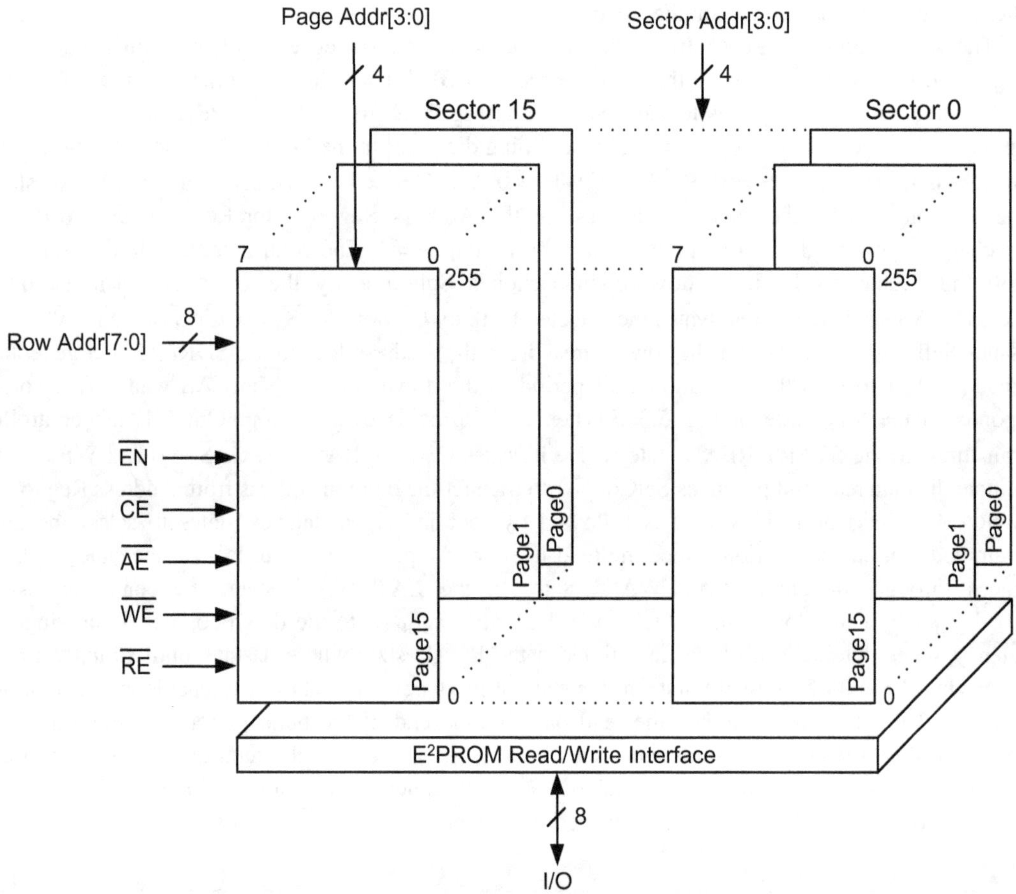

Fig. 5.31 A typical E^2PROM organization

Typical E^2PROM architecture consists of a memory core, address decoder, output data buffer, status, address and command registers, and control logic as shown in Fig. 5.32. Prior to any operation, command and address registers are programmed. When the operation starts, the control logic enables the address decoder, the data buffer and the memory core using the active-high Enable Address (ENA), Enable Data Buffer (END), and Write Enable Core (WEC) or Read Enable Core (REC) signals depending on the operation. The address stored in the address register is decoded to point the location of data. If the read operation needs to be performed, the required data is retrieved from the E^2PROM core and stored in the data buffer before it is delivered to the I/O bus. If the operation is a write (or program), the data is stored in the data buffer first before it is uploaded to the designated E^2PROM address. In all cases, $\overline{EN}$ needs to be at logic 0 to place E^2PROM into standby mode before starting an operation. The table in Fig. 5.33 describes all major operation modes. Hibernate mode disables the address decoder, memory core and data buffer to reduce power dissipation, and puts the device into sleep.

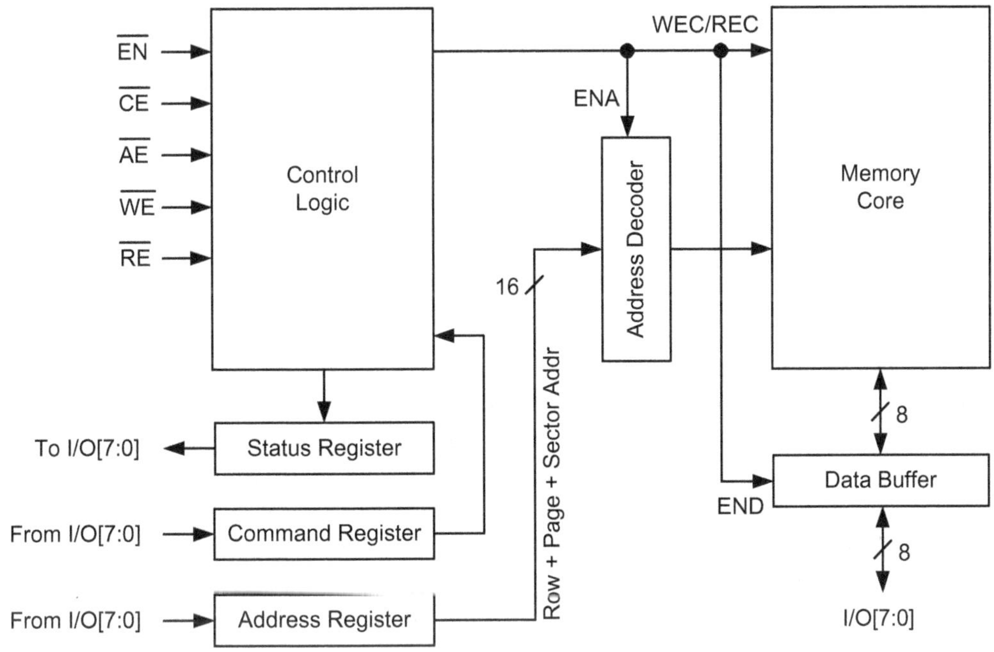

Fig. 5.32 A typical E^2PROM architecture

$\overline{EN}$	$\overline{WE}$	$\overline{RE}$	
0	1	1	Standby
0	0	1	Write
0	1	0	Read
1	X	X	Hibernate

Fig. 5.33 E^2PROM major operation modes

The E^2PROM cell shown in Fig. 5.34 is basically an N-channel MOS transistor with an additional floating gate layer sandwiched between its control gate terminal (Wordline) and the channel where the electronic conduction takes place. This device has also drain (Bitline) and source (Sourceline) terminals for connecting the cell to the neighboring circuitry.

To write logic 0 into the memory cell, a high voltage is applied between Wordline and Bitline terminals while the Sourceline node remains connected to ground. This configuration generates hot carriers in the transistor channel which tunnel through the gate oxide and reach the floating gate, raising the threshold voltage of the transistor. The raised threshold voltage prevents the programmed device to be turned on by the standard gate-source voltage used during normal circuit operations, and causes the value stored in the device to be interpreted as logic 0. An unprogrammed device with no charge on the floating gate, on the other hand, exhibits low threshold voltage characteristics and can be turned on by the standard gate-source voltage, producing a channel current. In this state, the value stored in the device is interpreted as logic 1.

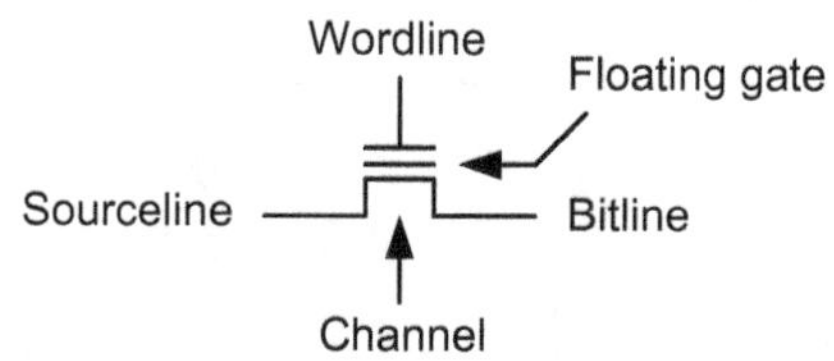

Fig. 5.34 E^2PROM cell

Figure 5.35 shows a typical command input sequence. There are four basic commands for this E^2PROM example: read, write (program), page-erase and status register read. Write and program commands will be used interchangeably in E^2PROM or Flash memories as they mean the same operation. The operation sequence is always the command input followed by the address and data entries. To issue a command input, $\overline{EN}$ is lowered to logic 0, $\overline{AE}$ is raised to logic 1 (because the entry is not an address), and $\overline{CE}$ is lowered to logic 0, indicating that the value on the I/O bus is a command input. Since the command input is written to the command register $\overline{WE}$ is also lowered to logic 0 some time after the negative edge of $\overline{CE}$ signal. This delay is called the setup time (t_S) as shown in Fig. 5.35. The low phase $\overline{WE}$ signal lasts for a period of t_{LO}, and transitions back to logic 1 some time before the positive edge of $\overline{CE}$. This time interval is called the hold time, t_H. Prior to the positive edge of $\overline{WE}$, a valid command input is issued, satisfying the data setup, t_{DS}, and the data hold, t_{DH}, times as shown in Fig. 5.35.

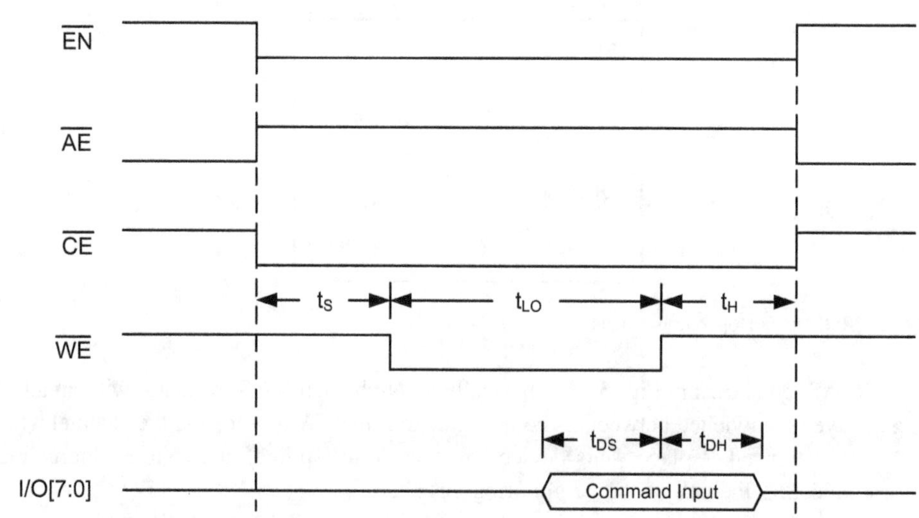

Fig. 5.35 Command input timing diagram

The address input timing shown in Fig. 5.36 has the same principle as the command input timing described above: $\overline{EN}$ needs to be at logic 0 to enable the device, $\overline{AE}$ must be at logic 0 for the address entry, and $\overline{CE}$ needs to be at logic 1 because this operation is not a command entry. During the low phase of $\overline{EN}$ signal, $\overline{WE}$ signal must be lowered to logic 0 twice to locate data in the E^2PROM. The first time $\overline{WE} = 0$, an eight-bit row address is entered at the first positive edge of $\overline{WE}$. This is followed by the combination of four-bit page address and four-bit sector address at the next $\overline{WE} = 0$. The $\overline{WE}$ signal must be lowered to logic 0 after a period of t_S, and then back to logic 1 for a period of t_H before the positive edge of $\overline{EN}$. The $\overline{WE}$ signal must also be at the low phase for a period of t_{LO} and at the high phase for a period of t_{HI} (or longer) during the address entry. Valid address values are issued at each positive edge of $\overline{WE}$ within the t_{DS} and t_{DH} setup and hold time periods.

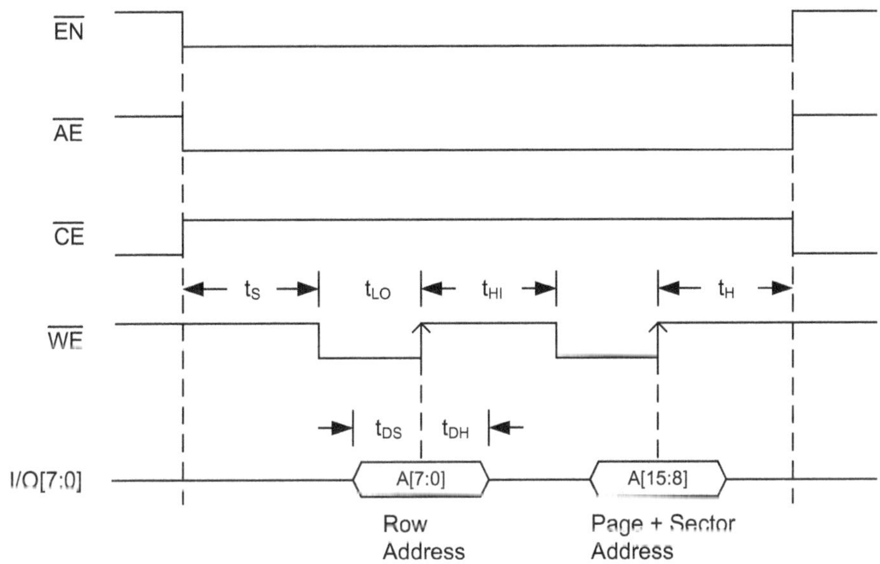

Fig. 5.36 Address input timing diagram

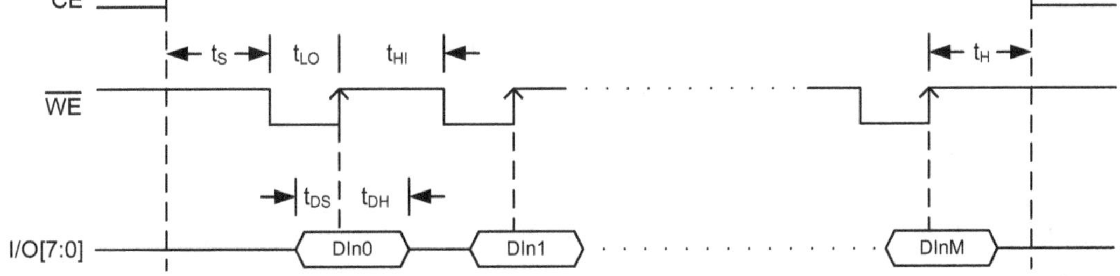

Fig. 5.37 Data input (write or program) timing diagram

Figure 5.37 describes data entry sequence where (M + 1) number of data packets are written to the E^2PROM core. During the entire write cycle $\overline{AE}$ signal must be at logic 1, indicating that the operation is a data entry but not an address. Data packets are written at each positive edge of $\overline{WE}$ signal.

During a read the active-low control signals, $\overline{AE}$ and $\overline{CE}$, are kept at logic 1. The Read Enable signal, $\overline{RE}$, enables the E^2PROM to read data from its memory core at each negative edge as shown in Fig. 5.38. The time delay between the negative edge of $\overline{RE}$ and the actual availability of data from the memory is called the access time, t_A, as shown in the same timing diagram. The $\overline{RE}$ signal must have the specified t_S, t_H, t_{LO} and t_{HI} time periods to be able to read data from the memory core.

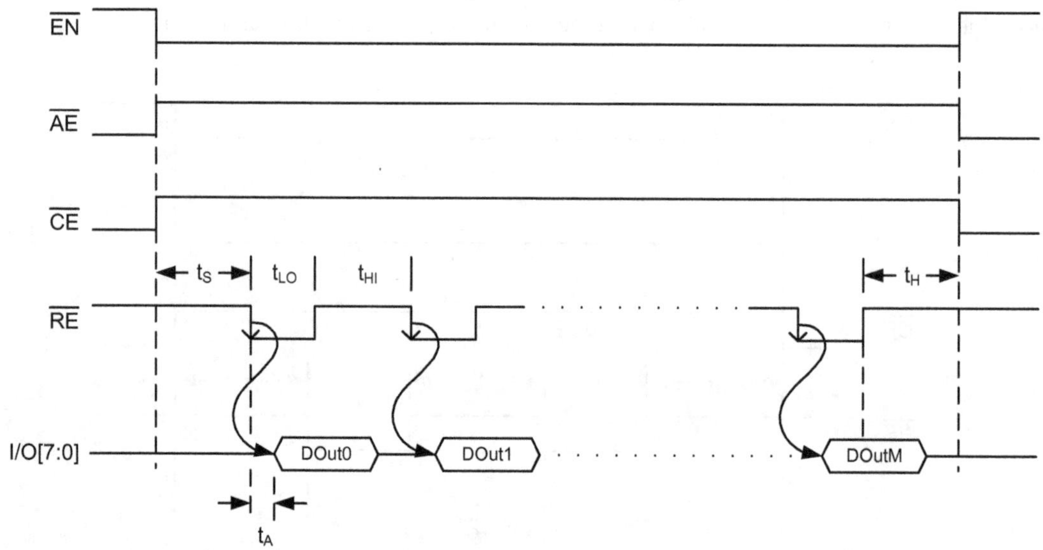

Fig. 5.38 Data output (read) timing diagram

Reading data from the Status Register is a two-step process. The first step involves entering the command input, Status Register Read, at the positive edge of $\overline{WE}$ signal. The contents of the register are subsequently read sometime after the negative edge of $\overline{RE}$ (t_A) as shown in Fig. 5.39. Note that $\overline{CE}$ signal is initially kept at logic 0 when entering the command input, but raised to logic 1 when reading the contents of the Status Register.

Full-page data write entry consists of the combination of four tasks as shown in Fig. 5.40. The first task is entering the Write into Data Buffer command at the positive edge of $\overline{WE}$ while keeping $\overline{CE}$ at logic 0. The second task is entering the page and sector addresses at the positive edge of $\overline{WE}$ while $\overline{AE}$ is at logic 0. The third task is entering the full-page of data from D(0) to D(255) into the data buffer at each positive edge of $\overline{WE}$ signal. Both $\overline{AE}$ and $\overline{CE}$ are kept at logic 1 during this phase. The last task is entering the Write to Core Memory command in order to transfer all 256 bytes of data from the data buffer to the memory core. The last cycle needs a relatively longer time period, t_{WRITE}, to complete the full-page write.

The read operation is composed of three individual tasks similar to the write operation as shown in Fig. 5.41. The first task is entering the Read from Memory command at the positive edge of $\overline{WE}$ while $\overline{CE}$ is at logic 0. The second step is entering the starting address by specifying the row, page and sector address values at each positive edge of $\overline{WE}$ while $\overline{AE}$ is at logic 0. The third task is to read data from the memory core at each negative edge of $\overline{RE}$ while $\overline{CE}$ and $\overline{AE}$ signals are at logic 1.

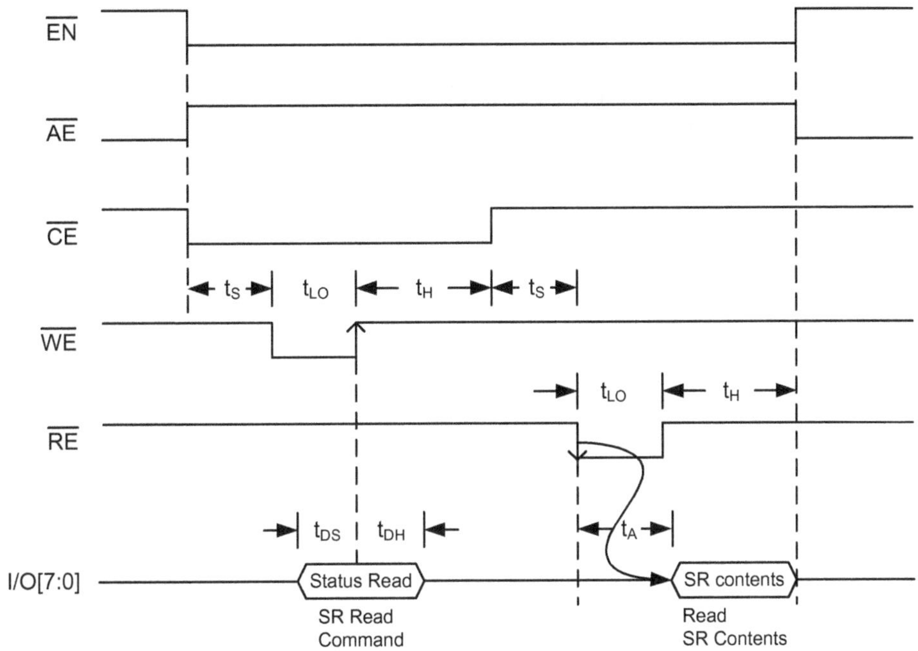

Fig. 5.39 Timing diagram for reading status register

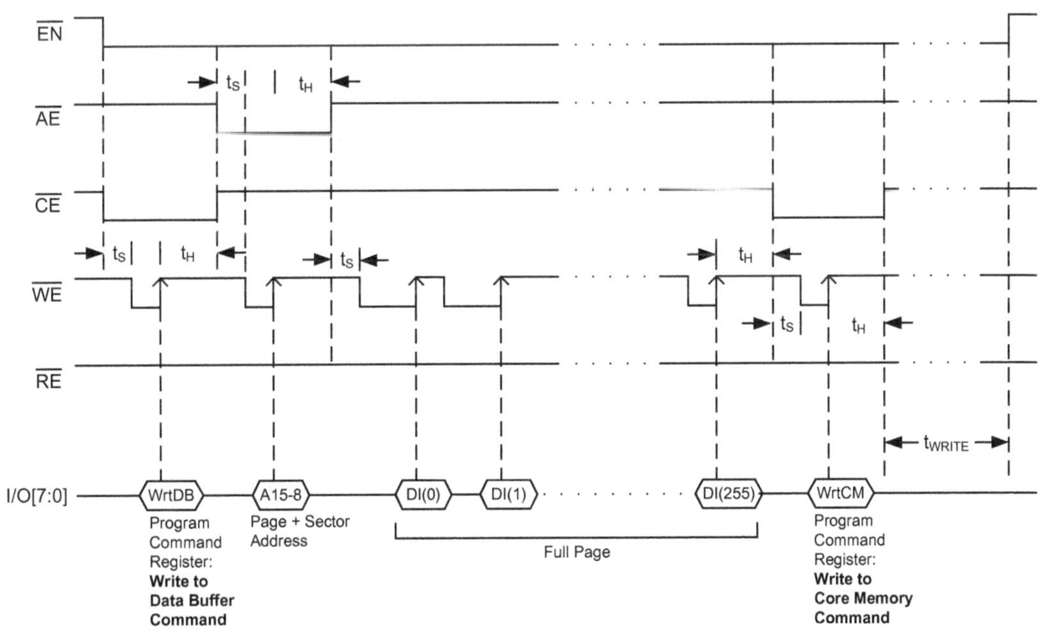

Fig. 5.40 Timing diagram for full-page write (program)

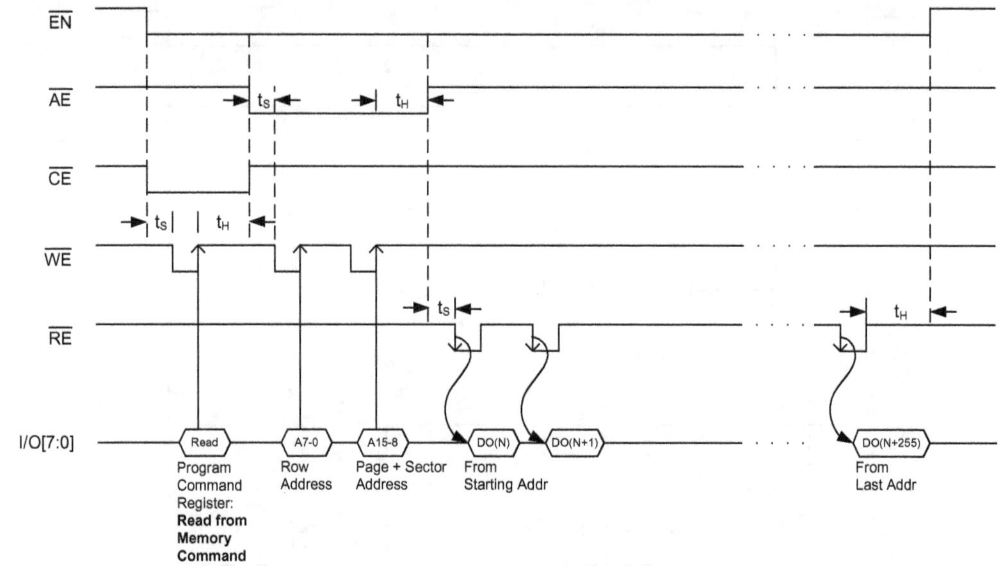

Fig. 5.41 Timing diagram for full-page read

A typical full-page erase is described in Fig. 5.42. In this figure, the Erase Full Page command is entered first at the positive edge of $\overline{WE}$ while $\overline{CE}$ is at logic 0. The memory address composed of page

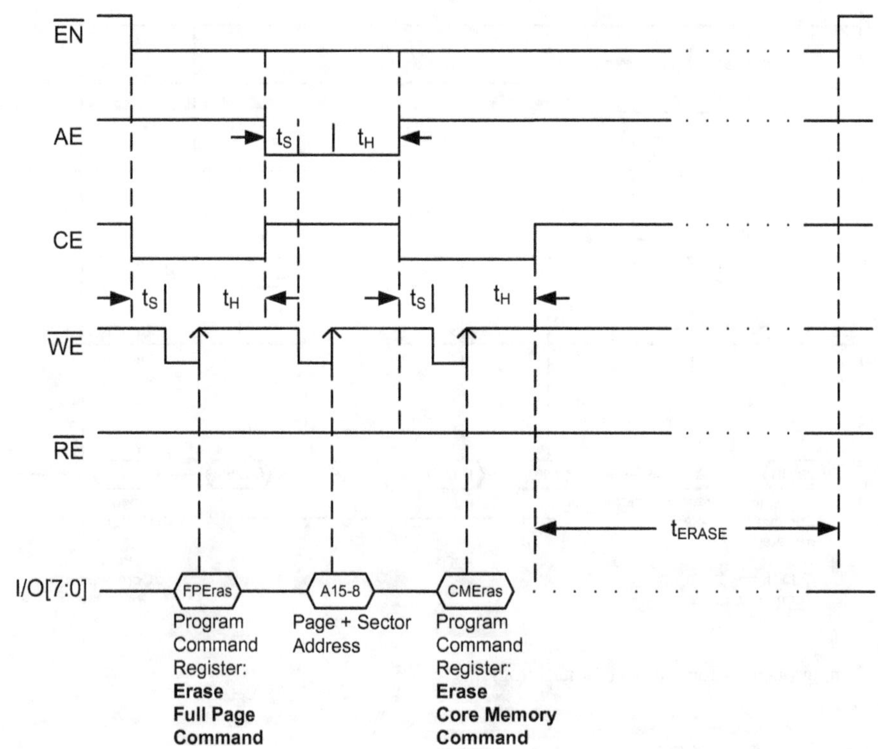

Fig. 5.42 Timing diagram for full-page erase

and sector addresses is entered next while $\overline{AE}$ is at logic 0. The Erase Core Memory command is entered following the address while $\overline{CE}$ is at logic 0. Full-page erase time period, t_{ERASE}, must be employed to complete the operation.

5.4 Flash Memory

Flash memory is the successor of the Electrically-Erasable-Programmable-Read-Only-Memory, and as its predecessor it has the capability of retaining data after power is turned off. Therefore, it is ideal to use in hand-held computers, cell phones and other mobile platforms.

A typical Flash memory is composed of multiple sectors and pages as shown in Fig. 5.43. An eight-bit word can be located in a Flash memory by specifying the sector, the page and the row addresses. To be compatible with the E^2PROM architecture example given in the previous section, this particular Flash memory also contains 16 sectors and 16 pages. Each page contains 256 bytes. The sector address constitutes the most significant four bits of the 16-bit Flash address, namely Addr [15:12]. Each page in a sector is addressed by Addr[11:8], and each byte in a page is addressed by Addr[7:0]. There are five main control signals in Flash memory to perform basic read, write (program), erase, protect and reset operations. Write and program commands are equivalent to each other, and used interchangeably throughout the manuscript when describing Flash memory

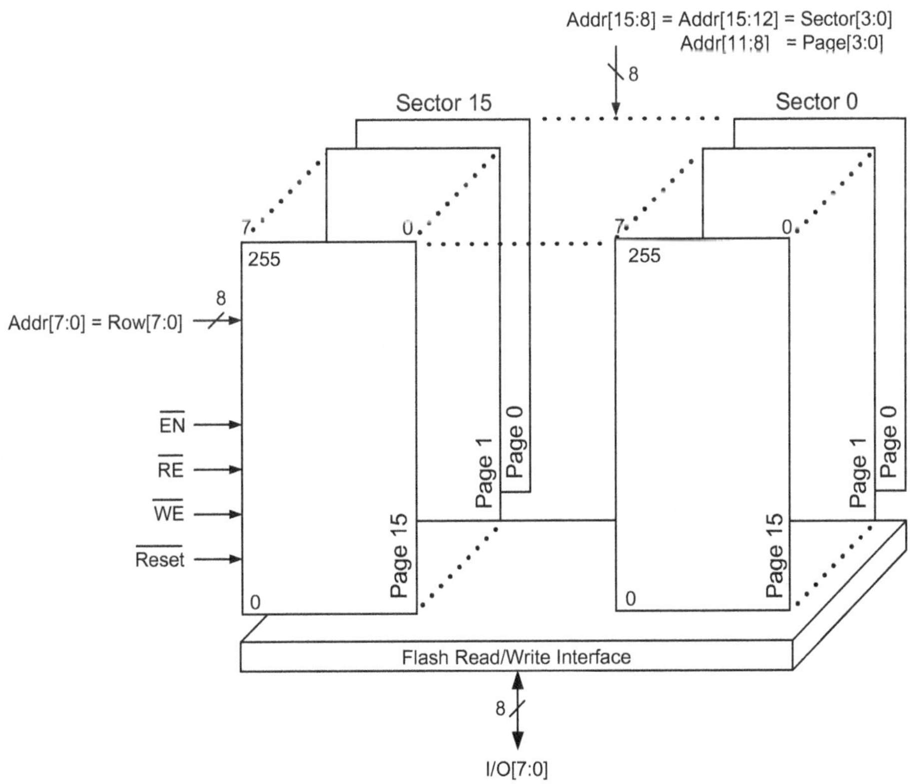

Fig. 5.43 Flash memory organization

operations. Many Flash datasheets use the term, program, to define writing a byte or a block of data to Flash memory.

The active-low Enable input, $\overline{\text{EN}}$, activates a particular page in the Flash memory to prepare it for an upcoming operation. The active-low Read Enable input, $\overline{\text{RE}}$, activates the Read/Write interface to read data from the memory. The active-low Write Enable input, $\overline{\text{WE}}$, enables to write (program) data to the memory. The active-low Reset input, $\overline{\text{Reset}}$, is used for resetting the hardware after which the Flash memory automatically goes into the read mode.

Typical Flash memory architecture, much like the other memory structures we have examined earlier, consists of a memory core, address decoder, sense amplifier, data buffer and control logic as shown in Fig. 5.44. When a memory operation starts, the control logic enables the address decoder, the address register, and the appropriate data buffers in order to activate the read or the write data-path. The address in the address register is decoded to point the location of data in the memory core. If a read operation needs to be performed, the retrieved data is first stored in the data buffer, and then released to the bus. If the operation calls for a write, the data is stored in the data buffer first, and then directed to the designated address in the memory core. The standby mode neither writes to the memory nor reads from it. The hibernation mode disables the address decoder, memory core and data buffer to reduce power dissipation. The main Flash operation modes are tabulated in Fig. 5.45.

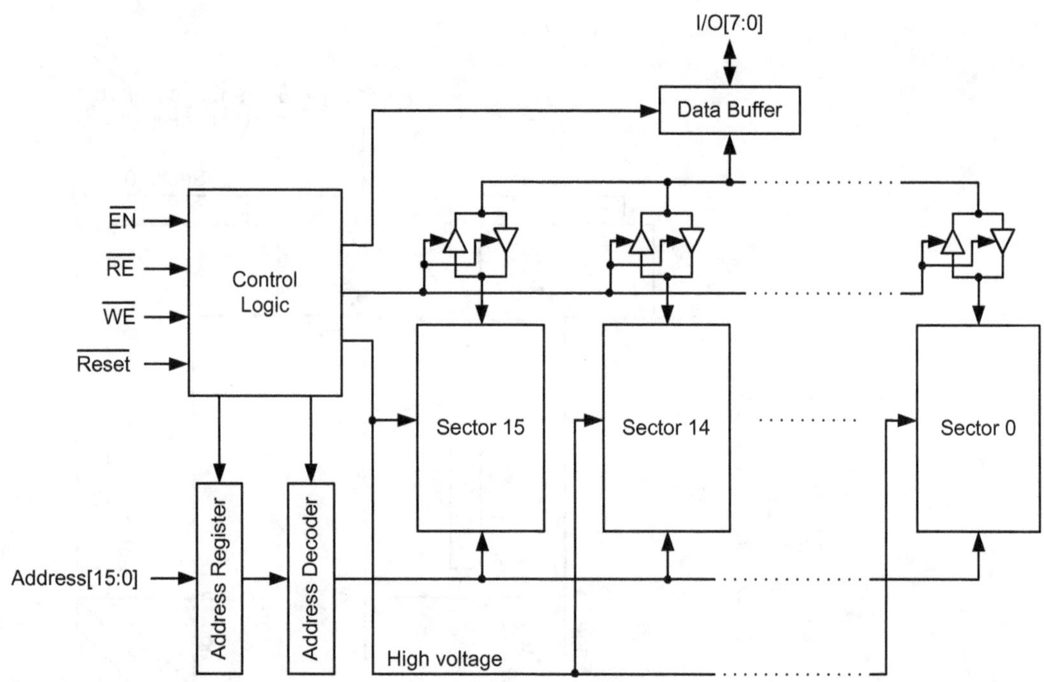

Fig. 5.44 Flash memory architecture

$\overline{EN}$	$\overline{RE}$	$\overline{WE}$	$\overline{reset}$	MODE
0	0	1	1	Read
0	1	0	1	Write
0	1	1	1	Standby
1	X	X	1	Hibernate
X	X	X	0	Hardware reset

Fig. 5.45 Main modes of Flash memory

The memory cell shown in Fig. 5.34 is the basic storage element in the Flash memory core. It is an N-channel MOS transistor with a floating gate whose sole purpose is to store electronic charge. The device needs high voltages well above the power supply voltage to create and transfer electrons to the floating gate. In order to obtain a much higher DC voltage from power supply for a short duration, the control logic in Fig. 5.44 contains a charge pump circuit composed of a constant current source and a capacitor. As the constant current charges the capacitor, the voltage across the capacitor rises linearly with time, ultimately reaching a high DC potential to create electrons for the floating gate. The mechanism of electron tunneling to the floating gate requires time. Therefore, a write or erase operation may take many consecutive clock cycles compared to simple control operations such as suspend or resume.

Figure 5.46 shows the basic read operation provided that data has already been transferred from the memory core to the data buffer. Once a valid address is issued, data is produced at the I/O terminal some time after the falling edge of the Read Enable signal, $\overline{RE}$. Data is held at the I/O port for a period of hold time, t_h, following the rising edge of $\overline{RE}$ as shown in the timing diagram below. The actual read operation takes about four clock cycles as the entire data retrieval process from the memory core takes time. This involves sensing the voltage level at the Flash cell, amplifying this value using the sense amplifier, and propagating the data from the sense amplifier to the data buffer.

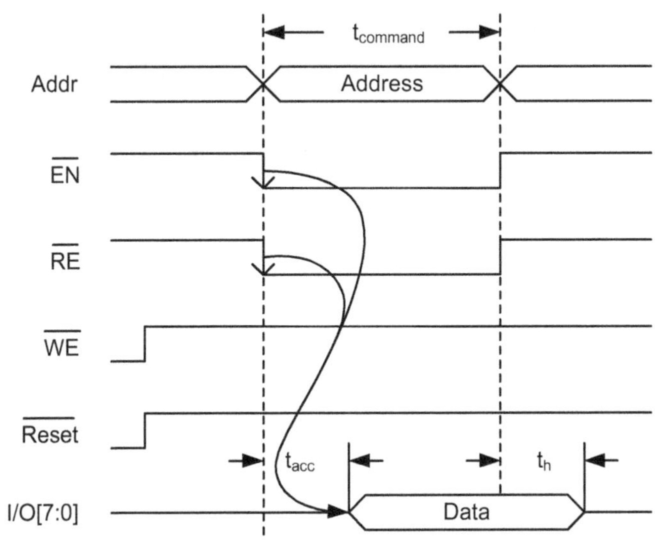

Fig. 5.46 Basic read operation timing diagram

In contrast to read, the basic write operation follows the timing diagram of Fig. 5.47. In this figure, a valid address must be present at the address port when the Enable and Write Enable signals, $\overline{EN}$ and $\overline{WE}$, are both at logic 0. Valid data satisfying the setup and hold times, t_s and t_h, is subsequently written to the data buffer. The actual write process can take up to four clock cycles due to the data propagation from the I/O port to the data buffer, and then from the data buffer to the Flash core.

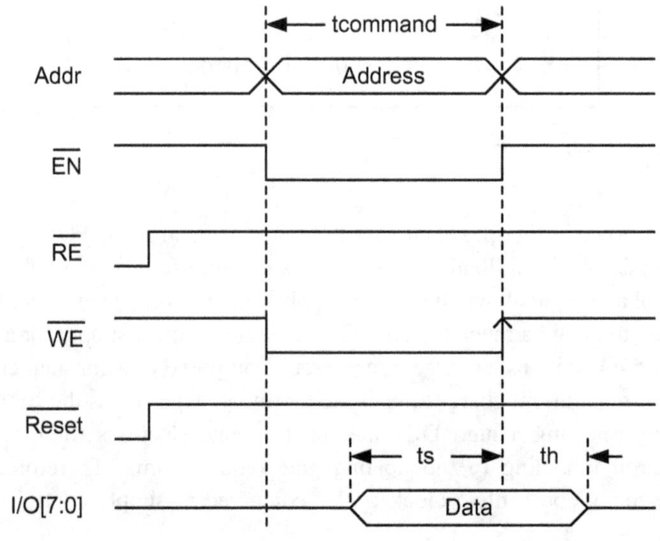

Fig. 5.47 Basic write (program) operation timing diagram

Disabling the I/O port for read or write, and therefore putting the device in standby mode requires $\overline{EN}$ signal to be at logic 0 as shown in Fig. 5.48. The I/O port will float and show high impedance (Hi-Z).

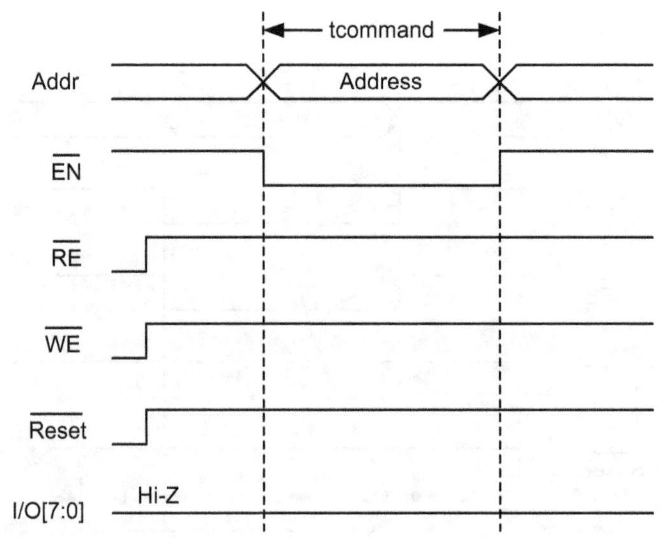

Fig. 5.48 Basic standby operation

Hardware reset requires only lowering $\overline{\text{Reset}}$ signal during the command cycle. The actual reset operation takes three bus cycles and resets the entire Flash memory as shown in 5.49.

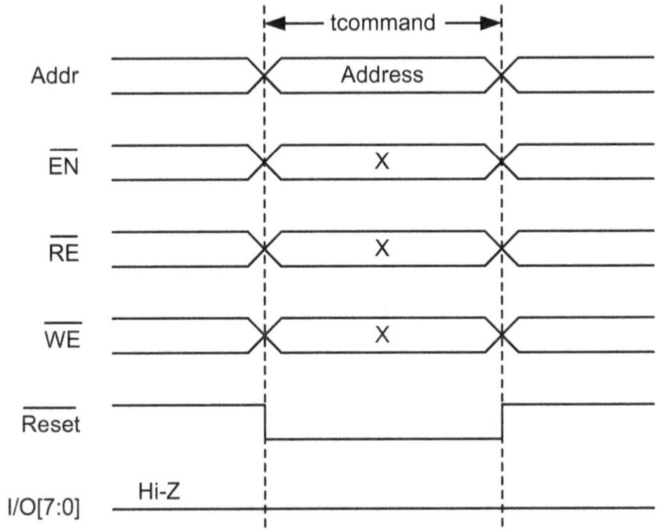

Fig. 5.49 Basic hardware reset operation timing diagram

Basic Flash memory operations are tabulated in Fig. 5.50. In actuality, there are a lot more commands in commercially available Flash memories than what is shown in this table. This section considers only essential byte-size operations in a Flash memory. Word-size operations, very specific Flash command sequences, such as hidden ROM programs, query and verification commands and boot protection processes are avoided in order to emphasize the core Flash memory operations for the reader. Address and data entries indicating a specific command in Fig. 5.50 are also modified compared to the actual datasheets to simplify the read, write (program) and erase sequences. The number of clock cycles, the address and data preamble values in each cycle, and the operational codes to perform read, write, page erase, chip erase, page protect, fast write and other modes of operation may be different from the actual datasheets.

The first task of Fig. 5.50 is the Flash memory read sequence which takes four clock cycles. The first three clock cycles of this sequence represents the waiting period to prepare the read path from the memory core. During this period, address and data values in the form of alternating combinations of 1s and 0s, such as 0x5555/0xAA and then 0xAAAA/0x55, are introduced to the address and data ports as shown in Fig. 5.51. Once the read command, 0x00, is issued in the third clock cycle, a byte of data becomes available shortly after the negative edge of the $\overline{\text{RE}}$ signal in the fourth and final clock cycle.

Figure 5.52 shows an example of the read operation which extracts the manufacturer's ID and device ID from the Flash memory. The first three clock cycles of this sequence are the same as the normal read operation, but with the exception of the ID read code, 0x10. The next two cycles deliver the manufacturer's ID and the device ID following the negative edge of the $\overline{\text{RE}}$ signal.

MAIN COMMANDS	CYCLE 1		CYCLE 2		CYCLE 3		CYCLE 4		CYCLE 5		CYCLE 6	
	Addr	Data	Addr	Data	Addr	Data	Addr	Data				
Read	0x5555	0xAA	0xAAAA	0x55	0x5555	0x00	Read Addr	Read Data				
ID Read	0x5555	0xAA	0xAAAA	0x55	0x5555	0x10	Manuf. Addr	Manuf. Data	Device Addr	Device Data		
Write	0x5555	0xAA	0xAAAA	0x55	0x5555	0x20	Write Addr	Write Data				
Write suspend	Page Addr	0x30										
Write resume	Page Addr	0x40										
Chip erase	0x5555	0xAA	0xAAAA	0x55	0x5555	0x50	0x5555	0xAA	0xAAAA	0x55	0x5555	0x60
Page erase	0x5555	0xAA	0xAAAA	0x55	0x5555	0x50	0x5555	0xAA	0xAAAA	0x55	Page Addr	0x70
Page erase suspend	Page Addr	0x80										
Page erase resume	Page Addr	0x90										
Page protect	Page Addr	0xA0	Page Addr	0xA0	Page Addr	0xA0	Page Addr	Verif. Code				
Fast write set	0x5555	0xAA	0xAAAA	0x55	0x5555	0xB0						
Fast write	0xXXXX	0xC0	Write Addr	Write Data								
Fast write reset	0xXXXX	0xD0	0xXXXX	0xE0								
Reset	0x5555	0xAA	0xAAAA	0x55	0x5555	0xF0						

Fig. 5.50 Flash memory commands with required clock cycles

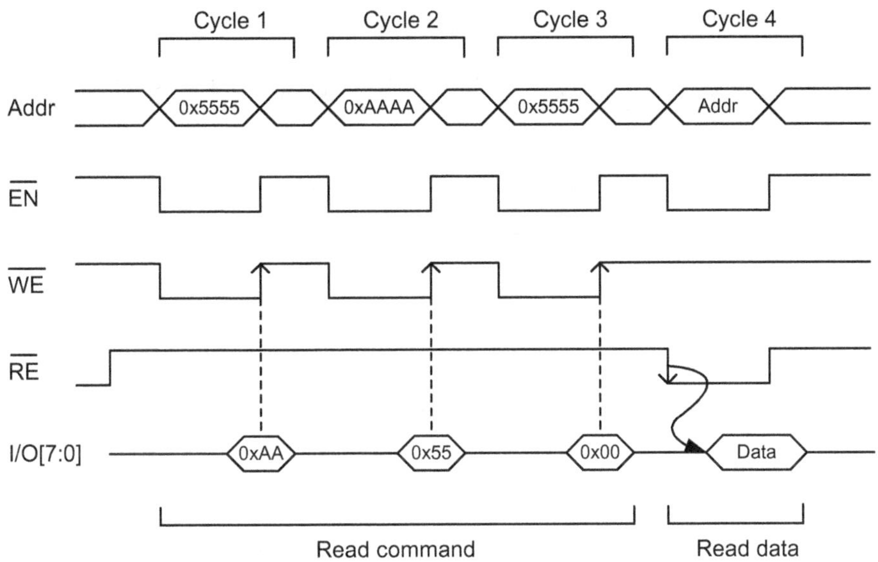

Fig. 5.51 Timing diagram for read operation

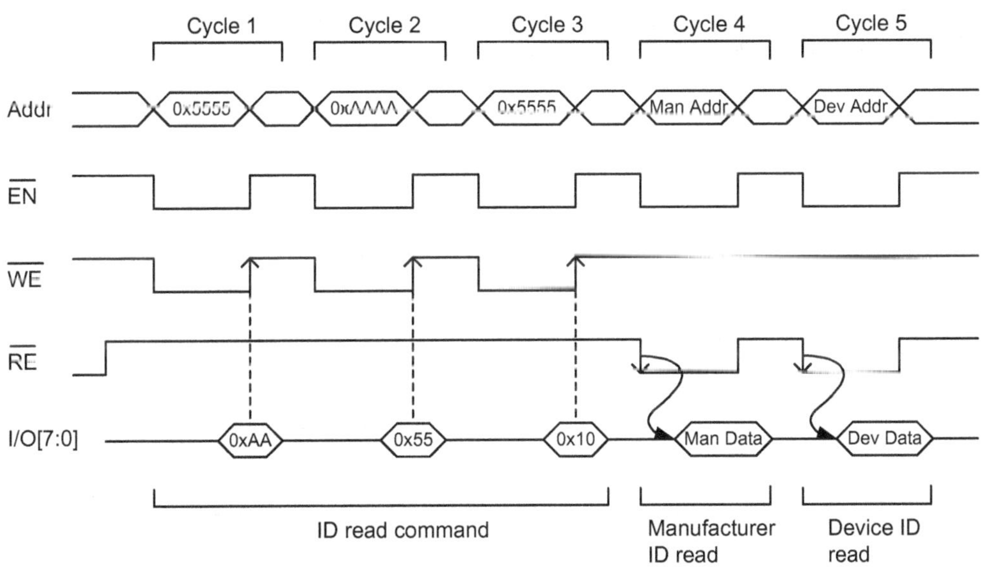

Fig. 5.52 Timing diagram for ID read operation

There are basically two types of write (program) operations for the Flash memory: auto write (program) and fast write (program). Figure 5.53 explains the auto write sequence where the first three cycles are the same as the read sequence with the exception of the auto write command code, 0x20, in the third clock cycle. In the fourth cycle, a valid address and a data are entered to the device when $\overline{EN}$ and $\overline{WE}$ are both lowered to logic 0. The valid data is subsequently written to the specified address at the positive edge of $\overline{WE}$. The data written to the Flash memory can be retrieved in the following cycle without going through a separate read sequence. This is called the auto write verification step, and the most recent written data becomes available at the I/O port as soon as $\overline{RE}$ is lowered to logic 0.

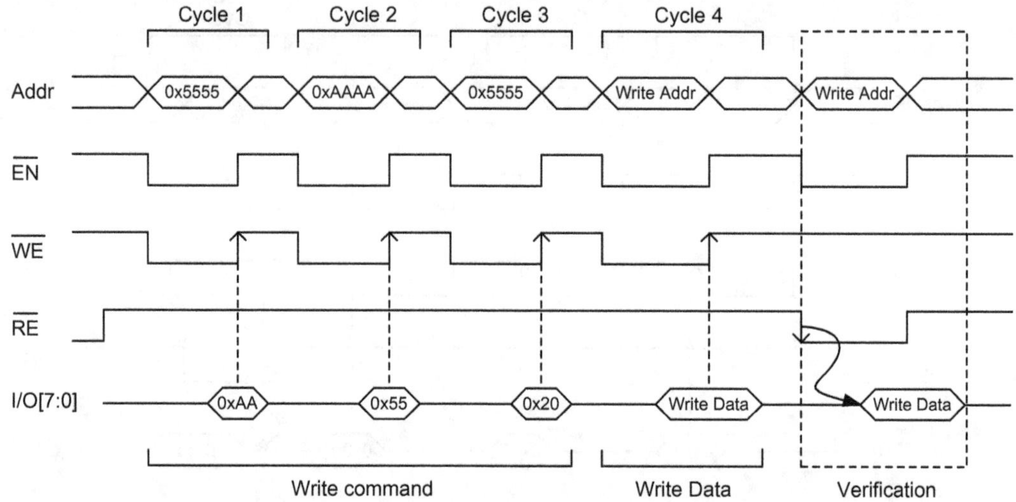

Fig. 5.53 Timing diagram for write (program) operation with the optional verification cycle

The write sequence can be suspended or resumed depending on the need. Both operations take only one clock cycle with the appropriate suspend and resume codes as shown in Fig. 5.54 and Fig. 5.55, respectively. The write suspend and resume codes can be read in the second cycle as a verification step.

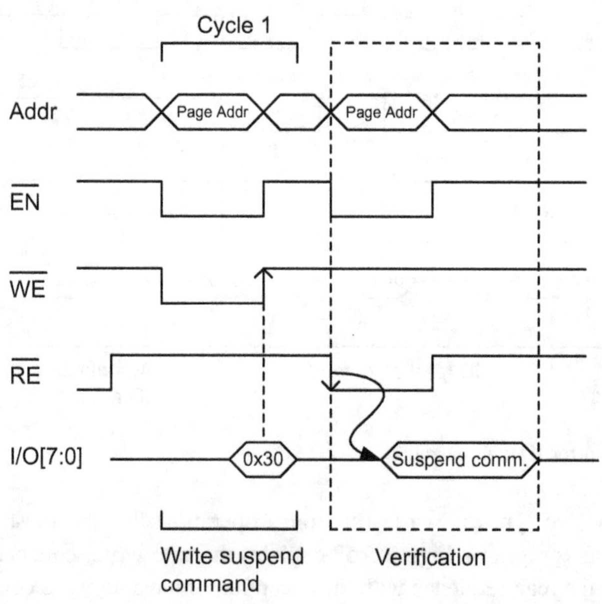

Fig. 5.54 Timing diagram for write (program) suspend operation with the optional verification cycle

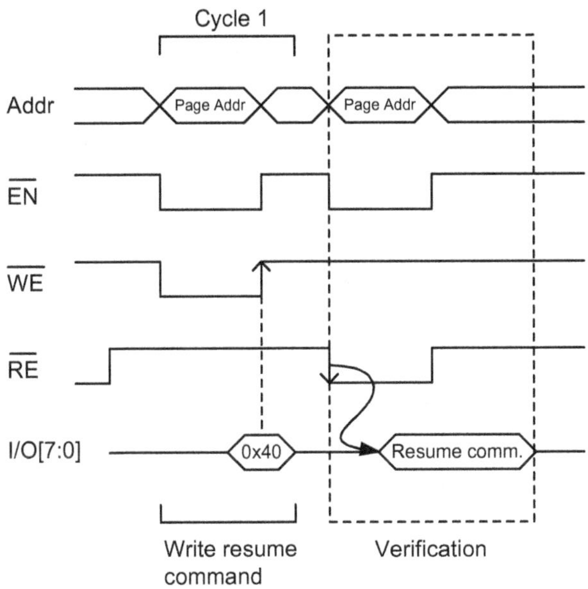

Fig. 5.55 Timing diagram for write (program) resume operation with the optional verification cycle

The erase operation can be applied either to the entire chip or to a particular page. Both sequences take six clock cycles because of the lengthy nature of erase process. In auto chip erase, the first three and the last three cycles are almost identical except two new codes, 0x50 and 0x60, are issued in the third and the sixth cycles. These codes signify the command to erase data in the entire chip as shown in Fig. 5.56.

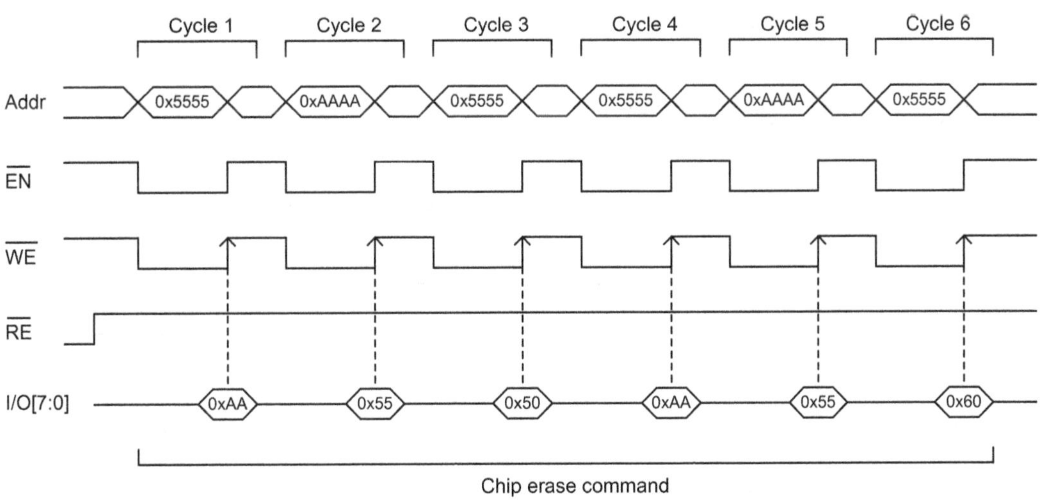

Fig. 5.56 Timing diagram for chip erase operation

In auto page erase, the first five clock cycles are identical to the auto chip erase as shown in Fig. 5.57. The page address to be erased is supplied with the page erase command, 0x70, in the sixth cycle.

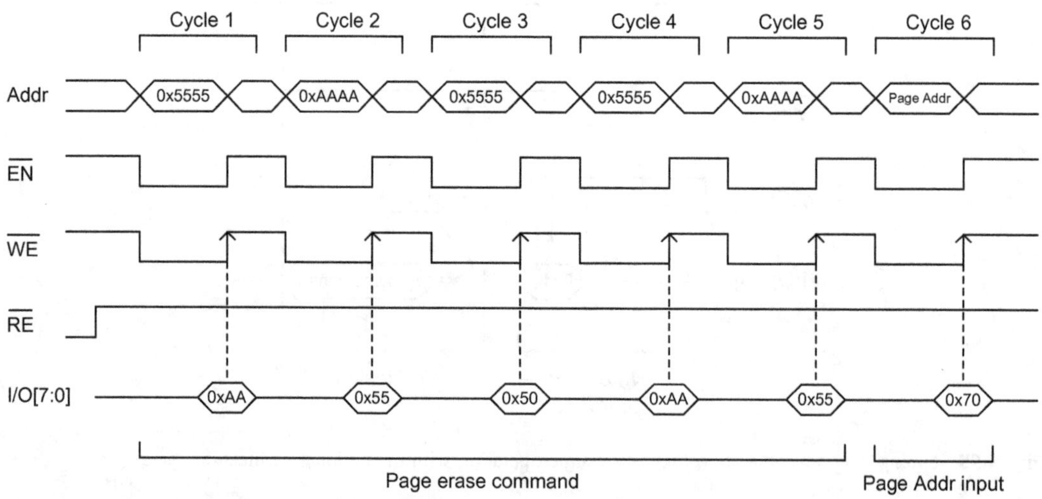

Fig. 5.57 Timing diagram for page erase operation

A certain Flash memory page can be protected from being overwritten or erased by issuing a page protect operation. This is a three-cycle operation as shown in Fig. 5.58. In all three cycles, the page address and the page protect code, 0x0A, have to be specified.

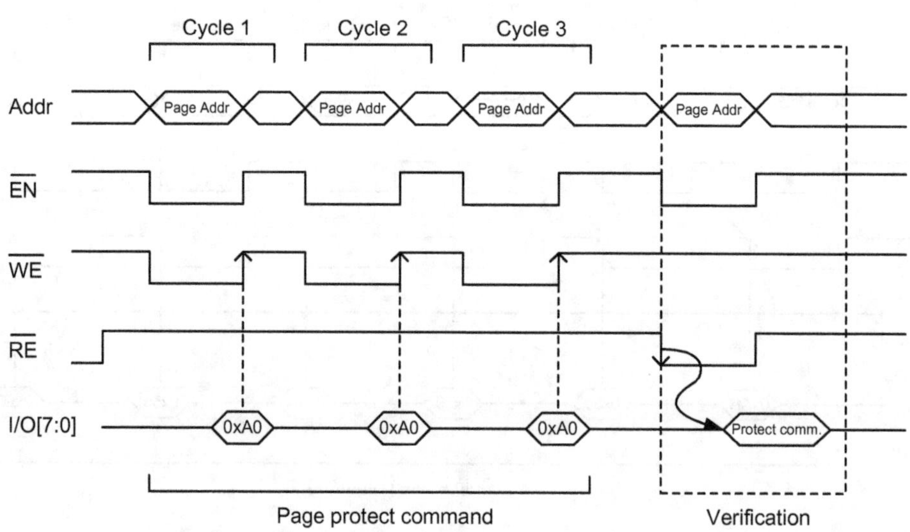

Fig. 5.58 Timing diagram for page protect operation with the optional verification cycle

If faster writing speed is required from the Flash memory, the fast write (program) sequence can be used. This sequence is composed of three parts: fast write set, fast write and fast write reset. The fast write set and reset codes are entered at the beginning and at the end of a write sequence. Figure 5.59 shows the timing diagram for the fast write set sequence where the set code, 0xB0, is entered in the third clock cycle.

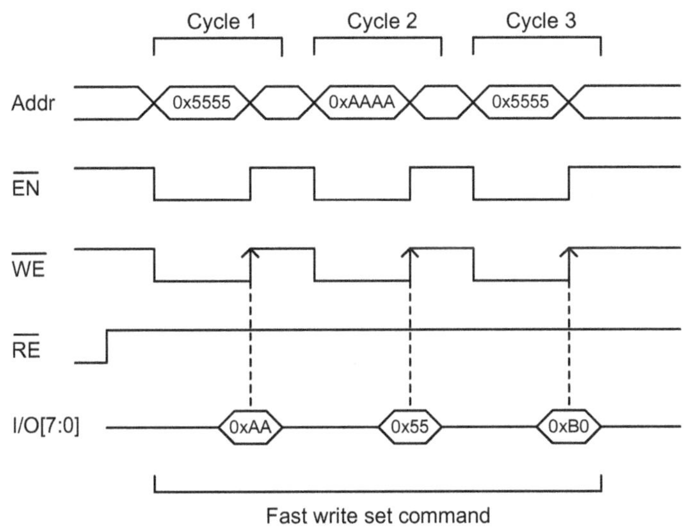

Fig. 5.59 Timing diagram for fast write (program) set operation

The timing diagram for the fast write is a two-cycle sequence as shown in Fig. 5.60. In the first cycle, the fast write code, 0xC0, is entered. In the second cycle, a valid address and a data are entered at the positive edge of $\overline{WE}$.

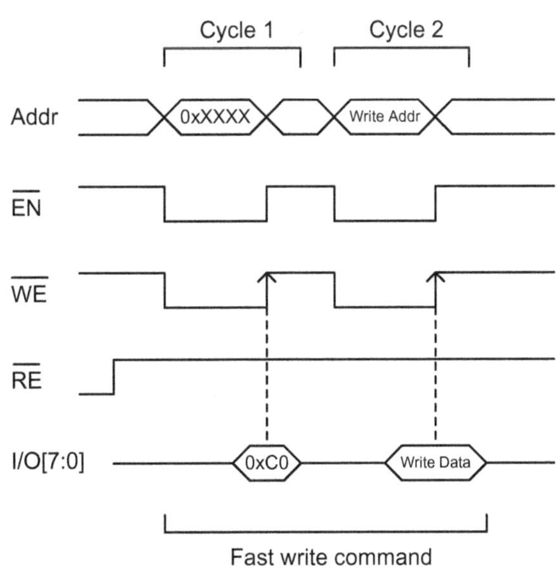

Fig. 5.60 Timing diagram for fast write (program) operation

The fast write reset sequence shown in Fig. 5.61 is also a two-cycle process with two fast write termination codes, 0xD0 and 0xE0, entered in two consecutive clock cycles.

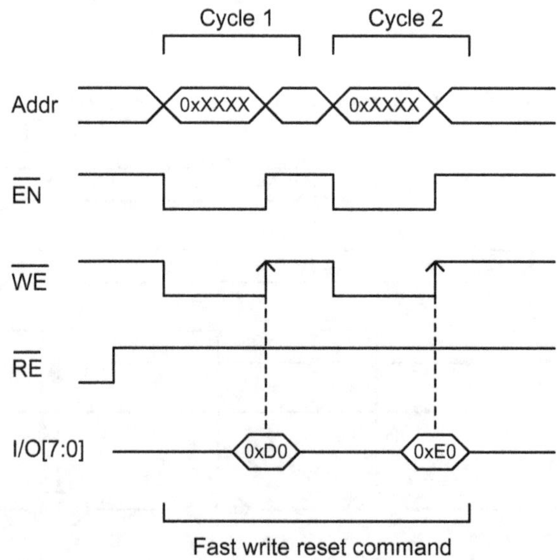

Fig. 5.61 Timing diagram for fast write (program) reset operation

Device reset can be initiated either by $\overline{\text{Reset}}$ input in Fig. 5.44 or by entering the reset code, 0xF0, in the third clock cycle of Fig. 5.62.

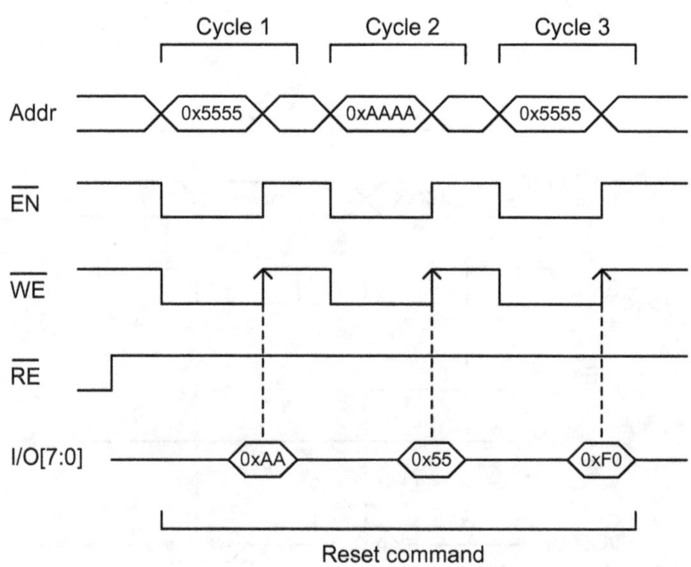

Fig. 5.62 Timing diagram for Reset operation

Both reading and writing (programming) data are cyclic processes. This means that a loop has to be established in the user program to generate a series of memory addresses to read or write data.

Figure 5.63 shows the auto write (program) flow chart where a loop is created to generate the next write address. Each box in the flow chart corresponds to a clock cycle. The first three boxes of the flow chart are responsible for preparing the memory core prior to write. The preparation period terminates with the auto write code, 0x20, as mentioned earlier in Fig. 5.53. After entering the first write address and data, the memory address is incremented. The same process repeats itself prior to issuing the next address and data. When the final address is reached, the write process simply terminates.

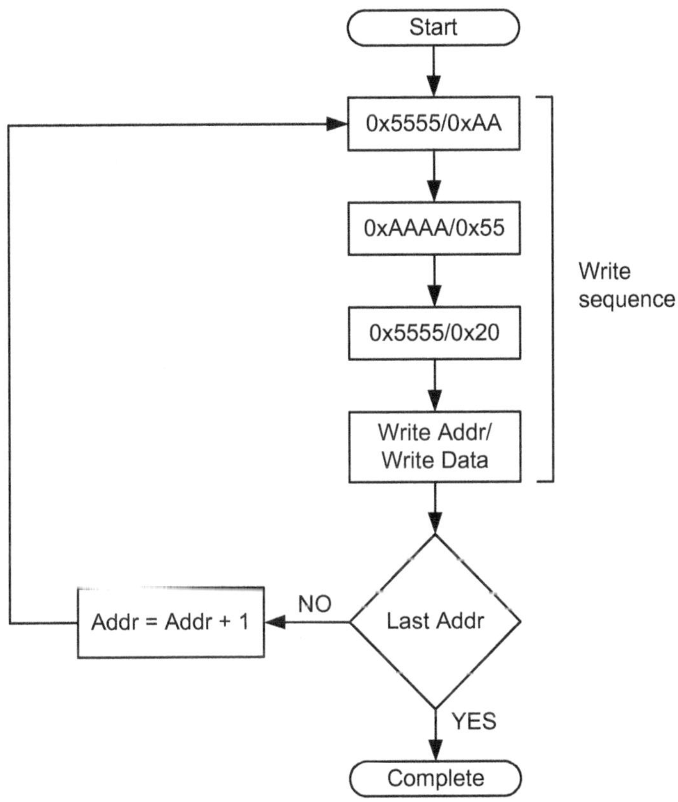

Fig. 5.63 Flow chart for write (program)

The sequence of events is a little different for the fast write (program) in Fig. 5.64. The fast write phase starts with the three-cycle long fast write set sequence followed by the two cycle long fast write sequence. Memory address keeps incrementing until the last data byte is written to the core. The fast write process ends with the two-cycle long fast write reset sequence.

Auto write and fast write processes can be interrupted or resumed by issuing one-cycle long suspend and resume commands anytime during the write process.

In the following sections, we will demonstrate how to design I^2C bus interfaces with Flash memory in order to perform read, write and erase operations. In each design, we will assume only one mode of operation to simplify the design process. The reader is encouraged to design a single

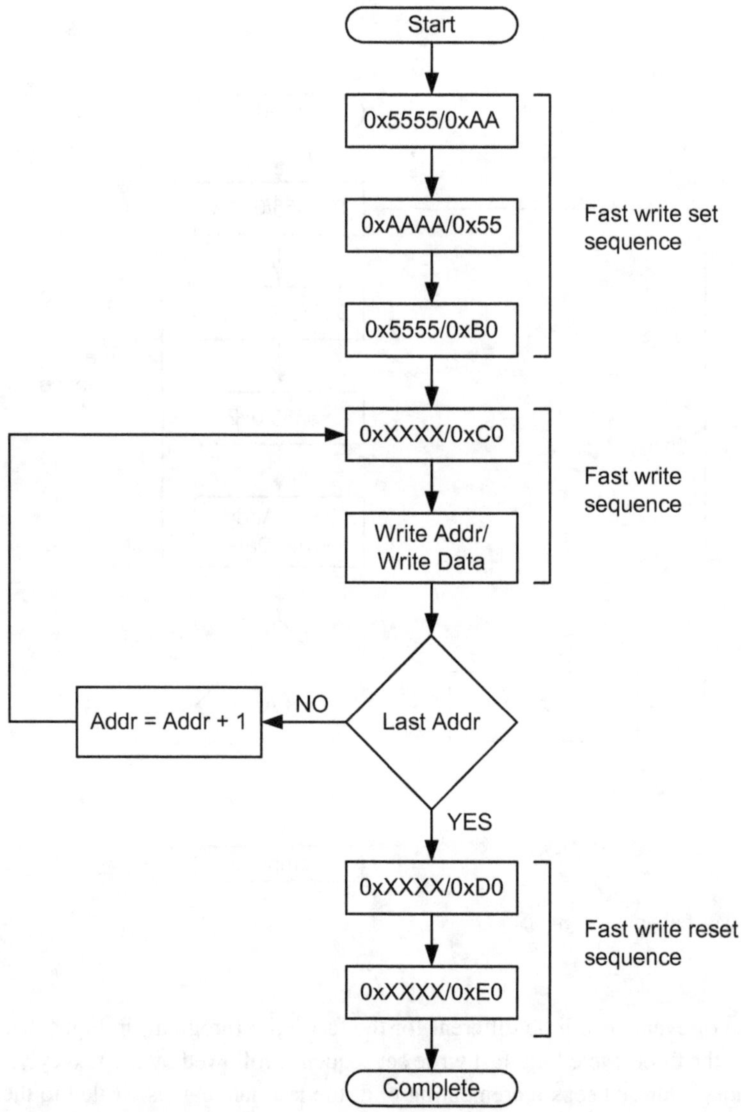

Fig. 5.64 Flow chart for fast write (program)

interface that combines the read, write, erase and other Flash memory operations after studying all three design examples.

Example 5.1: I^{2}C fast write (program) interface for Flash memory

The following design example constructs only the I^{2}C fast write (program) interface for a Flash memory that has parallel address and data ports as shown in Fig. 5.44 using a modified seven-bit address mode. No read, auto write, erase, page protect, reset or other modes are included in this design for the sake of simplicity.

Before dealing with the design details and methodology, it may be prudent to review the timing diagram of I^{2}C write sequence using the seven-bit addressing mode. Although Fig. 5.65 includes only one byte of data, it describes the entire write protocol for the seven-bit address mode in Fig. 4.22. This diagram also includes the start and the stop conditions in Fig. 4.23, and when data (or address) is allowed to change in Fig. 4.24. After generating the start condition, the bus master delivers a seven-bit slave address, starting from the most significant bit, A6. The address sequence is followed by the write bit at logic 0. Once the slave receives the seven-bit address and the write (or the read) command, it produces an acknowledgment, ACK, by lowering the SDA bus to logic 0. The master detects the ACK signal, and sends out an eight-bit data starting from the most significant bit, D7. Once the entire byte is received, the slave responds with another ACK. More data packets follow the same routine until the master generates the stop condition.

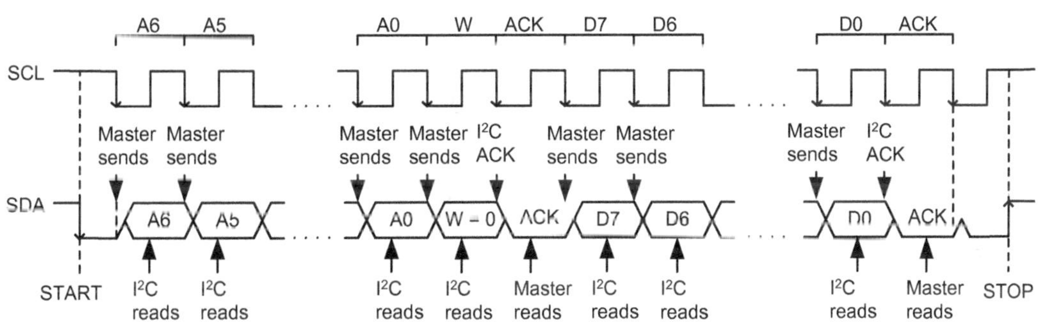

Fig. 5.65 A simple timing diagram for I^{2}C write (program) using seven-bit address mode

In Fig. 5.65, the names that appear on top of each SCL cycle describe a distinct state. If a state machine needs to be constructed from this timing diagram, we simply assign an independent state that corresponds to each name in Fig. 5.65 and produce a state diagram in Fig. 5.66. In this diagram, the start condition activates the state machine, which goes through the address and the command sequences before the data. As long as the state machine does not detect any stop condition, it constantly traces the data states D7 to D0. However, when there is a stop condition, the state machine goes to the IDLE state and waits for another start condition to emerge.

Even though this example only shows the fast write interface, it sets up a solid foundation of how to design a typical I^{2}C interface between a bus master and a Flash memory. The first step of the

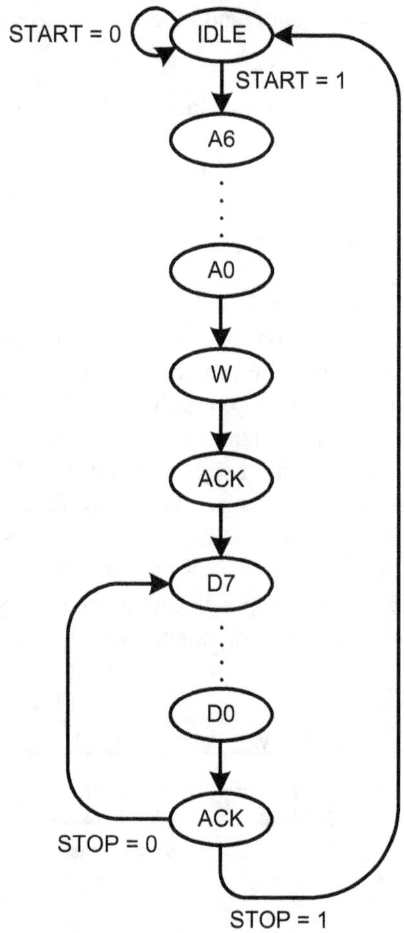

Fig. 5.66 The state diagram for the simple I^2C write (program) in Fig. 5.65

design process is to create a rough interface block diagram showing all the major I/O ports between the Flash memory and the I^2C bus as shown in Fig. 5.67. For the fast write sequence, the address and data packets are serially transferred to the interface through the SDA port. However, the Flash memory needs the address and data fields all at once. Therefore, the write operation requires the interface to perform serial-to-parallel conversion of incoming data. The interface also needs to produce two control signals, $\overline{EN}$ and $\overline{WE}$, for the fast write sequence, and the control signal, EnDataOut, for writing an eight-bit data to a designated Flash memory address.

Figure 5.68 shows the architectural block diagram of the Flash memory interface for the fast write operation. As mentioned earlier when designing SRAM and SDRAM memory interfaces, creating a complete data-path for an interface is not a single-step process. The design methodology requires building a simple data-path with all of its functional units and a corresponding timing diagram showing the flow of data in each clock cycle, the start and stop conditions, address and data formations. However, as more detail is added to the architecture, the initial timing diagram also becomes

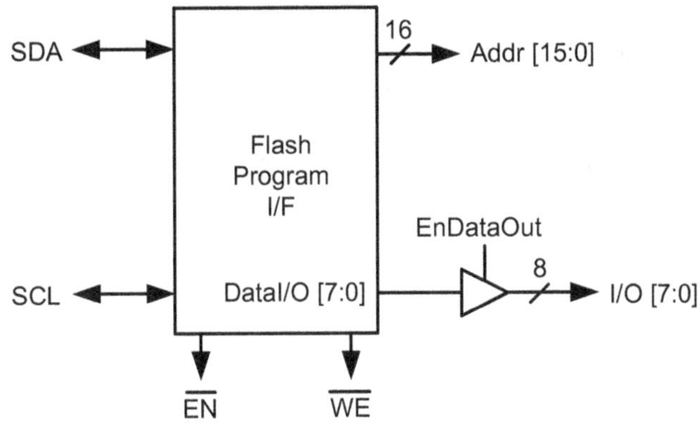

Fig. 5.67 Simplified page diagram of the I^{2}C fast write (program) interface

more complex to match the architecture. Figure 5.69 through Fig. 5.71 show a set of timing diagrams related to the architecture in Fig. 5.68. These diagrams describe the start and stop conditions, preamble sequence, address and data formations, repacking and delivery of serial address and data in a systematic manner.

Once the cycle-by-cycle nature of address and data entries are accurately described in the timing diagram, the control signals responsible for routing the address and data can be added to the diagram. The final step of the design process is to assign distinct states to each clock cycle of the timing diagram that contains different sets of control signals in order to generate a Moore-type state machine.

To start, we need to include four functional units in Fig. 5.68 to be able to handle a simple fast write operation. The first functional unit is an eight-bit shift register whose sole purpose is to convert the incoming serial data from the SDA port into a parallel form for the Flash memory. If the fast-write process requires an authentication step prior to data exchange, the first eight-bit packet coming to this interface must be delivered to the device ID register, which is considered the second functional unit. The second and the third eight-bit data packets arriving at the interface belong to the most and the least significant bytes of the 16-bit starting Flash memory address, respectively, and they are stored in the address counter. The address counter, which constitutes the third functional unit, uses this initial address to generate subsequent addresses for programming the Flash memory. All eight-bit data that follows the address entry is routed directly to the data port of the Flash memory. There are also several fixed-value registers connected to the inputs of the address and data MUXes in Fig. 5.68. These registers contain the preamble data for setting and resetting the fast write modes for the Flash memory prior to address and data sequences. The write controller, which is considered to be the fourth functional unit, generates all the control signals necessary for storing data, incrementing the address, and routing the address and data to the output ports of the interface. The host processor delivers all address, control and data signals to the interface at the negative edge of the SCL, which requires all the registers to operate at the positive edge of SCL in Fig. 5.68.

The START condition in Fig. 5.69 is produced by the bus master lowering the SDA signal to logic 0 while keeping the SCK signal at logic 1 in cycle 1. Once the START condition is detected, the serial data on the SDA port is transferred to an eight-bit shift register which converts this data into a parallel form before sending it to different registers in Fig. 5.68. In clock cycles 2 to 8, the seven-bit Flash memory ID is loaded to the shift register starting from the most significant bit if device authentication is required prior to data transmission. In cycle 9, the bus master sends the write bit, W, stored in the shift register. In cycle 10, a number of events take place simultaneously. First, the write bit at the least significant bit position of the shift register activates the write controller. Second, the device ID and the write bit are transferred from the shift register to a special device ID register. Third, the Flash memory

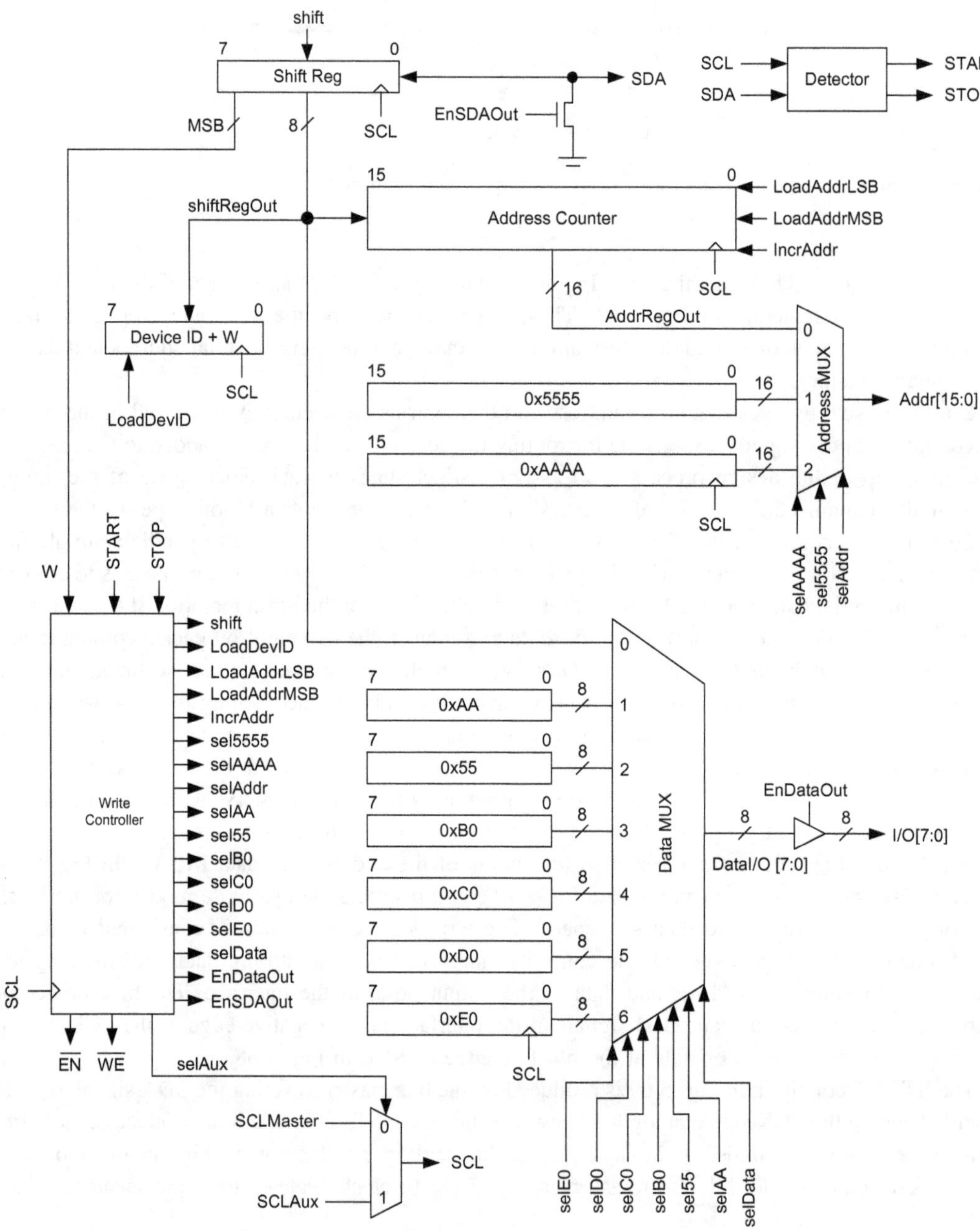

Fig. 5.68 I^{2}C fast write (program) interface data-path

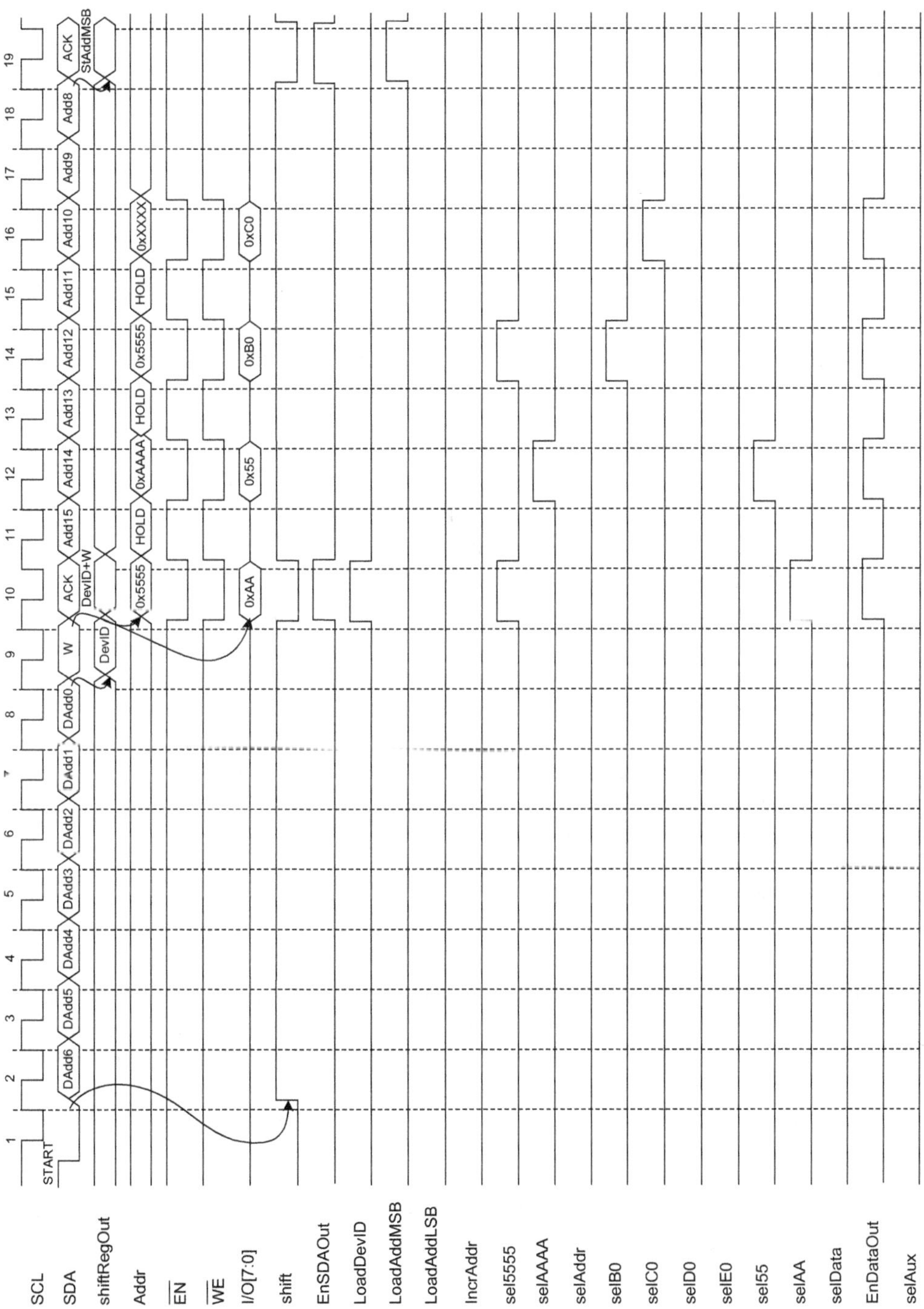

Fig. 5.69 I^2C fast write (program) set sequence, device ID and Flash memory address formations

interface produces the first acknowledge signal, ACK, by EnSDAOut = 1, thus lowering the SDA bus to logic 0. Finally, the interface sends the first preamble which consists of the address, 0x5555, and the data, 0xAA, to the Flash memory and lowers $\overline{EN}$ and $\overline{WE}$ to logic 0 as the first step of the fast write set.

In cycle 11, the most significant bit of the 16-bit initial Flash memory address, Add15, is received from the SDA bus and stored in the shift register. This cycle is considered a hold period for $\overline{EN}$ and $\overline{WE}$ signals. In cycle 12, the second most significant address bit, Add14, is stored in the shift register. In this cycle, the second preamble that contains the address, 0xAAAA, and the data, 0x55, are sent to the Flash memory as the second step of the fast write set. In cycle 14, the third address and command preamble, 0x5555 and 0xB0, are sent to the Flash memory, completing the fast write set sequence. The fast write sequence starts at cycle 16 where the fast write command, 0xC0, is sent to the Flash memory. In cycle 19, the most significant byte of the 16-bit starting Flash memory address, StAddMSB, is transferred from the shift register to the address register which resides inside the address counter. In this cycle, the interface also generates the second ACK signal by EnSDAOut = 1.

From cycles 20 to 27 in Fig. 5.70, the least significant byte of the starting Flash address, StAddLSB, is received by the shift register. In cycle 28, this byte is transferred to the least significant byte of the address register in order to form the 16-bit starting Flash address. In this cycle, the interface generates the third ACK signal. From cycles 29 to 36, the first set of data bits starting from the most significant bit, DF7, to the least significant bit, DF0, are received by the shift register. In cycle 37, the first eight-bit data packet, Data0, is transferred directly to the bidirectional I/O port of Flash memory. The control signal, EnDataOut, enables the tri-state buffers in Fig. 5.68 to write this data packet to the Flash memory.

Cycles 38 through 45 in Fig. 5.70 and Fig. 5.71 are used to store the second eight-bit data packet in the shift register. Cycle 46 transfers this data packet, Data1, to the I/O port and generates an ACK signal for receiving the second data byte from the bus master. If the STOP condition is detected during the next clock cycle, the fast write process halts. The write controller goes into the fast write reset mode and asynchronously produces selAux = 1 to engage the auxiliary clock, SCLAux, instead of using the main SCL clock, SCLMaster, generated by the bus master. This is because the Flash memory needs two more preambles that contain 0xD0 and 0xE0 commands to complete the fast write reset sequence. Therefore, starting from cycle 48, SCL resumes with three more cycles. In cycle 48, the first preamble that contains 0xD0, and in cycle 50 the second preamble that contains 0xE0 are sent to the Flash memory by setting selD0 and then selE0 to logic 1, respectively. In the next cycle, selAux becomes logic 0. Therefore, SCL switches back to the SCLMaster input which is permanently raised to logic 1.

The Moore machine in Fig. 5.72 implements the write controller in Fig. 5.68. At the onset of the START condition, the controller wakes up and goes into the device ID retrieval mode. From cycle 2 to cycle 8 in Fig. 5.69, the serial device ID is received by the shift register on the SDA bus. These cycles correspond to the states DAdd6 to DAdd0 in Fig. 5.72 where the shift signal is constantly kept at logic 1, and writing data to the Flash memory is disabled. In cycle 9, the write bit is also stored in the shift register. This corresponds to the W state in the state machine. In cycle 10, numerous events take place simultaneously. First, shifting serial data into the shift register stops by shift = 0. Second, the seven-bit device ID and the write bit are delivered to the device ID register by LoadDevID = 1. Third, the first address and data preamble, 0x5555 and 0xAA, is delivered to the Flash memory through port 1 of the address MUX by sel5555 = 1 and port 1 of the data MUX by selAA = 1. Fourth, the control signals, $\overline{EN}$ and $\overline{WE}$, are lowered to logic 0 in order to write the address and data

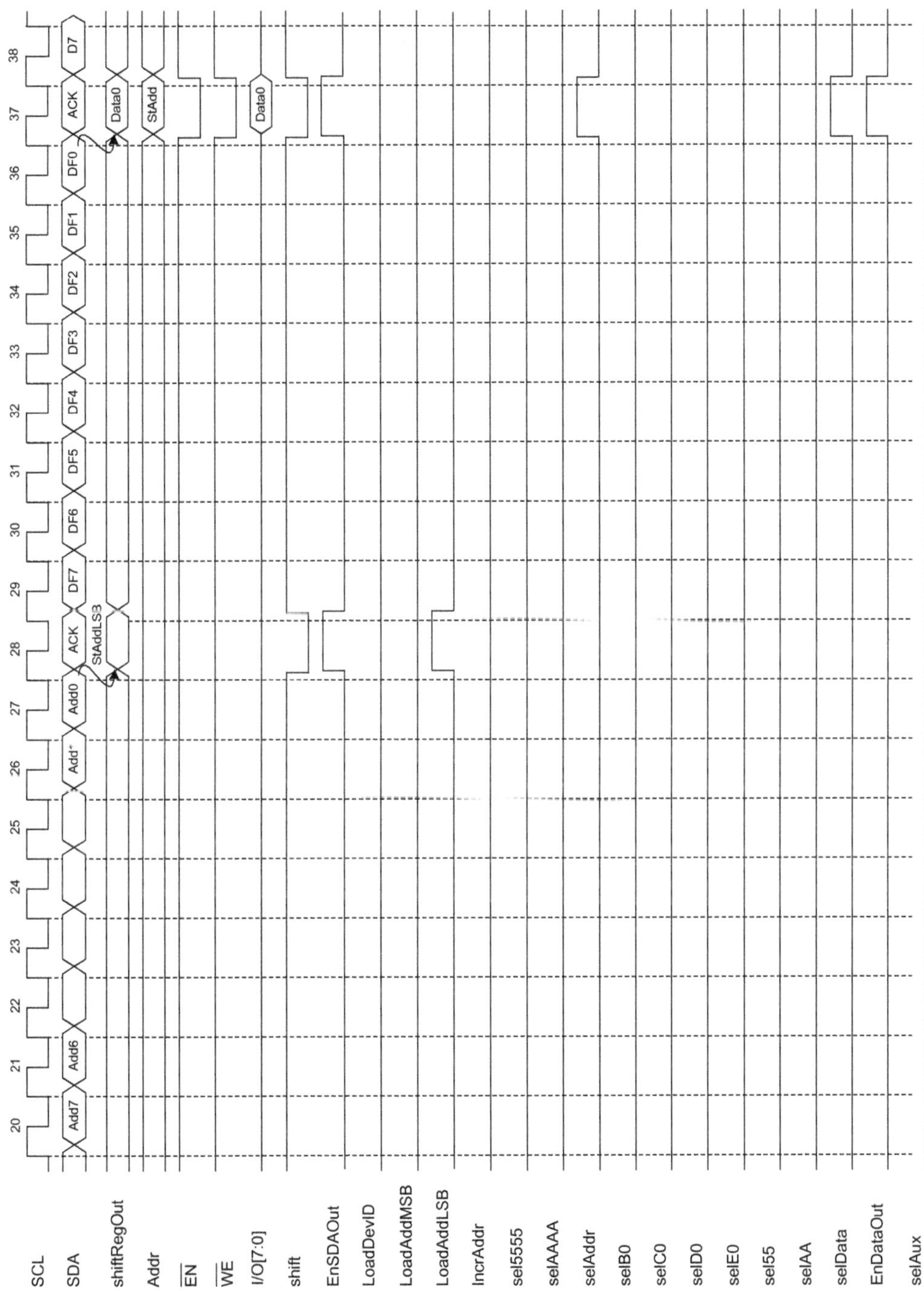

Fig. 5.70 I^2C fast write (program) sequence

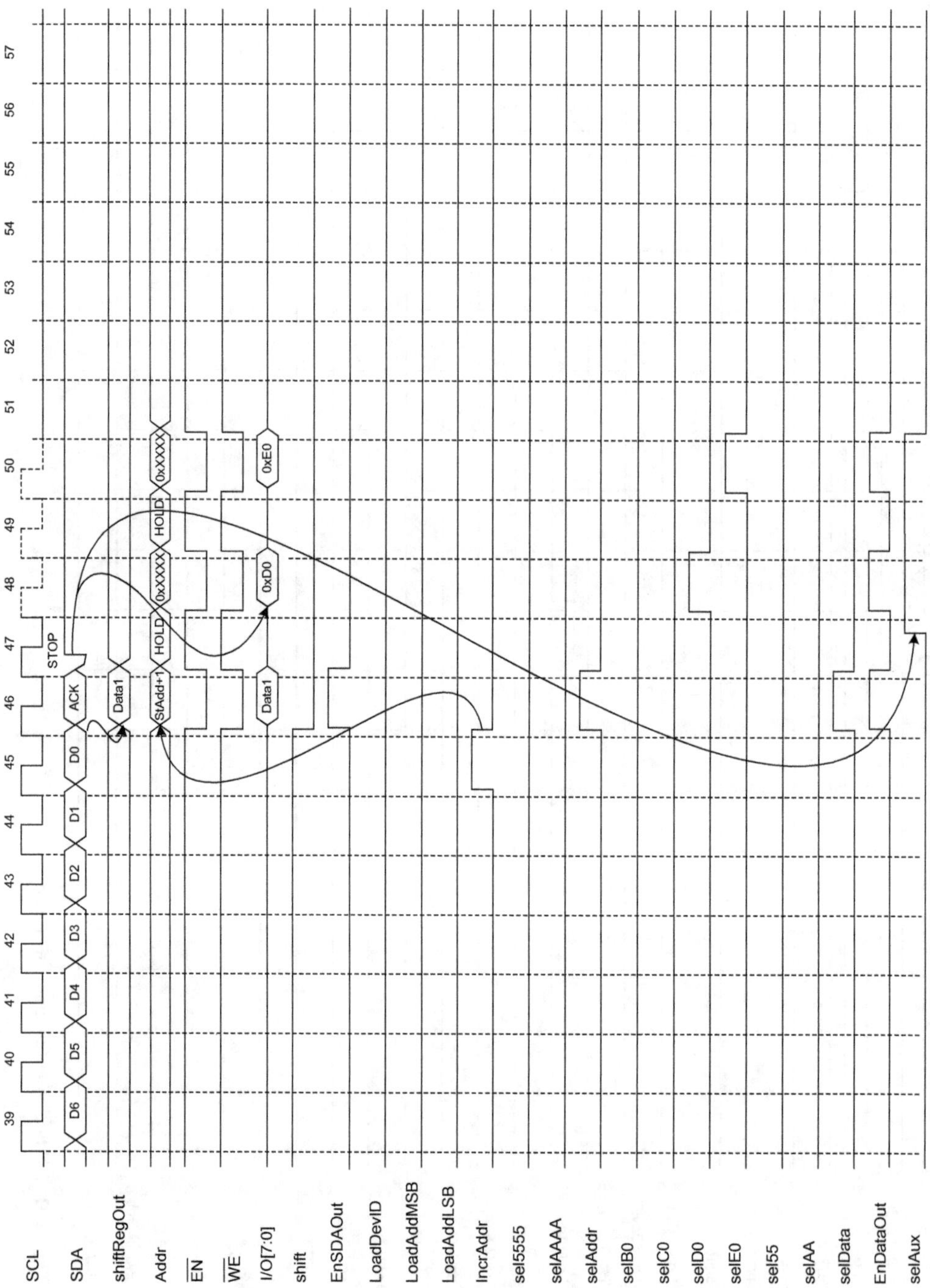

Fig. 5.71 I^2C fast write (program) reset sequence

preamble to the Flash memory. Finally, an ACK signal is generated by EnSDAOut = 1. This cycle corresponds to the DevID ACK state of the write controller.

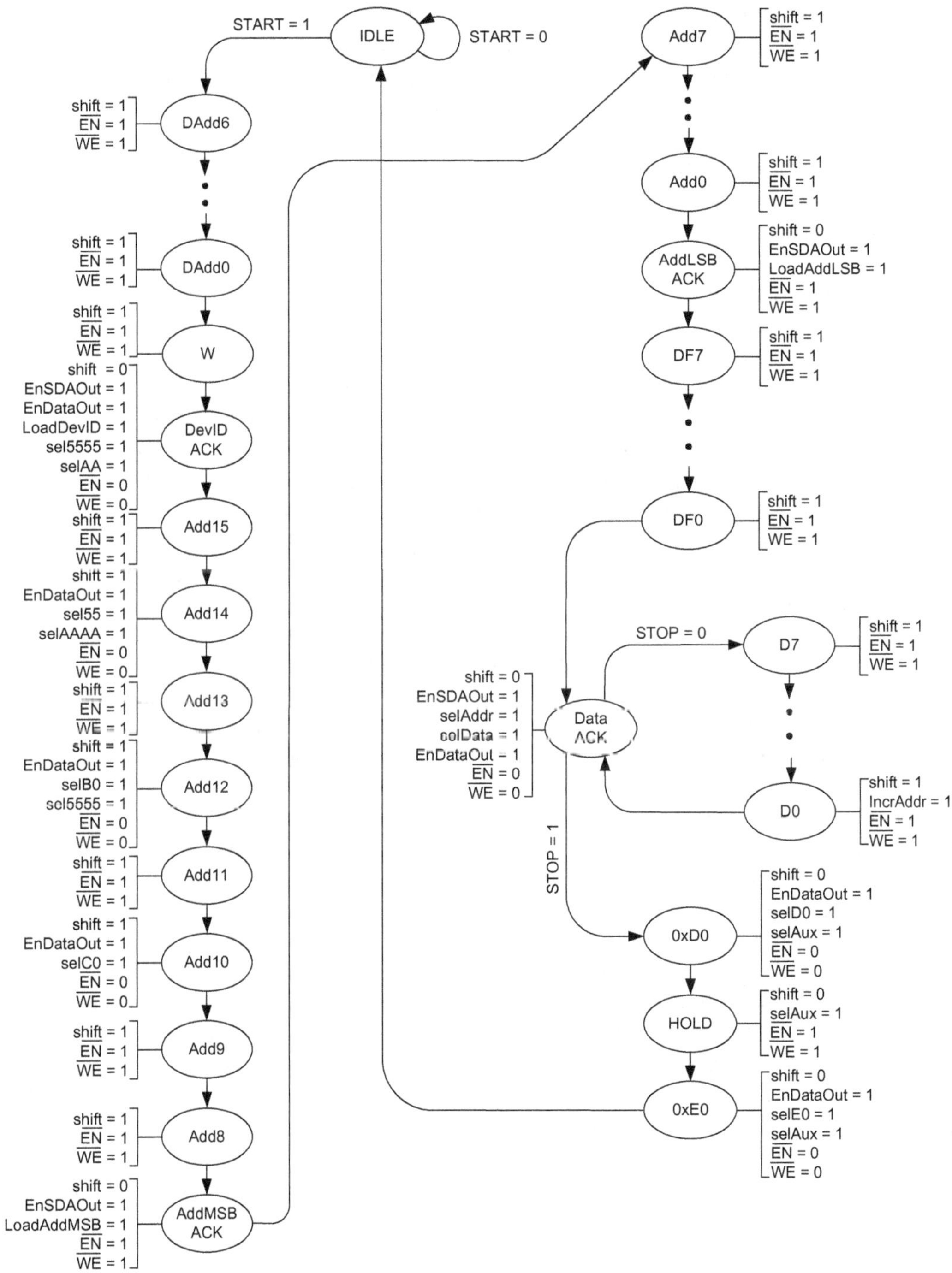

Fig. 5.72 I^2C fast write (program) interface controller (only the essential control signals are indicated in each state to avoid complexity)

In cycle 11, shifting data resumes, and the shift register receives the most significant bit of the initial Flash address, Add15, while in the Add15 state. In cycle 12, corresponding to the Add14 state, the second most significant address bit, Add14, is latched in the shift register by shift = 1. In the same cycle, the second address and data preamble, 0xAAAA and 0x55, is delivered to the Flash memory through port 2 of the address MUX by selAAAA = 1 and port 2 of the data MUX by sel55 = 1. The control signals, $\overline{\text{EN}}$ and $\overline{\text{WE}}$, are also lowered to logic 0 in order to write this preamble to the Flash memory. In cycle 13, Add13 is stored in the shift register. This cycle corresponds to the Add13 state. In cycle 14, the third address and data preamble, 0x5555 and 0xB0, is written to the Flash memory through port 1 of the address MUX by sel5555 = 1 and port 3 of the data MUX by selB0 = 1. In this cycle, the control signals, $\overline{\text{EN}}$ and $\overline{\text{WE}}$, are lowered to logic 0 in order to write the last address and data preambles. This cycle corresponds to the Add12 state. Cycle 15 designates the end of the fast write set cycle, and corresponds to the Add11 state when the address bit, Add11, is loaded to the shift register.

Cycle 16 enters the fast write command mode and writes the address and data preambles, 0xXXXX and 0xC0, through port 4 of the data MUX by selC0 = 1. In this cycle, $\overline{\text{EN}}$ and $\overline{\text{WE}}$ signals are lowered to logic 0 to accommodate the write operation, and Add10 is latched in the shift register. This clock cycle corresponds to the Add10 state. Storing the higher byte of the initial Flash address becomes complete by the end of cycle 18. In cycle 19, the higher byte of the initial address is transferred to the address register by LoadAddMSB = 1, and an acknowledge signal is generated by EnSDAOut = 1. This cycle corresponds to the AddMSB ACK state in the state diagram. Similar events take place when storing the least significant byte of the starting address in the shift register. These states are marked as Add7 to Add0 in the state diagram, and correspond to the cycles 20 to 27, respectively. The cycle 28, which corresponds to the AddLSB ACK state, generates the third acknowledge for the bus master by EnSDAOut = 1, and transfers the least significant byte of the starting Flash memory address to the address register by LoadAddLSB = 1.

From cycles 29 to 36, the first set of data bits are delivered to the shift register starting from the most significant data bit, DF7. This sequence is shown as the states DF7 to DF0 in the state diagram. In cycle 37, an acknowledgement is sent to the bus master by EnSDAOut = 1 shown as the Data ACK state. During this period, the initial 16-bit address and eight-bit data are delivered to the Flash memory through port 0 of the address MUX by selAddr = 1 and port 0 of the data MUX by selData = 1. Tri-state buffer at the I/O port is also enabled by EnDataOut = 1. The second data byte is received during cycles 38 to 45, which correspond to the states D7 to D0, respectively. Cycle 45 is also the cycle to increment the Flash memory address by issuing IncrAddr = 1. In Cycle 46, the write controller goes into the Data ACK state once again and issues an acknowledgement for receiving the second data packet by EnSDAOut = 1. In this cycle, the second data packet is delivered to the incremented Flash memory address, StAdd + 1, by selAddr = 1, selData = 1 and EnDataOut = 1. As long as the STOP condition is not detected, data packets are delivered to the Flash memory at each incremented address. However, if the bus master issues a STOP condition, the auxiliary SCL generator, SCLAux, is asynchronously enabled within the same cycle by selAux = 1. The write controller goes into the fast write reset mode in the next clock cycle and keeps the auxiliary SCL generator enabled by selAux = 1. For the next three clock cycles, the 0xD0 and 0xE0 command codes, corresponding to the 0xD0 and 0xE0 states in the state diagram, are delivered to the Flash memory through port 5 of the data MUX by selD0 = 1 and port 6 of the data MUX by selE0 = 1.

Example 5.2: I^2C read interface for Flash memory

The following design example constructs only the I^2C read interface for a Flash memory that has parallel address and data ports as shown in Fig. 5.44 using a modified seven-bit address mode. No other modes are included in this design except the read.

The timing diagram for I^2C read with seven-bit address mode is given in Fig. 5.73 where eight-bit data packets are serially read from a slave after issuing an address. The address sent by the bus master requires slave's acknowledgment (ACK). In contrast, data packets sent by the slave require the master's acknowledgment. If the bus master chooses not to acknowledge the receipt of data (NACK), the data transfer stops in the next cycle. Figure 5.74 shows the sequence of events taking place in Fig. 5.73 in the form of a state diagram where the logic level in Master ACK/NACK state determines the continuation or the end of the data transfer.

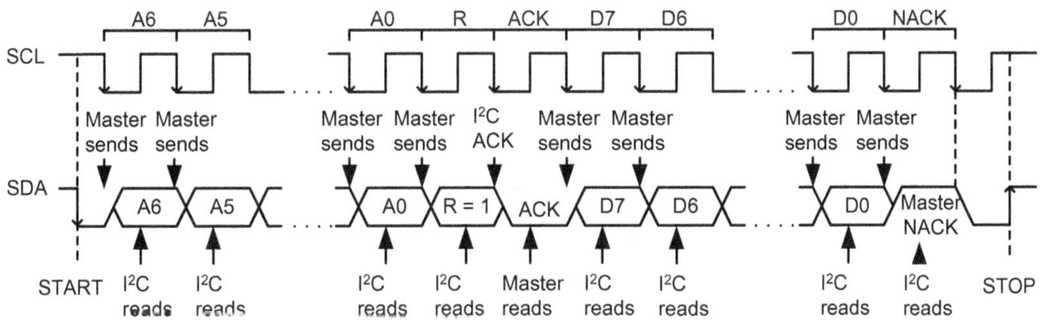

Fig. 5.73 A simple timing diagram for the I^2C read operation using seven-bit address mode

The Flash memory read sequence described in Fig. 5.75 is a four-cycle process as mentioned earlier in Fig. 5.50. The bus master sends the address and data preambles, 0x5555/0xAA and 0xAAAA/0x55, in the first two cycles. This is followed by the 0x5555/0x00 preamble containing the read command code in the third cycle. All three cycles can be considered a preparation period for a read operation which takes place in the fourth cycle. Following the read operation, the address is incremented either by one or a predefined value according to the Flash memory address generation protocol before the next data read sequence takes place.

To read data from the Flash memory, the address and command entries are serially sent by the host processor to the I^2C interface through the SDA port. The Flash memory requires a 16-bit address all at once in order to read an eight-bit data, and this necessitates an interface to perform both serial-to-parallel and parallel-to serial conversions. The interface has to produce three active-low control signals, $\overline{CE}$, $\overline{WE}$ and $\overline{OE}$, to be able to read data from the Flash memory. It also needs to produce the control signals, EnDataIn and EnDataOut, to route the incoming and outgoing data.

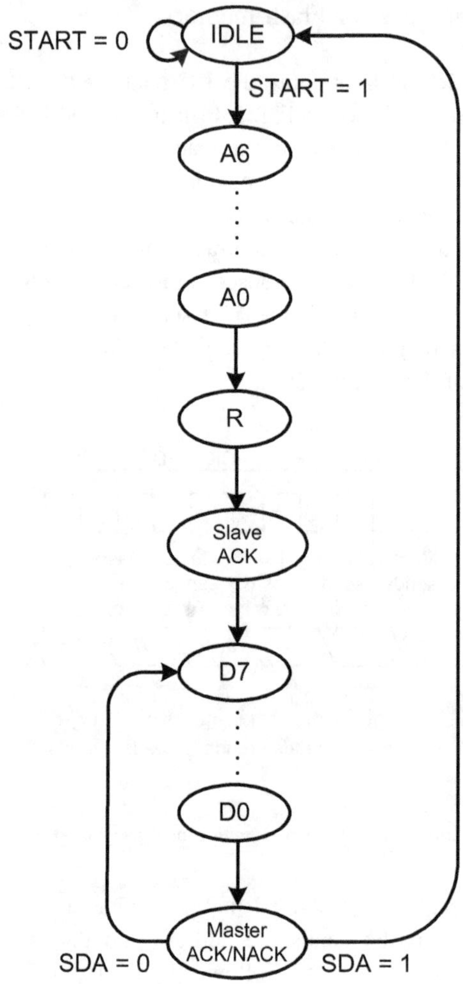

Fig. 5.74 The state diagram for the simple I^2C read operation in Fig. 5.73

Figure 5.76 shows the architectural diagram of the Flash memory read interface. Figures 5.77, 5.78 and 5.79 show the timing diagrams related to this architecture. These waveforms describe a complete picture of the preamble formation, device ID creation, address generation and serializing the read data from the Flash memory.

The architecture in Fig. 5.76 still contains four functional units as in the fast write (program) data-path. The first functional unit is an eight-bit shift register which stores the serial address, command and data, and converts the serial data on the SDA bus into a parallel form and the parallel data from the Flash memory into a serial form. The second functional unit stores the device ID if the Flash memory requires an identification process prior to data exchange, and the command bit. The third unit stores the initial 16-bit Flash memory address and generates the subsequent memory addresses using an up-counter. The fourth unit is the read controller, which is responsible for storing the device ID, the command bit, forming and incrementing the initial address, and handling the proper

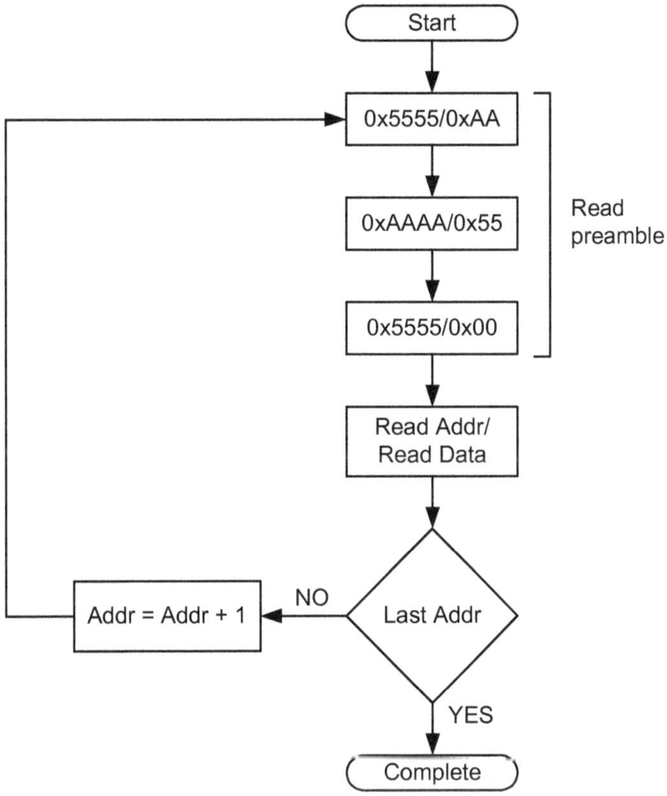

Fig. 5.75 Flow chart for the read sequence

data-flow that complies with the timing diagrams in Figs. 5.77, 5.78 and 5.79. There are also several fixed-value registers, each of which contains the address and data preambles for retrieval of data from the Flash memory. The host processor dispatches all address and control signals at the negative edge of the SCL in order to read data from the Flash memory, and therefore it requires all the registers in Fig. 5.76 to operate at the positive edge of SCL clock.

Figures 5.77, 5.78 and 5.79 describe the complete picture of reading data from the Flash memory. Figure 5.77 shows the device ID and the command bit formations followed by the generation of the most significant byte of the initial Flash memory address. Figure 5.78 describes the formation of the least significant byte of the initial Flash memory address and the first data byte read from the memory. Figure 5.79 includes two additional bytes of data sent to the bus master and the termination of data transfer.

The bus master initiates the data transfer by issuing the START condition in Fig. 5.77. Between cycles 1 and 8, the bus master sends the device ID followed by the read command on the SDA bus, both of which are serially loaded to an eight-bit shift register by ShiftIn = 1. These cycles are represented by the states DAdd6 to DAdd0 followed by the R state, which corresponds to the read command, in the state diagram in Fig. 5.80. In cycle 9, the Flash memory interface responds to the bus master with an acknowledgement by issuing EnSDAOut = 1, but pauses shifting data by ShiftIn = 0. In the same cycle, the interface also transfers data from the shift register to the device ID register by LoadDevID = 1. This cycle corresponds to the first slave-acknowledgement state, DevID SACK, in Fig. 5.80. In cycle 10, the interface starts sending the preamble to the Flash memory

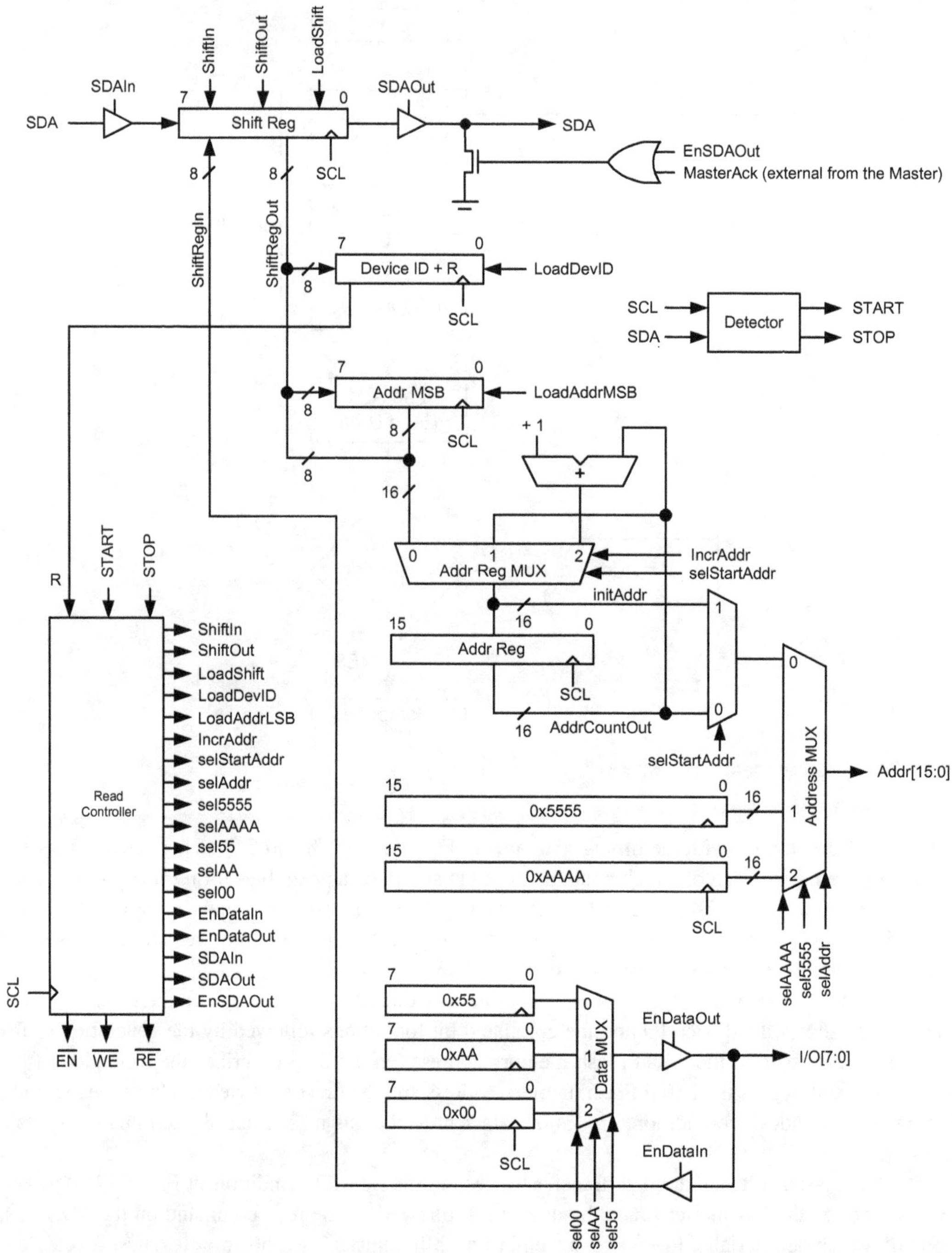

Fig. 5.76 I^{2}C read interface data-path

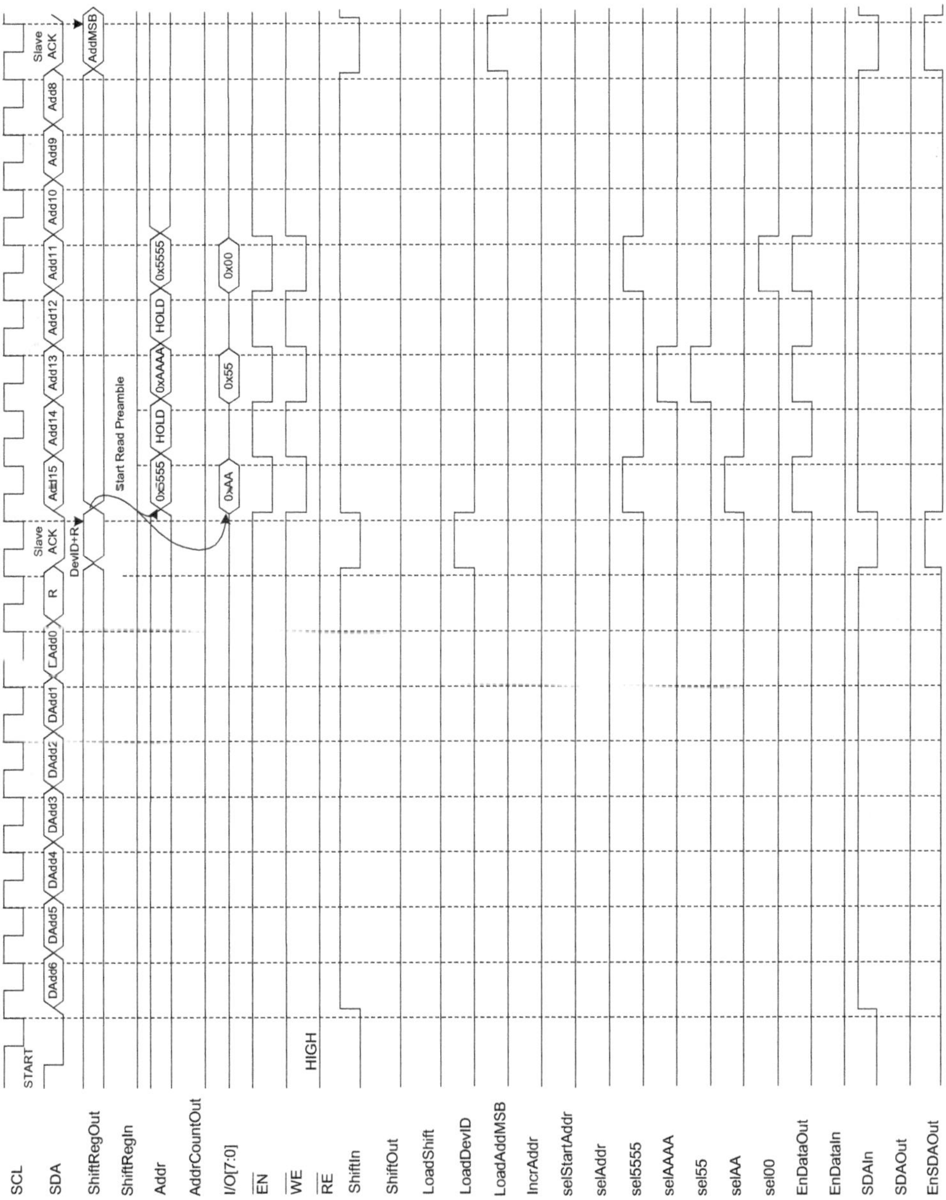

Fig. 5.77 I^2C interface timing diagram emphasizing device ID and LSB of the starting Flash memory address formations during the read operation

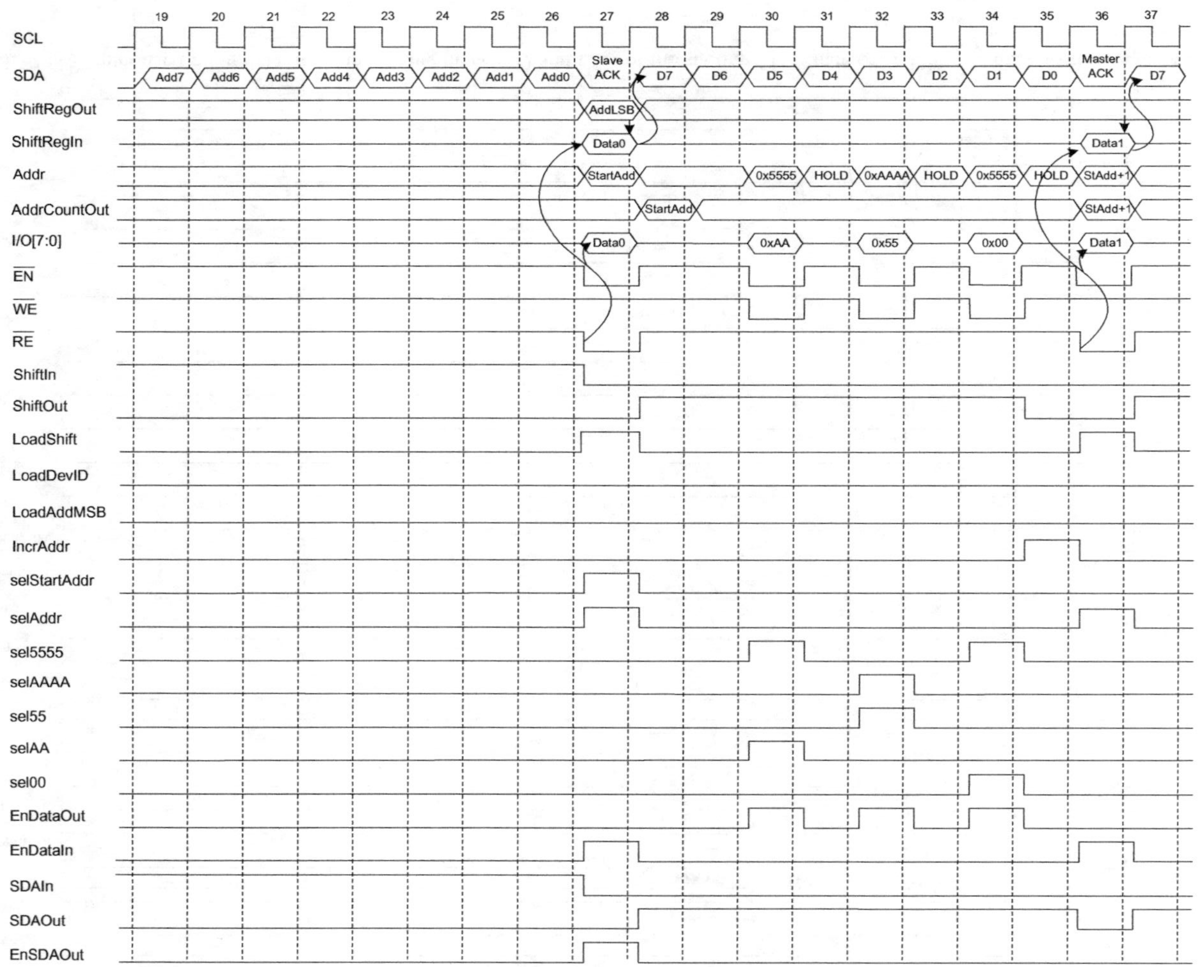

Fig. 5.78 I^{2}C interface timing diagram emphasizing the MSB of the starting Flash memory address and data formations during the read operation

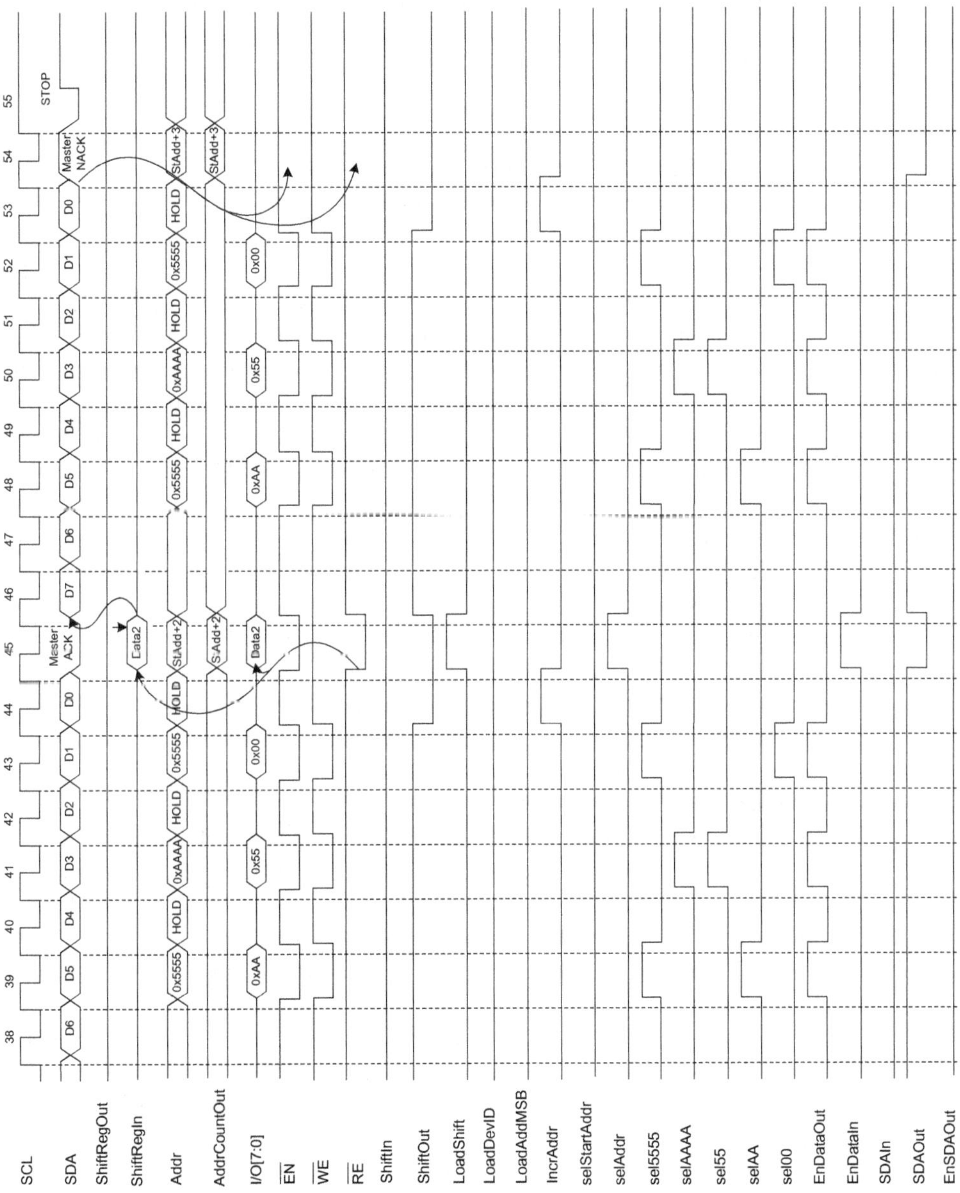

Fig. 5.79 I^{2}C interface timing diagram emphasizing the data formations and the end of the read cycle

for a read operation. In this cycle, the first address and data preamble, 0x5555/0xAA, is fetched from the fixed-data registers, 0x5555 and 0xAA in Fig. 5.76. This preamble is subsequently sent to the address port of the device through port 1 of the address MUX by sel5555 = 1, and to the data port of the device through port 1 of the data MUX by selAA = 1. In this cycle, the bus master also sends the most significant bit of the 16-bit Flash memory address, Add15, on the SDA bus.

In cycle 12, the interface sends the second address and data preamble, 0xAAAA/0x55, through port 2 of the address MUX by selAAAA = 1 and port 0 of the data MUX by sel55 = 1. In cycle 14, the last address and data preamble, 0x5555/0x00, containing the read command, is sent to the Flash memory address and data ports. The cycles 10 to 17 correspond to storing the most significant byte of the starting Flash memory address, Add15 to Add8, in the shift register by ShiftIn = 1. In cycle 18, the interface sends an acknowledgement to the bus master by EnSDAOut = 1 to indicate that it has received the higher byte of the starting Flash memory address. Within the same cycle, this higher byte is stored in the 16-bit address register that resides in the address counter by LoadAddMSB = 1. This cycle represents the second slave-acknowledgement state, AddLSB SACK, in Fig. 5.80.

In cycles 19 to 26, the bus master sends the least significant byte of the starting Flash memory address by ShiftIn = 1. These cycles correspond to the states Add7 to Add0 in Fig. 5.80, respectively. Cycle 27 constitutes the third slave-acknowledgement state, AddLSB SACK, in Fig. 5.80. There are numerous events that take place during this clock cycle, and they are all inter-related. The first event concatenates the least significant byte of the starting Flash memory address in the shift register with the most significant byte in the Addr MSB register to form the complete 16-bit starting Flash memory address. This address is subsequently sent to the Addr[15:0] terminal of the Flash memory through port 0 of the address register MUX by selStartAddr = 1 and port 0 of the address MUX by selAddr = 1. The second event lowers $\overline{\text{EN}}$ and $\overline{\text{RE}}$ control signals to logic 0 and produces EnDataIn = 1 in order to fetch the first data byte from the Flash memory, Data0, since the read preamble has already been sent between cycles 10 and 14. The third event stores Data0 in the shift register through its ShiftRegIn port by LoadShift = 1. Finally, the last event sends an acknowledgement signal to the bus master by EnSDAOut = 1, signifying the least significant byte of the starting address has been received so that the bus master can start receiving serial data on the SDA bus in the next cycle.

In cycle 28, the starting address, which could not be registered due to time limitations in the earlier cycle, is now registered in the address register, and the address counter output, AddrCountOut, becomes equal to the starting address, StartAdd. In the same cycle, the most significant bit of Data0, D7, becomes available on the SDA bus by SDAOut = 1. Starting from cycle 30, the read preamble associated with the second data is sent to the Flash memory. The read preamble could have been issued as early as cycle 28 or 29 since the address counter still held StartAdd at the AddrCountOut node during these periods. In cycle 35, the interface increments the starting address by IncrAddr = 1 and uses port 2 of the address register MUX to feed through the result. Until the beginning of cycle 36, all eight bits of Data0, D7 to D0, are serially sent to the bus master by SDAOut = 1. Therefore, cycles 28 to 35 correspond to the states D7 to D0 in Fig. 5.80, respectively. In cycle 36, while the bus master acknowledges the reception of Data0 by lowering the SDA bus to logic 0, and the interface sends the incremented Flash memory address, StAdd + 1, to the Addr[15:0] terminal through port 0 of the address MUX by selAddr = 1. In the same cycle, the interface lowers $\overline{\text{EN}}$ and $\overline{\text{RE}}$ signals to logic 0, fetches Data1 from the I/O port of the Flash memory by EnDataIn = 1, and stores this value in the shift register by LoadShift = 1. This particular cycle corresponds to the master-acknowledge state, MACK, in Fig. 5.80, where the state machine continues fetching data from the Flash memory.

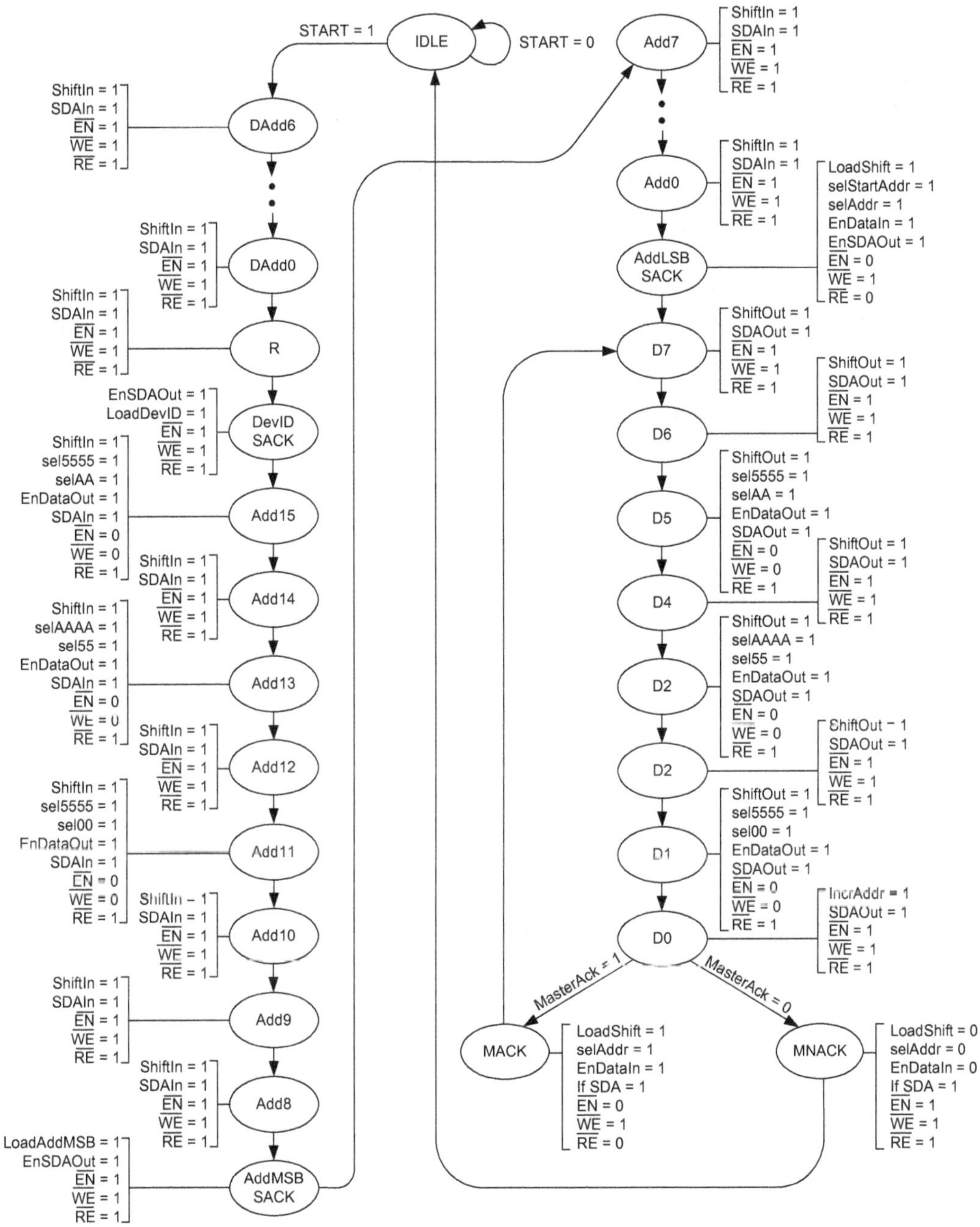

Fig. 5.80 I²C read interface controller (only the essential control signals are indicated in each state to avoid complexity)

Cycles 37 to 44 and cycles 46 to 53 in Figs. 5.78 and 5.79 are comprised of identical events to the one between cycles 28 and 35. They both correspond to the states D7 to D0 in Fig. 5.80. In cycle 54, the bus master decides not to issue any more acknowledgements by keeping the MasterAck signal at logic 0. As a result, the interface does not lower $\overline{EN}$ and $\overline{RE}$ signals to logic 0, and no data reading takes place from the Flash memory. In the next cycle, the bus master terminates the SCL activity and issues the STOP condition, signifying the end of data transfer.

Example 5.3: I^{2}C page erase interface for Flash memory

The following design example constructs only the I^{2}C page erase interface for a Flash memory that has parallel address and data ports as shown in Fig. 5.44 using a modified seven-bit address mode. No other Flash memory mode is implemented in this design except the erase.

The Flash memory page erase is a six-cycle sequence as described earlier in Fig. 5.50. The flow chart for this process is shown in Fig. 5.81. In the first five cycles, the bus master sends fixed address/data combinations to the Flash memory as a preamble to prepare the memory to erase a block of data at a specified memory location. The page erase command is the 0x50 entry in the third cycle followed by the page address and the second erase command, 0x70, in the sixth cycle to initiate the process.

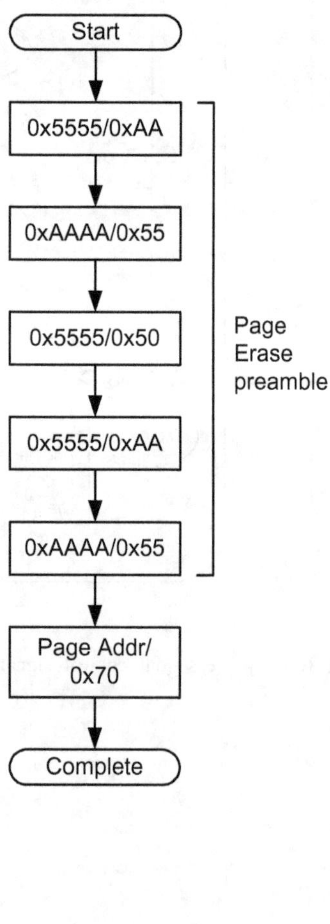

Fig. 5.81 Flow chart for Page Erase

Figure 5.82 shows the data-path for the I^{2}C page erase interface. The shift register acquires the device ID (if the Flash memory requires any type of device authentication prior to page erase) and the page address from the SDA bus, and transfers them to the device ID register and the page address register, respectively. There are also address and data registers that store only fixed values, and they are routed to the address and data ports of the Flash memory in order to produce the correct preamble and page erase command in Fig. 5.81.

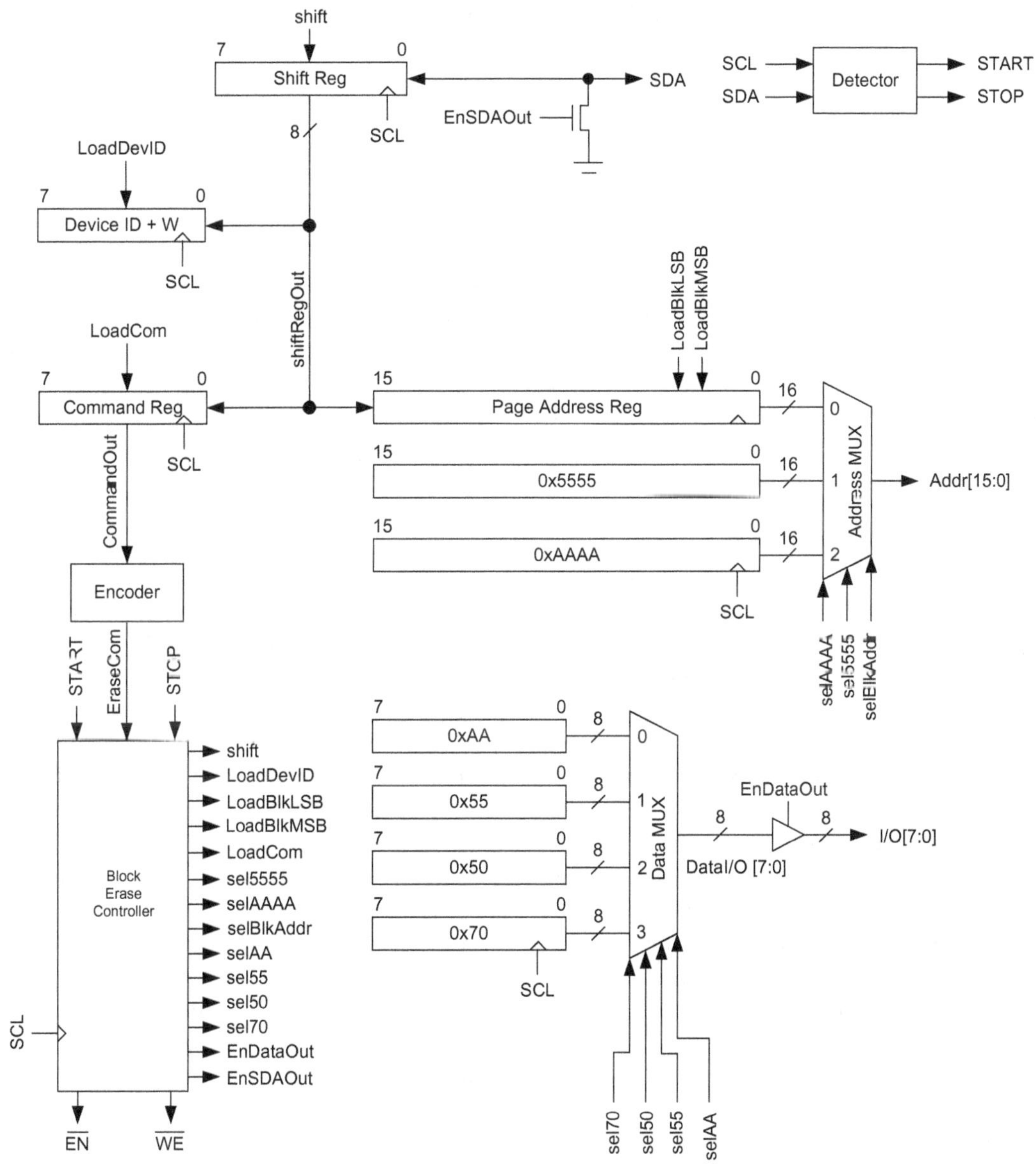

Fig. 5.82 I^{2}C Page Erase interface data-path

The page erase process is described in the timing diagrams of Figs. 5.83, 5.84 and 5.85. The process starts with the bus master generating the START condition in Fig. 5.83. In cycles 2 to 9, the bus master sends the seven-bit device ID and the write bit, starting with the most significant device ID bit, DAdd6. Even though the write bit is considered a command bit, it does not have any significance in the page erase preamble. The bus master sends this bit only to comply with the I^2C protocol. All these bits are temporarily stored in the shift register and correspond to the states DAdd6 to W in the state diagram in Fig. 5.86. In cycle 10, the interface generates an acknowledgement, ACK, to signify that it has received the first eight bits from the bus master by EnSDAOut = 1, and transfers the device ID stored in the shift register to the device ID register by LoadDevID = 1. This cycle corresponds to the DevID ACK state in Fig. 5.86.

From cycles 11 to 18, which correspond to the states Add15 to Add8 in the state diagram, the bus master sends the most significant byte of the Flash memory page address to the interface. These bits are received by the shift register and immediately transferred to the page address register in cycle 19 by LoadBlkMSB = 1. In this cycle, the interface also sends a second acknowledgment to the bus master by EnSDAOut = 1, which is represented by the AddMSB ACK state in the state diagram.

From cycles 20 to 27, the interface receives the least significant byte of the page address. It stores this byte in cycle 28 by LoadBlkLSB = 1, and sends a third acknowledgement to the bus master by EnSDAOut = 1. These events are shown by the states Add7 to Add0 and the state AddLSB ACK in the state diagram, respectively. Starting in cycle 29, the complete page address becomes available at the Addr[15:0] terminal in Fig. 5.82 even though the page erase process has not been initiated. This cycle is also the starting point for the bus master to send the erase command, 0x50, to the Flash memory interface. Without this step, the interface will not be able to recognize if the ongoing process is actually about erasing a block of data.

From cycle 29 to 36 that correspond to the states 0x50-0 to 0x50-7, the interface receives all eight bits of the command code, 0x50, in the shift register. Then in cycle 37, it generates the fourth acknowledgment by EnSDAOut = 1, and transfers the contents of the shift register, 0x50, to the command register by LoadCom = 1. Later on, the interface uses this value to be able to generate the correct preamble for the page erase operation. Cycle 37 corresponds to the 0x50 ACK state in the state diagram. While the bus master sends the second command code, 0x70, from cycle 38 to 45 to initiate the page erase, the interface, now aware of the page erase operation, sends the first address and data preamble, 0x5555/0xAA, to the Flash memory in cycle 39. In this cycle, the fixed register value, 0x5555, is routed through port 1 of the address MUX by sel5555 = 1. The fixed register data, 0xAA, is also sent to the I/O[7:0] port through port 0 of the data MUX by selAA = 1 and EnDataOut = 1. In cycle 41, the second address and data preamble, 0xAAAA/0x55, is sent. This is followed by sending the third preamble (including the first page erase command), 0x5555/0x50, in cycle 43, and then the fourth preamble, 0x5555/0xAA, in cycle 45. The interface pauses for one cycle after dispatching each address and data combination to comply with the Flash memory protocol of writing data.

Cycle 38 to cycle 45 are represented by the states 0x70-0 to 0x70-7 in the state diagram, respectively. The interface sends the fifth acknowledgment to the bus master in cycle 46 by EnSDAOut = 1 while in the 0x70 ACK state. In cycle 47, the interface sends the fifth address and data preamble, 0xAAAA/0x55, and finally in cycle 49, it sends the page address with the second page erase command, 0x70, to erase the entire block of data. Cycles 47 to 49 are represented by the DontCare-0, DontCare-1 and DontCare-2 states in Fig. 5.86, respectively.

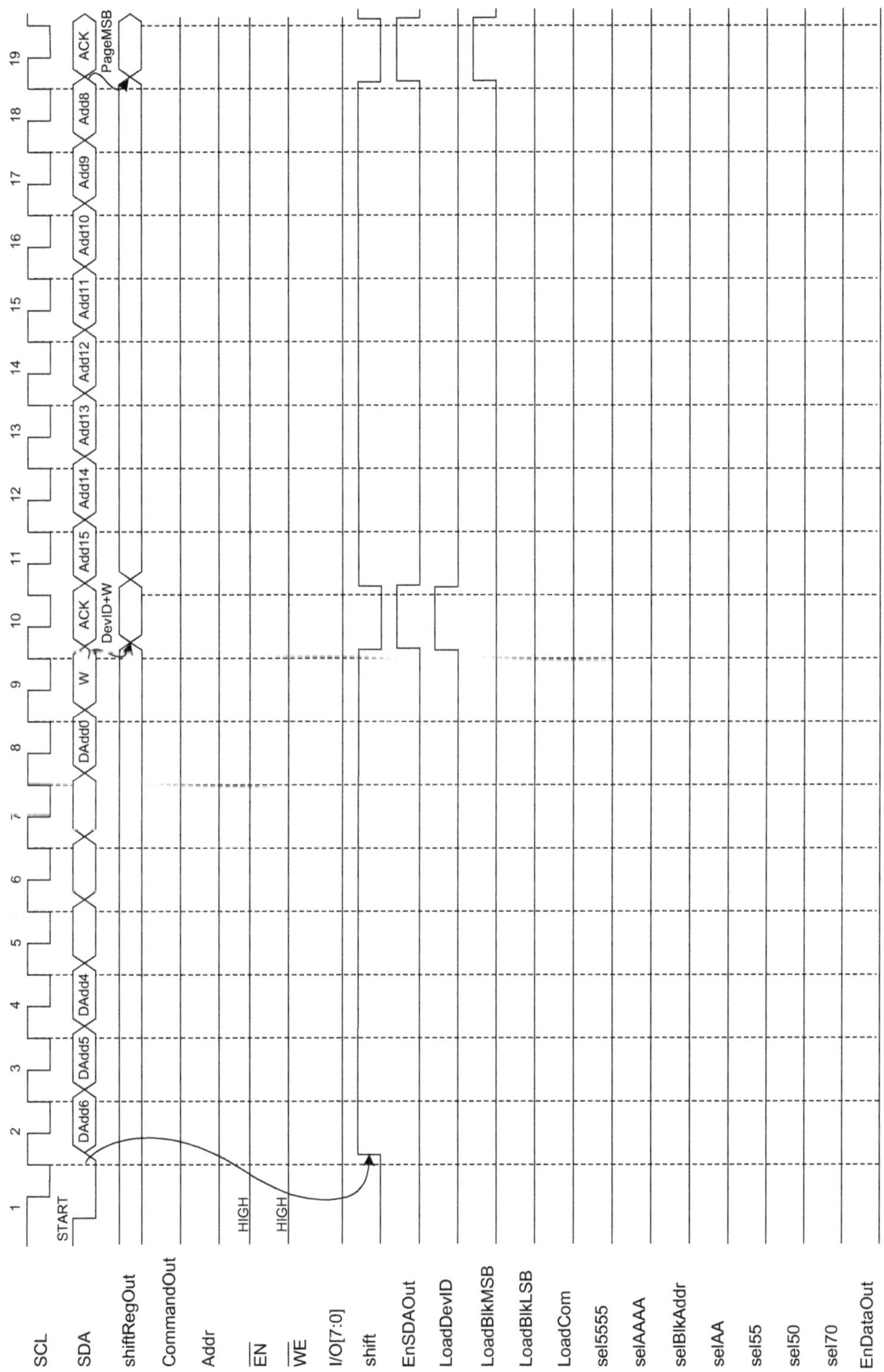

Fig. 5.83 I^2C page erase sequence with device ID and the LSB of page address

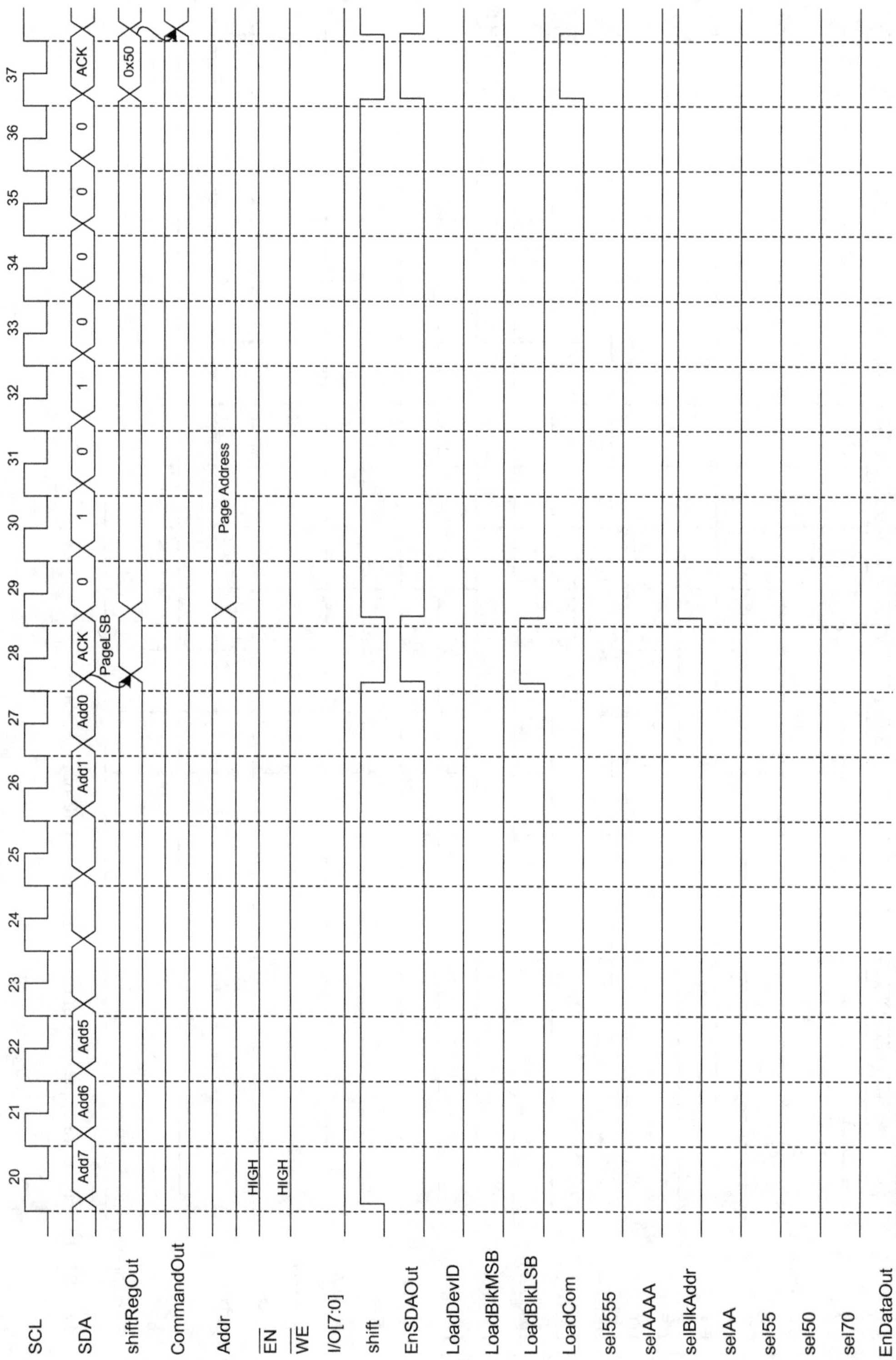

Fig. 5.84 I^2C page erase sequence with the MSB of page address and the erase command

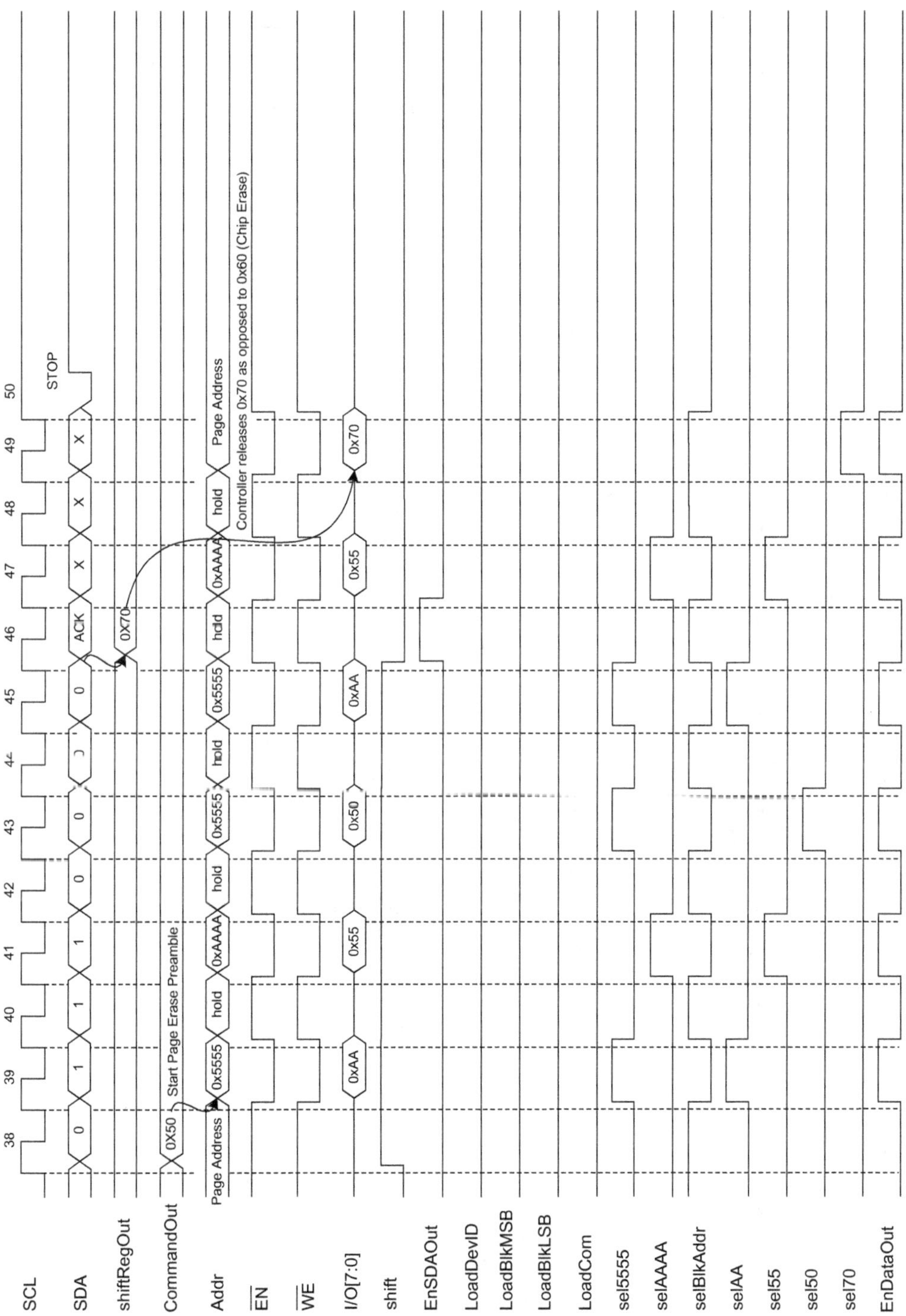

Fig. 5.85 I²C page erase preamble and the page erase command

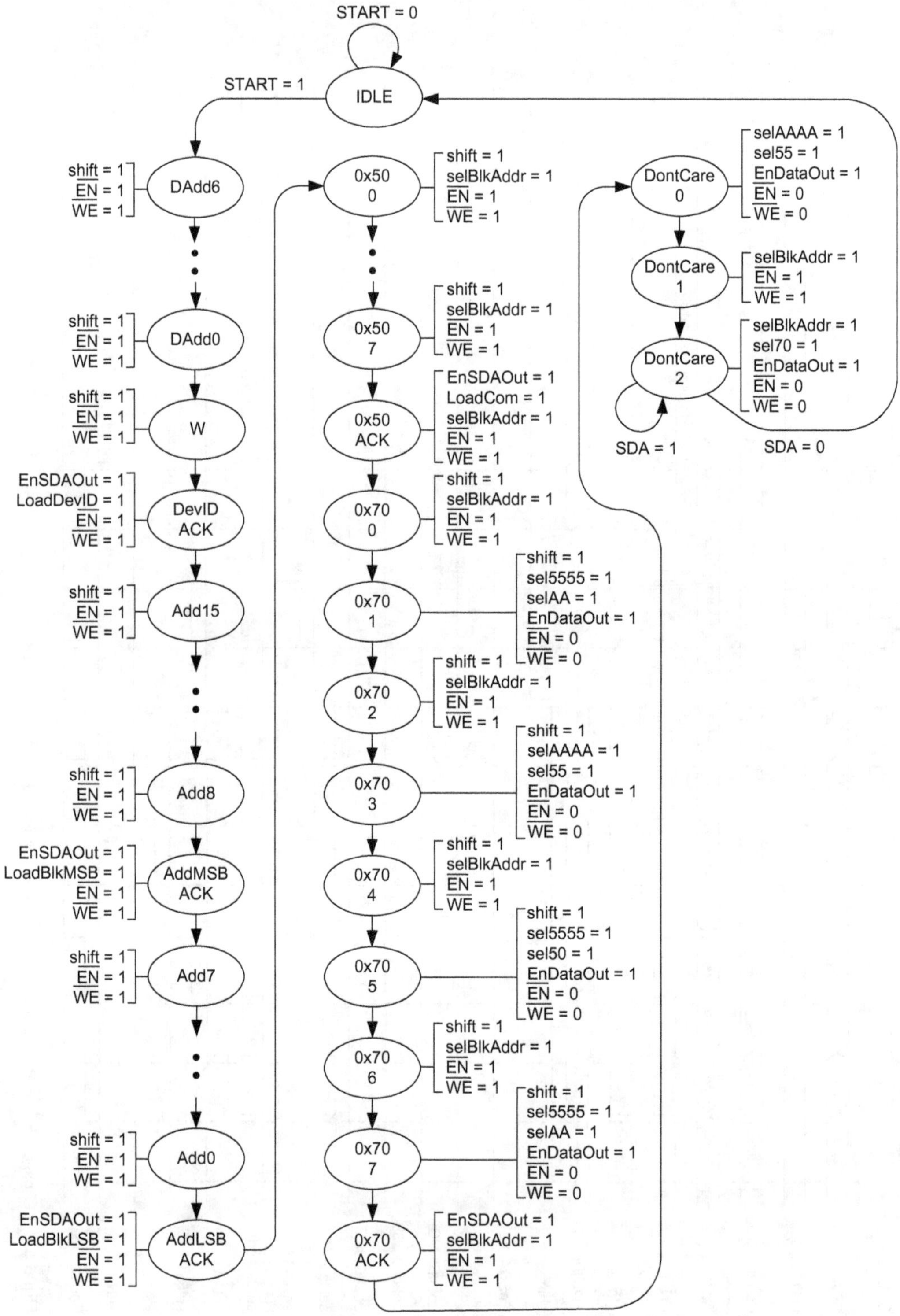

Fig. 5.86 I^{2}C page erase interface controller (only the essential control signals are indicated in each state to avoid complexity)

An architecture combining the read and fast write interfaces can be implemented by a data-path shown in Fig. 5.87. A shift register can be used to receive the incoming device address and command bit from the SDA bus, which is subsequently is stored in an auxiliary register as shown in this figure. The seven-bit device ID field can be used to activate one of the maximum 128 Flash memory chips. The command bit, R or W, is used to enable either the read interface or the fast write interface depending on its value. The Flash memory address residing in the shift register is then forwarded to the eight-bit shift register in either the read or the fast write interface to prepare the Flash memory for a data transfer.

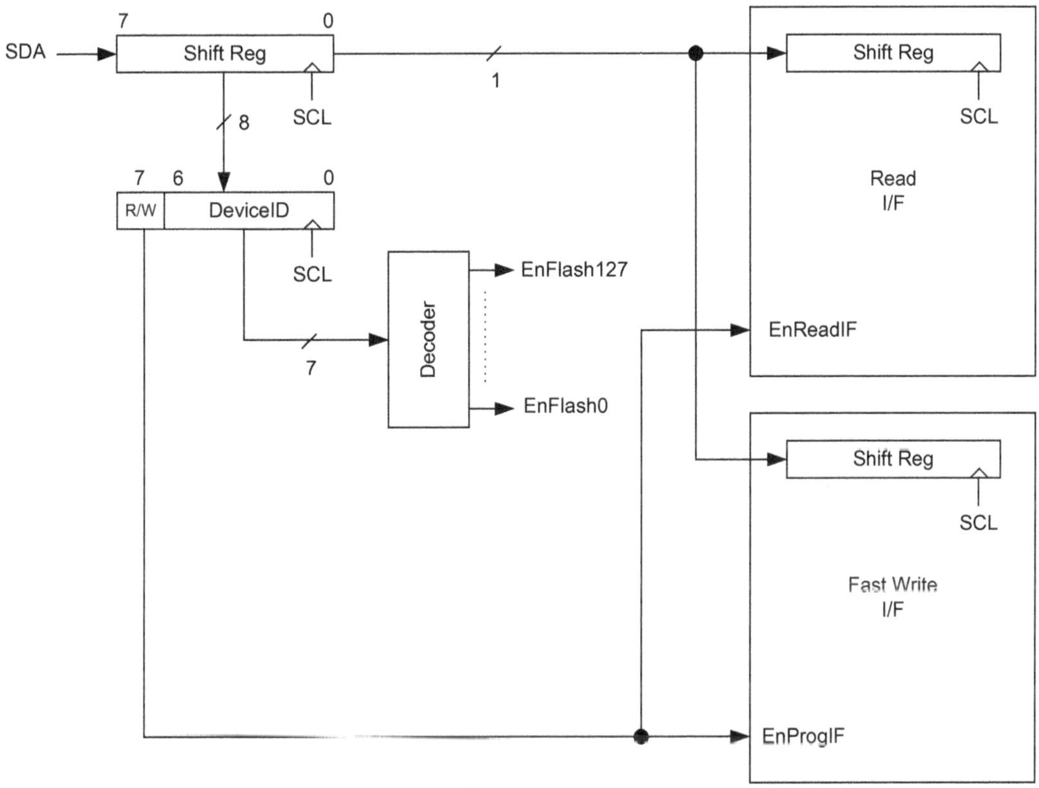

Fig. 5.87 I^2C read and fast write (program) interface topologies

The reader should be quite familiar with all three I^2C interface designs shown above to be able to integrate them in one interface to achieve a complete design.

5.5 Serial Flash Memory

Recent Flash memory chips already include I^2C or SPI interface to interact with a host processor or another bus master. The user does not have to deal with preambles, waiting periods or other complexities of the serial bus, but simply write an I^2C or SPI-compliant embedded program to initiate a read, write or erase operation with the Flash memory.

This section examines the operation of a typical Flash memory with an SPI interface. Figure 5.88 shows the basic internal architecture of the Flash memory where an external active-low Slave Select control signal, $\overline{SS}$, is applied to enable the memory. The clock is supplied through the SCK port. The serial data comes into the memory through Serial-Data-In (SDI) port and departs from Serial-Data-Out (SDO) port. Once a serial address is retrieved from the SPI bus, it is stored in the address register. The address decoder uses the contents of the address register to access the Flash memory core and read the data to an internal data buffer. The serial data is subsequently delivered to the bus master through the SDO port. If the operation is a write, the bus master sends serial data to the SDI port, which is then transferred to an internal data buffer, and subsequently to the memory core.

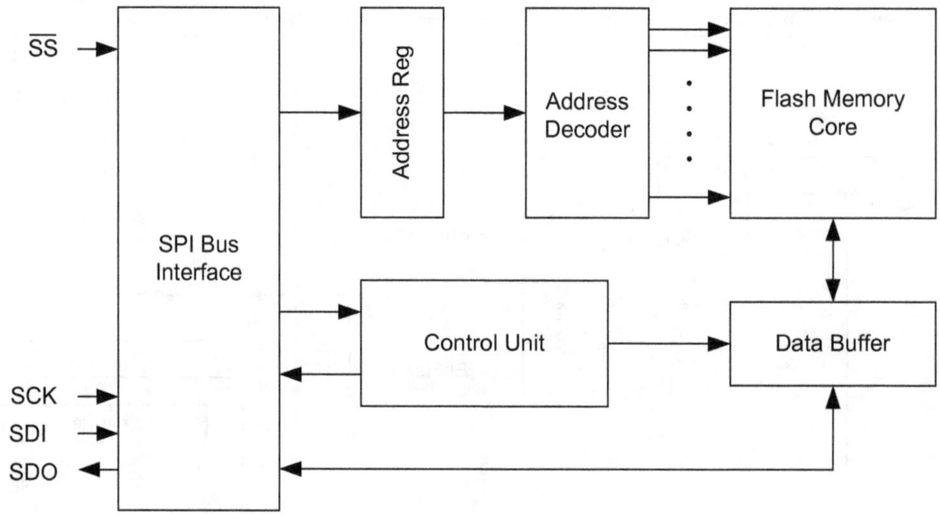

Fig. 5.88 Serial Flash memory architecture with SPI interface

Figure 5.89 describes a typical partitioning scheme of a 1 MB memory core that requires a 20-bit address for each byte of data. To be consistent with the E^2PROM and Flash memory organizations discussed in previous sections, the entire memory block in this example is initially divided into sixteen 64 KB sectors. Each sector is subdivided into sixteen 4 KB blocks, and each block is further subdivided into 16 pages. Each page contains 256 bytes, any of which is accessible through the SPI bus. Figure 5.90 shows the detailed address mapping of Block 0 in Sector 0 to further illustrate the internal memory organization.

Figure 5.91 shows nine basic modes of operation for this Flash memory. Some of the modes in this table are further divided into sub-modes according to the complexity of the main mode. For example, in the write (program) mode, the opcode 0x20 assumes to write between 1 and 255 bytes to the memory core while the opcode 0x23 writes 64 KB of data to a sector. Similarly, the erase mode can be configured to erase a page, a block or the entire chip. The protect operation prevents over-writing to a sector or the chip. The write enable feature is a security measure for the serial Flash

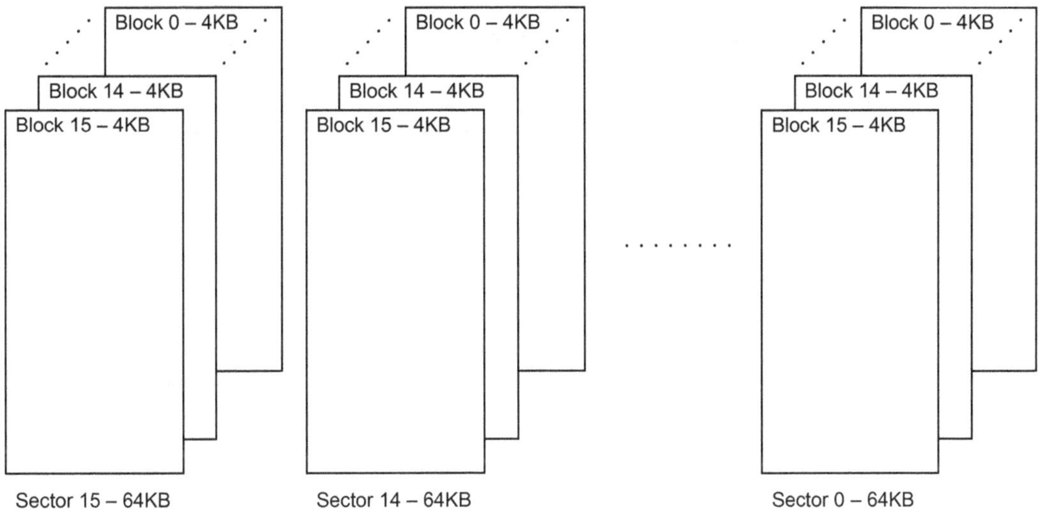

Fig. 5.89 A serial Flash core memory organization: 16 Sectors and 16 Blocks in each sector

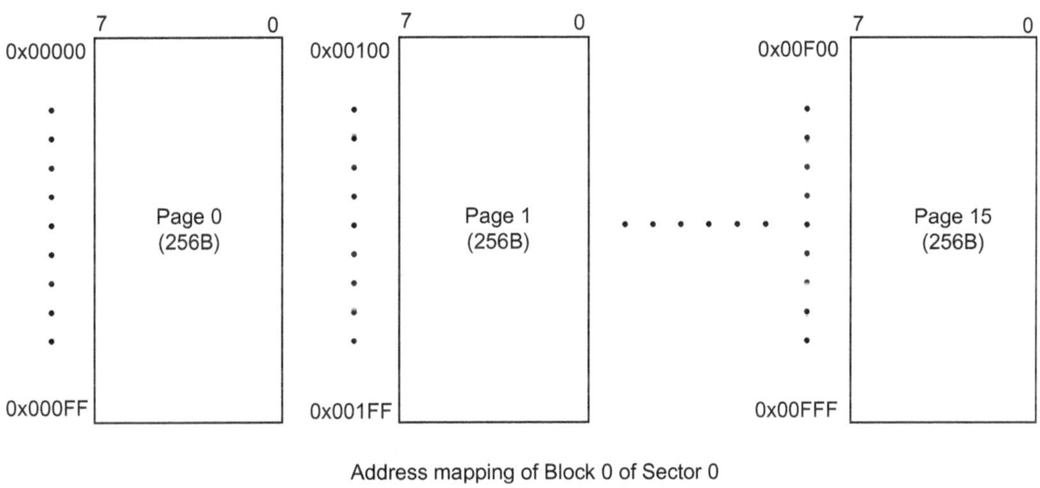

Fig. 5.90 Memory organization of Block 0 of Sector 0: 16 pages, 256 bytes per page

memory, and it is used prior to an actual write or an erase operation. Once the write enable command is issued, any kind of data alteration in the memory core becomes possible. Both the write enable and protect features are registered in the status register and can be read on demand. The status register also indicates whether the device is busy, such as in the middle of a write or read operation, write enable is engaged or not, and which sector is protected. The Flash memory can be placed into a long term hibernation mode to save power. The modes in Fig. 5.91 are at minimum compared to a typical serial Flash memory to emphasize only the primary modes of operation. The opcode value for each mode is also randomly selected. Actual serial Flash memory datasheets contain many more operational modes with different opcode values assigned to each mode.

FLASH MEMORY COMMANDS		OPCODES
Read		0x10
Write	Write Byte (1-255)	0x20
	Write Page	0x21
	Write Block	0x22
	Write Sector	0x23
Erase	Erase Page	0x30
	Erase Block	0x31
	Erase Chip	0x32
Protect/Unprotect	Protect Sector	0x40
	Unprotect Sector	0x41
	Protect Chip	0x42
	Unprotect Chip	0x43
Write Enable		0x55
Write Disable		0x66
Read Status Register		0x77
Hibernate		0x88
Wake up		0x99

Fig. 5.91 Main serial Flash memory commands

This particular Flash memory operates in both mode 0 (SCK is initially at logic 0) and mode 3 (SCK is initially at logic 1) of the SPI protocol. However, most of the timing diagrams in this section will refer to mode 0 when explaining different commands in Fig. 5.91.

Figures 5.92 and 5.93 explain the basic write protocols in mode 0 and mode 3, respectively. Once $\overline{SS}$ signal is lowered to logic 0, data bits at the SDI port can be written to the Flash memory's data buffer at the positive edge of SCK. The data transaction stops when $\overline{SS}$ is raised to logic 1. The entire data buffer is subsequently transferred to the memory core within a write period of t_{WRITE}.

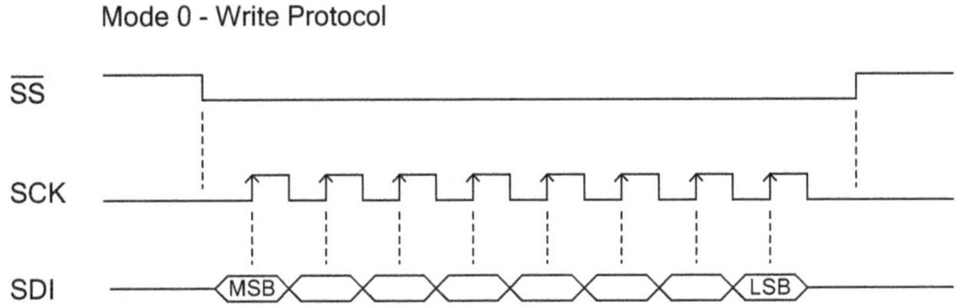

Fig. 5.92 Serial Flash memory mode 0 SPI write (program) protocol

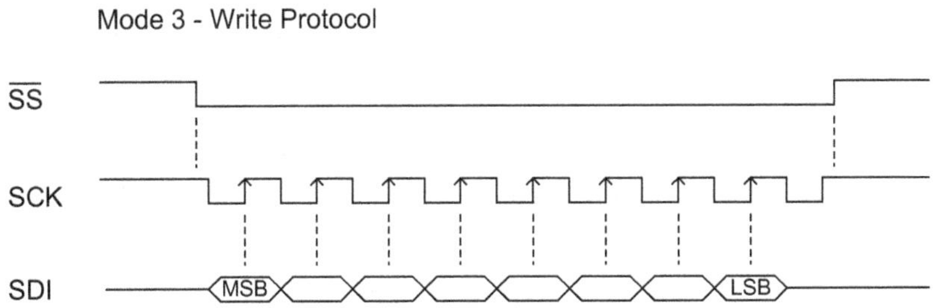

Fig. 5.93 Serial Flash memory mode 3 SPI write (program) protocol

Similarly, Figs. 5.94 and 5.95 explain the basic read protocols in mode 0 and mode 3, respectively. When $\overline{SS}$ signal is at logic 0, data is delivered from the memory core to the data buffer, and then from the data buffer to SDO terminal at the negative edge of each SCK cycle. The memory access is equal to t_{READ} with respect to the negative edge of SCK. When $\overline{SS}$ signal is raised to logic 1, SCK is no longer allowed to change, and the read process terminates.

In both write and read operations, the most significant data bit is delivered first and the least significant data bit is delivered last.

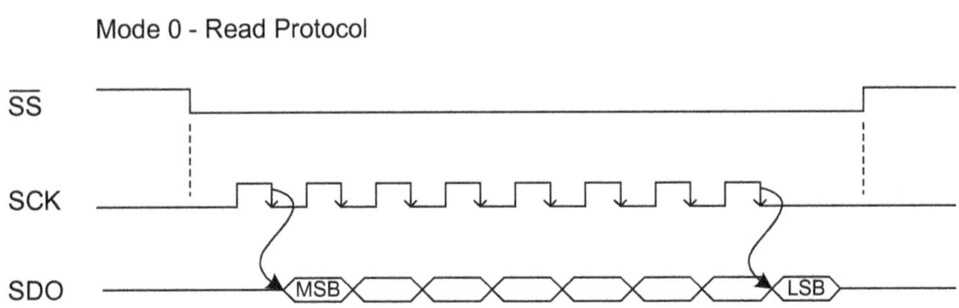

Fig. 5.94 Serial Flash memory mode 0 SPI read protocol

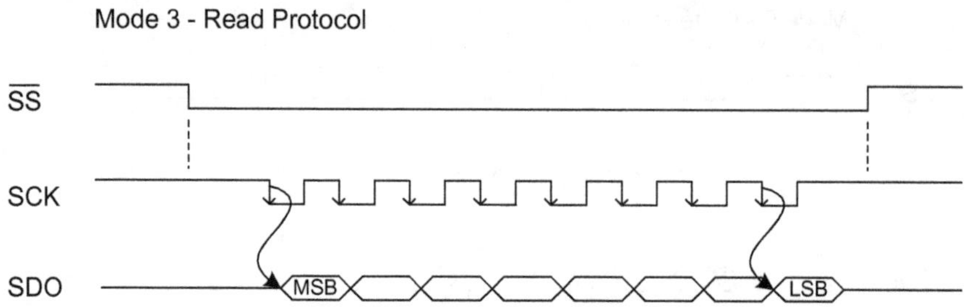

Fig. 5.95 Serial Flash memory mode 3 SPI read protocol

A typical Flash memory byte read is shown in Fig. 5.96. The process starts with sending the opcode, 0x10, corresponding to a read operation according to the table in Fig. 5.91. A 20-bit address follows the opcode with the most significant address bit, A19, first, and the least significant address bit, A0, last. The first data bit, D7 (also the most significant bit of data), is delivered to the SDO terminal at the negative edge of SCK according to Fig. 5.96. The remaining seven data bits of the data are sequentially delivered at each negative edge of SCK until SCK signal stabilizes at logic 0 and $\overline{SS}$ transitions to logic 1.

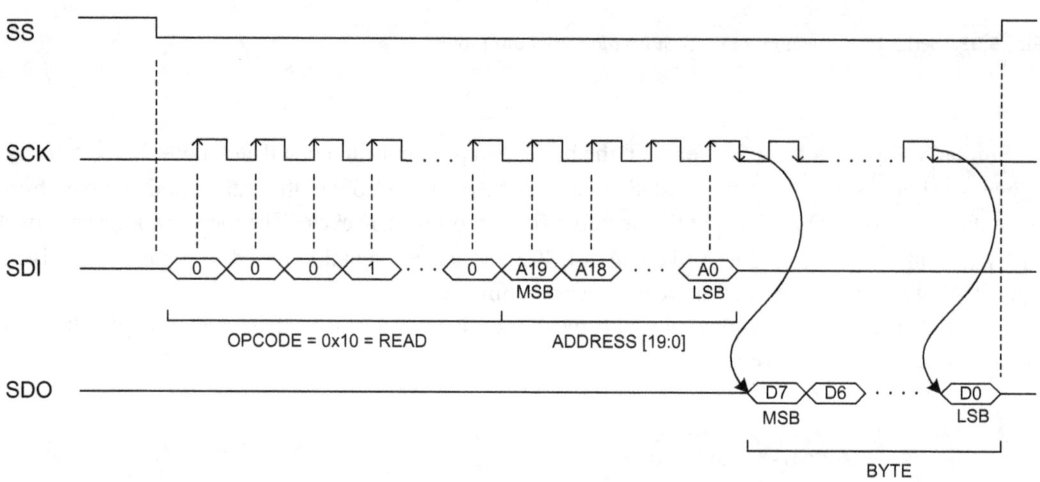

Fig. 5.96 Serial Flash memory byte read in mode 0

Once a starting 20-bit address is issued, a number of bytes, ranging from one byte to the contents of the entire memory (1 MB), can be read from the SDO terminal as long as the $\overline{SS}$ signal is kept at logic 0, and the SCK activity is present. Terminating SCK and raising $\overline{SS}$ to logic 1 ceases the read process as shown in Fig. 5.97.

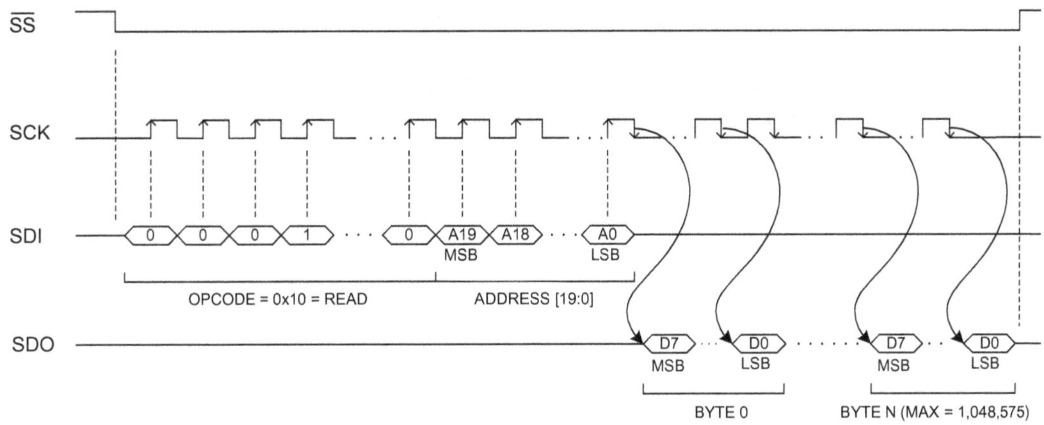

Fig. 5.97 Serial Flash memory in a burst of byte read in mode 0

The Flash memory write (program) mode has four sub modes. In the byte write mode, bytes ranging between 1 and 255 can be written to a page following the opcode, 0x20, and a 20-bit memory address as shown in Fig. 5.98. After data is written to the last address of the page, subsequent bytes at SDI terminal are considered invalid and will be ignored even though there may still be SCK activity and/or $\overline{\text{SS}}$ may still be at logic 0. In some serial Flash memory chips, excess data is not ignored but written to the memory core starting from the first address of the page (address looping).

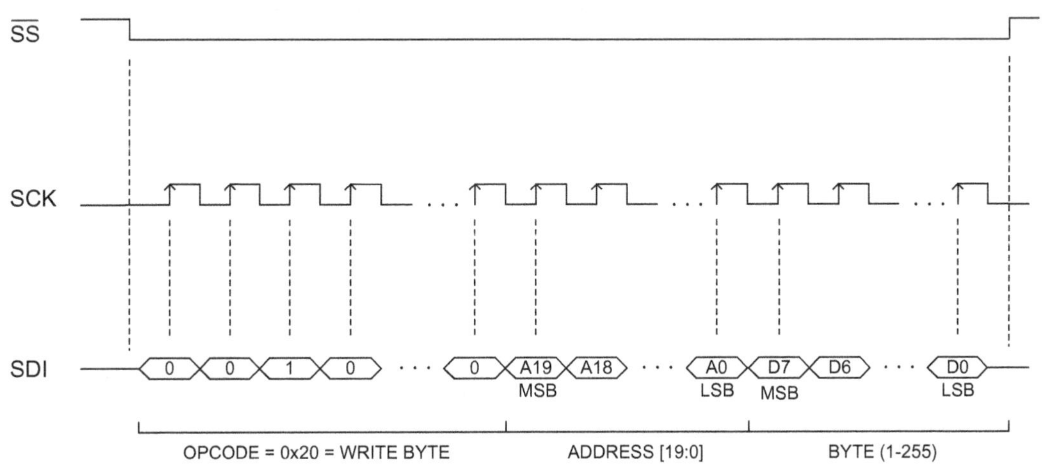

Fig. 5.98 Serial Flash memory byte write in a burst (1 to 255 bytes) in mode 0

Figure 5.99 describes the page write mode. After issuing the write page opcode, 0x21, and a 20-bit page address, 256 bytes of data are sequentially written to the memory core starting from the top of the page. Any data beyond 256 bytes will be ignored by the device. It is vital that the 20-bit starting address aligns with the first address of the page. For example, if page 0 of block 0 in sector 0 needs to be accessed to write data, the starting address has to be 0x00000 according to Fig. 5.90. Similarly, the starting address has to be 0x00100 for page 1 or 0x00F00 for page 15 if the contents of either page need to be written.

Writing to a block or a sector is not any different from writing to a page. In both instances, the starting 20-bit address needs to align with the topmost address of the block or the sector. For example, writing a 4 KB of data to block 0 of sector 0 requires the starting address to be 0x00000. Similarly, the starting address of block 1 of sector 0 is 0x01000 if a 4 KB data needs to be written to this block.

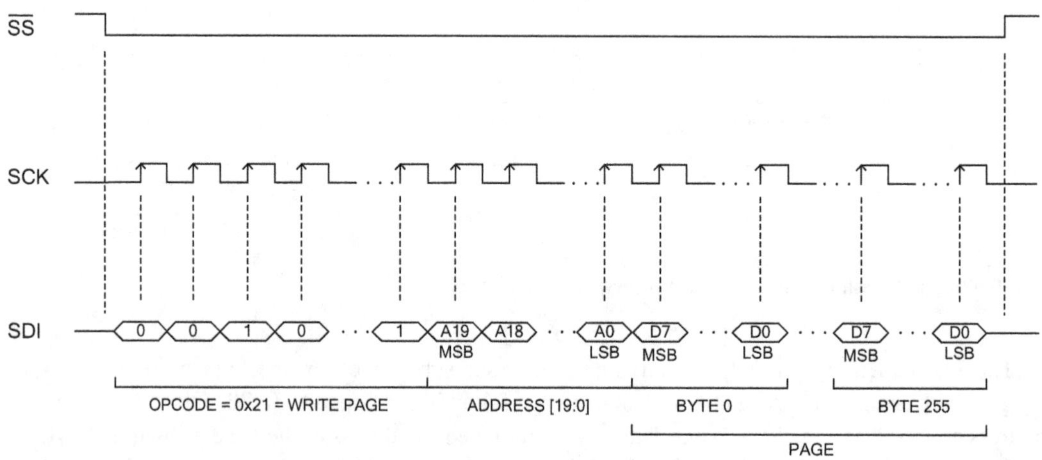

Fig. 5.99 Serial Flash memory page write in mode 0

Erase can be performed on a page, a block or the entire chip according to the table in Fig. 5.91. The page erase requires the opcode, 0x30, followed by the topmost address of the page as shown in Fig. 5.100. Erasing the entire chip only requires the opcode, 0x32, as shown in Fig. 5.101.

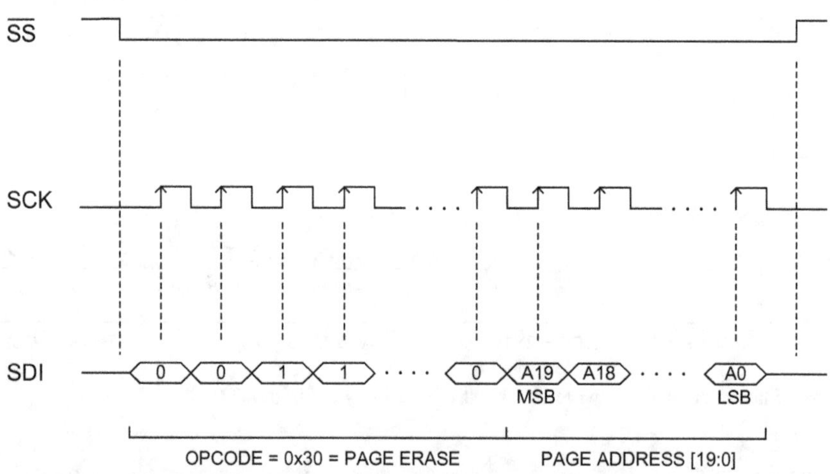

Fig. 5.100 Serial Flash memory page erase in mode 0

Accidentally altering the contents of the Flash memory is a non-reversible process. Therefore, many manufacturers formulate a security measure, such as a write enable command, prior to a write or an erase operation. The write enable command requires issuing an opcode, 0x55, according to Fig. 5.91, and it is implemented in Fig. 5.102. This code changes the write enable bit in the status register which then enables the Flash memory for write or erase. For example, in Fig. 5.103 the write enable opcode, 0x55, is issued prior to the write byte opcode, 0x20, to allow any number of bytes to be written to a page. If the write enable opcode is omitted prior to a byte, a page, a block or a sector write, the data delivered to the memory core becomes invalid and is ignored.

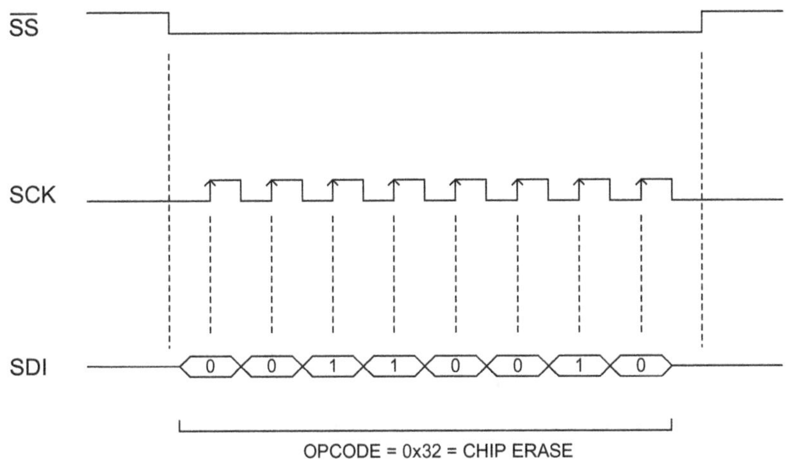

Fig. 5.101 Serial Flash memory chip erase in mode 0

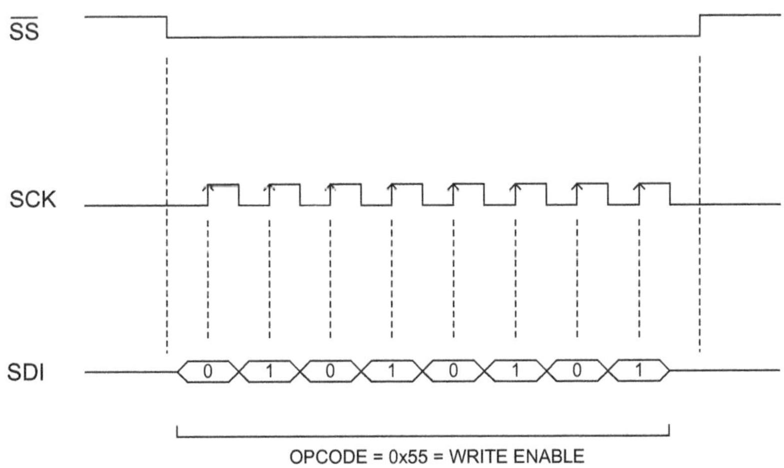

Fig. 5.102 Serial Flash memory write enable operation in mode 0

Protecting a sector or the entire chip is also a vital security measure for the Flash memory. For example, if a Flash memory contains BIOS data in specific sectors, accidentally accessing these sectors for write or erase becomes fatal. Therefore, such accesses need to be prevented at all costs. The opcode, 0x40, is issued with a specific sector address to protect the data in this sector as shown in Fig. 5.104. However, as with the write and erase modes, the write enable opcode, 0x55, must accompany the sector protect opcode, 0x40, to make the sector protect a valid entry as shown in the timing diagram in Fig. 5.105.

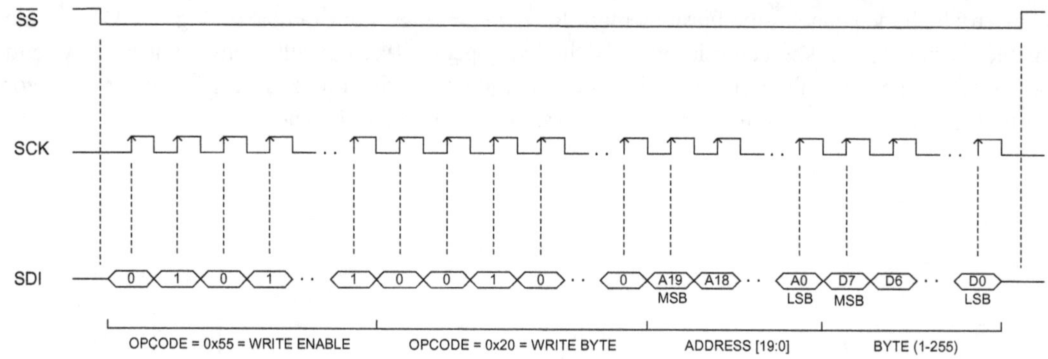

Fig. 5.103 Serial Flash memory write (program) burst (1 to 255 bytes) followed by write enable

The user may reverse the write enable status of the device by issuing a write disable command, 0x66, as shown in the timing diagram in Fig. 5.106.

Status register constitutes an important part of Flash memory programming. For this particular Flash memory, there are four entries in the status register that contain vital operational information as shown in Fig. 5.107. The SP0, SP1 and SP2 bits identify which sectors are protected. The Write Enable Latch bit, WEL, signifies if the device has already been write-enabled or not. The Write-In-Progress, WIP, bit defines if the device is busy with a write process.

The user can access the contents of the status register at any time by issuing the read status register command, 0x77, as shown in Fig. 5.108. After executing this command, the contents of the status register bits become available at the SDO port.

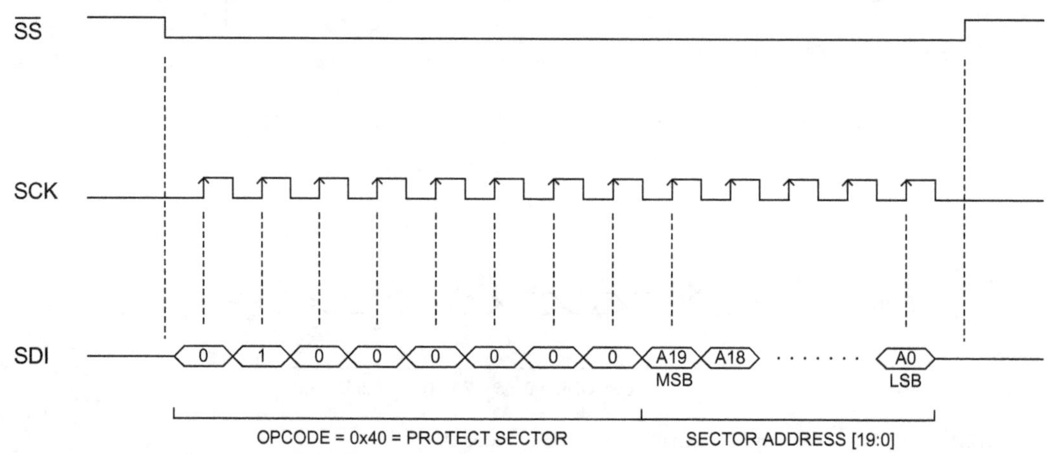

Fig. 5.104 Serial Flash memory protect sector operation in mode 0

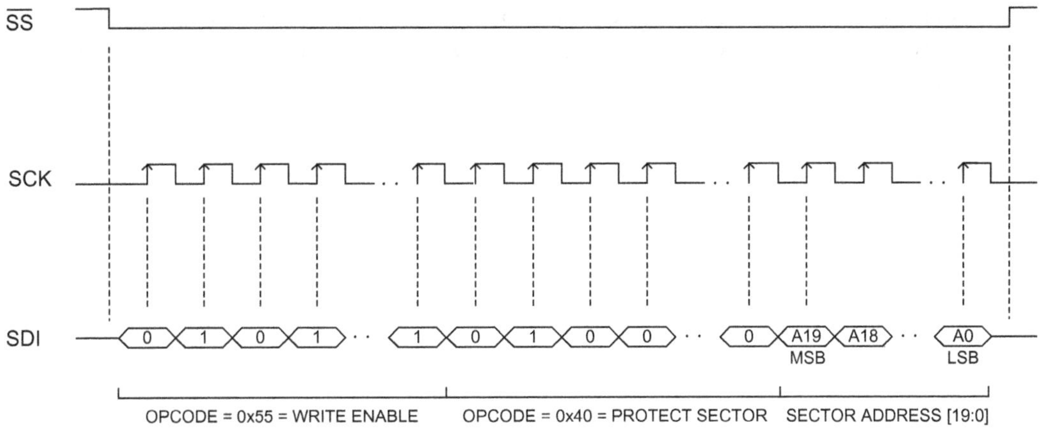

Fig. 5.105 Serial Flash memory write enable operation followed by protect sector in mode 0

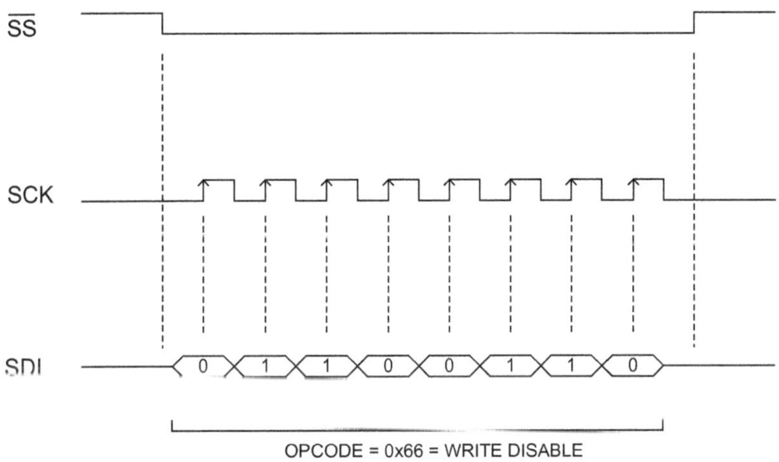

Fig. 5.106 Serial Flash memory write disable operation in mode 0

7	6	5	4	3	2	1	0
Reserved			SP2	SP1	SP0	WEL	WIP

WIP = Write In Progress = 1 Device is busy with write
0 Device is not busy with write

WEL = Write Enable Status = 1 Write Enable is active
0 Write Enable is inactive

SP2	SP1	SP0	
0	0	0	No sector is protected
0	0	1	Address 0x00000 to 0x0FFFF is protected
0	1	0	Address 0x00000 to 0x1FFFF is protected
0	1	1	Address 0x00000 to 0x3FFFF is protected
1	0	0	Address 0x00000 to 0x7FFFF is protected
1	0	1	Address 0x00000 to 0xFFFFF is protected
1	1	0	Address 0x00000 to 0xFFFFF is protected
1	1	1	Address 0x00000 to 0xFFFFF is protected

Fig. 5.107 Serial Flash memory status register

The user can also place the Flash memory into the sleep mode to conserve power by issuing hibernate opcode, 0x88, as shown in Fig. 5.109. The hibernation mode can be reversed by issuing the wake-up opcode, 0x99, as shown in Fig. 5.91.

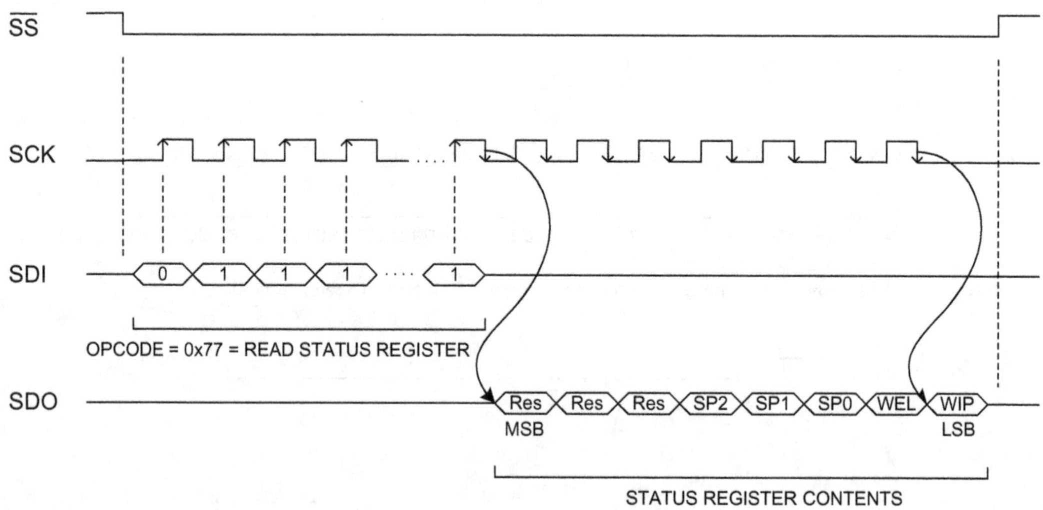

Fig. 5.108 Serial Flash memory status register read operation in mode 0

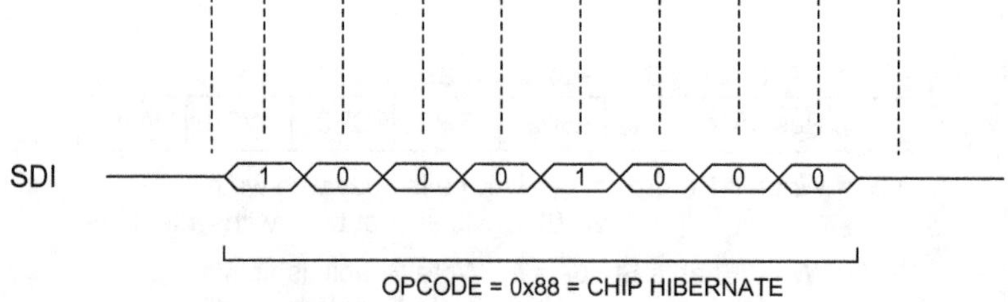

Fig. 5.109 Serial Flash memory chip hibernate operation in mode 0

Review Questions

1. An SDRAM is composed of two16-bit wide banks, bank 0 and bank 1, as shown below.

SDRAM Bank0 address	15 … 0	SDRAM Bank1 address	15 … 0
	⋮		⋮
0xAB00	0000	0xCD00	AA00
0xAB01	1111	0xCD01	BB11
0xAB02	2222	0xCD02	CC22
0xAB03	3333	0xCD03	DD33
0xAB04	4444	0xCD04	EE44
0xAB05	5555	0xCD05	FF55
0xAB06	6666	0xCD06	AA11
0xAB07	7777	0xCD07	BB22
0xAB08	8888	0xCD08	CC33
0xAB09	9999	0xCD09	DD44
0xAB0A	AAAA	0xCD0A	EE55
	⋮		⋮

The truth table below defines precharge, activate and read cycles.

Operation	$\overline{CS}$	$\overline{RAS}$	$\overline{CAS}$	$\overline{WE}$
Precharge	0	0	1	0
Activate	0	0	1	1
Read	0	1	0	1

Each 16-bit SDRAM address is composed of two parts: the most significant byte corresponds to the row address, and the least significant byte to the column address as shown below.

Address = {Row Address, Column Address}

The wait period between the precharge and activate commands is one clock cycle. Similarly, the wait period between the activate and read commands is also one cycle. The read burst from the specified address starts after a latency of two cycles. The waiting period between the last data packet and the next precharge command is also one cycle if the read repeats from the same bank.

(a) Show two read sequences in sequential addressing mode from Bank 0. Each burst contains four data packets: the first burst from address 0xAB03 and the next from address 0xAB07.

(b) Show the two read sequences in sequential addressing mode from different banks with no delay in between. Each burst contains four data packets: the first burst from Bank 0 with the starting address 0xAB03 and the next from Bank 1 with the address 0xCD06.

2. A Flash memory is composed of two byte-addressable sectors. It has an eight-bit bidirectional I/O port for reading and writing data, and a 16-bit unidirectional address port. The upper eight bits of the address field are allocated for the sector address and the lower eight bits for the program address. The three active-low inputs, $\overline{EN}$, $\overline{RE}$ and $\overline{WE}$, control the Flash memory according to the following chart:

Operations	$\overline{EN}$	$\overline{RE}$	$\overline{WE}$
Read	0	0	1
Write	0	1	0
Standby	0	1	1
Off	1	x	x

The initial data contents in this memory are shown below:

Sector 0

15	0	
0xEE	0xFF	0x00
0xCC	0xDD	0x02
0xAA	0xBB	0x04
0x88	0x99	0x06

Sector 1

15	0	
0x34	0x12	0x00
0x78	0x56	0x02
0xBC	0x9A	0x04
0xF0	0xDE	0x06

The Flash command chart is as follows:

		Cycle 1		Cycle 2		Cycle 3	
		Addr	Data	Addr	Data	Addr	Data
Sector protect	word	0x5555	0xAA	0xAAAA	0x55	Sector Addr	0x01
Fast write	word	0x5555	0xAA	0xAAAA	0x02	Write Addr	Write Data
	byte	0xAAAA		0x5555		Write Addr	Write Data
Read	word	0x5555	0xAA	0xAAAA	0x03	Read Addr	Read Data
	byte	0xAAAA		0x5555		Read Addr	Read Data

In this diagram, the sector protect is a three-cycle sequence where the sector protect code, 0x01, is provided in the third cycle along with the sector address.

Both the fast write and the read processes are initially three cycles. However, once the process starts, additional reads or writes are reduced to two-cycle operations as shown in the flow chart below.

According to this chart, for each additional data to be read or written, the command code must be employed in the second cycle, and the address/data combination in the third cycle.

The flow chart for fast write and read is as follows:

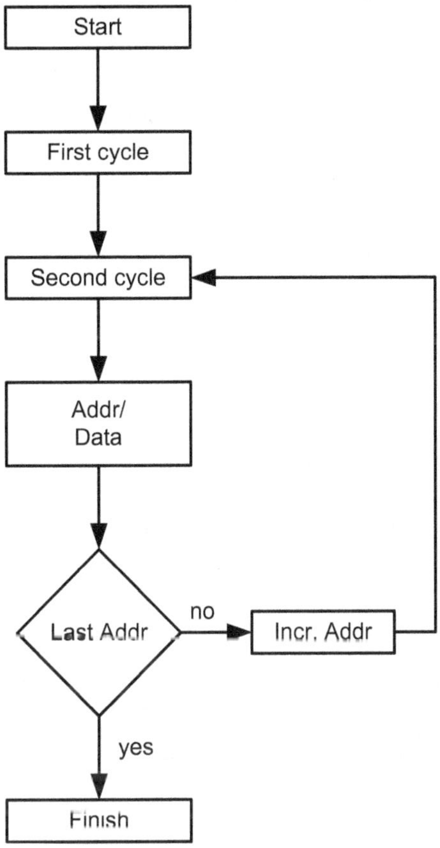

(a) Protect sector 1. Show the timing diagram with control inputs, address and data.
(b) Fast write to sector 0 with four bytes of data, 0x11, 0x22, 0x33, 0x44, starting from address 0x04 and incrementing the address to write each byte of data. Show the timing diagram with control inputs, address and data.
(c) Read four bytes from sector 0 at addresses 0x00, 0x02, 0x04 and 0x06. Show the timing diagram with control inputs, address and data.

3. Two reads can be accomplished from a 16-bit wide SDRAM by a single CPU instruction.

SDRAM is organized by four banks with data shown below.

Bank0 address	Bank0 [15:0]	Bank1 address	Bank1 [15:0]	Bank2 address	Bank2 [15:0]	Bank3 address	Bank3 [15:0]
	⋮		⋮		⋮		⋮
AA00	1111	BB00	AAAA	CC00	FFFF	DD00	8888
AA01	2222	BB01	BBBB	CC01	EEEE	DD01	7777
AA02	3333	BB02	CCCC	CC02	DDDD	DD02	6666
AA03	4444	BB03	DDDD	CC03	CCCC	DD03	5555
AA04	5555	BB04	EEEE	CC04	BBBB	DD04	4444
AA05	6666	BB05	FFFF	CC05	AAAA	DD05	3333
AA06	7777	BB06	0000	CC06	1111	DD06	2222
AA07	8888	BB07	1111	CC07	0000	DD07	1111
	⋮		⋮		⋮		⋮

Each SDRAM address is composed of an eight-bit wide row address, RA[7:0], and an eight-bit wide column address, CA[7:0], as in the following format:

SDRAM Address = {RA[7:0], CA[7:0]} where the row address occupies higher bits.

To control SDRAM, the following controls are available:

Operation	$\overline{CS}$	$\overline{RAS}$	$\overline{CAS}$	$\overline{WE}$
Precharge	0	0	1	0
Activate	0	0	1	1
Read	0	1	0	1

The wait period between the precharge and activate commands is one clock cycle long. Similarly, the wait period between the activate and read commands is also one clock cycle. The precharge command for the next read operation can be issued one clock cycle after the last data is read out from SDRAM. BS[1:0] = 0 selects Bank0, BS[1:0] = 1 selects Bank1, BS[1:0] = 2 selects Bank2, and BS [1:0] = 3 selects Bank3 in the timing diagrams.

(a) Assuming that the mode register is pre-programmed with sequential mode addressing, a burst length of four and a CAS latency of two, construct a timing diagram to show the two reads from SDRAM addresses, 0xAA00 and 0xAA04. Start from the precharge cycle to accomplish each read.

(b) With the same mode register contents in part (a), construct a timing diagram such that the two reads from the SDRAM addresses, 0xAA00 and 0xBB02, take place in the shortest possible time. Again start from the precharge cycle to accomplish each read. In this part, the bus master can extend the wait period more than one clock cycle between the precharge, activate and read commands in such a way that no two commands overlap with each other.

(c) With the same mode register contents in part (a), accomplish one read from the SDRAM address 0xCC00 with a burst length of two, and one read from the SDRAM address 0xDD02 with a burst length of eight. Start from the precharge cycle to accomplish each read.

4. Subsequent write and read operations are performed on an SDRAM that consists of two banks. Both banks have eight-bit wide I/O data ports.

The first SDRAM operation is a write operation that writes 0x11 to a starting address of 0xAB in bank 0. This is followed by writing data values, 0xEE, 0x00 and 0xFF, to bank 0 in sequential mode.

The read operation takes place from bank 1 without any interruption. This means the first read data is delivered to the data bus immediately after the last write data, 0xFF. The first read address is defined as 0x12. Four data packets are read from this starting address in sequential mode with a latency of two cycles.

Both write and read operations require $t_{PRE} = t_{CAS} = 1$ cycle.

Construct a timing diagram with control, address and data values to achieve these two consecutive operations. Assume all initial data values in Bank 0 are 0x00. Make sure to mark each precharge, activate, write and read cycle on the timing diagram. Indicate where latency happens.

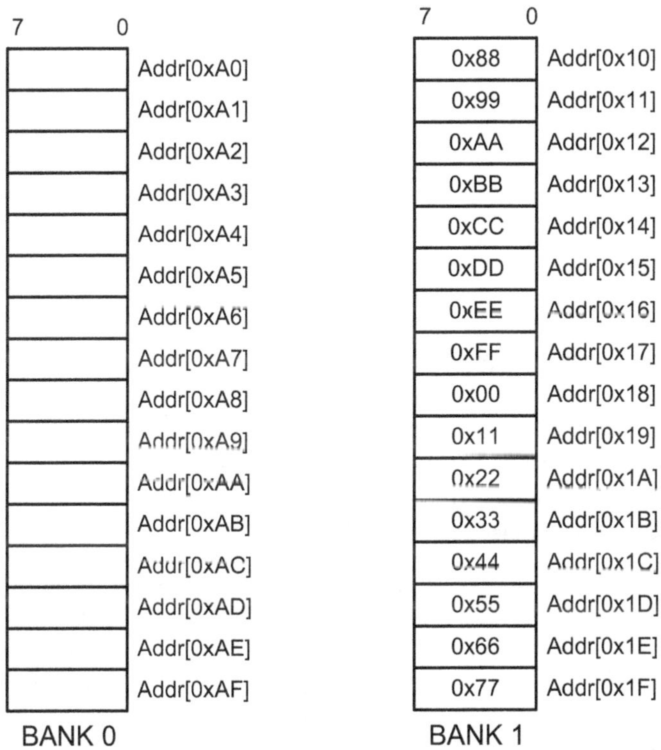

5. An E^2PROM memory is organized in four sectors. There are eight columns in each sector but no pages. The existing data in this memory is shown below.

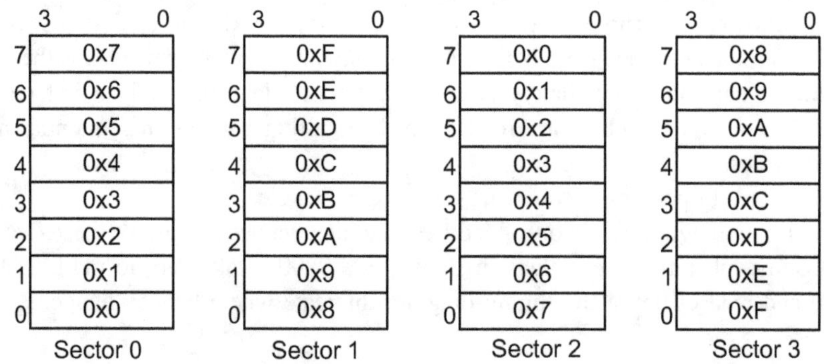

The command truth table is given below.

Function	Command code
SR Read	0x0
Read	0x1
Write buffer	0x2
Write core	0x3

The memory has five control pins:
$\overline{EN}$ is an active-low signal that activates the sector
AE is an active-high signal that accepts address
CE is an active-high signal that enables command function
$\overline{WE}$ is an active-low signal that enables write
$\overline{RE}$ is an active-low signal that enables read

Writing to the memory takes place at the rising edge of $\overline{WE}$. At the falling edge of $\overline{RE}$, reads take place from the memory. The sequence of write starts with the command function followed by an address and then data. A read sequence follows a similar fashion: it starts with the read command, then an address and data. Assume all AE, $\overline{WE}$ and $\overline{RE}$ set-up times are 0 s. The setup and hold times for command, address and data are all different from 0 s. It takes t_{WRITE} to transfer data from the buffer to the memory core.

(a) Draw a timing diagram to read data from column address = 0 and sector address = 3.

(b) Draw a timing diagram to write 0xA, 0xB, 0xC, 0xD, 0xE, 0xF, 0x7, 0x6 starting from the column address = 2 and the sector address = 2 in the following manner: the first data, 0xA, to column address 2; the second data, 0xB, to address 0x3 and so forth. Draw the contents of the memory after the write sequence is complete.

6. A Flash memory has an eight-bit address, and all reads and writes are achieved on an eight-bit bidirectional data bus. The Flash memory write sequence contains a preamble, a write command, and an address and data combination as shown in the flow chart below.

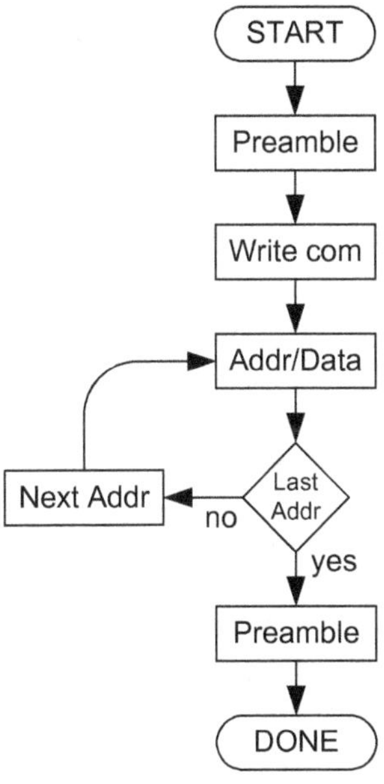

Once the write command is issued, address and data are generated continuously until the last write. The sequence ends with the same preamble that started the write.

In the read sequence, the bus master starts fetching data once the preamble and read command are issued. The sequence has the same exit preamble as shown below.

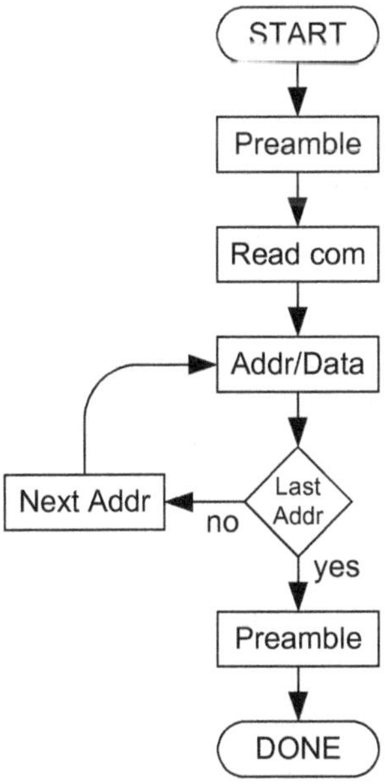

START and DONE do not have any significance in timing diagrams other than that they indicate the start and the end of the sequence, respectively.

The preamble, write and read commands are issued with hexadecimal values shown in the truth table below.

COMMANDS	Address	Data
Preamble	FF	00
Write com	AA	FF
Read com	BB	FF

The state of the Flash memory before any read or write operation is shown below. The leftmost column in this figure shows the Flash memory address in hexadecimal.

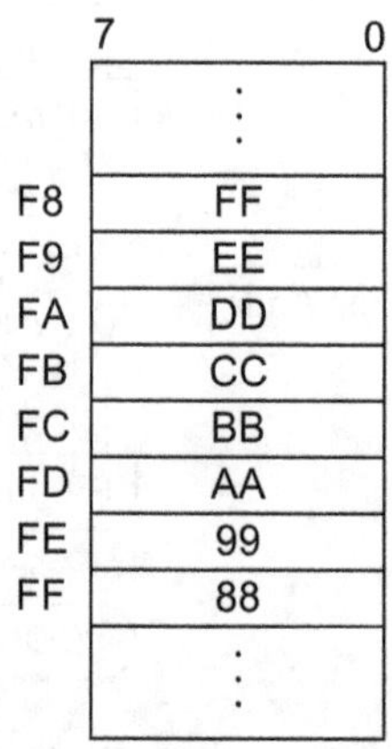

The bus master produces three data transmissions for the Flash memory. In the first transmission, four data packets below are written into the Flash memory.

Packet no	Address	Data
1	F8	00
2	F9	11
3	FA	22
4	FB	33

In the second transmission, the bus master reads two data packets from the following addresses below.

Packet no	Address	Data
1	FA	
2	FB	

In the third transmission, the bus master reads two more data packets from the following addresses.

Packet no	Address	Data
1	FE	
2	FF	

Construct a timing diagram with Address, $\overline{WE}$, $\overline{RE}$ and Data values. Note that the Flash memory requires a hold period which coincides with the high phase of $\overline{EN}$ signal. However, in the low phase, when the Flash memory is active, the device either writes or reads depending on the value of $\overline{WE}$ and $\overline{RE}$ signals, respectively.

7. A serial on-chip SPI bus described in Chap. 4 is used to program an SDRAM register file that consists of five registers (see the SDRAM bus interface architecture).

Assume that each register in the register file has an eight-bit long address. Data in each register is also assumed to be eight bits long.

Wait register receives the number of clock periods which is equivalent to t_{WAIT}, Latency register to t_{LAT}, Burst register to t_{BURST}, CAS register to t_{CAS}, and Precharge register to t_{PRE}.

The SPI bus uses its SDI terminal and sends out an eight-bit address (starting with the most significant bit) followed by an eight-bit data (again starting with the most significant bit) at the positive edge of SCK until all five registers are programmed while $\overline{SS}$ is at logic 0. Once the programming is finished, $\overline{SS}$ node is pulled to logic 1.

Design the interface between the SPI bus and the register file. Make sure to show each I/O port of the SPI bus (such as SCK, SDI, $\overline{SS}$ etc.), the internal address, data and control signals of the interface on the timing diagram. The functionality of the interface must be identical between the timing diagram and the data-path.

Start building the timing diagram that includes only the address and the data. Then form the corresponding data-path that matches the timing diagram. Increase the complexity of the design by including the control signals in the timing diagram, governing the flow of data. Lastly, draw the state diagram of the Moore type controller for the interface.

Projects

1. Implement and verify the SRAM bus interface unit with the unidirectional bus designed in Chap. 4. Use Verilog as the hardware design language for the module implementation and functional verification.
2. Implement and verify the SDRAM bus interface unit with the unidirectional bus designed in Chap. 4. Use Verilog as the hardware design language for the module implementation and functional verification. Produce the hardware to program the SDRAM register file. Assume a serial bus such as SPI or I^2C to distribute the program data to the registers.
3. Implement and verify the I^2C fast write interface in the first design example in this chapter using Verilog.
4. Implement and verify the I^2C read interface in the second design example using Verilog.
5. Combine the read and the fast write interfaces into a single unit. Design and verify the complete interface using Verilog.

Note that for projects 3 through 5, write a behavioral Verilog code that mimics the bus master in order to send data on the I^2C bus.

References

1. Toshiba datasheet TC59S6416/08/04BFT/BFTL-80, -10 Synchronous Dynamic RAM
2. Toshiba datasheet TC58DVM72A1FT00/TC58DVM72F1FT00 128Mbit E^2PROM
3. Toshiba datasheet TC58256AFT 256Mbit E^2PROM
4. Toshiba datasheet TC58FVT004/B004FT-85, -10, -12 4MBit CMOS Flash memory
5. Toshiba datasheet TC58FVT400/B400F/FT-85, -10, -12 4MBit CMOS Flash memory
6. Toshiba datasheet TC58FVT641/B641FT/XB-70, -10 64MBit CMOS Flash memory
7. Atmel datasheet AT26DF161 16Mbit serial data Flash memory

Central Processing Unit

6

This chapter describes a basic Central Processing Unit (CPU), operating with a simplified Reduced Instruction Set (RISC). The chapter is divided into four parts.

In the first part, fixed-point instructions are described. This section first develops dedicated hardware (data-path) to execute a single RISC instruction, and then groups several instructions together in a set and designs a common data-path to be able to execute user programs that use this instruction set. In each step of the process, the instruction field is partitioned into several segments as the instruction flows through the data-path, and the necessary hardware is formed to execute the instruction and generate an output. In this section, fixed-point-related structural, data and program control hazards are described, and the methods how to prevent each type of hazard is explained.

The second part of this chapter is dedicated to the IEEE single and double-precision floating-point formats, leading to the simplified designs of floating-point adder and multiplier. These designs are then integrated with the fixed-point hardware to obtain a RISC CPU capable of executing both fixed-point and floating-point arithmetic instructions. In the same section, floating-point-related data hazards are described. A new floating-point architecture is proposed based on a simplified version of the Tomasula algorithm in order to reduce and eliminate these hazards.

In the third part, various techniques to increase the program execution efficiency are discussed. The trade-offs between static and dynamic pipelines, single-issue versus dual and triple-issue pipelines are explained with examples. Compiler enhancement techniques, such as loop unrolling and dynamic branch prediction methods, are illustrated to reduce overall CPU execution time.

The last section of this chapter explains different types of cache memory architectures, including direct-mapped, set-associative and fully-associative caches, their operation and the trade-off between each cache structure. The write-through and write-back mechanisms are discussed, and compared with each other, using various design examples.

6.1 Fixed-Point Unit

Instruction Formats

In a RISC CPU, all instructions include an Operation Code (OPC) field which instructs the processor what to do with the rest of the fields in the instruction, and when to activate different hardware components in the CPU to be able to execute the instruction. The OPC field is followed by one or more operand fields. Each field either corresponds to a register address in the Register File (RF) or contains immediate user data to process the instruction.

A. Bindal, *Fundamentals of Computer Architecture and Design*,
https://doi.org/10.1007/978-3-030-00223-7_6

There are three types of instructions in a RISC CPU: register-to-register-type, immediate-type and jump-type.

A register-to-register-type instruction contains an OPC followed by three operands: two source register addresses and one destination register address pointing the RF, namely RS1, RS2 and RD. The format of this instruction is shown below.

OPC RS1, RS2, RD

This type of instruction fetches the contents of the first and second source registers, Reg[RS1] and Reg[RS2], from the RF, processes them according to the OPC, and writes the result to the destination register, Reg[RD] in the RF. This operation is described below.

Reg[RS1] (OPC) Reg[RS2] $\rightarrow$ Reg[RD]

An immediate-type instruction contains an OPC followed by three operands: one source register address, RS, one destination register address, RD, and an immediate data as shown below.

OPC RS, RD, Imm Value

This type of instruction combines the contents of the source register, Reg[RS], with a sign-extended immediate value according to the OPC, and writes the result to the destination register, Reg[RD], in the RF. This operation is shown below.

Reg[RS] (OPC) Immediate Value $\rightarrow$ Reg[RD]

The jump-type instruction contains an OPC followed by a single immediate value shown below.

OPC Imm Value

This type of instruction uses the immediate field to modify the contents of the Program Counter (PC) for the instruction memory. The operation of this instruction is given below.

Immediate Value $\rightarrow$ PC

All three instruction types fit in a 32-bit wide instruction memory as shown Fig. 6.1. In this figure, the numbers on top of each field correspond to the bit positions of the instruction memory, defining the borders of the OPC or a particular operand field.

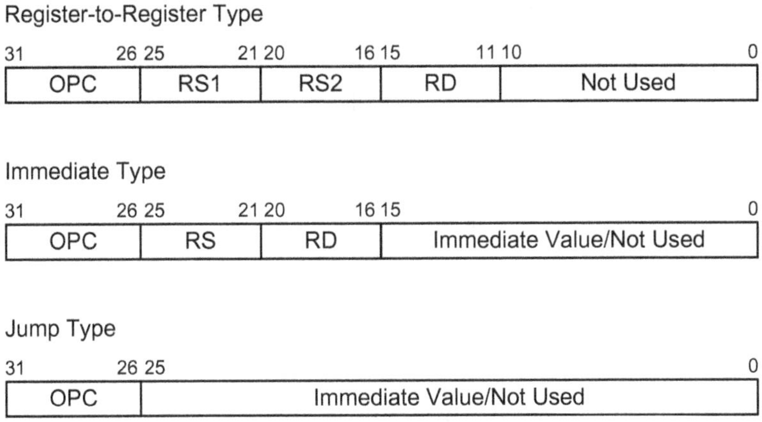

Fig. 6.1 Instruction field formats

CPU Data-Path

A modern RISC CPU is composed of an Arithmetic Logic Unit (ALU) to execute operands, a small memory in the form of registers (RF) to store temporary data, and two large size memories operating as instruction and data caches. A Program Counter (PC) generates the instruction memory address for each instruction in the instruction memory as shown in Fig. 6.2. Each instruction is fetched from this memory and separated into OPC and operand fields. The OPC field guides the data-flow through the rest of the CPU. The operand field contains either a number of RF addresses or the user data or the combination of the two. Once the source and destination RF addresses become available at the output of the instruction memory, the corresponding data is fetched from the RF and processed in the ALU according to the OPC. The processed data is consequently written back to the RF at a destination address. On the other hand, if a particular instruction needs to fetch data from the data memory instead of the RF, the ALU calculates the effective memory address of the data memory. When the data becomes available at the data memory output, it is routed back to a destination address in the RF. Sometimes, instructions contain user-defined immediate values. These are separated from the rest of the operand fields and combined with the contents of a source register, Reg[RS], in the ALU. The processed data is subsequently written back to the RF.

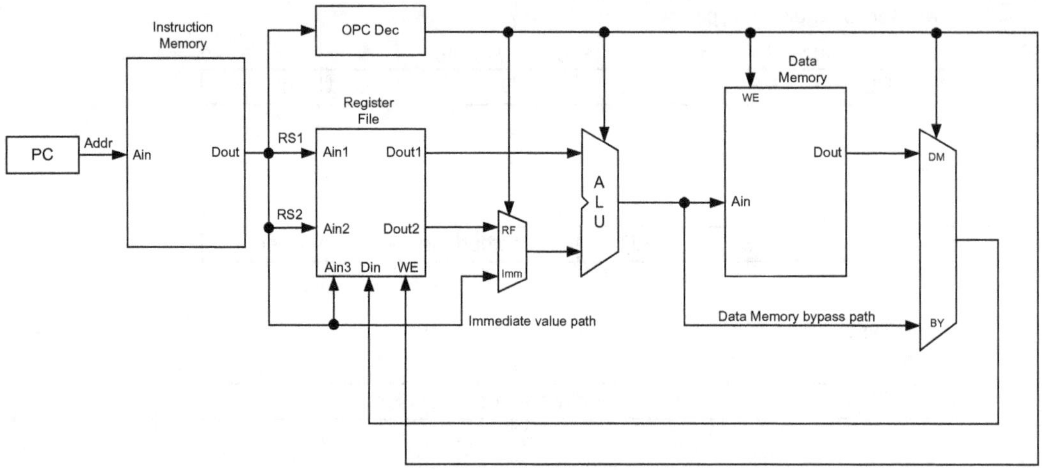

Fig. 6.2 A non-pipelined CPU

Instructions can be executed in RISC CPUs in two ways. In a non-pipelined CPU architecture, instructions are fetched from the instruction memory and processed through the remaining four stages of the CPU in Fig. 6.2 before the CPU takes the next instruction. This is shown in Fig. 6.3. In this figure, IF corresponds to the Instruction Fetch stage, RF to the Register File stage, A to the ALU stage, DM to the Data Memory stage, and finally WB to the Write-Back stage.

NON-PIPELINED	cycle 1	cycle 2	cycle 3
Instruction 1	IF RF A DM WB		
Instruction 2		IF RF A DM WB	
Instruction 3			IF RF A DM WB

Fig. 6.3 A non-pipelined CPU timing table

Non-pipelined structures are inefficient in terms of data throughput but require lower clock frequencies to operate. The CPU becomes more efficient in terms of throughput if the architecture in Fig. 6.2 is subdivided into smaller functional stages where each stage is separated from its neighboring stage by a flip-flop boundary that stores data (or address) only for one clock cycle as shown in Fig. 6.4. In this figure, the CPU data-path consists of five stages where individual tasks are executed in each stage within one clock cycle. According to this scheme, the clock frequency becomes five times higher compared to the architecture in Fig. 6.2.

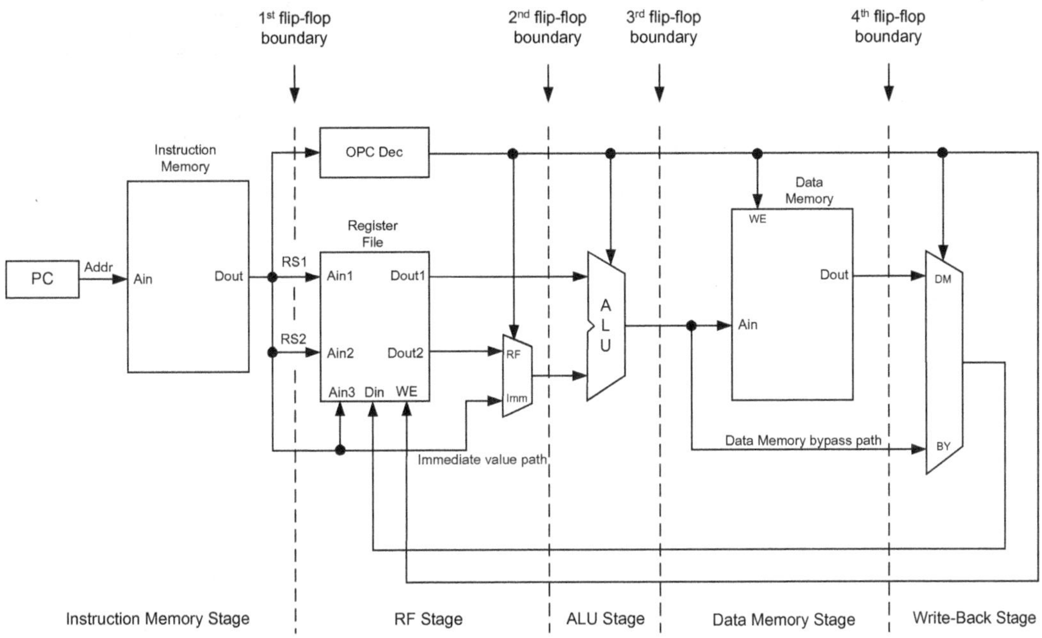

Fig. 6.4 A pipelined five-stage CPU

The first stage of this new pipeline in Fig. 6.4 is the instruction memory access as mentioned earlier. In this stage, each program instruction is fetched from the instruction memory and stored in the instruction register at the first flip-flop boundary. This architecture supports a word-addressable instruction memory, and therefore requires the PC to increment by one.

The next pipeline stage is the RF stage where the instruction OPC is separated from its operands. The OPC is decoded in order to generate control signals to route the address and data in the rest of the CPU. Operand fields are either source register addresses to access the data in the RF or immediate data supplied by the user as mentioned earlier. If the operand corresponds to an RF address, the data fetched from this address is loaded to the register that resides at the second flip-flop boundary. If the operand is an immediate data, it is sign extended to 32 bits before it is loaded to a register in the second flip-flop boundary.

The third stage of the CPU pipeline is the ALU stage. The data from the source registers in the RF or the immediate data are processed in this stage according to the OPC and loaded to the register at the third flip-flop boundary.

The fourth stage is the data memory stage. This stage accesses the data memory contents or bypasses the data memory completely. If the instruction calls for a load or a store operation, the ALU calculates the data memory address to access its contents. Otherwise, the ALU result simply bypasses the data memory to be stored at the fourth flip-flop boundary.

The last stage of the CPU pipeline is the write-back stage. In this stage, data is either routed from the output of the data memory or from the bypass path to a designated destination address in the RF.

A pipelined RISC CPU's efficiency and speed are shown in the timing table in Fig. 6.5. This figure extends to 15 high frequency clock cycles, which is the equivalent to three low frequency clock cycles in Fig. 6.3. The number of completed instructions in this new pipeline is almost 12, which is four times larger than the number of instructions in Fig. 6.3. The difference between non-pipelined and pipelined CPU efficiency only gets better as the number of instructions increases.

PIPELINED	cycle 1	cycle 2	cycle 3	cycle 4	cycle 5	cycle 6	cycle 7	cycle 8	cycle 9	cycle 10	cycle 11	cycle 12	cycle 13	cycle 14	cycle 15	cycle 16
Instruction 1	IF	RF	A	DM	WB											
Instruction 2		IF	RF	A	DM	WB										
Instruction 3			IF	RF	A	DM	WB									
Instruction 4				IF	RF	A	DM	WB								
Instruction 5					IF	RF	A	DM	WB							
Instruction 6						IF	RF	A	DM	WB						
Instruction 7							IF	RF	A	DM	WB					
Instruction 8								IF	RF	A	DM	WB				
Instruction 9									IF	RF	A	DM	WB			
Instruction 10										IF	RF	A	DM	WB		
Instruction 11											IF	RF	A	DM	WB	
Instruction 12												IF	RF	A	DM	WB

Fig. 6.5 A pipelined CPU timing table

The next section of this chapter examines the hardware requirements of register-to-register-type, immediate-type and jump-type RISC instructions.

Register-to-Register Type ALU Instructions

Fixed-point register-to-register type ALU instructions interact with the ALU only. The most fundamental instruction in this category is the Add (ADD) instruction that contains the ADD opcode, two source register addresses, RS1 and RS2, and a destination register address, RD, as shown below.

ADD RS1, RS2, RD

This instruction fetches data from the source addresses, RS1 and RS2, adds them, and returns the result to the destination address, RD, according to the equation below.

$$\text{Reg[RS1]} + \text{Reg[RS2]} \rightarrow \text{Reg[RD]}$$

The field format of this instruction in the instruction memory is shown in Fig. 6.6. The numbers on top of each field represent the OPC and operand bit boundaries.

31 – 26	25 – 21	20 – 16	15 – 11	10 – 0
ADD	RS1	RS2	RD	Not Used

Fig. 6.6 Fixed-point ADD instruction field format

The required hardware for the ADD instruction is shown in Fig. 6.7. In this figure, the PC generates an instruction memory address, and loads the contents of the ADD instruction from the instruction memory to the instruction register at the end of the first clock cycle. Since the instruction contains two source register addresses, RS1 and RS1, the contents of these registers, Reg[RS1] and Reg[RS2], are fetched from the RF and stored at the second flip-flop boundary. In the third clock cycle, the ALU adds Reg[RS1] and Reg[RS2], and stores the result at the third flip-flop boundary. In the fourth clock cycle, the ALU result is written back to the destination register address, RD, in the RF. No processing is applied to RD. The destination address simply propagates from one flip-flop stage to another and points where the ALU result needs to go in the RF.

Since the ADD instruction does not require any data to be stored or fetched from the data memory, the data memory stage is omitted from this data-path which reduces the number of stages from five to four.

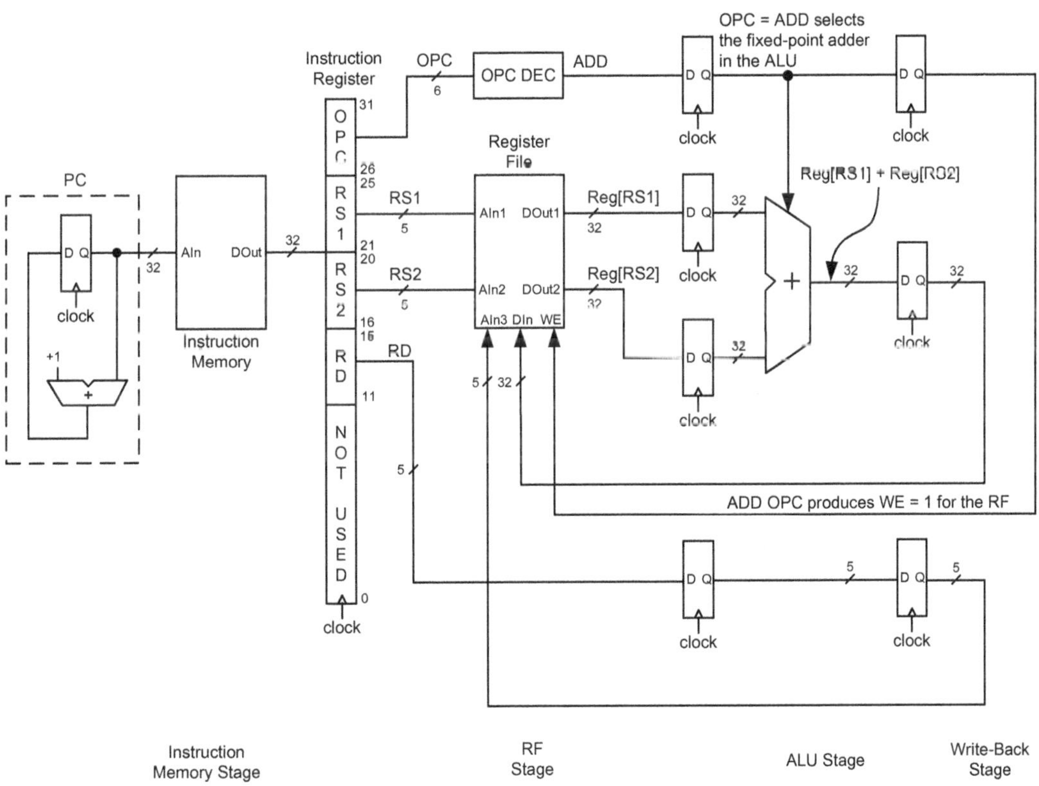

Fig. 6.7 ADD instruction data-path

Similar to the ADD instruction, the Subtract (SUB) instruction subtracts the 32-bit data at RS2 from the 32-bit data at RS1 and returns the result to RD. The instruction and its operational equation are shown below.

SUB RS1, RS2, RD
Reg[RS1] − Reg[RS2] → Reg[RD]

The field format of the SUB instruction in the instruction memory is the same as the ADD instruction in Fig. 6.6 except the ADD OPC is replaced by SUB. This instruction also follows the same data-path as the ADD instruction except the OPC selects the subtractor in the ALU instead of the adder.

The fixed-point Multiplication (MUL) instruction requires four operands as shown below. This instruction multiplies the contents of RS1 and RS2 and generates a 64-bit result. The lower and upper 32 bits of the result in curly brackets are written to RD1 and RD2, respectively.

MUL RS1, RS2, RD1, RD2
Reg[RS1] * Reg[RS2] → {Reg[RD2], Reg[RD1]}

The field format of this instruction in the instruction memory is shown in Fig. 6.8.

31 26	25 21	20 16	15 11	10 6	5 0
MUL	RS1	RS2	RD2	RD1	Not Used

Fig. 6.8 Fixed-point MUL instruction field format

There are two ways to generate a data-path for this particular instruction. The first method executes this instruction using four pipeline stages and requires two RF write-back ports as shown in Fig. 6.9. In this figure, the lower 32 bits of the multiplication result, MUL[31:0], are written to the RF address RD1 while the higher 32 bits, MUL[63:32], are written to the address RD2.

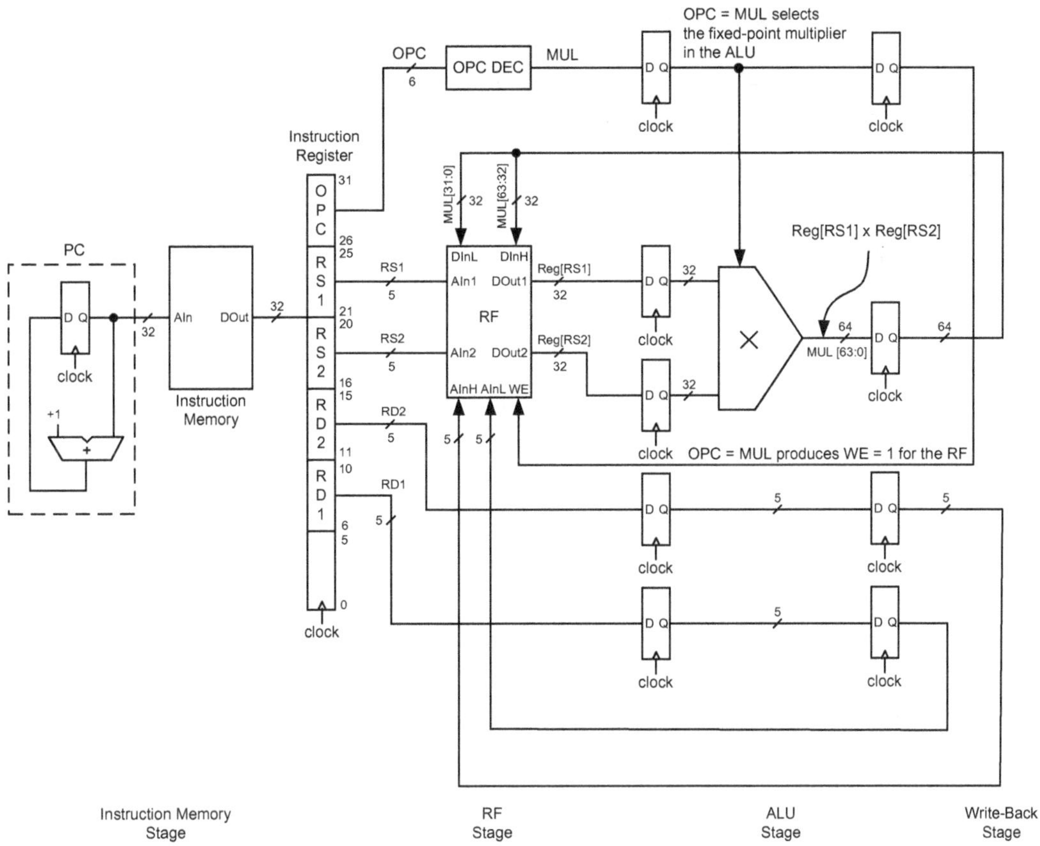

Fig. 6.9 Data-path for fixed-point multiplication using 2 write-back ports

The second method writes the multiplication result back to the RF in two successive clock cycles instead of one but does not require the RF to have two write-back ports. In this scheme, the 64-bit multiplication result is divided between two distinct paths in the ALU as shown in Fig. 6.10. The lower 32 bits, MUL[31:0], are written back to the RF at the end of fourth cycle while the higher 32 bits, MUL[63:32], are stored at an additional flip-flop boundary. The higher 32 bits are subsequently written back to the RF at the end of the fifth cycle.

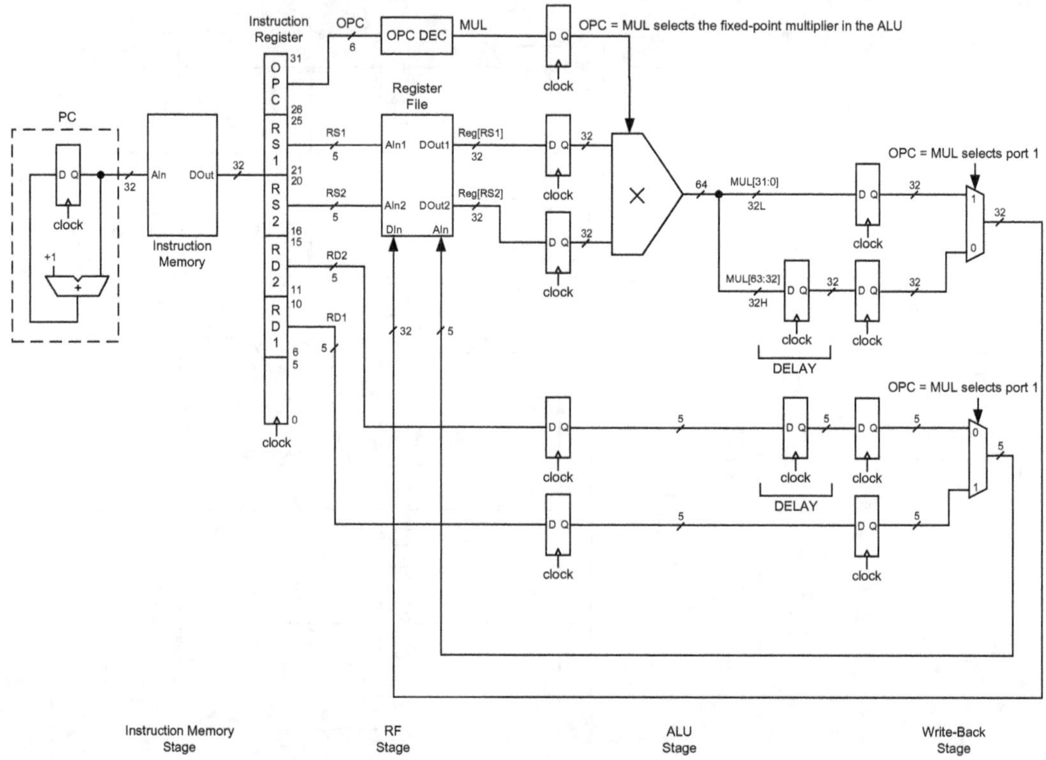

Fig. 6.10 Data-path for fixed-point multiplication using a single write-back port (WE = 1 to RF is not shown for clarity)

Logical operations utilize the logical functional units in the ALU. The And (AND) instruction below "bitwise ANDs" the contents of RS1 and RS2, and returns the result to the RF at the address, RD. The operational equation of this instruction contains the "&" sign to indicate that this is an AND operation. The field format of this instruction is shown in Fig. 6.11.

AND RS1, RS2, RD
Reg[RS1] & Reg[RS2] → Reg[RD]

31 — 26	25 — 21	20 — 16	15 — 11	10 — 0
AND	RS1	RS2	RD	Not Used

Fig. 6.11 Fixed-point AND instruction field format

The AND instruction data-path in Fig. 6.12 is identical to the ADD or SUB instruction data-paths except for the ALU which requires 32 sets of two-input AND gates to carry out the instruction.

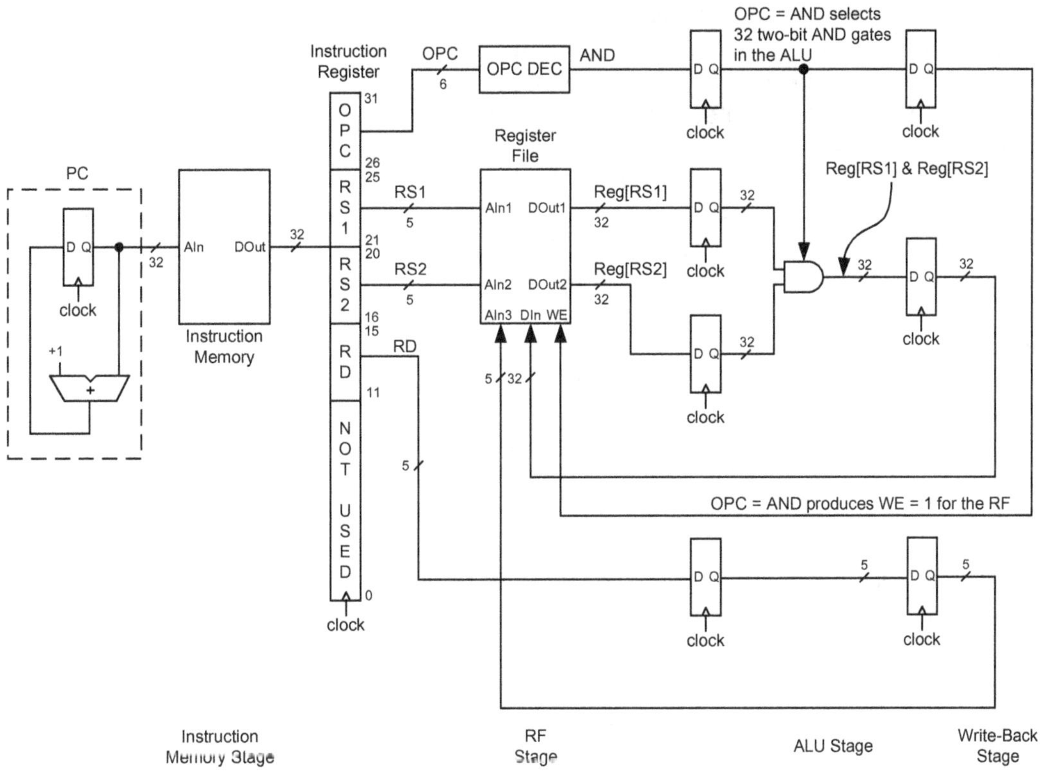

Fig. 6.12 AND instruction data-path

The Or (OR), Exclusive Or (XOR), Nand (NAND), Nor (NOR) and Exclusive Nor (XNOR) instructions have identical instruction formats except the opcode field. These operations are shown below.

OR RS1, RS2, RD
Reg[RS1] | Reg[RS2] → Reg[RD]

XOR RS1, RS2, RD
Reg[RS1] ^ Reg[RS2] → Reg[RD]

NAND RS1, RS2, RD
Reg[RS1] ~ & Reg[RS2] → Reg[RD]

NOR RS1, RS2, RD
Reg[RS1] ~ | Reg[RS2] → Reg[RD]

XNOR RS1, RS2, RD
Reg[RS1] ~^ Reg[RS2] → Reg[RD]

Here, "|" and "^" signs refer to the OR and XOR operations, respectively. The "∼" sign corresponds to negation and generates a complemented value. Therefore, "∼&", "∼|" and "∼^" operations denote the bitwise-NAND, NOR and XNOR, respectively.

The OR, XOR, NAND, NOR and XNOR instructions follow the same, four-stage data-path as the AND instruction in Fig. 6.12. However, each logical instruction requires different types of logic gates in the ALU stage, and the OPC field selects which to use.

Another important register-to-register type instruction is the shift instruction. The Shift Left (SL) instruction shifts the contents of RS1 to the left by the amount stored at the address RS2, and returns the result to RD. The format and the operation of this instruction are shown below. The "<<" sign indicates left-shift operation. This instruction's field format is similar to the previous register-to-register-type instructions as shown in Fig. 6.13.

31 … 26	25 … 21	20 … 16	15 … 11	10 … 0
SL	RS1	RS2	RD	Not Used

Fig. 6.13 Fixed-point Shift-Left (SL) instruction field format

SL RS1, RS2, RD
Reg[RS1] << Reg[RS2] → Reg[RD]

The Shift Right (SR) instruction is similar to the SL instruction except the contents of RS1 are shifted to the right by the amount indicated in RS2. The ">>" sign corresponds to the SR operation.

SR RS1, RS2, RD
Reg[RS1] >> Reg[RS2] → Reg[RD]

Both the SL and SR instructions require linear shifters in the ALU. These units are large combinational logic blocks that are predominantly made out of multiplexers as examined in Chap. 1. Both of these instructions follow the same data-path as any other register-register-type instructions with three operands. Figure 6.14 shows the combined data-path for the SL and SR instructions. The ALU stage contains both a left and a right linear shifter. The first input to either shifter, Reg[RS1], represents the value to be shifted to the right or left. The second input, Reg[RS2], specifies the amount to be shifted in number of bits. Even though the ALU executes both the left and right-shifted versions of Reg[RS1] simultaneously, only one result is selected by the OPC and written back to the RF. If the instruction is a shift-right (SR) instruction, then OPC selects the right (R) port of the 2-1 MUX, and the SR result is written to the RF. Otherwise, the OPC selects the left (L) port, and the shift-left (SL) result is written to the RF. Both the SL and SR instructions require a four-stage CPU pipeline.

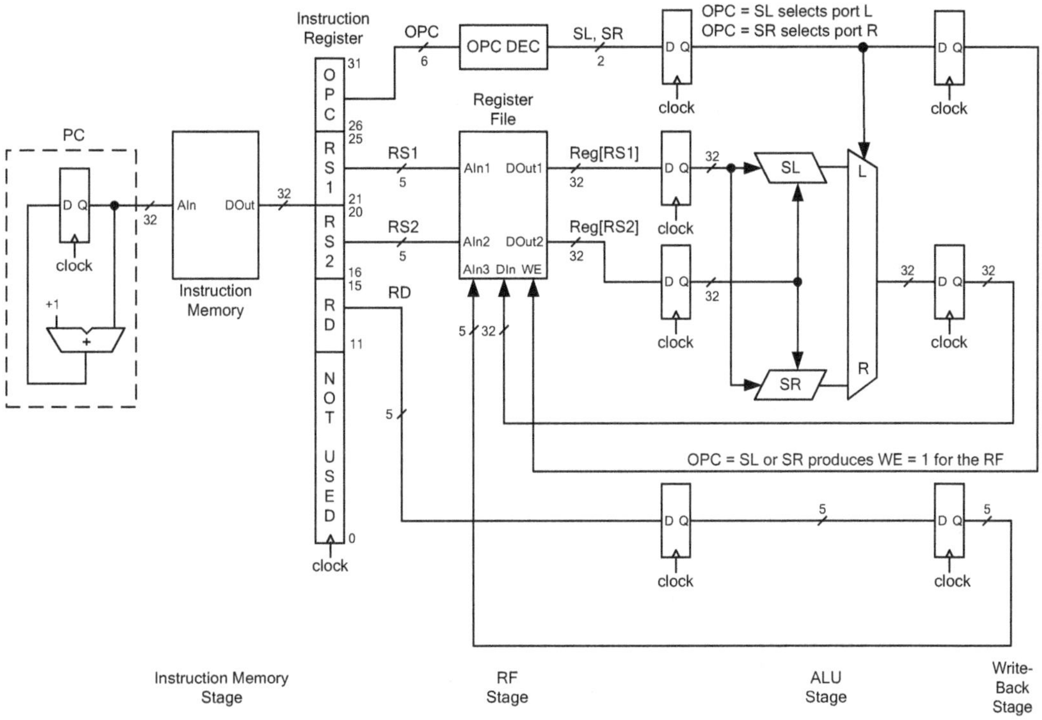

Fig. 6.14 SL and SR instruction data-paths

As an example, let us combine the individual data-paths for ADD, SUB, AND, NAND, OR, NOR, XOR, XNOR, SL and SR instructions in a single CPU to execute a user program. The architecture in Fig. 6.15 shows eight individual functional units in the ALU followed by an 8-1 MUX to select the desired ALU output. In this figure, there is only one adder, and it is able to execute a two's complement addition to perform subtraction. The left and right linear shifters are also combined in a single unit. The SL or SR opcode selects the output of either the left shifter or the right shifter for the destination register. The outputs of all logical units are selected by an 8-1 MUX and forwarded to the RF.

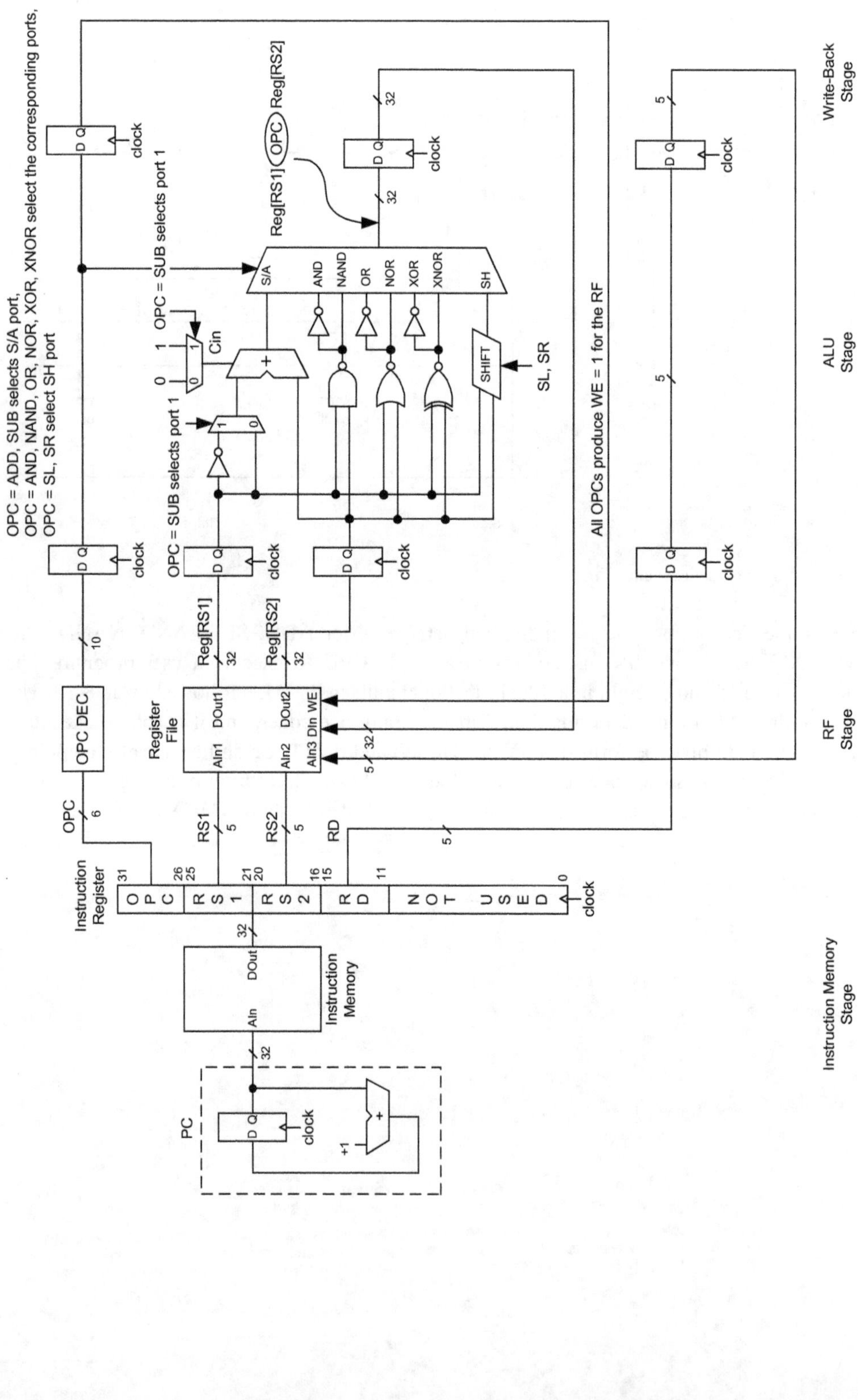

Fig. 6.15 Combined register-to-register ALU instruction data-paths

The next category of register-register-type instructions are the Set instructions used in decision-making situations where two source register values are compared with each other prior to a branch statement in a program.

The Set-Greater-than-or-Equal (SGE) instruction below describes setting the contents of RD to 0x0000001 if the contents of RS1 are found to be greater than or equal to the contents of RS2. The data in RS1 and RS2 registers are considered unsigned integers. If the comparison fails, then the contents of RD are set to 0x0000000. The field format of this instruction is given in Fig. 6.16.

SGE RS1, RS2, RD
If Reg[RS1] $\geq$ Reg[RS2] then 1 $\rightarrow$ Reg[RD] else 0 $\rightarrow$ Reg[RD]

31 26	25 21	20 16	15 11	10 0
SGE	RS1	RS2	RD	Not Used

Fig. 6.16 Fixed-point Set-Greater-than-or-Equal (SGE) instruction field format

The data-path for the SGE instruction in Fig. 6.17 tests if the contents of RS1 are greater than or equal to the contents of RS2 using a subtractor in the ALU. Again, the data in RS1 and RS2 registers are considered unsigned integers. To perform this test, Reg[RS1] is subtracted from Reg[RS2], and the sign bit of the result is used to make the decision. If Reg[RS1] is greater than or equal to Reg [RS2], the sign bit becomes zero. The complemented sign bit is then forwarded to the 2-1 MUX at the

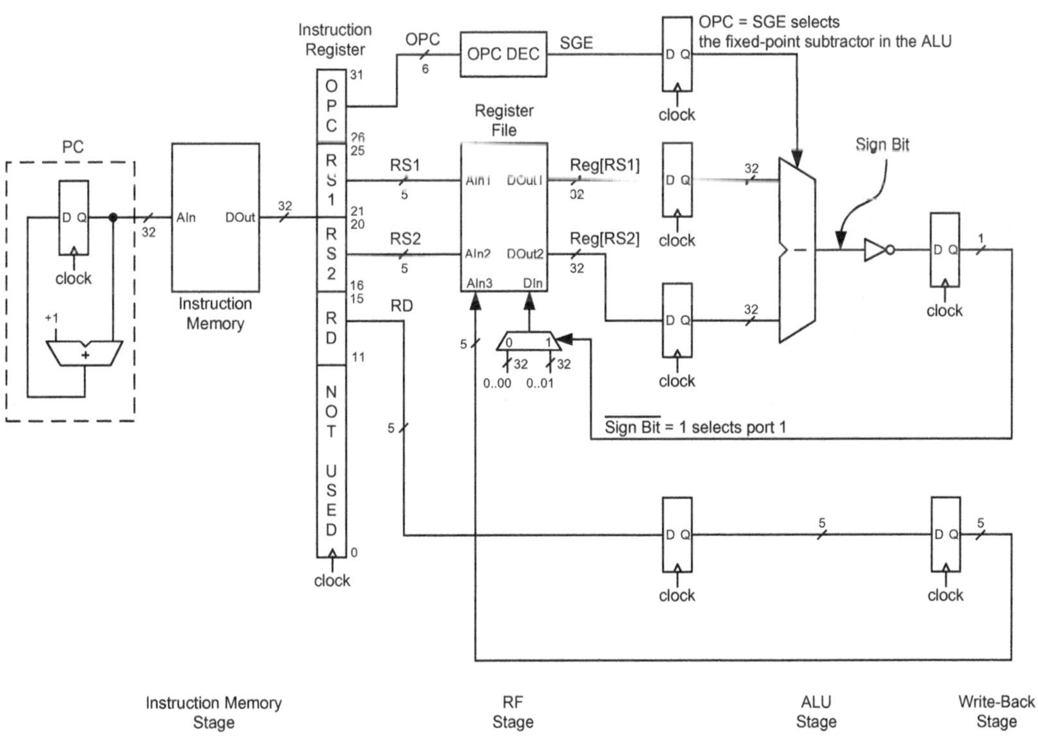

Fig. 6.17 SGE instruction data-path (WE signal to RF is not shown for clarity)

write-back port of the RF to store 0x00000001 in the destination register. If the subtraction yields a negative number, the complemented sign bit selects 0x00000000 to be stored in the RD.

The Set-Greater-Than (SGT) instruction is another instruction that tests if Reg[RS1] is greater than Reg[RS2]. If the comparison is successful, the instruction stores 0x00000001 in the RD. Otherwise, the instruction stores 0x00000000 as described below.

SGT RS1, RS2, RD
If Reg[RS1] > Reg[RS2] then 1 $\rightarrow$ Reg[RD] else 0 $\rightarrow$ Reg[RD]

The data-path of the SGT instruction is shown in Fig. 6.18. In this figure, two tests are performed in the ALU stage. The first test checks if Reg[RS1] is greater than or equal to Reg[RS2] using the sign bit of the subtractor as it was applied to the SGE instruction. The second test checks if Reg[RS1] is not equal to Reg[RS2] using 32 sets of two-input XNOR gates followed by a single 32-input NAND gate. The condition, Reg[RS1] $\neq$ Reg[RS2], is then logically isolated from the condition, Reg[RS1] $\geq$ Reg[RS2], in the form of a two-input AND gate to determine if the final condition, Reg[RS1] > Reg[RS2], is met. If Reg[RS1] > Reg[RS2], then port 1 of the 2-1 MUX is selected to store 0x00000001 in RD. Otherwise, port 0 is selected to store 0x00000000.

Similar to the SGE and SGT instructions, there are four other set instructions that compare the contents of the two source registers in a variety of different ways to set or reset the destination register. The instructions, Set-Less-than-or-Equal (SLE), Set-Less-Than (SLT), Set-Equal (SEQ) and Set-Not-Equal (SNE), and how they operate inside the CPU are listed below.

SLE RS1, RS2, RD
If Reg[RS1] $\leq$ Reg[RS2] then 1 $\rightarrow$ Reg[RD] else 0 $\rightarrow$ Reg[RD]

SLT RS1, RS2, RD
If Reg[RS1] < Reg[RS2] then 1 $\rightarrow$ Reg[RD] else 0 $\rightarrow$ Reg[RD]

SEQ RS1, RS2, RD
If Reg[RS1] = Reg[RS2] then 1 $\rightarrow$ Reg[RD] else 0 $\rightarrow$ Reg[RD]

SNE RS1, RS2, RD
If Reg[RS1] $\neq$ Reg[RS2] then 1 $\rightarrow$ Reg[RD] else 0 $\rightarrow$ Reg[RD]

All Set instructions require four clock cycles to write the result to the RF like all the other register-to-register-type instructions.

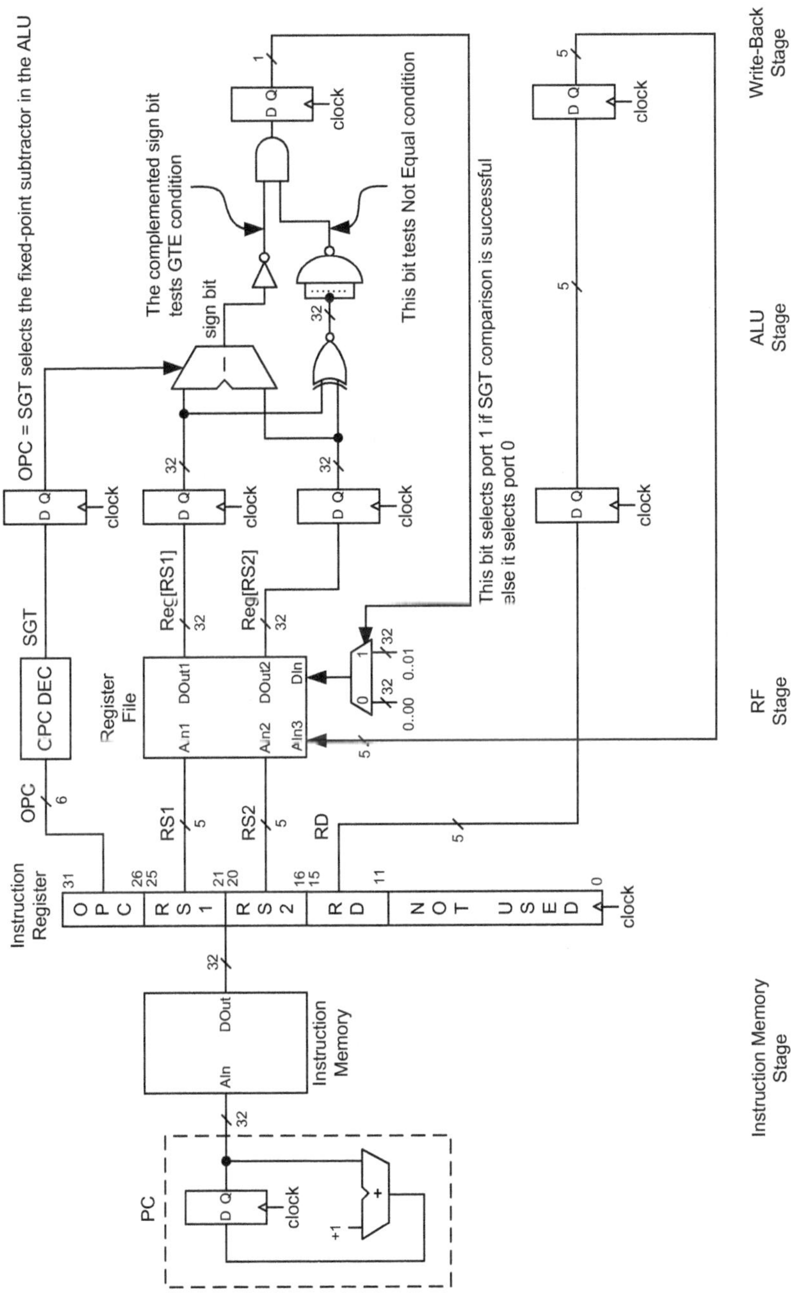

Fig. 6.18 SGT instruction data-path (WE signal to RF is not shown for clarity)

Immediate Type ALU Instructions

Immediate ALU instructions allow user data to be included in the instruction. However, these instructions still fetch register data from the RF to be combined with the user data.

The Add-Immediate instruction (ADDI) adds the contents of RS to the user-supplied 16-bit immediate value and returns the result to RD in the RF. This instruction and how it operates in the CPU are shown below. The field format of this instruction in the instruction memory is given in Fig. 6.19.

ADDI RS, RD, Imm Value
Reg[RS] + Immediate Value → Reg[RD]

31 — 26	25 — 21	20 — 16	15 — 0
ADDI	RS	RD	Immediate Value

Fig. 6.19 Fixed-point ADD Immediate (ADDI) instruction field format

The ADDI instruction data-path is shown in Fig. 6.20. In this figure, the contents of the instruction are transferred from the instruction memory to the instruction register at the end of the first clock

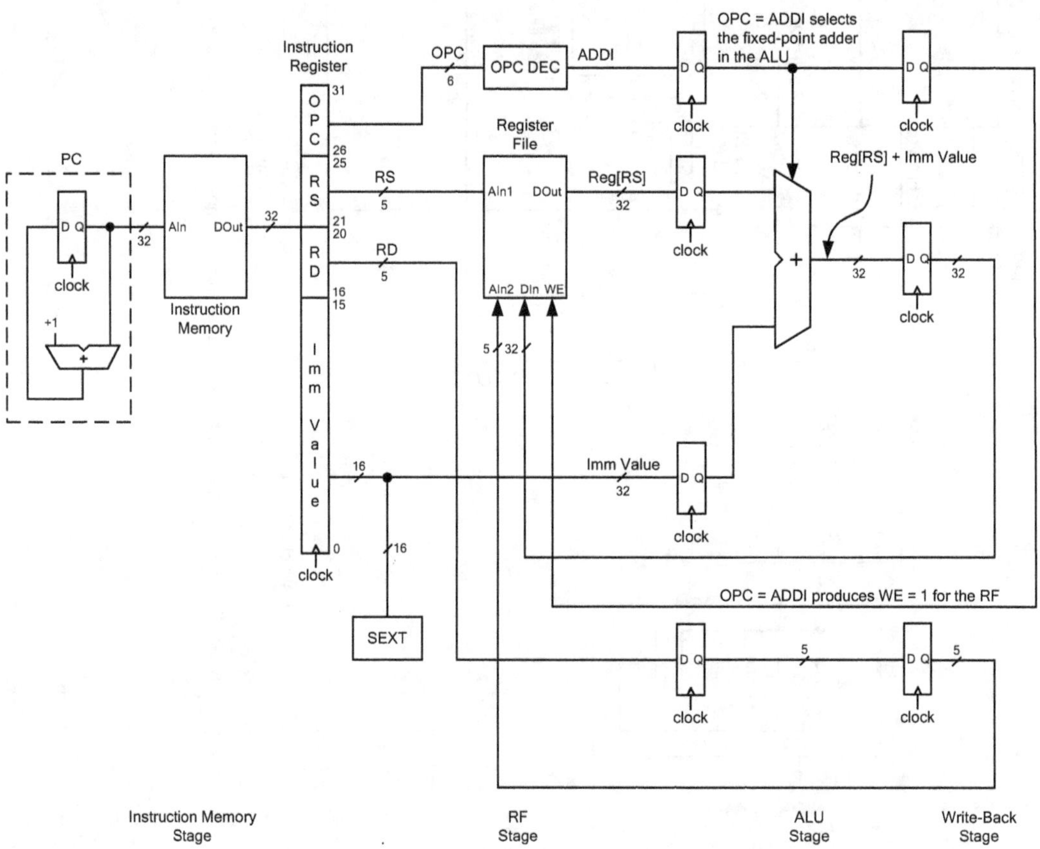

Fig. 6.20 ADDI instruction data-path

cycle. In the second clock cycle, the 16-bit immediate value in the instruction is sign extended to 32 bits while the contents of RS are fetched from the RF. In the third clock cycle, the two values are added in the ALU. At the end of the fourth cycle, the processed data is written back to the RF at the address RD. Therefore, the ADDI instruction requires only four clock cycles to execute.

The Subtract-Immediate instruction (SUBI) behaves similar to the ADDI, but it subtracts the immediate value from the contents of RS, and returns the result to RD as illustrated below.

SUBI RS, RD, Imm Value
Reg[RS] − Immediate Value → Reg[RD]

There are also immediate logical instructions that operate with the user data. The AND-Immediate (ANDI) instruction, for example, bitwise ANDs the contents of RS with the immediate value and returns the result to RD as shown below. The field format of this instruction is given in Fig. 6.21.

ANDI RS, RD, Imm Value
Reg[RS] & Immediate Value → Reg[RD]

31 26	25 21	20 16	15 0
ANDI	RS	RD	Immediate Value

Fig. 6.21 Fixed-point AND Immediate (ANDI) instruction field format

The ANDI instruction uses a similar data-path as the ADDI, but replaces the fixed-point adder with 32 two-input AND gates as shown in Fig. 6.22. The contents of RS and the sign-extended immediate value are combined using the AND gates, and the result is returned to RD in the RF.

Similar to the ANDI instruction, the ORI, XORI, NANDI, NORI, and XNORI instructions operate with the immediate data and follow the same data-path as the ANDI instruction. These instructions and how they operate inside the CPU are listed below.

ORI RS, RD, Imm Value
Reg[RS] | Immediate Value → Reg[RD]

XORI RS, RD, Imm Value
Reg[RS] ^ Immediate Value → Reg[RD]

NANDI RS, RD, Imm Value
Reg[RS] ~& Immediate Value → Reg[RD]

NORI RS, RD, Imm Value
Reg[RS] ~| Immediate Value → Reg[RD]

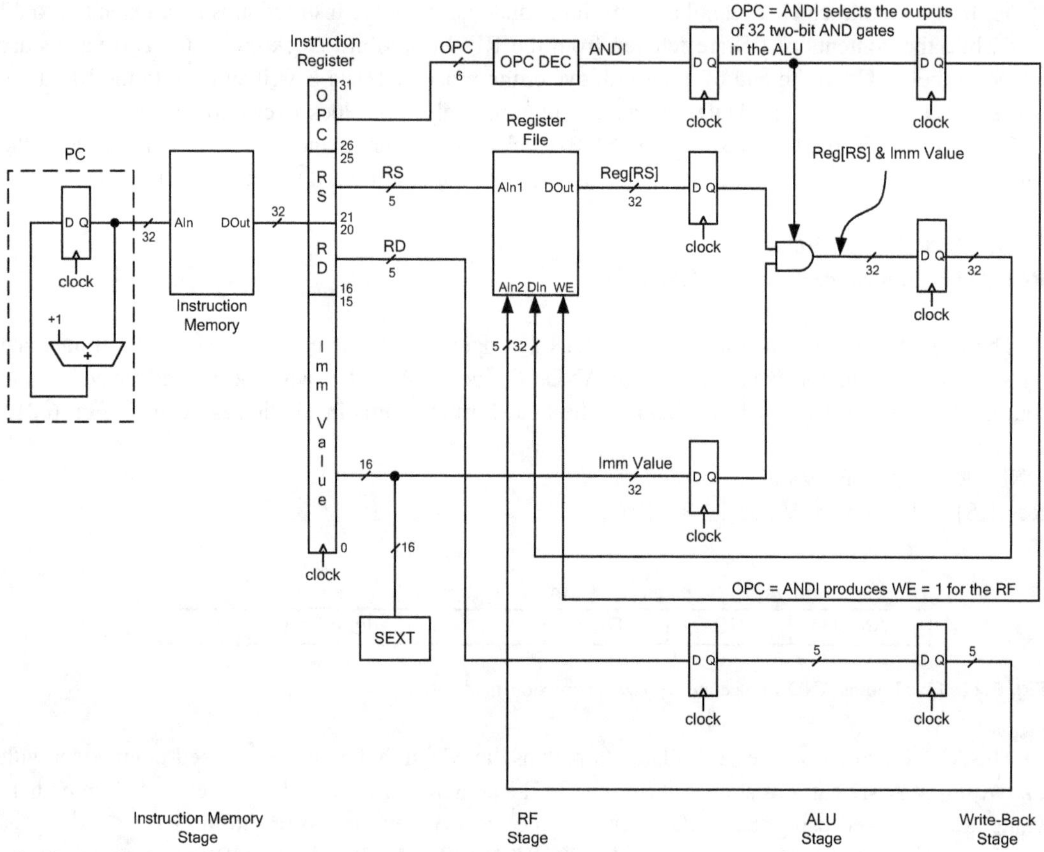

Fig. 6.22 ANDI instruction data-path

XNORI RS, RD, Imm Value
Reg[RS] ~^ Immediate Value → Reg[RD]

All logical immediate instructions require four cycles to form the result.

The Shift Left Immediate (SLI) and Shift Right Immediate (SRI) instructions use the same functional units as the SL and SR instructions in the ALU.

The SLI instruction fetches the contents of RS from the RF, shifts it to the left by an immediate value, and stores the result in RD as indicated below. Figure 6.23 shows the field format of this instruction.

SLI RS, RD, Imm Value
Reg[RS] << Immediate Value → Reg[RD]

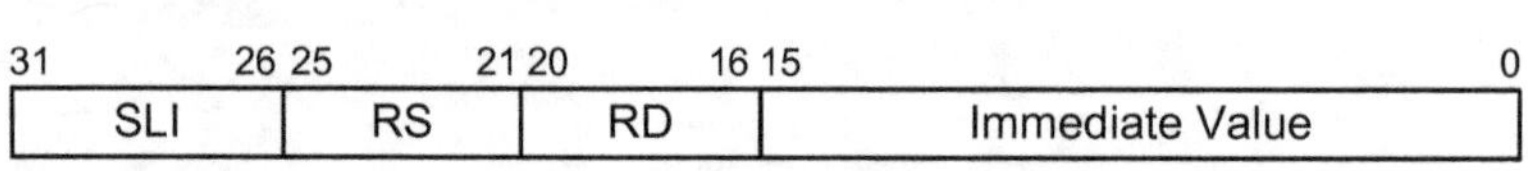

Fig. 6.23 Fixed-point SL Immediate (SLI) instruction field format

Similar to the SLI instruction, the SRI instruction shifts the contents of RS to the right by an amount equal to the immediate value, and stores the result in RD as shown below.

SRI RS, RD, Imm Value
Reg[RS] >> Immediate Value → Reg[RD]

The SLI and SRI instruction data-paths in Fig. 6.24 still require the left and right linear shifters in the ALU stage. In these instructions, one shifter input receives an immediate data that specifies the number of bits to be shifted to the left or to the right while the other input receives the contents of RS.

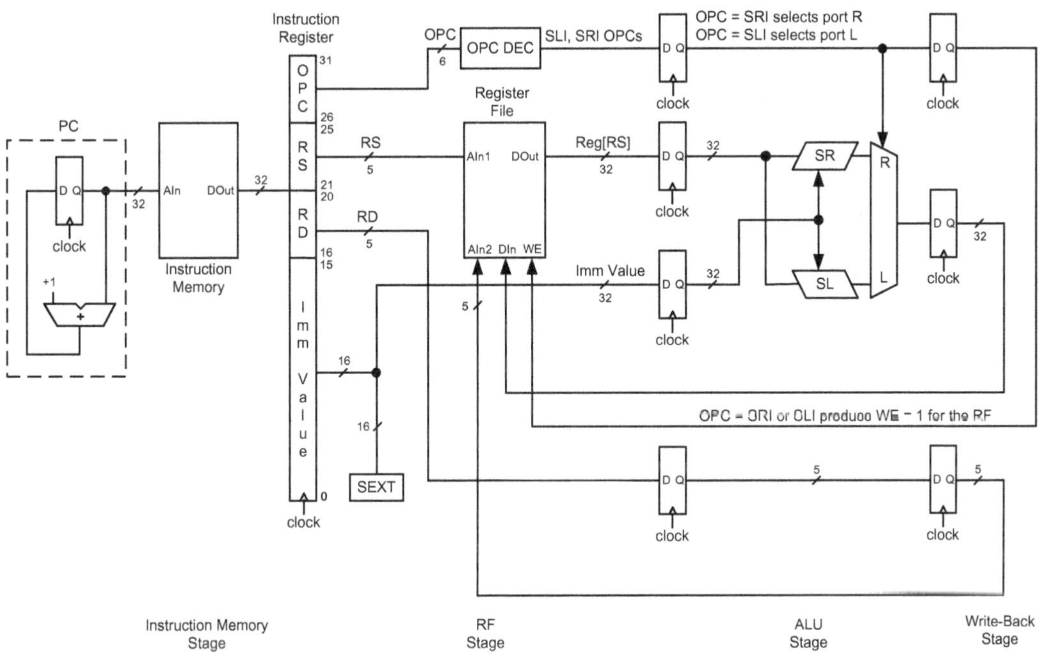

Fig. 6.24 SLI and SRI instruction data-paths

Figure 6.25 combines all the immediate ALU instructions examined so far in one schematic. These include the ADDI, SUBI, SLI, SRI, ANDI, ORI, XORI, NANDI, NORI and XNORI instructions with their common sign-extended inputs. The port selection process at the ALU MUX is as follows. If the OPC is ADDI or SUBI, the adder/subtractor output is routed through the S/A port. For the SLI and SRI OPCs, the shifter outputs are routed through the SH port. For all other OPCs, the processed data is routed through the corresponding MUX port, carrying the same OPC name, and becomes the ALU output.

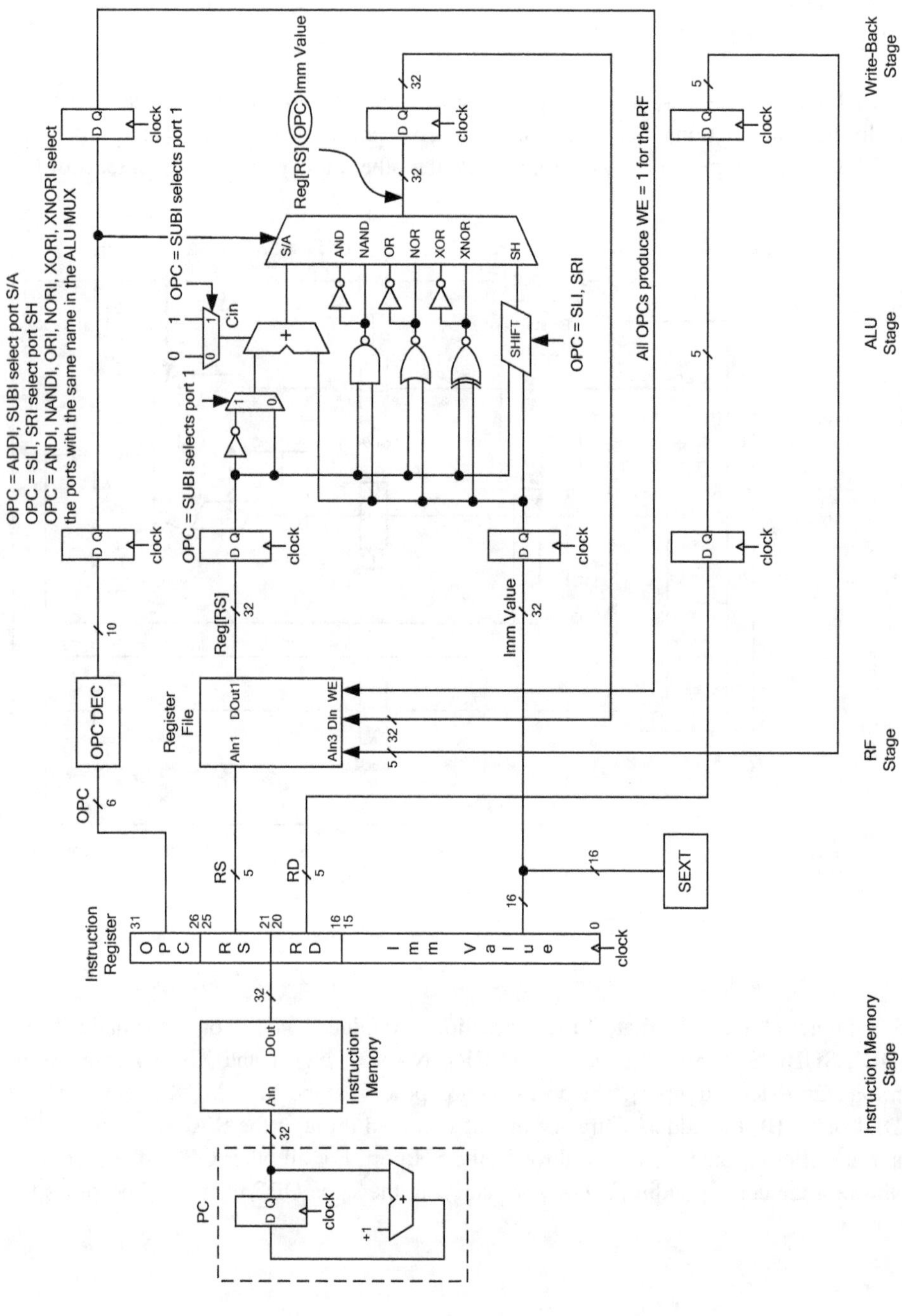

Fig. 6.25 Combined immediate ALU instruction data-paths

The immediate-set instructions compare the contents of RS with an immediate value for setting or resetting the register RD.

The Set-Greater-than-or-Equal-Immediate (SGEI) instruction sets the contents of RD if the instruction finds the contents of RS to be greater than or equal to the immediate value. This instruction and how it operates inside the CPU is shown below. The field format is given in Fig. 6.26.

SGEI RS, RD, Imm Value
If Reg[RS] $\geq$ Immediate Value then 1 $\rightarrow$ Reg[RD]

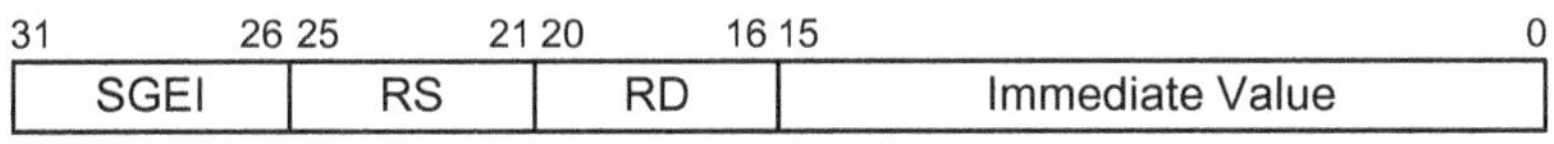

Fig. 6.26 Fixed-point SGE Immediate (SGEI) instruction field format

The SGEI instruction data-path in Fig. 6.27 tests the relative magnitudes of Reg[RS] and the sign-extended immediate value to make a decision about the contents of RD. If Reg[RS] is larger than the immediate value or equal to it, the sign bit of the ALU result becomes zero, which in turn, stores 0x00000001 in RD. Otherwise, RD becomes 0x00000000.

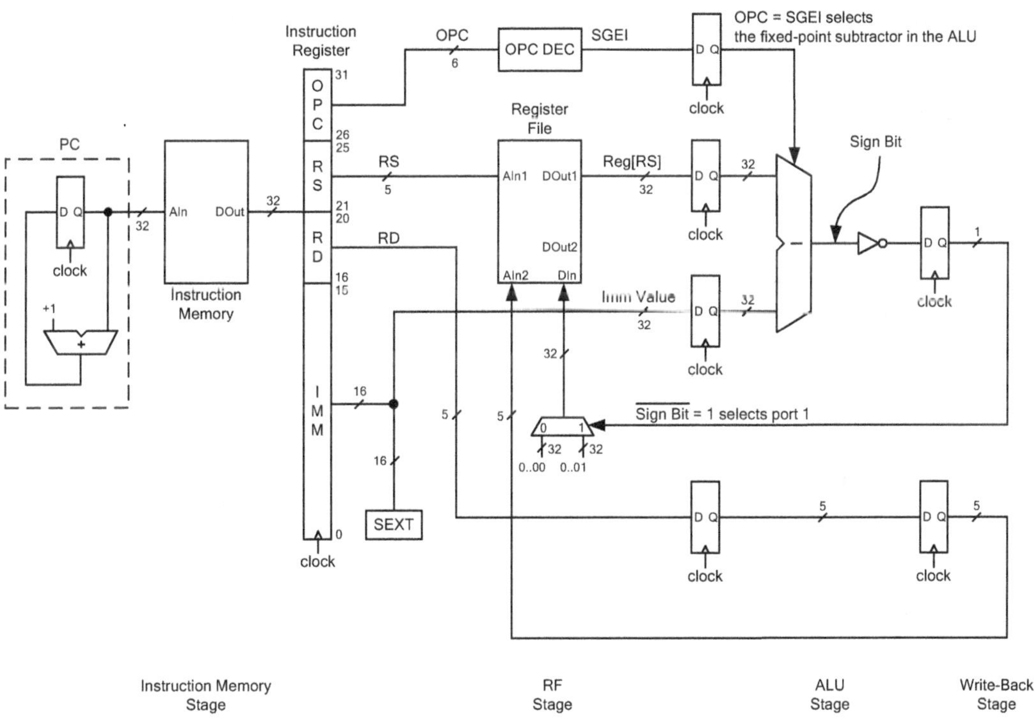

Fig. 6.27 SGEI instruction data-path (WE = 1 to RF is not shown for clarity)

Similarly, the formats for Set-Greater-Than-Immediate (SGTI), Set-Less-than-or-Equal-Immediate (SLEI), Set-Less-Than-Immediate (SLTI), Set-Equal-Immediate (SEQI), Set-Not-Equal-Immediate (SNEI) instructions, and how they operate inside the CPU are given below.

SGTI RS, RD, Imm Value
If Reg[RS] > Immediate Value then 1 $\rightarrow$ Reg[RD] else 0 $\rightarrow$ Reg[RD]

SLEI RS, RD, Imm Value
If Reg[RS] $\leq$ Immediate Value then 1 $\rightarrow$ Reg[RD] else 0 $\rightarrow$ Reg[RD]

SLTI RS, RD, Imm Value
If Reg[RS] < Immediate Value then 1 $\rightarrow$ Reg[RD] else 0 $\rightarrow$ Reg[RD]

SEQI RS, RD, Imm Value
If Reg[RS] = Immediate Value then 1 $\rightarrow$ Reg[RD] else 0 $\rightarrow$ Reg[RD]

SNEI RS, RD, Imm Value
If Reg[RS] $\neq$ Immediate Value then 1 $\rightarrow$ Reg[RD] else 0 $\rightarrow$ Reg[RD]

All Set Immediate instructions require four clock cycles to store the result in RD.

Data Movement Instructions

The first data movement instruction that relocates data from the data memory to a register in the RF is the Load (LOAD) instruction. This instruction first adds the contents of RS to a user-defined immediate value to form an effective data memory address. It then fetches the data at this address and moves it to a destination register, RD, in the RF. This instruction and how it operates inside the CPU is shown below. The term, Reg[RS] + Imm Value, defines the effective data memory address, and mem {Reg[RS] + Imm Value} corresponds to the data at this address. The field format of this instruction while it is in the instruction memory is given in Fig. 6.28.

LOAD RS, RD, Imm Value
mem {Reg[RS] + Immediate Value} $\rightarrow$ Reg[RD]

31 – 26	25 – 21	20 – 16	15 – 0
LOAD	RS	RD	Immediate Value

Fig. 6.28 Fixed-point LOAD instruction field format

The LOAD instruction data-path in Fig. 6.29 adds the contents of RS to the sign-extended immediate value and uses this sum as an address for the data memory. The OPC selects the adder in the ALU to calculate the effective data memory address and enables the data memory for read. The data from the data memory is subsequently written back to the RF at the address, RD. This instruction requires five clock cycles to complete, and it traces through all five stages of the CPU.

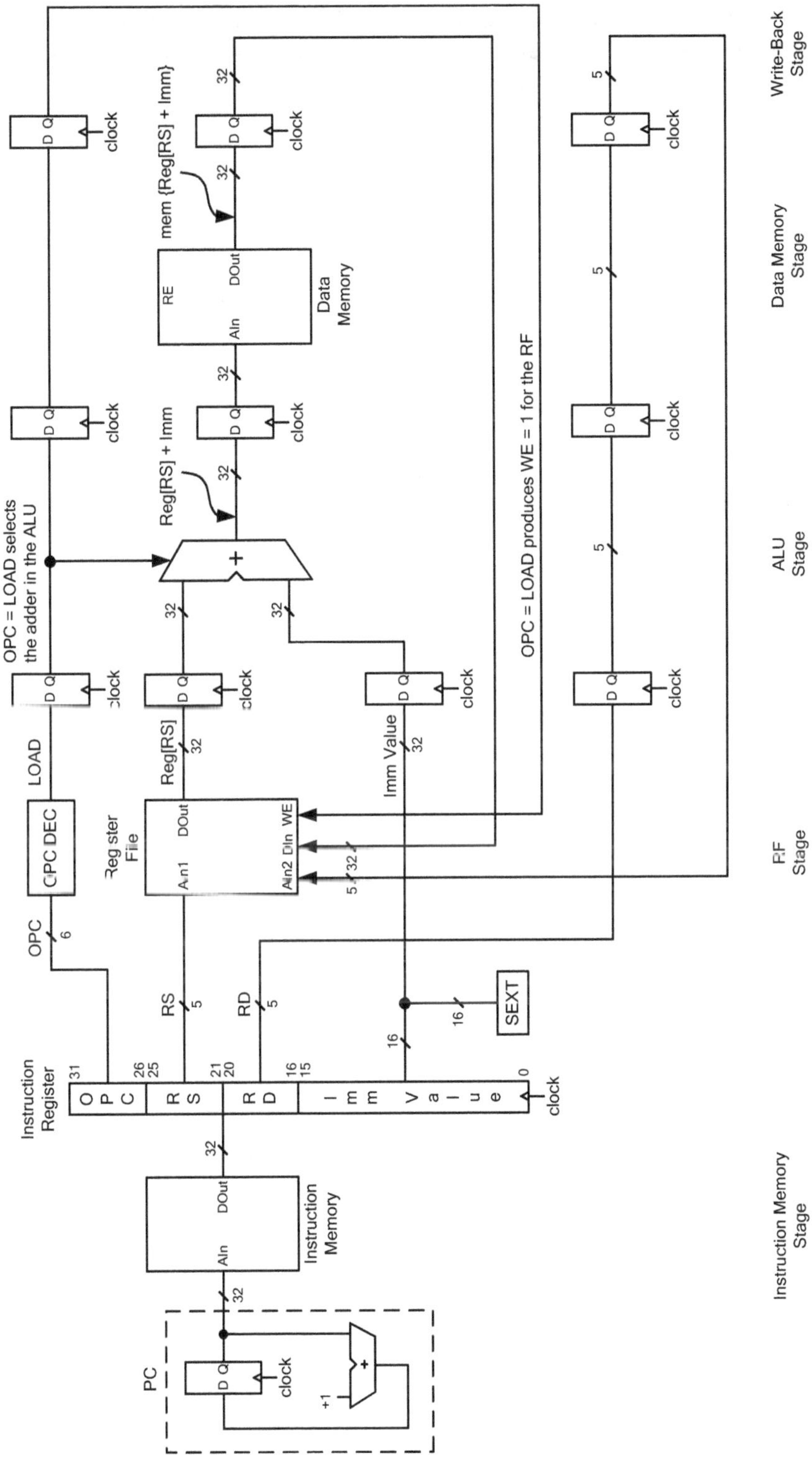

Fig. 6.29 LOAD instruction data-path

The Store (STORE) instruction moves data in the opposite direction of the LOAD instruction. This instruction uses the contents of RD and the immediate value to form a data memory address, and moves the contents of RS to this address as described below. Figure 6.30 shows this instruction's field format.

STORE RS, RD, Imm Value
Reg[RS] → mem {Reg[RD] + Immediate Value}

31 26	25 21	20 16	15 0
STORE	RS	RD	Immediate Value

Fig. 6.30 Fixed-point STORE instruction field format

The data-path for the STORE instruction is shown in Fig. 6.31. In this figure, the OPC selects the adder in the ALU to perform an add operation between the sign-extended immediate value and Reg[RD]. The contents of RS are then written to the data memory at this calculated address. The STORE instruction needs only four clock cycles to complete.

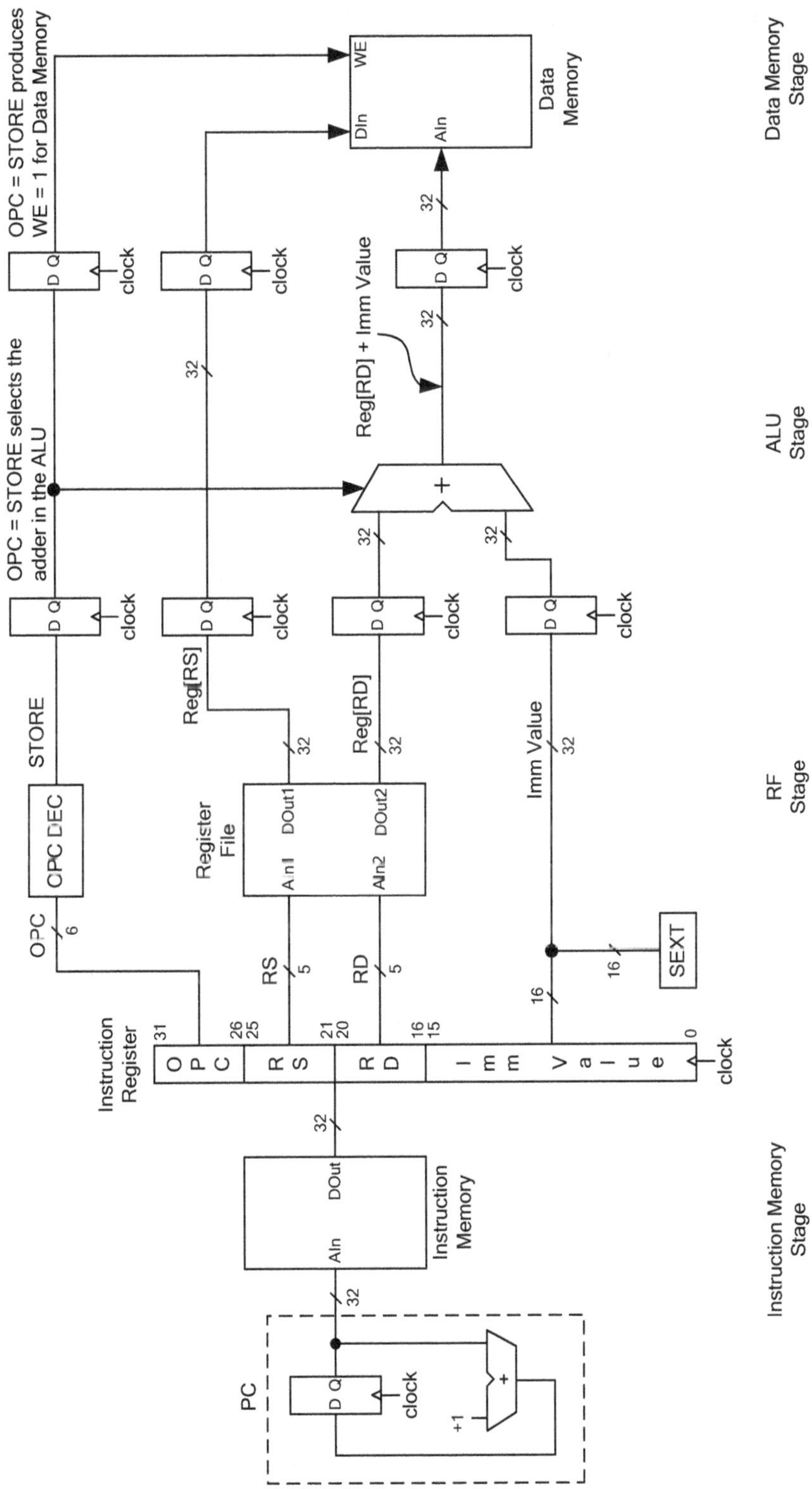

Fig. 6.31 STORE instruction data-path

The Move (MOVE) instruction does not interact with the data memory. Nevertheless, it moves data from one register to another in the RF. The instruction and how it operates inside the CPU are given below. The field format for MOVE instruction is shown in Fig. 6.32.

MOVE RS, RD
Reg[RS] $\rightarrow$ Reg[RD]

31–26	25–21	20–16	15–0
MOVE	RS	RD	Not Used

Fig. 6.32 Fixed-point MOVE instruction field format

The Move Immediate (MOVEI) instruction moves an immediate value to a destination register, RD, in the RF as shown below. The field format of this instruction is given in Fig. 6.33.

MOVEI Imm Value, RD
Immediate Value $\rightarrow$ Reg[RD]

31–26	25–21	20–16	15–0
MOVEI	Not Used	RD	Immediate Value

Fig. 6.33 Fixed-point MOVEI instruction field format

The schematic in Fig. 6.34 combines the data-paths of LOAD, STORE and MOVE instructions. The data memory address is configured by adding the sign-extended immediate value to either Reg[RS] or Reg[RD] depending on the OPC. The write-back stage selects either the contents of RS for the MOVE instruction or the contents of data memory for the LOAD instruction, and writes the result back to the RF.

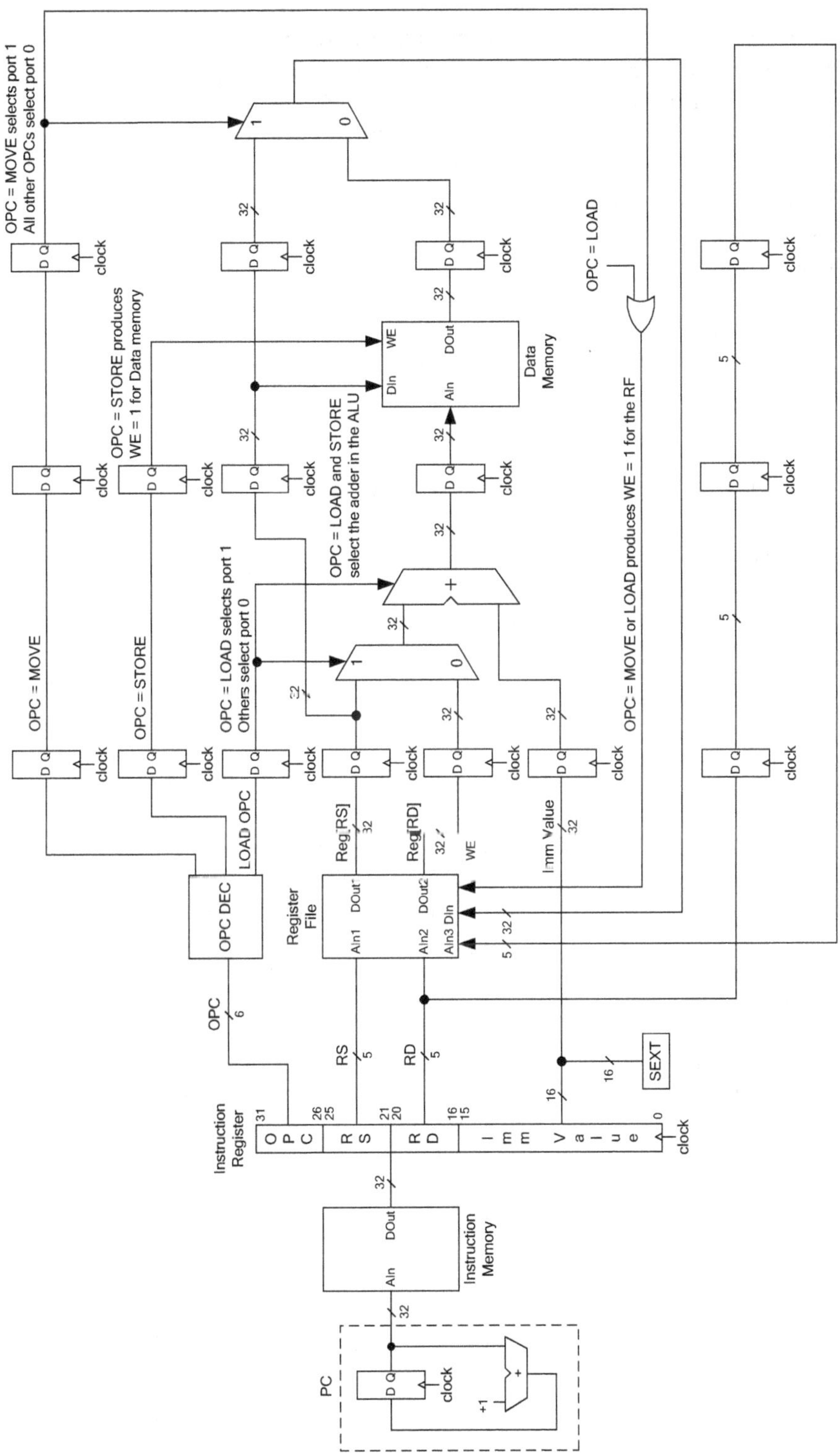

Fig. 6.34 Data movement instruction data-paths (MOVE, LOAD and STORE)

Program Control Instructions

To be able to make decisions in a program, we need program control instructions.

The Branch (BRA) instruction is one of the most essential instructions in a program as it controls direction of the program flow. This instruction first compares the contents of RS with a five-bit RS value specified by the user. If the comparison is successful, the BRA instruction redirects the program to fetch an instruction from a different PC address as shown below. This new PC address is calculated by incrementing the old PC address by an immediate value specified by the user. If the comparison is not successful, the program skips the next instruction but executes the instruction after. This instruction's field format is shown in Fig. 6.35.

BRA RS, RS Value, Imm Value
If Reg[RS] = RS Value then PC + Immediate Value $\rightarrow$ PC
else PC + 2 $\rightarrow$ PC

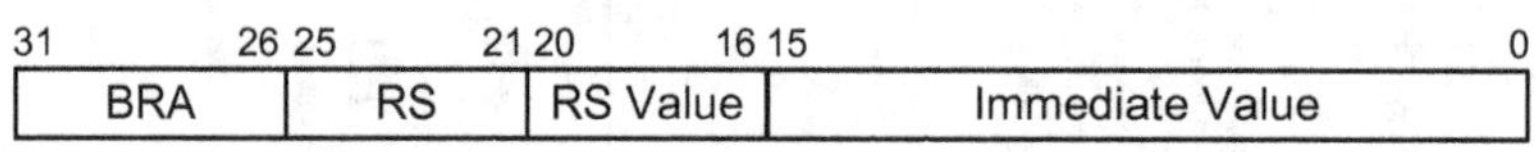

Fig. 6.35 Fixed-point BRANCH instruction field format

The BRA instruction usually depends on a previously stored data value in the RF as a result of a Set or Set-Immediate instruction. To be able to carry out the BRA instruction, first the contents of RS is compared with the RS Value. The comparator in Fig. 6.36 is composed of 32 two-input XNOR

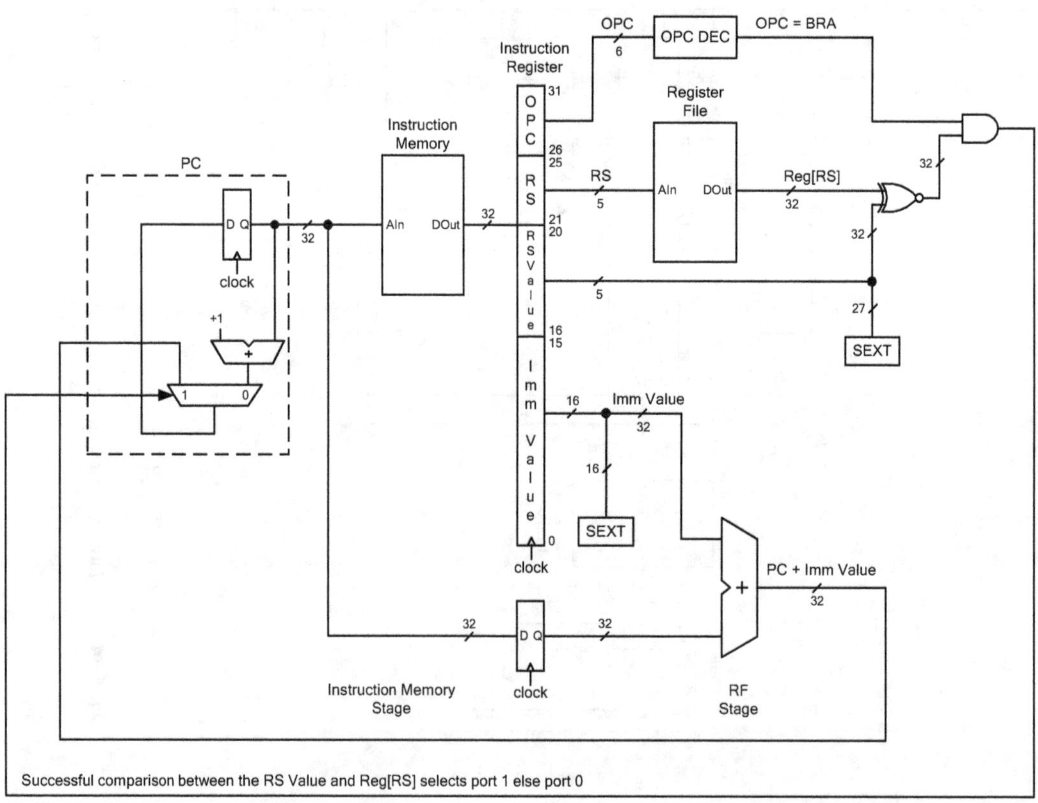

Fig. 6.36 BRA instruction data-path

gates to perform a bitwise comparison between Reg[RS] and the sign-extended RS Value. All 32 XNOR outputs are then fed to a 32-input AND gate to make a decision for the new PC value. If the comparison is successful, the current PC value is replaced with (PC + Imm Value). If the comparison is unsuccessful, however, the PC value is incremented by (PC + 2). The reason for (PC + 2) is because the BRA data-path has to transverse two flip-flop boundaries by the time a new PC value forms. In actuality, a hazard forms when the PC increments to (PC + 1), and the compiler either inserts a No Operation (NOP) instruction right after the BRA instruction or finds an unrelated instruction in the program and inserts it into this slot in order to remove the hazard.

Unconditional decisions in the program do not need a comparison. The programmer can simply change the flow of the program by using a jump-type instruction.

The first unconditional jump-type instruction is the Jump (JUMP) instruction which simply replaces the current PC value with an immediate value as shown below. Its field format is shown in Fig. 6.37.

JUMP Imm Value
Immediate Value → PC

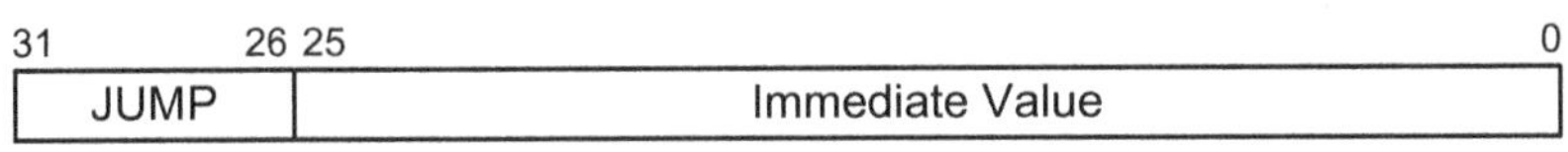

Fig. 6.37 Fixed-point JUMP instruction field format

The data-path for the JUMP instruction is shown in Fig. 6.38. In this data-path, the 26 bit jump value is an unsigned (positive) integer extended to 32 bits before being forwarded to the PC. However, the PC value has already incremented twice and equals to (PC + 2) right before it is replaced by a jump value. Therefore, this instruction also creates a control hazard when the PC is at (PC + 1) and requires compiler's intervention to remove the hazard.

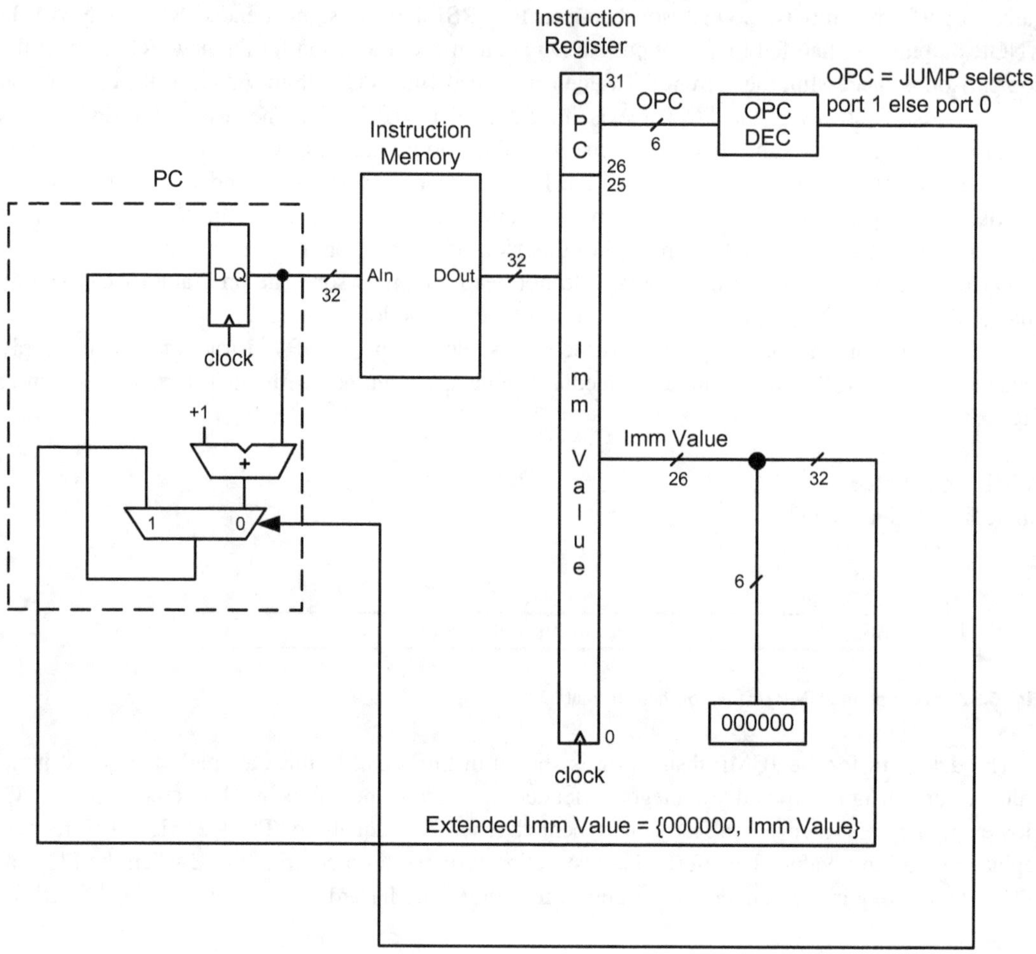

Fig. 6.38 JUMP instruction data-path

The second unconditional jump-type instruction is the Jump Register (JREG) instruction, which is similar to the JUMP instruction, but uses the contents of RS to replace the current PC value as shown below. Figure 6.39 describes this instruction's field format.

JREG RS
Reg[RS] → PC

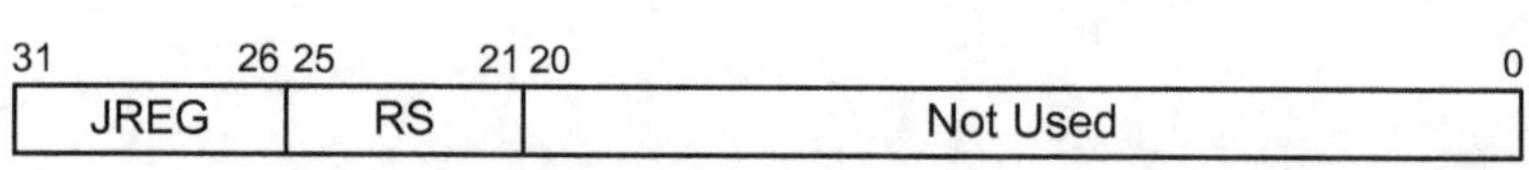

Fig. 6.39 Fixed-point Jump Register (JREG) instruction field format

The data-path for the JREG instruction is shown in Fig. 6.40. This instruction reads the contents of RS and loads it to the PC as the jump value. The program control hazard also exists for this instruction. The hazard can be removed either by inserting a NOP instruction or another unrelated instruction in the program after the JREG instruction.

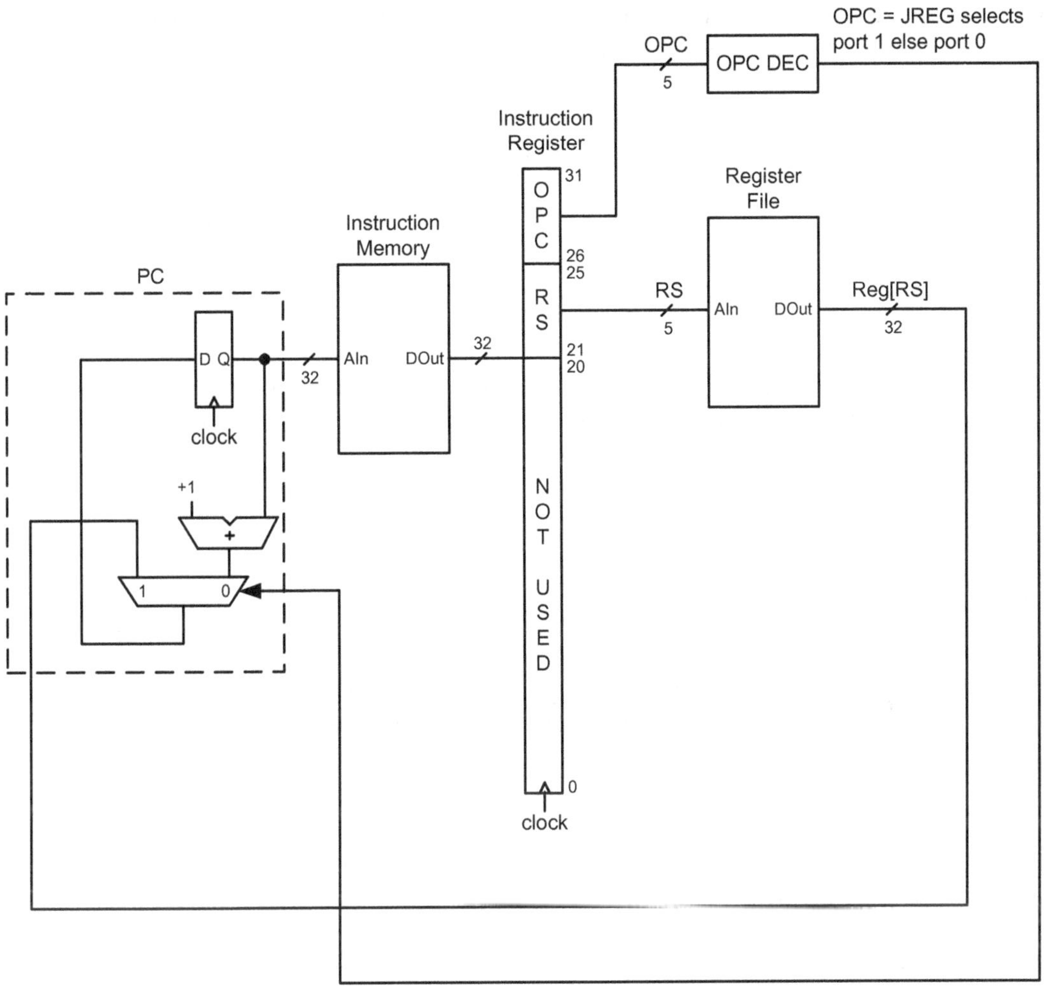

Fig. 6.40 JREG instruction data-path

The Jump-And-Link (JAL) instruction is the third unconditional jump-type instruction, and it requires two steps to operate. In the first step, the PC address, (PC + 2), following the JAL instruction is stored in the register R31. In the second step, the current PC value is replaced with an immediate value as shown below. Its field format is given in Fig. 6.41.

JAL Imm Value
(PC + 2) → Reg[R31] followed by Immediate Value → PC

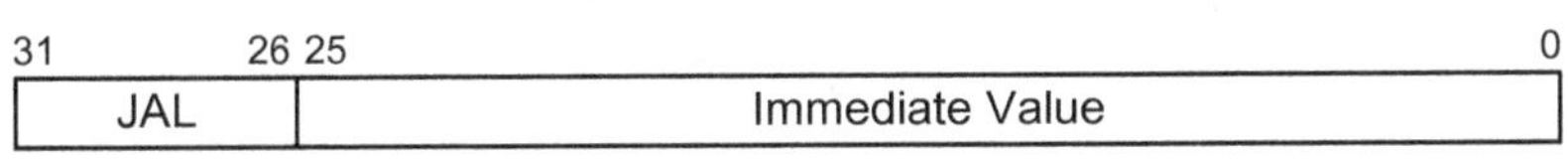

Fig. 6.41 Fixed-point Jump-And-Link (JAL) instruction field format

The last unconditional jump-type instruction is the Jump-And-Link Register (JALR) instruction. This instruction also requires a two-step process. In the first step, the PC address, (PC + 2), is stored in the register R31. In the second step, the PC is loaded with the contents of RS as shown below. The field format of this instruction is shown in Fig. 6.42.

JALR RS
(PC + 2) → Reg[R31] followed by Reg[RS] → PC

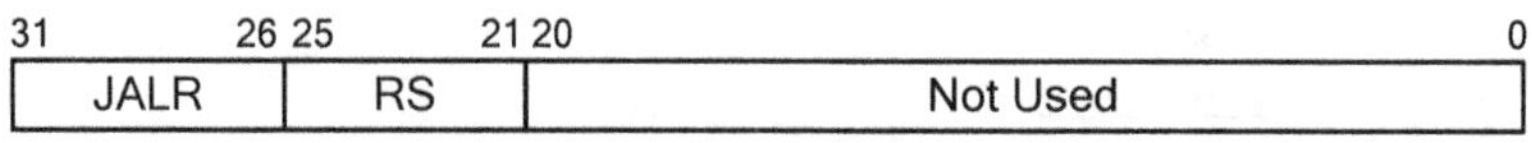

Fig. 6.42 Fixed-point Jump-And-Link Register (JALR) instruction field format

The Return instruction (RET) works with the JAL or JALR instruction. It retrieves the old program address stored in the register R31, and replaces the current PC value with the contents of R31 in order to go back to the original program location as described below. This instruction's field format is given in Fig. 6.43.

RET
Reg[R31] → PC

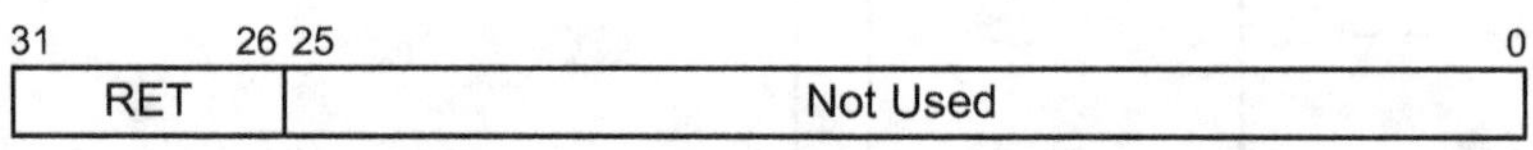

Fig. 6.43 Fixed-point Return (RET) instruction field format

6.2 Stack Pointer and Subroutines

Subroutines

A subroutine is a small program within the main program. Its structure is shown in Fig. 6.44. In this figure, the JAL 10 instruction first stores the (PC + 2) = 5 in R31 (the instruction at PC = 4 must be either a NOP or an unrelated instruction to the JAL instruction due to the Jump-type hazard), then

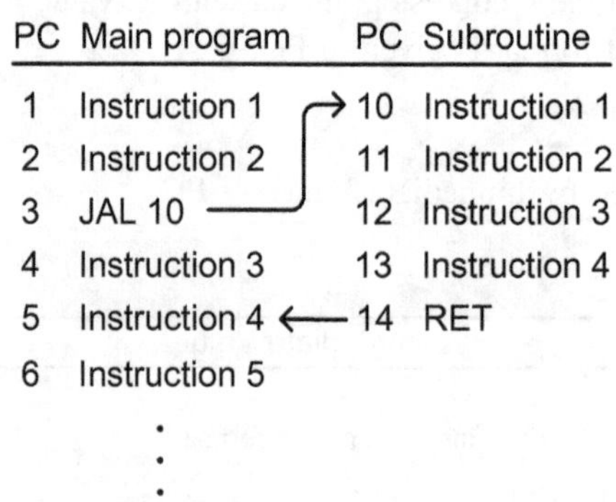

Fig. 6.44 Subroutine (function call) structure

jumps to the instruction at PC = 10. Once there, the program continues to execute the subroutine until it reaches the RET instruction. The RET instruction retrieves the return value in R31 and jumps back to PC = 5 to resume the rest of the program.

Although storing the return value in R31 works in small programs which execute one subroutine at a time, this strategy will not work in bigger programs which may contain numerous "nested" subroutines (one subroutine inside another). To solve this dilemma, programs use the stack architecture.

The Stack and the Stack Pointer

The stack is part of the data memory dedicated to store the return values of (nested) subroutines in a user program. This process is shown in Fig. 6.44. The operation of the stack somewhat resembles to First-In-Last-Out (FILO) memory.

Suppose there are three data values, A, B and C, which need to be pushed onto the data stack. Assuming that the top of the stack is empty, PUSH A instruction stores data A in the stack with nothing else on top of it as shown in the first column of Fig. 6.45.

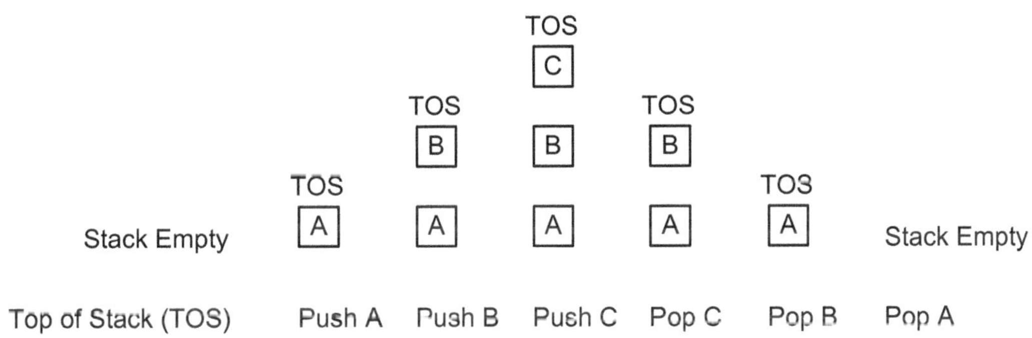

Fig. 6.45 Operation of PUSH and POP instructions

PUSH B and PUSH C instructions further push data values B and C on top of data A as shown in the second and third columns. POP instruction operates the opposite way. POP C instruction removes data C from the top of the stack as shown in the fourth column. POP B and POP A instructions remove the remaining data in the stack, leaving the stack empty.

Once a number of data packets are stored in the stack by PUSH instructions, the removal of the data by POP instructions has to follow the reverse order. In other words, issuing a POP A instruction while the stack has data values A, B and C as in the third column in Fig. 6.45 cannot remove data A from the bottom of the stack. Therefore, data C has to be removed first, data B second and data C third in order to successfully empty the stack as mentioned above.

Data values are pushed and popped onto the data stack or from the stack by the Stack Pointer (SP), which acts as an address pointer. R31 can be defined to store the stack address and named as the Stack Pointer.

To activate the stack operation, a variety of PUSH and POP instructions can be added on top of the integer-point instructions. These instructions either push or pop the contents of data memory or the contents of RF as shown below:

PUSH f

mem [f] = Imm. Value → mem [SP] = mem {Reg [R31]}

SP + 1 → SP

PUSH RS

Reg [RS] → mem [SP] = mem {Reg [R31]}

SP + 1 → SP

POP f

SP − 1 → SP

mem [SP] → mem [f]

POP RD

SP − 1 → SP

mem [SP] → Reg [RD]

Here, RS and RD stand for a source and destination register addresses in the RF, and f corresponds to an address in the data memory.

Example 6.1 Assume the contents of the data memory and the RF are shown in Fig. 6.46 before PUSH 800 instruction.

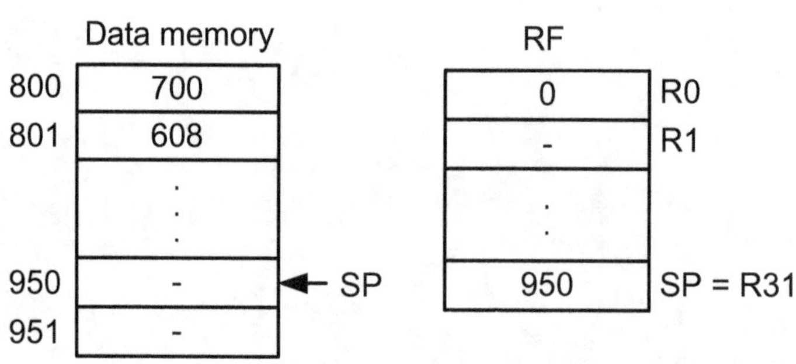

Fig. 6.46 Contents of the data memory and the RF before PUSH 800 instruction

After executing PUSH 800 instruction, and pushing the contents of the data memory at the address, 800, onto the stack, the data memory and the RF become:

mem [800] = 700 → mem [SP] = mem [950]
SP + 1 → SP

This is shown in Fig. 6.47.

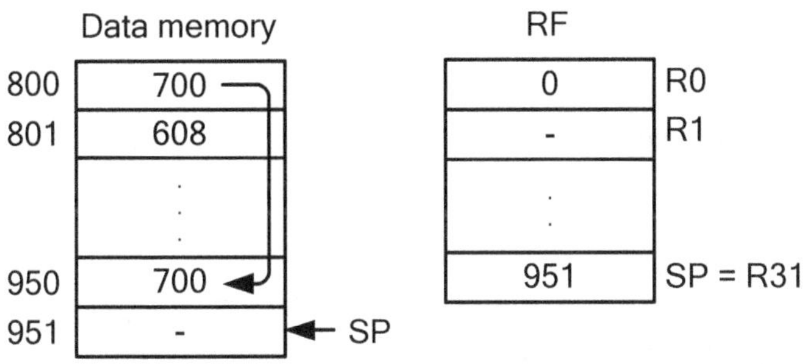

Fig. 6.47 Contents of the data memory and the RF after PUSH 800 instruction

Example 6.2 Assume the contents of the data memory and the RF are shown in Fig. 6.48 before PUSH R1 instruction.

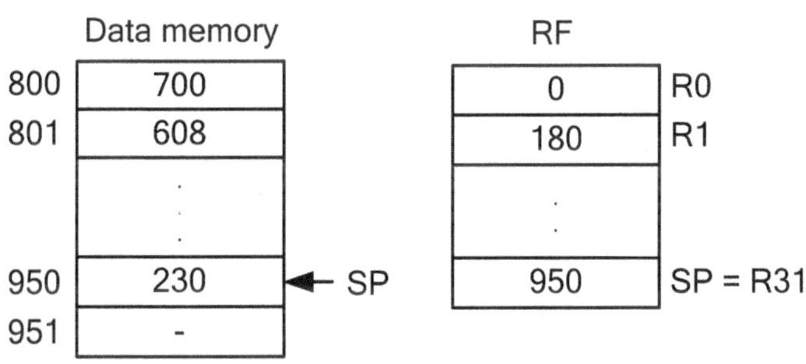

Fig. 6.48 Contents of the data memory and the RF before PUSH R1 instruction

After executing PUSH R1 instruction, and pushing the contents of R1 onto the stack, the data memory and the RF become:

Reg[R1] = 180 → mem [SP] = mem [950]
SP + 1 → SP

This is shown in Fig. 6.49.

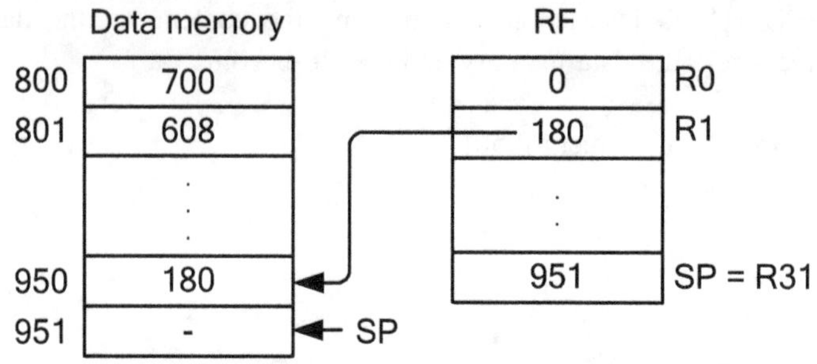

Fig. 6.49 Contents of the data memory and the RF after PUSH R1 instruction

Example 6.3 Assume the contents of the data memory and the RF are shown in Fig. 6.50 before POP 800 instruction.

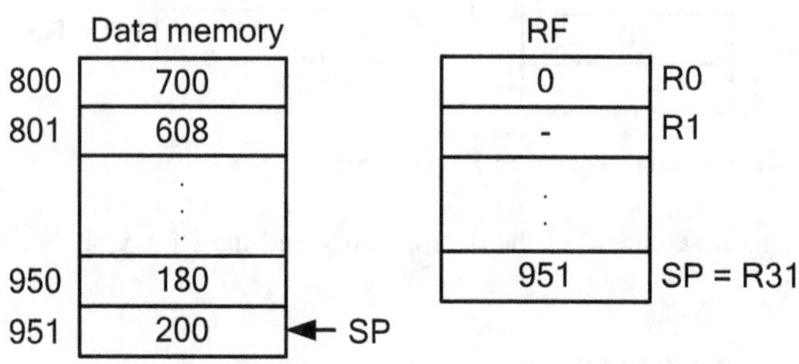

Fig. 6.50 Contents of the data memory and the RF before POP 800 instruction

After executing POP 800 instruction, and popping the contents of the stack to a destination address, 800, in the data memory, the data memory and the RF become:

SP – 1 = 951 – 1 = 950 → SP then
mem [SP] = mem [950] = 180 → mem [800]

This is shown in Fig. 6.51.

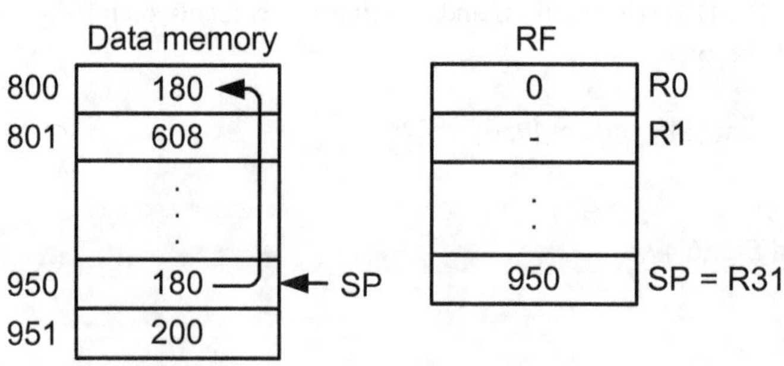

Fig. 6.51 Contents of the data memory and the RF after POP 800 instruction

Example 6.4 Assume the contents of the data memory and the RF are shown in Fig. 6.52 before POP R1 instruction.

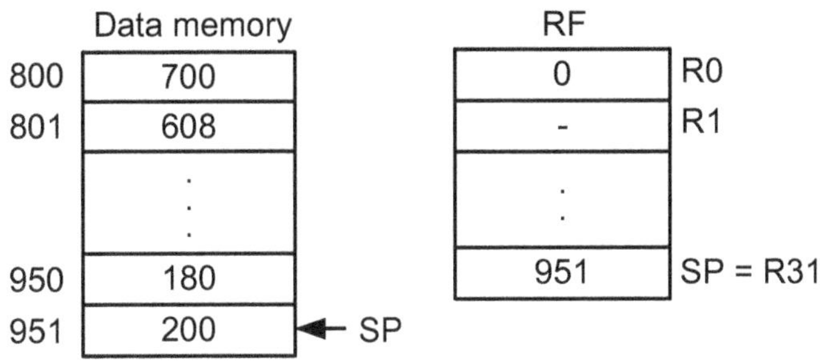

Fig. 6.52 Contents of the data memory and the RF before POP R1 instruction

After executing POP R1 instruction, and popping the contents of the stack to a destination address, R1, at the register file, the data memory and the RF become:

SP − 1 = 951 − 1 = 950 → SP then
mem [SP] = mem [950] = 180 → Reg [R1]

This is shown in Fig. 6.53.

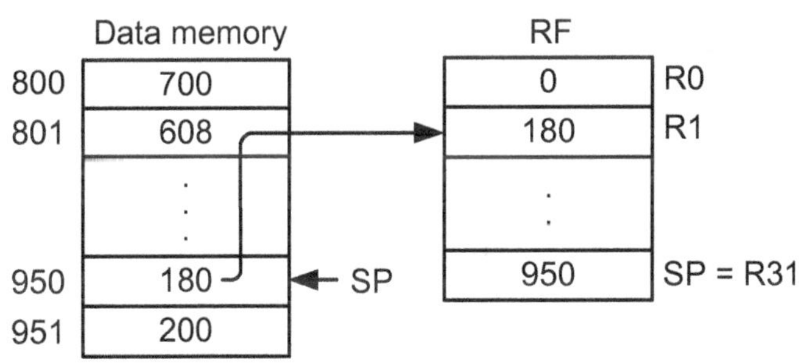

Fig. 6.53 Contents of the data memory and the RF after POP R1 instruction

Call and Return Instructions and the Use of Data Stack

As opposed to using different versions of the Jump-And-Link instructions discussed earlier, it may be more prudent to replace them with a CALL instruction that includes a label or a register address as the target value. The CALL instruction shown below eliminates the need for retracing and renumbering the operand value for each JAL and JALR instruction every time an instruction is added to the program or a modification is made. The CALL instruction functions the same as the JAL or JALR instruction, and stores the return address in the data stack before branching to a target address. The format of this instruction is shown below.

CALL <target>

The RET instruction acts the opposite way of the CALL instruction, and returns the program to its original location after executing a subroutine call. The format of this instruction is shown below.

RET

The following describes the operation of the CALL and RET instructions that use the data stack.

CALL <label>
PC + 2 (return address) → mem [SP]
SP + 1 → SP
<label> → PC where SP = Reg [R31]

RET
SP − 1 → SP
mem [SP] → PC where SP = Reg [R31]

Example 6.5 The program in Fig. 6.54 contains three subroutine calls which need to interact with the stack. Determine the contents of the data memory and the RF at each step. Assume that SP is initialized at the address 900.

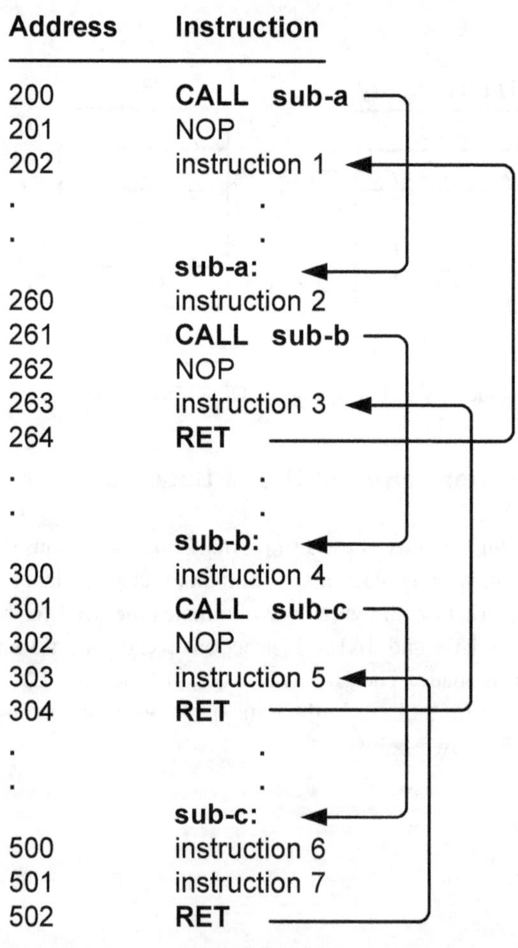

Fig. 6.54 An arbitrary program with three subroutine calls

Before **CALL sub-a**, the contents of the stack and R31 are shown in Fig. 6.55.

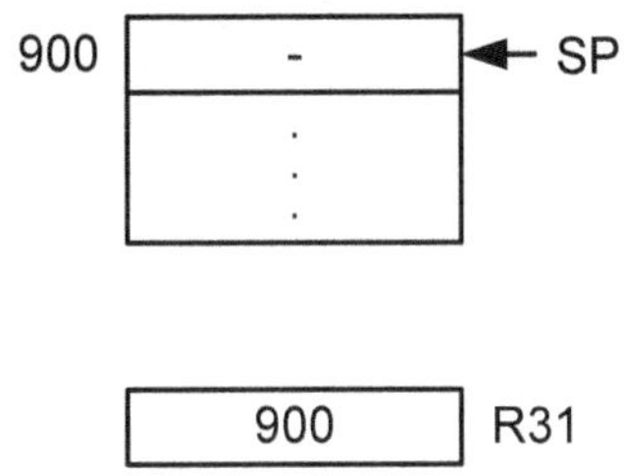

Fig. 6.55 Contents of the stack before the first CALL instruction

After **CALL sub-a**:

PC + 2 = 200 + 2 = 202 → mem [SP] = mem [900] (return address is stored)
SP + 1 = 900 + 1 = 901 → SP = Reg [R31]
sub-a = 260 → PC

Thus, the contents of the stack and R31 become as in Fig. 6.56.

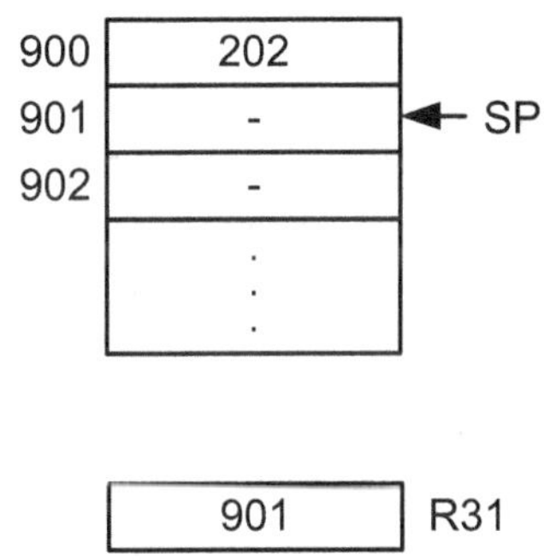

Fig. 6.56 Contents of the stack after the first CALL instruction

The program then executes instruction 2.
After **CALL sub-b**:

PC + 2 = 261 + 2 = 263 → mem [SP] = mem [901] (return address is stored)
SP + 1 = 901 + 1 = 902 → SP = Reg [R31]
sub-b = 300 → PC

Thus, the contents of the stack and R31 become as in Fig. 6.57.

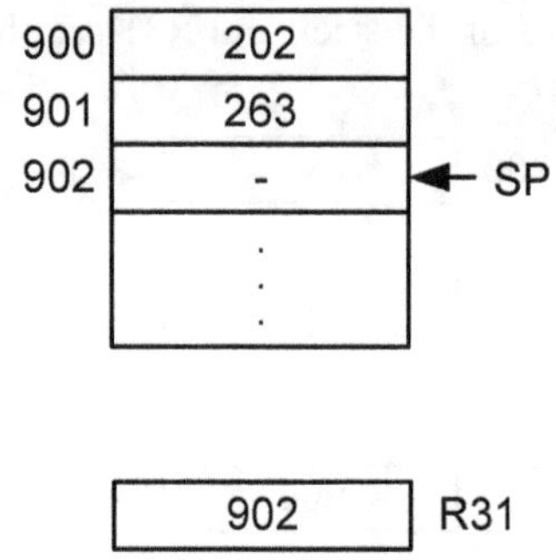

Fig. 6.57 Contents of the stack after the second CALL instruction

The program then executes instruction 4.
After **CALL sub-c**:

PC + 2 = 301 + 2 = 303 → mem [SP] = mem [902] (return address is stored)
SP + 1 = 902 + 1 = 903 → SP = Reg [R31]
sub-c = 500 → PC

Thus, the contents of the stack and R31 become as in Fig. 6.58.

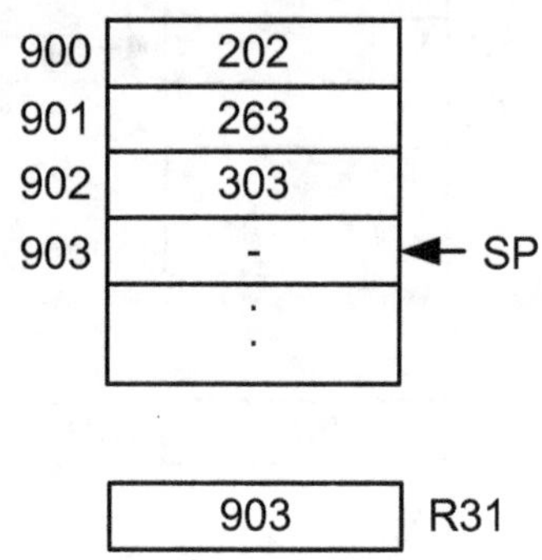

Fig. 6.58 Contents of the stack after the third CALL instruction

The program then executes instructions 6 and 7.
After the first return, **502 RET**:

SP − 1 = 903 − 1 = 902 → SP = Reg [R31]
mem [SP] = mem [902] = 303 → PC

Thus, the contents of the stack and R31 become as in Fig. 6.59.

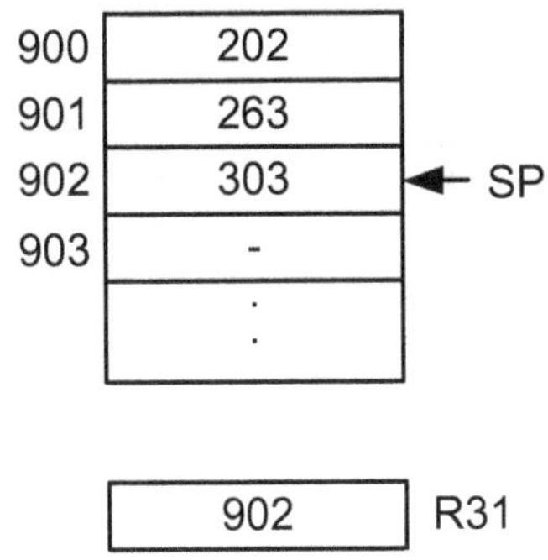

Fig. 6.59 Contents of the stack after the first RET instruction

The program then executes instruction 5.
After the second return, **304 RET**:

SP − 1 = 902 − 1 = 901 → SP = Reg [R31]
mem [SP] = mem [901] = 263 → PC

Thus, the contents of the stack and R31 become as in Fig. 6.60.

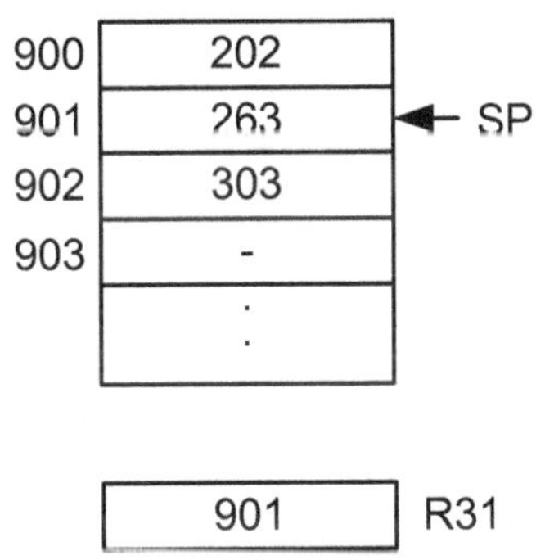

Fig. 6.60 Contents of the stack after the second RET instruction

The program then executes instruction 3.
After the third and final return, **264 RET**:

SP − 1 = 901 − 1 = 900 → SP = Reg [R31]
mem [SP] = mem [900] = 202 → PC

Thus, the contents of the stack and R31 become as in Fig. 6.61.

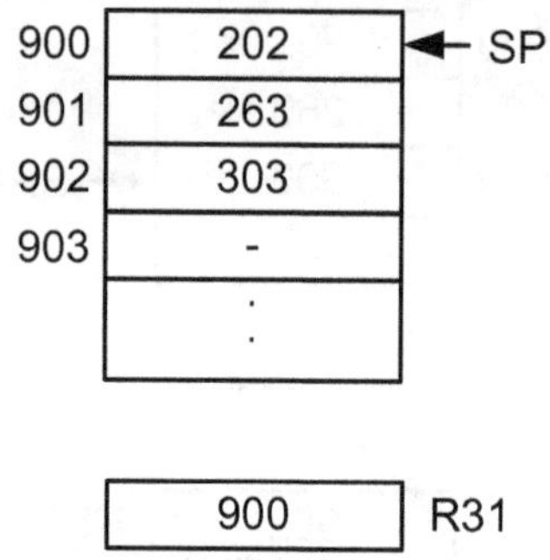

Fig. 6.61 Contents of the stack after the third RET instruction

6.3 Fixed-Point Design Examples

Example 6.6 Design a fixed-point CPU data-path to execute an instruction set whose opcodes are listed below:

ADD, LOAD, STORE and MOVE

The first step of the design process is to start with a single instruction in the instruction list and build its data-path. Each new instruction brings new hardware requirements to the design, and they are added incrementally to the existing data-path. Once the data-path reaches its final form so that it is able to execute a set of instructions, the second step is to build the OPC decoders to control each pipeline stage and guide the data.

In this design, we start with the data-path for the ADD instruction as shown in Fig. 6.7. The next step is to introduce additional hardware for the LOAD instruction. To accommodate this requirement, the first modification is to place a 2-1 MUX in the ALU stage to select between the immediate value required by the LOAD instruction and Reg[RS2] required by the ADD instruction as shown in Fig. 6.62. The second modification is to add a bypass path in the data memory stage so that the adder result bypasses the data memory if the OPC is ADD, or it is used as an effective address for the data memory if the OPC is LOAD. Finally, a third modification is to place a 2-1 MUX in the write-back stage in order to select either the contents of the data memory for the LOAD instruction or the adder output for the ADD instruction before writing the result back to the destination register, RD.

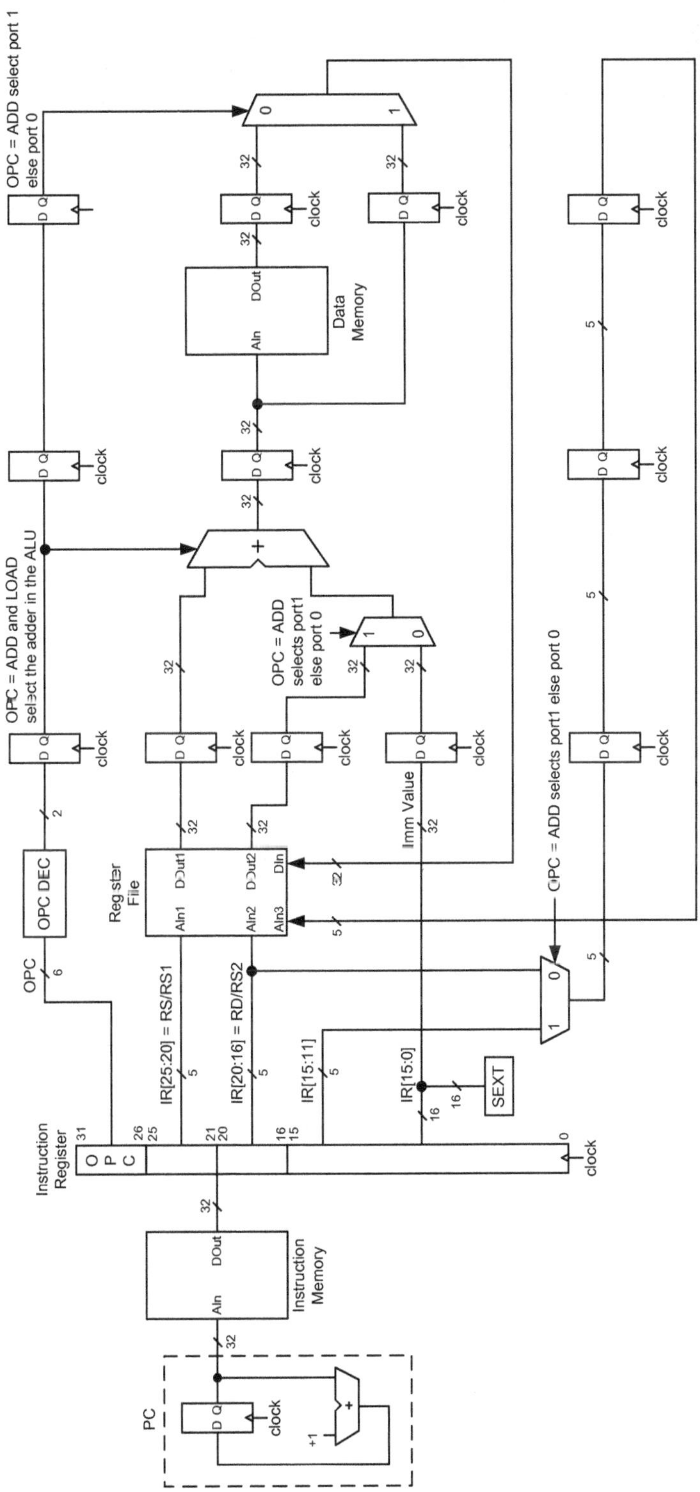

Fig. 6.62 CPU data-path with ADD and LOAD instructions (WE = 1 to the RF and Data Memory are omitted for clarity)

When the STORE instruction is introduced as a third instruction, it prompts a change in the calculation of the data memory address from {Reg[RS] + Imm Value} to {Reg[RD] + Imm Value}. This change requires a secondary 2-1 MUX to be placed in the ALU stage to guide the effective memory address to the AIn port of the data memory, and an additional path to transfer Reg[RS] to the DIn port of the data memory as shown in Fig. 6.63. The modifications for the STORE instruction, however, should not alter the existing data-paths for the ADD and LOAD instructions. If the ADD and LOAD instruction data-paths are individually traced after adding the hardware to support the STORE instruction, both Reg[RS1] + Reg[RS2] and the memory contents at {Reg[RS] + Imm Value} should still be available to write back to a destination register in the RF.

Introducing the MOVE instruction requires another write-back path to be integrated with the three existing write-back paths in the architecture. Therefore, a 3-1 MUX needs to placed in the write-back stage to pass the contents of RS to the RF as shown in Fig. 6.64. While the MOVE and the LOAD instructions use port 0 and port 1 of the 3-1 MUX, respectively, the rest of the instructions use port 2 for the write-back path.

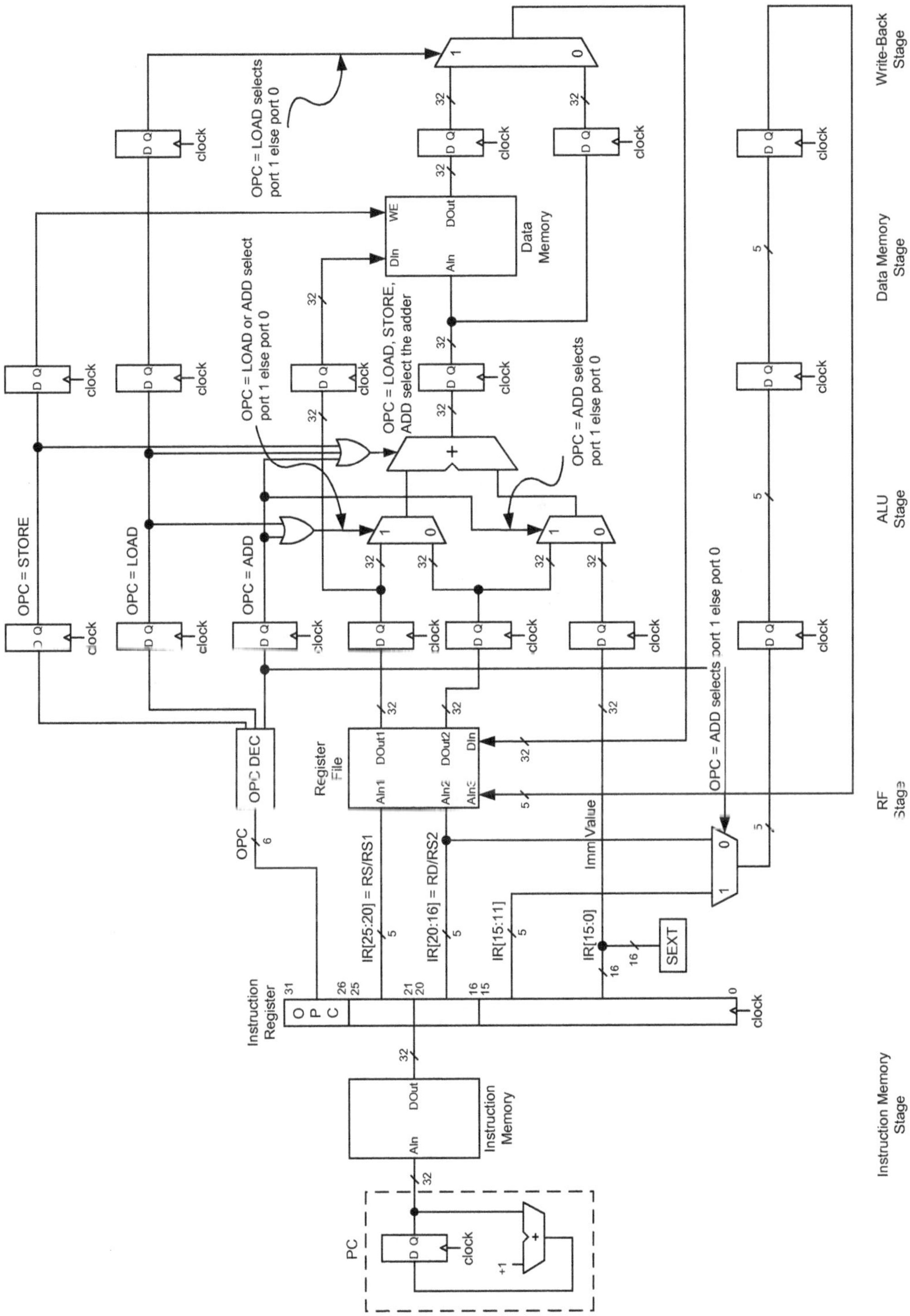

Fig. 6.63 CPU data-path with ADD, LOAD and STORE instructions (WE = 1 to the RF and Data Memory are omitted for clarity)

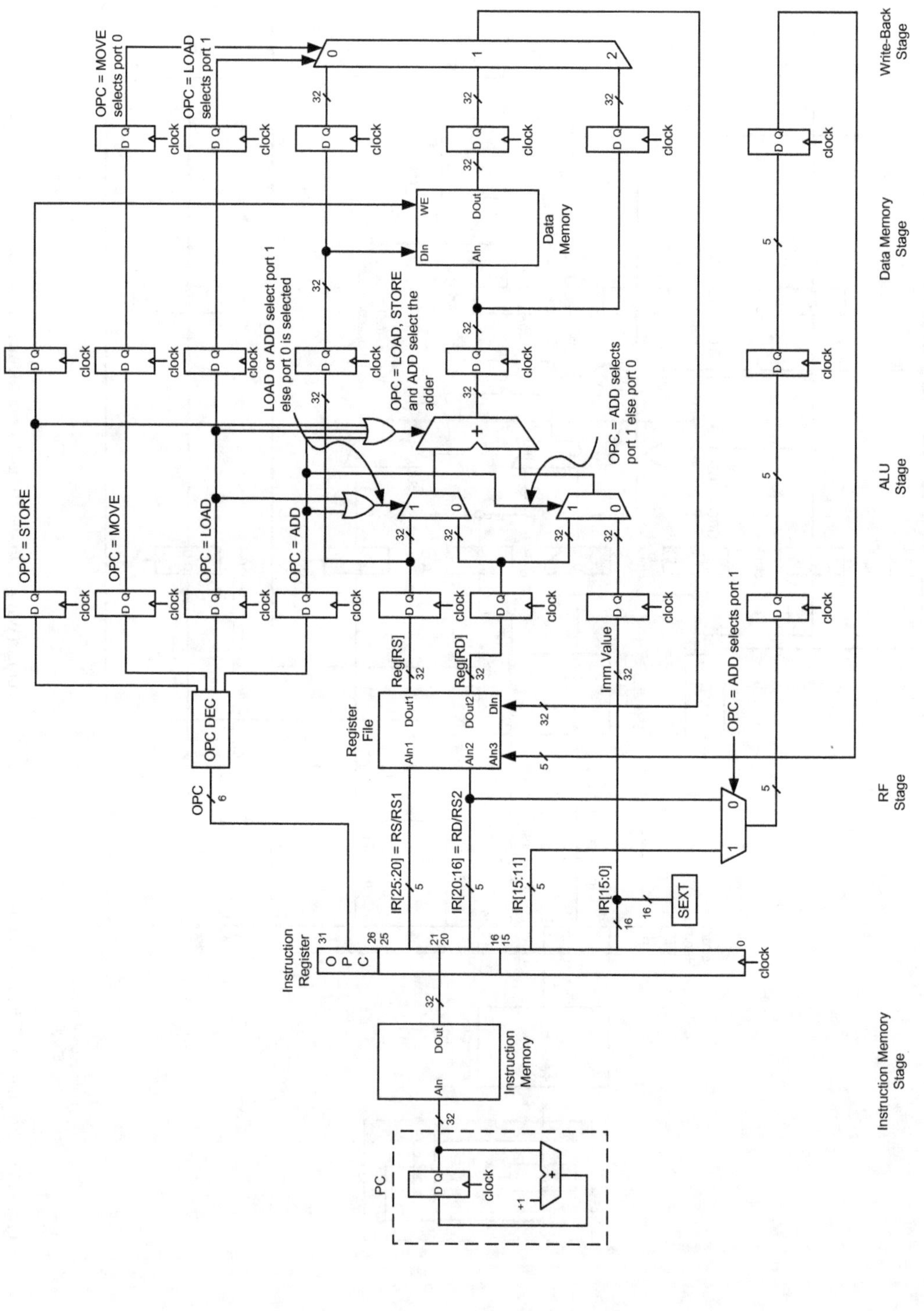

Fig. 6.64 CPU data-path with ADD, LOAD, STORE and MOVE instructions (WE = 1 to the RF and Data Memory are omitted for clarity)

The controller for each stage of the data-path is OPC-dependent and completely combinational as shown in Fig. 6.65. Since the OPC propagates from one stage to another with the data, it can effectively be used as a control input to guide data and to activate the required hardware in each stage. Four instructions need only two input bits stemming from the instruction register, IR[27:26], to design the OPC decoder in Fig. 6.65. The more significant four OPC bits, IR[31:28], are considered zero for an instruction set of four.

OPC	IR[27]	IR[26]
ADD	0	0
LOAD	0	1
STORE	1	0
MOVE	1	1

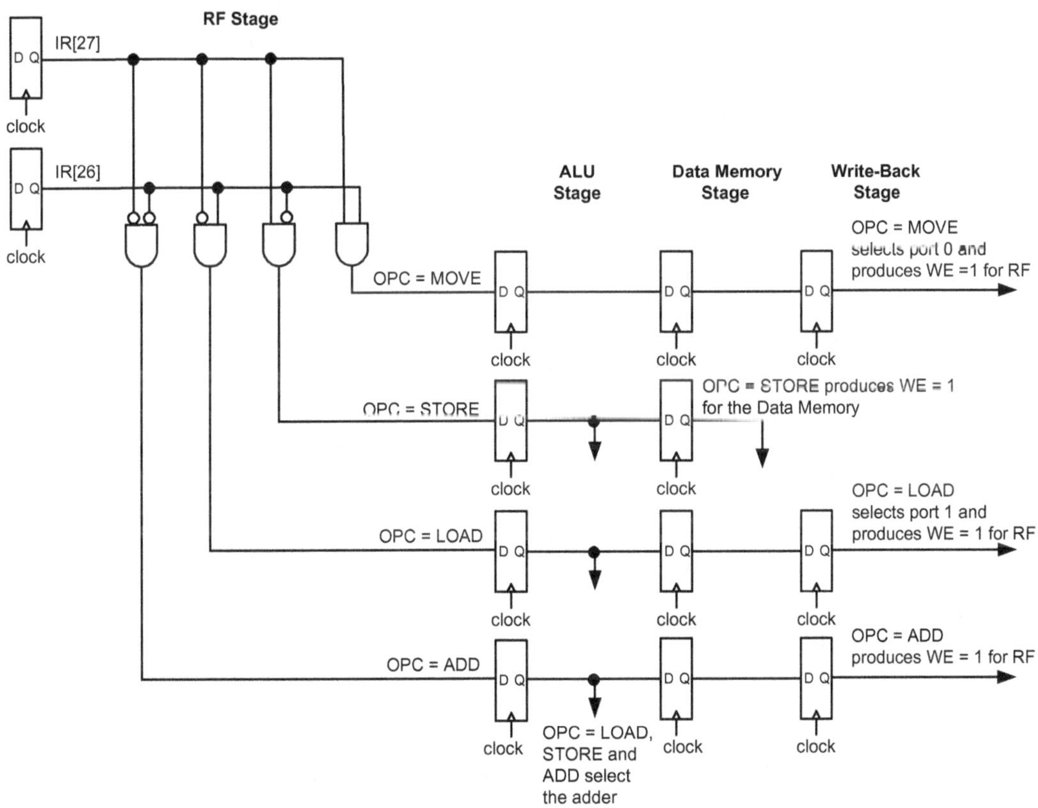

Fig. 6.65 OPC table for ADD, LOAD, STORE and MOVE instructions and the control circuitry

To generate the ADD selector input, both IR[27] and IR[26] are complemented and then ANDed according to the table in Fig. 6.65. The selector inputs for LOAD, STORE and MOVE instructions are also generated using the same OPC table.

The STORE selector input is connected to Write Enable (WE) bit for the data memory since the STORE instruction is the only instruction that writes data to the data memory. All other instructions write back the results to the RF, and therefore the WE input to the RF should be enabled. However, this signal is not shown in Figs. 6.64 or 6.65 to avoid complexity.

Example 6.7 Design a fixed-point CPU data-path to execute an instruction set whose opcodes are listed below:

ADD, LOAD, STORE, MOVE, SLI, SRI, JUMP and BRA

The design methodology used in this example is the same as in the previous example. First, as additional instructions are introduced to the design, new hardware for each instruction is incrementally added to the existing data-path. Second, an OPC truth table should be constructed from the instruction set. Third, controller outputs should be generated from the OPC truth table to guide the data in each CPU stage.

We start with the ADD instruction data-path given in Fig. 6.7 to form the base platform. The SLI and SRI instructions are implemented next. Both of these instructions require left and right linear shifters in the ALU stage where each shifter can individually be selected by the SLI or SRI inputs as shown in Fig. 6.66. These instructions also require a bypass path in the RF stage that connects IR [15:0] to the ALU input as explained earlier.

When the LOAD instruction is introduced as the fourth instruction, the existing 32-bit adder in the ALU is used to calculate the effective address for the data memory. The 2-1 MUX in the write-back stage is replaced by a 3-1 MUX to be able to write the contents of the data memory back to the RF.

The STORE instruction is the fifth instruction added to this design. This instruction requires two separate paths to calculate the data memory address and to write the contents of RS to the data memory. The STORE instruction also necessitates a secondary 2-1 MUX in the ALU stage so that an immediate value is added to the contents of RD instead of RS.

The MOVE instruction is the sixth instruction which requires a continuous bypass path to the RF, bypassing both the ALU and the data memory. This path passes through port 0 of the 3-1 MUX in the write-back stage to transfer the contents of RS back to the RF.

The BRA instruction is the seventh instruction in this set and requires a path to compare Reg[RS] with RS Value, IR[20:16]. The bitwise comparison is done by 32 two-input XNOR gates followed by a single 32-input AND gate, and produces a single bit that selects between (PC + Imm Value) and (PC + 2). The selected target address is loaded to the PC to redirect the program to a different PC address. This instruction also requires a special 32-bit adder in the RF stage to calculate (PC + Imm Value) as shown earlier.

The JUMP instruction simply forwards the jump value, IR[25:0], extended to 32 bits to the PC. It requires a second 2-1 MUX that decides between (PC + 2) and the jump value before loading the result to the PC. The ALU adder is selected by the ADD, LOAD and STORE instructions. However, this selection is not shown in Fig. 6.66 to avoid complexity.

Once the data-path for all eight instructions is complete, the OPC truth table in Fig. 6.67 is formed. Since there are eight instructions in this set, only IR[28:26] are used for designing the OPC decoder. The upper three bits of the OPC field become equal to zero.

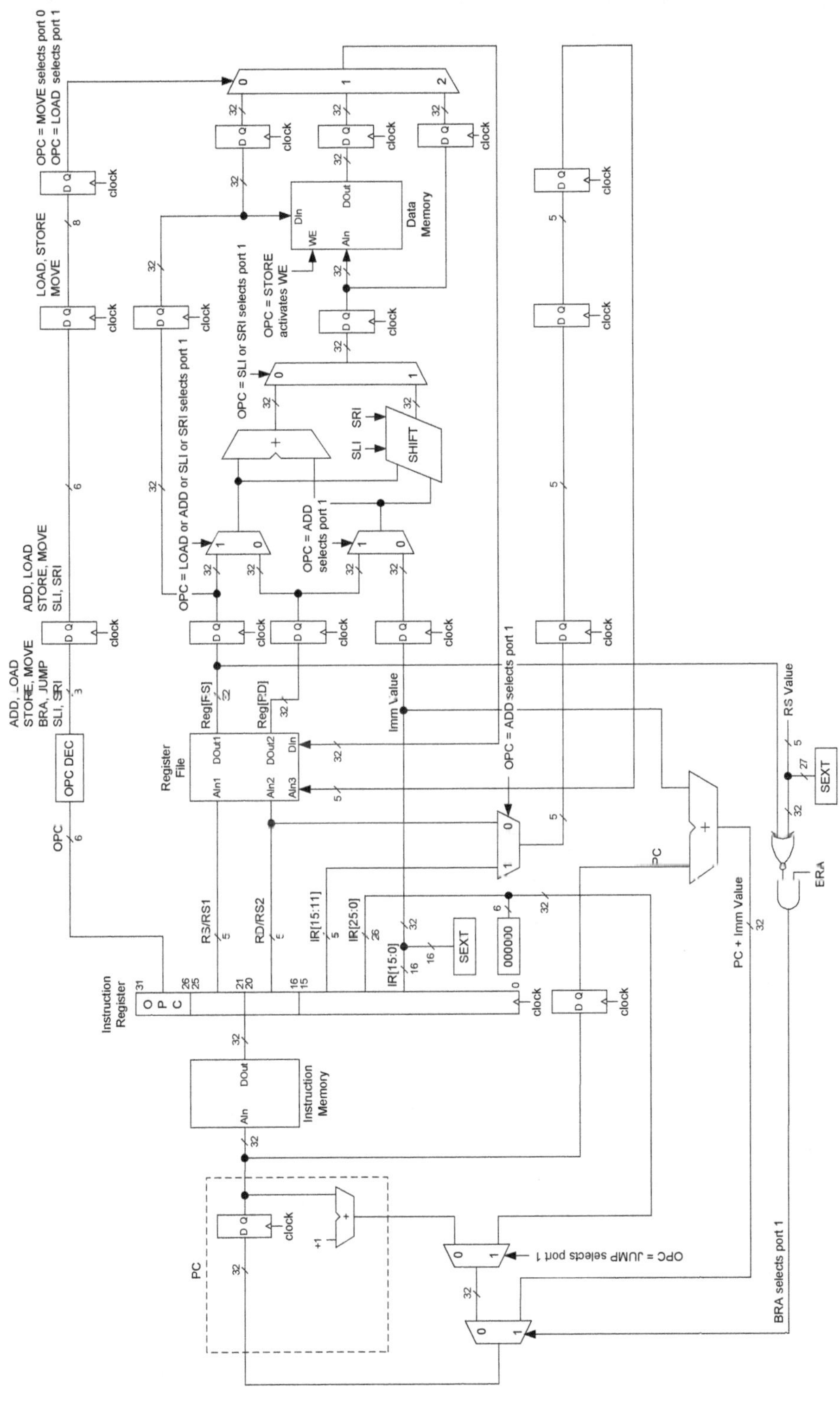

Fig. 6.66 CPU data-path with ADD, LOAD, STORE, MOVE, SLI, SRI, JUMP and BRA (WE = 1 for RF and Data Memory are omitted for clarity)

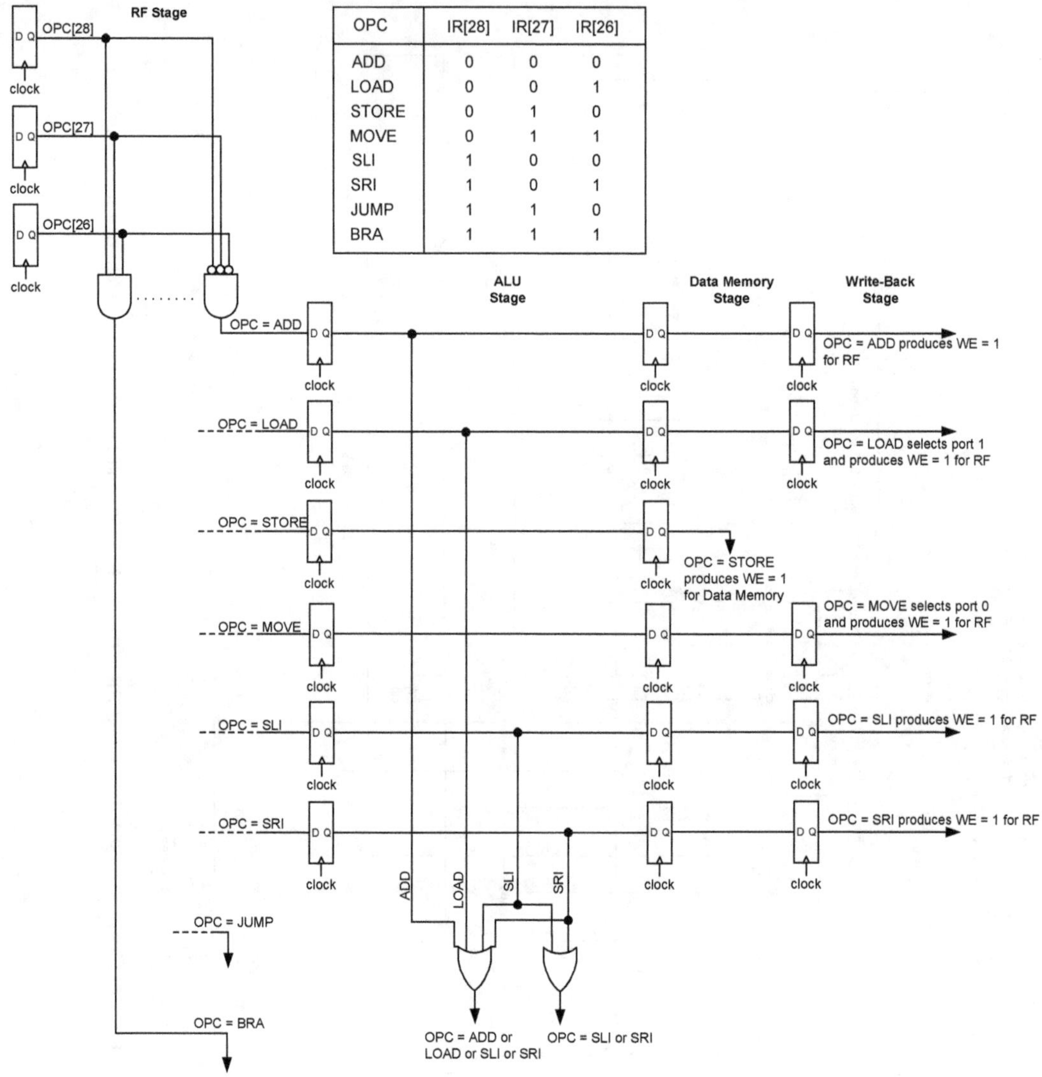

OPC	IR[28]	IR[27]	IR[26]
ADD	0	0	0
LOAD	0	0	1
STORE	0	1	0
MOVE	0	1	1
SLI	1	0	0
SRI	1	0	1
JUMP	1	1	0
BRA	1	1	1

Fig. 6.67 OPC table for ADD, LOAD, STORE, MOVE, SLI, SRI, JUMP and BRA and the control circuitry

To generate the ADD selector input (used in the RF and ALU stages), the first row of the OPC table is implemented. This requires IR[28], IR[27] and IR[26] to be complemented and ANDed. The LOAD, STORE, MOVE, SLI, SRI, JUMP and BRA selector inputs are also generated similarly using the same OPC table. The ALU stage requires the ADD, LOAD, SLI and SRI selector inputs to be ORed to select Reg[RS] for the adder input. Similarly, SLI and SRI inputs are ORed to select the shifter outputs. The STORE selector input is connected to the WE input of the data memory since STORE instruction is the only instruction in the instruction set that writes data to the data memory. The WE for the RF is omitted to avoid complexity.

6.4 Fixed-Point Hazards

Structural Hazards

When instructions are fetched from the instruction memory and introduced to the CPU pipeline, they follow each other by one clock cycle as shown in Fig. 6.68. In this figure, when the first instruction transitions to the RF stage in cycle 2, the second instruction starts its instruction fetch cycle. In cycle 3, the first instruction enters the ALU stage, the second the RF stage, and the third the instruction fetch stage.

CYCLES	1	2	3	4	5	6	7	8	9
INSTRUCTION 1	I	R	A	D	W				
INSTRUCTION 2		I	R	A	D	W			
INSTRUCTION 3			I	R	A	D	W		
INSTRUCTION 4				I	R	A	D	W	
INSTRUCTION 5					I	R	A	D	W

Fig. 6.68 Structural hazards in a five-stage CPU

The fourth cycle in Fig. 6.68 should be viewed with a particular importance because it creates a structural hazard. In this cycle, the first instruction accesses the data memory while the fourth instruction is fetched from the instruction memory. If there is only one memory block with a single port (to read instructions and to store data), this configuration will create a structural hazard because the first and the fourth instructions will try to access the memory in the same cycle. This is the primary reason to have separate instruction and data memories in a RISC CPU.

Cycles 5 and 6 create another type of structural hazard. In both of these cycles, data has to be read from the RF and written to the RF in the same cycle. If the data read from the RF depends on the data written to the RF, this condition will create a hazard because the write needs to take place before the read in order to produce correct results. Therefore, RF should be designed in such a way that all the writes have to take place at the high phase of the clock and all the reads at the low phase.

Data Hazards

Data hazards create a situation where the required data is unavailable when it is needed by an instruction. There are four common data hazards in this architecture. The examples below illustrate each data hazard type and propose solutions in the CPU architecture to avoid them.

The first type of data hazard is shown in the example in Fig. 6.69. In this example, the ADD instruction adds the contents of R1 and R2, and then writes the result to a destination register, RD.

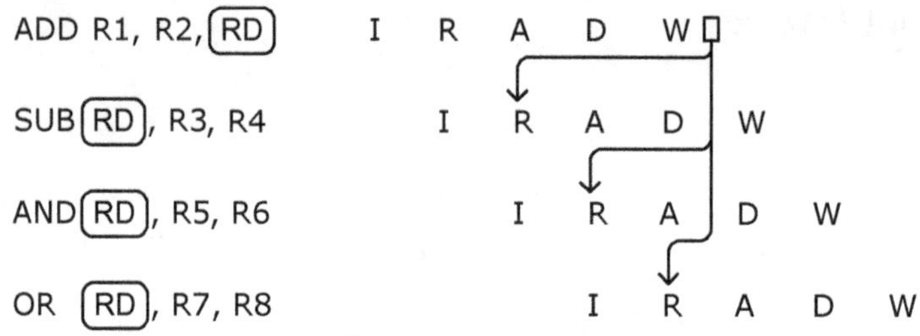

Fig. 6.69 Data hazard: a register-type instruction followed by other register-type instruction(s)

The second, third and fourth instructions all require the contents of RD to proceed. The only hazard-free case here is the data exchange between the ADD and OR instructions since the RF permits writes during the high phase of the clock and reads during the low phase. However, the SUB and AND instructions need to fetch the contents of RD before they become available. Therefore, executing any of these instructions without any protection in the CPU pipeline would produce two separate hazards. In order to circumvent this problem, a technique called "data-forwarding" is applied to the CPU pipeline. This method requires a special data route in the CPU data-path so that partially processed data is immediately transferred from a particular pipeline stage to the next when it is needed. Figure 6.70 shows the two forwarding paths to remove the data hazards associated with the SUB and AND instructions. The first path transfers data from the ALU output to the ALU input when the SUB instruction needs the ADD instruction's ALU result to proceed. The second path transfers data from the output of the data memory to the ALU input when the AND instruction needs the ADD instruction's ALU output.

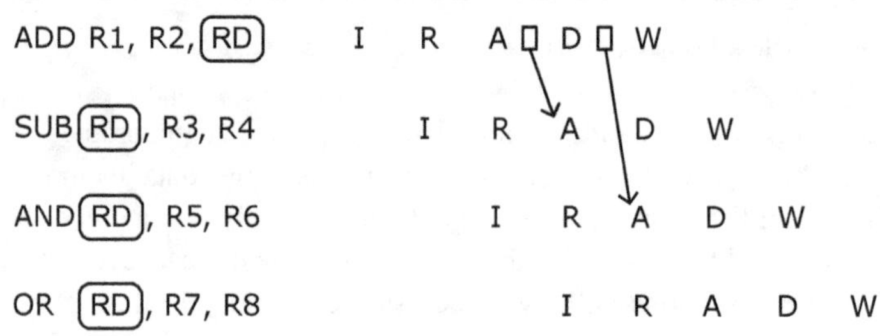

Fig. 6.70 Forwarding paths to remove the data hazards caused by a register-type instruction followed by other register-type instruction(s)

Figure 6.71 shows the first forwarding path from the ALU output to the ALU input to remove the data hazard caused by the ADD-SUB instruction pair in Fig. 6.69. In this figure, the source RF address of the SUB instruction is compared with the destination RF address of the ADD instruction. If there is a match, then the ALU output to ALU input forwarding path(s) is activated by selecting port 1 of the 2-1 MUX at the input of the ALU.

The second forwarding path shown in Fig. 6.72 feeds back the output of the data memory stage to the input of the ALU and removes the data hazard caused by the ADD-AND instruction pair in Fig. 6.69. Again, the source RF address of the AND instruction is compared against the destination RF address of the ADD instruction. The path that connects the data memory output at the end of the bypass path to the ALU input is activated by selecting port 1 of the 2-1 MUX if there is a match.

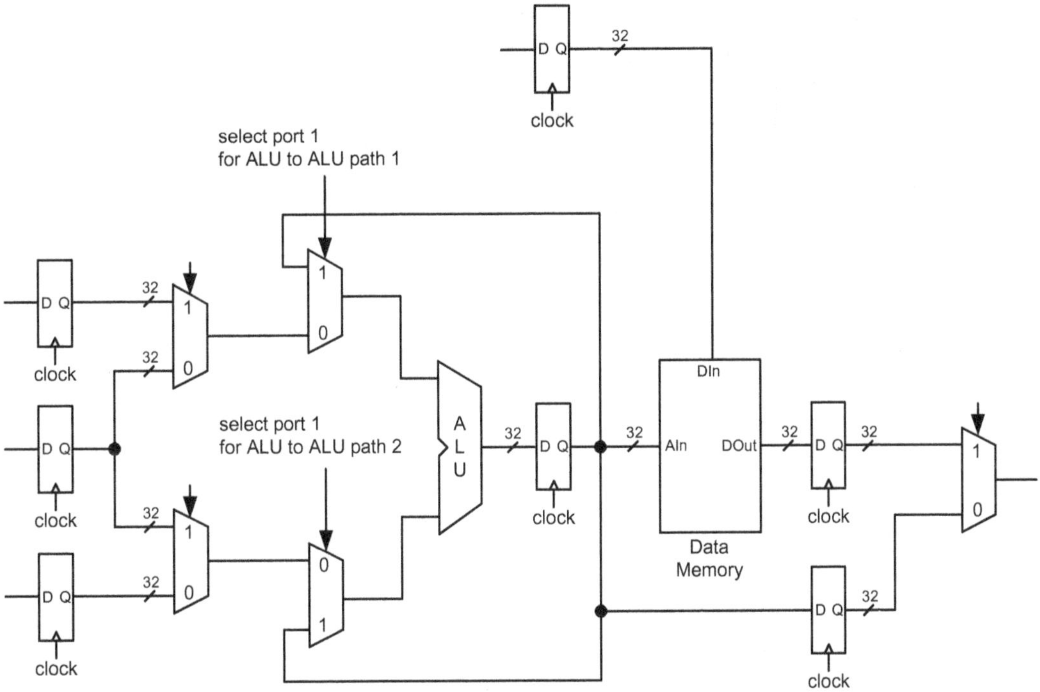

Fig. 6.71 ALU output to ALU input forwarding path (ALU to ALU path 1 or path 2 depends on if the forwarded data needs to replace Reg[RS1] or Reg[RS2], respectively)

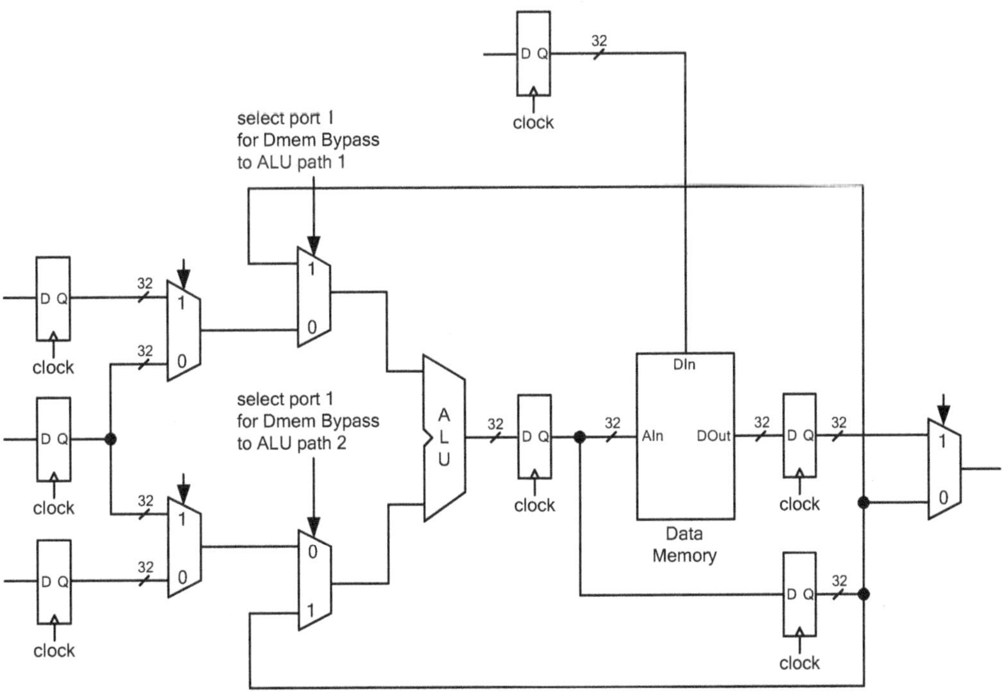

Fig. 6.72 Data memory bypass output to ALU input forwarding path (Data memory bypass to ALU path 1 or path 2 depends on if the forwarded data needs to replace Reg[RS1] or Reg[RS2], respectively)

Another type of data hazard is shown in Fig. 6.73. This hazard originates from an instruction that requires the contents of the data memory in the next clock cycle. A forwarding path that connects the output of the data memory to the input of the ALU may not be sufficient to remove this data hazard as shown in Fig. 6.74. However, if a No Operation (NOP) instruction is inserted between the LOAD and ADD instructions, the one cycle delay created by this instruction can avoid this hazard as shown in Fig. 6.75. With the NOP instruction in place, the LOAD instruction can now forward the contents of the data memory as a source operand for the ADD instruction as it enters the ALU stage.

Fig. 6.73 Data hazard created by the LOAD instruction followed by a register-type instruction

Fig. 6.74 A forwarding path to remove the data hazard caused by the LOAD instruction followed by a register-type instruction

Figure 6.76 shows the hardware implementation to remove the particular data hazard in Fig. 6.73. The data memory output to the ALU input path is activated by implementing a logic block that compares the destination RF address of the LOAD instruction with the source RF address of the ADD instruction, and enables port 1 of the 2-1 MUX.

Fig. 6.75 A forwarding path and a NOP instruction to remove the data hazard caused by the LOAD instruction followed by a register-type instruction

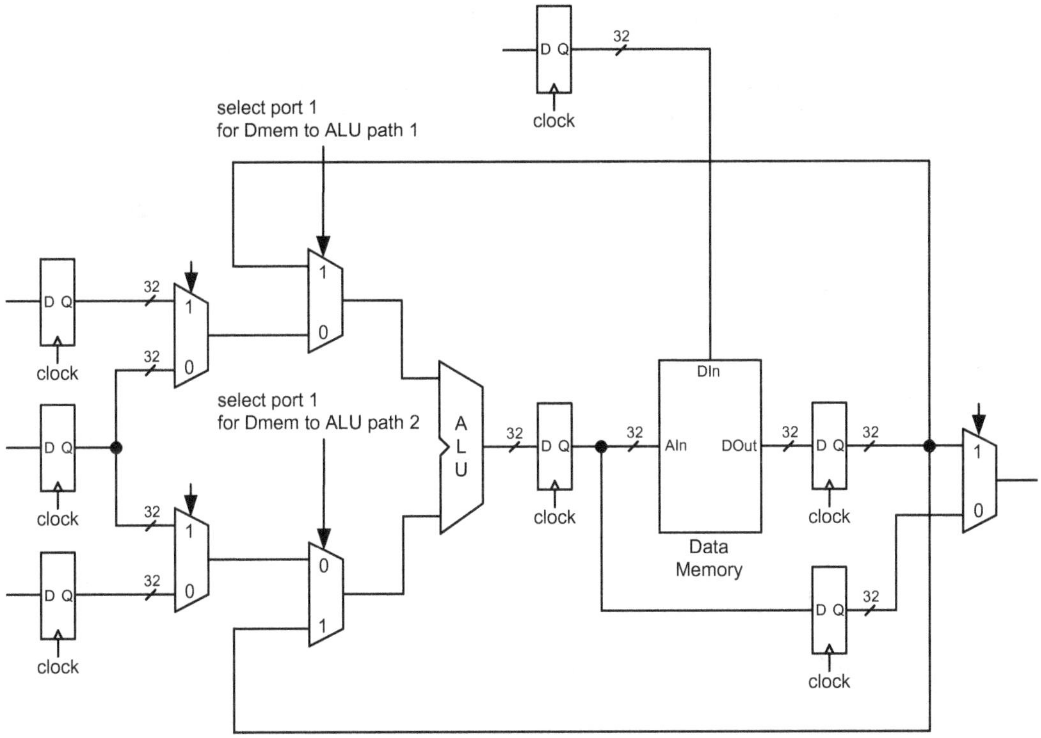

Fig. 6.76 Data memory output to ALU input forwarding path (Data memory output to ALU path 1 or path 2 depends on if the forwarded data needs to replace Reg[RS1] or Reg[RS2], respectively)

The final data hazard shown in Fig. 6.77 stems from a LOAD instruction followed by a STORE instruction. Here, the destination RF address of the LOAD instruction is the same as the source RF address of the STORE instruction. This results in a situation where the data written back to the RF has to be stored in the data memory within the same clock cycle. A forwarding path that connects the output of the data memory to the input of the data memory removes this hazard as shown in Figs. 6.78 and 6.79. Once again, we need a comparator that compares the contents of the destination RF address of the LOAD instruction against the source RF address of the STORE instruction to activate this forwarding path.

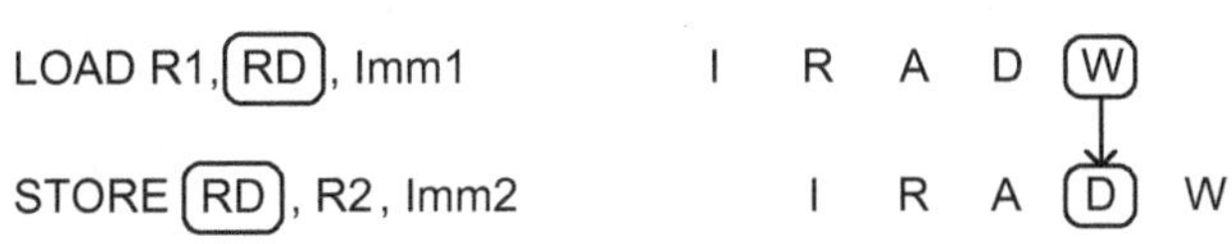

Fig. 6.77 Data hazard: a STORE instruction followed by a register-type instruction

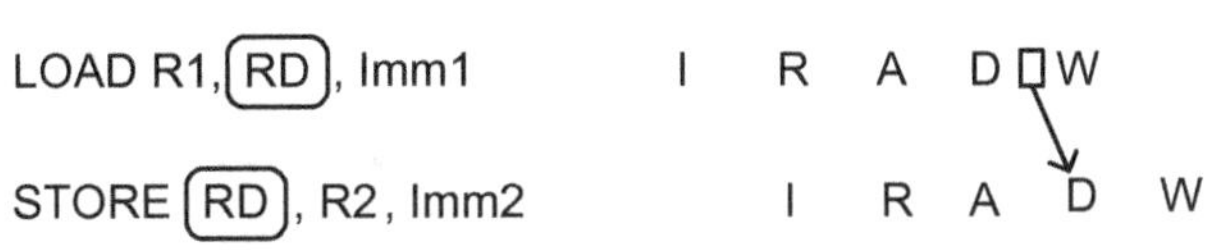

Fig. 6.78 A forwarding path to remove the data hazard caused by a STORE instruction followed by a register-type instruction

Figure 6.80 shows the combination of all four forwarding paths to remove the common data hazards from the CPU pipeline as discussed above. The selector inputs in this figure originate from individual comparators that compare the destination RF address of an instruction with the source RF address of the subsequent instruction.

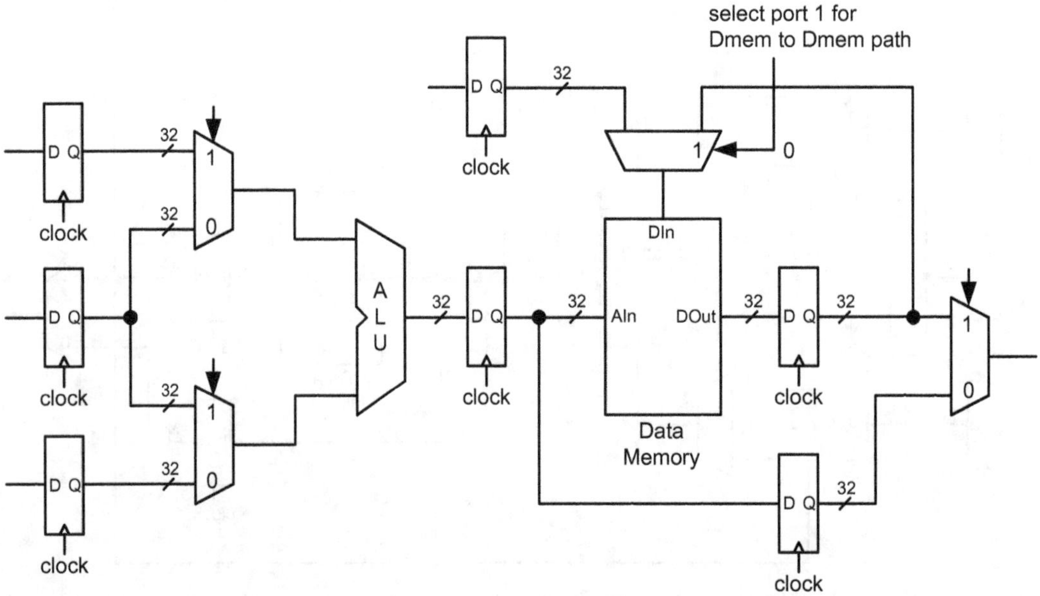

Fig. 6.79 Data memory output to data memory input forwarding path

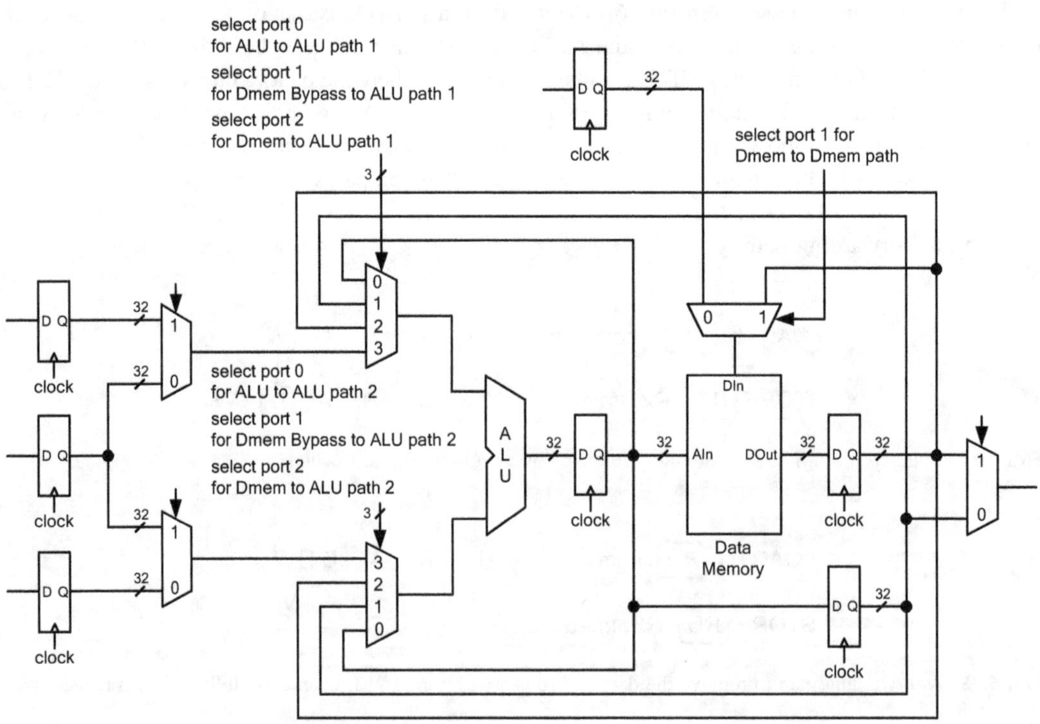

Fig. 6.80 CPU schematic containing all data hazard corrections

Program Control Hazards

Branch and jump instructions also create hazards. In the program shown in Fig. 6.81, the earliest time for the BRA instruction to produce a branch target address is when this instruction is at the RF stage. Therefore, the instruction following the BRA instruction cannot be fetched from the instruction memory in the next clock cycle but it requires one cycle delay. This delay can be implemented either by inserting an unrelated instruction to the branch, or using a NOP instruction following the branch instruction (branch delay slot) as shown in Fig. 6.82.

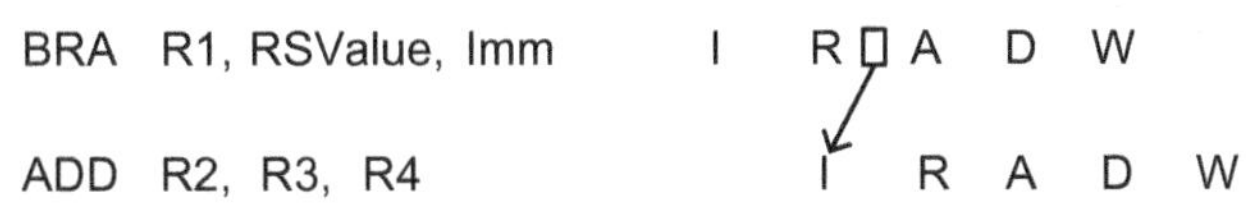

Fig. 6.81 Control hazard: a BRA instruction

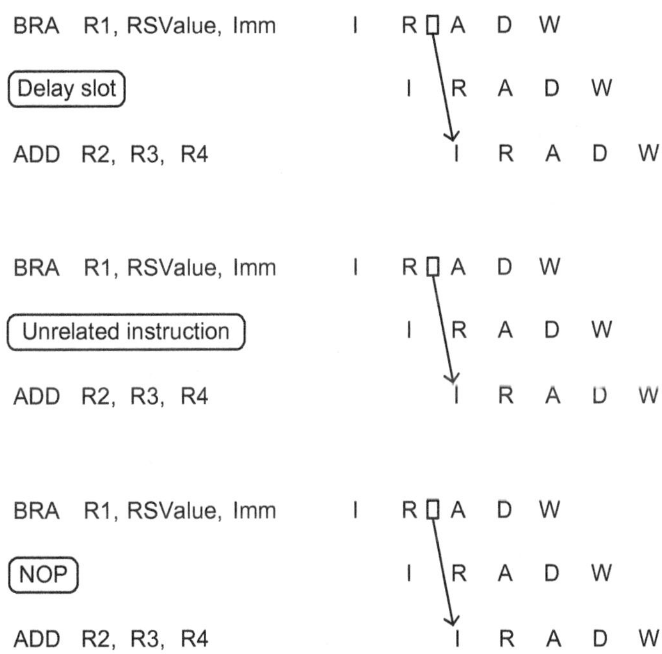

Fig. 6.82 Removal of BRA control-hazard using an unrelated instruction in the program or a NOP instruction

Similar to the BRA instruction, the jump-type (JUMP, JREG, JAL and JALR) instructions also create control hazards since they can only define the address of the next instruction when they are in the RF stage as shown in Fig. 6.83. Inserting a NOP instruction or an unrelated instruction following the jump-type instruction (jump delay slot) removes the pending control hazard as shown in Fig. 6.84.

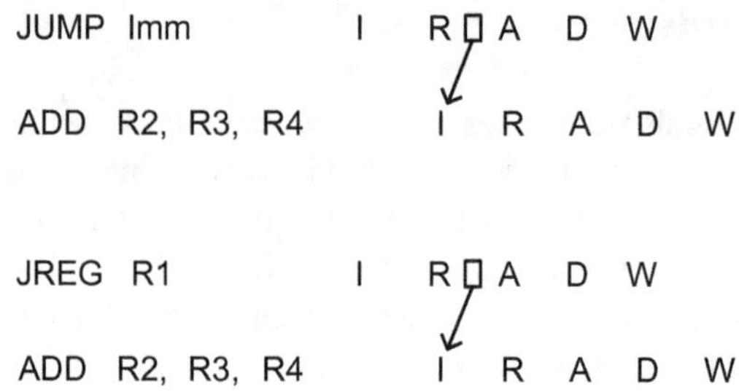

Fig. 6.83 Control hazard: JUMP and JREG instructions

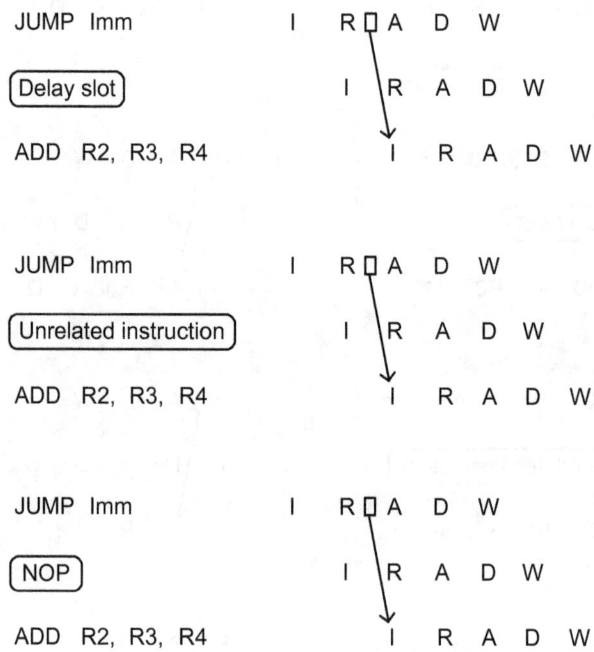

Fig. 6.84 Removal of JUMP and JREG control-hazards using an unrelated instruction in the program or a NOP instruction

Even though the jump-type and branch-type instructions update the contents of the PC, the branch instruction requires a comparator and a special adder in the RF stage to calculate the target address.

Example 6.8 Construct an instruction chart and write a small program according to the flow chart in Fig. 6.85. Show how to handle data and control hazards for a five-stage RISC CPU using forwarding paths. Use NOP instruction(s) when forwarding paths become insufficient to remove a particular hazard.

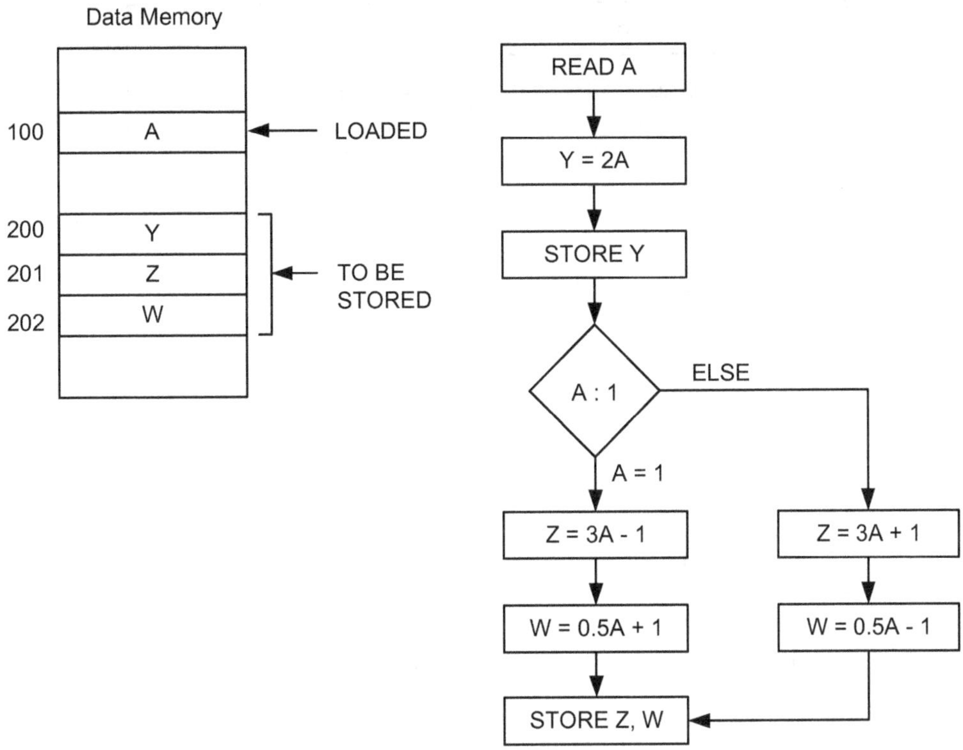

Fig. 6.85 Flow-chart of an example program and data memory contents

Figure 6.85 describes the flow chart of a small program and shows the contents of the instruction memory before this program is executed. Data A in the flow chart is read from the memory address 100. Data Y, Z and W are stored at the memory addresses 200, 201 and 202, respectively.

After Y is stored, the flow chart comes to a decision box where the value of A is compared against one. At this point, the program splits into two where each branch performs calculations to determine the values of Z and W before they are stored in the data memory.

Now, let us convert this flow chart into a program. The instruction, LOAD R0, R1, 100, in Fig. 6.86 adds the contents of R0, which contains only zero, to 100 to calculate the data memory address. It then fetches data A from the data memory address, 100, and writes it to the RF address, R1.

The SLI R1, R2, 1 instruction shifts the contents of R1 to the left by one digit to produce 2A, and writes the result, Y = 2A, to R2. The ADD R1, R2, R3 and SRI R1, R4, 1 instructions compute the values 3A and 0.5A, respectively. Both values will be used later in the program.

The BRA R1, 1, 5 instruction compares the contents of R1, which currently holds A, with the RS Value = 1. If the comparison is successful, the program branches off to fetch the next instruction at the instruction memory location, PC = 4 + 5 = 9, to execute the SUBI R3, R5, 1 that computes Z = 3A – 1. Otherwise, the program fetches the next instruction, ADDI R3, R5, 1 at PC = 4 + 2 = 6 to compute Z = 3A + 1.

PC	Instruction	Comments
0	LOAD R0, R1, 100	A→ Reg [R1]
1	SLI R1, R2, 1	2A → Reg [R2]
2	ADD R1, R2, R3	3A→ Reg [R3]
3	SRI R1, R4, 1	0.5A → Reg [R4]
4	BRA R1, 1, 5	If A = 1 then PC + 5→ PC
5	STORE R2, R0, 200	2A → mem [200]
6	ADDI R3, R5, 1	Z = 3A + 1→ Reg [R5]
7	SUBI R4, R6, 1	W = 0.5A - 1→ Reg [R6]
8	JUMP 11	11→ PC
9	SUBI R3, R5, 1	Z = 3A - 1 → Reg [R5]
10	ADDI R4, R6, 1	W = 0.5A + 1→ Reg [R6]
11	STORE R5, R0, 201	Z→ mem [201]
12	STORE R6, R0, 202	W → mem [202]

Fig. 6.86 Instruction memory contents of the example and explanation of each instruction

The STORE R2, R0, 200 instruction is an unrelated instruction to the branch. Therefore, it is used in the branch delay slot following the branch instruction. This instruction stores the contents of R2, which contains 2A, to the data memory location 200 in Fig. 6.85.

The JUMP 11 instruction at PC = 8 changes the value of the PC with an immediate value of 11. Therefore, the program skips both the SUBI and ADDI instructions following the JUMP instruction to execute the STORE R5, R0, 201 and STORE R6, R0, 202 instructions. As a result, the values of Z and W are stored at the data memory addresses 201 and 202, respectively, regardless of the branch outcome.

Figure 6.87 shows the instruction chart of the program in Fig. 6.86. The LOAD R0, R1, 100 instruction causes a data hazard as explained in Fig. 6.73, but it is corrected by a combination of a NOP instruction and a forwarding path from the data memory stage (D) of the LOAD instruction to the ALU stage (A) of the SLI instruction.

The ADD R1, R2, R3 instruction also requires data forwarding from the A stage of the SLI instruction to the A stage of the ADD instruction.

The target address in the branch instruction has changed from (PC+5) to (PC+2) according to the new branch instruction, BRA R1, 1, 2, in Fig. 6.87 which computes the branch target while it is in the RF (R) stage. An instruction unrelated to the branch instruction, such as STORE R2, R0, 200 is used in the branch delay slot so that the CPU has enough time to fetch ADDI R3, R5, 1 if the branch comparison is unsuccessful or SUBI R3, R5, 1 if the comparison is successful.

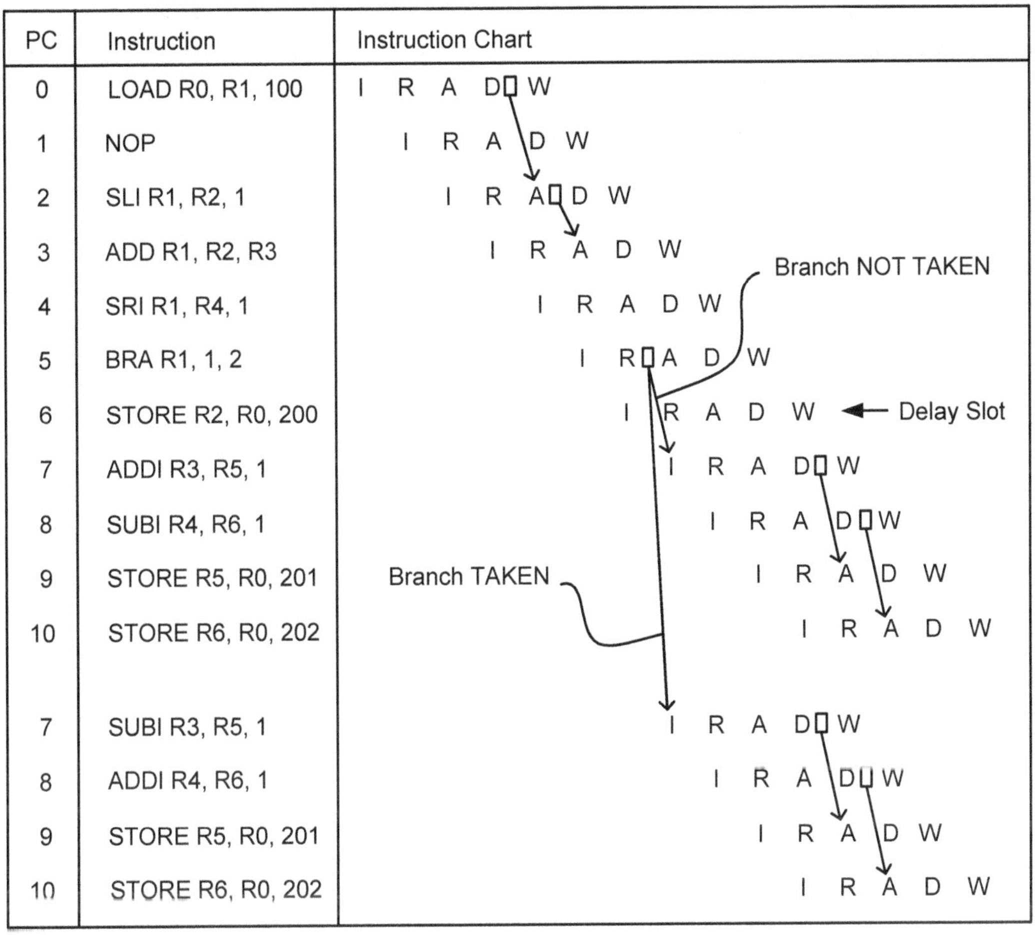

Fig. 6.87 Five-stage CPU instructional chart

From this point forward, the program follows two separate instruction charts. The first chart shows the case where the branch comparison is unsuccessful, and contains two forwarding paths required by two STORE instructions. However, this section does not contain any JUMP instruction because the program has two separate copies after the branch instruction. The second chart follows the case where the branch comparison is successful, and also contains two forwarding paths required by the STORE instructions. Both of these forwarding paths direct data from the D stage of the immediate ALU instructions to the A stage of the STORE instructions to avoid data hazards.

The entire program takes 15 clock cycles to complete whether the branch is successful or unsuccessful.

Can the program in Fig. 6.86 be executed more efficiently in a shorter amount of time and with fewer forwarding paths if the number of pipeline stages is reduced? To answer this question, two additional CPU pipelines are implemented: one with four pipeline stages and the other with three.

Example 6.9 Construct an instruction chart for the same flow chart in Fig. 6.85. This time, show how to handle data and control hazards for a four-stage RISC CPU using forwarding paths. Use NOP instruction(s) when forwarding paths become insufficient to remove a particular hazard.

Figure 6.88 shows a four-stage RISC CPU where the ALU and data memory stages are combined into a single stage.

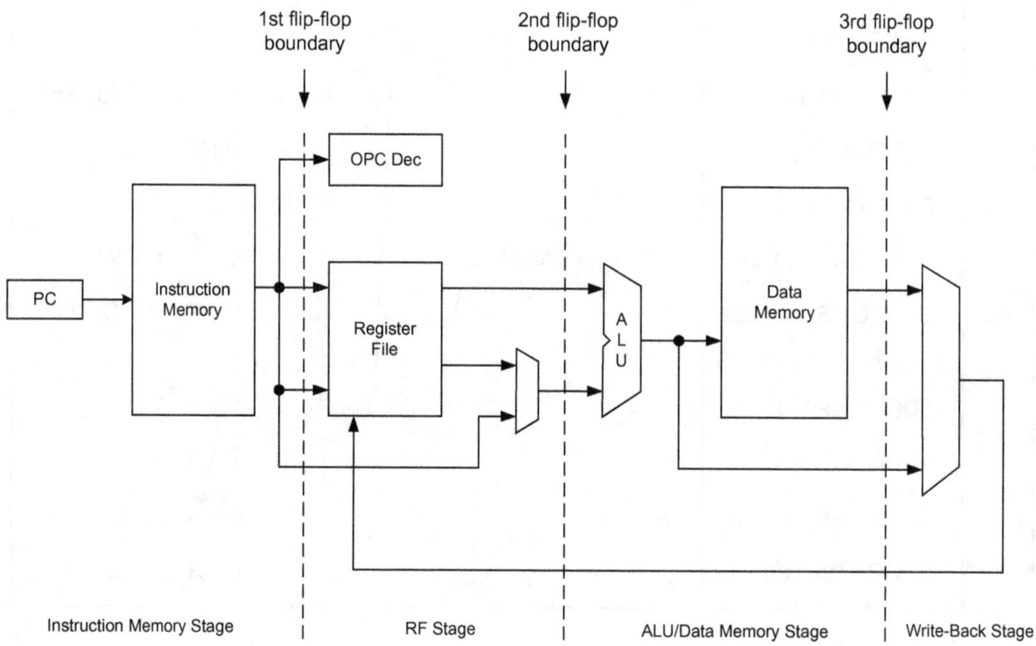

Fig. 6.88 Four-stage fixed-point CPU data-path

If the instructional chart is reconstructed to execute the program in Fig. 6.86 in a four-stage CPU, it will produce a chart in Fig. 6.89.

The first observation in this figure is that there are fewer NOP instructions. For example, the NOP instruction that follows LOAD R0, R1, 100 is eliminated because the memory contents become available during the combined ALU/Data Memory stage (AD). Also, the forwarding paths from the immediate ALU to the STORE instructions in Fig. 6.87 are eliminated because the write-back stage (W) of the immediate ALU instruction lines up with the RF access stage (R) of the STORE instruction in this new CPU pipeline.

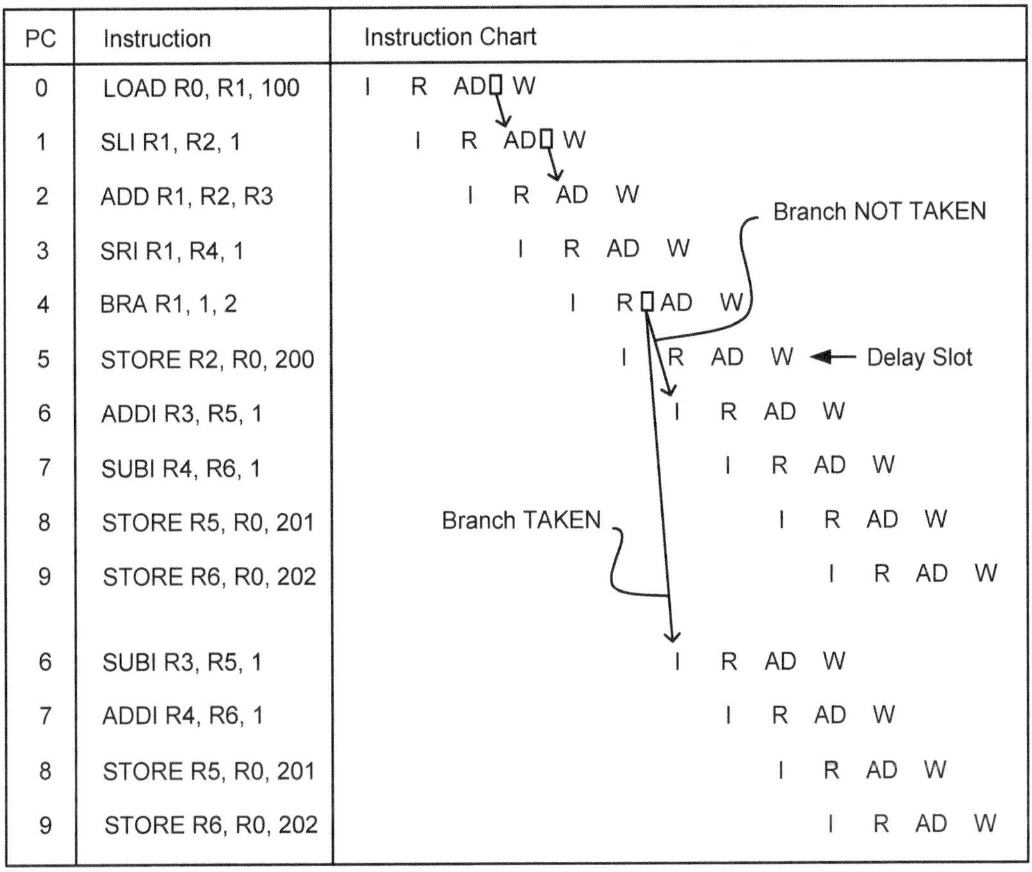

Fig. 6.89 Four-stage CPU instructional chart

Branch-related hazard still exists in the chart in Fig. 6.89, and it is removed by inserting STORE R2, R0, 200 instruction in the branch delay slot.

The new CPU pipeline executes the program in a shorter time: 13 clock cycles whether the branch is successful or unsuccessful.

Example 6.10 Reconstruct the instruction chart for the flow chart in Fig. 6.85, and show how the data and control hazards in a three-stage RISC CPU are handled using forwarding paths. Use NOP instruction(s) when forwarding paths become insufficient to remove a particular hazard.

Can there be a continuing improvement in the program efficiency and the overall execution time if the number of pipeline stages is reduced further? To answer this question, the CPU pipeline in Fig. 6.88 is repartitioned into three stages. Its third stage combines the ALU, data memory and write-back stages in a single stage shown in Fig. 6.90.

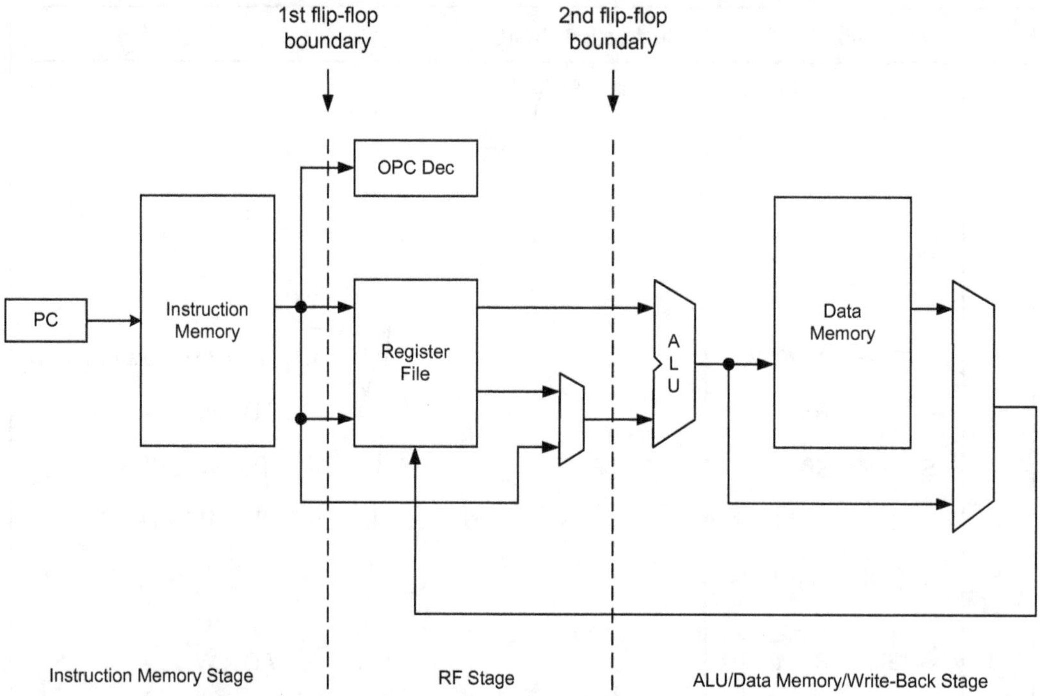

Fig. 6.90 Three-stage fixed-point CPU data-path

When the instructional chart is reconstructed for the three-stage CPU in Fig. 6.90, we see an immediate improvement of this new CPU configuration for reducing the number of forwarding paths to one. This is shown Fig. 6.91. However, the combined ALU, Data Memory and Write-Back stages (AW) create a practical engineering problem: calculating the data memory address, accessing the data memory, and writing the results back to the RF may not fit in half a clock cycle.

For the SLI R1, R2, 1 and ADD R1, R2, R3 instruction pair, half a cycle may be sufficient to shift the contents of R1, and store the result in R2 before the ADD instruction accesses this data. Therefore, no NOP instruction is inserted between the SLI and ADD instructions.

Branch-related delay slot still cannot be avoided in Fig. 6.91. STORE R2, R0, 200 is inserted in the branch delay slot just as in Figs. 6.87 and 6.89.

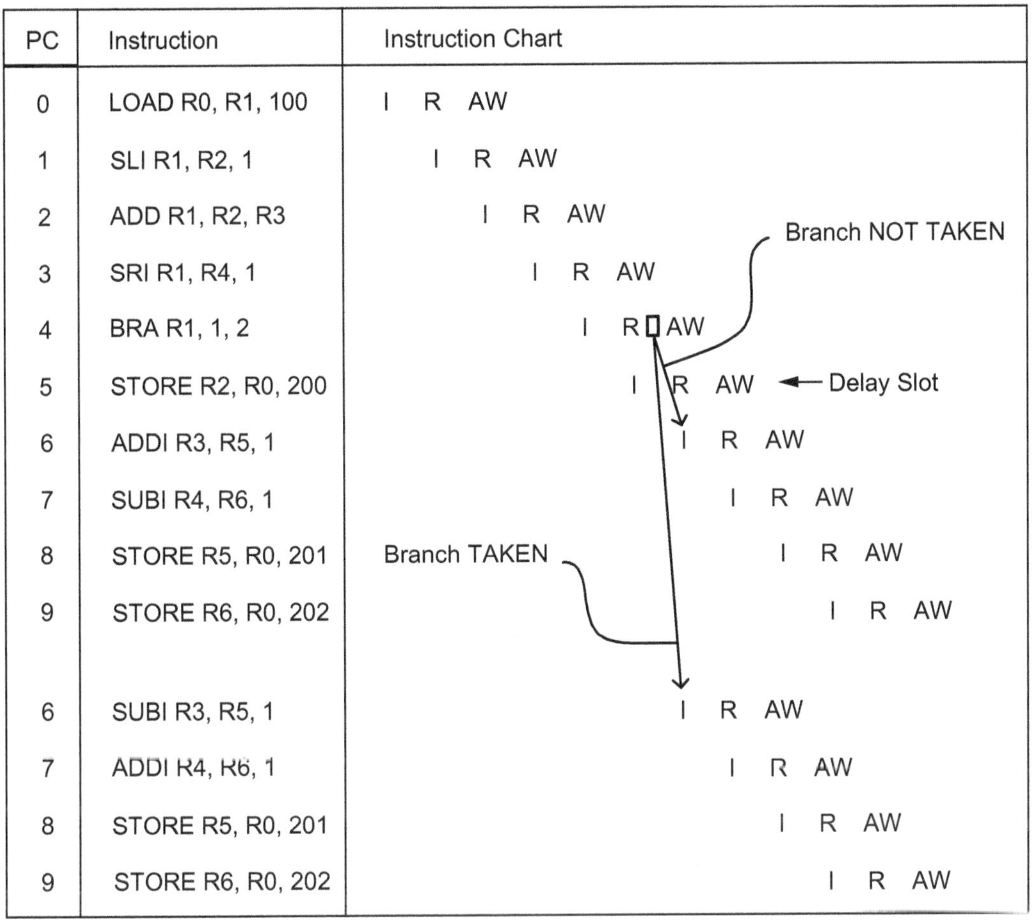

Fig. 6.91 Three-stage CPU instructional chart

The three-stage CPU executes the program in 12 clock cycles whether the branch is successful or unsuccessful. Therefore, there is almost no gain in speed compared to the four-stage CPU. However, this may not be true for larger programs, which are likely to contain more branch instructions or instructions that access data memory.

6.5 Floating-Point Unit

Floating-Point Instructions

This RISC instruction set contains six floating-point (FP) instructions: floating-point add (ADDF), floating-point subtract (SUBF), floating-point multiply (MULF), floating-point divide (DIVF), floating-point load (LOADF) and floating-point store (STOREF). The ADDF, SUBF, MULF and DIVF instructions use the IEEE single-precision floating-point format which specifies the most significant bit to be the sign, the lesser eight most significant bits to be the exponent and the least significant 23 bits to be the fraction. The LOADF and STOREF use the same integer field format as the fixed-point LOAD and STORE instructions, and they require a secondary register file (besides the fixed-point register file) to store floating-point data.

The ADDF instruction adds two single precision floating-point numbers at the registers, FS1 and FS2, and returns the result to the register FD as described below. The bit field format of this instruction is given in Fig. 6.92.

ADDF FS1, FS2, FD
Reg[FS1] + Reg[FS2] → Reg[FD]

31 26	25 21	20 16	15 11	10 0
ADDF	FS1	FS2	FD	Not Used

Fig. 6.92 Floating-point add (ADDF) instruction field format

The SUBF instruction subtracts the contents of FS2, Reg[FS2], from the contents of FS1, Reg [FS1], and returns the result to FD as described below. The bit field format of the instruction is shown in Fig. 6.93.

SUBF FS1, FS2, FD
Reg[FS1] − Reg[FS2] → Reg[FD]

31 26	25 21	20 16	15 11	10 0
SUBF	FS1	FS2	FD	Not Used

Fig. 6.93 Floating-point subtract (SUBF) instruction field format

The MULF instruction multiplies two floating-point numbers in registers FS1 and FS2, and returns the result to FD as shown below. This instruction's field format is described in Fig. 6.94.

MULF FS1, FS2, FD
Reg[FS1] * Reg[FS2] → Reg[FD]

31 26	25 21	20 16	15 11	10 0
MULF	FS1	FS2	FD	Not Used

Fig. 6.94 Floating-point multiply (MULF) instruction field format

The DIVF instruction divides two floating-point numbers in registers FS1 and FS2, and returns the result to FD as shown below. This instruction's field format is described in Fig. 6.95.

DIVF FS1, FS2, FD
Reg[FS1]/Reg[FS2] → Reg[FD]

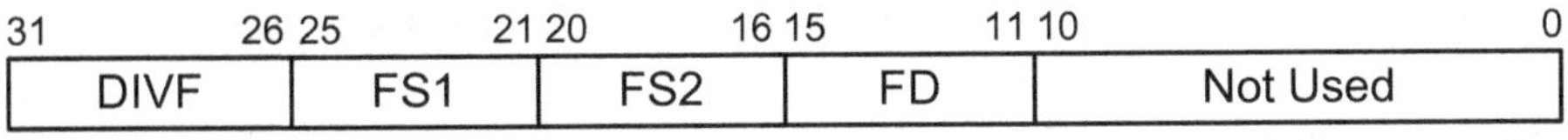

Fig. 6.95 Floating-point divide (DIVF) instruction field format

The LOADF instruction loads the contents of register FD, Reg[FD], from the data memory to the floating-point register file. The data memory address is calculated by adding the contents of RS, Reg[RS], to the immediate value. This instruction's field format is described in Fig. 6.96.

LOADF RS, FD, Imm. Value
mem {Reg[RS] + Imm. Value} → Reg[FD]

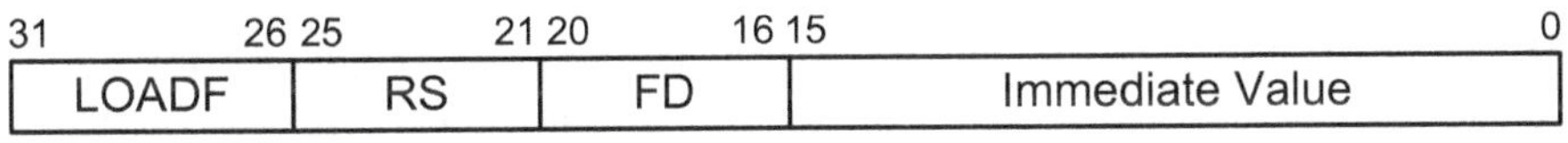

Fig. 6.96 Floating-point load (LOADF) instruction field format

Similar to the LOADF instruction, the STOREF instruction stores the contents of register FS, Reg[FS], to the data memory. The data memory address is calculated by adding the contents of RD, Reg[RD], to the immediate value. This instruction's field format is described in Fig. 6.97.

STOREF FS, RD, Imm. Value
Reg[FS] → mem {Reg[RD] + Imm. Value}

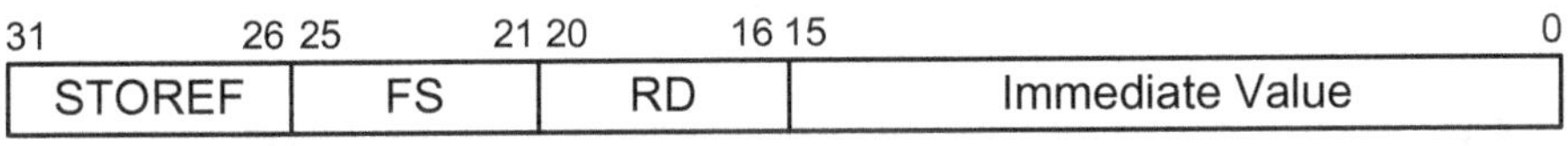

Fig. 6.97 Floating-point store (STOREF) instruction field format

Floating-Point Bit Field Formats

Single and double-precision floating-point bit field formats are designed to operate with 32 or 64-bit buses, respectively, and each format consists of the sign, exponent and fraction entries.

An IEEE single-precision (SP) floating-point number shown in Fig. 6.98 has an eight-bit field for the exponent and 23-bit field for the fraction. The most significant bit constitutes the sign bit for the fraction.

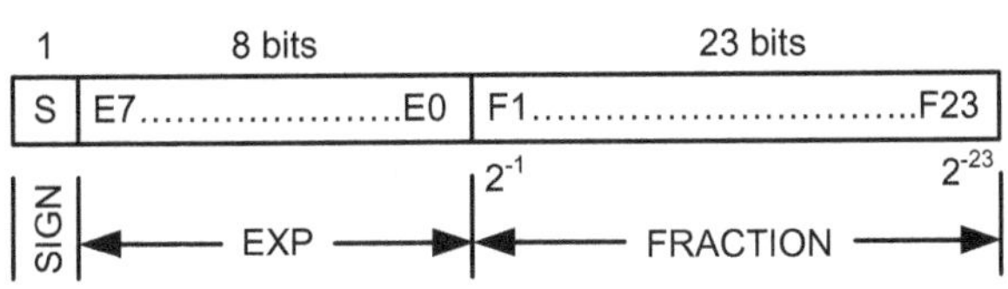

Fig. 6.98 IEEE single-precision floating-point format

Mathematically, a single-precision floating-point number is expressed as follows:

Single-precision FP number = $(-1)^S \times (1 + F1 \times 2^{-1} + F2 \times 2^{-2} + \cdots + F23 \times 2^{-23})\ 2^{REXP}$

Here, s = 1 indicates the fraction is negative, and s = 0 indicates the fraction is positive. F1 to F23 are the 23 fraction bits that range from the most significant to the least significant bit positions of the fraction field as shown in Fig. 6.98, respectively.

Since the exponent field does not possess a sign bit, a biasing system is employed to distinguish the negative exponents from the positive ones. The biased exponent (EXP) field shown in Fig. 6.98 is the combination of the real exponent (REXP) and the BIAS as shown below.

EXP = REXP + BIAS

The BIAS is calculated by substituting the lowest and the highest eight-bit numbers in the EXP field to calculate the most negative and the most positive real exponent values. The most negative REXP is equal to −MAX when EXP is at its minimum value of zero. Thus,

0 = −MAX + BIAS

Similarly, when EXP reaches its maximum value of 255, the most positive REXP becomes +MAX. Thus,

255 = MAX + BIAS

Hence substituting MAX = BIAS into 255 = MAX + BIAS yields:
BIAS = 127 for single-precision numbers.

Example 6.11 Represent -0.75_{10} as a single-precision floating-point number.

$-0.75_{10} = -0.11_2 = (-1)^S\ (1 + F1 \times 2^{-1} + F2 \times 2^{-2} + F3 \times 2^{-3} + \cdots + F23 \times 2^{-23})\ 2^{REXP}$

$-1.1 \times 2^{-1} = (-1)^S\ (1 + F1 \times 2^{-1} + F2 \times 2^{-2} + F3 \times 2^{-3} + \cdots + F23 \times 2^{-23})\ 2^{REXP}$

Thus,

s = 1
F1 = 1
F2 through F23 = 0
REXP = −1 = EXP − BIAS = EXP − 127
EXP = 127 − 1 = 126

Filling the fraction and the exponent fields in Fig. 6.99 yields:

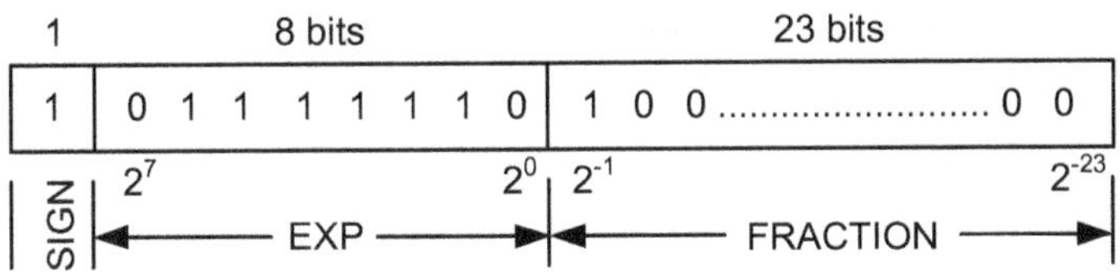

Fig. 6.99 Single-precision floating-point number in Example 6.11

Example 6.12 Represent -527.5_{10} as a single-precision floating-point number.

$$-527.5_{10} = (-1)^1\ (1 + F1 \times 2^{-1} + F2 \times 2^{-2} + F3 \times 2^{-3} + \cdots + F23 \times 2^{-23})\ 2^{REXP}$$

The closest real exponent to 527.5 is $512 = 2^9$. Thus,

$$-527.5_{10} = (-1)^1\ (1 + F1 \times 2^{-1} + F2 \times 2^{-2} + F3 \times 2^{-3} + \cdots + F23 \times 2^{-23})\ 2^9$$

where,

$$(1 + F1 \times 2^{-1} + F2 \times 2^{-2} + F3 \times 2^{-3} + \cdots + F23 \times 2^{-23}) = 527.5/512 = 1.03027_{10}$$
$$\approx 1.00001_2 = 1.03125_{10}$$

$$\text{Error in the fraction} = \frac{1.03125 - 1.03027}{1.03125} = 0.01\%$$

The biased exponent is calculated as follows:

$$REXP = 9 = EXP - BIAS = EXP - 127$$

$$EXP = 127 + 9 = 136$$

After entering the fraction and exponent fields in Fig. 6.98, the single-precision floating-point number for -527.5_{10} produces the format as shown in Fig. 6.100.

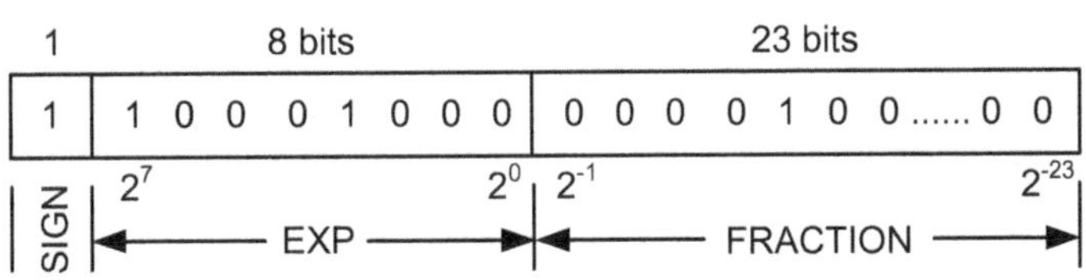

Fig. 6.100 Single-precision floating-point number in Example 6.12

Example 6.13 Convert the single-precision floating-point number in Fig. 6.101 to a decimal number.

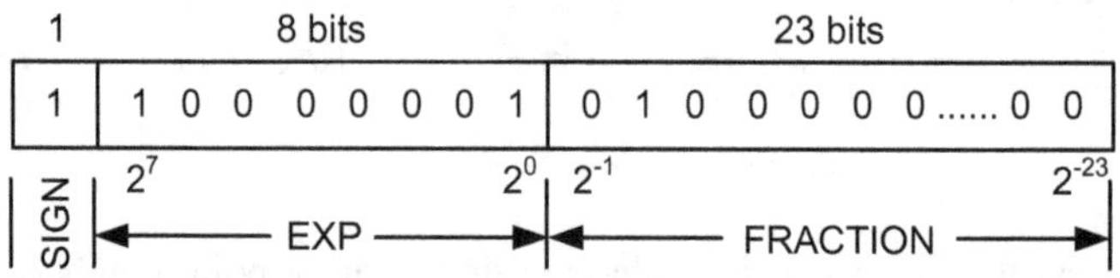

Fig. 6.101 Single-precision floating-point number in Example 6.13

Here, s = 1 corresponds to a negative fraction. The fraction and biased exponent fields yield 0.25 and 129, respectively. Thus,

REXP = 129 − 127 = 2

The decimal number = $(-1)^1\ (1 + 0.25)\ 2^2 = -1.25 \times 4 = -5$

Example 6.14 Convert the single-precision floating-point number in Fig. 6.102 to a decimal number.

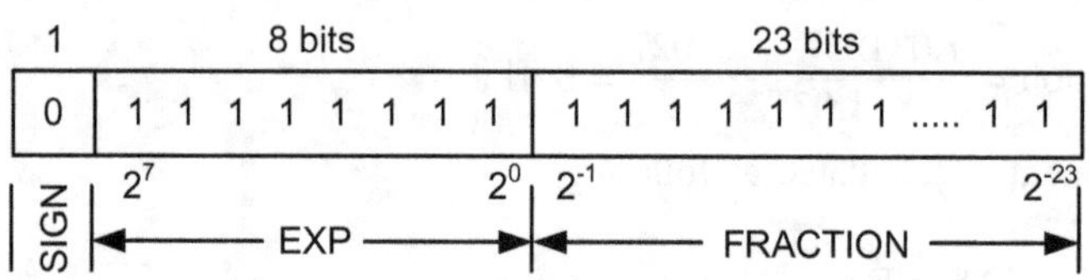

Fig. 6.102 Single-precision floating-point number in Example 6.14

Here, s = 0, which corresponds to a positive fraction. The fraction field yields approximately 1. The biased exponent produces EXP = 255. Therefore,

REXP = 255 − 127 = 128

The decimal number = $(-1)^0\ (1 + 1)\ 2^{128} = 2 \times 10^{38}$

The IEEE double-precision (DP) floating-point number in Fig. 6.103 has 11 bits for the exponent and 52 bits for the fraction for better accuracy. Once again, the most significant bit corresponds to the sign bit for the fraction.

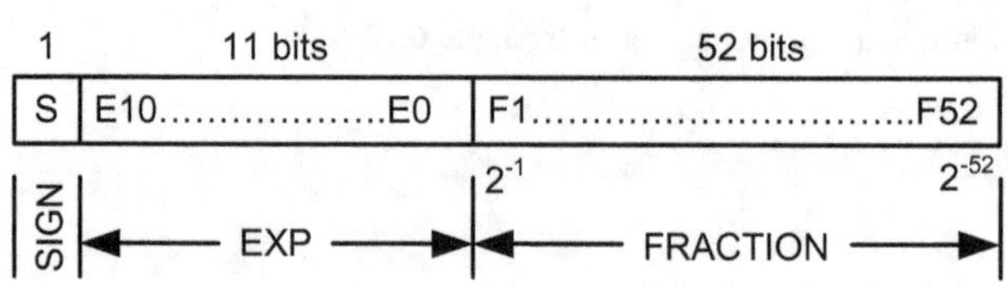

Fig. 6.103 IEEE double-precision floating-point format

The double-precision floating-point number is expressed as follows:

Double-precision FP Number = $(-1)^S \times (1 + F1 \times 2^{-1} + F2 \times 2^{-2} + \cdots + F52 \times 2^{-52})\, 2^{REXP}$

Here, s = 1 corresponds to a negative, and s = 0 corresponds to a positive fraction. Bits F1 to F52 are the 52 fraction bits from the most significant bit position to the least significant bit position, respectively.

The bias system to represent single-precision floating-point exponents can also be applied to the double-precision numbers. Thus,

EXP = REXP + BIAS

Here, EXP is the 11-bit biased exponent field in Fig. 6.103, and REXP is the real exponent. The BIAS is calculated by substituting the lowest and the highest biased exponents in the equation, respectively.

Therefore, for the most negative real exponent of −MAX, EXP becomes equal to zero:

0 = −MAX + BIAS

Similarly, EXP value becomes 2047 for the most positive real exponent of +MAX:

2047 = MAX + BIAS

Substituting MAX = BIAS into 2047 = MAX + BIAS yields BIAS = 1023 for double-precision floating-point numbers.

Example 6.15 Represent -0.75_{10} as a double-precision floating-point number.

$-0.75_{10} = -0.11_2 = (-1)^S\, (1 + F1 \times 2^{-1} + F2 \times 2^{-2} + F3 \times 2^{-3} + \cdots + F52 \times 2^{-52})\, 2^{REXP}$
$-1.1 \times 2^{-1} = (-1)^S\, (1 + F1 \times 2^{-1} + F2 \times 2^{-2} + F3 \times 2^{-3} + \cdots + F52 \times 2^{-52})\, 2^{REXP}$

Thus,

s = 1
F1 = 1
F2 through F52 = 0
REXP = −1 = EXP − BIAS = EXP − 1023
EXP = 1023 − 1 = 1022

Therefore, entering the fraction and the exponent fields in Fig. 6.104 yields:

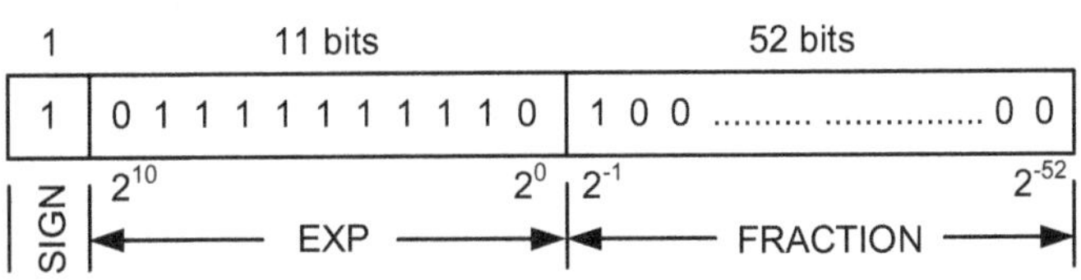

Fig. 6.104 Double-precision floating-point in Example 6.15

Example 6.16 Represent 4.0_{10} as a double-precision floating-point number.

$$4_{10} = (-1)^0\,(1 + F1 \times 2^{-1} + F2 \times 2^{-2} + F3 \times 2^{-3} + \cdots + F52 \times 2^{-52})\,2^{REXP}$$

The closest exponent to 4 is $4 = 2^2$. Thus,

$$4_{10} = (-1)^0\,(1 + F1 \times 2^{-1} + F2 \times 2^{-2} + F3 \times 2^{-3} + \cdots + F52 \times 2^{-52})\,2^2$$

where,

$$(1 + F1 \times 2^{-1} + F2 \times 2^{-2} + F3 \times 2^{-3} + \cdots + F52 \times 2^{-52}) = 1.0_2$$

Therefore,

F1 through F52 = 0
REXP = 2 = EXP−BIAS = EXP − 1023
EXP = 1025

Thus, entering the fraction and exponent fields in Fig. 6.105 yields:

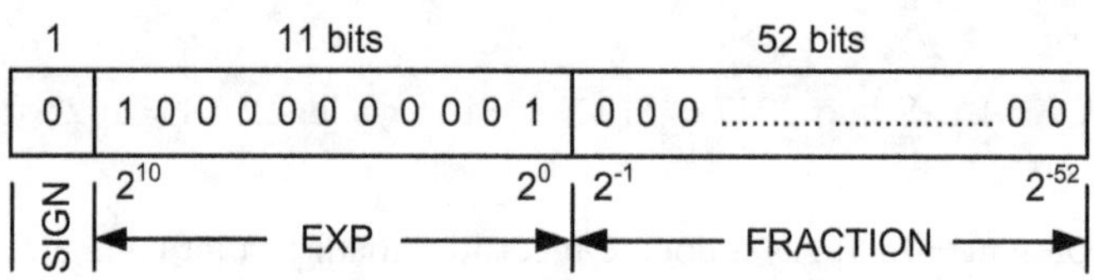

Fig. 6.105 Double-precision floating-point in Example 6.16

Example 6.17 Convert the double-precision floating-point number in Fig. 6.106 to a decimal number.

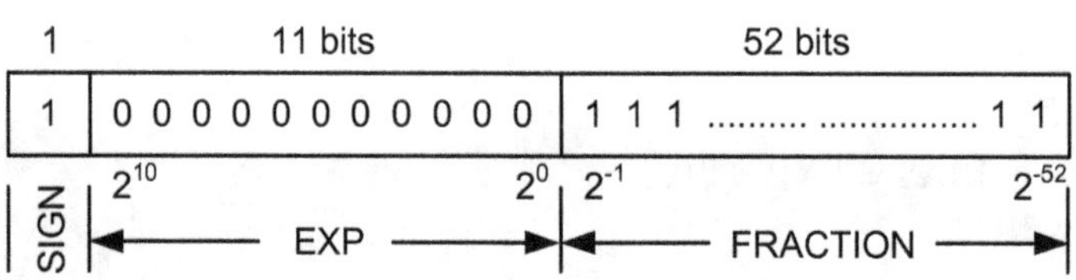

Fig. 6.106 Double-precision floating-point in Example 6.17

Here, s = 1, which corresponds to a negative fraction. The fraction and biased exponent fields yield 1 and 0, respectively. Therefore,

REXP = 0 − 1023 = −1023

Thus, the decimal number = $(-1)^1\,(1 + 1)\,2^{-1023} = -2 \times 10^{-308}$.

Floating-Point Adder

Floating-point addition requires equating the exponents before adding the fractions. There are two ways to equate exponents. The first method is to shift the fraction of the floating-point number with greater exponent to the left until both exponents become equal. The second method is to shift the fraction of the floating-point number with lesser exponent to the right until the exponents are equal.

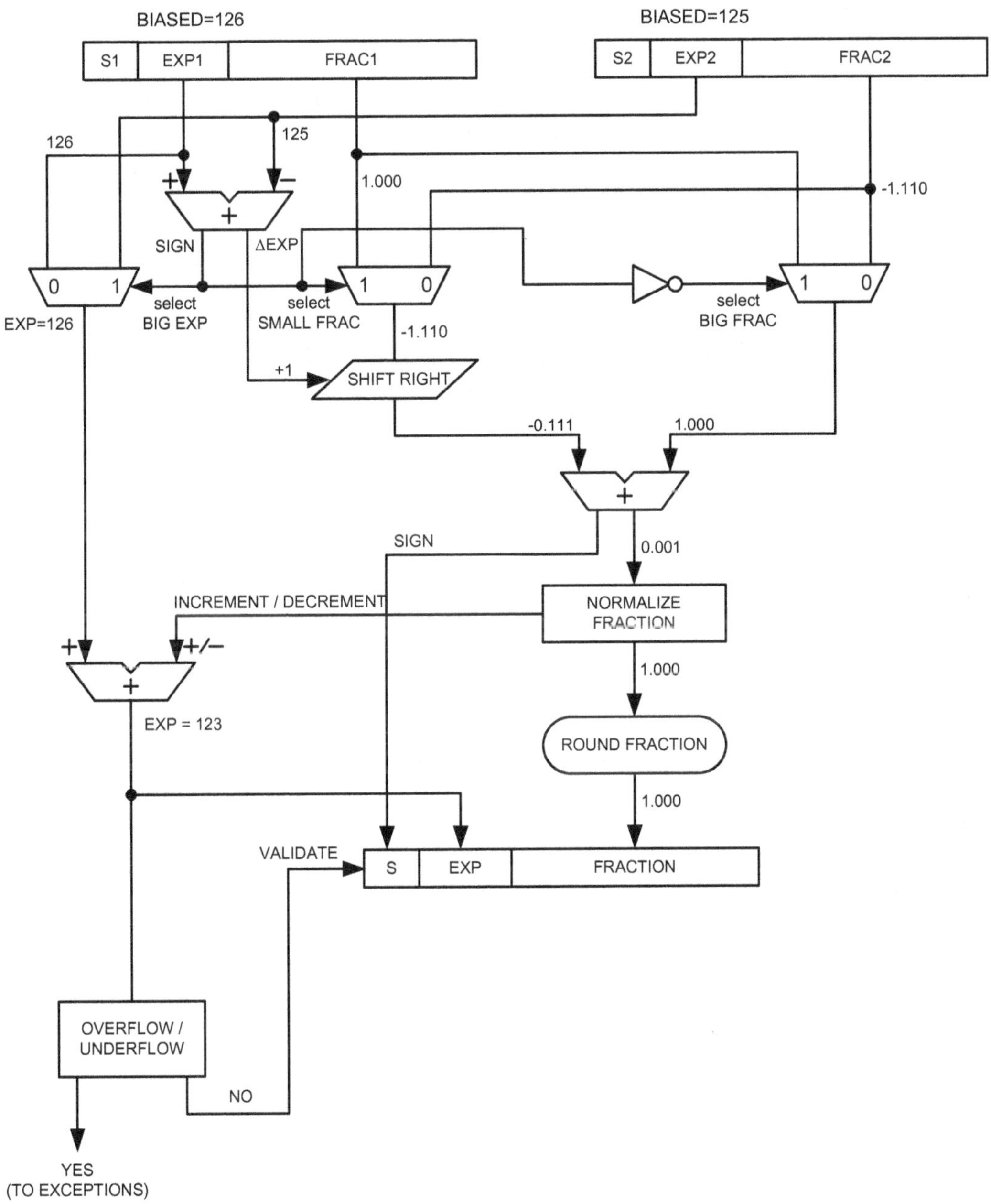

Fig. 6.107 A floating-point adder implementation

The floating-point adder in Fig. 6.107 implements the second method. The first step of this method is to determine which of the two floating-point numbers has a smaller exponent. This leads to subtracting the two exponents from each other and examining the sign bit of the result. In the schematic in Fig. 6.107, the second number's exponent, EXP2, is subtracted from the first number's exponent, EXP1, to obtain the difference, ΔEXP = EXP1 − EXP2. If the sign bit of ΔEXP becomes zero, it indicates EXP1 is larger than EXP2. Therefore, the fraction of the second number, FRAC2, is shifted to the right by an amount equal to ΔEXP before adding the fractions. If, on the other hand, the sign bit of ΔEXP becomes one, FRAC1 is shifted to the right by ΔEXP before adding FRAC1 to FRAC2.

The numerical example in Fig. 6.107 describes the process of adding the fraction fields of two floating-point numbers according to ΔEXP. In this figure, 126, 1.000, 125 and −1.110 are assigned to the EXP1, FRAC1, EXP2 and FRAC2 fields, respectively. Initially, EXP2 is subtracted from EXP1, which yields ΔEXP = +1. The sign of ΔEXP becomes zero, and routes the larger exponent, EXP1 = 126, to an adder that calculates the real exponent. The sign bit also directs the fraction field of the larger exponent, FRAC1 = 1.000, to a separate adder to compute the fraction field.

FRAC2 = −1.110, on the other hand, is moved to the input of the right shifter which shifts this value by ΔEXP = 1, and produces −0.111. This shifted fraction, −0.111, is then added to FRAC1 = 1.000, producing 0.001 while the exponent stays at 126. The normalization mechanism takes place next. During this process, the result, 0.001, is shifted to the left until the binary value, 1, in the 0.001 field is detected. The normalizer output becomes 1.000, but the shifted amount, −3, is added to the current exponent, 126, yielding 123 at the output of the adder that computes the exponent.

The normalized fraction goes through a rounding process and truncates the fraction field to 23 bits. The result is stored in a 23-bit register at the output of the floating-point adder. The output of the exponent adder is similarly stored in an eight-bit register along with the sign bit for further processing in the CPU.

This floating-point adder also deals with overflow and underflow conditions in case the exponent values become higher than 255 or smaller than 0, both of which generate exceptions for the CPU.

Floating-Point Multiplier

The processing complexity of comparing two exponent fields in the floating-point adder does not take place in the floating-point multiplier. The algorithm of multiplying two floating-point numbers simply consists of multiplying the fractions and adding the exponents.

The floating-point multiplier architecture in Fig. 6.108 computes the fraction and the exponent fields with a numerical example. In this example, 126, 1.000, 125 and 1.110 values are assigned to the EXP1, FRAC1, EXP2 and FRAC2 fields of the multiplier, respectively.

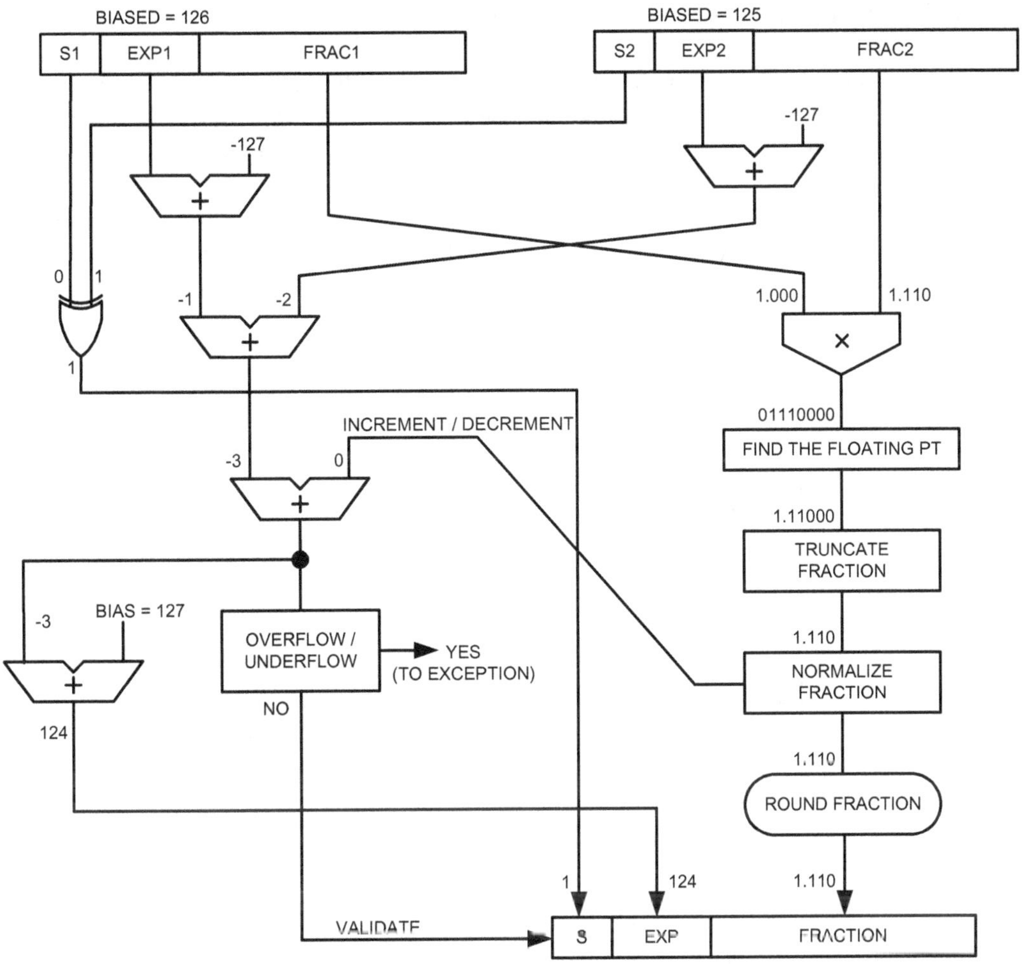

Fig. 6.108 A floating-point multiplier implementation

The first step of the multiplication process is to calculate the real exponent values of the two floating-point numbers. Therefore, both EXP1 = 126 and EXP2 = 125 are subtracted from the single-precision bias, 127, yielding −1 and −2, respectively. The real exponents are then added, producing −3, which becomes the input of an adder that increments or decrements the real exponent after multiplying the fractions. The fractions, FRAC1 = 1.000 and FRAC2 = 1.110, are multiplied, producing 1110000. The floating point is assigned immediately after locating the leading 1 in the 1110000 field which results in 1.11000. This result is subsequently truncated to 1.110. Since no normalization needs to be performed on 1.110, this step effectively produces a zero increment/decrement value, and the fraction is stored in the output register after rounding takes place. The real exponent, −3, on the other hand, is added to the bias, 127, and the result is also stored in the output register. Finally, sign bits of the two floating-point numbers are XORed and stored in the same register.

As in the floating-point adder, the underflow and overflow conditions in the exponent field of the floating-point multiplier cause the CPU to generate exceptions.

RISC CPU with Fixed and Floating-Point Data-Paths

The floating-point adder, FP Adder, and the floating-point multiplier, FP Multiplier, can be included in the existing fixed-point five-stage CPU data-path as shown in Fig. 6.109. In this figure, the fixed-point register file outputs are connected to the fixed-point ALU, and the floating-point register file outputs are connected to the floating-point adder and the multiplier, forming two completely isolated data-paths from each other. The OPC field for the ADDF (SUBF) instruction selects port 1 of the 2-1 MUX, and routes the floating-point adder (subtractor) result back to the floating-point register file. Similarly, the OPC field of the MULF instruction selects port 0 of the 2-1 MUX, and routes the floating-point multiplier output to the floating-point register file.

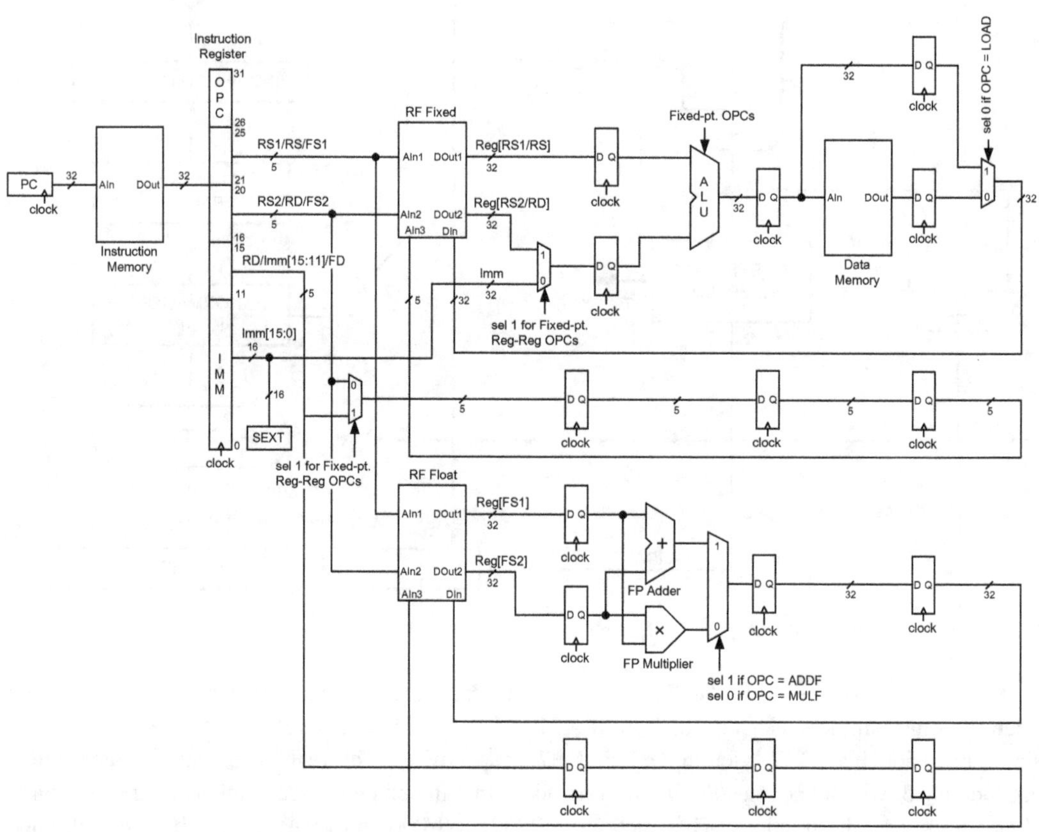

Fig. 6.109 Floating and fixed-point execution data-paths

Figure 6.110 shows four distinct data-paths that belong to the fixed and floating-point load and store instructions, namely LOAD, STORE, LOADF and STOREF. These data-paths are deliberately avoided from Fig. 6.109 to prevent complexity in this figure. LOAD and STORE instruction data-paths in this figure have already been described earlier in this chapter. For LOADF instruction, the contents of RS from the fixed-point register file and the sign-extended immediate value from the instruction are added in the ALU stage, forming an effective address for the data memory. The data at this address is subsequently routed back to the floating-point register file at the address FD. Similarly, the STOREF instruction calculates the data memory address by adding the sign-extended immediate value to the contents of RD, and stores the contents of FS to this address.

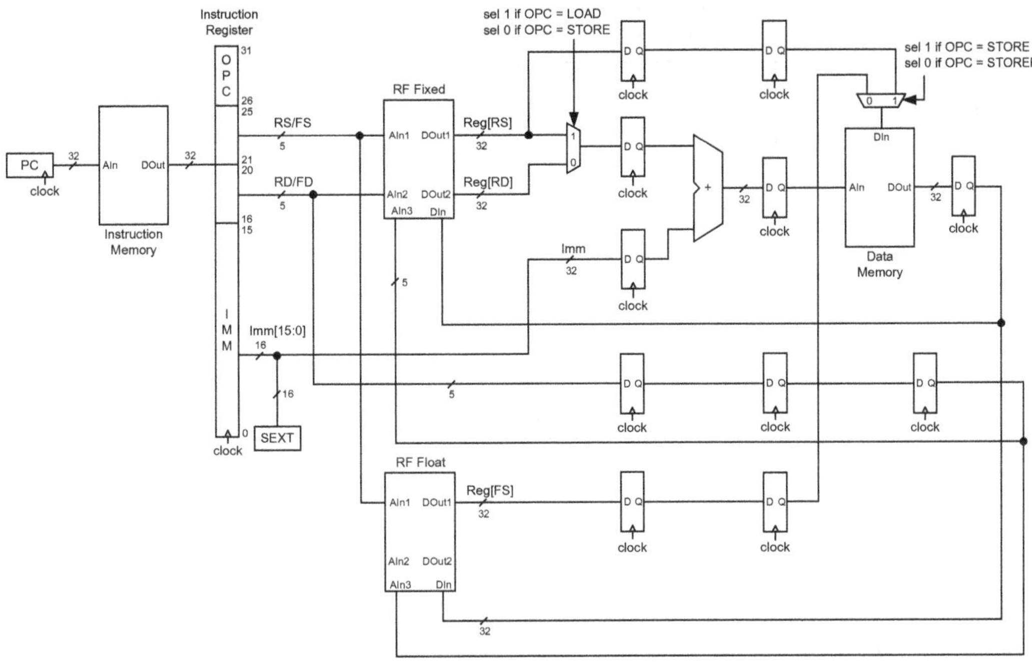

Fig. 6.110 LOAD, STORE, LOADF and STOREF instruction data-paths

In reality, Figs. 6.109 and 6.110 must be combined together to form a single figure to enable a RISC CPU to execute all fixed and floating point-ALU instructions except DIVF.

Floating-Point Data Hazards

There are three types of floating-point hazards, all culminating from variable execution times in the floating point data-path.

The first hazard is Read After Write (RAW)-type hazard. This hazard is shown as an example in Fig. 6.111, illustrating its mechanics. In this example, all floating-point instructions go through three stages of execution: Instruction Fetch (I), Execution (E) and Write-back (W) stages. Both the ADDF and SUBF instructions are assumed to take only one cycle in the E stage, whereas MULF takes eight consecutive cycles to produce a result. In the instruction chart in Fig. 6.111, the FPU produces the MULF result in the ninth cycle, and writes it to the floating-point register file in the tenth cycle. Instead of waiting for the result from the MULF operation, the SUBF instruction prematurely fetches an old value of F2 and combines it with its other source operand in the fourth cycle, producing an invalid data for F7 in the fifth cycle. Therefore, this scenario outlines the case where a latter instruction (SUBF) reads an invalid data entry "before" an earlier instruction (MULF) is able to produce valid data, creating a RAW hazard.

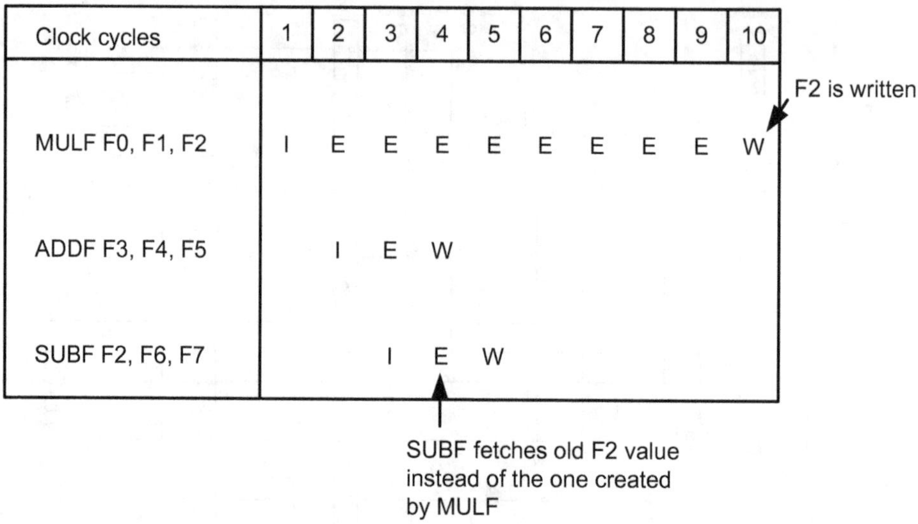

Fig. 6.111 A RAW hazard

The next floating-point hazard is the Write After Read (WAR)-type. Again, this hazard is shown as an example in Fig. 6.112, explaining how it forms. In this instruction chart, MULF produces a result, Reg[F2], in the ninth cycle, and forwards it as an operand to the ADDF instruction in the tenth cycle. In the same cycle, instead of fetching the original operand from the register F3, the ADDF instruction fetches the operand produced by a latter instruction, SUBF, in the fifth cycle, and uses it to produce Reg[F4], resulting in an invalid data. In other words, a latter instruction (SUBF) writes data to the register file "before" an earlier instruction (ADDF) is able to read and execute, creating a WAR hazard.

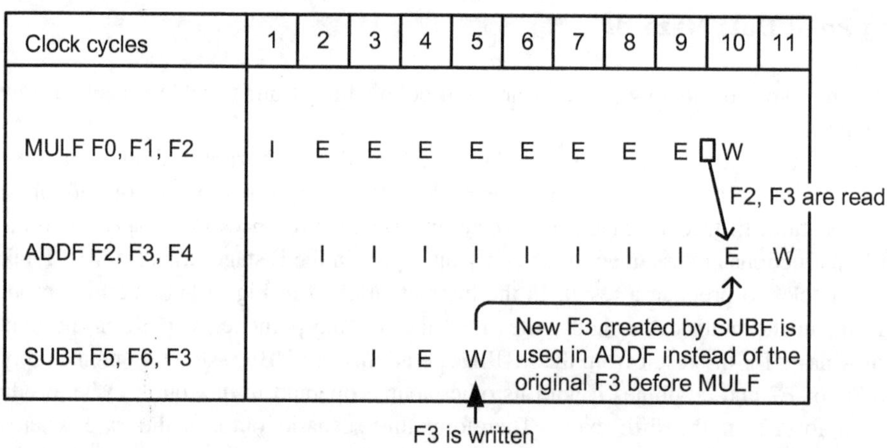

Fig. 6.112 A WAR hazard

The last floating-point hazard, Write After Write (WAW), is illustrated as an example in Fig. 6.113. In this example, a latter instruction, ADDF, writes to the register address, F2, in the fourth cycle while an earlier instruction, MULF, writes to the same register address in the sixth cycle, overwriting the valid data entry by ADDF. The final instruction, SUBF F2, F11, F12, uses the data written by the earlier instruction (MULF) instead of using the data from the latter instruction (ADDF), resulting in invalid data in F12. In other words, a latter instruction (ADDF) writes data to the register file "before" an earlier instruction (MULF) is able to write, causing a WAW hazard.

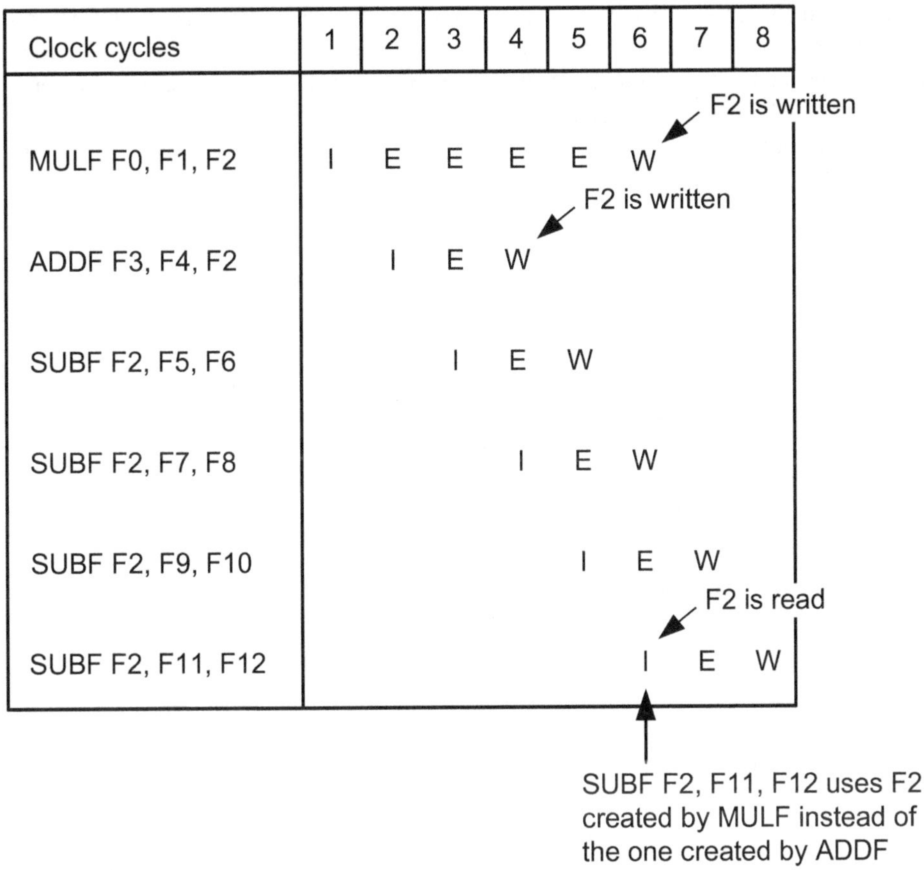

Fig. 6.113 A WAW hazard

Figure 6.114 presents many forms of floating-point hazards in one program. The RAW hazard indicates if a source operand is prematurely read before an earlier instruction has a chance to write it back to the register file. In this figure, MULF reads the contents of F8 in cycle 4 before SUBF completes the operation in cycle 5. Therefore, forwarding the contents of F8 becomes necessary from the E-stage of SUBF to the first E-stage of MULF to eliminate the hazard.

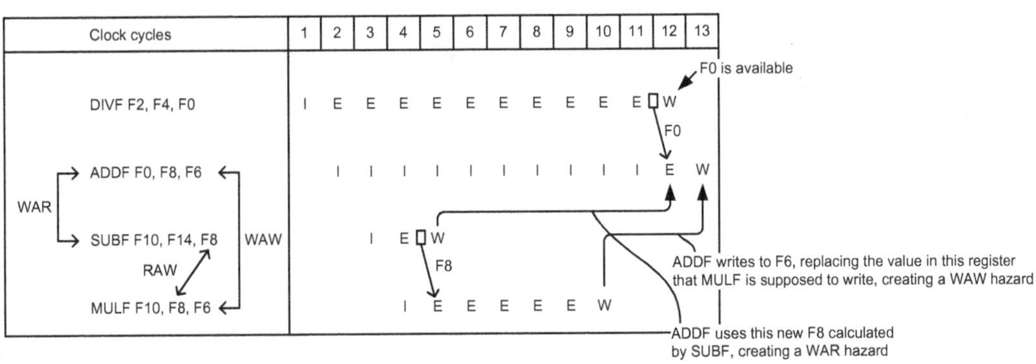

Fig. 6.114 Mixture of RAW, WAW and WAR hazards

A WAR hazard also presents itself between the ADDF and SUBF instructions in this program. The SUBF instruction writes the contents of F8 in cycle 5 while an earlier instruction, ADDF, tries to use this operand in cycle 12. Therefore, the ADDF instruction will most likely read the contents of F8 written by the SUBF instruction instead of an earlier instruction.

The last hazard in this program is a WAW hazard between the ADDF and MULF instructions. In this case, the MULF instruction writes the contents of F6 in cycle 10 while an earlier instruction, ADDF, writes the same operand much later in cycle 13, creating this hazard.

Out-of-Order Execution and the Need for Register-Renaming

Executing floating-point instructions may require quite a number of clock cycles if instructions are executed in order. The instruction chart in Fig. 6.115 removes potential RAW, WAR and WAW hazards from the program in Fig. 6.114 by deliberately delaying the execution stages of the ADDF, SUBF and MULF instructions, and executes all four floating-point instructions in order. However, the penalty from the delays adds six more clock cycles to the chart compared to Fig. 6.114.

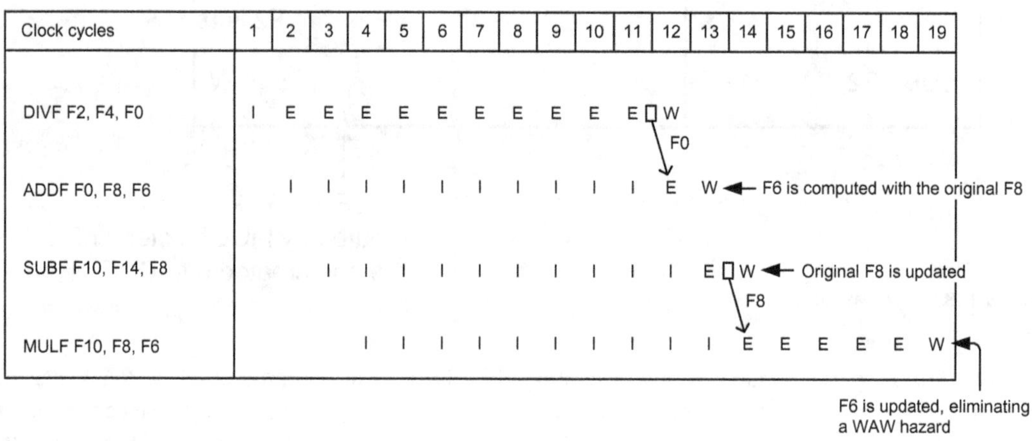

Fig. 6.115 Hazard-free program from Fig. 6.114, executing in-order

Fortunately, there are ways to reduce the number of clock cycles to complete the program while having a hazard-free instruction chart. One of these methods is called register-naming shown in Fig. 6.116 where the subsequent instructions following DIVF can be executed out of order. According to this figure, the destination registers in the ADDF and SUBF instructions are renamed as X and Y instead of F6 and F8, respectively. Also, the source register for MULF is changed to Y to be able to forward data to this instruction properly. The resultant program requires only 13 clock cycles instead of 19 and free of any floating-point hazards.

In this new chart, the ADDF instruction fetches F8 along with the forwarded F0 from the DIVF instruction, and writes the result to a renamed register, X, which is physically a different register than F6. This mechanism removes the possibility of the WAW hazard between the ADDF and MULF instructions in Fig. 6.114 because the former writes to X and the latter writes to F6.

In a similar fashion, the SUBF instruction uses F10 and F14, and writes the result to a renamed register, Y. Having this separate destination register for SUBF prevents the ADDF instruction to use this register as a source register much later, preventing a potential WAR hazard.

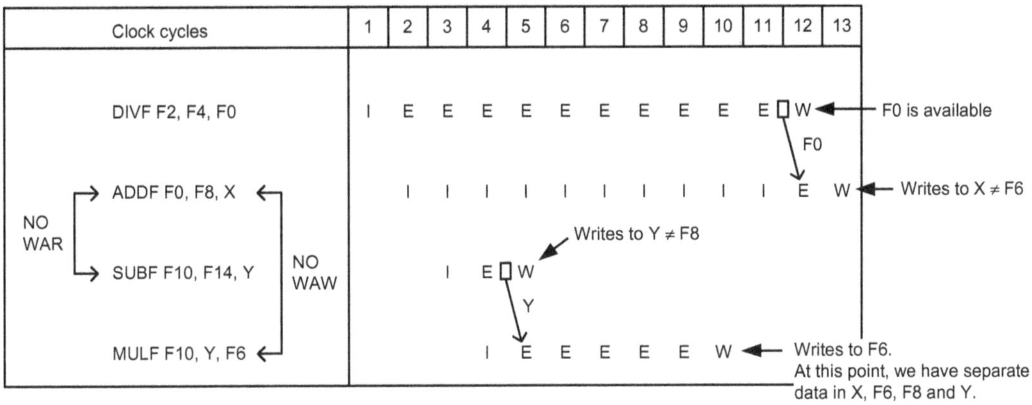

Fig. 6.116 Hazard-free program from Fig. 6.114, executing out-of-order

The MULF instruction, on the other hand, uses F10 and the forwarded Y operands, and writes the result to the register, F6, which is physically a different register than register, X, as mentioned above.

In the end, the program contains true values in registers, F0, F6, X and Y, which are ready to be used by subsequent programs.

Effects of Tomasula Algorithm on Floating-Point Hazards

To be able to obtain hazard-free instruction charts where programs can be executed in minimum number of clock cycles, we need to examine a special algorithm, called Tomasula algorithm, which brings forth a new CPU architecture. To simplify matters in this architecture, assume both the fixed-point and floating-point instructions go through three stages in a CPU pipeline: Instruction Fetch (I), Execution (E), and Write-Back (W) as outlined earlier. The instruction fetch stage is also the stage where the register file is accessed and the instruction is decoded. Similarly, the execution stage is where the instruction combines its source operands according to the opcode, and the length of this process can vary from one cycle to many cycles. This stage also combines data memory access if the opcode is a fixed or floating-point load or a store. Finally, the write-back stage is where the result from the execution stage is written back to the fixed or floating-point register file.

The process of partitioning typical fixed and floating point instructions in each stage of this new CPU is shown in Fig. 6.117.

In this figure, the LOADF instruction requires the memory address calculation to be complete and in the Load Address Buffer in the I-stage, the data to be extracted from the data memory and loaded to the Floating-Point Input Register in the E-stage and written back to the floating-point register file in the W-stage. Similarly, the STOREF instruction calculates the data memory address and writes it to the Store Address Buffer in the I-stage, fetches data from the floating-point register file and writes it to the Store Data Buffer in the E-stage, and transfers this data from the Store Data Buffer to the data memory in the W-stage. However, STORE and STOREF instructions can also be executed in two cycles. In the first cycle, a data memory address is calculated and written to the Store Address Buffer while the corresponding data is fetched from the register file and written to the Store Data Buffer. These two events can be accomplished in the same clock cycle because the address and data paths in the Tomasula CPU are independent of each other. In the second cycle, the data in the Store Data Buffer is simply written to the address in the data memory.

Instruction Type	Instruction Fetch / RF Access Stage	Execution / Data Memory Access Stage	Write-Back Stage
LOAD RS, RD, Imm	Reg [RS] + Imm → Load Address Buffer	Data Memory → Fixed RF Reg	Fixed RF Reg → Reg [RD]
LOADF RS, FD, Imm	Reg [RS] + Imm → Load Address Buffer	Data Memory → Float RF Reg	Float RF Reg → Reg [FD]
STORE RS, RD, Imm	Reg [RD] + Imm → Store Address Buffer	Reg [RS] → Store Data Buffer	Store Data Buffer → Data Memory
STOREF FS, RD, Imm	Reg [RD] + Imm → Store Address Buffer	Reg [FS] → Store Data Buffer	Store Data Buffer → Data Memory
ADD RS1, RS2, RD	Reg [RS1], Reg [RS2] → Res. Station	Reg [RS1] + Reg [RS2] → Fixed RF Reg	Fixed RF Reg → Reg [RD]
ADDI RS, RD, Imm	Reg [RS] → Res. Station	Reg [RS] + Imm → Fixed RF Reg	Fixed RF Reg → Reg [RD]
ADDF FS1, FS2, FD	Reg [FS1], Reg [FS2] → Res. Station	Reg [RS1] + Reg [RS2] → Float RF Reg	Float RF Reg → Reg [FD]

Fig. 6.117 Partitioning typical fixed and floating-point instructions of a three-stage CPU for a simplified Tomasula algorithm

A typical floating-point instruction, such as ADDF, accesses the contents of source operands, FS1 and FS2, in the I-stage, adds the contents of these registers and stores the result in a general area, called the Reservation Station, in the E-stage, and writes the result back to the floating-point register file at a destination address in the W-stage. As mentioned earlier, some floating-point instructions may take more than one clock cycle to execute, and the result may not be immediately available for the write-back. In these instances, the subsequent instructions, which are dependent on this result, usually wait in the Reservation Station until the result becomes available.

The sample instruction chart in Fig. 6.118 explains the usage and functionality of the Reservation Station where subsequent instruction(s) wait and use the result of a particular instruction(s).

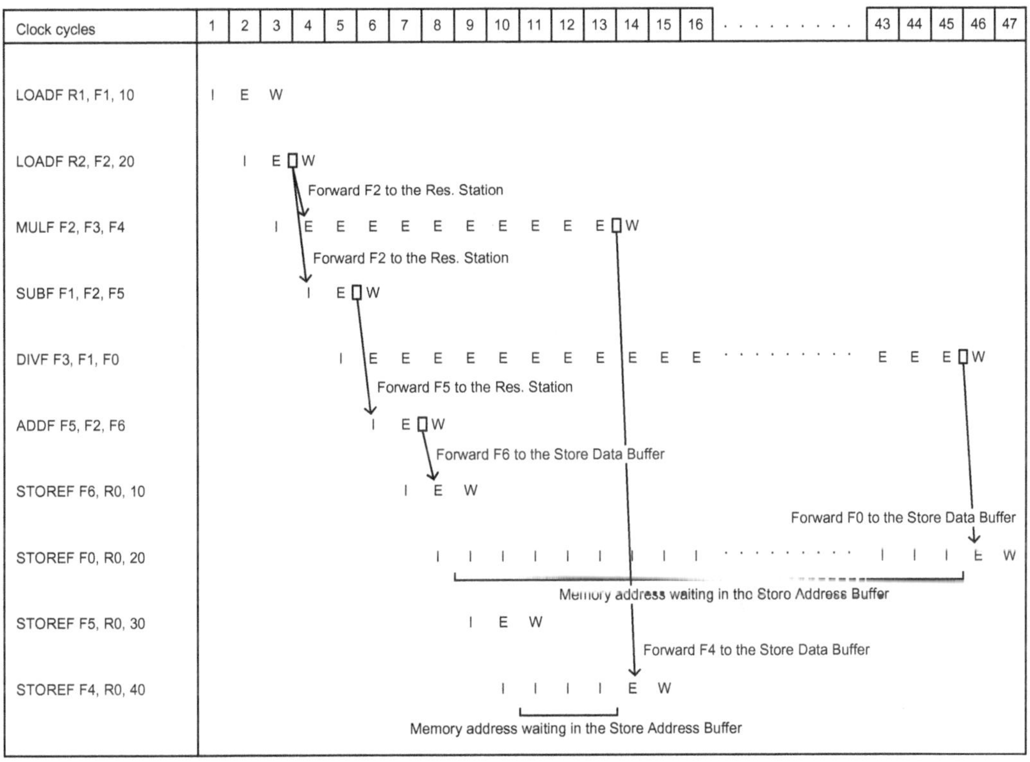

Fig. 6.118 A sample program executing the Tomasula algorithm

In this chart, the second instruction, LOADF R2, F2, 20, normally writes the data fetched from the data memory address, Reg[R2] + 20, to the floating-point register file in cycle 4 according to Fig. 6.117. However, the two subsequent instructions, MULF F2, F3, F4 and SUBF F1, F2, F5, also need Reg[F2] in cycle 4. Therefore, the data, waiting to be written to F2, but presently residing at the data memory output in cycle 3, is immediately forwarded to the Reservation Station so that the subsequent instructions, MULF F2, F3, F4 and SUBF F1, F2, F5, can use this data in cycle 4 when they are in the E and I-stages, respectively.

The fourth instruction, SUBF F1, F2, F5, also forwards the floating-point subtraction result to the Reservation Station at the end of cycle 5 because the sixth instruction, ADDF F5, F2, F6, uses the contents of F5 in its I-stage during cycle 6.

Not all cases need to be forwarded to the Reservation Station. For example, the sixth instruction, ADDF F5, F2, F6, forwards the floating-point adder result to the Store Data Buffer instead of the Reservation Station at the end of cycle 7 so that STOREF F6, R0, 10 can use it during cycle 8.

There are also instances in the program where an instruction forwards the result to the Reservation Station only to be used by a subsequent instruction scheduled many instructions later. Such a case occurs to the third instruction, MULF F2, F3, F4. At the end of its floating-point multiplication stage, the result is forwarded to the Reservation Station in cycle 13. This result is subsequently used by the tenth instruction, STOREF F4, R0, 40, in cycle 14. Another example is the fifth instruction, DIVF F3, F1, F0. According to the instruction chart, the floating-point division result is available in the register file in cycle 46. But, the eighth instruction, STOREF F0, R0, 20, also needs the division result in cycle 46. In order to avoid a potential data hazard, the DIVF instruction simply forwards its result to the Store Data Buffer at the end of cycle 45 so that the STOREF instruction can use it in cycle 46.

The program in Fig. 6.118 and similar floating-point-based programs require the CPU architecture in Fig. 6.119 where instructions are queued initially, and then separated into two different data-paths while they are in execution stage.

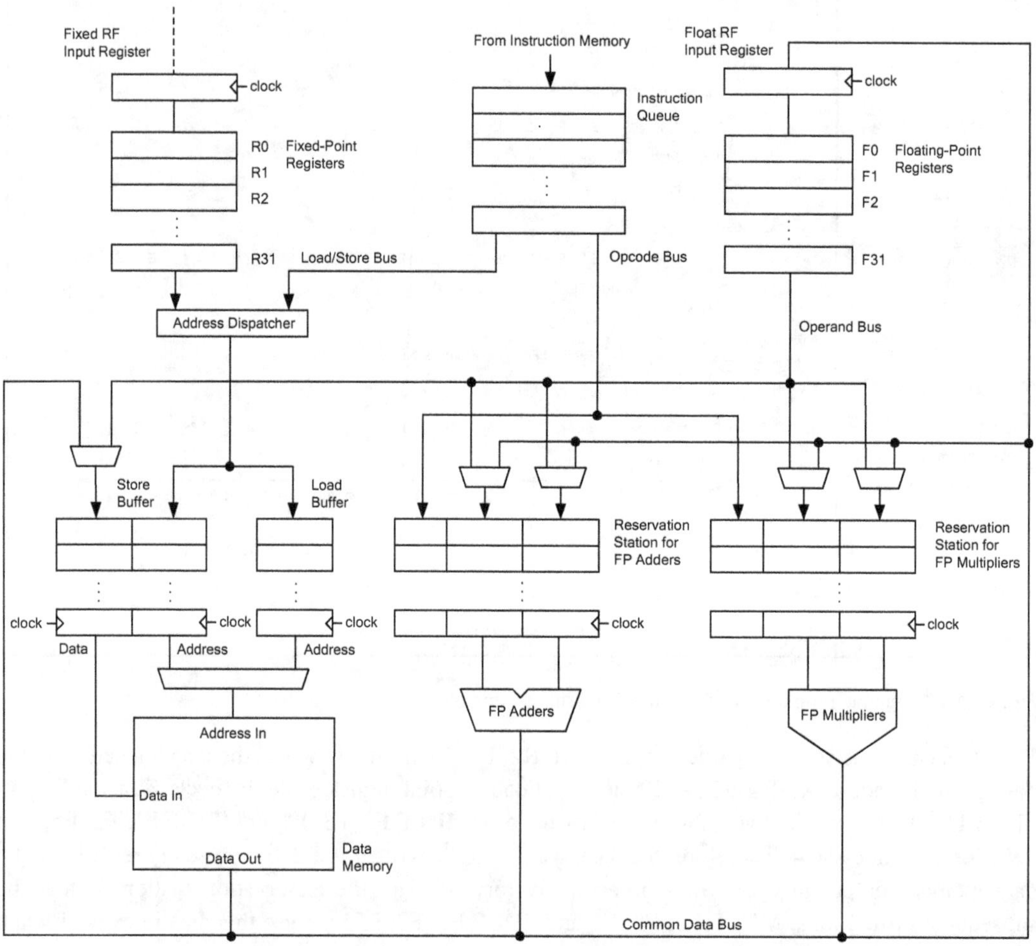

Fig. 6.119 A three-stage Tomasula CPU with floating-point execution and load/store units

The data movement instructions, such as LOADF and STOREF, fetch the operand values for data memory address from the fixed-point register file, form an effective address, and place this address in the Load or Store Address buffers, respectively. This constitutes the I-stage activities of these two instructions shown in Fig. 6.117. In the E-stage, the LOADF instruction downloads contents of the data memory, and stores this data in the Floating-Point Input Register file. The STOREF instruction, on the other hand, fetches the source operand, FS, from the floating-point register file, and places it in the Store Data Buffer. In the W-stage, the LOADF instruction stores the data memory contents to the destination register, FD, in the floating-point register file. The STOREF instruction stores the contents of FS in the data memory.

In this architecture, there are two Reservation Stations to hold temporary data for the floating-point add (subtract) and multiply (divide) operations. The data at the memory output can be moved either of these Reservation Stations in case a subsequent floating-point instruction(s) requires this data immediately. Because of this need, a path exists between the data memory and the Reservation Stations. There are two examples in the instruction chart in Fig. 6.118 to relate this case. The first instance is between LOADF R2, F2, 20 and MULF F2, F3, F4 where the MULF instruction needs the contents of F2 in the E-stage. The second instance is between LOADF R2, F2, 20 and SUBF F1, F2, F5 where the SUBF instruction requires the contents of F2 in its I-stage.

The floating-point arithmetic instructions, such as ADDF, SUBF, MULF and DIVF, fetch their source operands from the floating-point register file and place them in the Reservation Station during the I-stage of execution. In the E-stage, the source data are sent to respective floating-point unit(s) to start execution, and the result is stored either in the Floating-Point RF Input Register or in the Reservation Station(s) in case a subsequent instruction requires this data immediately. The latter necessitates a feedback path between the floating point adder/multiplier outputs and the Reservation Station. The association between ADDF F5, F2, F6 and SUBF F1, F2, F5 in Fig. 6.118 where the ADDF instruction uses the contents of F5 in the I-stage proves this need. In the W-stage, the result is simply written back to the floating-point register file.

The results of floating-point arithmetic instructions may also be needed immediately by a subsequent STOREF instruction, and therefore a path becomes necessary between the floating-point add/multiply units and the Store Data Buffer. There are three such instances in Fig. 6.118. In the first instance, STOREF F6, R0, 10 uses the contents of F6 produced by ADDF F5, F2, F6 in the E-stage. In the second instance, STOREF F0, R0, 20 uses the contents of F0 produced by DIVF F3, F1, F0 in the E-stage. Finally in the third instance, STOREF F4, R0, 40 uses the contents of F4 produced by MULF F2, F3, F4 in the E-stage.

Figure 6.120 through Fig. 6.129 explains how the data flows for each instruction in the program in Fig. 6.118. These figures contain three floating-point adders and two floating-point multipliers. There is a dedicated register for each floating-point unit, containing the Opcode and the source Operand entries.

Figure 6.120 shows the data activity of the first instruction, LOADF R1, F1, 10, while it enters the I-stage. In this cycle, LOADF R1, F1, 10 is fetched from the Instruction Queue, and separated into opcode and operand fields. The opcode, LOADF, is loaded to the Opcode bus. The contents of R1 from the fixed-point register file and the immediate value, 10, are loaded to the Operand bus, and then added in order to calculate an effective memory address. This address is subsequently loaded to the Load Address Buffer.

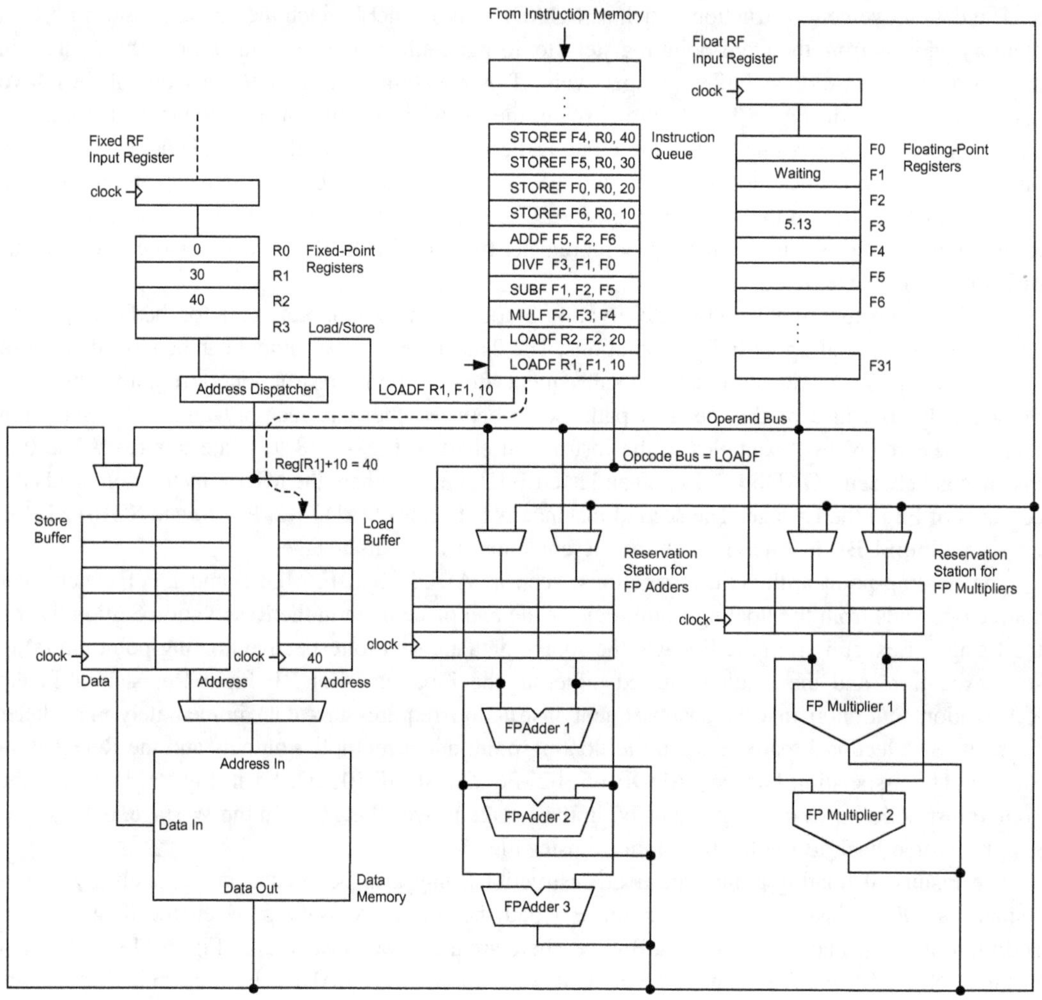

Fig. 6.120 Tomasula FPU starts executing LOADF R1, F1, 10

Figure 6.121 shows the data activity of the second instruction, LOADF R2, F2, 10. Similar to the events which took place in the previous clock cycle, the CPU calculates another memory address by adding the contents of R2 to 20, and stores this value in the Load Address Buffer.

The memory contents at address, 40, now become available at the output of the data memory due to the first LOADF instruction. This data, 0.83, is subsequently stored in the Floating-Point RF Input Register.

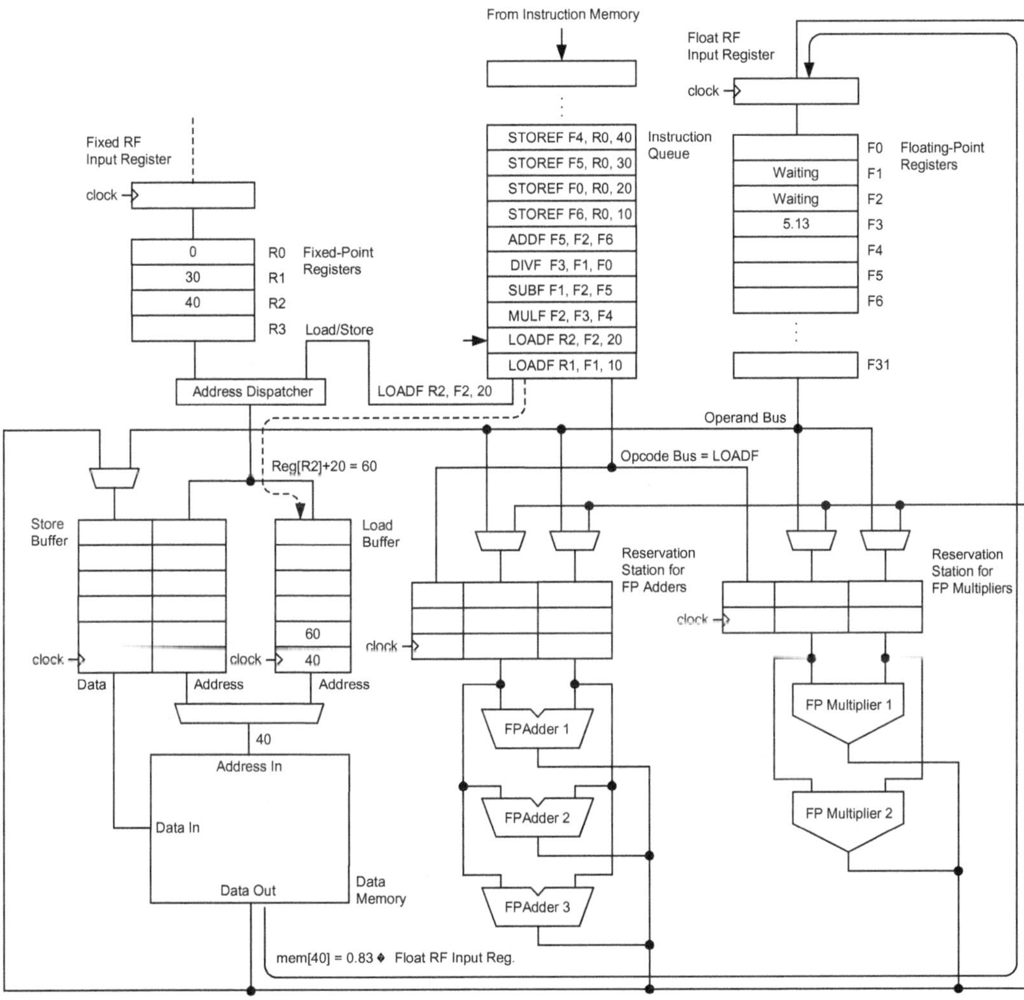

Fig. 6.121 Tomasula FPU starts executing LOADF R2, F2, 20

Figure 6.122 shows the data activity of the third instruction, MULF F2, F3, F4. In this clock cycle, the Opcode bus changes its value to the new opcode, MULF. The contents of the first source operand, Reg[F3] = 5.13, are assumed to be available in the Floating-Point Register File, and directly routed to the Reservation Station. The contents of the second source operand, Reg[F2] = 1.23, on the other hand, do not reside in the Floating-Point Register File, but available at the output of the data memory. This value is also written to the Reservation Station for the MULF instruction. Because the next instruction, SUBF F1, F2, F5, also uses F2 as the source operand, the CPU forwards the contents of F2 to the Reservation Station as well.

In the same cycle, the contents of F1, Reg[F1] = 0.83, due to LOADF R1, F1, 10 is transferred from the Floating-Point RF Input Register to the Floating-Point Register File.

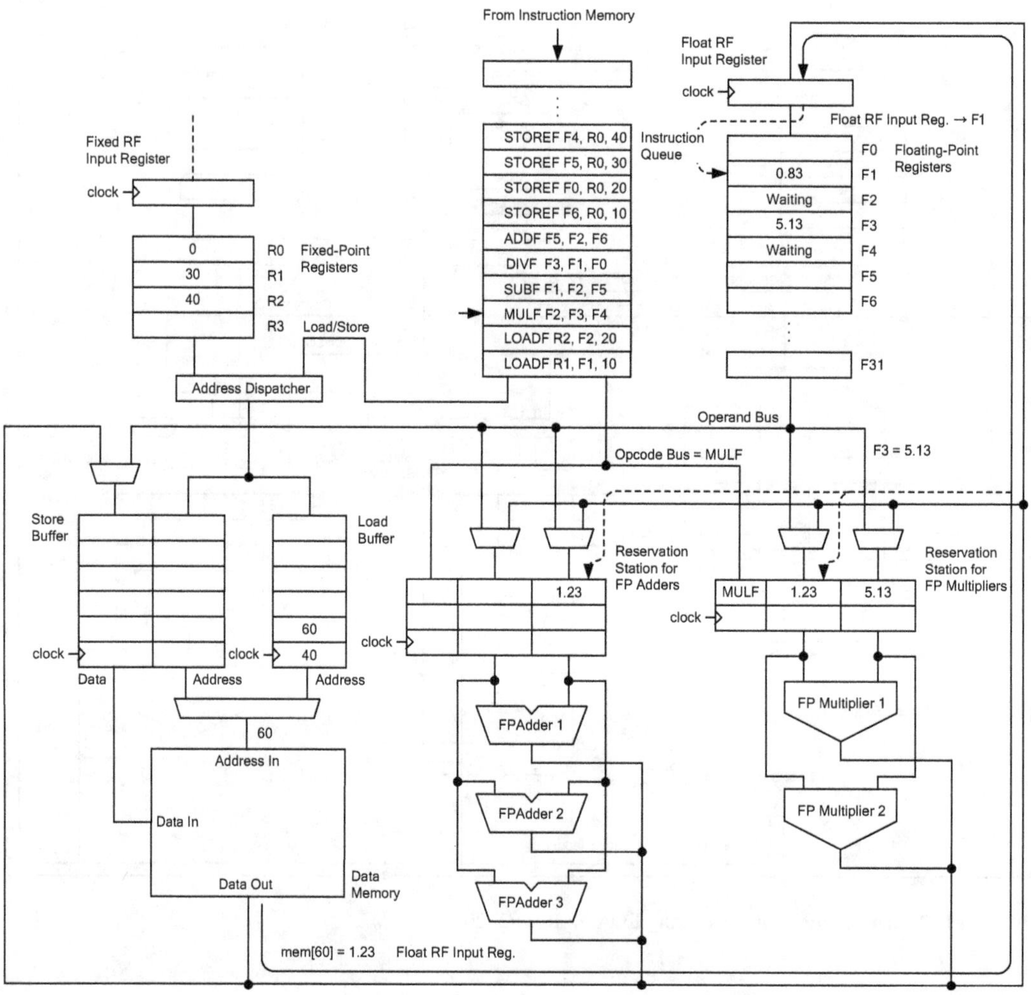

Fig. 6.122 Tomasula FPU starts executing MULF F2, F3, F4

Similar to the data activity of MULF, the SUBF instruction, SUBF F1, F2, F5, follows a similar path in Fig. 6.123. This instruction starts the instruction fetch cycle by loading the Opcode bus with SUBF and the Operand bus with the contents of Reg[F1] = 0.83. Both of these values are subsequently stored in the Reservation Station. The second operand, Reg[F2] = 1.23, was already written to the Reservation Station a cycle before, therefore the MULF instruction starts its first execution stage in the floating-point multiplier, which will last another nine cycles according to Fig. 6.118.

The third event in this cycle is writing the contents of the Floating-Point RF Input Register back to F2 in the Floating-Point Register File.

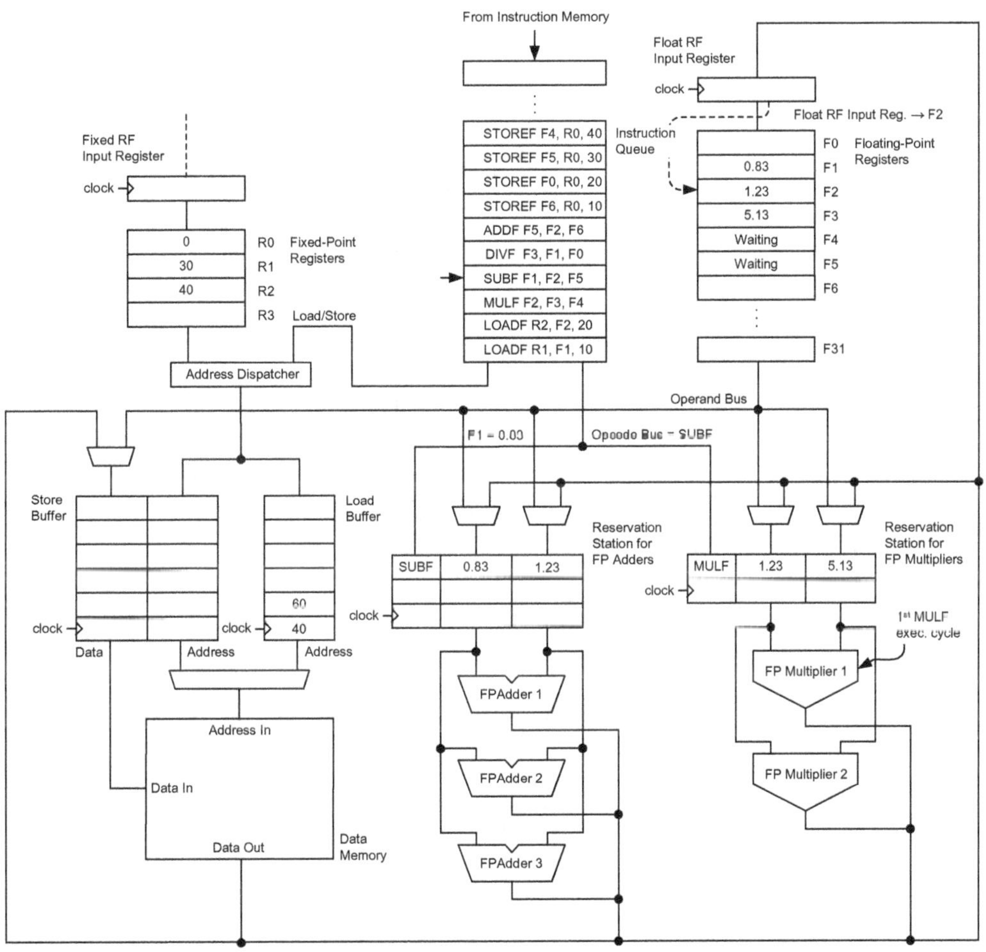

Fig. 6.123 Tomasula FPU starts executing SUBF F1, F2, F5

The next instruction, DIVF F3, F1, F0, starts its I-stage according to Fig. 6.124. The Opcode bus is updated with DIVF. The Operand bus is loaded with the contents of F3, Reg[F3] = 5.13, and the contents of F1, Reg[F1] = 0.83, which are stored in the Reservation Station.

In this cycle, two other events take place. In the first event, the floating-point multiplier enters the second execution stage. In the second event, the floating point adder completes the subtraction operation due to SUBF, and routes the result to the Floating-Point RF Input Register and the Reservation station. The latter is for the ADDF instruction which needs the contents of F5, Reg[F5] = −0.4, in its E-stage, and is scheduled next.

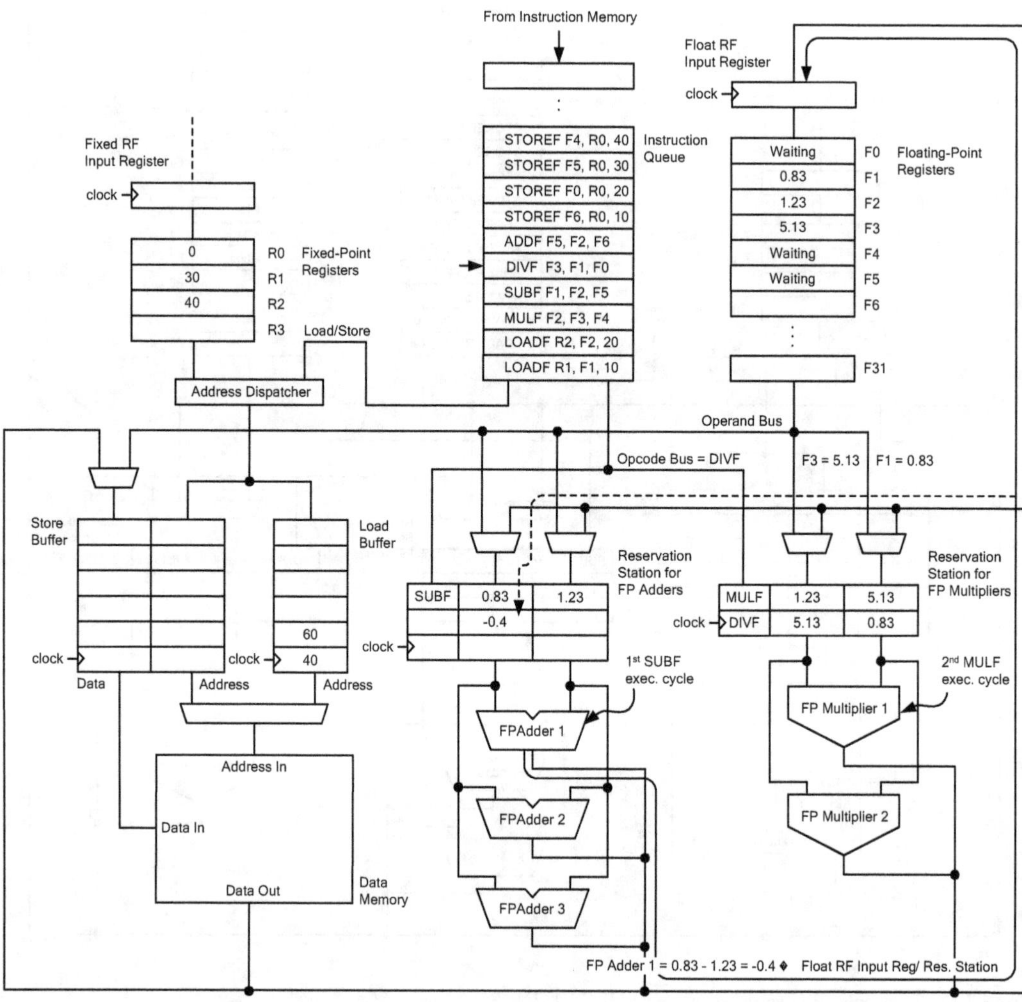

Fig. 6.124 Tomasula FPU starts executing DIVF F3, F1, F0

In Fig. 6.125, the CPU starts executing ADDF F5, F2, F6. In this cycle, the Opcode bus changes to ADDF. The contents of the first source operand, Reg[F2] = 1.23, is fetched from the Floating-Point Register File and directed to the Reservation Station. The contents of the second source operand, Reg [F5] = −0.4, has already been stored in the Reservation Station in the previous clock cycle, and therefore ready to be used.

Within the same cycle, three other events take place. In the first event, the CPU transfers the contents of the Floating-Point RF Input Register, −0.4, to F5 in the Floating-Point Register File. In the second event, the floating point-multiplier enters the third execution stage for MULF. In the last event, the DIVF instruction starts its first execution cycle in the second floating-point multiplier. There will be 39 more clock periods until the DIVF instruction determines the floating point quotient according to Fig. 6.118 since the floating-point division algorithm successively approximates a quotient value by using the floating-point multiplier at each approximation.

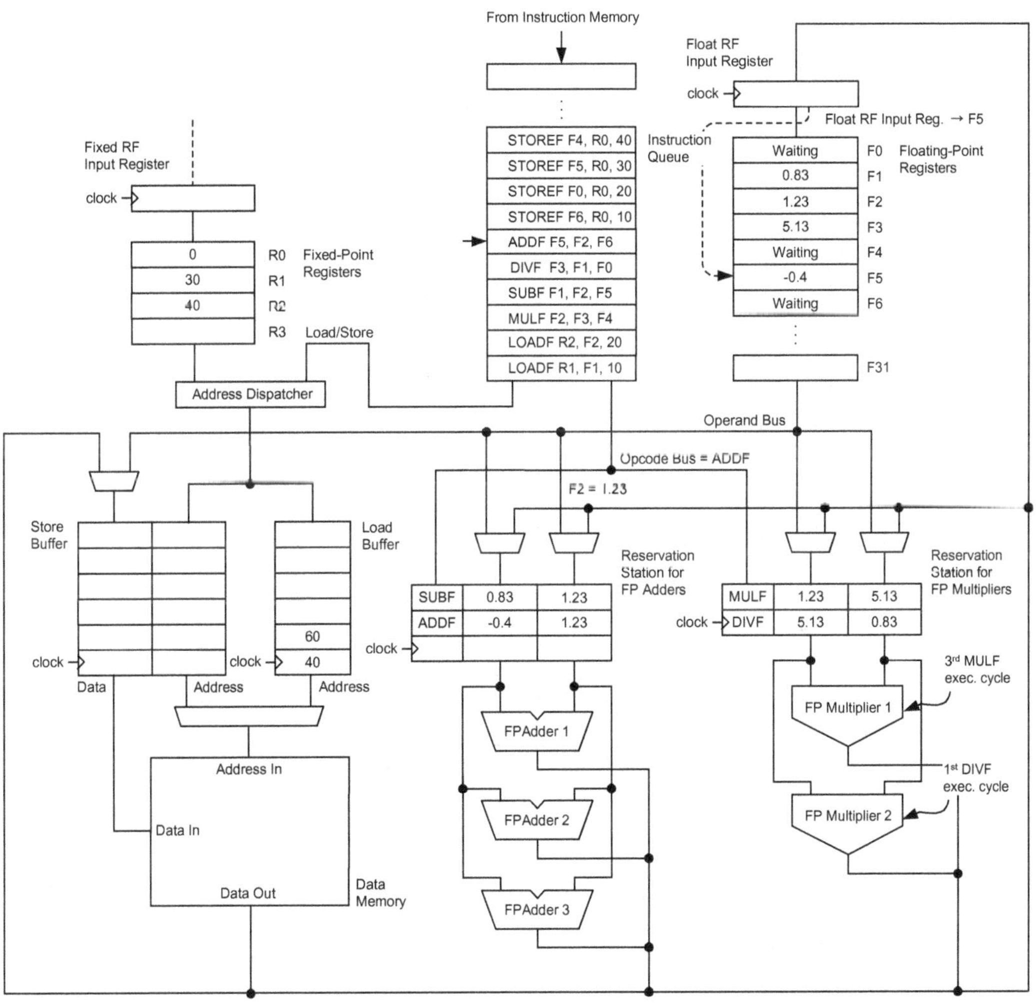

Fig. 6.125 Tomasula FPU starts executing ADDF F5, F2, F6

In this cycle, the CPU executes the first floating-point store instruction, STOREF F6, R0, 10, according to the program in Fig. 6.118. All data activities are shown in Fig. 6.126. The address dispatcher calculates the store address, Reg[R0] + 10 = 10, and writes it to the Store Address Buffer while the Opcode bus is loaded with the STOREF opcode.

Three other data activities take place in this cycle. In the first event, the floating-point adder completes executing the ADDF instruction, and the CPU transfers the result, 0.83, to the Floating-Point RF Input Register. In the second and third events, the first floating-point multiplier, executing the MULF instruction, now enters its fourth execution stage while the second multiplier, executing the DIVF instruction, starts the second execution stage.

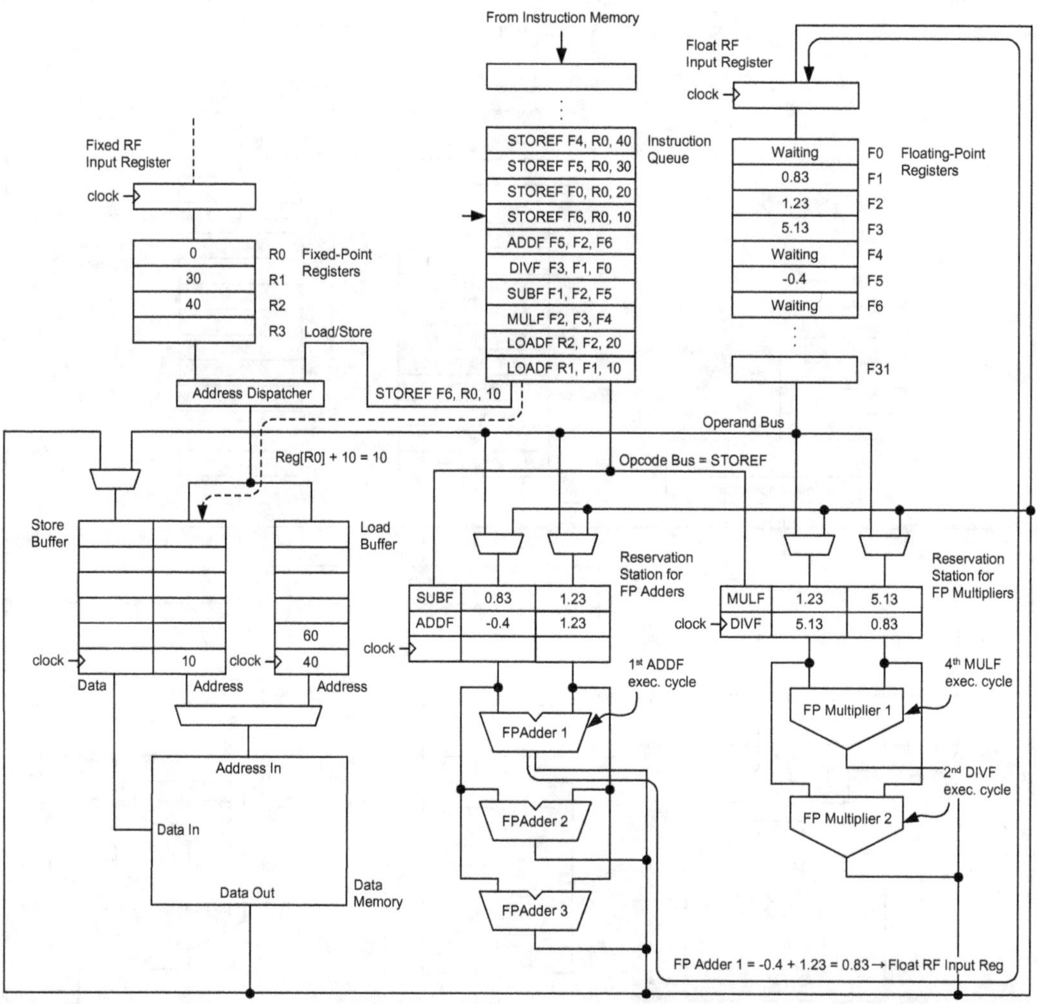

Fig. 6.126 Tomasula FPU starts executing STOREF F6, R0, 10

The second floating-point store instruction, STOREF F0, R0, 20, in the program starts in this cycle. All data activities are shown in Fig. 6.127. According to this figure, the address dispatcher calculates a new store address, Reg[R0] + 20 = 20, and stores it in the Store Address Buffer. The Opcode bus keeps the STOREF opcode.

In the same cycle, the floating point multipliers enter the fifth and the third execution cycles for the MULF and DIVF instructions, respectively.

Lastly, the result from the first floating-point adder is transferred to the Store Data Buffer because the first STOREF instruction, STOREF F6, R0, 10, needs to store this data at the data memory address of 10 according to Fig. 6.118.

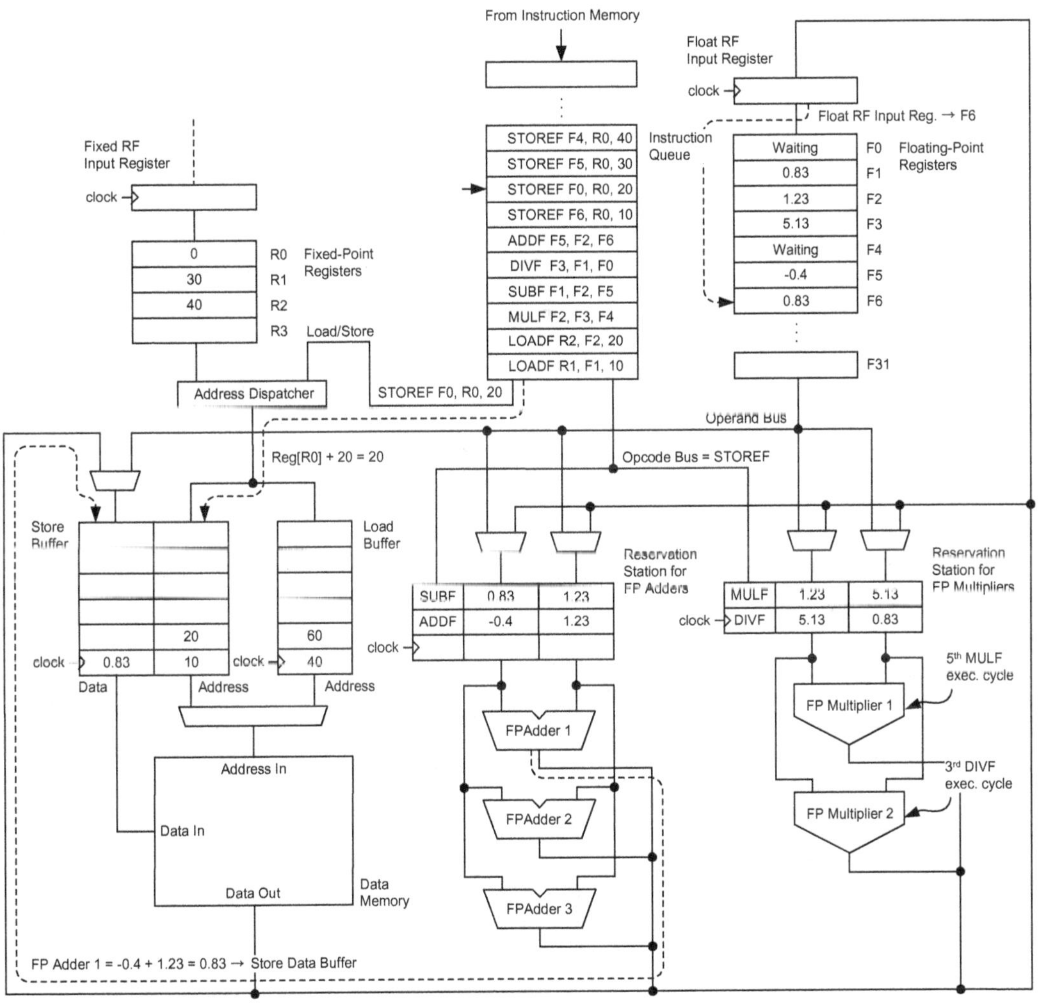

Fig. 6.127 Tomasula FPU starts executing STOREF F0, R0, 20

In this cycle, the third floating-point store instruction, STOREF F5, R0, 30, enters its instruction fetch stage according to Fig. 6.128. The store address, Reg[R0] + 30 = 30, is calculated and stored in the Store Address Buffer. The Opcode bus still contains the STOREF opcode. The floating-point multipliers enter their sixth and fourth execution stages for the MULF and DIVF instructions, respectively.

Two other operations take place in this clock cycle. In the first operation, the CPU transfers the contents of F0 from the Floating-Point Register File to the Store Data Buffer. However, since the CPU is still waiting for the results of the DIVF instruction to determine the contents of F0, this transfer does not actually materialize. Instead, the operation goes into a pending stage where the data entry in the Store Data Buffer is tagged with WAIT F0. In the second operation, the contents of F6, Reg[F6] = 0.83, is written to the data memory address, 10, due to the first STOREF instruction, STOREF F6, R0, 10.

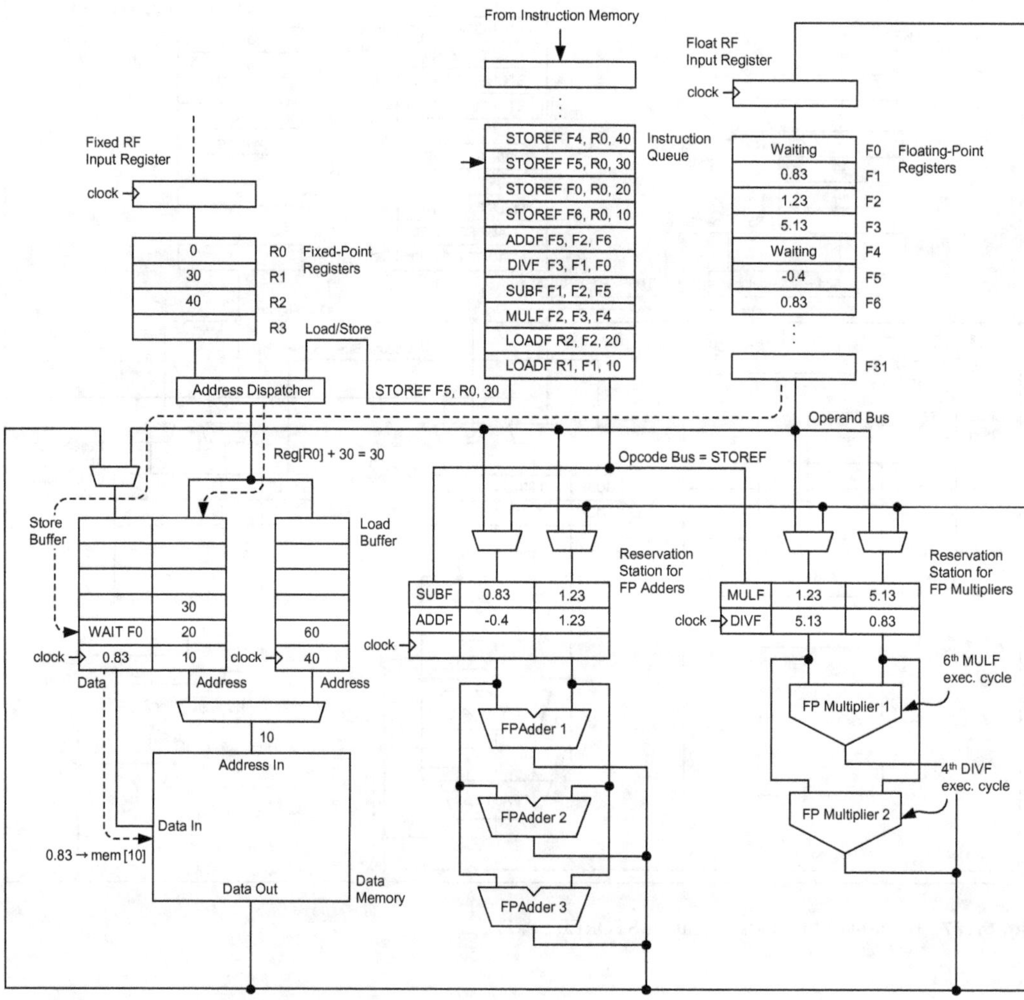

Fig. 6.128 Tomasula FPU starts executing STOREF F5, R0, 30

The data activities in this cycle are pretty much the same as in the previous clock period. The fourth floating-point store instruction, STOREF F4, R0, 40, enters its instruction fetch stage as shown in Fig. 6.129. A new store address, Reg[R0] + 40 = 40, is calculated and stored in the Store Address Buffer. The floating-point multipliers enter their seventh and fifth execution stages for the MULF and DIVF instructions, respectively.

Two other operations take place in this clock period. In the first one, the CPU moves the contents of F5, Reg[F5] = −0.4, from the Floating-Point Register File to the Store Data Buffer due to the STOREF F5, R0, 30 instruction. In the second, the CPU transfers the contents of F0 from the Store Data Buffer to the data memory at the address of 20. However, since the CPU is still waiting for a valid data entry for F0, the actual write to the data memory does not materialize. Instead, this operation stays in the pending stage to be executed later when the DIVF instruction finishes its execution stage.

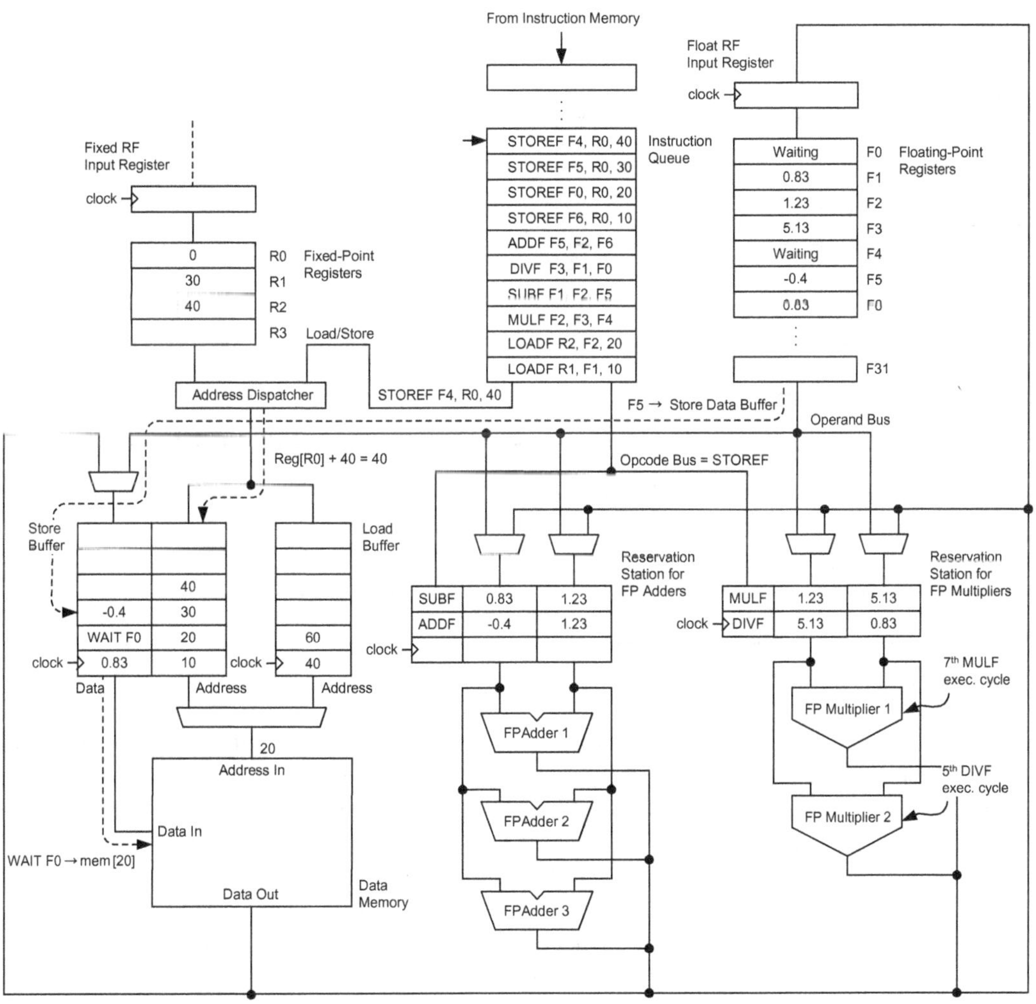

Fig. 6.129 Tomasula FPU starts executing STOREF F4, R0, 40

A curious reader may question how the CPU knows when to forward an operand from the output of the floating-point adder, multiplier or data memory to either the Store Data Buffer or the Reservation Station. There may be several implementation schemes to realize these data transfers. But, the simplest scheme may reside inside the compiler. This scheme may include a "virtual" instruction chart that compares the source and destination register addresses for each instruction in the program such as in Fig. 6.130. In this chart, the first LOADF instruction has no data dependency, so it is tagged with a "No" tag. The second LOADF instruction has data two data dependencies: one with the MULF instruction scheduled in the next clock cycle, and the other with the SUBF instruction scheduled two cycles later as shown in Fig. 6.118. Therefore, this LOADF instruction is tagged with a "Yes" tag, and it will forward the data both to the Reservation Station for the floating-point adder and to the Reservation Station for the floating-point multiplier as soon as the data memory contents become available. The subsequent instructions, MULF, SUBF, DIVF and ADDF, also show where the floating-point data needs to be forwarded from the floating-point adder/multiplier output(s). For the STOREF instructions, the destination register is R0, which always is zero. Therefore, these four instructions have "Invalid" tags for data dependency.

Prog. Instructions	Dest. match?	N	Y	I	ADD Res. Stat.	MUL Res. Stat.	Store Data Buf.
LOADF R1, F1, 10	F1	✓			-	-	-
LOADF R2, F2, 20	F2		✓		✓	✓	-
MULF F2, F3, F4	F4		✓		-	-	✓
SUBF F1, F2, F5	F5		✓		✓	-	-
DIVF F3, F1, F0	F0		✓		-	-	✓
ADDF F5, F2, F6	F6		✓		-	-	✓
STOREF F6, R0, 10	R0	-	-	✓	-	-	-
STOREF F0, R0, 20	R0	-	-	✓	-	-	-
STOREF F5, R0, 30	R0	-	-	✓	-	-	-
STOREF F4, R0, 40	R0	-	-	✓	-	-	-

Fig. 6.130 Virtual instruction chart

This chart may also include other entries depending on how sophisticated the compiler needs to be for specific applications. For example, additional tags may indicate which subsequent instruction's source operand is dependent on the destination operand, and when the subsequent instruction is scheduled to be executed in case the Reservation Station(s) or the Store Data Buffer has a limited capacity to sustain long programs and cannot hold operand values for long periods of time.

Figure 6.131 contains the same Tomasula floating-point topology shown in Fig. 6.120 with two separate register files and dedicated Reservation Stations for the fixed and floating-point units. The inclusion of the reservation station to the fixed point unit eliminates the forwarding hardware between the ALU output and the ALU input, and between the data memory output and the ALU input as discussed earlier in this chapter.

Both fixed-point and floating-point instructions go through three stages of execution in this architecture: the combined instruction fetch and register file access stage (I), the execution or data memory

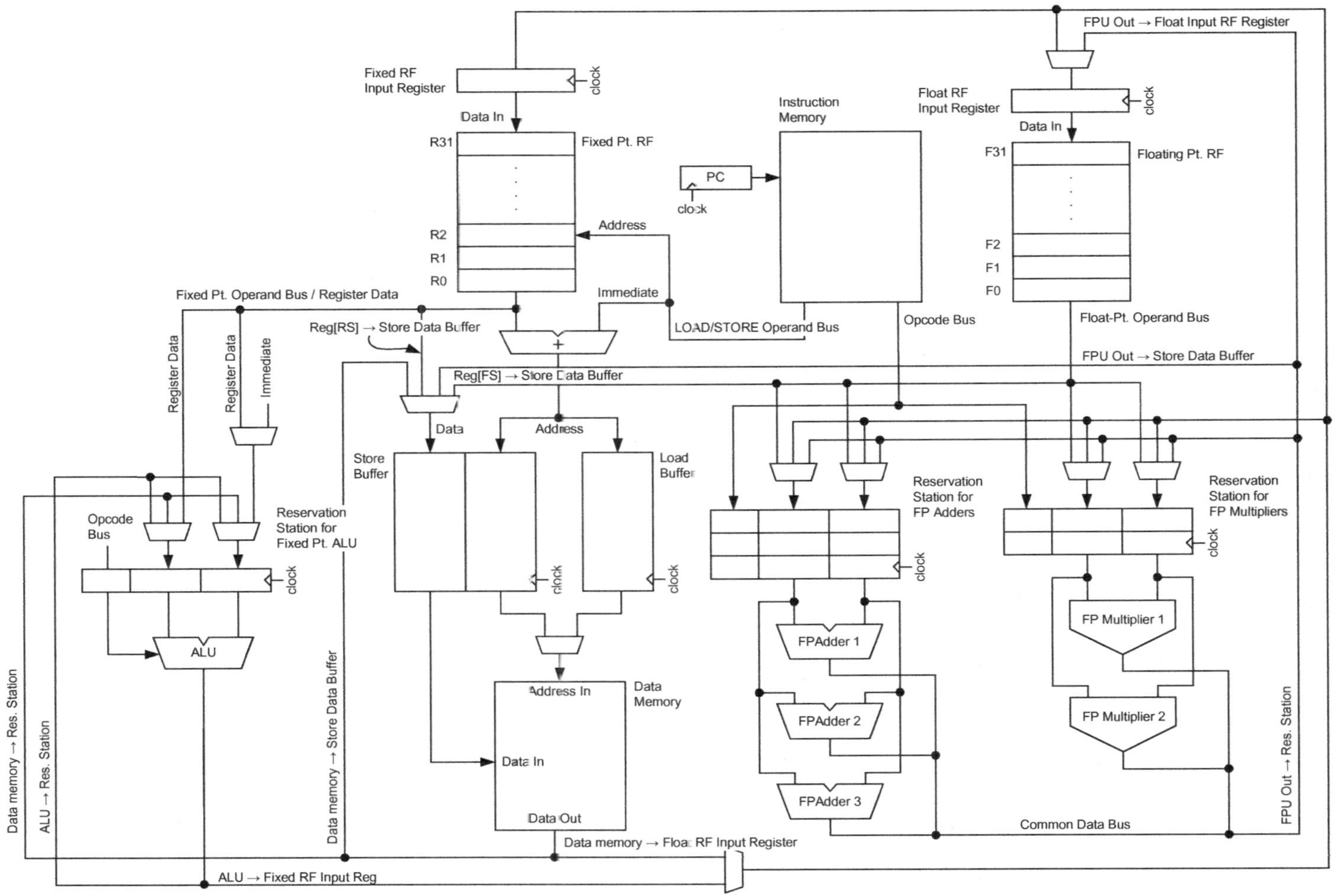

Fig. 6.131 A three-stage Tomasula CPU with fixed and floating-point execution and load/store units

access stage (E) and the write-back stage (W). Although the execution stage of a floating-point add or subtract may take as little as one clock cycle, a floating-point multiply or divide operation may take as many as ten clock cycles or more as demonstrated in the sample program in Fig. 6.118.

6.6 Increasing Program Execution Efficiency

Static Versus Dynamic Pipelines

In the previous sections of this chapter, we described a CPU architecture that issued fixed and floating-point instructions into a single pipeline without any consideration of parallelism to reduce the overall program execution time. Program instructions, however, can be separated from each other and guided into several pipelines, each pipeline specializing to handle certain types of instructions, so that different segments of the program can be executed simultaneously. To demonstrate this approach, consider a program with fixed-point instructions in Fig. 6.132. From the instruction chart, this program is assumed to be executed in a four-stage pipeline where the ALU and data memory stages are combined.

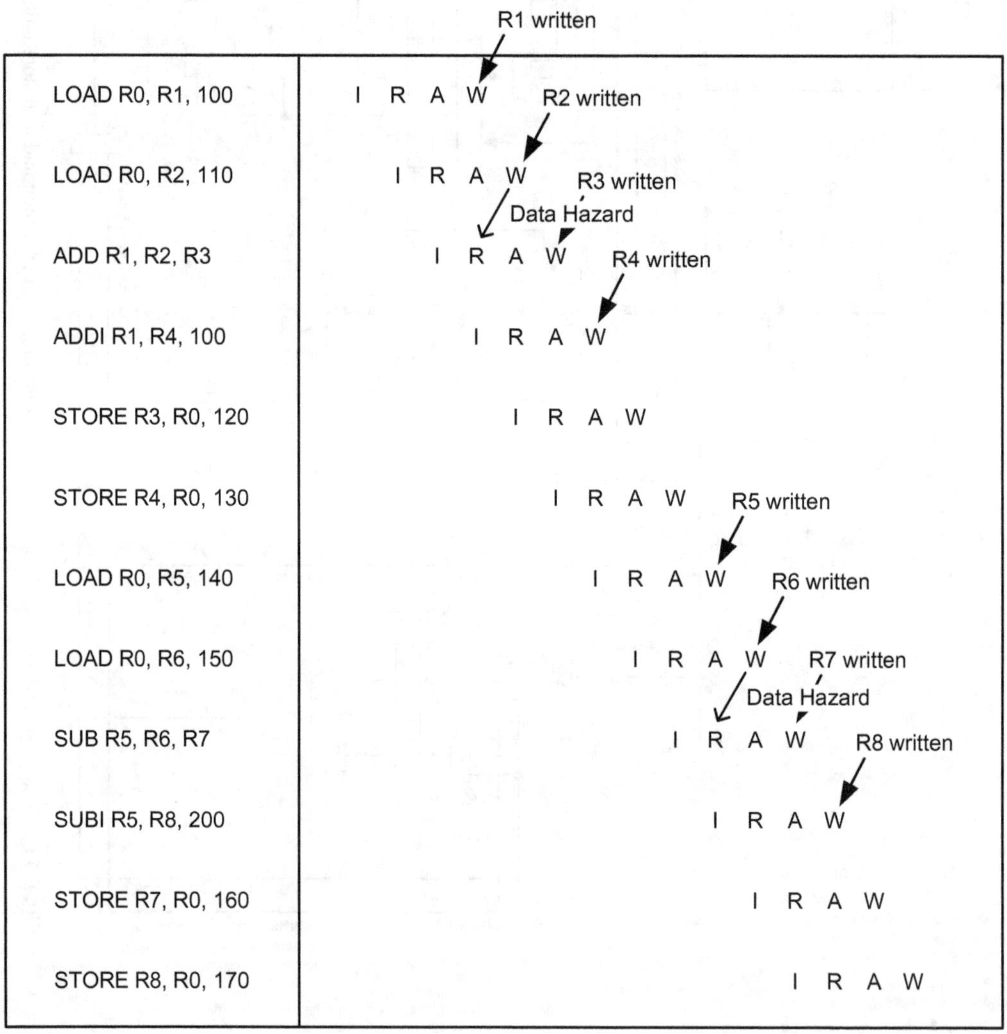

Fig. 6.132 A single-issue program with fixed-point instructions

During the program execution, instructions may present data dependencies with one another and create data hazards. To eliminate a data hazard, the CPU may use existing forwarding path(s) and/or choose to stall the pipeline by adding NOP instruction(s), but it always keeps the program instructions in order. This type of CPU is called the static-issue CPU.

Two data hazards exist in the instruction chart in Fig. 6.132. The first one is between LOAD R0, R2, 110 and ADD R1, R2, R3 where the ADD instruction needs the contents of R2, but cannot get it immediately. The second one is between LOAD R0, R6, 150 and SUB R5, R6, R7 where the LOAD instruction cannot supply the contents of R6 to the SUB instruction on time.

Therefore, the CPU has two choices to eliminate these hazards. The first choice is to issue two NOP instructions: one NOP is placed before the ADD instruction and the other one before the SUB instruction. This way, both the ADD and the SUB instructions will have time to fetch their operands from the corresponding LOAD instructions properly. However, this approach increases the overall CPU execution time by two clock cycles, and not optimal. The second choice is to employ two

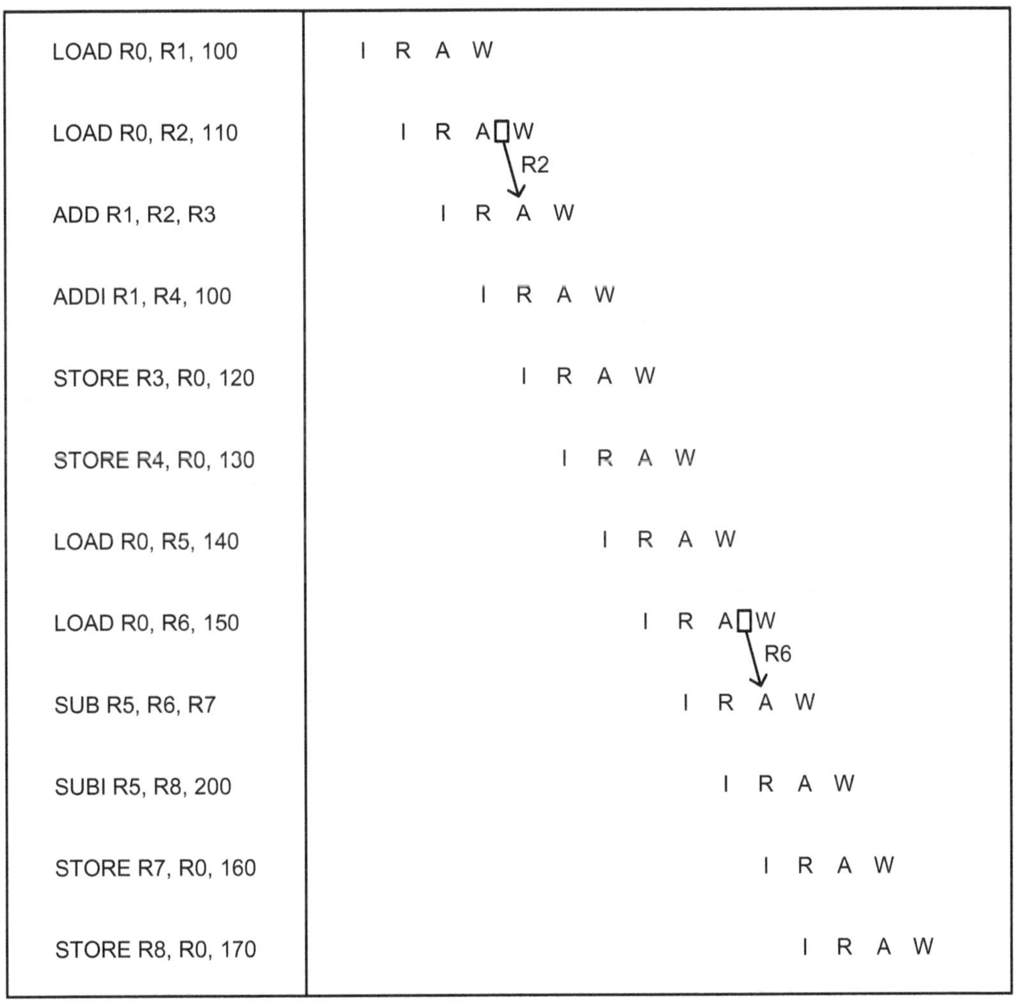

Fig. 6.133 Hazard-free single-issue program from Fig. 6.132

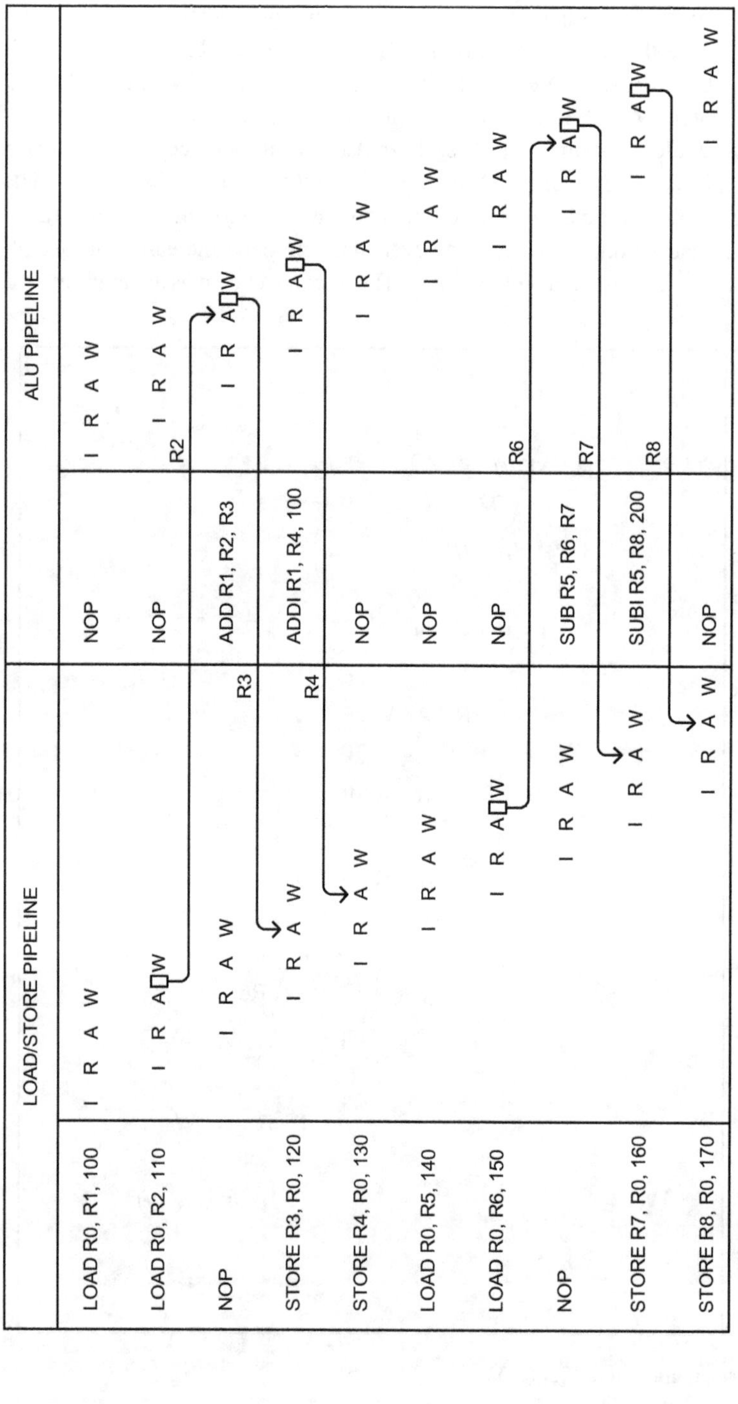

Fig. 6.134 Scheduling the static dual-issue program in Fig. 6.133 for two pipelines

forwarding paths as shown in Fig. 6.133. The first forwarding path moves the contents of R2 from the ALU output to the ALU input, and the second path repeats the same process for R6.

Now, let us take the hazard-free program in Fig. 6.133, and create two static-issue pipelines instead of one as shown in Fig. 6.134. In this new design, the first pipeline is designated to handle only the load/store operations, and the second pipeline operates only on the ALU-type register-to-register or immediate instructions. Because the pipeline is assumed static, there should be no deviation from the original program order shown in Fig. 6.133. This means that the load/store pipeline schedules LOAD R0, R1, 100 first and STORE R8, R0, 170 last to match the program order in Fig. 6.133. In a similar fashion, the ALU pipeline schedules ADD R1, R2, R3 first and SUBI R5, R8, 200 last, again according to Fig. 6.133. Because the first data dependency occurs between LOAD R0, R2, 110 and ADD R1, R2, R3, the CPU issues two NOP instructions prior to the ADD instruction in the ALU pipeline, and forwards the contents of R2 from the load/store pipeline to the ALU pipeline. Next, the CPU issues a NOP instruction in the load/store pipeline instead of scheduling STORE R3, R0, 120 because the STORE instruction needs the contents of R3 from the ADD instruction to proceed. The CPU issues five more NOP instructions and employs four more data forwarding paths to eliminate all the data hazards in each pipeline before the program comes to an end. As a result, the CPU executes the entire program in 13 clock cycles using two static pipelines instead of 15 with a single pipeline.

This time, let us remove the static-issue restriction, and allow the compiler to change the program order in order to avoid forwarding paths and/or NOP penalties. With this new approach, we introduce a dynamic behavior to the program and execute the instructions out of order. After the compilation phase is complete, the original program in Fig. 6.132 becomes hazard-free, needing no forwarding paths for a single-issue CPU as shown in Fig. 6.135. In this new program, the compiler reschedules two instructions: the ADDI instruction is scheduled before the ADD instruction; the SUBI instruction is moved in front of the SUB instruction. Since the W and R-stages can take place in the same clock cycle, overlapping these two stages do not impose any data hazard as shown in the figure.

As a second step, let us also allow the compiler to distribute these instructions between the same load/store and the ALU pipelines in such as way that the resultant instruction chart contains minimum number of forwarding paths and NOP instructions. With these new rules, the instruction chart in Fig. 6.136 no longer follows the original program order in Fig. 6.132, but rather contains four instructions to be executed simultaneously in both pipelines. As a result, the program finishes in 11 clock cycles with two forwarding paths compared to 13 clock cycles and six forwarding paths as in Fig. 6.134.

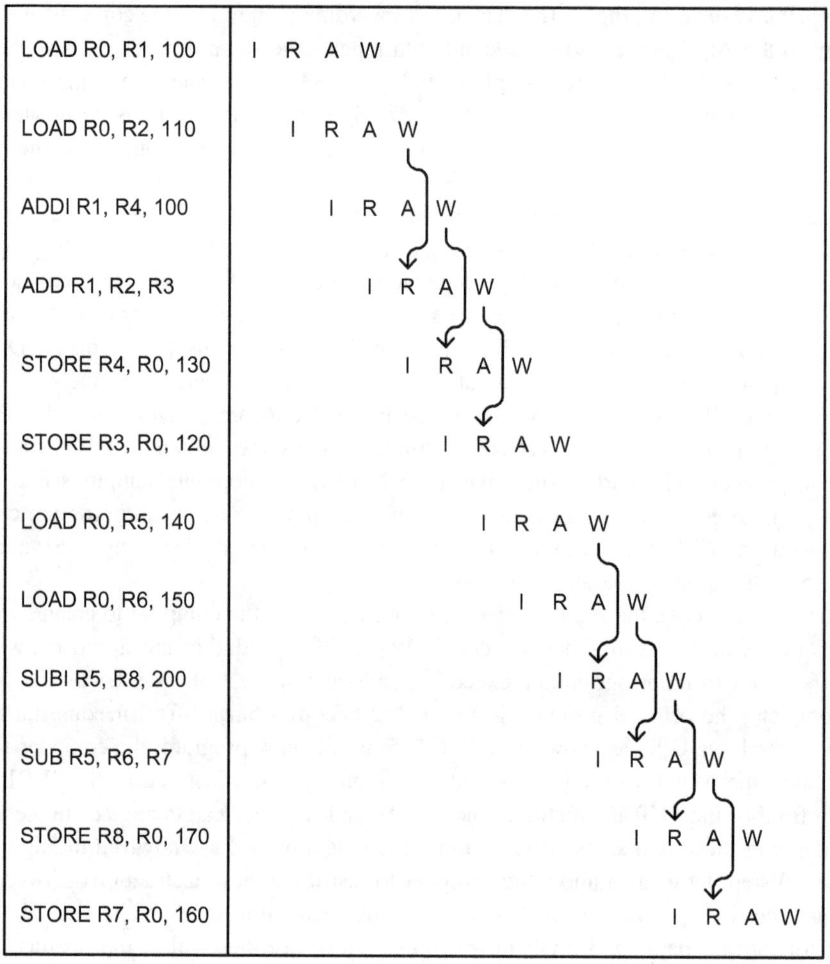

Fig. 6.135 Reordering the program instructions in Fig. 6.132 and achieving an out-of-order program execution

The resultant dual-issue, fixed-point CPU architecture is shown in Fig. 6.137, containing all four stages to match the instruction charts in Figs. 6.134 or 6.136, regardless the program instructions are executed in-order or out-of-order formation. In this new architecture, the CPU dispatches the compiled instructions to either the load/store or the ALU pipelines according to the opcode. The dispatcher hardware is not shown in this figure, but it is a part of the RF access-stage. Since there may be two simultaneous write-backs from the LOAD and the ALU instructions within the same clock cycle, the register file has two sets of write-back ports. Note that the forwarding paths between the load/store and the ALU pipelines in this figure are also not shown to avoid complexity.

LOAD/STORE PIPELINE		ALU PIPELINE	
LOAD R0, R1, 100	I R A W	NOP	I R A W
LOAD R0, R2, 110	I R A W	NOP	R2 I R A W
LOAD R0, R5, 140	I R A W	ADD R1, R2, R3	I R A W
LOAD R0, R6, 150	I R A W	ADDI R1, R4, 100	R6 I R A W
STORE R3, R0, 120	I R A W	SUB R5, R6, R7	I R A W
STORE R4, R0, 130	I R A W	SUBI R5, R8, 200	I R A W
STORE R7, R0, 160	I R A W	NOP	I R A W
STORE R8, R0, 170	I R A W	NOP	I R A W

Fig. 6.136 Scheduling the out-of-order execution program in Fig. 6.135 for two pipelines

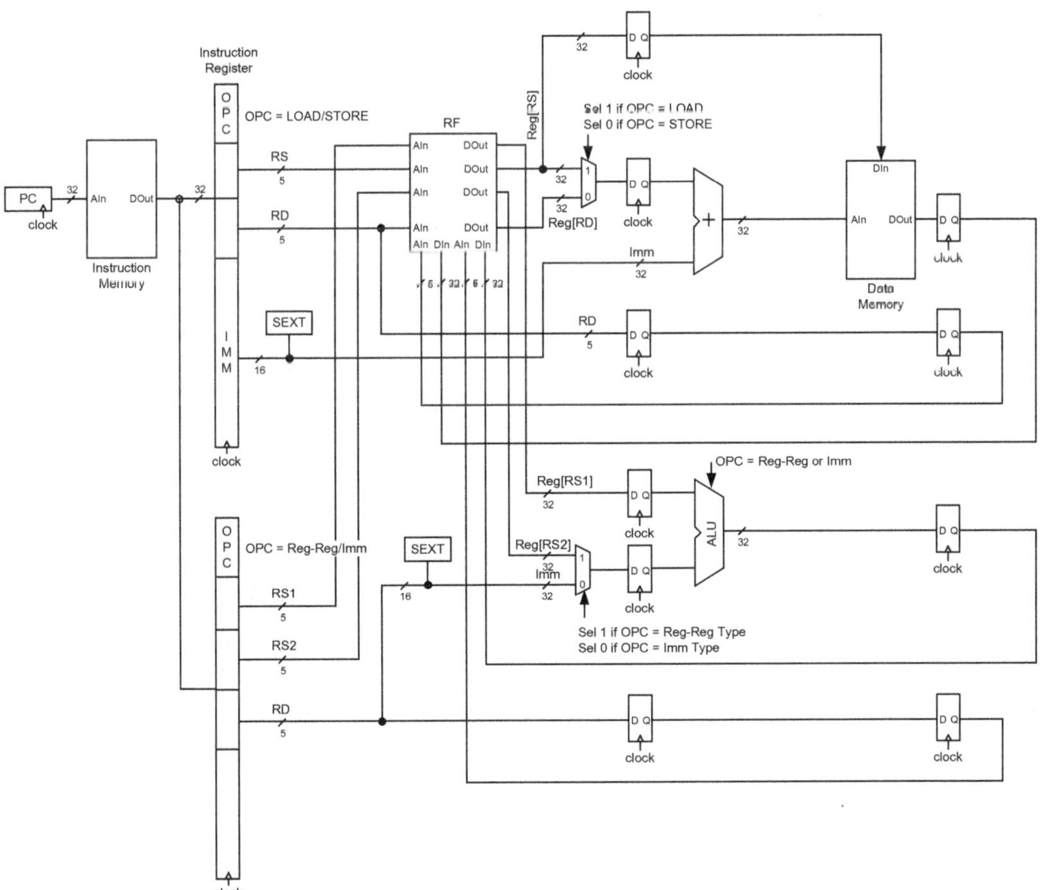

Fig. 6.137 Dual-issue, four-stage CPU (forwarding paths are not shown for clarity)

We can use the same argument in the dual-issue CPU and install more pipelines to achieve even more parallelism. For example, the single-issue program in Fig. 6.138 contains three types of instructions: the load/store, the fixed-point and the floating-point. Each instruction type can be branched to an associated pipeline and executed simultaneously to reduce the overall execution time. This approach is used in Fig. 6.139 where instructions are separated from each other according to

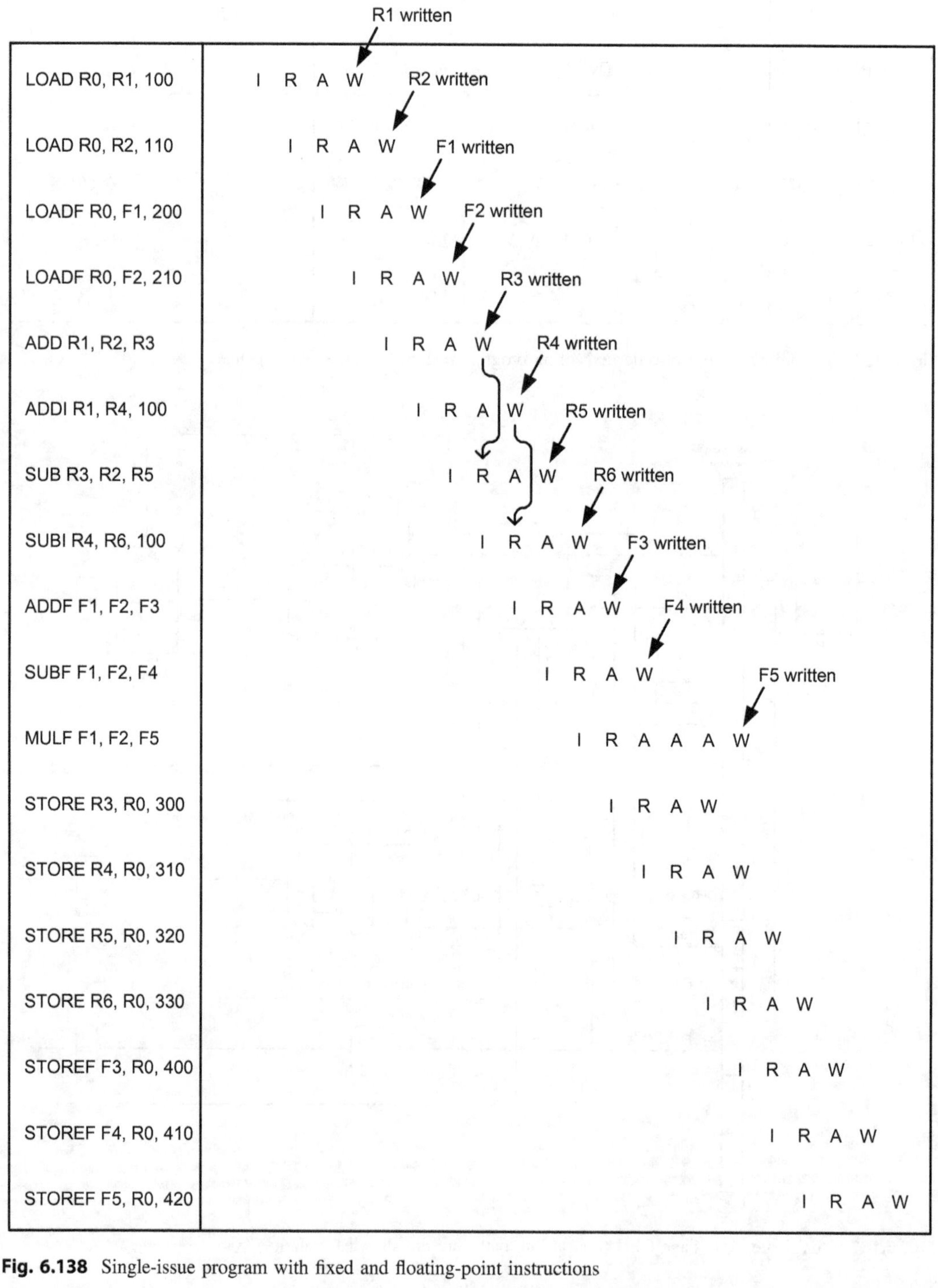

Fig. 6.138 Single-issue program with fixed and floating-point instructions

their opcodes types, compiled and scheduled for execution. In the process, the instruction order is changed in the load/store and the fixed-point ALU pipelines compared to the instruction order in Fig. 6.138 to make sure that the entire program is free of data hazards, and maximum parallelism is achieved among instructions. According to this new chart, the distributed program executes in 14 clock cycles instead of 21 as in Fig. 6.138.

LOAD/STORE PIPELINE		FIXED PT. ALU PIPELINE		FLOAT. PT. FPU PIPELINE	
LOAD R0, R1, 100	I R A W	NOP		NOP	
LOAD R0, R2, 110	I R A W	NOP		NOP	
LOADF R0, F1, 200	I R A W	ADDI R1, R4, 100	I R A W	NOP	
LOADF R0, F2, 210	I R A W	ADD R1, R2, R3	I R A W	NOP	
STORE R4, R0, 310	I R A W	SUBI R4, R6, 100	I R A W	NOP	
STORE R3, R0, 300	I R A W	SUB R3, R2, R5	I R A W	ADDF F1, F2, F3	I R A W
STORE R6, R0, 330	I R A W	NOP		SUBF F1, F2, F4	I R A W
STORE R5, R0, 320	I R A W	NOP		MULF F1, F2, F5	I R A A A W
STOREF F3, R0, 400	I R A W	NOP		NOP	
STOREF F4, R0, 410	I R A W	NOP		NOP	
STOREF F5, R0, 420	I R A W	NOP		NOP	

Fig. 6.139 Scheduling the program in Fig. 6.138 for three pipelines, each operating in out-out-order fashion

Figure 6.140 shows the triple-issue CPU architecture with a three-stage floating-point multiplier to match the instruction chart in Fig. 6.139. Both the fixed and the floating-point register files have two write-back ports to accommodate LOAD/LOADF and fixed (or floating-point) instructions in the same cycle. The address and data forwarding paths are omitted from this figure for clarity.

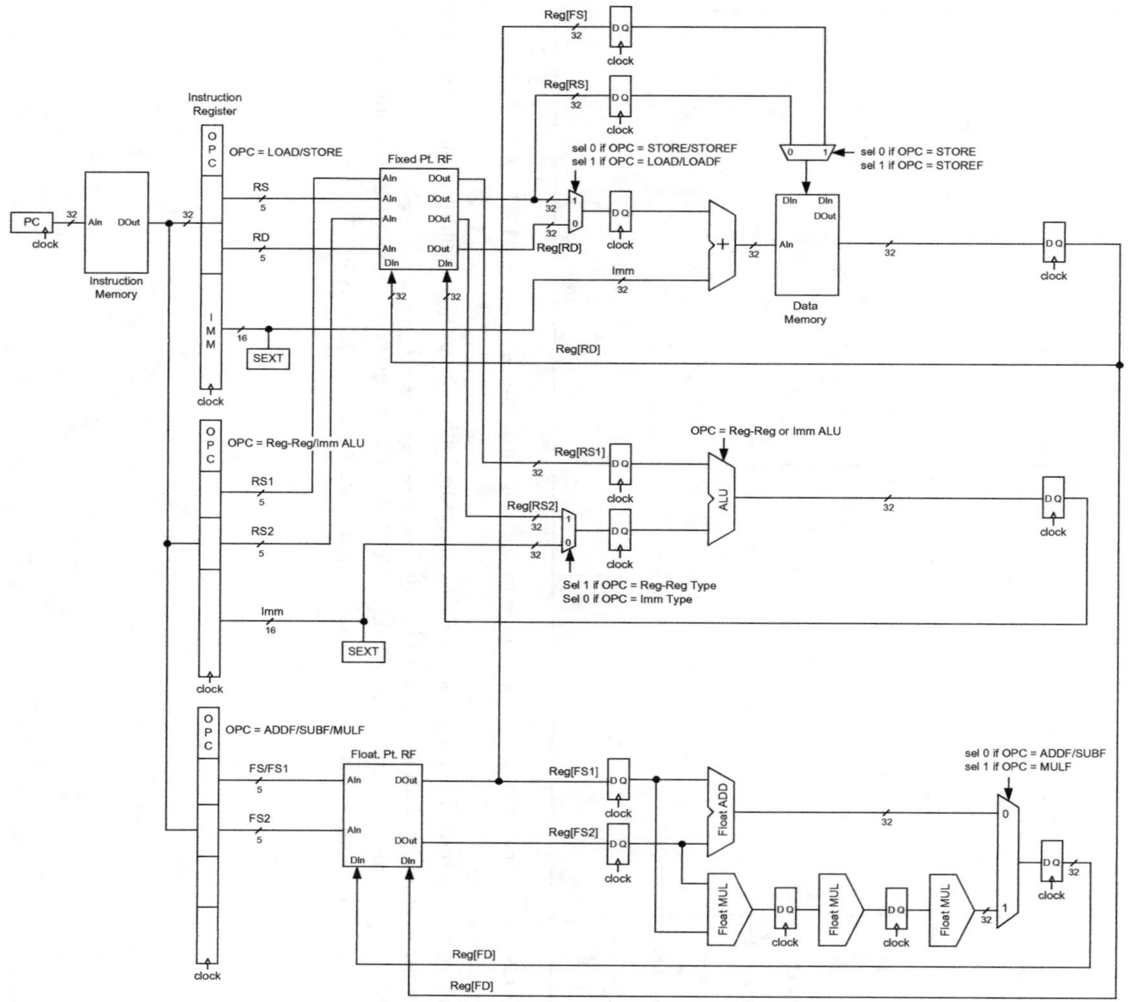

Fig. 6.140 Triple-issue four-stage CPU (address and forwarding paths are not shown for clarity)

Loop Unrolling

The second method to shorten the overall program execution time is to unroll a loop when the program encounters a loop function. The only drawback in this method is that the user needs to duplicate the program segment contained in the loop at every iteration. However, a compiler can also be designed to duplicate the instructions in a loop, assigning index(s) to each instruction if necessary, and rescheduling them properly for execution.

The flow chart in Fig. 6.141 moves the data memory contents to F0, Reg[F0], at the index, i (the index corresponds to an address in data memory, and initially assumed to be greater than zero), adds Reg[F0] to the contents of F2, Reg[F2], to form the contents of F4, Reg[F4], and stores this result back to the data memory at the same index, i. The program repeats the same operations until the index, i, decrements to zero.

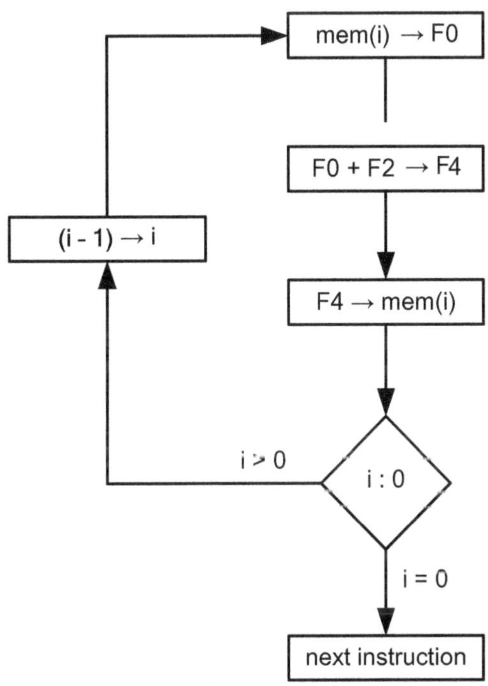

Fig. 6.141 Flow chart of a load/store program with memory address decrementer

The instruction chart in Fig. 6.142 is formed based on the flow chart in Fig. 6.141, and it is assumed to have five CPU stages.

PC	Clock cycles	1	2	3	4	5	6	7	8	9	10	11	12	13	14	15
0	LOADF R1, F0, 0	I	R	A	D	W										
1	ADDF F0, F2, F4		S	I	R	A	A	A	D	W						
2	STOREF F4, R1, 0				S	S	I	R	A	D	W					
3	BRA R1, 0, 5							I	R	A	D	W				
4	NOP								I	R	A	D	W			
5	SUBI R1, R1, 1									I	R	A	D	W		
6	JUMP 0										I	R	A	D	W	
7	NOP											I	R	A	D	W
8	Next instruction									I	R	A	D	W		

Fig. 6.142 Program instructions of the flow chart in Fig. 6.141

The first instruction, LOADF R1, F0, 0, assumes the index, i, is stored in the register, R1. Therefore, the CPU uses Reg[R1] to access the data memory, and downloads its contents to the register, F0, in the floating-point register file. The second instruction, ADDF F0, F2, F4, adds the contents of F0 to the contents of F2, forming the contents of F4. However, the data from the memory becomes available in clock cycle four. Therefore, the CPU immediately forwards the memory contents from the memory-stage (D) to the ALU-stage (A) after inserting a NOP instruction (or stalling the pipeline for one cycle as shown by the bold letter, **S**) in order to prevent a potential data hazard. The floating-point addition in this example is assumed to take three clock cycles; therefore, the CPU forwards the floating-point adder result to the ALU-stage of the floating-point store instruction, STOREF F4, R1, 0, to store the result at the memory address, Reg[R1]. In the process, the CPU stalls the pipeline twice prior to the STOREF instruction in order to avoid a second data hazard.

The branch instruction, BRA R1, 0, 5, compares the index value in the register, R1, with zero, and branches the program to the next instruction at PC = 8 if the compare is successful. If the compare is not successful, the index is decremented by one by SUBI R1, R1, 1, and the program counter goes back to the LOADF instruction.

In this instruction chart, the CPU needs to insert two NOP instructions to prevent control-related hazards: one after the BRA instruction and the other after the JUMP instruction.

The instruction chart in Fig. 6.143 includes only three iterations, and shows how the loop instructions in Fig. 6.141 are executed in each iteration. At the PC values, 5, 13 and 21, the contents of R1 is decremented by one, updating the index value, i. However, the instructions in the loop do not change their nature and perform the same tasks as mentioned earlier. According to this chart, the CPU completes executing all 24 instructions in 29 clock cycles.

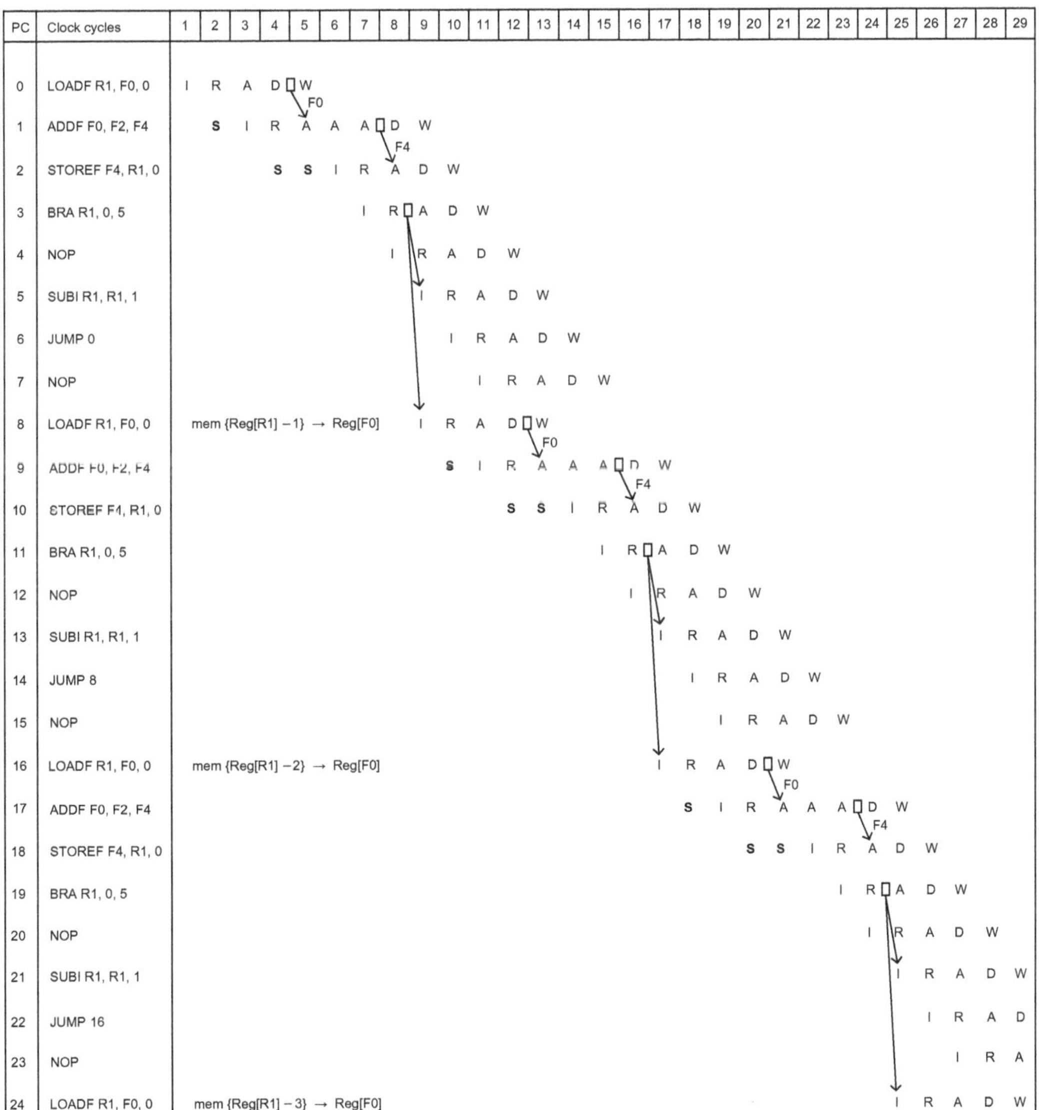

Fig. 6.143 First three iterations of the program in Fig. 6.142 with the BRA and JUMP instructions

When the loop is unrolled, and the branch and jump instructions are eliminated from Fig. 6.143, the program shortens almost by half of its original size as shown in Fig. 6.144. However, this does not mean that the number of clock cycles after the first three iterations also decrease by half as shown in this figure. The end savings in the overall execution time is only three clock cycles from 29 to 26, which is hardly a gain. However, as user programs become larger, containing many more loop functions, each employing larger number of iterations, the end savings in program execution time that uses loop unrolling technique will be significant.

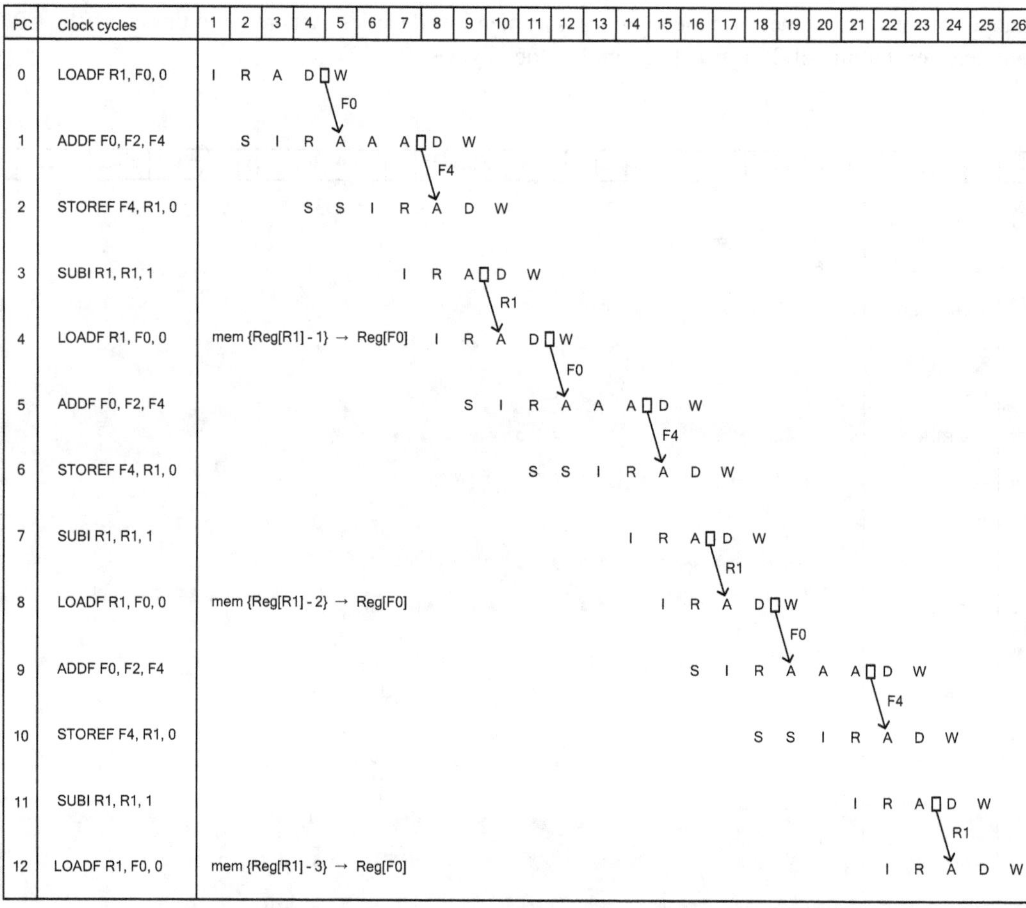

Fig. 6.144 First three iterations of the unrolled loop in Fig. 6.142 without the BRA and JUMP instructions

Figure 6.145 illustrates a modified flow chart where the decrementing index has been relocated from the path that signifies an unsuccessful branch to the main body of the program after memory contents are downloaded to the register, F0. The resultant program in Fig. 6.146 now decrements after the LOADF instruction, and effectively uses the SUBI instruction instead of stalling the pipeline for one clock cycle. The rest of the instructions, including the BRA and the JUMP instructions exactly remain the same as in Fig. 6.142. This instruction chart, therefore, produces exactly the same results as the earlier chart in Fig. 6.142, but gains one clock cycle as expected.

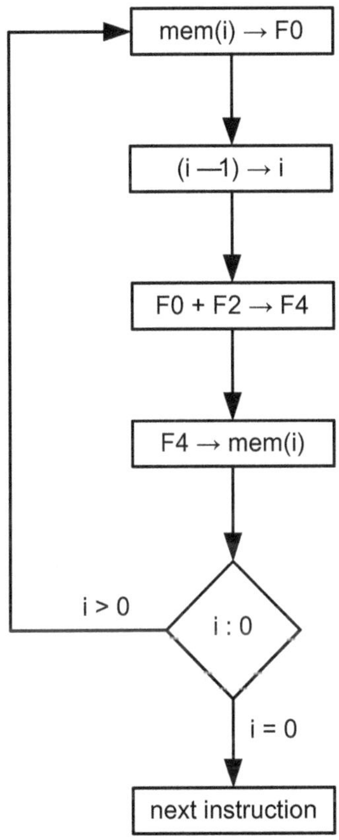

Fig. 6.145 Relocating (i − 1) → i in the flow chart in Fig. 6.141

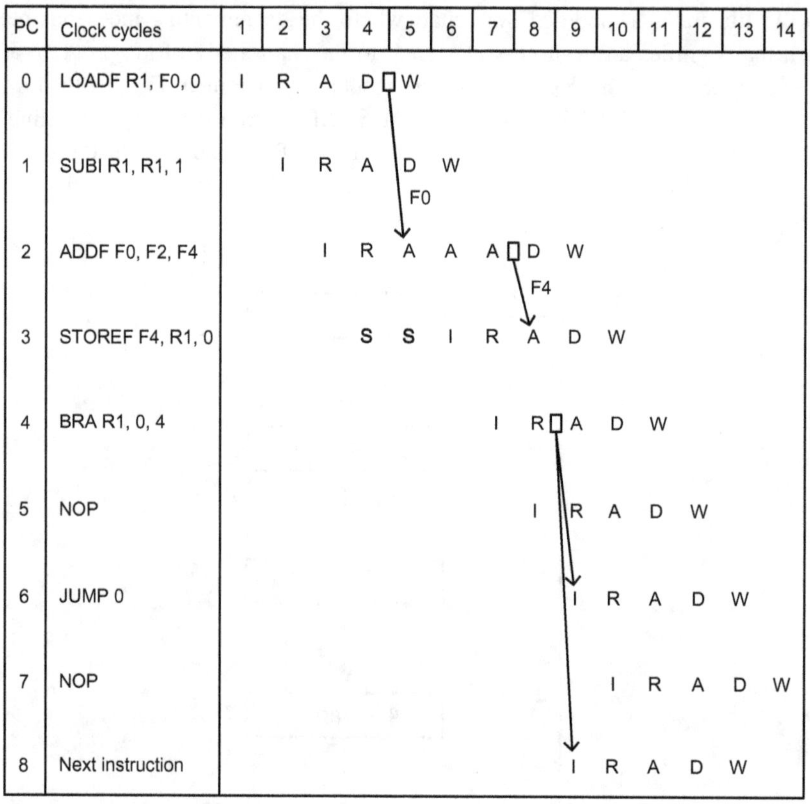

Fig. 6.146 Program instructions of the flow chart in Fig. 6.145 with the SUBI instruction relocated

When we unroll the loop function in the instruction chart in Fig. 6.146, but still keep the BRA and the JUMP instructions inside the loop in each iteration, we end up producing the chart in Fig. 6.147 after the first three iterations. This chart also updates the contents of R1 three times at the PC values, 1, 9 and 17, and completes the program in 29 clock cycles.

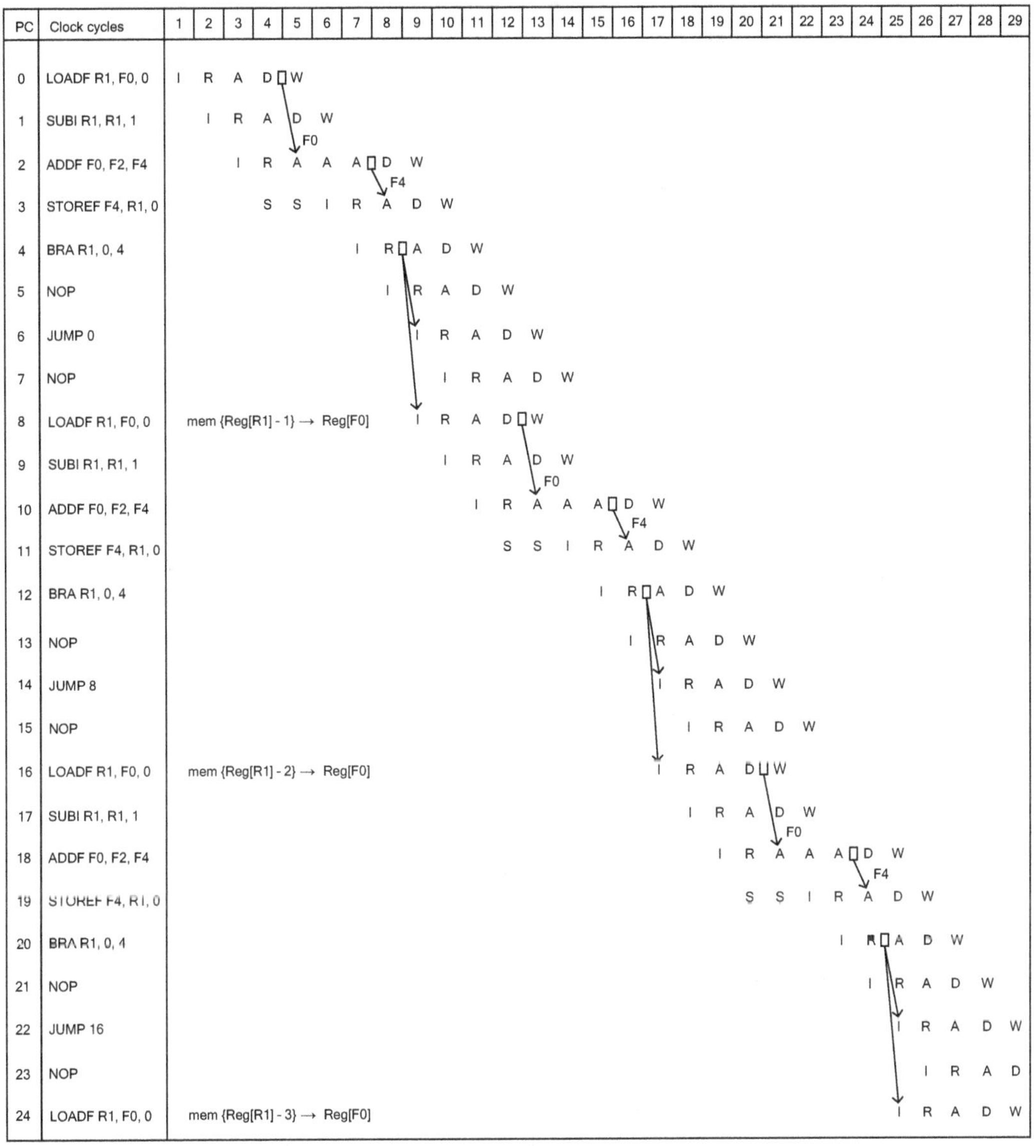

Fig. 6.147 First three iterations of the program in Fig. 6.146 with the BRA and JUMP instructions

However, if we unroll the loop, and eliminate BRA and JUMP instructions for duration of three iterations, we observe the number of instructions in the resultant program is reduced by half as shown in Fig. 6.148. The number of clock cycles is also reduced from 29 to 23. Again, this is not a considerable time savings in this example, but as programs get larger and contain many more loops, each loop controlled by a large index, then we may expect a dramatic reduction in the overall program execution time if we use this technique.

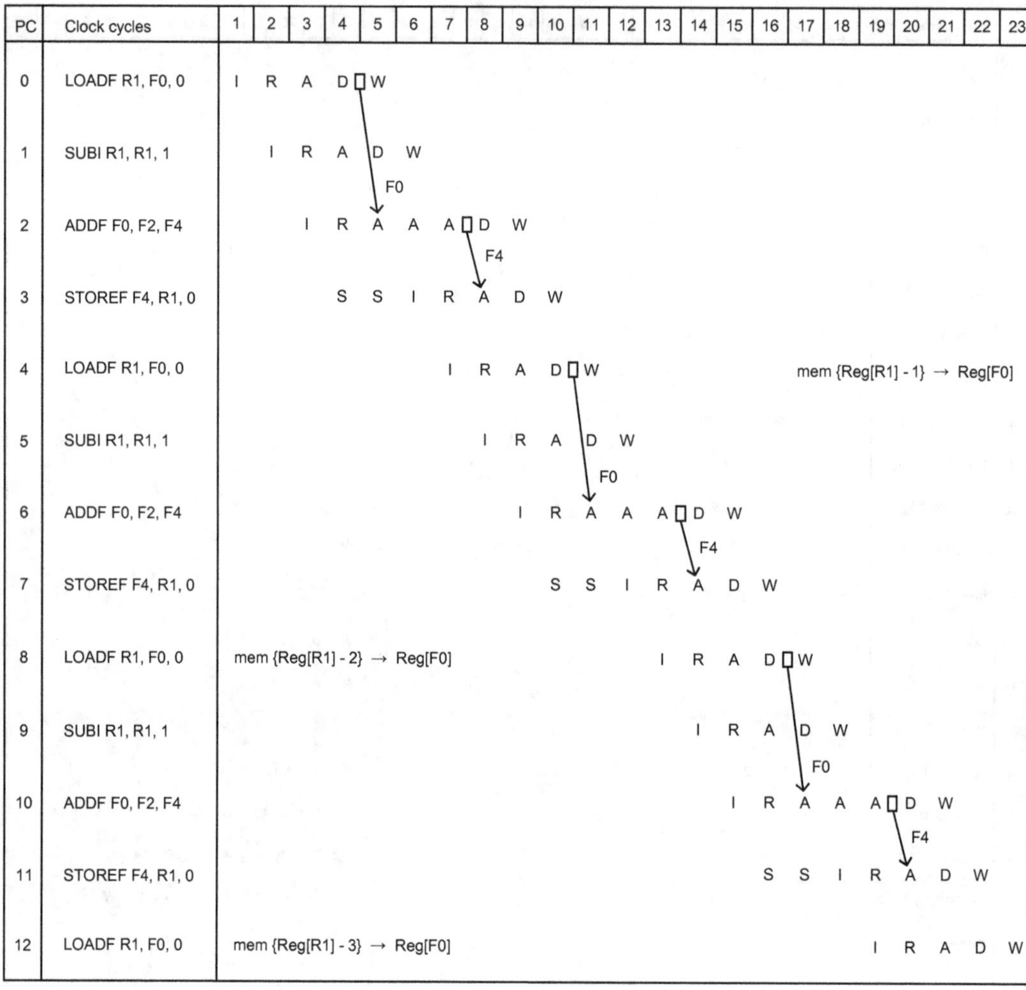

Fig. 6.148 First three iterations of the unrolled loop in Fig. 6.146 without the BRA and JUMP instructions

The instruction chart in Fig. 6.149 employs a different technique than the previous two loop unrolling techniques to shorten the overall execution time even further. In this case, the reduced index values are distributed to three different registers, R1, R2 and R3 (the register, R4, is also included in this chart to supply an index for the fourth LOADF instruction in Fig. 6.148) instead of keeping them in one register, and data memory contents are also kept in three separate floating-point registers, F0, F1 and F3, instead of only F0. After three floating-point add operations, the results are stored in the floating-point registers, F4, F5 and F6, and written back to the data memory at the addresses, Reg [R1], Reg[R2] and Reg[R3], respectively. Even though the number of instructions did not change too much in this chart compared to the one in Fig. 6.148, the overall CPU execution time is reduced from 23 cycles to 16. Even though this technique produces a gain of seven clock cycles for such a small program, it is not efficient because it tends to occupy large number of registers in both the fixed and floating-point register files. However, for programs with small index values, this method is preferable compared to the two loop unrolling methods described earlier.

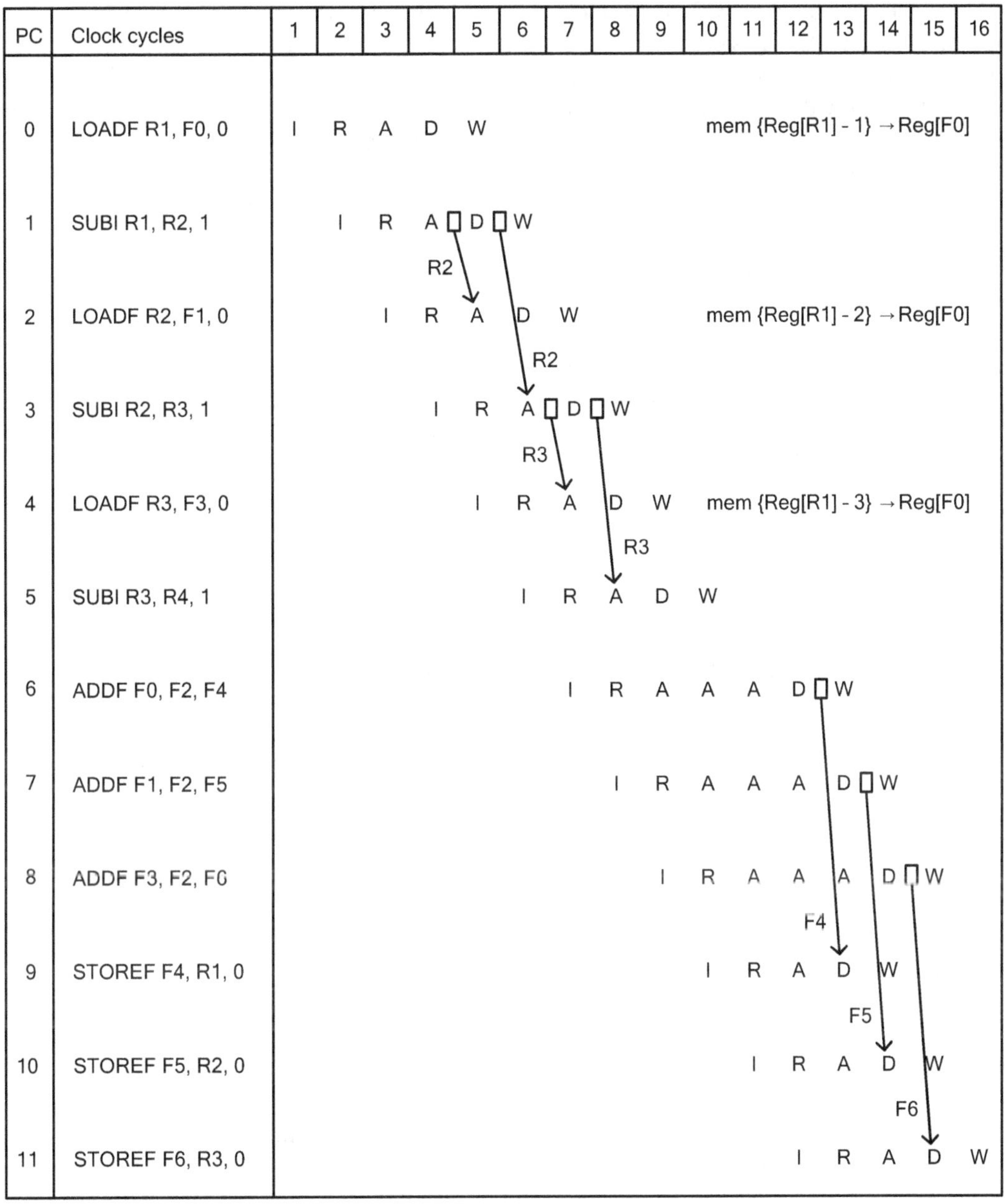

Fig. 6.149 First three iterations of the unrolled loop in Fig. 6.146 with rescheduling and renaming

Static and Dynamic Branch Prediction

The third method to reduce the program execution time is to predict if the branch will take place while the instruction is at the instruction memory stage rather than waiting until it arrives at the RF stage. We know that every time the program encounters a branch (or a jump) instruction, the overall program execution time increases by one clock cycle. Employing a branch (or a jump) instruction in a program loop becomes especially expensive if the loop index is set to a high value.

Static and dynamic-issue CPUs use different types of branch prediction mechanisms to eliminate the branch (or jump) delay slot in case the compiler cannot find an unrelated instruction to the branch (or jump), and it ends up issuing a NOP instruction.

In static-issue CPUs, the simplest method to combat against the clock loss when the program encounters a branch instruction is to assume the branch is taken. This approach especially applies to programs with loops. Another method is to construct a branch history profile that belongs to earlier program runs, and predict if the branch will be taken according to the accumulated history.

In dynamic-issue CPUs, the branch prediction mechanisms are more involved. One method is to use a buffer in the instruction fetch stage that stores the branch history. The buffer may be addressed by the lower portion of the branch address in the instruction. Buffer contents at each address also contain a bit to indicate if the branch is recently taken or not. However, if there are two branch instructions whose lower address bits are equal to each other, this bit may produce prediction in the wrong direction. To improve the accuracy of this method, a two-bit system may be employed in the Branch History Buffer as shown in Fig. 6.150. In this system, the branch prediction must miss twice before an assignment is made to indicate if the next branch is taken or not.

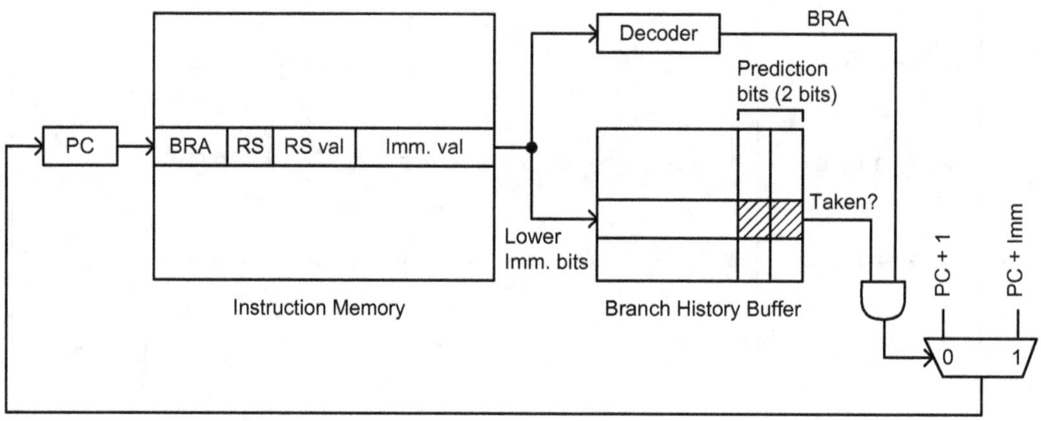

Fig. 6.150 Simplified dynamic branch prediction data-path

The state machine employed in the two-bit prediction scheme is shown in Fig. 6.151. According to this figure, the state of the branch predictor starts with the state 00, indicating branch is "Not Taken". If the next branch is taken, the state transitions to 01, but still keeps the "Not Taken" prediction status. This means that the next branch instruction fetched from the instruction memory will still be classified as not taken, and the program counter value will be incremented by one as shown in Fig. 6.150. Only the second taken branch will be able to transition the current state to 11, and change the prediction status to "Taken". While in the state 01, a "Not Taken" branch switches the prediction decision back to the state 00, and starts building the branch history one more time from the beginning. Similarly, there have to be two consecutive "Not Taken" branches from the state 11 in order to change the prediction decision to "Not Taken".

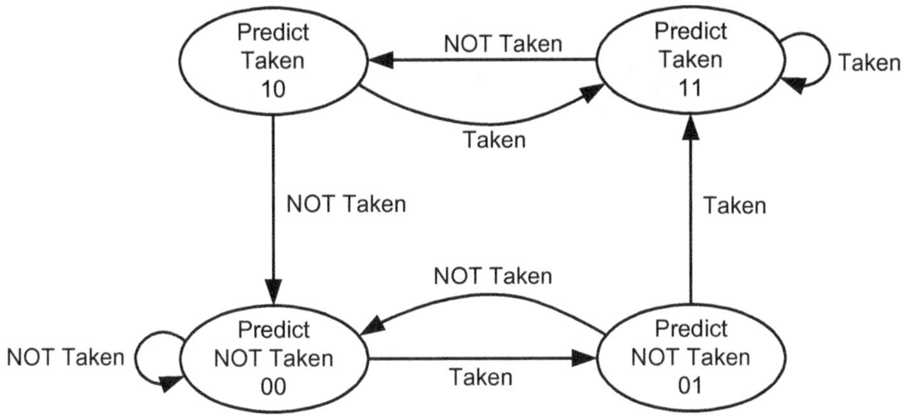

Fig. 6.151 Dynamic branch prediction state-diagram

Compared to the one-bit prediction scheme with 50–50 chance in making branch decisions, the two-bit system in Fig. 6.151 is more robust in deciding if the next branch will take place or not.

6.7 Multi-core Architectures and Parallelism

We can quantify parallelism both in software and hardware by examining four different computing platforms as shown in Fig. 6.152.

1. *Single instruction input stream, single data output stream* (SISD)

This category belongs to a single CPU at the top part of Fig. 6.152.

2. *Single instruction input stream, multiple data output streams* (SIMD)

This type corresponds to a single instruction executed by multiple processors to create different data outputs as shown in the middle part of Fig. 6.152. SIMD computers exploit *data-level parallelism* by applying the same operations to produce multiple data in parallel. Each processor has its own data memory, but the system contains a single instruction memory from which instructions are fetched and dispatched to the rest of the pipeline. CPUs specializing in multimedia extensions produce a form of SIMD parallelism where a single instruction can produce multiple forms of image data. Vector processors used in pixel manipulation and creation is another application of SIMD approach. SIMD protocols are virtually used in every arena of graphics where performance is a priority to create three dimensional, real-time virtual environments.

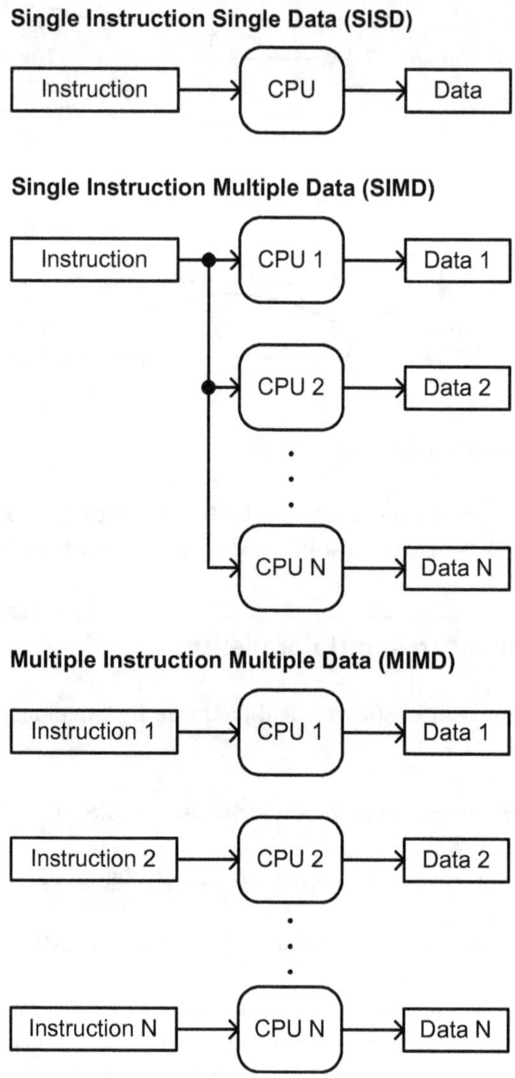

Fig. 6.152 Multi-core architectures

3. *Multiple instruction input streams, single data output stream* (MISD)

No commercial application is available for this type of architecture.

4. *Multiple instruction input streams, multiple data output streams* (MIMD)

In this category, each processor fetches its own instructions and uses on its own data. MIMD computers exploit parallelism mostly in hardware since multiple tasks must run in independent CPU pipelines connected in parallel as shown at the bottom part of Fig. 6.152.

In a MIMD environment, each processor is assigned to execute a code composed of multiple processes. A "process" is a segment of the code that may run independently, and each process is typically independent of the other processes. It is useful to be able to have multiple processors executing a single program. Multiple processes share the same program and address space, and they are often called "threads". A thread may also refer to multiple programs running on different processors, even when they do not share the same address space. Therefore, it is possible to mold a multi-threaded architecture into a form that allows either simultaneous execution of multiple processes with separate address spaces or multiple threads that share the same address space. Because MIMD structures can generally promote parallelism in hardware, they are considered the architecture of choice for general-purpose multiprocessor platforms explained in this chapter. MIMD platforms also offer flexibility and execute one application when high performance is needed, or they can run multiple tasks simultaneously. MIMD structures either employ a single centralized main memory shared by every CPU, or contain a distributed memory architecture where each CPU has it own memory.

The first group with centralized, shared memory architecture has usually much less than a few dozen CPU cores as shown in Fig. 6.153. For multiprocessors with small processor counts, it is possible to share a single centralized memory without increasing the waiting period to access the memory and/or decreasing the memory bandwidth.

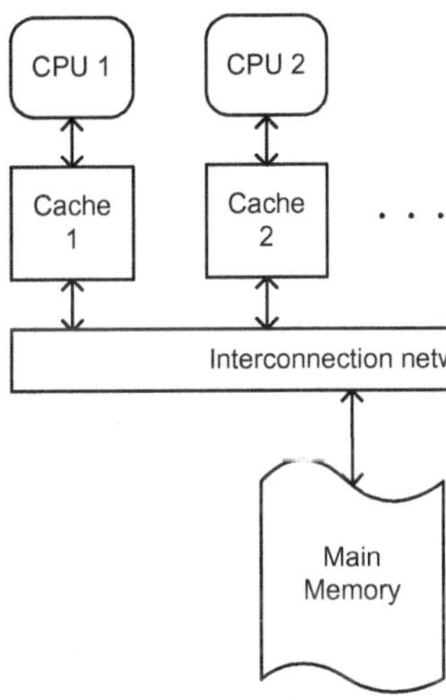

Fig. 6.153 Centralized shared memory architecture

The second group consists of multiprocessors with distributed memories as shown in Fig. 6.154. To support larger processor counts, the entire system memory must be distributed among the processors. Otherwise, the single memory system will not be able to support the memory bandwidth demand from a large group of CPUs, and as a result it will produce excessively long memory access times. However, in recent years with rapid advancements in processor performance and memory bandwidth, the need for the multi-processor system with distributed memories has shrunk. Larger number of processors in MIMD platforms also raises the need for high bandwidths in interconnection networks.

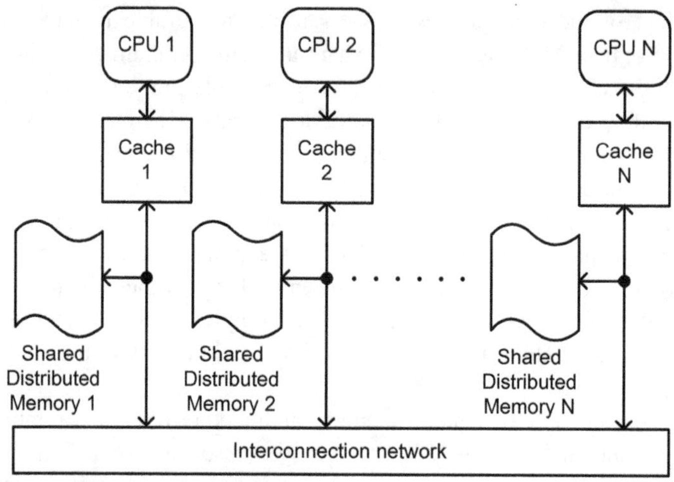

Fig. 6.154 Distributed shared memory architecture

Memory Coherency in Multi-core Architectures

One of the most challenging issues in systems with multiple CPU cores is to keep the central memory (or distributed memories) coherent with each CPU's cache memory. In a single CPU system, we go to great lengths to maintain the data coherency between the cache and the main memory in the shortest possible time. To reach this goal, we often formulate different types of cache architectures from write-through to write-back to minimize a read or a write cycle, and use many types of data replacement policies to be able to preserve the same data in both memories.

Multiple CPU cores take this challenge to the next level. A simple explanation of memory coherency is illustrated in Fig. 6.155. Let us assume that the main memory contains data A at the address X in a single memory system. When a single CPU writes data B to this address, it will read the same data back at a later time as shown at the bottom left hand side of Fig. 6.155. However, in a system with two CPUs complications arise rather quickly if precautions are not taken in the data exchange policy as shown at the bottom right hand side of Fig. 6.155. Suppose CPU 2 writes data A at the address X initially, and wants to read it back when the program calls for it. If CPU 1 writes a new data B at this address before CPU 2 reads the data, CPU 2 will read data B instead of data A, resulting in a computation error. So, how do we maintain data coherency while we deal with multiple CPUs to employ parallelism? We need to approach this problem in two ways: (1) by designing smart compilers to detect data hazards caused by multiple CPUs, (2) by employing fail-safe data exchange algorithms for memories.

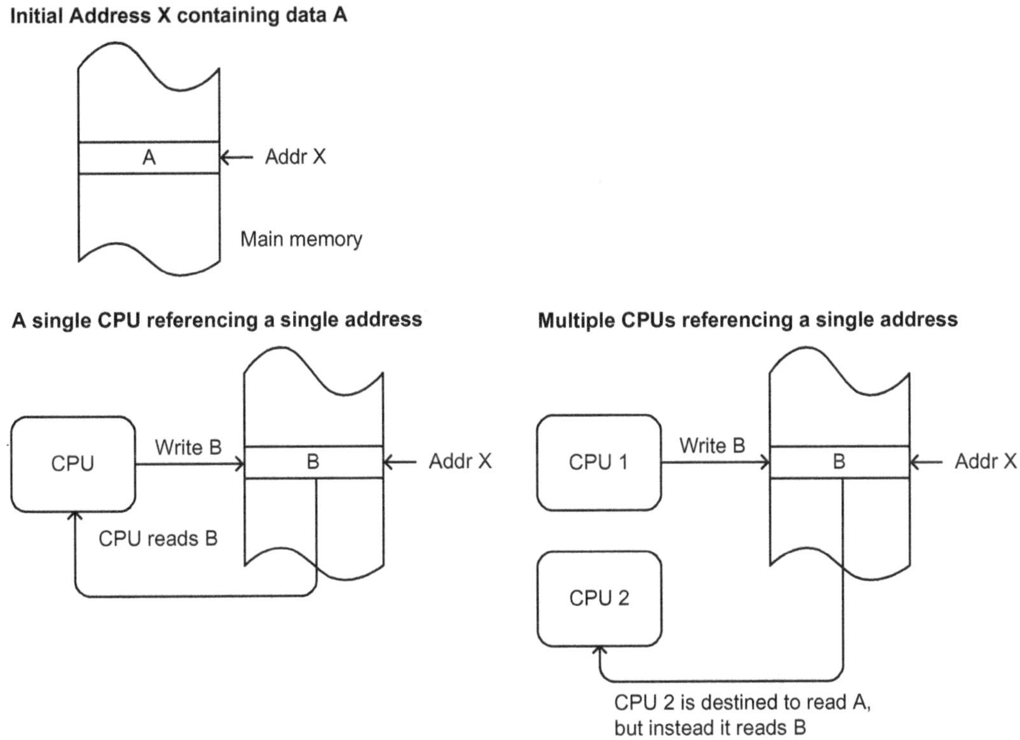

Fig. 6.155 Memory coherency

Parallelism in Software

One good example to understand how parallelism in software works is through image processing. Suppose the image composed of black and white pixels in Fig. 6.156 is modified such that the resultant image must contain the original pixels and the interstitial pixels as a result of averaging the black and white pixels.

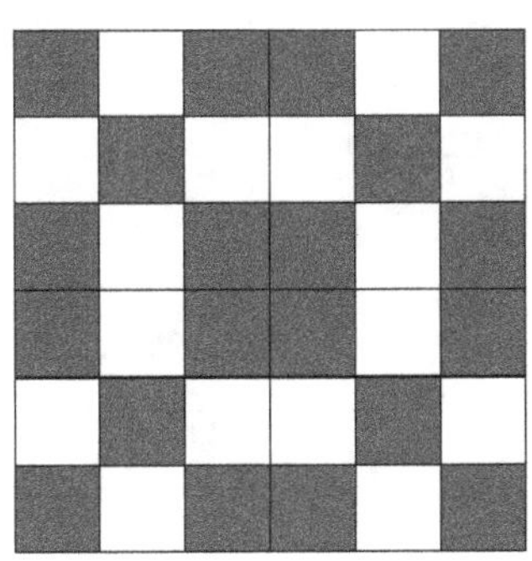

Fig. 6.156 Unmodified image

If we employ a single CPU to complete this process, we obtain the image in Fig. 6.157. In this image, every gray interstitial pixel is the result of averaging neighboring black and white pixels in Fig. 6.156.

Fig. 6.157 Modified image

However, if we employ four independent processors working simultaneously on each quadrant of Fig. 6.157, the time that takes to complete the final image shortens considerably. The result is shown in Fig. 6.158.

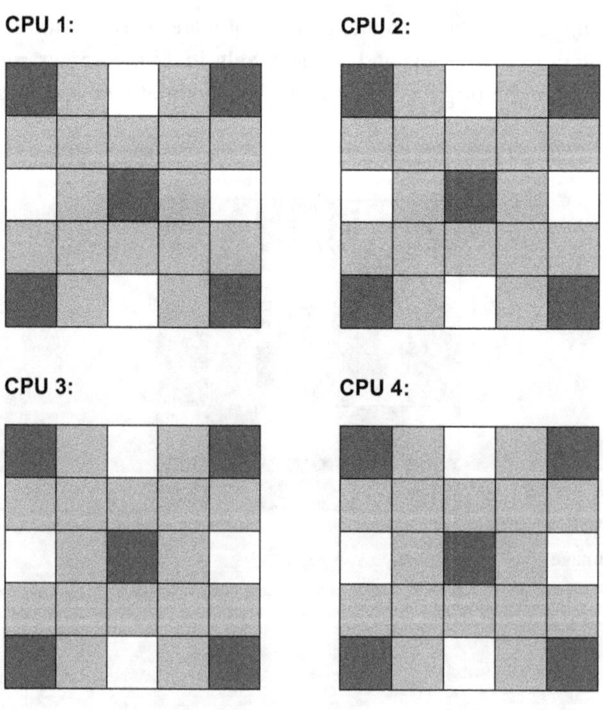

Fig. 6.158 Partitioning the work among four CPU cores

Once each quadrant of Fig. 6.158 is complete, then one of the four processors can be assigned to produce the pixels at the quadrant boundaries, resulting in an image shown at the top portion of Fig. 6.159. If the total processing time is still a concern, the task can again be divided among the four processors shown at the bottom part of Fig. 6.159. Finally, to form the final pixel in the middle can be assigned to any one of the four processors (here it is assigned to CPU 1).

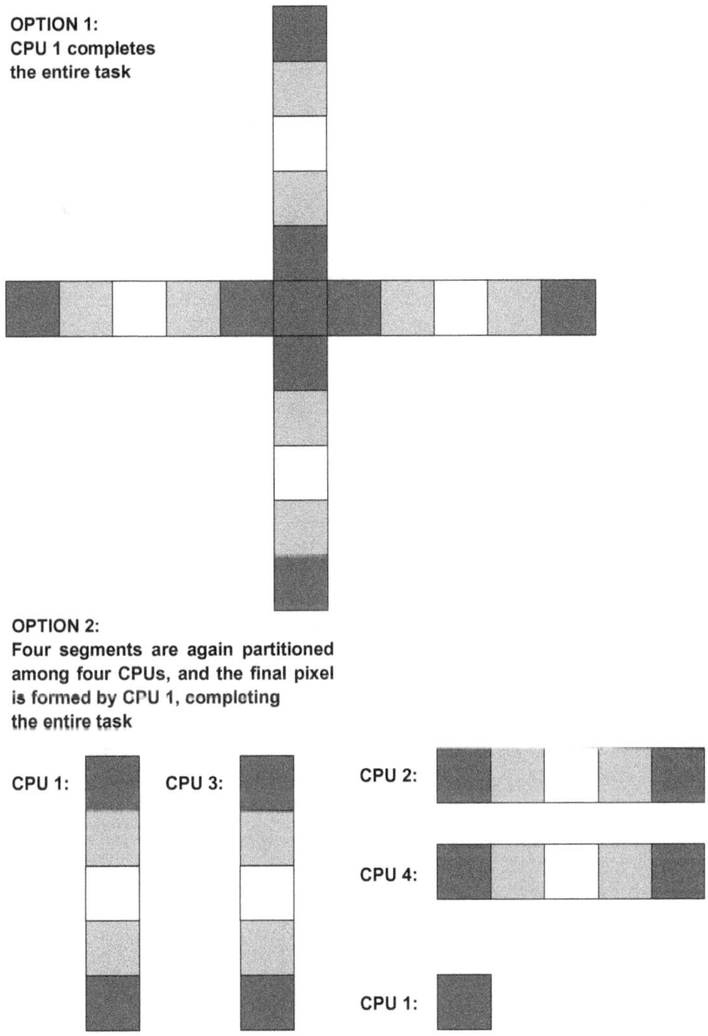

Fig. 6.159 Evaluating pixels at boundaries

To produce the quadrants in Fig. 6.158, each of the four processors executes the algorithm shown in Fig. 6.160 simultaneously. In this algorithm, the original pixels in Fig. 6.156 are defined as op(x, y), and the interstitial ones created by the algorithm are defined as p(i, j). The first loop in Fig. 6.160 creates the pixels in the first row of the quadrant as long as $x < 2$. When $x \geq 2$, the algorithm goes into the second loop and creates the pixels in the next row. This process continues until the pixels in the third row are created as long as $y < 2$. When $y \geq 2$, the algorithm first goes into the third, and then into the fourth loop, creating the pixels between row 0 and row1, and between row 1 and row 2. The process ends when the algorithm detects the row value, j, to be greater than or equal to 4.

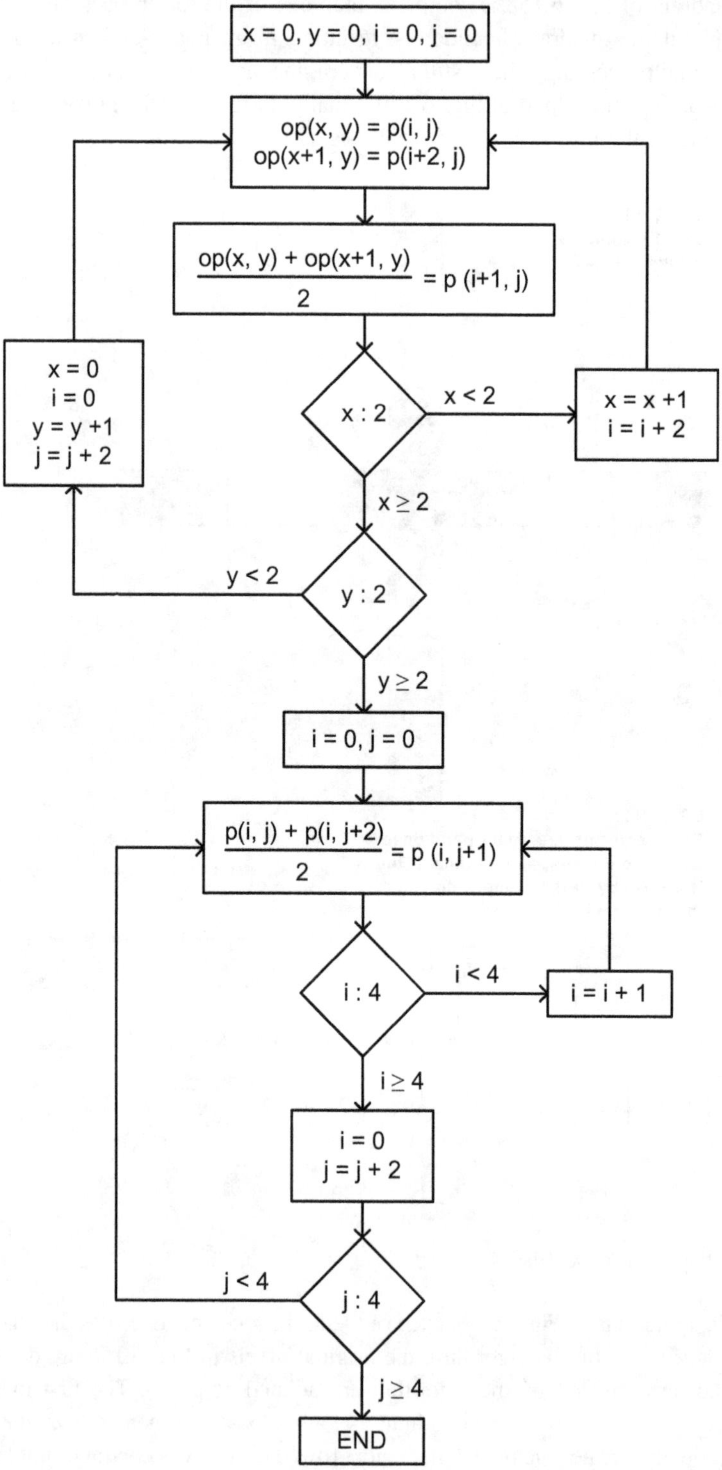

Fig. 6.160 Pixel averaging algorithm

Multi-core Systems with a Central Memory and Parallelism in Hardware

In order to understand the basic principles of multi-processor systems with a central memory or distributed memories, the reader should have "some" familiarity in cache memory operation and its interaction with the main memory under different conditions. Fundamental terms, such as cache miss and cache hit during a memory read or a write, will often be mentioned in this section with a central memory, and also in the next section with distributed memories. Therefore, in case of insufficient background or understanding, it is important for the reader to refer to the section in this chapter about cache memories before studying multi-core systems and parallelism in hardware.

To be able to realize parallelism in systems with multiple CPU cores, one method is to have all CPU cores share the same memory. This platform, however, cannot sustain high-speed data rate if the number of cores exceeds a few dozen as mentioned earlier. It also assumes that every CPU has its own cache memory that has to be coherent with a central main memory in an interconnection network as shown in Fig. 6.153.

The basic idea behind this platform is to have a broadcasting system to invalidate data in all CPU cache memories as soon as one of the CPU cores updates its cache (and the main memory). This approach produces a unique cache controller design for every CPU to manage data and be coherent with the central memory. In this new design, the local cache controller needs to define the data state in its own cache in one of the three forms depending on a transaction involving another CPU.

The invalid state is the first data state in this design. This state describes a case where the data is either present in the cache, but in invalid form, or not in the cache. The shared state describes a situation where the data is in a local cache (perhaps in the other caches as well), but also resides in the main memory. Finally, the modified state describes a state where the data resides only in a local cache, but not in other caches, and not in the main memory. The local cache controller transitions between these three states according to the transactions taking place in the local CPU as well as in other CPUs.

In the following section, we will examine the types of event(s) that make the local cache controller stay in its current state or transition to another state.

Case 1: Transition from the Invalid to the Invalid state

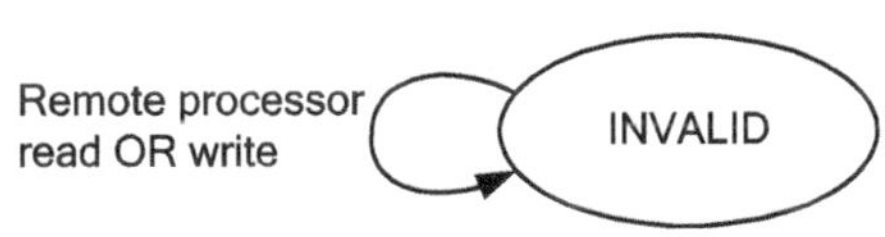

The local cache controller stays in the Invalid state if a remote CPU issues an address where the data at this address is initially invalid (or does not exist) in the local cache according to the state diagram above. This can be triggered by one of the two independent events:

(a) A remote processor write

This event is shown on the left hand side of Fig. 6.161 and goes through the following steps:

- A remote processor places a write request on the bus for other processors to snoop, and issues a write address.
- Assume this address produces a write miss for the remote CPU. This scenario prompts the remote cache controller to look for an address match in all other cache memories, including the local cache. If the remote cache controller finds an address match in the local cache with an invalid data entry, then it aborts the write.
- Thus, the local cache controller stays in the Invalid state.

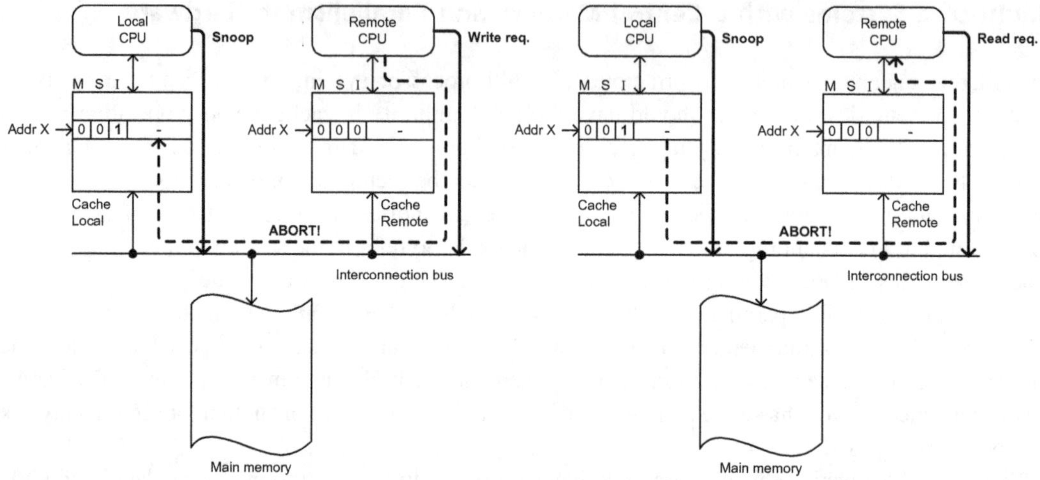

Fig. 6.161 Transitions from the Invalid to the Invalid state

(b) A remote processor read

This event is shown on the right hand side of Fig. 6.161 and goes through the following steps:

- A remote processor places a read request on the bus for other processors to snoop, and issues a read address.
- Assume this address produces a read miss for the remote CPU. Therefore, the remote cache controller searches an address match in all other cache memories, including the local cache. If the remote cache controller finds an address match in the local cache with an invalid data entry, then it aborts the read.
- Thus, the local cache controller stays in the Invalid state.

Case 2: Transition from the Invalid to the Modified state

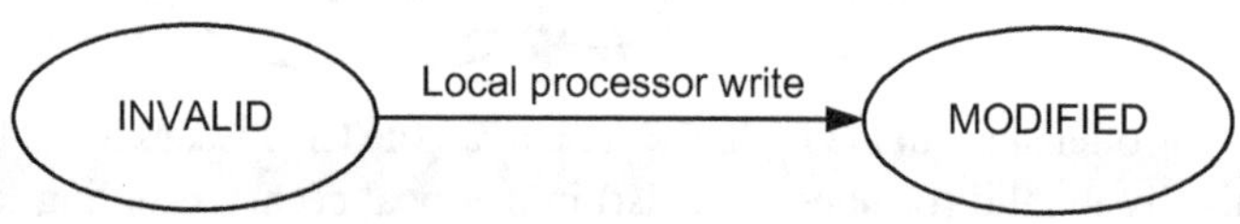

The local cache controller transitions from the Invalid to the Modified state if the local CPU issues an address, which resides in the local cache, but has an invalid data entry. This scenario causes a write hit, which prompts the local processor to write data in its own cache, changing the current state to the Modified state, according to the state diagram above.

This event is shown in Fig. 6.162, and goes through the following steps:

- The local processor places a write request on the bus for other processors to snoop, and issues a write address.
- Assume this address produces a write hit for the local CPU, and enables the local cache controller to write data directly to its own local cache.
- Since the written data does not exist anywhere else in the system, the local processor invalidates all data entries in every remote cache at this address.
- The local cache controller changes its state from Invalid to Modified due to this written block while the remote cache controller changes its state to Invalid at this address.

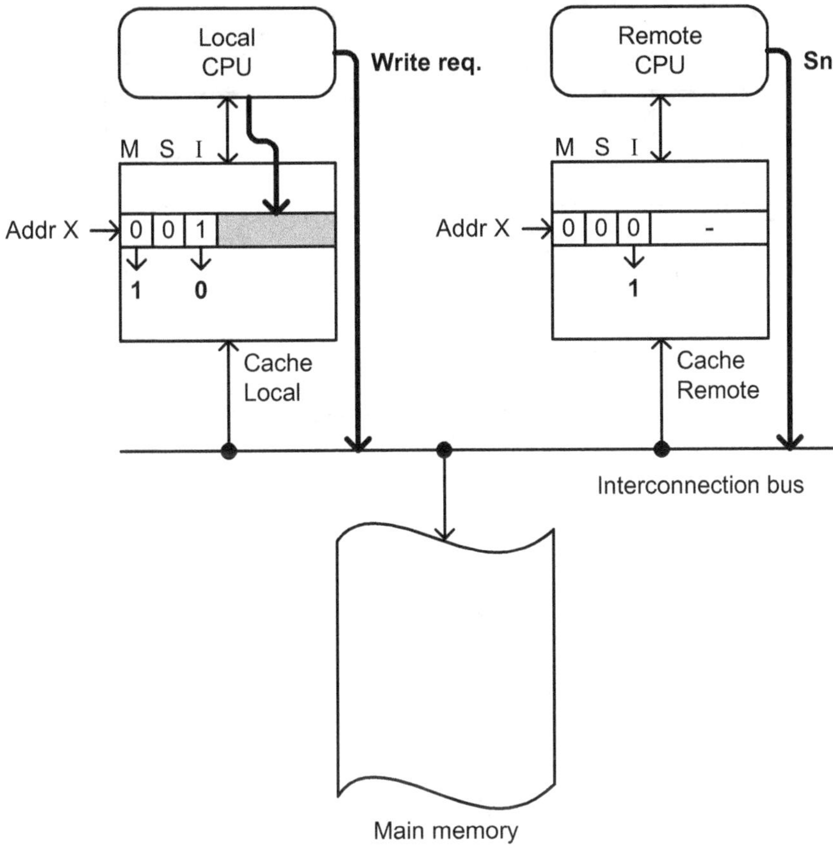

Fig. 6.162 Transition from the Invalid to the Modified state

Case 3: Transition from the Invalid to the Shared state

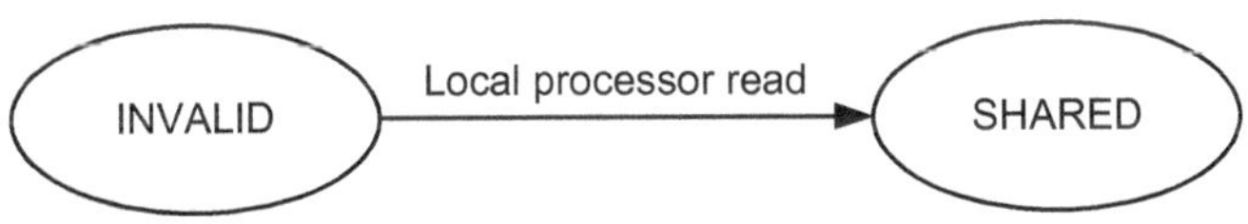

The local cache controller transitions from the Invalid to the Shared state if the local CPU issues an address where the data entry at this address is originally invalid, and causes a read miss. Thus, the local cache controller fetches data from the main memory, changing the current state from Invalid to Shared according to the state diagram above.

This event is shown in Fig. 6.163, and goes through the following steps:

- The local processor places a read request on the bus for other processors to snoop, and issues a read address.
- Assume this address produces a read miss for the local CPU due to an invalid entry. Therefore, the local cache controller searches an address match in all other caches and the main memory.
- When the local cache controller cannot find an address match in other cache memories, it fetches this data from the main memory as shown in Fig. 6.163. It then updates the state of this data block from Invalid to Shared because the data exists both in the local cache and the main memory.

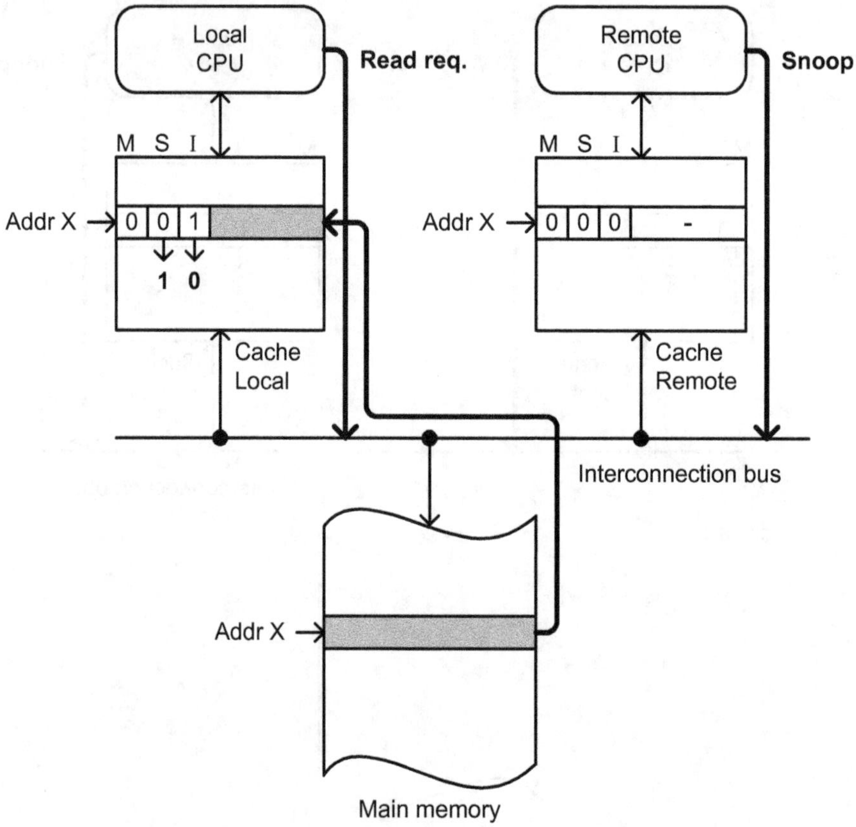

Fig. 6.163 Transition from the Invalid to the Shared state

After integrating these three cases above, we finally obtain Fig. 6.164, showing all possible state transitions from the Invalid state.

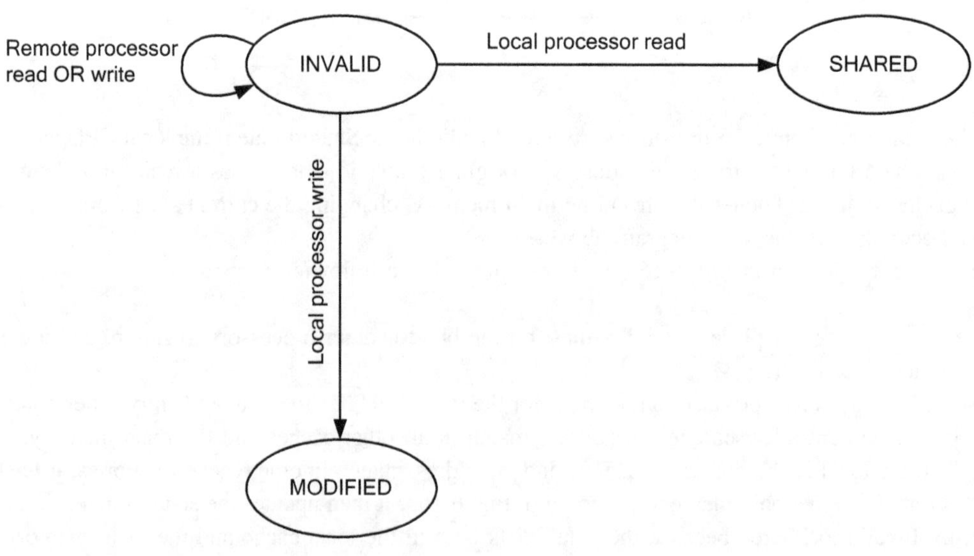

Fig. 6.164 Transitions from Invalid to Shared and Modified states

Case 4: Transition from the Modified to the Modified state

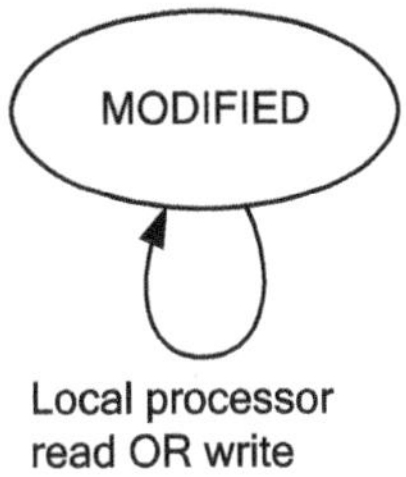

The local cache controller stays in the Modified state if additional addresses issued by the local CPU result in write or read hits according to the state diagram above.

(a) A local processor write

This event is shown on the left hand side of Fig. 6.165, and goes through the following steps:

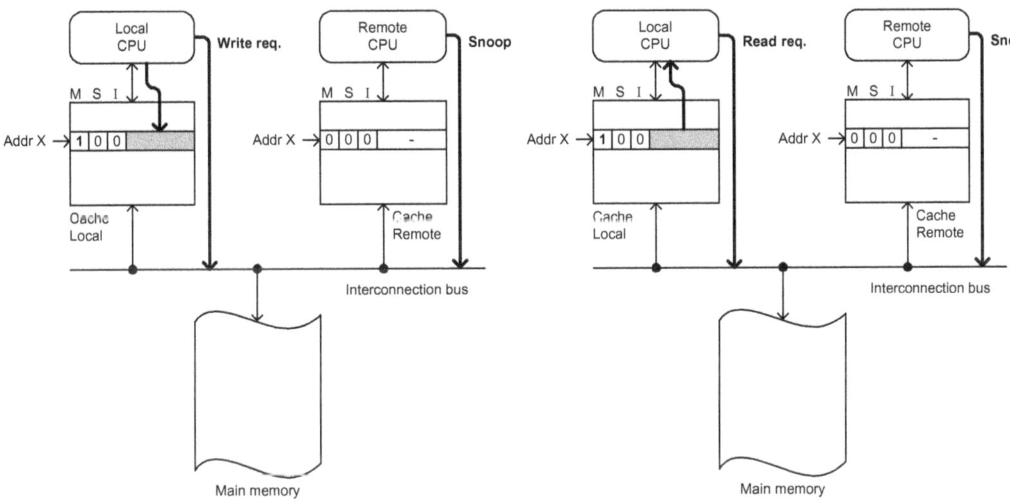

Fig. 6.165 Transition from the Modified to the Modified state

- The local processor places a write request on the bus for other processors to snoop, and issues a write address.
- Assume this address produces a write hit for the local CPU, and enables the cache controller to overwrite new data on top of the existing data at this address.
- Since the block is already in modified form in the local cache, any additional write(s) only updates the data contents, but does not alter the state.
- Therefore, the local cache controller stays in the Modified state.

(b) A local processor read

This event is shown on the right side of Fig. 6.165, and goes through the following steps:

- Local processor places read request on the bus for other processors to snoop and issues a read address.
- Assume this address produces a read hit for the local CPU, and enables the local cache controller to read directly from its own cache.

- Since the block is already in modified form in the local cache, any additional read(s) from the same address does not affect its state.
- Therefore, the local cache controller stays in the Modified state.

Case 5: Transition from the Modified to the Invalid state

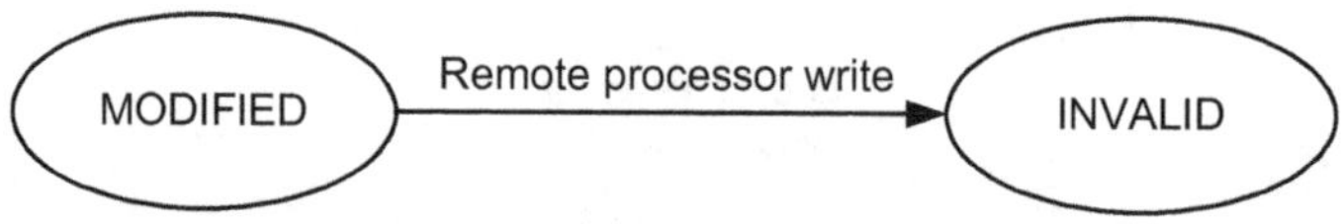

The local cache controller transitions from the Modified to the Invalid state according to the state diagram above if a remote CPU issues an address, which results in a write hit for the remote cache. This event is shown in Fig. 6.166, and goes through the following steps:

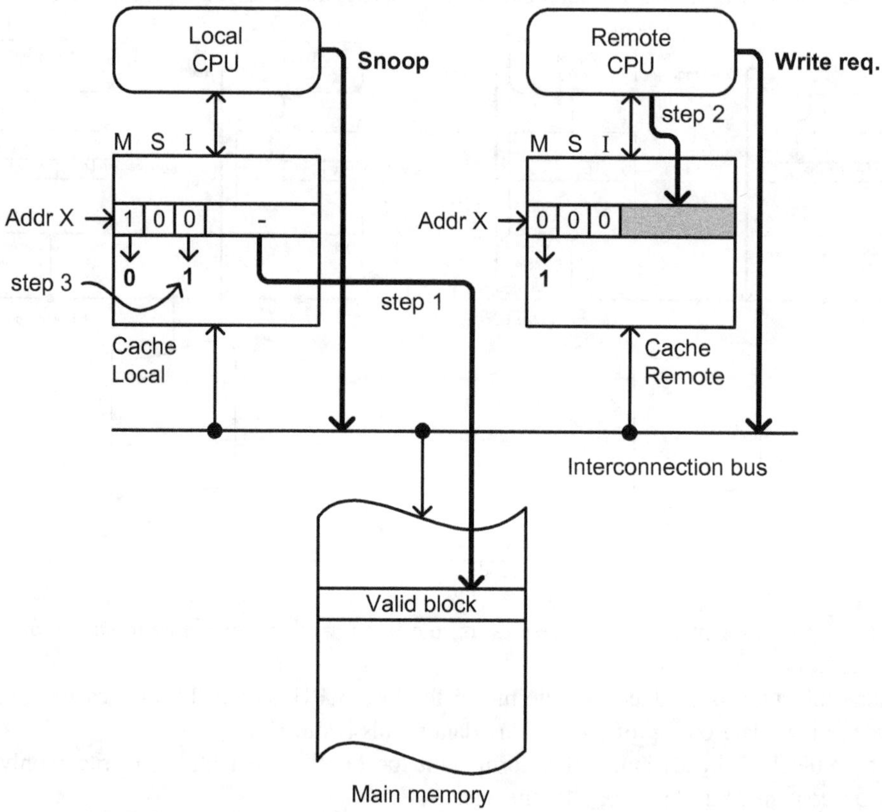

Fig. 6.166 Transition from the Modified to the Invalid state

- The remote processor places a write request on the bus for other processors to snoop, and issues a write address.
- Assume address matches an existing address in the remote cache, and produces a write hit. The remote cache controller is now enabled to overwrite data on top of the existing data at this address. The state in the remote cache changes to Modified because the data exists only in the remote cache but nowhere else.

- The remote cache controller, therefore, invalidates all data entries in all other caches at this address.
- Since the addressed block in the local cache is still valid, the local cache controller preserves the data block by writing it to the main memory (step 1).
- The remote processor then proceeds with writing data to its own cache (step 2).
- The local cache controller changes the state of the data block from Modified to Invalid, and invalidates the data block at this address (step 3).

Case 6: Transition from the Modified to the Shared state

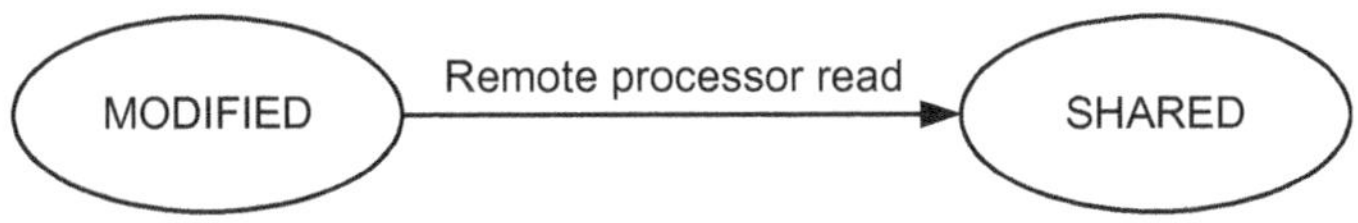

The local cache controller transitions from the Modified to the Shared state if a remote CPU issues a read address, but misses the remote cache. If this read address matches one of the addresses in the local cache, the local cache controller first transfers the block of data to the main memory for the remote processor to read, and then updates its state from Modified to Shared according to the state diagram above since this data block exists more than one place.

This event is shown in Fig. 6.167, and goes through the following steps:

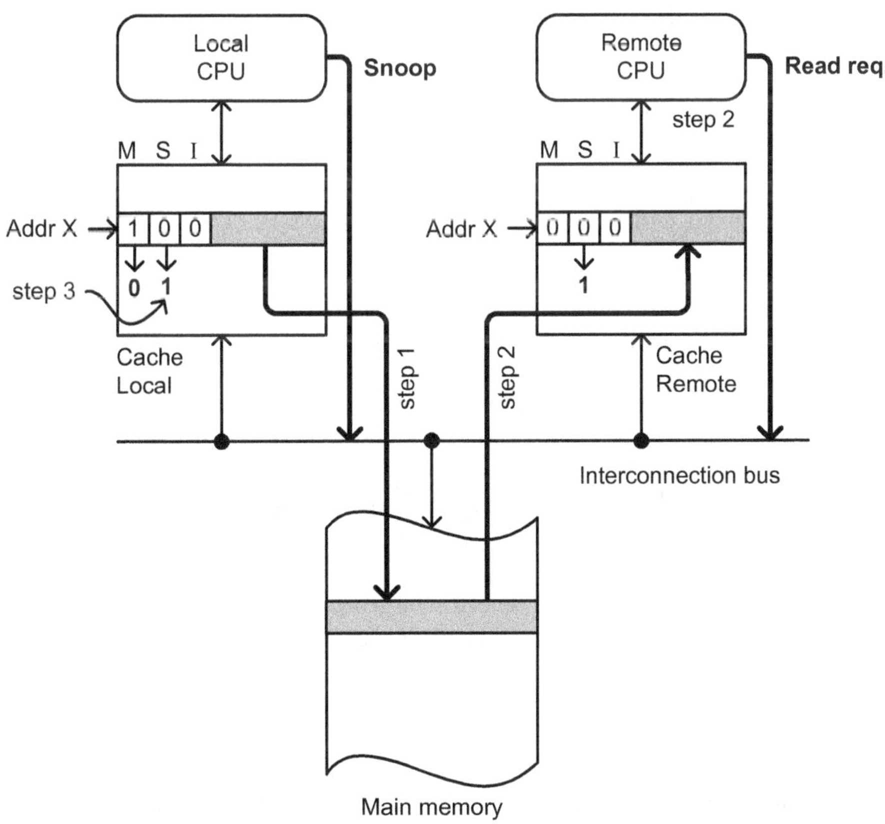

Fig. 6.167 Transition from the Modified to the Shared state

- The remote processor places read request on the bus for other processors to snoop, and issues a read address.
- Assume this address produces a read miss for the remote CPU, but the remote cache controller is able to locate the data in the local cache in modified form.
- Therefore, the local CPU first transfers the data block to the main memory (step 1).
- Then the remote CPU fetches this data from the main memory, and brings it to its own cache (step 2).
- The local cache controller changes the state from Modified to Shared (step 3).

Therefore, after integrating the cases from 4 to 6 above, we finally obtain Fig. 6.168, showing all possible state transitions from the Modified state.

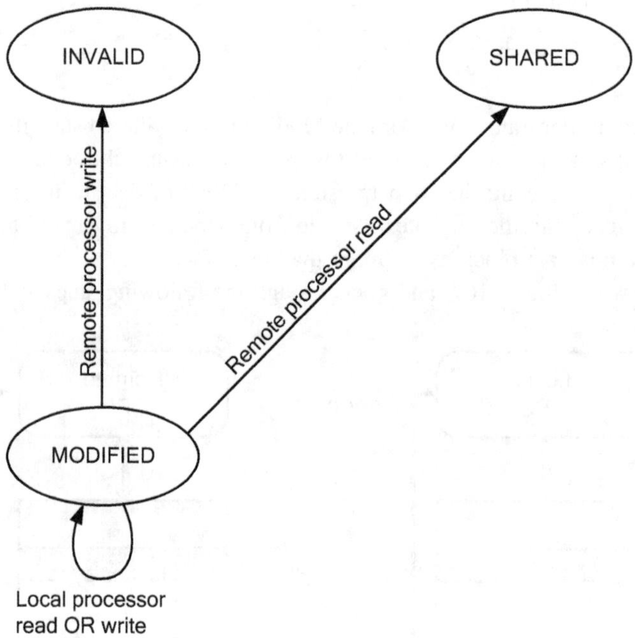

Fig. 6.168 Transitions from Modified to Invalid and Shared states

Case 7: Transition from the Shared to the Shared state

The local cache controller stays in the Shared state if the local or a remote CPU issues a read, which results in a hit as shown by the state diagram above. This can be triggered by one of the two independent events:

(a) A local processor read

This event is shown on the left side of Fig. 6.169, and goes through the following steps:

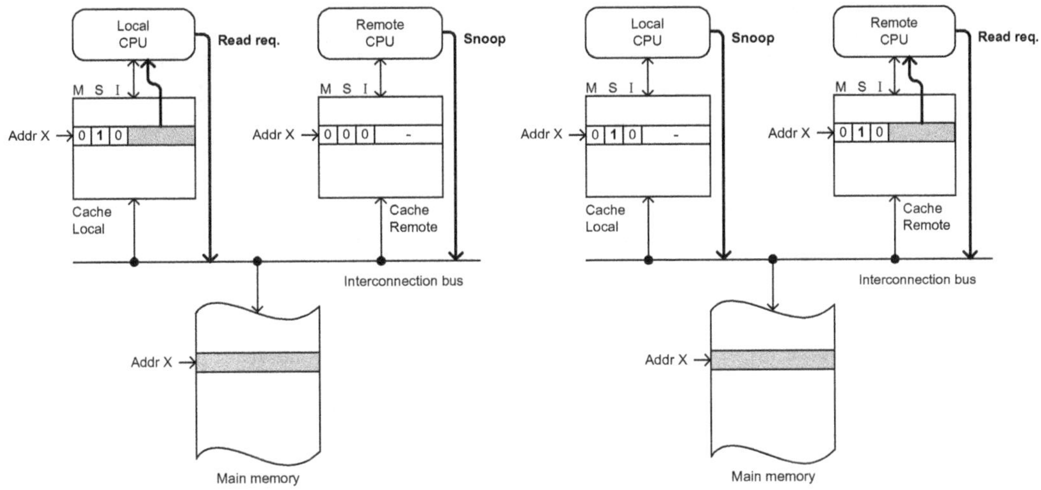

Fig. 6.169 Transition from the Shared to the Shared state

- Local processor places read request on the bus for other processors to snoop, and issues a read address.
- This address produces a read hit for the local cache. Since this data is shared, additional reads from the same address does not affect the cache state.
- As a result, the local cache controller stays in the Shared state.

(b) A remote processor read

This event is shown on the right side of Fig. 6.169, and goes through the following steps:

- The remote processor places read request on the bus for other processors to snoop, and issues a read address.
- This address produces a read hit for the remote cache. Since the data is already shared with other CPUs and the main memory, additional reads from the same address does not affect the cache state of the remote CPU, and it certainly does not change the shared state of the local cache at this address.
- Therefore, the local cache controller stays in the Shared state.

Case 8: Transition from the Shared to the Invalid state

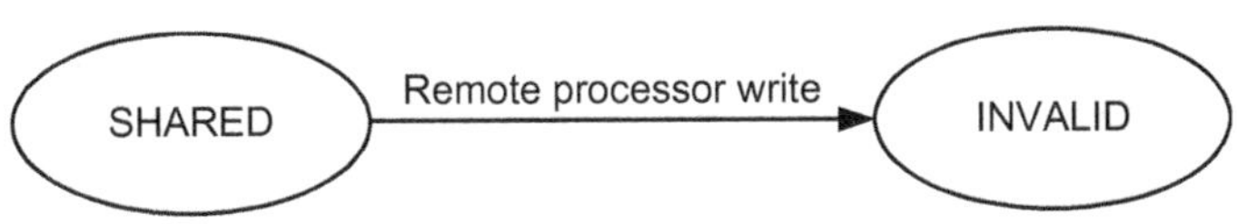

The local cache controller transitions from the Shared to the Invalid state if a remote CPU produces a write hit, modifies the data in its own cache, and invalidates the data entry at this address in all other caches, including the local cache, as shown by the state diagram above. This event is shown in Fig. 6.170, and goes through the following steps:

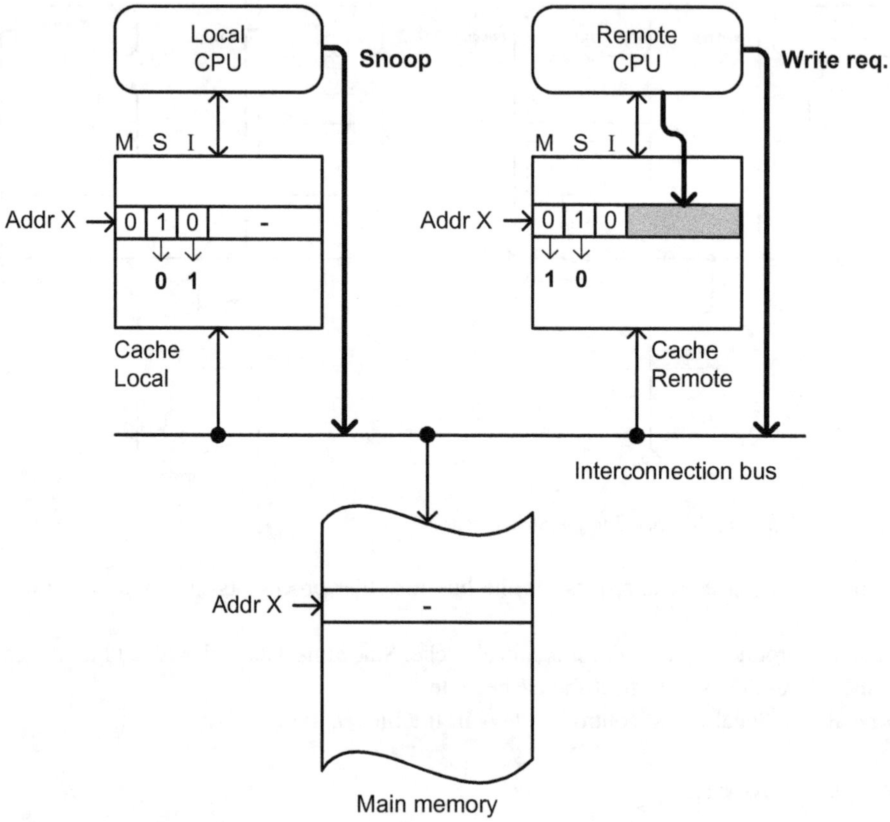

Fig. 6.170 Transition from the Shared to the Invalid state

- The remote processor places write request on the bus for other processors to snoop, and issues a write address.
- Assume this address produces a write hit for the remote cache. Therefore, the remote cache controller writes the data to its own cache, and invalidates the data entries in all other caches at this address.
- Thus, the local cache controller changes its cache state from Shared to Invalid at this address while the remote cache controller changes the state from Shared to Modified.

Case 9: Transition from the Shared to the Modified state

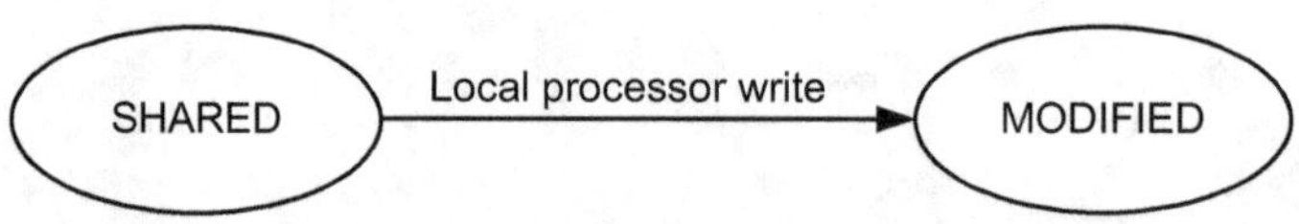

The local cache controller transitions from the Shared to the Modified state if the local CPU issues a write, which results in a hit, and modifies the data entry in its own cache as shown in the state diagram above. This event is shown in Fig. 6.171, and goes through the following steps:

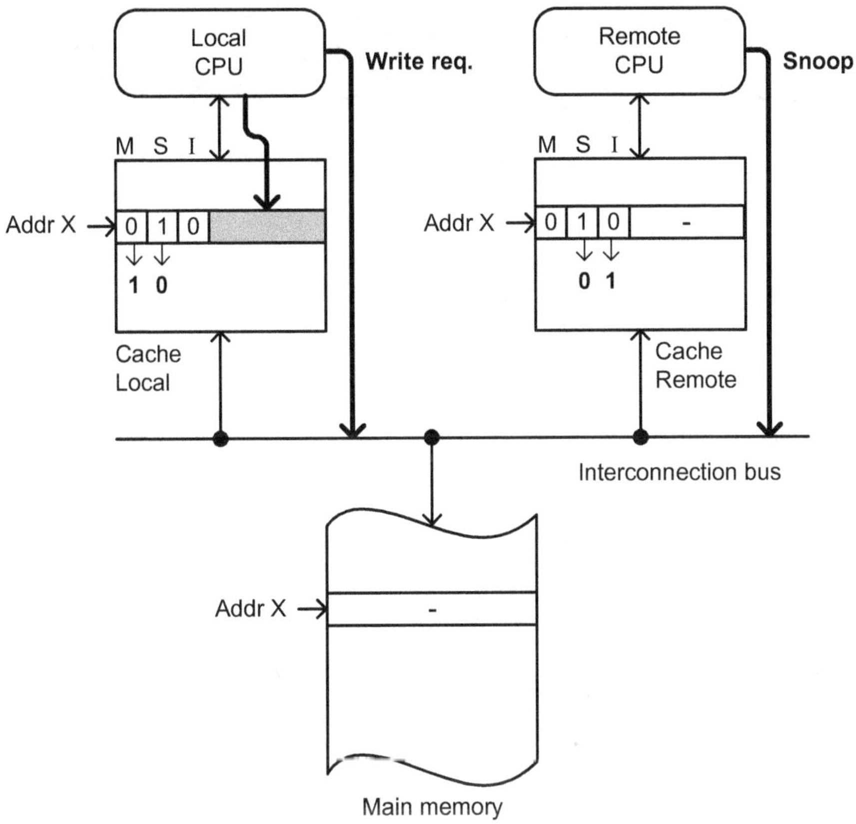

Fig. 6.171 Transition from the Shared to the Modified state

- The local processor places write request on the bus for other processors to snoop, and issues a write address.
- Assume this address produces a write hit for the local cache. Therefore, the local cache controller is now enabled to write new data to its own cache, invalidating the data entries in all other caches at this address.
- Since this new data block only exists in the local cache but nowhere else, the local cache controller changes the cache state from Shared to Modified while all remote cache controllers change their state to Invalid at this address.

Therefore, after grouping the cases 7, 8 and 9, we obtain Fig. 6.172, showing all possible state transitions from the Shared state.

Now, if we further integrate the state transitions from Figs. 6.164, 6.168 and 6.172, we obtain the final form of the state machine for the local cache controller in Fig. 6.173 that manages the data coherency in multi-core CPUs with a single, centralized memory.

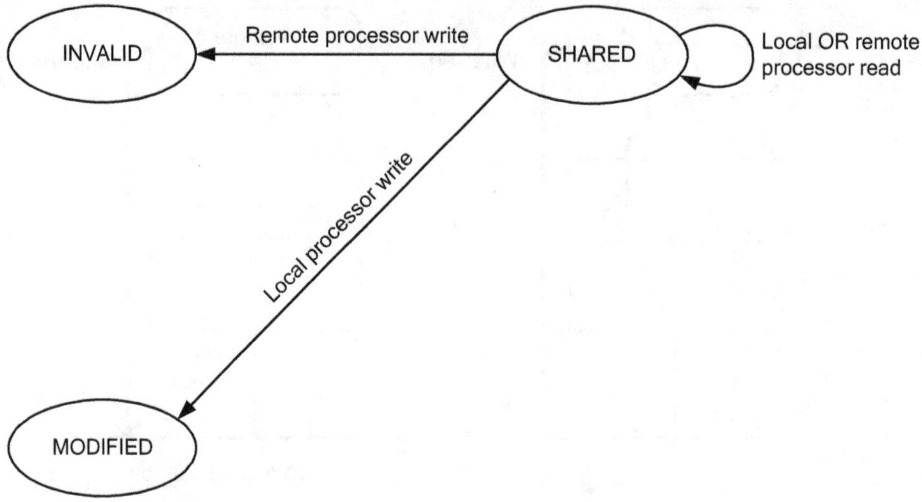

Fig. 6.172 Transitions from Shared to Invalid and Modified states

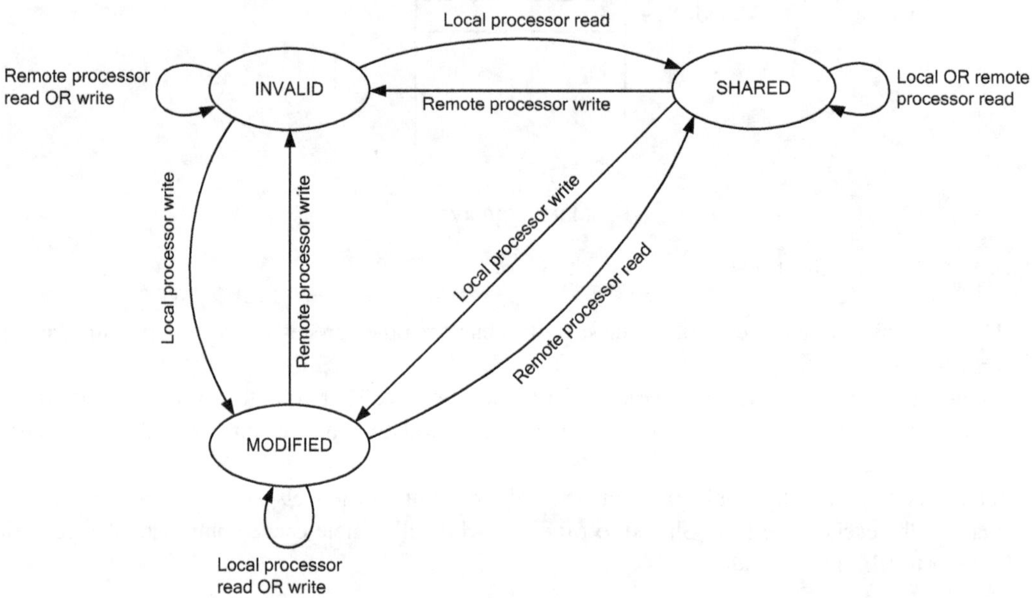

Fig. 6.173 Cache controller state machine for Snooper-based multi-core CPUs

Multi-core Systems with Distributed Memories and Parallelism in Hardware

When the number of CPU cores approaches to 100 and over, it may be prudent not to use a single, central main memory, but allow each CPU to have its own individual memory as shown in Fig. 6.174.

In this platform, every cache is still controlled by a local cache controller, often referred as the MSI controller (after Modified, Shared and Invalid states), and each distributed memory is managed by a directory controller. Individual directories contain a set of entries, called Sharers' IDs, indicating which CPU core(s) shares the local data block at a particular address.

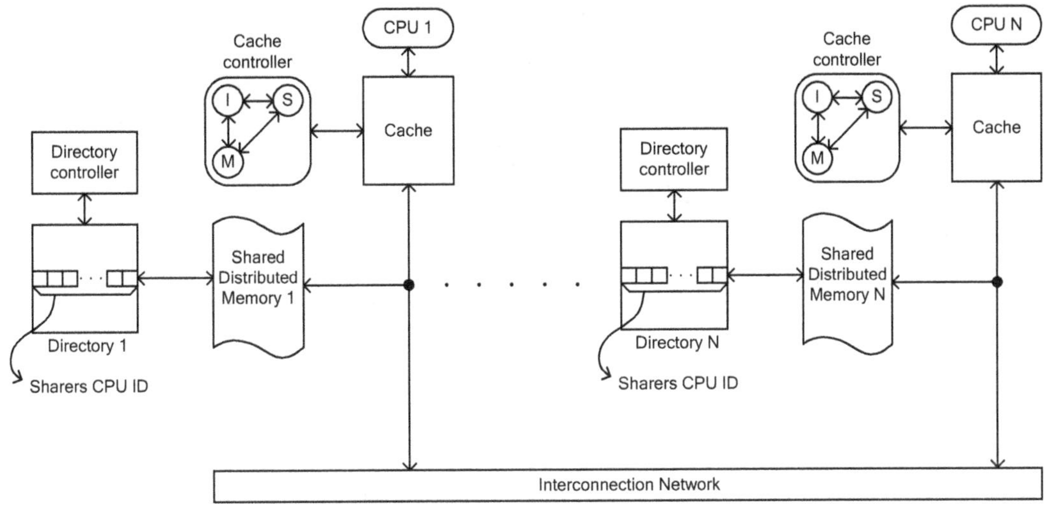

Fig. 6.174 Directory-based system with distributed memory

Each distributed memory is also managed by a directory system shown in the same figure. Each directory contains a set of entries, called Sharers' IDs, indicating which CPU core(s) shares the data block at a particular address.

The properties of this directory system are as follows:

- For each block in the local memory, there is a corresponding Sharers' ID entry in the directory.
- This entry lists all possible outcomes after a data transfer: (1) if a data block is shared by other CPU cores (the Shared state), (2) if a data block only exists in a local memory but nowhere else (the Modified state), (3) if a data block does not exist in a local memory (the Uncached state).
- On a cache miss, the directory controller searches and locates the wanted data block, and updates the Sharers' ID after each read/write operation.

There are basically three CPU cores that involve in every data transaction. Each CPU core in a transaction is named differently and called a "node".

- The CPU core that issues an address and requests data from other CPU core(s) is called the Requesting node.
- The CPU core that contains this issued address in its memory is called the Home node.
- The CPU core that actually has the requested data in its memory is called the Owner node.

Figure 6.175 shows these CPU nodes and their function. Assume there are three CPU cores connected to a network shown in this figure. In the first step, the requesting node generates an address and a read/write request. In the second step, the home node locates and fetches the data from the owner node, and it updates its directory. Finally, in the third step the home node forwards the data to the requesting node, completing the data transfer.

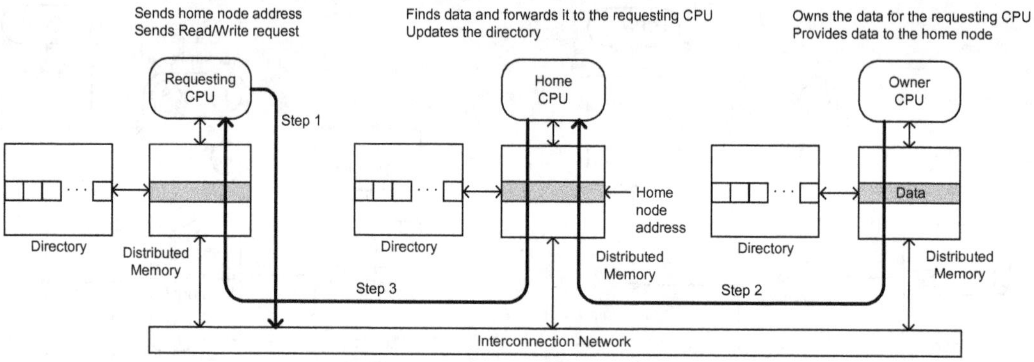

Fig. 6.175 CPU nodes in directory-based system with distributed memory

The cache controller (the MSI controller) of each CPU node is very similar to the cache controller for the snooper-based multi-core systems in Fig. 6.173, and is shown in Fig. 6.176.

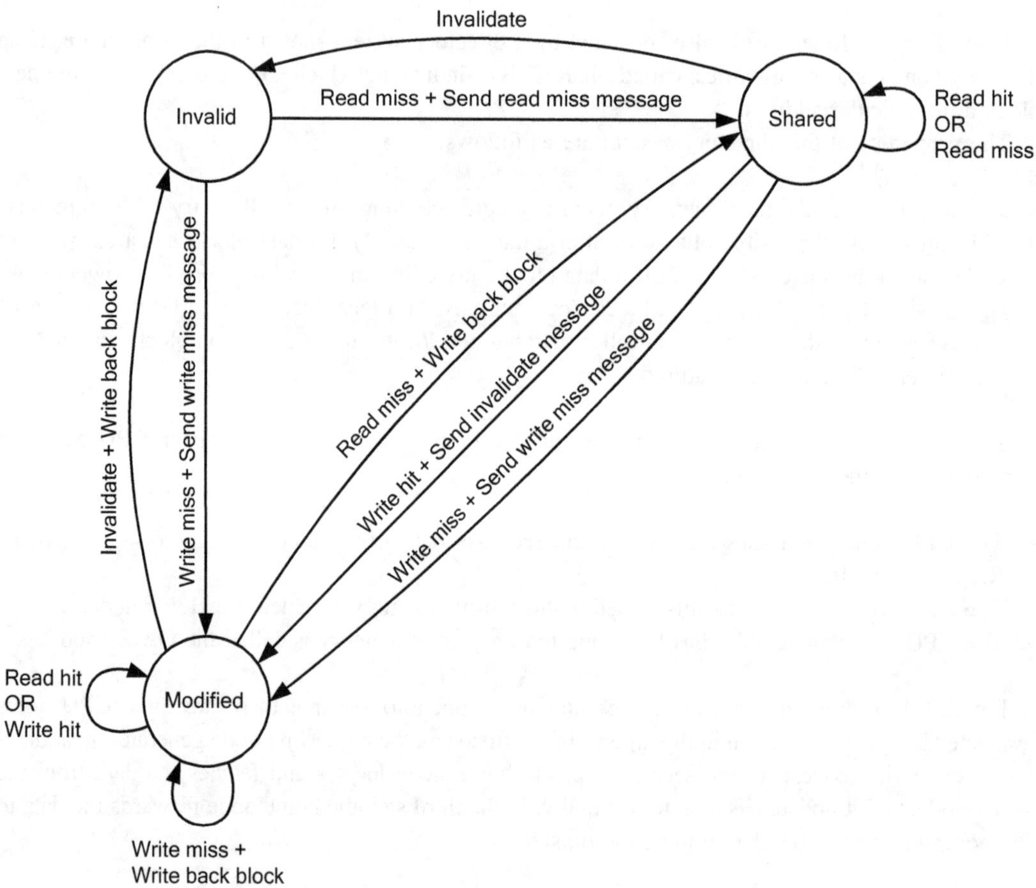

Fig. 6.176 Cache (MSI) controller in directory-based systems

In this figure, the transition from the Invalid state to the Shared state is due to a read request from a local processor. The local processor (or the requesting node) cannot locate the data in its own cache, which results in a read miss. At this point, the local processor sends a read miss message, and transitions to the Shared state because eventually the home node will find the data, and forward it to the requesting node. The transition from the Invalid state to the Modified state is because of a write request from a local processor, which results in a write miss. The local processor will eventually write the data to its own cache and transition to the Modified state because this data is unique and exists only in the local processor's cache.

Once in the Shared state, the local cache controller stays in this state whether the local processor's read request results in a read hit or a miss. While in the Shared state, if a remote processor issues a write request, and it results in a write hit, this scenario invalidates the data entry in the local cache, causing the local cache to transition to the Invalid state. Similarly, a local write causes the cache state to transition from Shared to Modified because the modified data exists only in the local cache.

A read hit or a write hit while the local cache is in the Modified state keeps the cache in the Modified state. A write miss while in the Modified state, on the other hand, means that the local cache controller needs to write data to its own cache and local memory. However, before this event takes place, the old data block, which is still valid in the local cache, is written back to the local memory. Then the local cache controller writes the new data to the local cache, transitioning the cache state from Modified to Modified. Finally, a read miss while in the Modified state causes the cache controller to locate and bring back the data to the local cache, transitioning the current cache state to Shared since the data will reside in more than one place. But, before this transaction takes place, the old data block, which is still valid in the local cache, has to be written back to the local memory.

Similar to the cache controller design in Fig. 6.176, the directory controller has also three distinct states. The Uncached state represents a block of data that does not yet exist in the directory-based system. The Modified state represents a block of data that exists only in one of the distributed memories, but nowhere else. Finally, the Shared state represents the same block of data existing in more than one distributed memory. The directory controller of the requesting node transitions among these three states depending on the nature of event(s) between the local memory and remote memories, and updates the Sharers' ID every time a data transaction is completed.

Case 1: Transition from the Uncached to the Shared state

In the first case, the directory controller transitions from the Uncached state to the Shared state as a result of a read miss shown in the state diagram above and in Fig. 6.177, and it goes through the following steps:

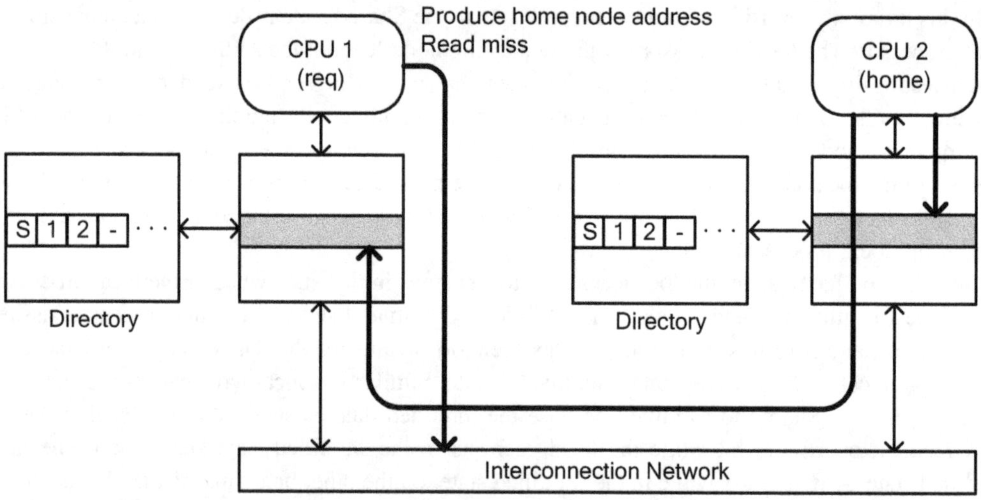

Fig. 6.177 Transition from the Uncached to the Shared state

- The requesting node places a read miss on the bus because the address that it needs to read from is not in the local memory.
- The home node finds the data either from its own memory or fetches it from some other CPU node (owner node), and forwards it to the requesting CPU.
- The directory controller changes the state of the requesting directory to Shared.
- The directory controller also updates the sharers' ID with CPU 1 and CPU 2, corresponding to the requesting node and the home node, respectively.

Case 2: Transition from the Uncached to the Modified state

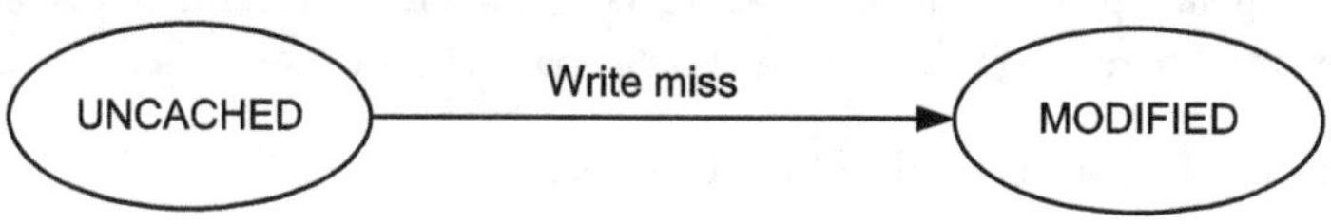

In the second case, the directory controller transitions from the Uncached state to the Modified state as a result of a write miss as shown in the state diagram above and in Fig. 6.178, and it goes through the following steps:

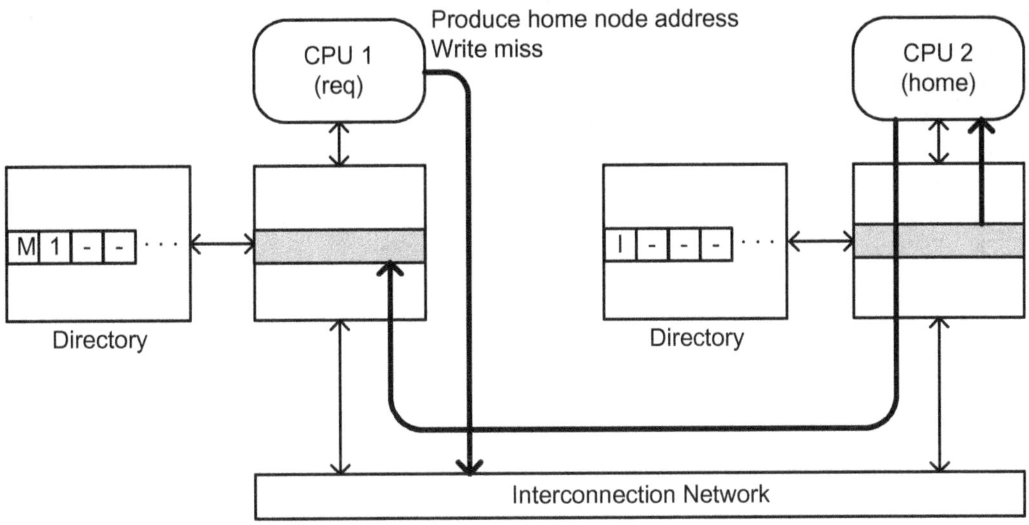

Fig. 6.178 Transition from the Uncached to the Modified state

- The requesting CPU places a write miss on the bus because the address it needs to write to is not in the local memory.
- The home node locates the data, and writes it directly to the requesting CPU's memory.
- Because this data exists only in the requesting CPU's memory, the directory controller changes the state of the local memory to Modified, and invalidates the data in all other memories at this address.
- The directory controller updates the sharers' ID with CPU 1, signifying the data exists at the requesting node only.

Case 3: Transition from the Shared to the Shared state

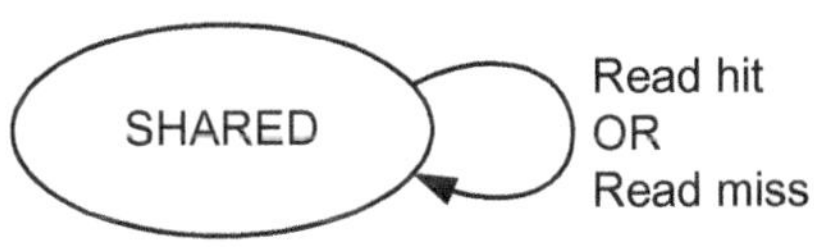

In the third case, the directory controller stays in the Shared state as a result of a read hit or a read miss as shown in the state diagram above and in Fig. 6.179, and it goes through the following steps:

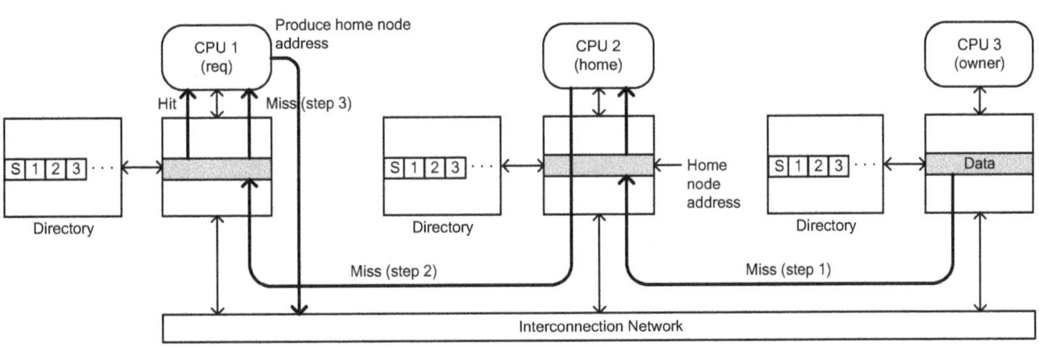

Fig. 6.179 Transition from the Shared to the Shared state

- If a block of data is in Shared state, it must reside in more than one memory in the system, including the requesting CPU's own memory.
- In case the data resides in another node and misses the local memory, the requesting CPU places a read miss on the bus while pointing the home node.
- Since the requested data is shared, the home node finds and fetches the data from the owner node, and forwards it to the requesting CPU (steps 1, 2 and 3).
- The directory controller keeps the state of the requesting CPU's memory in Shared.
- The directory controller also updates the sharers' ID with CPU 1, CPU 2 and CPU 3, corresponding to the requesting, home and owner nodes, respectively. However, the sharers' ID would only contain CPU 1 and CPU 2 if the data were located at the home node and only CPU 1 if it were at the requesting node.

Case 4: Transition from the Shared to the Modified state

In the fourth case, the directory controller transitions from the Shared state to the Modified state as a result of a write miss or a write hit as shown in the state diagram above and in Fig. 6.180, and it goes through the following steps:

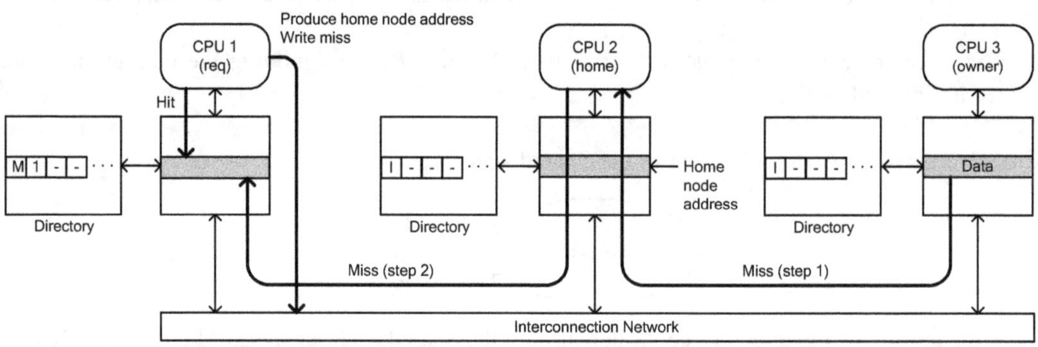

Fig. 6.180 Transition from the Shared to the Modified state

- If the block of data is in the Shared state, it must reside in more than one memory in the system, including the requesting CPU's own memory.
- When the requesting CPU produces a write hit and writes the data to its own memory, the directory controller changes the state of the memory to Modified, and updates the sharers' ID with CPU 1, indicating the data exists only in the local memory.
- When the requesting CPU misses, however, this time the home node fetches the data from the owner node, and forwards it to the requesting node (steps 1 and 2). The local directory controller changes the state of the memory to Modified while invalidating the data in all other memories. It also updates the sharers' ID with CPU 1.

Case 5: Transition from the Modified to the Modified state

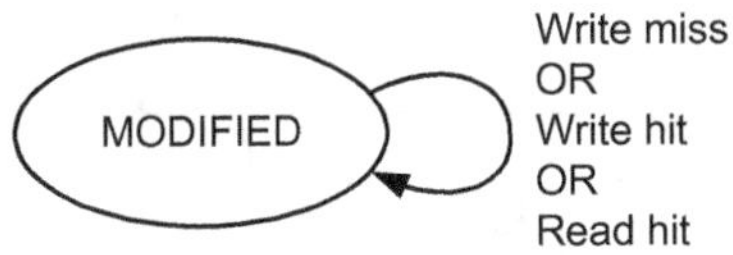

In the fifth case, the directory controller stays in the Modified state as a result of a write miss, a write hit or a read hit as shown by the state diagram above and in Fig. 6.181, and it goes through the following steps:

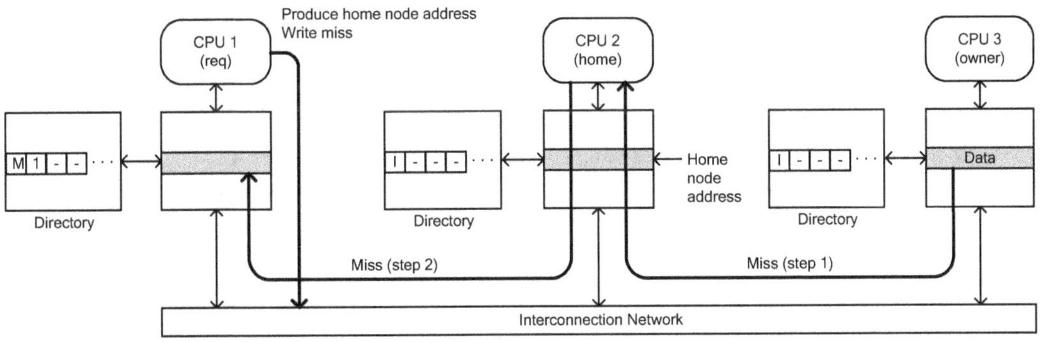

Fig. 6.181 Transition from the Modified to the Modified state

- If the requesting node produces a write hit, and writes new data directly to its own memory, the state of the memory remains in the Modified state. The directory controller keeps CPU 1 in the Sharers' ID after this event.
- If the requesting node produces a read hit, and reads directly from its own memory, the state of the memory again remains in the Modified state. The directory controller does not change the current entry in Sharers' ID, but keeps it at CPU 1 after this event.
- If the requesting node misses the write while in the Modified state, this time the home node fetches the data from the owner node, and forwards it to the requesting CPU (steps 1 and 2).
- The directory controller keeps the state of the requesting CPU's memory at the Modified state, and invalidates the data in all other memories at this address.
- The directory controller also keeps the sharers' ID at CPU 1.

Case 6: Transition from the Modified to the Shared state

In the sixth and final case, the directory controller transitions from the Modified state to the Shared state as a result of a read miss as shown in the state diagram above and in Fig. 6.182, and it goes through the following steps:

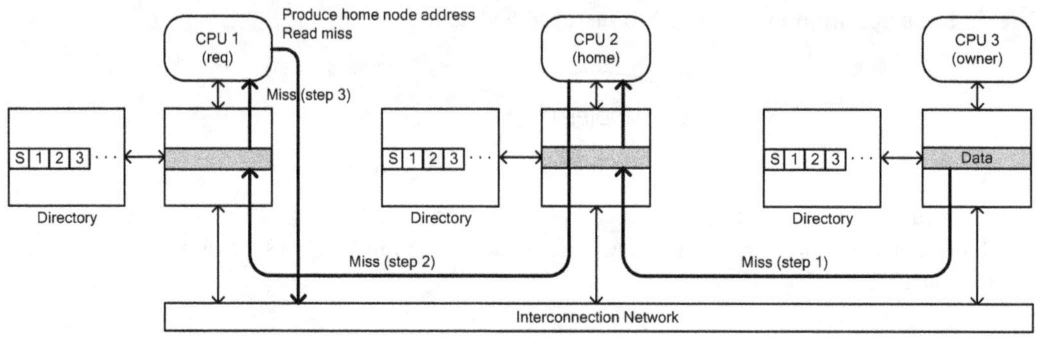

Fig. 6.182 Transition from the Modified to the Shared state

- While in the Modified state if the requesting CPU tries to read data from its own memory but misses, it places a read miss on the bus while the home node fetches the requested data from the owner node and writes it to the requesting CPU's memory (steps 1 and 2).
- The requesting CPU then proceeds with reading the data from its own memory (step 3).
- Since this data is now shared among the requesting, home and owner nodes, the directory controller changes the state of the requesting CPU's memory to Shared.
- The directory controller also updates the sharers' ID with CPU 1, CPU 2 and CPU 3.

When the state transitions from case 1 to case 6 are integrated, we obtain the state machine in Fig. 6.183 for the local directory controller.

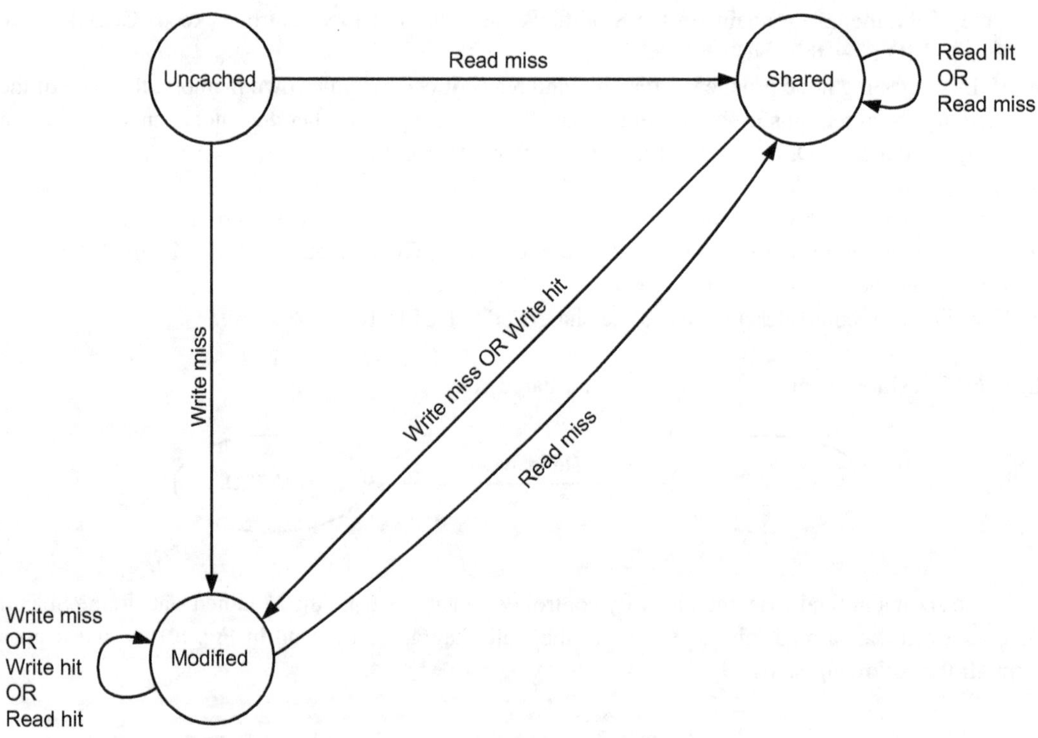

Fig. 6.183 Directory controller for the multi-core distributed memory system

6.8 Caches

Cache Topologies

Cache is a local memory to the CPU where the temporary blocks of data is kept until it is permanently stored in the system memory. No other bus master but the CPU is allowed to access the cache memory.

There are three types of cache architectures in modern CPUs: fully-associative, set-associative and direct-mapped.

Fully-associative cache protocol allows a block of data to be written (or read) anywhere in the cache as shown in Fig. 6.184. In this type of cache architecture, a block of data is searched in the entire cache before it is read. The range of cache memory addresses to store a block of data is called a set. In Fig. 6.184, the entire cache containing N number of blocks belongs to a single set.

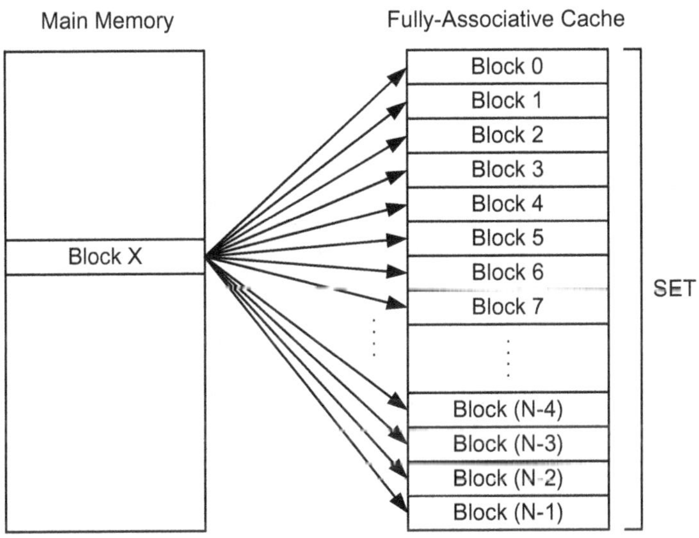

Fig. 6.184 Fully-associative cache topology

Set-associative cache protocol, in contrast, allows a block of data to be written (or read) only to a limited set of addresses in the cache. The top figure of Fig. 6.185 shows a two-way set-associative cache where a block of data from the main memory is written only to two possible cache addresses, which defines a set. Conversely, when data needs to be read from a two-way set-associative cache, data is searched only within a given set. Therefore, the time to search and locate data is reduced by a factor of 2/N in a two-way set associative cache compared to a fully-associative cache containing N number of blocks where N is assumed much greater than two.

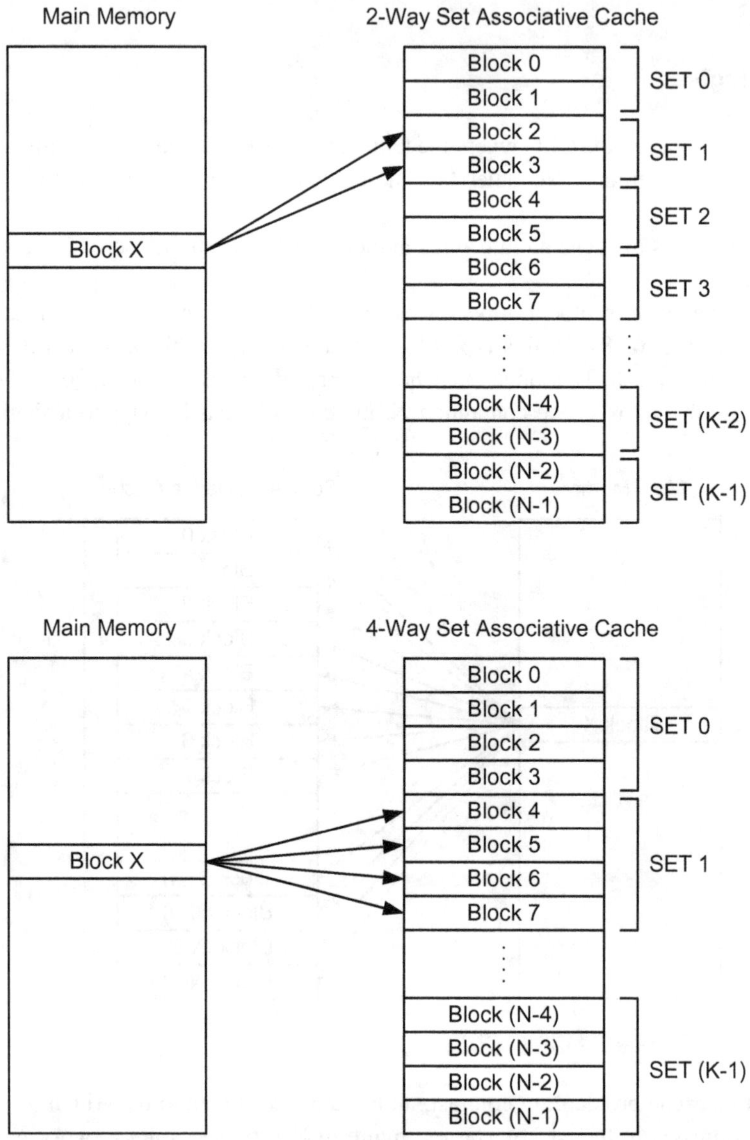

Fig. 6.185 Two-way and four-way set-associative cache topologies

Similar to the two-way set-associative cache, a block of data is searched in four possible addresses in a set when the CPU issues a cache read in a four-way set-associative cache as shown at the bottom part of Fig. 6.185. If data needs to be written to the cache, only four possible cache addresses in a set are considered according to the cache protocol.

The third cache type is the direct-mapped cache as shown in Fig. 6.186. Its organization is similar to an SRAM, and it maintains a one-to-one addressing scheme with the main memory. In other words, a block of data in the main memory can only be written to a specific location in the cache memory or vice versa.

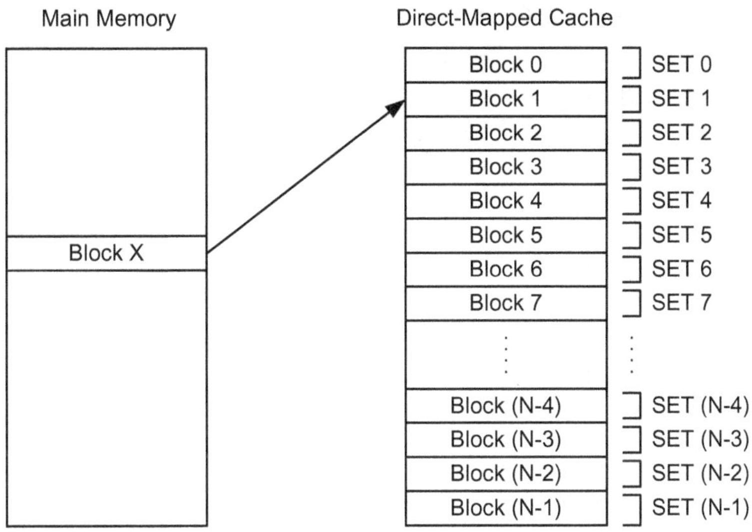

Fig. 6.186 Direct-mapped cache topology

All stored data blocks in the cache memory are tagged. Consequently, all data transactions between the cache and the main memory (or the CPU) require validation of the tag field before performing a cache read or write operation. Physically, tag fields are stored in a different memory block in the same cache structure. However, both the cache and tag memories retain one-to-one association with each other as shown in Fig. 6.187. In addition, the tag memory comes with valid bits. Each bit specifies if a block of data residing in the cache has an identical twin in the main memory or not. A valid bit equal to logic 0 means that the contents of the main memory have not been updated with the block of data residing in the cache at a certain address. When updating is complete and there is complete data coherency between the cache and main memories, the valid bit becomes logic 1.

The CPU address that references the cache memory consists of three separate fields as shown at the bottom section of Fig. 6.187: tag, index and block offset. The index field specifies the set address where the data block resides. Since a data block may contain many words, block offset selects the word in the block. Therefore, before a cache-related operation takes place, the tag field in a CPU address is compared against all the tag fields in a given set. If the tag comparison is successful, the block of data at the specified set address is transferred out of the cache memory or vice versa.

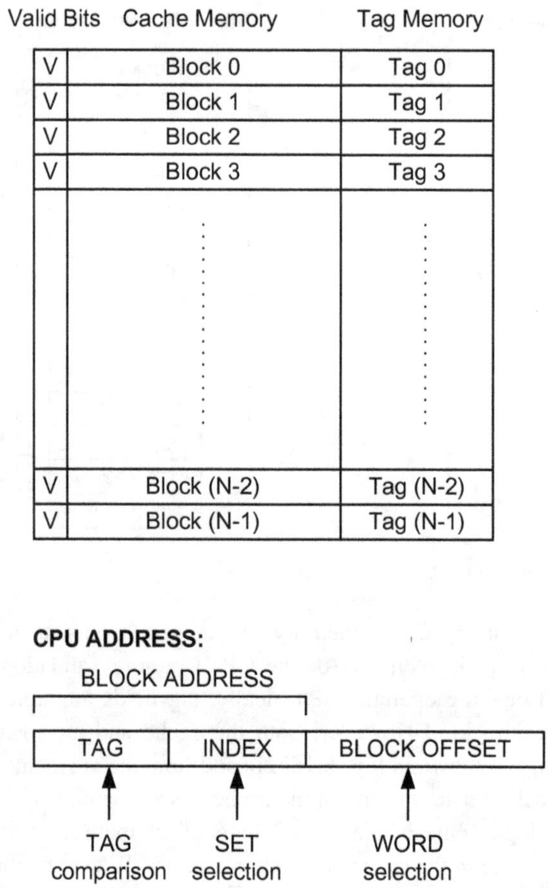

Fig. 6.187 Cache structure

Cache Write and Read Data-Paths

Cache write operation starts with comparing the tag field of the CPU address with all the tag fields of the addressed set in the tag memory. If the comparison is successful, this creates a hit signal, and prompts the CPU to write data to the specified set (and block offset) address in the cache memory.

Figure 6.188 shows the cache write operation to a 32-bit, four-way set-associative cache. This cache contains 22-bit tag fields, 256 sets due to the eight-bit index field, and four words in a block due to the two-bit block offset field. The write process starts with identifying the set address using the eight-bit index field. All four tag fields at this set address are individually compared against the CPU tag using XNOR-gates as shown at the output stage of the tag memory in Fig. 6.188. If one of the tags at the set address compares successfully with the CPU tag, it creates a hit signal for the CPU to write a block of data to the corresponding set address. The CPU data is routed through the tri-state buffers placed at the input stage of the cache memory, and written to the designated address via the index field and the block offset field.

The cache read operation is similar to the cache write operation except that the cache and the tag memories are accessed simultaneously to shorten the cache read access period.

Figure 6.189 shows the same four-way set-associative cache structure shown in Fig. 6.188. With the eight-bit index field defining the set address, four tag blocks from the tag memory and four data

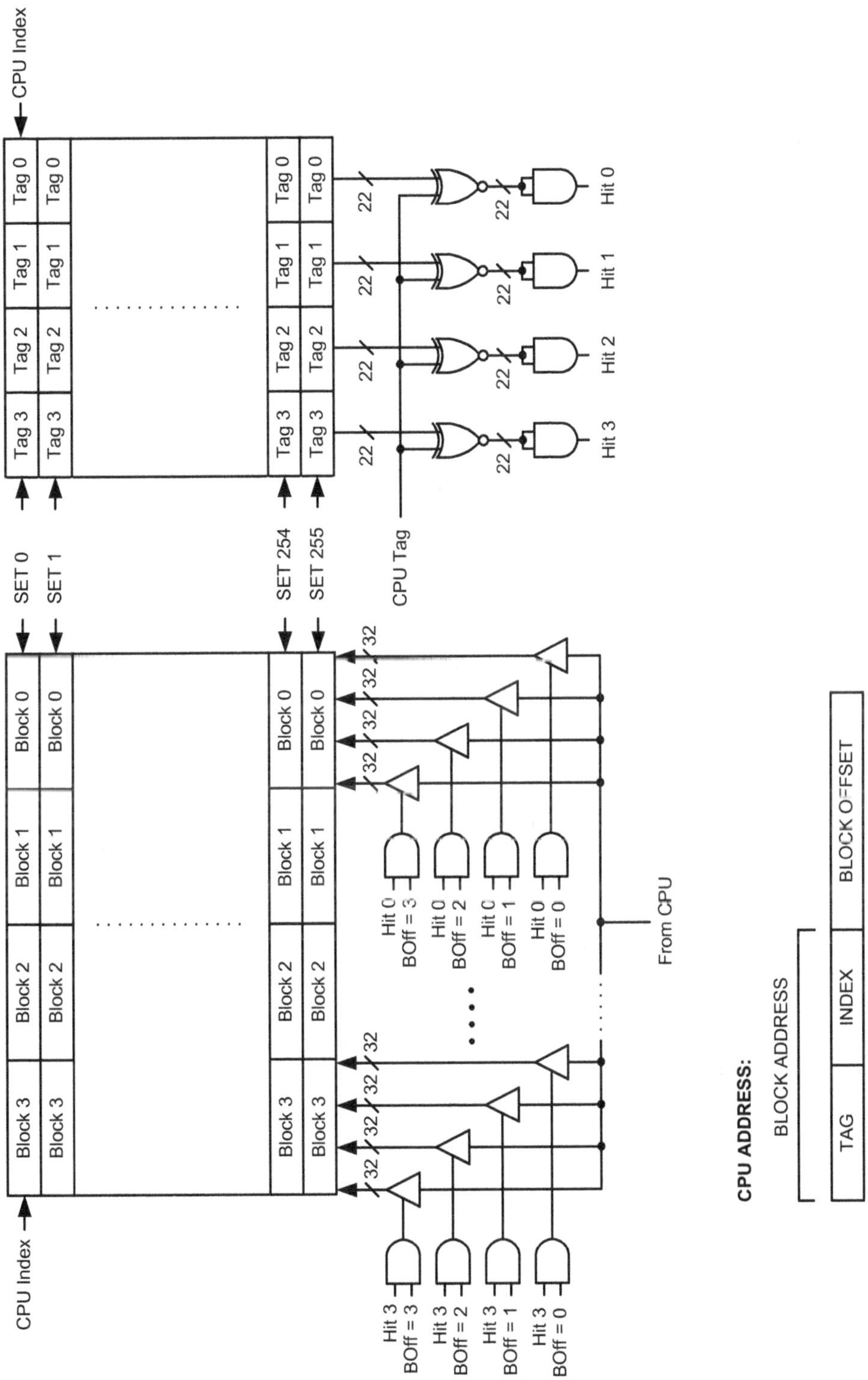

Fig. 6.188 Cache write data-path and operation

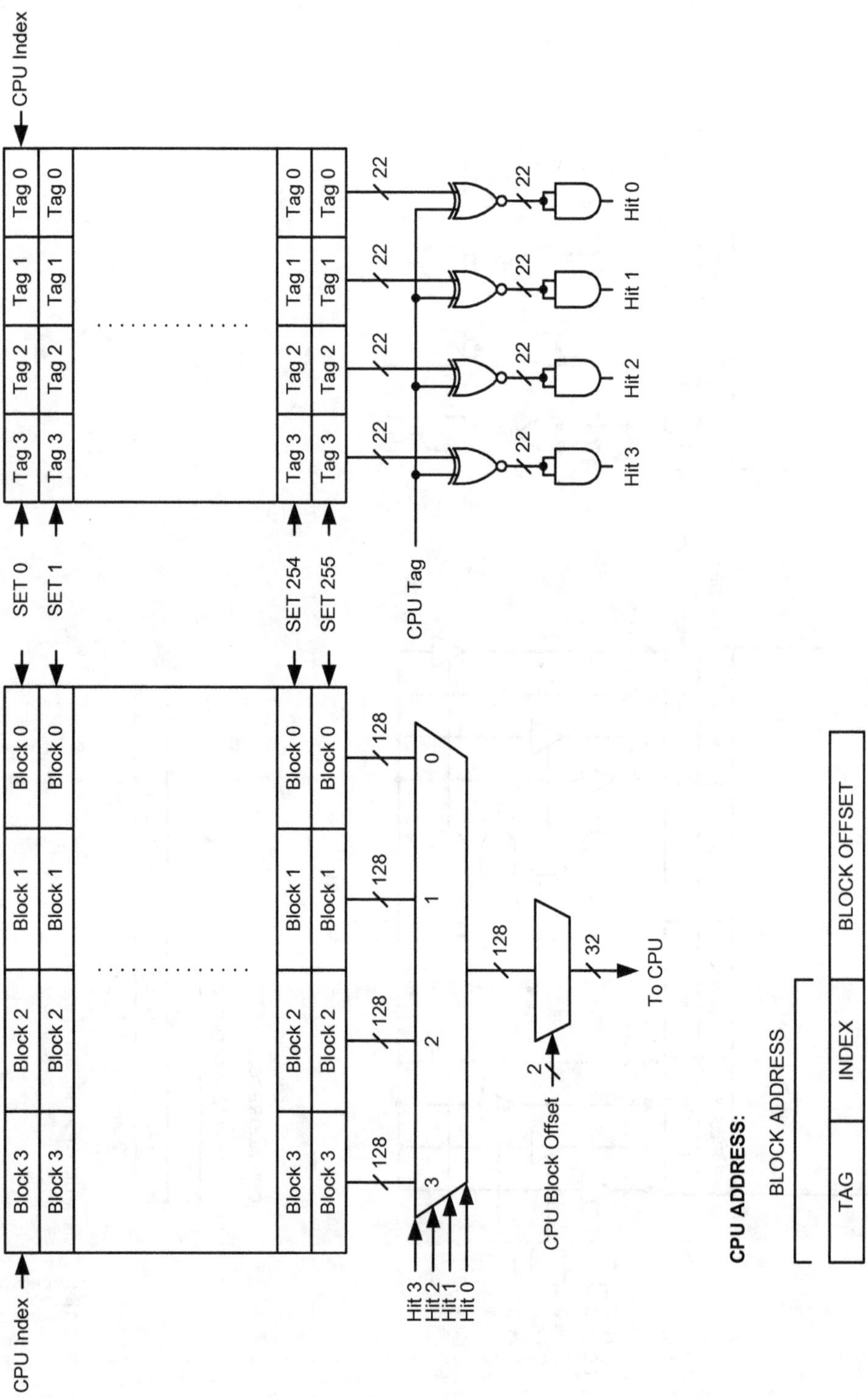

Fig. 6.189 Cache read data-path operation

blocks from the cache memory are read out simultaneously. Each tag is individually compared with the 22-bit tag field in the CPU address, and hit signals are generated by 22 sets of two-bit XNOR gates followed by a 22-bit AND gate. These hit signals are subsequently used as selector inputs for the 4-1 MUX at the output of the cache memory to select one of the cache blocks. The word from the chosen block is selected by the two-bit block offset field and directed to the CPU.

Example 6.18 Examine the memory transactions in Fig. 6.190 to understand the operation of a direct-mapped cache.

Each CPU address in this example contains a five-bit word with three-bit index field and a two-bit tag field. Therefore, the cache structure consists of eight sets.

Assume that the CPU issues cache reads from the following addresses: 10101, 10010, 10101, 01010, 10000, 10110, 10000, 10100 and 11111. Initially, the valid bit, the tag and the data fields of the cache memory are all zero. The index column in Fig. 6.190 is not an actual part of cache memory. Its sole purpose is simply to indicate the set address.

When the CPU issues the first read from the address 10101, the tag memory contents 00 at the set address 101 are compared against the tag field contents of 10 in the CPU address. Since the two values are different from each other, the cache controller issues a cache miss, fetches the data, mem(10101), from the main memory address of 10101, delivers this data to the CPU, and stores the same data in the cache. It also updates the tag contents with 10, and issues valid bit = 1.

Next, the CPU issues the second read from the address 10010. This time, the tag memory contents 00 at the set address 010 are compared with the tag field of 10 in the CPU address. The comparison fails and produces another miss. The cache controller fetches the data, mem(10010), from the main memory, writes this data to the set address 010 of the cache memory, delivers the same data to the CPU, updates the tag memory with 10, and produces valid bit = 1.

The CPU reissues another read from the address 10101. The tag contents 10 at the set address 101 compare successfully with the CPU tag field of 10. As a result, the cache controller issues a hit. The data, mem(10101), at the set address 101 is transferred directly from the cache memory to the CPU.

The next CPU address 01010 accesses the set, 010, in the tag memory, but finds the tag memory contents of 10 at this address is different from the tag field contents of 01 in the CPU address. Therefore, the cache controller issues a miss, fetches mem(01010) from the main memory, and delivers this data to the CPU and the cache memory. It also updates the tag contents with 01, and issues valid bit = 1.

The fifth address 10000 produces a fourth miss because the tag contents of 00 at the set address 000 do not compare with the tag contents of 10 in the CPU address. The cache controller transfers mem(10000) from the main memory to the set address 000 of the cache, delivers the same data to the CPU, updates the tag contents with 10, and produces valid bit = 1.

When the CPU issues the sixth address 10110, the cache controller finds the tag contents of 00 at the set address 110 to be different from the tag field contents of 10 in the CPU address, and issues a miss. Consequently, the cache controller fetches mem(10110) from the main memory, delivers it to the CPU and the cache memory. It also updates the tag memory with 10, and assigns valid bit = 1.

Next, the CPU reissues the address 10000. This time, the CPU and the tag memory contents match, and produce a cache hit. The cache controller simply delivers mem(10000) from the set address 000 of the cache to the CPU.

The next address 10100 creates another cache miss. The cache controller updates the set address contents with mem(10100), and delivers the same data to the CPU. It also updates the tag memory with a value of 10 and issues valid bit = 1.

When the last CPU address 11111 is issued, the cache controller finds the tag field contents of 11 in the CPU address to be different from the tag memory contents 00 at the set address 111, and issues a miss. Subsequently, it delivers mem(11111) to both the CPU and the cache. It updates the tag memory with 11 and assigns valid bit = 1.

CPU Address:

TAG	INDEX	BOFF
2 bits	3 bits	0 bits

(1) Initial state of the cache

INDEX	TAG	V	DATA
000	00	0	0
001	00	0	0
010	00	0	0
011	00	0	0
100	00	0	0
101	00	0	0
110	00	0	0
111	00	0	0

(2) After the **MISS** at address 10101

INDEX	TAG	V	DATA
000	00	0	0
001	00	0	0
010	00	0	0
011	00	0	0
100	00	0	0
101	10	1	mem (10101)
110	00	0	0
111	00	0	0

(3) After the **MISS** at address 10010

INDEX	TAG	V	DATA
000	00	0	0
001	00	0	0
010	10	1	mem (10010)
011	00	0	0
100	00	0	0
101	10	1	mem (10101)
110	00	0	0
111	00	0	0

(4) After the **HIT** at address 10101

INDEX	TAG	V	DATA
000	00	0	0
001	00	0	0
010	10	1	mem (10010)
011	00	0	0
100	00	0	0
101	10	1	mem (10101)
110	00	0	0
111	00	0	0

(5) After the **MISS** at address 01010

INDEX	TAG	V	DATA
000	00	0	0
001	00	0	0
010	01	1	mem (01010)
011	00	0	0
100	00	0	0
101	10	1	mem (10101)
110	00	0	0
111	00	0	0

(6) After the **MISS** at address 10000

INDEX	TAG	V	DATA
000	10	1	mem (10000)
001	00	0	0
010	01	1	mem (01010)
011	00	0	0
100	00	0	0
101	10	1	mem (10101)
110	00	0	0
111	00	0	0

(7) After the **MISS** at address 10110

INDEX	TAG	V	DATA
000	10	1	mem (10000)
001	00	0	0
010	01	1	mem (01010)
011	00	0	0
100	00	0	0
101	10	1	mem (10101)
110	10	1	mem (10110)
111	00	0	0

(8) After the **HIT** at address 10000

INDEX	TAG	V	DATA
000	10	1	mem (10000)
001	00	0	0
010	01	1	mem (01010)
011	00	0	0
100	00	0	0
101	10	1	mem (10101)
110	10	1	mem (10110)
111	00	0	0

(9) After the **MISS** at address 10100

INDEX	TAG	V	DATA
000	10	1	mem (10000)
001	00	0	0
010	01	1	mem (01010)
011	00	0	0
100	10	1	mem (10100)
101	10	1	mem (10101)
110	10	1	mem (10110)
111	00	0	0

(10) After the **MISS** at address 11111

INDEX	TAG	V	DATA
000	10	1	mem (10000)
001	00	0	0
010	01	1	mem (01010)
011	00	0	0
100	10	1	mem (10100)
101	10	1	mem (10101)
110	10	1	mem (10110)
111	11	1	mem (11111)

Fig. 6.190 A direct-mapped cache operation

Write-Through and Write-Back Cache Structures in Set Associative Caches

It is very common to see two types of cache structures when analyzing set-associative caches: write-through caches and write-back caches.

In write-through caches, the cache controller maintains data coherency between the cache and the main memory before issuing a new data transaction.

In write-back caches, the wait time for data coherency is an essence. For example, if the transaction requires writing to the main memory as well as the cache, the cache controller temporarily stores the data in the write-back buffer instead of waiting to write it to the main memory, and starts a new task. When the bus arbiter grants the bus access, the cache controller resumes transferring this data from the write-back buffer to the main memory.

Example 6.19 Examine the data transactions between the CPU and a two-way set-associative write-through cache in Fig. 6.191 to understand how write-through caches operate. The initial contents of the main memory are shown in the same figure. Each transaction is specified by a CPU address, the type of transaction and data.

Transaction No	CPU Address	Read/Write	CPU Data
1	010101	Read	-
2	110001	Read	-
3	100000	Read	-
4	100011	Read	-
5	110000	Read	-
6	010011	Read	-
7	000110	Write	0x0B
8	001110	Write	0x1B
9	110000	Write	0x2B
10	100000	Write	0x3B
11	100100	Write	0x0C
12	100100	Write	0x1C
13	111100	Write	0x2C
14	000000	Read	-

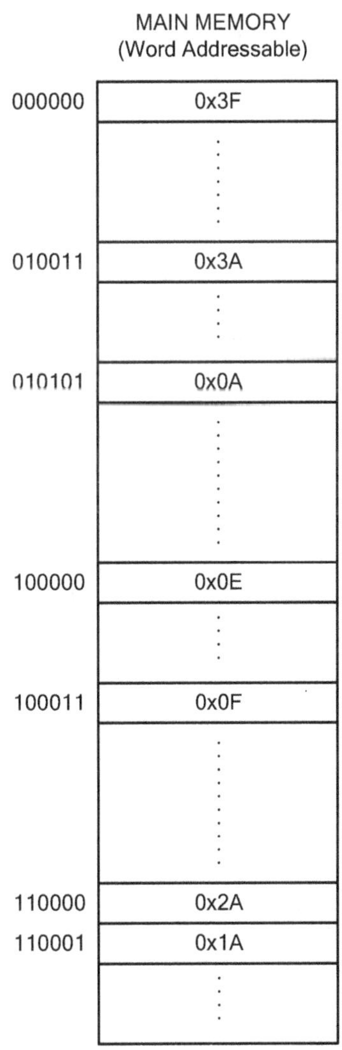

Fig. 6.191 A two-way set-associative write-through cache pending transactions and initial data memory contents

The CPU address in this example consists of six bits: the most significant four bits define the tag address; the least significant two bits indicate the set address as shown in Fig. 6.192. There are no bits for block offset, which indicates every data block consists of a single word with six bits of data. Since this is a two-way set-associative cache, the memory is organized as two adjacent data blocks for every set. The tag memory also consists of two adjacent tag fields with valid bits at every set as shown in Fig. 6.192.

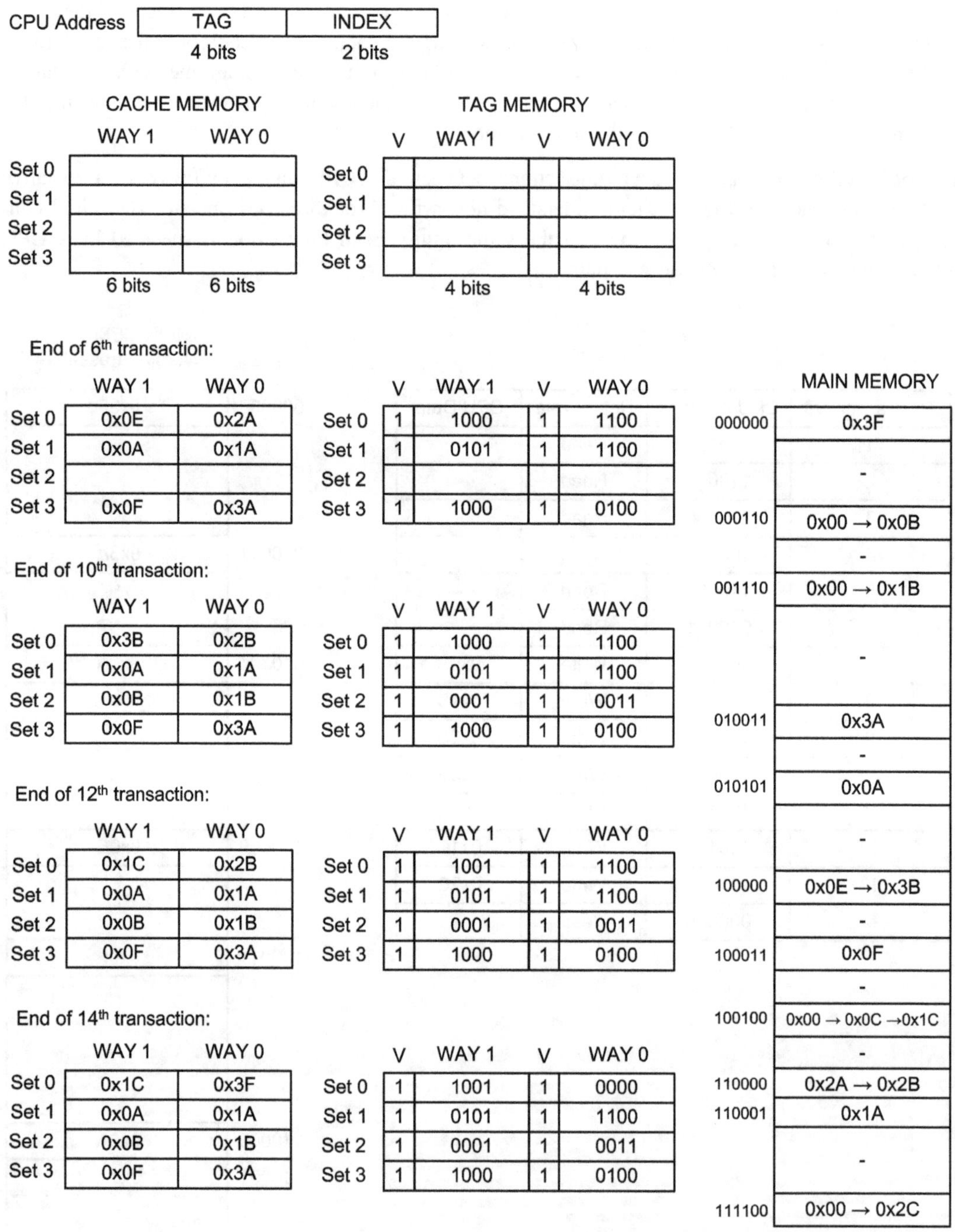

Fig. 6.192 A two-way set-associative write-through cache, tag and data memory contents after the sixth, tenth, twelfth and fourteenth transactions

The data from the main memory can either go to the most significant (WAY 1) or the least significant (WAY 0) block position of the two-way set-associative cache. Therefore, a data replacement policy must be defined when designing set-associative cache architecture. In this example, let us assume that the data replacement policy replaces old data with smaller number of memory references (the total number of reads and writes) with new data. If the number of memory references is the same at a particular set, the block of data at the most significant cache position is replaced with new data.

The first transaction reads from the memory address 010101. In this transaction, the cache controller compares the tag field contents 0101 in the CPU address with the tag memory contents 0000 at the set address 01, and issues a miss. Next, the cache controller fetches 0x0A from the main memory address of 010101, and delivers it to the CPU. Since the number of memory references at the most and the least significant cache positions are both zero at this point, the cache controller places 0x0A at the most significant cache position as the result of the data replacement policy. It also updates the tag memory with 0101, and assigns valid bit = 1.

The next five read transactions result in five consecutive cache misses. By the end of the sixth transaction, six new data entries from the main memory are written to both the cache and the tag memories as shown in Fig. 6.192.

The seventh CPU transaction is a write. The CPU issues to write 0x0B to the main memory address 000110. As for the read operations, the cache controller compares the tag memory contents of 0000 at the set address 10 with the tag field contents of 0001 in the CPU address, and issues a miss. The data, 0x0B, is written to both the main memory address 000110 and the most significant cache position at the set address 10. The tag memory is also updated with 0001, and valid bit = 1.

In the eight transaction, the cache controller again issues a miss for the set address 10 because the tag field comparison fails. Consequently, the cache controller stores the CPU data, 0x1B, in the main memory address 001110, and also writes this data to the least significant cache position at the set address 10. The tag memory is updated with 0011, and valid bit = 1.

The next write compares the tag field entry 1100 in the CPU address with the tag memory entries at the set address 00. Since the least significant tag memory contents are identical to the CPU tag entry, the cache controller issues a hit, but still replaces the contents of the main memory at the address 110000 and the contents of the cache memory at the set address 00 with 0x2B. This is because the architecture of this cache is a write-through which requires the cache and the main memory contents to be the same before the cache controller starts a new task. Therefore, there is really no difference between a cache write miss and cache write hit when it comes to updating the cache and the main memory contents. However, the tag memory contents require no updating in a cache write hit.

The tenth transaction is another CPU write which results in a cache hit. Like the previous write transaction, the cache controller has to replace the contents of the main memory at the address 100000 and the contents of the cache memory at the set address 00 with 0x3B. At the end of the tenth transaction, the cache and the tag memory contents are shown in Fig. 6.192 with two memory references for the set address 00, and one memory reference for the remaining set addresses.

The eleventh transaction creates a cache miss, and causes the cache controller to replace the data, 0x3B, at the most significant cache position with the CPU data, 0x0C, at the set address 00. The most significant tag entry 1000 is also replaced with 1001 at the same set address.

The twelfth transaction creates a cache hit because the CPU tag field 1001 compares successfully with the tag memory contents at the set address 00. However, the cache controller replaces 0x0C at the main memory address 100100 and at the set address 00 with the new CPU data, 0x1C.

In the thirteenth transaction, the CPU address 111100 causes a cache miss. The CPU data, 0x2C, is written to both the main memory address 111100 and the least significant cache position at the set address 00 per data replacement policy. The least significant tag memory contents at the set address 00 are also updated with 1111.

The fourteenth transaction is a memory read and causes another miss. The cache controller delivers 0x3F from the main memory address, 000000, to the CPU and writes the same data to the least significant cache position at the set address 00.

Example 6.20 Examine the data transactions between the CPU and a two-way set-associative write-back cache in Fig. 6.193 to understand how write-back caches operate. The initial contents of

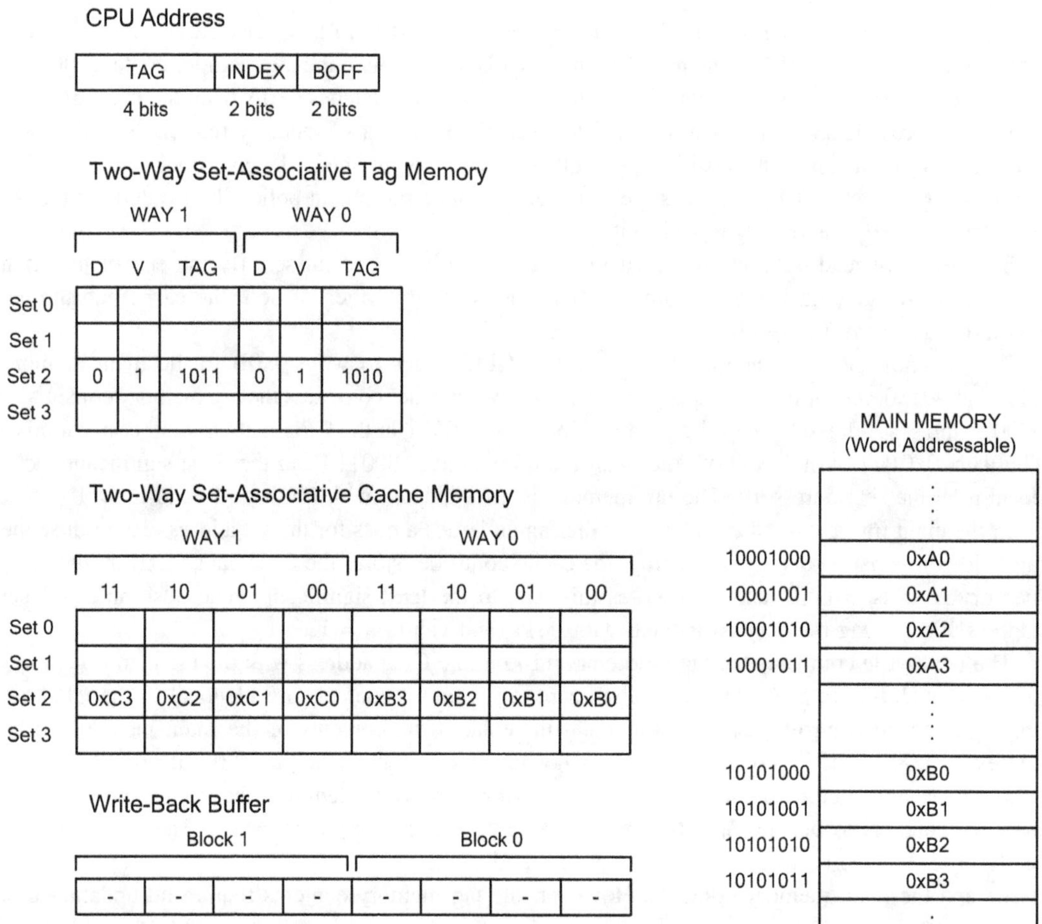

Transaction No	CPU Address	Read/Write	CPU Data
1	10001000	Read	-
2	10000000	Write	0xE0
	10000001	Write	0xE1
	10000010	Write	0xE2
	10000011	Write	0xE3
3	10000000	Write	0xF0
	10000001	Write	0xF1
	10000010	Write	0xF2
	10000011	Write	0xF3
4	11000000	Read	-

Fig. 6.193 A two-way set-associative write-back cache pending transactions and initial data memory

the main memory are shown in the same figure. Each transaction is specified by a CPU address, the type of transaction and data.

The CPU address in this example has eight bits. The most significant four bits are reserved for the tag field. The two-bit index field indicates that there are four sets in the cache memory. The two-bit block offset field signifies that there are four eight-bit wide words in each data block. The cache memory consists of two adjacent blocks at WAY 1 and WAY 0 positions, each of which contains four words at a given set. The tag memory has also two adjacent tag entries with valid bits, and each tag represents a data block in the cache memory. The data replacement policy in the cache memory assumes to replace the block of data with the least number of memory references. If the number of memory references is the same at a particular set, then the policy dictates to replace the old data at the least significant cache position with new data.

Since this is a write-back cache, its architecture enables the cache controller to store CPU data temporarily in a buffer to be written back to the main memory at a later time. Every time a block of data is written to this buffer, the dirty bit associated with this block becomes logic 1, designating that this block is waiting to be written to the main memory. Therefore, the cache coherency mechanism that presides over write-through caches is not applicable to write-back caches. Instead, the dirty bit attached to each tag entry determines if the same block of data exists in both the cache and the main memories.

Initially, identical data reside in both the main memory and the cache, and therefore both valid bit entries at the set address 10 in the tag memory are equal to logic 1. Since there is no data in the write-back buffer waiting to be written back to the main memory, the dirty bits at the set address 10 are equal to logic 0.

The first CPU transaction is to read data from the memory address, 10001000. The cache controller compares the tag entry 1000 in the CPU address with all the tags at the set address 10 and issues a miss. It then transfers the block of data from the main memory address 10001000 (0xA0, 0xA1, 0xA2 and 0xA3) to replace the old block (0xB0, 0xB1, 0xB2 and 0xB3) at the least significant cache position. The cache controller also updates the corresponding tag contents with 1000 as shown at the top portion of Fig. 6.194.

AFTER THE 1st TRANSACTION

Two-Way Set-Associative Cache Memory

	WAY 1 11	WAY 1 10	WAY 1 01	WAY 1 00	WAY 0 11	WAY 0 10	WAY 0 01	WAY 0 00
Set 0								
Set 1								
Set 2	0xC3	0xC2	0xC1	0xC0	0xA3	0xA2	0xA1	0xA0
Set 3								

Two-Way Set-Associative Tag Memory

	WAY 1 D	WAY 1 V	WAY 1 TAG	WAY 0 D	WAY 0 V	WAY 0 TAG
Set 0						
Set 1						
Set 2	0	1	1011	0	1	1000
Set 3						

Write-Back Buffer

Block 1				Block 0			
-	-	-	-	-	-	-	-

AFTER THE 2nd TRANSACTION

Two-Way Set-Associative Cache Memory

	WAY 1 11	WAY 1 10	WAY 1 01	WAY 1 00	WAY 0 11	WAY 0 10	WAY 0 01	WAY 0 00
Set 0					0xE3	0xE2	0xE1	0xE0
Set 1								
Set 2	0xC3	0xC2	0xC1	0xC0	0xA3	0xA2	0xA1	0xA0
Set 3								

Two-Way Set-Associative Tag Memory

	WAY 1 D	WAY 1 V	WAY 1 TAG	WAY 0 D	WAY 0 V	WAY 0 TAG
Set 0				1	1	1000
Set 1						
Set 2	0	1	1011	0	1	1000
Set 3						

Write-Back Buffer

Block 1				Block 0			
-	-	-	-	0xE3	0xE2	0xE1	0xE0

AFTER THE 3rd TRANSACTION

Two-Way Set-Associative Cache Memory

	WAY 1 11	WAY 1 10	WAY 1 01	WAY 1 00	WAY 0 11	WAY 0 10	WAY 0 01	WAY 0 00
Set 0					0xF3	0xF2	0xF1	0xF0
Set 1								
Set 2	0xC3	0xC2	0xC1	0xC0	0xA3	0xA2	0xA1	0xA0
Set 3								

Two-Way Set-Associative Tag Memory

	WAY 1 D	WAY 1 V	WAY 1 TAG	WAY 0 D	WAY 0 V	WAY 0 TAG
Set 0				1	1	1000
Set 1						
Set 2	0	1	1011	0	1	1000
Set 3						

Write-Back Buffer

Block 1				Block 0			
-	-	-	-	0xF3	0xF2	0xF1	0xF0

AFTER THE 4th TRANSACTION

Two-Way Set-Associative Cache Memory

	WAY 1 11	WAY 1 10	WAY 1 01	WAY 1 00	WAY 0 11	WAY 0 10	WAY 0 01	WAY 0 00
Set 0	0xD3	0xD2	0xD1	0xD0	0xF3	0xF2	0xF1	0xF0
Set 1								
Set 2	0xC3	0xC2	0xC1	0xC0	0xA3	0xA2	0xA1	0xA0
Set 3								

Two-Way Set-Associative Tag Memory

	WAY 1 D	WAY 1 V	WAY 1 TAG	WAY 0 D	WAY 0 V	WAY 0 TAG
Set 0	0	1	1100	1	1	1000
Set 1						
Set 2	0	1	1011	0	1	1000
Set 3						

Write-Back Buffer

Block 1				Block 0			
-	-	-	-	0xF3	0xF2	0xF1	0xF0

Fig. 6.194 A two-way set-associative write-back cache with tag and write-back buffers

The next CPU transaction is a write, which results in a cache miss. Consequently, the cache controller writes the contents of the CPU block that consists of 0xE0, 0xE1, 0xE2 and 0xE3 to the least significant cache position at the set address 00, and updates the tag memory contents with 1000. The cache controller also stores this block in the write-back buffer, and assigns dirty bit = 1 for the block because it exists only in the cache but not in the main memory. For this transaction, the valid bit is set at logic 1 because this block is considered a valid block from the CPU, not yet updated in the main memory. When the block is transferred to the main memory, the cache controller assigns dirty bit = 0, but still keeps the valid bit at logic 1.

The third CPU transaction is a write and results in a cache hit. The cache controller simply writes the new block of data, 0xF0, 0xF1, 0xF2 and 0xF3, to the least significant block position of the cache memory, replacing the old block, 0xE0, 0xE1, 0xE2 and 0xE3. It also writes the same data to the write-back buffer. Since this transaction does not require transferring the old block from the write-back buffer to the main memory, the write-back scheme creates a distinct speed advantage over the write-through scheme.

The last CPU transaction is a memory read that results in a miss. Consequently, the data block, 0xD0, 0xD1, 0xD2 and 0xD3, that resides at the main memory address 11000000 is transferred to the most significant block position of the cache memory at the set address 00 since this position has zero memory references compared to the least significant block position. The tag memory contents at this set address are updated with 1100 with the valid bit is set to logic 1, and the dirty bit is set to logic 0.

Exploring the Least-Recently-Used (LRU) Replacement Algorithm

In the previous sections, we proposed a simple algorithm to replace data blocks with minimum number of references to help the reader understand how the cache controller works when a cache miss occurs. The real-life data replacement policy in caches, however, is based on a slightly different algorithm than what we have proposed. The Least-Recently-Used (LRU) algorithm is based on a policy to replace a data block with least amount of usage and not referenced recently in a set.

Example 6.21 Examine the example in Fig. 6.195 to understand the LRU algorithm for write-through caches.

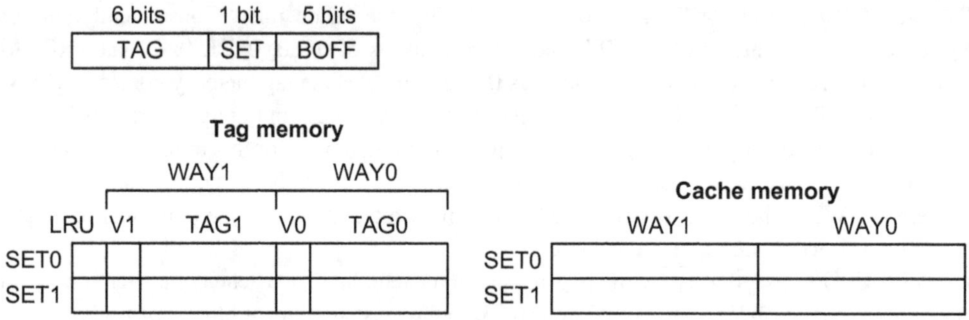

Transaction No	CPU Address	TAG	SET	BOFF
1	0x070	000001	1	10000
2	0x080	000010	0	00000
3	0x068	000001	1	01000
4	0x190	000110	0	10000
5	0x084	000010	0	00100
6	0x178	000101	1	11000
7	0x08C	000010	0	01100
8	0xF00	111100	0	00000
9	0x064	000001	1	00100

Fig. 6.195 A two-way set-associative write-through cache with LRU and pending transactions

This cache is assumed to be a two-way set-associative, write-through cache with a total memory capacity of 128 bytes. It has 32-byte block size and communicates with the CPU using a 12-bit address bus.

Therefore, we can determine the following entries in the CPU address:

Block offset = 5 bits to signify each byte of a 32-byte block size
Set = 128 total bytes/(32 bytes per block × 2 ways) = 128/64 = 2 sets
Tag = 12 − (5 + 1) = 6 bits

The most significant six bits in the CPU address are reserved for the tag field. The one-bit index field indicates that there are only two sets in the cache as shown above. The least significant five bits represent the block offset field. The cache memory consists of two adjacent blocks in WAY 1 and WAY 0 positions, each of which contains 32 words. The tag memory has also two adjacent tag entries with valid bits and a LRU bit for each set.

Since this example concentrates on how the LRU mechanism works, only the tag memory is shown after each transaction in Fig. 6.196, skipping the contents of the cache memory.

The first CPU transaction delivers a CPU address, 0x070, which produces a tag, 000001 and a block offset, 10000, referencing set 1. Since this first CPU tag does not match with the current tag of 000000, a miss occurs, and the least significant tag is replaced with 000001. This is because the LRU bit is initially set at zero, which shows the least significant tag position. The LRU bit in this set transitions to logic 1, signifying to replace the most significant tag contents next time another miss occurs.

The second transaction produces a tag value of 000010 with a block offset, 00000, pointing set 0. Just like the first transaction, this transaction produces a miss and replaces the least significant tag contents in set 0 with 000010. The LRU bit for this set is also set to logic 1.

Transaction 1: Cache miss

	LRU	V1	TAG1	V0	TAG0
SET0	0	0	-	0	-
SET1	1	0	-	1	000001

This bit indicates which block to replace next time

Transaction 2: Cache miss

	LRU	V1	TAG1	V0	TAG0
SET0	1	0	-	1	000010
SET1	1	0	-	1	000001

Transaction 3: Cache hit

	LRU	V1	TAG1	V0	TAG0
SET0	1	0	-	1	000010
SET1	1	0	-	1	000001

Transaction 4: Cache miss

	LRU	V1	TAG1	V0	TAG0
SET0	0	1	000110	1	000010
SET1	1	0	-	1	000001

Transaction 5: Cache hit

	LRU	V1	TAG1	V0	TAG0
SET0	0	1	000110	1	000010
SET1	1	0	-	1	000001

Transaction 6: Cache miss

	LRU	V1	TAG1	V0	TAG0
SET0	0	1	000110	1	000010
SET1	0	1	000101	1	000001

Transaction 7: Cache hit

	LRU	V1	TAG1	V0	TAG0
SET0	0	1	000110	1	000010
SET1	0	1	000101	1	000001

Transaction 8: Cache miss

	LRU	V1	TAG1	V0	TAG0
SET0	1	1	000110	1	111100
SET1	0	1	000101	1	000001

Transaction 9: Cache hit

	LRU	V1	TAG1	V0	TAG0
SET0	1	1	000110	1	111100
SET1	0	1	000101	1	000001

Fig. 6.196 A two-way set-associative write-through cache during transactions

The third transaction points set 1 with a tag value of 000001. This is a hit, and therefore no change takes place in the tag memory. The cache memory contents, however, would change if this transaction were a write.

The fourth transaction produces a new tag value, 000110, targeting set 0, and it creates a miss. The new tag is stored in the most significant tag position according to the current LRU bit, which promptly transitions to logic 0, indicating the least significant tag to be replaced next time a miss occurs.

Transaction five produces another hit and does not change the tag contents. Transaction six stores a new tag value of 000101 in the most significant tag position in set 1, producing LRU = 0. Transaction seven is another hit just like transactions three and five, and does not change the tag contents. Transaction eight creates a miss and points at set 0. The new tag value in this set, 111100, replaces the old value, 000010, switching the LRU bit to logic 0. The last transaction is another hit, and the tag contents remain the same.

Example 6.22 Examine the data transactions between the CPU and a set-associative write-back cache to understand how the LRU-based write-back cache operates.

Assume the following four-way set associative cache with a write back buffer in Fig. 6.197. This cache has four sets, and each set contains two eight-bit words which define the data block size. The CPU address is, therefore, divided into three segments as discussed before. The most significant five bits in the address field belongs to the tag. The index or the set is shown by the middle two bits followed by the block offset bit at the least significant bit position.

Assume the CPU carries out 15 data transactions with the main memory. The transactions details and the main memory contents prior to these transactions are shown in Fig. 6.198.

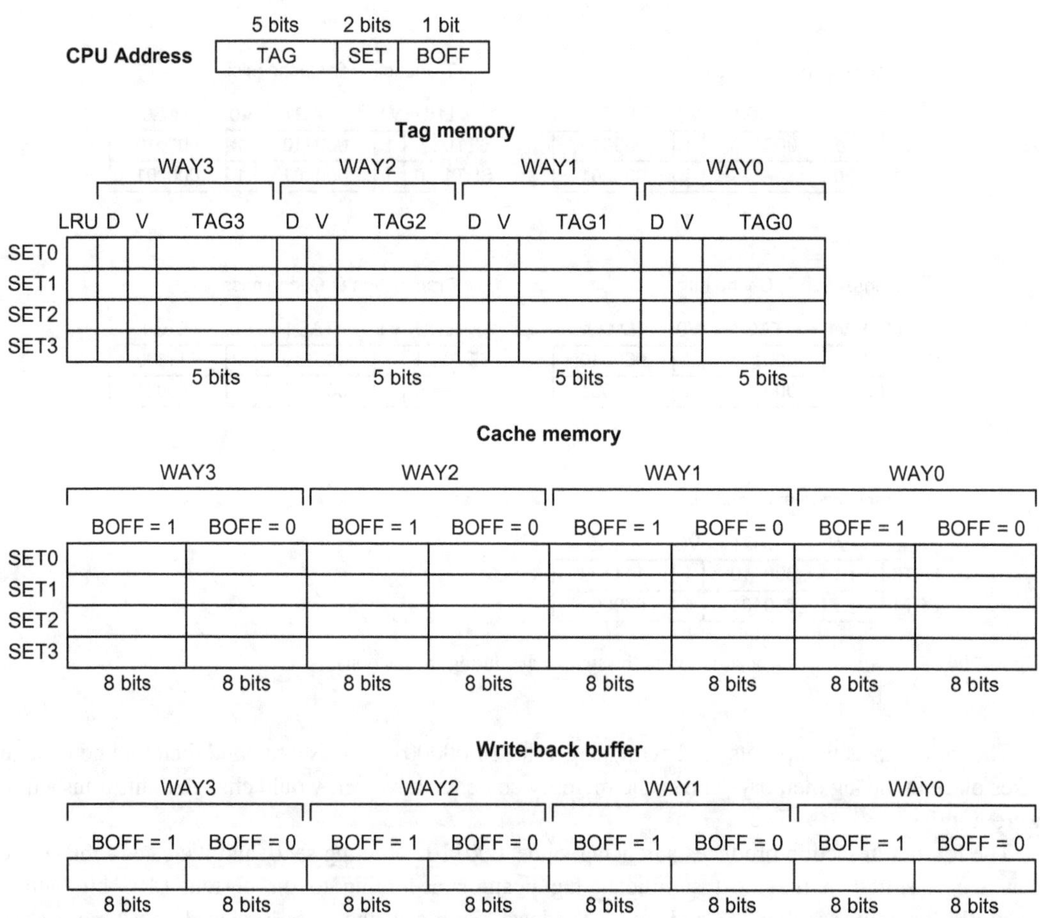

Fig. 6.197 A four-way set-associative write-back cache with LRU

Transaction No	R/W	CPU Address	CPU Data
1	R	00001100	
2	R	10000100	
3	R	11000100	
4	R	11111100	
5	R	00001110	
6	R	10100110	
7	R	11100110	
8	R	00010110	
9	W	00010100	0xA1
		00010101	0xA0
10	W	11110100	0xB1
		11110101	0xB0
11	W	11110100	0xC1
		11110101	0xC0
12	W	00001100	0xD1
		00001101	0xD0
13	W	11011100	0xE1
		11011101	0xE0
14	W	11111100	0xF1
		11111101	0xF0
15	W	11110100	0xF3
		11110101	0xF2

CPU Address	Main Memory
. . .	. . .
00001100	0X11
00001101	0X10
00001110	0X51
00001111	0X50
. . .	. . .
00010110	0X81
00010111	0X80
. . .	. . .
10000100	0X21
10000101	0X20
. . .	. . .
10100110	0X61
10100111	0X60
. . .	. . .
11000100	0X31
11000101	0X30
. . .	. . .
11100110	0X71
11100111	0X70
. . .	. . .
11111100	0X41
11111101	0X40
. . .	. . .

Fig. 6.198 Pending CPU transactions (left) and main memory contents (right)

The first two read transactions result in two consecutive cache misses. The cache controller, therefore, retrieves the data block, 0x11 and 0x10, from the main memory address, 00001100, and then the data block, 0x21 and 0x20, from the main memory address, 10000100, and brings them to the cache. The contents of the tag and the cache memory are shown in Fig. 6.199.

Tag memory

	WAY3				WAY2			WAY1			WAY0		
	LRU	D	V	TAG3	D	V	TAG2	D	V	TAG1	D	V	TAG0
SET0	0	0	0	-	0	0	-	0	0	-	0	0	-
SET1	0	0	0	-	0	0	-	0	0	-	0	0	-
SET2	2	0	0	-	0	0	-	0	1	10000	0	1	00001
SET3	0	0	0	-	0	0	-	0	0	-	0	0	-
				5 bits			5 bits			5 bits			5 bits

Cache memory

	WAY3		WAY2		WAY1		WAY0	
	BOFF = 1	BOFF = 0	BOFF = 1	BOFF = 0	BOFF = 1	BOFF = 0	BOFF = 1	BOFF = 0
SET0	-	-	-	-	-	-	-	-
SET1	-	-	-	-	-	-	-	-
SET2	-	-	-	-	0x20	0x21	0x10	0x11
SET3	-	-	-	-	-	-	-	-
	8 bits	8 bits	8 bits	8 bits	8 bits	8 bits	8 bits	8 bits

Write-back buffer

WAY3		WAY2		WAY1		WAY0	
BOFF = 1	BOFF = 0	BOFF = 1	BOFF = 0	BOFF = 1	BOFF = 0	BOFF = 1	BOFF = 0
-	-	-	-	-	-	-	-
8 bits	8 bits	8 bits	8 bits	8 bits	8 bits	8 bits	8 bits

Fig. 6.199 After the second transaction

The next six read transactions also end up with cache misses. The cache controller brings the blocks, 0x31 and 0x 30, 0x41 and 0x40, 0x51 and 0x50, 0x61 and 0x60, 0x71 and 0x70, and finally the block, 0x81 and 0x80, from the main memory, and places them in the cache. The updated tag and cache memory contents are shown in Fig. 6.200.

Tag memory

	WAY3				WAY2			WAY1			WAY0		
	LRU	D	V	TAG3	D	V	TAG2	D	V	TAG1	D	V	TAG0
SET0	0	0	0	-	0	0	-	0	0	-	0	0	-
SET1	0	0	0	-	0	0	-	0	0	-	0	0	-
SET2	0	0	1	11111	0	1	11000	0	1	10000	0	1	00001
SET3	0	0	1	00010	0	1	11100	0	1	10100	0	1	00001
				5 bits			5 bits			5 bits			5 bits

Cache memory

	WAY3		WAY2		WAY1		WAY0	
	BOFF = 1	BOFF = 0	BOFF = 1	BOFF = 0	BOFF = 1	BOFF = 0	BOFF = 1	BOFF = 0
SET0	-	-	-	-	-	-	-	-
SET1	-	-	-	-	-	-	-	-
SET2	0x40	0x41	0x30	0x31	0x20	0x21	0x10	0x11
SET3	0x80	0x81	0x70	0x71	0x60	0x61	0x50	0x51
	8 bits	8 bits	8 bits	8 bits	8 bits	8 bits	8 bits	8 bits

Write-back buffer

WAY3		WAY2		WAY1		WAY0	
BOFF = 1	BOFF = 0	BOFF = 1	BOFF = 0	BOFF = 1	BOFF = 0	BOFF = 1	BOFF = 0
-	-	-	-	-	-	-	-
8 bits	8 bits	8 bits	8 bits	8 bits	8 bits	8 bits	8 bits

Fig. 6.200 After the eighth transaction

The first two write transactions also result in two cache misses as shown in Fig. 6.201. After the first write miss, the cache controller places the block, 0xA1 and 0xA0, at the least significant cache position (WAY 0) because the LRU bit was equal to zero at the end of the eighth transaction. Once the cache controller replaces the old block, the LRU bit increments by one, and the dirty bit is set to logic 1 because the new data block, 0xA1 and 0xA0, does not yet exists in the main memory. This block is also written to the write-back buffer.

The next block, 0xB1 and 0xB0, is similarly placed at the WAY 1 position in the cache and in the write-back buffer because the LRU bit was equal to one after the ninth transaction. Since this block

Tag memory

	WAY3				WAY2			WAY1			WAY0		
	LRU	D	V	TAG3	D	V	TAG2	D	V	TAG1	D	V	TAG0
SET0	0	0	0	-	0	0	-	0	0	-	0	0	-
SET1	0	0	0	-	0	0	-	0	0	-	0	0	-
SET2	2	0	1	11111	0	1	11000	1	1	11110	1	1	00010
SET3	0	0	1	00010	0	1	11100	0	1	10100	0	1	00001
				5 bits			5 bits			5 bits			5 bits

Cache memory

	WAY3		WAY2		WAY1		WAY0	
	BOFF = 1	BOFF = 0	BOFF = 1	BOFF = 0	BOFF = 1	BOFF = 0	BOFF = 1	BOFF = 0
SET0	-	-	-	-	-	-	-	-
SET1	-	-	-	-	-	-	-	-
SET2	0x40	0x41	0x30	0x31	0xB0	0xB1	0xA0	0xA1
SET3	0x80	0x81	0x70	0x71	0x60	0x61	0x50	0x51
	8 bits	8 bits	8 bits	8 bits	8 bits	8 bits	8 bits	8 bits

Write-back buffer

WAY3		WAY2		WAY1		WAY0	
BOFF = 1	BOFF = 0	BOFF = 1	BOFF = 0	BOFF = 1	BOFF = 0	BOFF = 1	BOFF = 0
-	-	-	-	0xB0	0xB1	0xA0	0xA1
8 bits	8 bits	8 bits	8 bits	8 bits	8 bits	8 bits	8 bits

Fig. 6.201 After the tenth transaction

does not exist in the main memory, the associated dirty bit is also set to logic 1. The LRU bit increments by one after this transaction, and equals to two as shown in this figure.

In transaction 11, the CPU attempts to write the block, 0xC1 and 0xC0, to the main memory as shown in Fig. 6.202. However, one of the tag values in set 2 matches the CPU tag of 11110 in this transaction, resulting a write hit. Consequently, the cache controller overwrites the new block, 0xC1 and 0xC0, on top of the old one, 0xB1 and 0xB0, both in the cache and the write-back buffer.

Note that the LRU bit stays at two because there was no data replacement in the cache.

Tag memory

	WAY3				WAY2			WAY1			WAY0		
	LRU	D	V	TAG3	D	V	TAG2	D	V	TAG1	D	V	TAG0
SET0	0	0	0	-	0	0	-	0	0	-	0	0	-
SET1	0	0	0	-	0	0	-	0	0	-	0	0	-
SET2	2	0	1	11111	0	1	11000	1	1	11110	1	1	00010
SET3	0	0	1	00010	0	1	11100	0	1	10100	0	1	00001
				5 bits			5 bits			5 bits			5 bits

Cache memory

	WAY3		WAY2		WAY1		WAY0	
	BOFF = 1	BOFF = 0	BOFF = 1	BOFF = 0	BOFF = 1	BOFF = 0	BOFF = 1	BOFF = 0
SET0	-	-	-	-	-	-	-	-
SET1	-	-	-	-	-	-	-	-
SET2	0x40	0x41	0x30	0x31	0xC0	0xC1	0xA0	0xA1
SET3	0x80	0x81	0x70	0x71	0x60	0x61	0x50	0x51
	8 bits	8 bits	8 bits	8 bits	8 bits	8 bits	8 bits	8 bits

Write-back buffer

WAY3		WAY2		WAY1		WAY0	
BOFF = 1	BOFF = 0	BOFF = 1	BOFF = 0	BOFF = 1	BOFF = 0	BOFF = 1	BOFF = 0
-		-	-	0xC0	0xC1	0xA0	0xA1
8 bits	8 bits	8 bits	8 bits	8 bits	8 bits	8 bits	8 bits

Fig. 6.202 After the eleventh transaction

In transactions 12 and 13, the CPU again attempts to write two new blocks to the main memory. Both of these writes result in cache misses as shown in Fig. 6.203. The first block, 0xD1 and 0xD0, is placed at the WAY 2 position of the cache because the LRU bit was equal to two at the end of the eleventh transaction. The second block, 0xE1 and 0xE0, is similarly placed at the WAY 3 position of the cache because the LRU bit was incremented to three at the end of the twelfth transaction.

Tag memory

	WAY3				WAY2			WAY1			WAY0		
	LRU	D	V	TAG3	D	V	TAG2	D	V	TAG1	D	V	TAG0
SET0	0	0	0	-	0	0	-	0	0	-	0	0	-
SET1	0	0	0	-	0	0	-	0	0	-	0	0	-
SET2	0	1	1	11011	1	1	00001	1	1	11110	1	1	00010
SET3	0	0	1	00010	0	1	11100	0	1	10100	0	1	00001
				5 bits			5 bits			5 bits			5 bits

Cache memory

	WAY3		WAY2		WAY1		WAY0	
	BOFF = 1	BOFF = 0	BOFF = 1	BOFF = 0	BOFF = 1	BOFF = 0	BOFF = 1	BOFF = 0
SET0	-	-	-	-	-	-	-	-
SET1	-	-	-	-	-	-	-	-
SET2	0xE0	0xE1	0xD0	0xD1	0xC0	0xC1	0xA0	0xA1
SET3	0x80	0x81	0x70	0x71	0x60	0x61	0x50	0x51
	8 bits	8 bits	8 bits	8 bits	8 bits	8 bits	8 bits	8 bits

Write-back buffer

WAY3		WAY2		WAY1		WAY0	
BOFF = 1	BOFF = 0	BOFF = 1	BOFF = 0	BOFF = 1	BOFF = 0	BOFF = 1	BOFF = 0
0xE0	0xE1	0xD0	0xD1	0xC0	0xC1	0xA0	0xA1
8 bits	8 bits	8 bits	8 bits	8 bits	8 bits	8 bits	8 bits

Fig. 6.203 After the thirteenth transaction

In the fourteenth and the fifteenth transactions, the CPU attempts to write two more data blocks, 0xF1 and 0xF0, and then, 0xF3 and 0xF2, to the main memory, respectively.

The first transaction results in a cache miss, and forces the cache controller to replace the old data block at the WAY 0 position because the LRU bit was set to zero after the last transaction in Fig. 6.204. Since the old block, 0xA1 and 0xA0, is a valid data block and has no duplicate in the main memory, the cache controller first transfers this block to the main memory. Once this transfer is complete, the cache controller writes the new block, 0xF1 and 0xF0, on top of the old one, 0xA1 and 0xA0, both in the cache and the write-back buffer. At the end of this transaction, the LRU bit increments by one, and the dirty bit is also set to logic 1 because the new block does not exist in the main memory.

Tag memory

	WAY3				WAY2			WAY1			WAY0		
	LRU	D	V	TAG3	D	V	TAG2	D	V	TAG1	D	V	TAG0
SET0	0	0	0	-	0	0	-	0	0	-	0	0	-
SET1	0	0	0	-	0	0	-	0	0	-	0	0	-
SET2	1	1	1	11011	1	1	00001	1	1	11110	1	1	11111
SET3	0	0	1	00010	0	1	11100	0	1	10100	0	1	00001
				5 bits			5 bits			5 bits			5 bits

Cache memory

	WAY3		WAY2		WAY1		WAY0	
	BOFF = 1	BOFF = 0	BOFF = 1	BOFF = 0	BOFF = 1	BOFF = 0	BOFF = 1	BOFF = 0
SET0	-	-	-	-	-	-	-	-
SET1	-	-	-	-	-	-	-	-
SET2	0xE0	0xE1	0xD0	0xD1	0xF2	0xF3	0xF0	0xF1
SET3	0x80	0x81	0x70	0x71	0x60	0x61	0x50	0x51
	8 bits	8 bits	8 bits	8 bits	8 bits	8 bits	8 bits	8 bits

Write-back buffer

WAY3		WAY2		WAY1		WAY0	
BOFF = 1	BOFF = 0	BOFF = 1	BOFF = 0	BOFF = 1	BOFF = 0	BOFF = 1	BOFF = 0
0xE0	0xE1	0xD0	0xD1	0xF2	0xF3	0xF0	0xF1
8 bits	8 bits	8 bits	8 bits	8 bits	8 bits	8 bits	8 bits

Fig. 6.204 After the fifteenth transaction

The last write transaction ends up with a write hit because the CPU tag matches one of the tag entries in set 2. As a result, the block, 0xF3 and 0xF2, is written both to the WAY 1 position of the cache memory and the write-back buffer as shown in the same figure.

Writing to Cache Memory After a Cache Hit

This section explains how to write data blocks to different types of cache structures, which can be often confusing to a new reader.

In the first example, we will write to a direct-mapped cache with eight sets (index = 3) with no block offsets if the transaction produces a hit.

Let us assume that the CPU produces a memory address of 11010110, and attempts to write a decimal value of 21700 to the set, 110 (the least significant three bits of the memory address). Also assume the tag memory contents at this address are already 11010 (the most significant five bits of the memory address) as shown in Fig. 6.205, and a decimal value of 42000 is stored both in the cache and the main memory, producing the valid bit, V = 1.

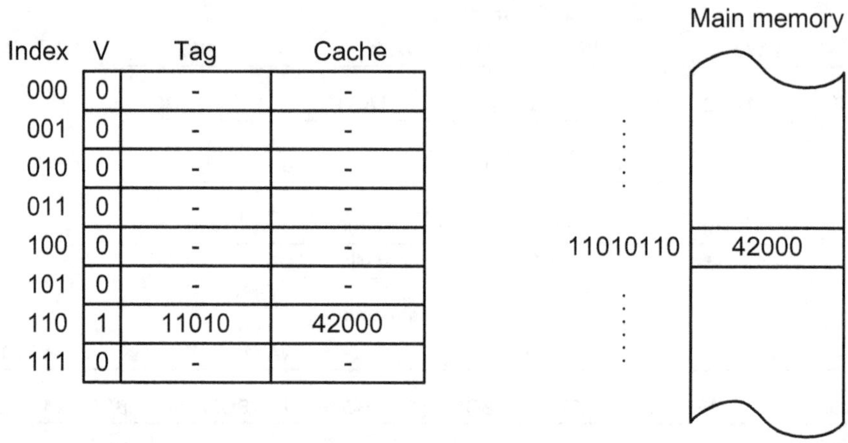

Fig. 6.205 Initial contents of a cache, tag and main memory prior to writing data

If the cache organization is a write-through type as shown in Fig. 6.206, the CPU data, 21700, is simply written to the cache and the main memory at the same time following a hit.

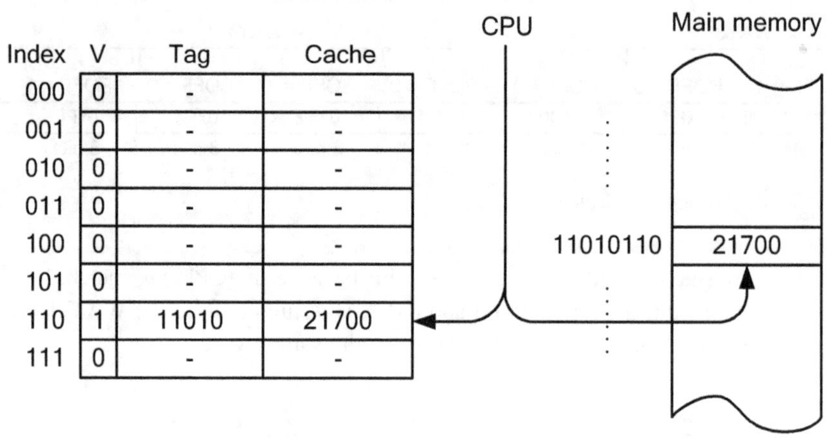

Fig. 6.206 Writing data to a write-through type cache following a hit

However, if the cache is a write-back type, the CPU data, 21700, is first written to the cache memory and its write-back buffer to shorten the wait time to complete the write process as shown in Fig. 6.207. To signify that the new data only exists in the cache, the dirty bit transitions to logic 1. The cache controller transfers the data from the write-back buffer to the main memory only when the CPU tries access data (read or write) at the same cache address, but misses as shown by dashed lines. Until this event takes place, the cache and the main memory will have different data values. After the data transfer, the dirty bit transitions back to logic 0, completing the write cycle.

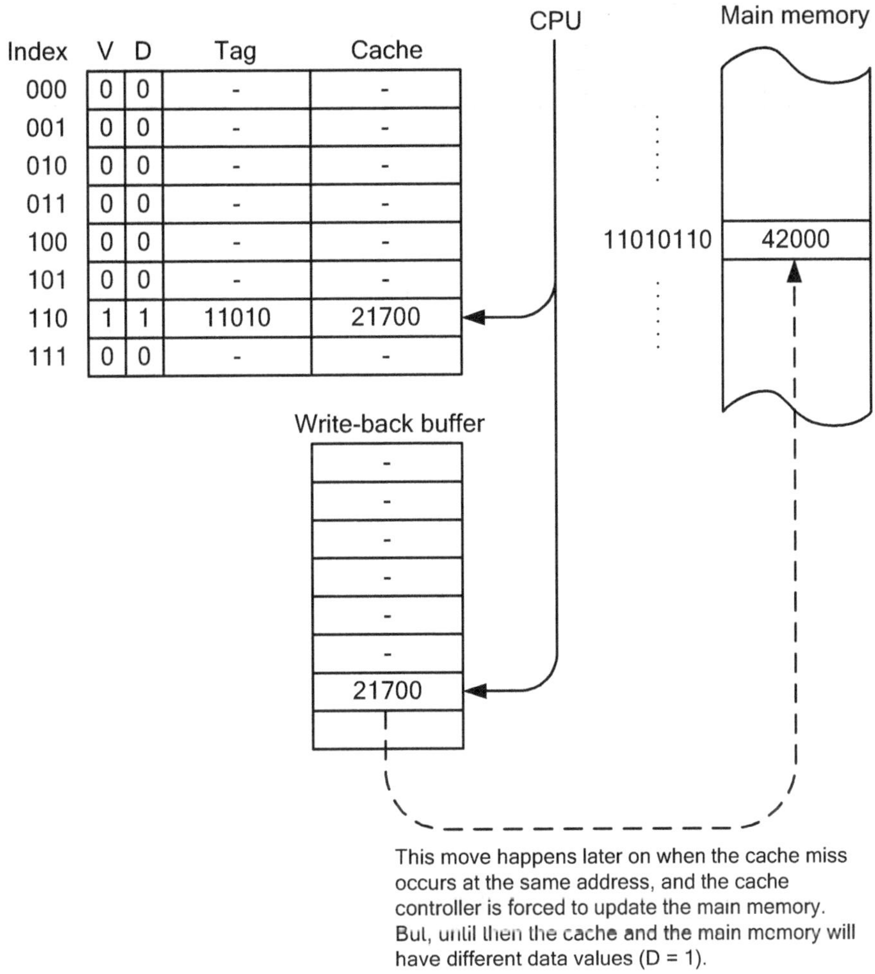

Fig. 6.207 Writing data to a write-back type cache following a hit

The second example in Fig. 6.208 clarifies the write-back operation when the cache misses a read transaction.

In this example, assume that the set, 110, has a tag value of 11010 and a cache value of 21700, which also resides in the write back buffer due a prior write miss. Also assume that the CPU wants to read from the memory location, 10001110, corresponding to the same set, 110. Since the CPU tag, 10001, does not match the existing tag, 11010, in the cache memory, the cache controller issues a read miss. Now, the data, 12000, that resides at the memory address, 10001110, has to be fetched from this location and brought to the cache. But, before this transfer takes place the cache controller first transfers the valid data, 21700, from the write-back buffer to the main memory address, 11010110, overwriting the existing data, 42000. Once this step is complete, then the cache controller copies the data, 12000, from the memory address, 10001110, and overwrites it over 21700 in the cache.

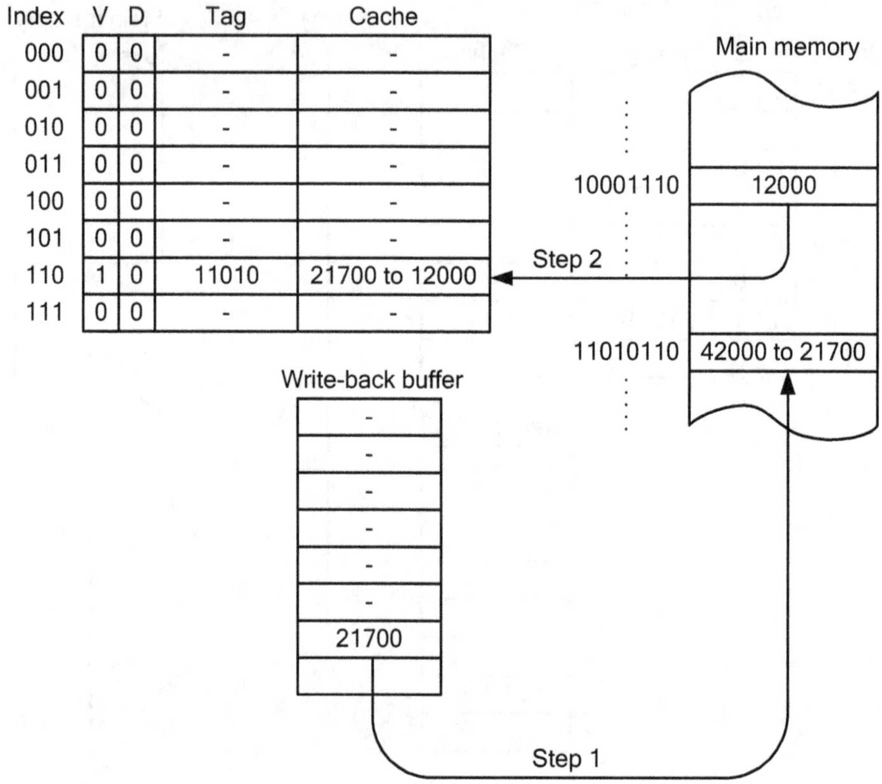

Fig. 6.208 Write-back action in sequence

Writing to Cache Memory After a Cache Miss

How does the CPU write a block of data to the cache after a miss?

To answer this question, let us assume the cache has eight sets with no block offsets as in the previous examples, and the initial values in the tag and the cache memories are 00010 and 36000 at the set 110, respectively. Let us also assume that the CPU intends to write the data, 21700, to the memory address, 11010110, but produces a write miss due to tag mismatch.

If this is a write-through cache, the cache controller first writes the new data, 21700, to the main memory address, 11010110, and then transfers the same data to the set location, 110, in the cache as shown in Fig. 6.209. The data transfer from the CPU to the main memory and the cache may also take place simultaneously depending on the cache architecture as shown in the same figure. This style of writing is called write allocate protocol.

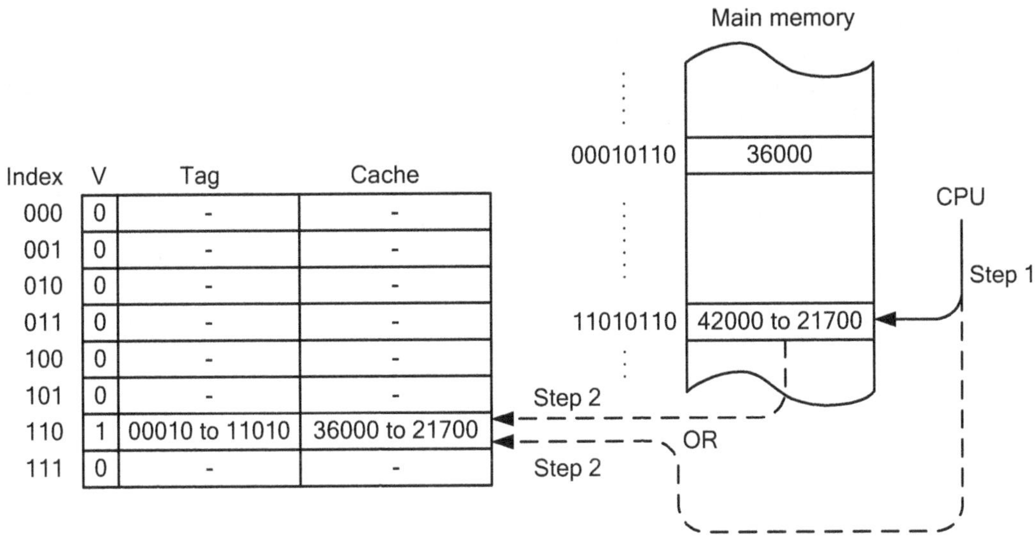

Fig. 6.209 Writing data to a write-through type cache following a miss with write-allocate protocol

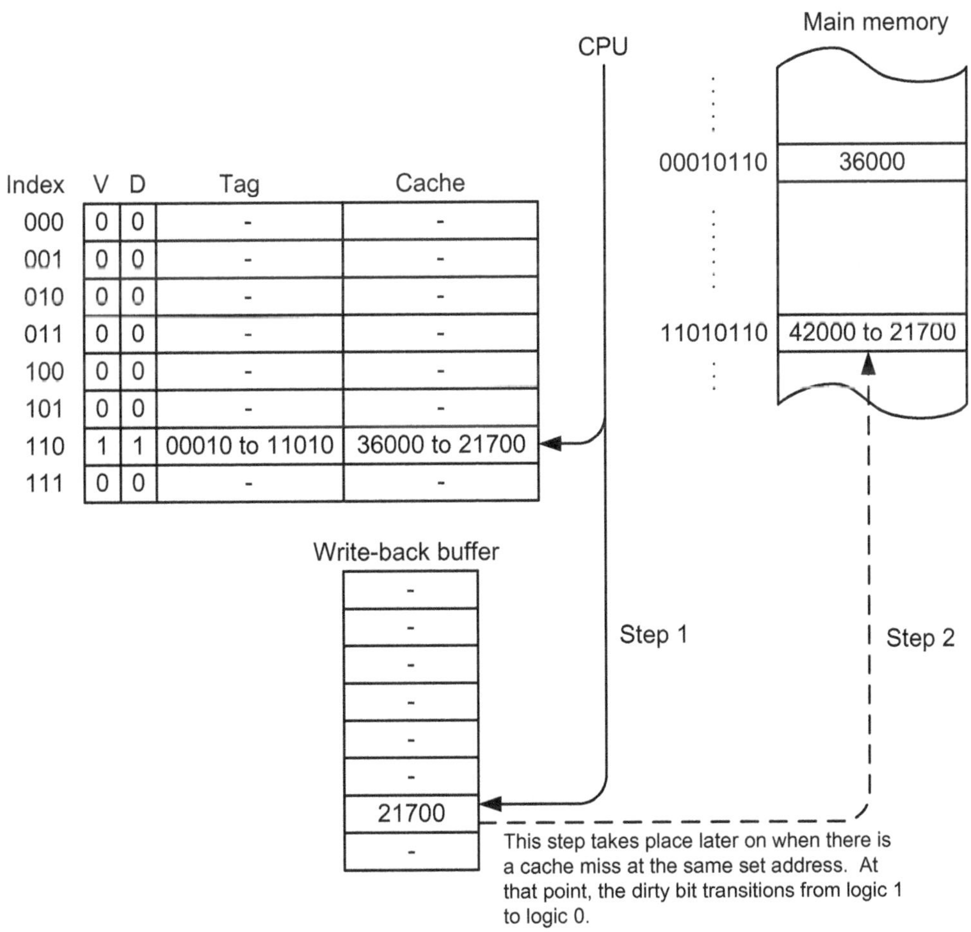

Fig. 6.210 Writing data to a write-back type cache following a miss with write-allocate protocol

If the cache is a write-back cache, the first step of the write-allocate protocol after a miss consists of writing the new tag, 11010, to the set address, 110, and at the same time writing the CPU data, 21700, to the cache and the write-back buffer as shown in Fig. 6.210. Updating the main memory, and therefore overwriting the old data, 42000, at the address 11010110 takes place later on when there is a cache miss at the same set. Until that point, the data memory and the cache retain different values with the dirty bit set to logic 1. When the write-back buffer contents are finally transferred to the main memory, the dirty bit transitions from logic 1 to logic 0, completing the write cycle.

The write-around protocol is a different approach to reduce the amount of data storage in the cache memory. According to this protocol, the CPU updates only the main memory in a write-through cache, but skips updating the cache altogether following a miss. Therefore, the CPU writes the new data, 21700, over the old data, 42000, at the main memory address 11010110 as shown in Fig. 6.211, but no cache update takes place.

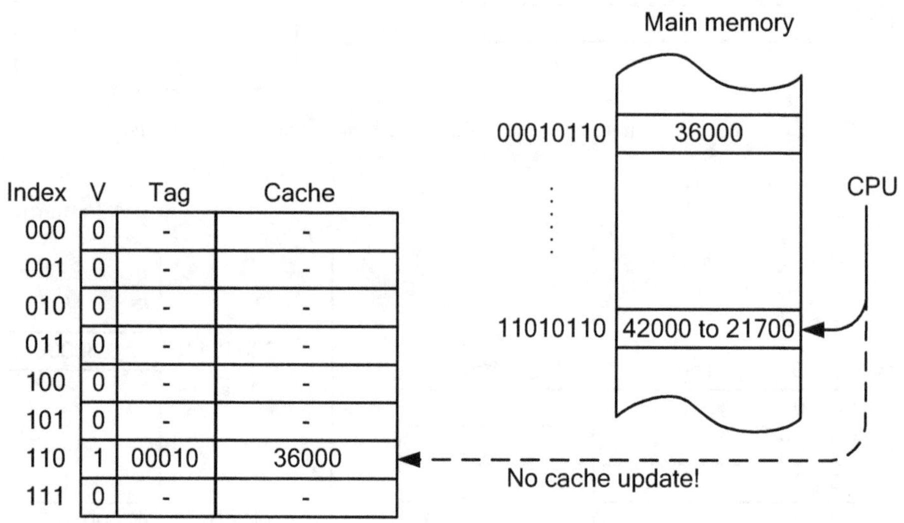

Fig. 6.211 Writing data to a write-through type cache following a miss with write-around protocol

To protect the cache memory from filling up too quickly, some processors with write-allocate type caches are equipped with special store instructions to activate the write-around mechanism. The for-loop function below fills up the cache memory rather quickly.

```
for (i = 0; i < n; i ++)
x[i] = i;
```

Here, n is assumed to be a large integer. Therefore, to prevent congesting the cache, the CPU instruction set may offer a store instruction to omit updating the cache.

Appendix: Iterative Fixed-Point Multiplication

The MUL instruction and its operational equation shown earlier in this chapter is rewritten below:

MUL RS1, RS2, RD1, RD2

Reg[RS1] * Reg[RS2] → {Reg[RD2], Reg[RD1]}

According to the equation above, the contents of the source registers, RS1 and RS2, are multiplied by a 32-bit fixed-point multiplier, and the 64-bit result is divided into two parts. The most significant 32-bit of the result is returned to RD2, and the least significant 32-bit result to RD1.

The iterative method suggests an algorithm that uses only SLI and ADD instructions. This algorithm does not require any additional hardware for the CPU; however, it takes many clock cycles to complete. The algorithm for a four-bit multiplication is shown in Figs. 6.212 and 6.213. The same method can be extended to 32 bits.

					c3	c2	c1	c0	← c
				×	r3	r2	r1	r0	← r
					c3 r0	c2 r0	c1 r0	c0 r0	← pp0
				c3 r1	c2 r1	c1 r1	c0 r1		← pp1
			c3 r2	c2 r2	c1 r2	c0 r2			← pp2
+		c3 r3	c2 r3	c1 r3	c0 r3				← pp3
	s7	s6	s5	s4	s3	s2	s1	s0	← result

i = 0

If r0 = 1 then c3 c2 c1 c0 = pp0 → pp

If r0 = 0 then 0 0 0 0 = pp0 → pp

i = 1

STEP 1 If r1 = 1 then c3 c2 c1 c0 = pp1

If r1 = 0 then 0 0 0 0 = pp1

STEP 2 pp1 << 1 → pp1 thus: c3 c2 c1 c0 0 = pp1 if r1 = 1

0 0 0 0 0 = pp1 if r1 = 0

STEP 3 pp1 + pp → pp thus, if r1 = 1

	c3	c2	c1	c0	0
+ pp →	0	p3	p2	p1	p0
	s4	s3	s2	s1	s0

if r1 = 0

	0	0	0	0	0
+ pp →	0	p3	p2	p1	p0
	s4	s3	s2	s1	s0

Fig. 6.212 An iterative fixed-point multiplication algorithm

```
i = 2

STEP 1    If r2 = 1      then   c3 c2 c1 c0 = pp2
          If r2 = 0      then    0  0  0  0 = pp2

STEP 2    pp2 << 2 → pp2    thus:   c3 c2 c1 c0 0 0 = pp2  if r2 = 1
                                     0  0  0  0 0 0 = pp2  if r2 = 0

STEP 3    pp2 + pp → pp     thus, if r2 = 1            c3 c2 c1 c0  0  0
                                               pp →     0 p4 p3 p2 p1 p0
                                             +  ------------------------
                                                       s5 s4 s3 s2 s1 s0

                                  if r2 = 0             0  0  0  0  0  0
                                               pp →     0 p4 p3 p2 p1 p0
                                             +  ------------------------
                                                       s5 s4 s3 s2 s1 s0
i = 3

STEP 1    If r3 = 1      then   c3 c2 c1 c0 = pp3
          If r3 = 0      then    0  0  0  0 = pp3

STEP 2    pp3 << 3 → pp3    thus:   c3 c2 c1 c0 0 0 0 = pp3  if r3 = 1
                                     0  0  0  0 0 0 0 = pp3  if r3 = 0

STEP 3    pp3 + pp → pp     thus, if r3 = 1            c3 c2 c1 c0  0  0  0
                                               pp →     0 p5 p4 p3 p2 p1 p0
                                             +  ---------------------------
                                                       s6 s5 s4 s3 s2 s1 s0

                                  if r3 = 0             0  0  0  0  0  0  0
                                               pp →     0 p5 p4 p3 p2 p1 p0
                                             +  ---------------------------
                                                       s6 s5 s4 s3 s2 s1 s0
```

Fig. 6.213 An iterative fixed-point multiplication algorithm

Assume that the quantities, r = {r3 r2 r1 r0} and c = {c3 c2 c1 c0}, represent a four-bit multiplier and a four-bit multiplicand, respectively. Also assume that the term, pp, corresponds to a partial product sum that adds a newly generated partial product to the old partial product in each iteration. The variable, i, represents the iteration index bounded by 0 and (n-1) where n signifies the number multiplier and multiplicand bits.

The first iteration (i = 0) generates the partial product pp0, which is equal to pp0 = {c3 c2 c1 c0} if r0 = 1; otherwise, pp0 becomes equal to {0 0 0 0} as shown in Fig. 6.212.

The second iteration is composed of three steps. The first step evaluates pp1 much like pp0. If r1 = 1, then pp1 = {c3 c2 c1 c0} else pp1 = {0 0 0 0}. In the second step, pp1 is shifted one bit to the left before adding it to pp0. This step produces pp1 = {c3 c2 c1 c0 0} if r1 = 1; otherwise, pp1 = {0

0 0 0 0}. The third step adds pp1 to the old partial product, pp0, and forms the partial product sum, pp = {s4 s3 s2 s1 s0}, as mentioned earlier.

The third and fourth iterations are also composed of three steps as shown in Fig. 6.213. The difference between them is that pp2 and pp3 are shifted to the left by two bits and three bits, respectively. This way, they will be in the correct bit position before they are added to the old partial product.

Finally, generating, left shifting and adding steps of partial products result in a compact flow chart shown in Fig. 6.214. Note that each r-term in this figure is indexed by the variable i. Therefore, r = {r0 r1 r2 r3} is identical to {r[0] r[1] r[2] r[3]}. Similarly, each pp-term uses indexed representations. Therefore, pp = {pp0, pp1, pp2, pp3} is identical to {pp[0], pp[1], pp[2], pp[3]}.

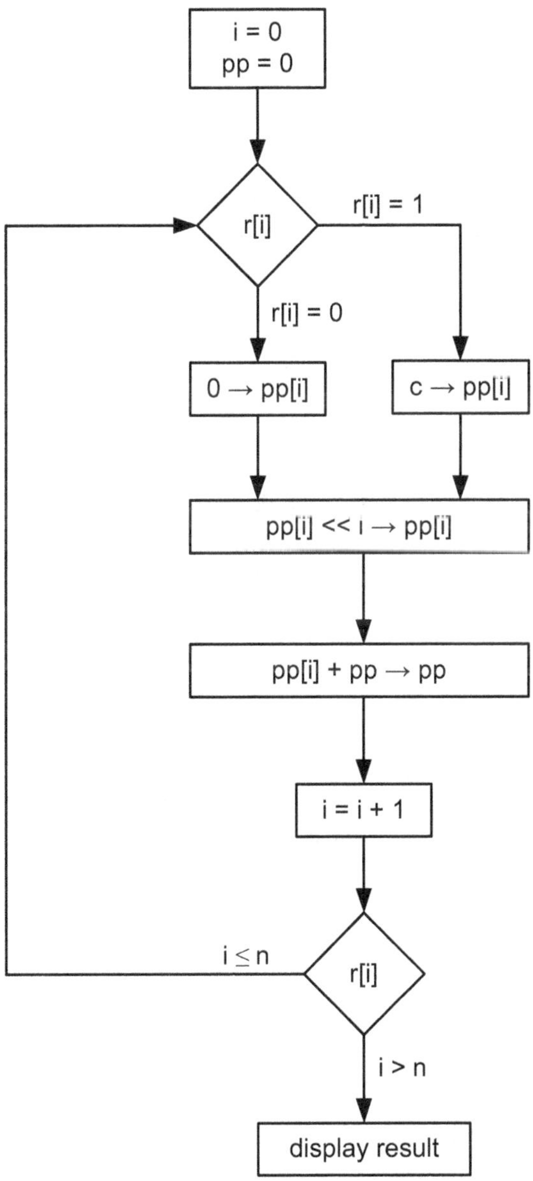

Fig. 6.214 The flow chart as a result of the iterative fixed-point multiplication algorithm

Review Questions

1. A 32-bit RISC CPU organized in Big Endian format has three pipeline stages to execute only the following two instructions:

ADDI RS, RD, Imm :	Reg[RS] + Imm → Reg[RD]
XOR RS1, RS2, RD :	Reg[RS1] ^ Reg[RS2] → Reg[RD]

Draw the detailed ALU and the CPU schematic that executes these two instructions. Label all interconnections, bus widths and control signals.

2. The following specification is given for implementing a 32-bit RISC processor that executes only integer multiply-add (MADD) and add (ADD) instructions:

   (i) Data, a, b, c and d are read at the same time from DOut1, DOut2, DOut3 and DOut4 ports of a 32-bit RF with 32 general purpose registers.
   (ii) There are four stages in the processor. The ALU consists of two stages.
   (iii) Multiplication is the first ALU stage for the MADD instruction between a and b, and between c and d. It takes one clock cycle to produce results which are eventually written to DinH (for higher 32 data input bits) and DinL (for lower 32 data input bits) ports of the RF simultaneously. This stage can be bypassed if addition is performed between a and c.
   (iv) Addition is the second ALU stage, and it also takes one clock cycle to produce results.
   (v) For MADD instruction, RS1 is the first source address that contains a, RS2 is the second source address that contains b, RS3 is the third source address that contains c, and RS4 is the fourth source address that contains d. RD1 is the first destination address that stores the lower 32 bits of the result, and RD2 the second destination address that stores higher 32 bits. For the ADD instruction, RS1 is the first source address that contains a, RS3 is the second address that contains c, and RD1 is the destination address that stores the result.

   (a) This CPU executes only these two instructions. Draw the instruction bit field format, indicating the opcode and operand fields for MADD and ADD instructions.
   (b) Draw the architectural diagram of the processor that executes ADD and MADD, indicating all the necessary hardware such as the required memories, the RF, the detailed ALU with all the port names and bit widths. Show how the opcode decoder enables multiplexers and other hardware in each stage.

Note: The reader should also attempt to implement the hardware that executes the integer multiply (MUL) instruction and superimpose it on top of the data-path that executes ADD and MADD instructions.

3. The area under y = x is calculated until the area equals 18. Here, x increments by one as shown in the figure below.

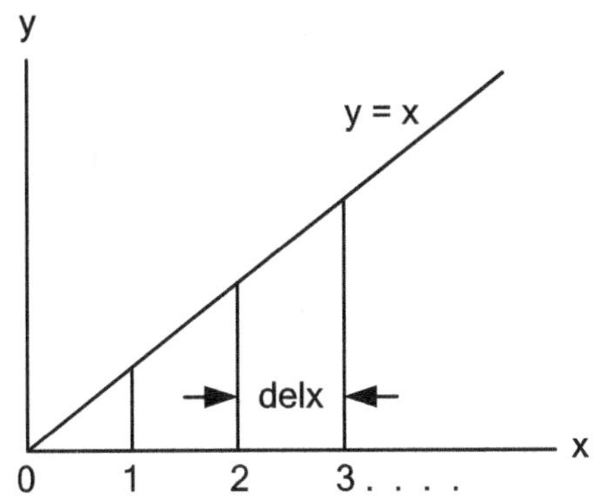

The incremental area is calculated by the flow chart given below.

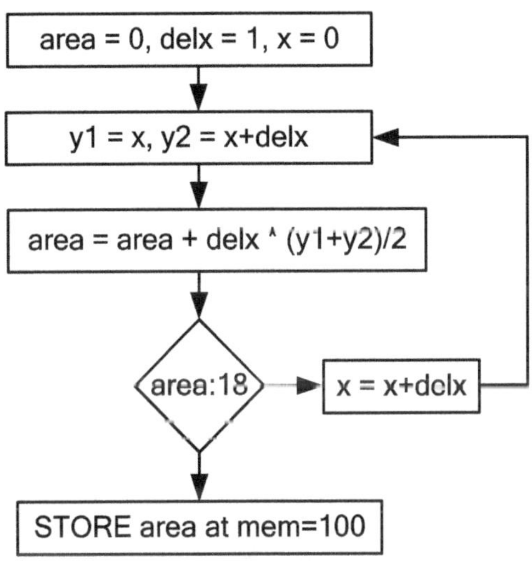

(a) Assuming Reg[R0] = 0, write a program using the instruction set given in Chapter 6. Make comments next to each instruction in the program.

(b) Form an instruction chart for this program, executing in a five-stage CPU, and show all the data dependencies that require forwarding loops. Stall the pipeline using the NOP instruction if necessary. Consider the branch or jump delay penalty to be 1 cycle.

4. A RISC CPU computes the following:

$X = 2\,A^2 + 1$

A is located at the data cache address 100. X needs to be stored at the address 200. All instructions take one cycle except multiply, which takes three cycles. The RF contains only R0 and R1. Reg [R0] = 0.

Make sure to have only 16-bit values in source registers, RS1 and RS2, in order to avoid the overflow condition in the destination register, RD, when the MUL instruction is used.

(a) Write an assembly code to compute and store the value of X. Make sure to write comments next to each instruction to keep track of the register values.
(b) Rewrite the assembly code with an instruction chart. Indicate all stalls caused by NOP instructions and forwarding loops on this chart.

5. Design a four-way set-associative write-through cache for an eight-bit CPU. The cache is organized in Little Endian format. It has four sets, and each data block in the set contains two eight-bit words.

The replacement policy on a cache miss is as follows:

(i) An entire block of data is transferred between the CPU and the cache
(ii) The block of with the fewest amount of references is replaced
(iii) The least significant block is replaced if all the memory references are the same in a set

The CPU transactions and the contents of the main memory before these transactions are shown below:

Address	Contents (bits 7 … 0)
00001010	0x33
00001011	0x44
00001100	0x55
00001101	0x66
	⋮
00100010	0x11
00100011	0x22
	⋮
01111100	0xEE
01111101	0xFF
	⋮
10000000	0x00
10000001	0x00
10000010	0xAA
10000011	0xBB
	⋮
10000110	0xCC
10000111	0xDD
10001000	0x77
10001001	0x88
10001010	0x00
10001011	0x99
	⋮
11011010	0x00
11011011	0x00
	⋮
11111010	0xAB
11111011	0xCD

Transaction no.	CPU Address	Data	Write/Read
1	10000010	-	R
2	10000110	-	R
3	01111100	-	R
4	00100010	-	R
5	00001010	-	R
6	00001100	-	R
7	10001000	-	R
8	10001010	-	R
9	10000000	0xAB	W
	10000001	0xCD	W
10	00001010	0xAB	W
	00001011	0xCD	W
11	11011010	0xAB	W
	11011011	0xCD	W
12	11111010	-	R

(a) Draw the block diagram of the cache and tag memories. Show the field format of the CPU address in terms of tag, index and block offset.
(b) Show the cache and tag memory contents after the eighth, tenth and twelfth transactions by individually drawing the cache and tag contents. Update the main memory contents if there is any change.

6. A 32-bit, five-stage RISC CPU organized in Little Endian format executes the flow chart below. The CPU contains an integer RF with 32 registers where Reg[R0] = 0. The integer values, such as SUM = 0, are stored at the data memory address 100, i = 1 is stored at 101, and the compare value of 100 for i is stored at 102. The final SUM value needs to be stored in the data memory address of 200.

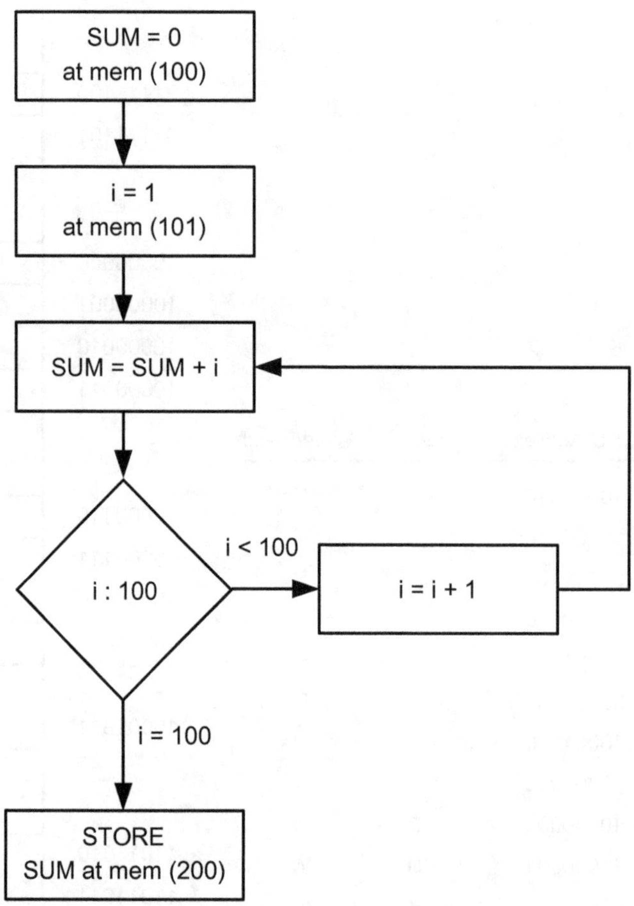

(a) Write an assembly program using the following instruction set. Accompany each instruction in the program with register data and comments.

Instruction Set	Instruction Definition	Instruction Bit Field
LOAD RS, RD, Imm Value	mem {Reg[RS] + Imm} → Reg[RD]	LOAD (31–26), RS (25–21), RD (20–16), Imm Value (15–0)
STORE RS, RD, Imm Value	Reg[RS] → mem {Reg[RD] + Imm}	STOR (31–26), RS (25–21), RD (20–16), Imm Value (15–0)
ADD RS1, RS2, RD	Reg[RS1] + Reg[RS2] → Reg[RD]	ADD (31–26), RS1 (25–21), RS2 (20–16), RD (15–11), Not used (10–0)
ADDI RS, RD, Imm Value	Reg[RS] + Imm Value → Reg[RD]	ADDI (31–26), RS (25–21), RD (20–16), Imm Value (15–0)
SEQ RS1, RS2, RD	If Reg[RS1] = Reg[RS2] then 1 → Reg[RD] else 0 → Reg[RD]	SEQ (31–26), RS1 (25–21), RS2 (20–16), RD (15–11), Not used (10–0)
BNEZ RS, Imm Value	If Reg [RS] ≠ 0 then PC + Imm → PC else PC + 2 → PC	BNEZ (31–26), RS (25–21), NU (20–16), Imm Value (15–0)
JUMP Imm Value	Imm → PC	JUMP (31–26), NU (25–21), NU (20–16), Imm Value (15–0)

(b) Draw the CPU schematic that executes the instructions in the flow chart above.

7. The function, $Y = \frac{5(A - B)}{32}$, needs to be executed using the instruction set below.

Instruction Set	Instruction Definition	Instruction Bit Field
LOAD RS, RD, Imm Value	mem {Reg [RS] + Imm} → Reg [RD]	LOAD (31–26), RS (25–21), RD (20–16), Imm Value (15–0)
STORE RS, RD, Imm Value	Reg [RS] → mem {Reg [RD] + Imm}	STOR (31–26), RS (25–21), RD (20–16), Imm Value (15–0)
ADD RS1, RS2, RD	Reg [RS1] + Reg [RS2] → Reg [RD]	ADD (31–26), RS1 (25–21), RS2 (20–16), RD (15–11), Not used (10–0)
SUB RS1, RS2, RD	Reg [RS1] - Reg [RS2] → Reg [RD]	SUB (31–26), RS1 (25–21), RS2 (20–16), RD (15–11), Not used (10–0)
SLI RS, RD, Imm Value	Reg [RS] << Imm → Reg [RD]	SLI (31–26), RS (25–21), RD (20–16), Imm Value (15–0)
SRI RS, RD, Imm Value	Reg [RS] >> Imm → Reg [RD]	SRI (31–26), RS (25–21), RD (20–16), Imm Value (15–0)

A is located at the memory address 100.
B is located at the memory address 101.
Y needs to be stored at the memory address 102.
Reg[R0] = 0.

(a) Write a program to compute Y.
(b) This program executes in a six-stage CPU. Two clock cycles are required to access data memory for a LOAD operation. Rewrite the program to accommodate this requirement. Show all forwarding loops and include all the necessary NOPs in the instruction chart.
(c) Indicate the minimum number of clock cycles to execute the program in part (b).

8. A 32-bit CPU organized in Big Endian format has 32 general purpose registers (R0 is also a general purpose register whose contents are not zero). This CPU executes the following flow chart:

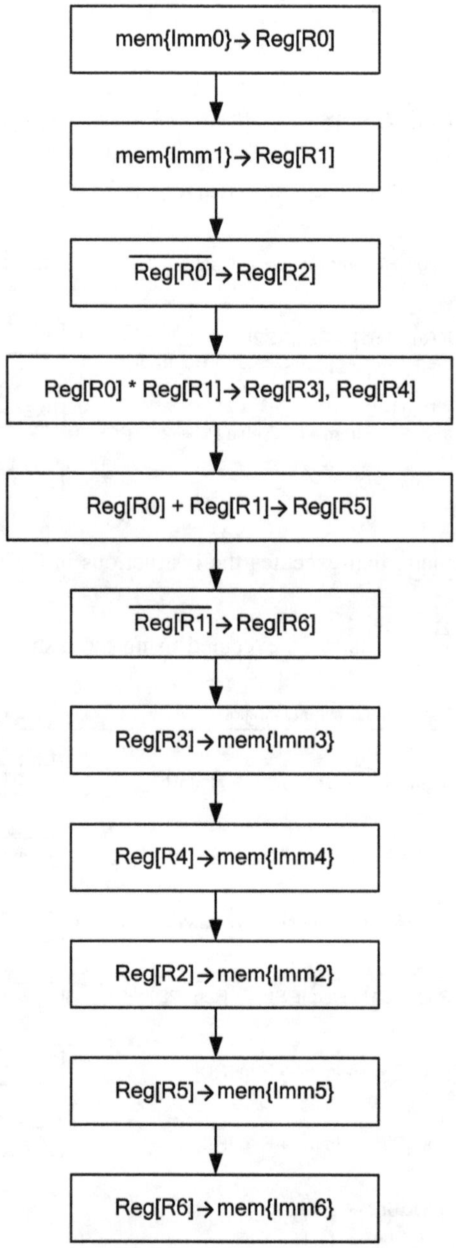

The instruction set and the bit-field format for each instruction are shown below.

Instruction	Definition	Bit field
LOAD Imm Value, RD	mem (Imm Value) → Reg[RD]	0–5: LOAD; 6–10: NU; 11–15: RD; 16–31: Imm Value
INVERT RS, RD	$\overline{\text{Reg[RS]}}$ → Reg[RD]	0–5: INVERT; 6–10: RS; 11–15: RD; 16–31: NU
MUL RS1, RS2, RD1, RD2	Reg[RS1] x Reg[RS2] → Reg[RD1] (lower), Reg[RD2] (upper)	0–5: MUL; 6–10: RS1; 11–15: RS2; 16–20: RD1; 21–25: RD2; 26–31: NU
ADD RS1, RS2, RD	Reg[RS1] + Reg[RS2] → Reg[RD]	0–5: ADD; 6–10: RS1; 11–15: RS2; 16–20: RD; 21–31: NU
STORE RS, Imm Value	Reg[RS] → mem (Imm Value)	0–5: STORE; 6–10: RS; 11–15: NU; 16–31: Imm Value

The CPU maintains the following rules:

(i) Every instruction is executed in a different number of clock cycles
(ii) No NOP instruction is allowed
(iii) LOAD does not have an ALU cycle but requires two data memory cycles
(iv) INVERT does not have a data memory cycle but requires one ALU cycle
(v) MUL does not have a data memory cycle but requires three ALU cycles
(vi) ADD does not have a data memory cycle but requires two ALU cycles
(vii) STORE does not have an ALU cycle but requires one data memory cycle

Construct the instruction chart to execute the flow chart above. Show all the necessary forwarding loops and possible data hazards. Show the cases in which there may be structural hazards and indicate how to prevent them.

9. The following instruction set needs to be executed in a 32-bit RISC CPU organized in Little Endian format. The CPU has three pipeline stages where the ALU and write-back stages are combined. The CPU is capable of executing the integer (ADDI, SLI and SRI) and floating-point (ADDF and MULF) instructions. The CPU stores the fixed and floating-point numbers in two separate register files, each containing 32 registers.

In the instruction set below, RS and RD are defined as the source and destination addresses for the integer registers, and FS1, FS2 and FD are the source and destination addresses for the floating-point registers, respectively.

Instruction Set	Instruction Definition	Instruction Bit Field
ADDI RS, RD, Imm Value	Reg[RS] + Imm → Reg[RD]	31–26: ADDI; 25–21: RS; 20–16: RD; 15–0: Imm Value
SLI RS, RD, Imm Value	Reg[RS] << Imm → Reg[RD]	31–26: SLI; 25–21: RS; 20–16: RD; 15–0: Imm Value
SRI RS, RD, Imm Value	Reg[RS] >> Imm → Reg[RD]	31–26: SRI; 25–21: RS; 20–16: RD; 15–0: Imm Value
ADDF FS1, FS2, FD	Reg[FS1] + Reg[FS2] → Reg[FD]	31–26: ADDF; 25–21: FS1; 20–16: FS2; 15–11: FD; 10–0: Not used
MULF FS1, FS2, FD	Reg[FS1] * Reg[FS2] → Reg[FD]	31–26: MULF; 25–21: FS1; 20–16: FS2; 15–11: FD; 10–0: Not used

Show a detailed data-path of this CPU, indicating all internal bus widths and port names. Include only the necessary functional units.

Projects

1. Implement a 32-bit four-stage RISC CPU that executes only ADD instruction using Verilog. On a timing diagram, trace through the data and control signals at the output ports of the instruction memory, RF, ALU and write-back stages.
2. Implement ADD, SUB, AND, NAND, OR, NOR, XOR, XNOR, SL and SR instructions in a 32-bit four-stage RISC CPU, and perform complete verification using Verilog.
3. Implement a 32-bit five-stage RISC CPU that executes LOAD, STORE, MOVE and MOVEI instructions using Verilog. Trace through the data and control signals at the output ports of the instruction memory, RF, ALU, data memory and write-back stages in a timing diagram.
4. Implement a 32-bit four-stage RISC CPU that executes only the BRA instruction using Verilog. Trace through the data and control signals at the output ports of the instruction memory and RF stages on a timing diagram.
5. Implement and verify the 32-bit floating-point adder using Verilog. Verify the validity of data at the outputs of every major stage using timing diagrams and perform functional verification for the entire adder.
6. Implement and verify the 32-bit floating-point multiplier using Verilog. Verify the validity of data at the outputs of every major stage using timing diagrams, and perform functional verification for the entire multiplier. Use behavioral Verilog to mimic the exponent adder and the integer multiplier.

System Peripherals 7

When the host processor or one of the co-processors executes a user program, it either exchanges data with system memories such as SRAM, SDRAM or Flash, or communicates with system peripherals using serial, and in some instances, parallel buses to perform various tasks.

A conventional computing system may consists of one or more CPU cores, co-processors such as hardware accelerators to perform specialized tasks, a Direct Memory Access (DMA) unit to do routine data transfers from one memory to another, a display adaptor to support a screen, and an interrupt controller to manage I/O-generated or user-generated interrupts. In most cases, data converters in charge of converting external analog signals into digital form or digital signals into analog form, timers to control the length of an event, and SPI or I^2C transceivers in charge of serially transmitting and receiving peripheral data are all interrupt-driven units and connected to the interrupt controller. The interrupt controller manages all event-driven or program driven tasks through a series of Interrupt Service Routines (ISR) that reside inside the program memory.

7.1 Overall System Architecture

A basic system architecture containing essential bus masters and slaves is shown in Fig. 7.1. In this figure, the CPU is a bus master that executes user programs. The Direct Memory Access (DMA) is another bus master in charge of transferring data between different system memories. Bus slaves are generally the system memories such as SRAM, SDRAM and Flash memory. However, other system devices that reside on the high speed bus such as the display adaptor or peripheral buffer memories connected to the low speed I/O bus are also considered bus slaves.

A. Bindal, *Fundamentals of Computer Architecture and Design*,
https://doi.org/10.1007/978-3-030-00223-7_7

The display adaptor is considered an essential high-speed peripheral that displays the results of a running program or application on the screen. Because of its bandwidth, this unit is usually connected to the parallel port of the CPU. However, there are times when the display adaptor can also be connected to the low speed I/O bus. This choice very much depends on how often the monitor needs to be used when running an application program.

Each type of memory mentioned in Chap. 5 serves a different purpose in a system. SRAM usually holds immediate data generated by the CPU or stores temporary data during a DMA transfer. Larger blocks of data are stored in SDRAM since this memory is many orders of magnitude larger in capacity compared to SRAM. Flash memory usually stores permanent data such as the Built-In Operating System (BIOS).

A bus adaptor translates commands, address and data signals between the parallel bus which operates at a high clock frequency and the low speed I/O bus which operates at a much lower clock frequency.

Sensors, electro-mechanical devices, human interface devices etc. that reside outside the main system commonly use SPI and I^2C bus protocols, and considered I/O devices. Ultimately, they are all connected to the interrupt controller, and memory-mapped due to their capability to store incoming or outgoing data.

The system can also be connected to other systems (or CPUs) with a network adaptor. The simplest connection protocol is Ethernet where many systems are serially connected to the same bus.

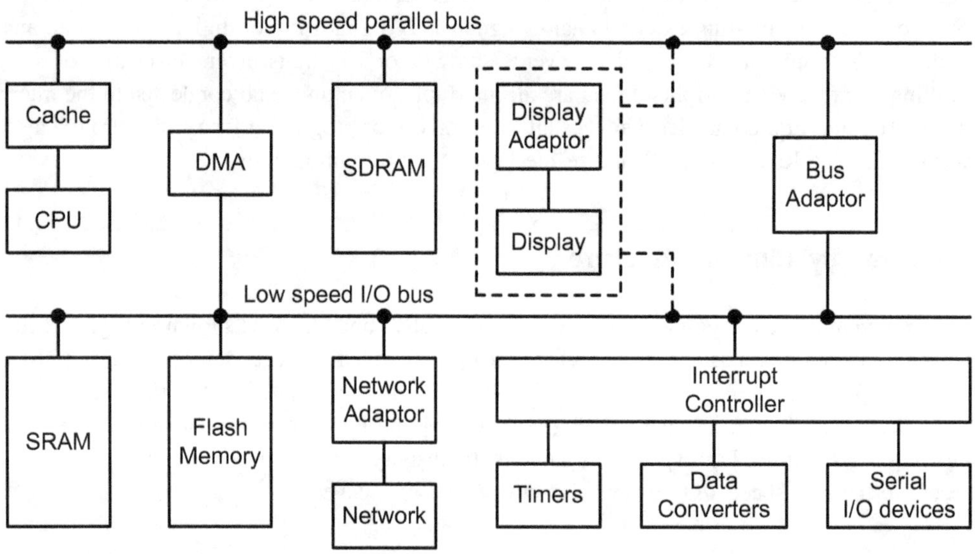

Fig. 7.1 A typical system architecture

7.2 Direct Memory Access Controller

The CPU assigns routine memory-to-memory data transfer operations to the Direct Memory Access (DMA) controller. Most of these transfers take place between two system memories or between the buffer memory of a peripheral device and a system memory. This section shows how to design the basic architecture of a DMA controller that transfers data from a source to a destination memory.

The DMA interface in Fig. 7.2 shows the I/O port description of a typical DMA controller. In this figure, the DMA controller interacts with the CPU through handshake signals, ReqM and AckM.

When the CPU initiates a DMA data transfer, it issues a request, ReqM, to the DMA controller. If the controller is not busy with another transfer, it then generates a request, ReqD, to the bus arbiter to use the bus. When the arbiter acknowledges the request by AckD, then the controller informs the CPU that it is ready to initiate the transfer by AckM, and at the same time it sends out its first address and control signals to the source memory. While the source memory is delivering data, the DMA controller issues the address and control signals for the destination memory within the same clock cycle. In order to accomplish this task, a direct data channel must exist between the source and the destination memories. This new configuration modifies the original bus structure in Fig. 5.1 in which all bus masters are assumed to have individual write data ports to be able to write data directly to a slave.

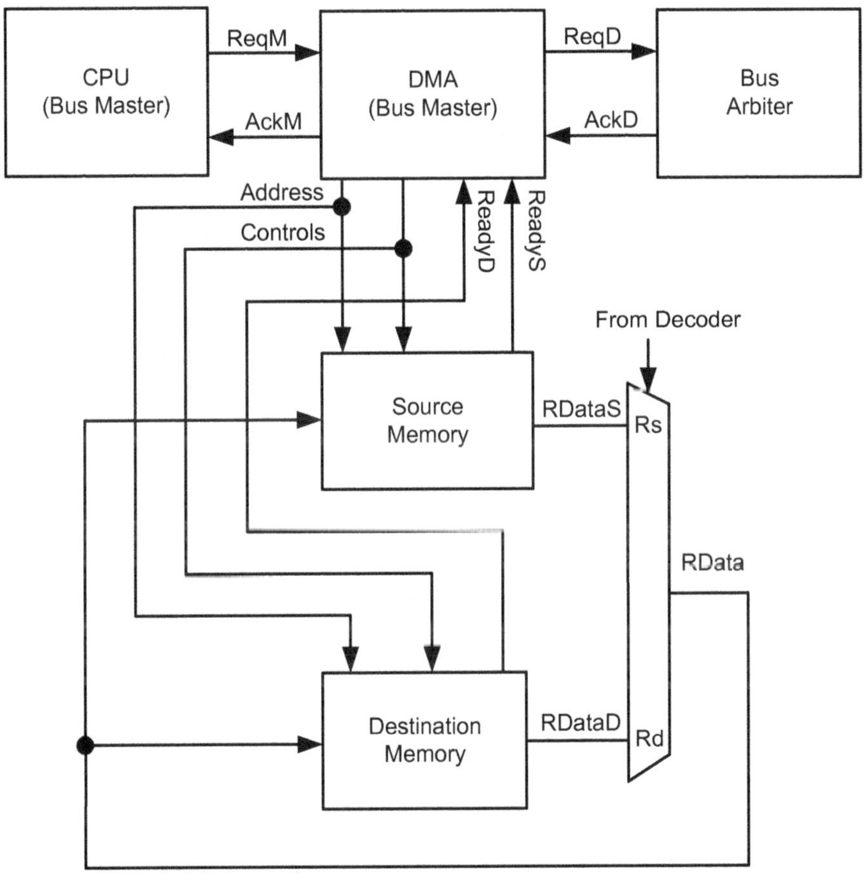

Fig. 7.2 The block diagram including a DMA, source and destination memories

Figure 7.3 shows a typical data transfer between the source and the destination memory until the last data packet, D4, is transmitted. The sequence starts with the DMA controller issuing Status = START and AddrS = AS1, indicating the beginning of the data transfer and the first source memory address, respectively. Since both memories are ready (ReadyS = ReadyD = 1) in the first clock cycle, the controller continues the data transfer by issuing Status = CONT, AddrS = AS2 (the second source memory address), and AddrD = AD1 (the first destination memory address) in the second cycle. In this cycle, the source memory also delivers the first data, D1, to the destination memory. The same process takes place in the third cycle, during which the DMA controller generates

AddrS = AS3, AddrD = AD2, and writes D2 to the destination memory. In the fifth clock cycle, as the DMA controller generates the last destination memory address, AddrD = AD4, and writes the last data, D4, to AD4. In this cycle, it also changes its status to Status = IDLE, indicating the end of the data transfer.

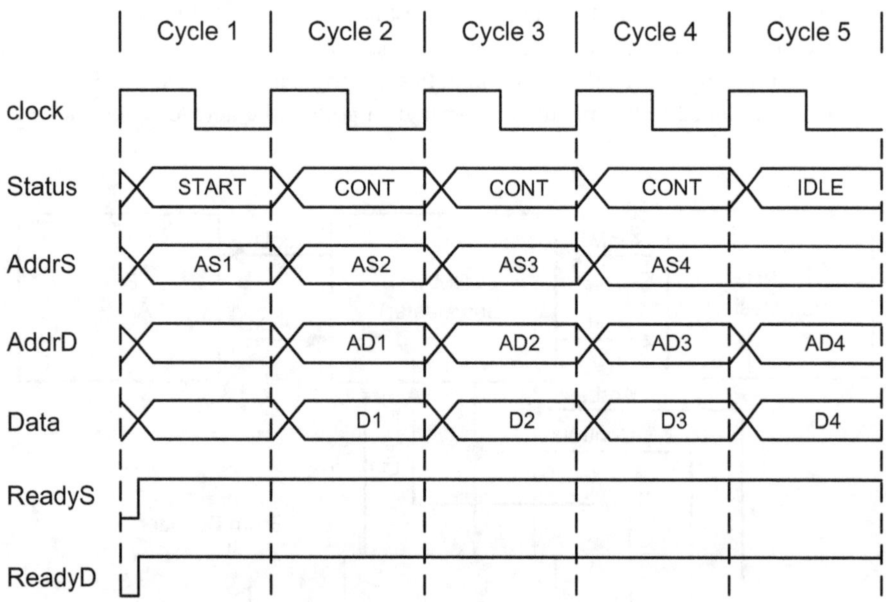

Fig. 7.3 Timing diagram showing a DMA-assisted data transfer

Any time one of the Ready signals from the source or the destination memories transition to logic 0, the DMA controller stalls the data transfer by repeating the control and address signals as long as ReadyS or ReadyD is at logic 0. Figure 7.4 shows a typical data transfer in which the destination memory is busy in the third cycle, and prompts the DMA controller to repeat AddrS = AS3 and AddrD = AD2 in the next cycle. The DMA controller stalls the bus again in the sixth cycle when it detects the source memory not ready in the fifth cycle.

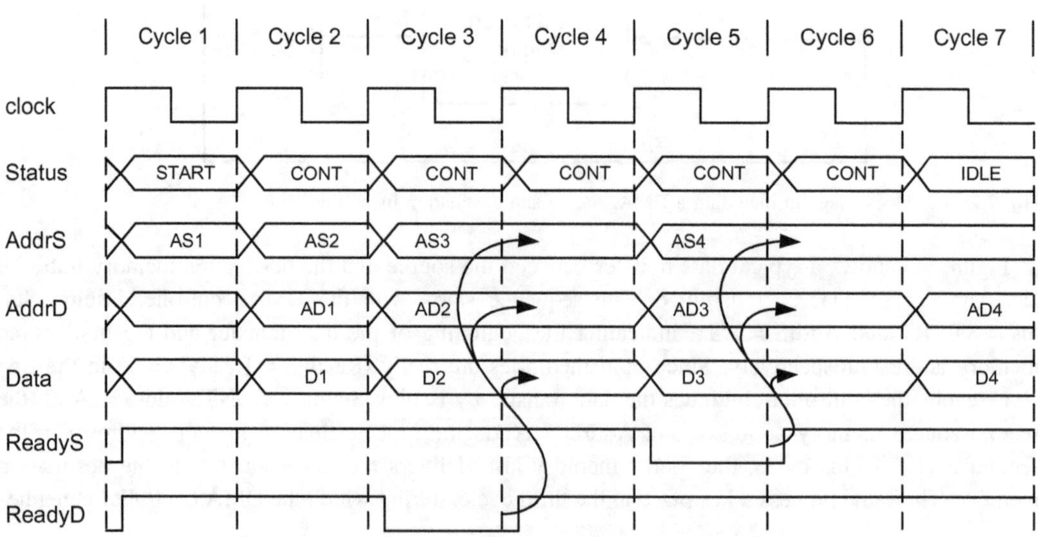

Fig. 7.4 DMA-assisted data transfer with varying Ready signals from memories

A basic DMA controller is shown in Fig. 7.5. There are three modules in this architecture. The first module is the DMA register file that stores the initial and incremental source address values, InitAddS, StepAddrS. Two other registers, InitAddrD and StepAddrD, store the initial and incremental destination addresses, respectively. This section also includes Size and Burst registers to store the data width and the burst length. A program data bus is used to store all six register entries before regular operations take place.

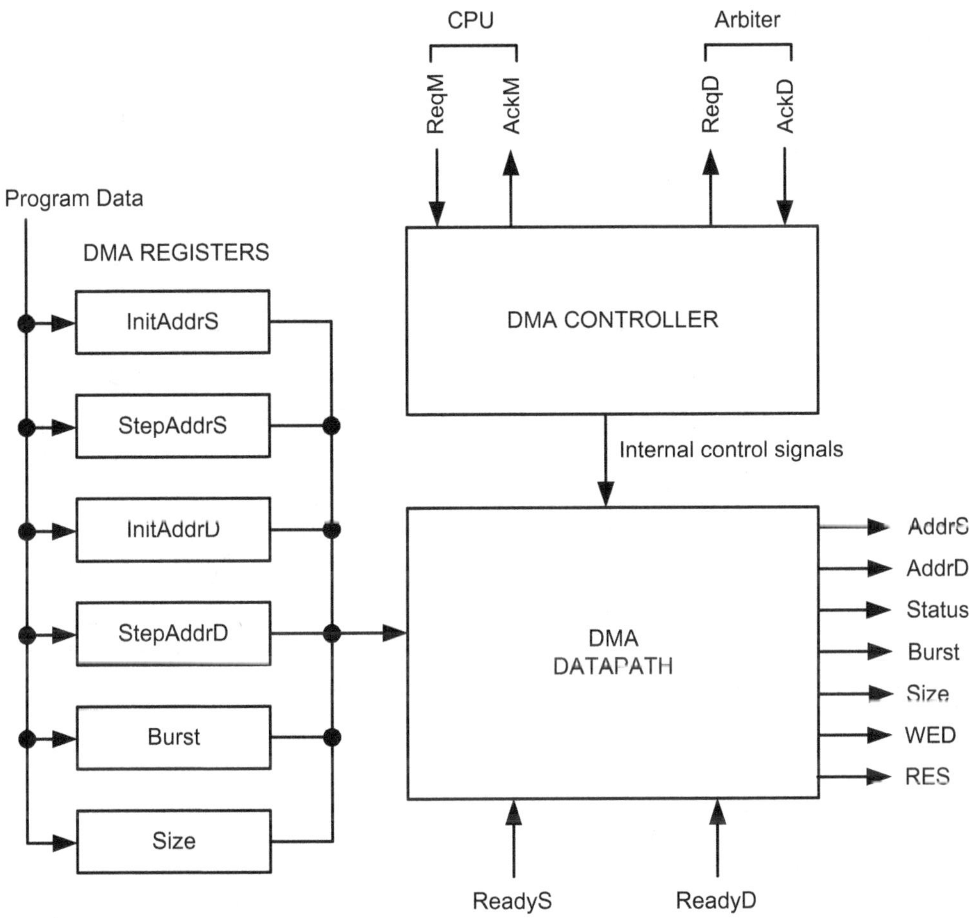

Fig. 7.5 A typical DMA architecture

The second section of the DMA controller manages the handshake mechanism with the CPU and the bus arbiter. This section also provides the internal control signals to the DMA data-path to guide the data flow.

In the third section, the DMA data-path produces source and destination addresses, AddrS and AddrD, and the bus master control signals, Status, Size, Burst, WED (write-enable for the destination memory) and RES (the read-enable for the source memory).

To be able to implement this architecture, three elements need to be examined in the design phase simultaneously: a timing diagram describing an entire data transfer process including the stall periods, a data-path that fully complies with the timing diagram, and a control logic that manages the data flow on the data-path.

As pointed out in previous chapters, the design always starts with forming a timing diagram that describes the complete data-flow in a logic block. The timing diagram generally includes a single clock (multiple clock domains or asynchronous event signals are also common but not relevant for a basic DMA design), address, data and control signals with respect to this clock. In order to generate an accurate timing diagram, a corresponding data-path must be developed simultaneously. The data-path generally consists of registers and logic gates. However, it can also contain mega cells such as complex arithmetic units or memories. As the design develops and more details are added to the original data-path, the corresponding timing diagram becomes more complex to accommodate the changes in the hardware. The design of the controller to govern the data flow is the last step in the design process. This step does not start until every internal and external control signal is defined, and the complete block functionality, including all corner cases, is described on the timing diagram.

A detailed timing diagram describing a typical DMA transfer is shown in Fig. 7.6. As we mentioned earlier in the memory-to-memory example in Chap. 2, this diagram is also developed in two

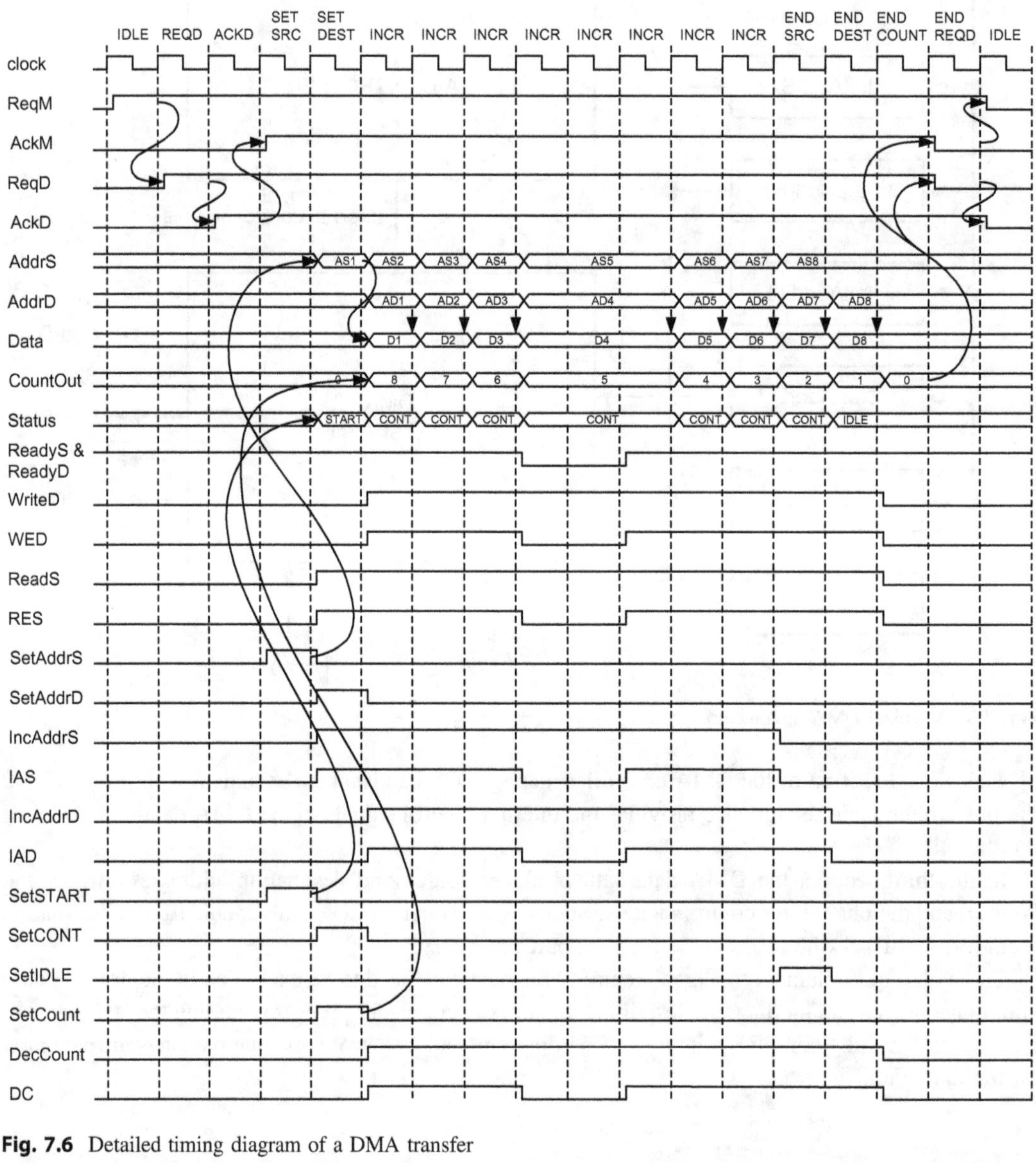

Fig. 7.6 Detailed timing diagram of a DMA transfer

phases. In the first phase, the main DMA signals, namely the handshake signals with the CPU and the arbiter (ReqM, AckM, ReqD and AckD), the source and the destination memory addresses (AddrS and AddrD), the data (Data), the bus master control signals (Status, Burst, Size, WriteD and ReadS), and the slave response signals (ReadyS and ReadyD) are included in the timing diagram. In the second phase, all internal control signals that support the address and data movement in the timing diagram are brought into the picture. This section also includes the control signals for an internal down-counter to keep track of the number of data packets transferred from one memory to the next.

In Fig. 7.6, the CPU initiates a DMA-assisted data transfer by issuing a request to the DMA controller, ReqM, in clock cycle 1. This request enables the DMA controller to generate a subsequent request, ReqD, to the arbiter in order to use the system bus in cycle 2. An acknowledgment from the arbiter, AckD, may come in the third cycle or many cycles later depending on the bus traffic and the other pending requests from higher priority bus masters. However, as soon as the DMA controller receives the acknowledgement from the arbiter, it notifies the CPU by issuing an acknowledgement signal, AckM, in cycle 4. This cycle also prepares the DMA for an upcoming data transfer by setting the SetAddrS and SetSTART signals to logic 1. That way, the first source memory address, AS1, can be fetched from the InitAddrS register and delivered to the AddrS port in Fig. 7.7, and similarly the START code can be produced at the Status port in cycle 5. The port selection guide for each 3-1 MUX in Fig. 7.7 is shown in Table 7.1. In cycle 5, the first data read command from AS1 is issued by ReadS = 1. This cycle also sets the control signals, SetAddrD, IncrAddrS and SetCONT, to logic 1, so that the first destination memory address can be retrieved from the InitAddrD register and transferred to the destination address port, AddrD, the second source memory address, (InitAddrS + StepAddrS), can be formed at the AddrS output, and the status code can be changed from START to CONT in the next clock cycle. Still in cycle 5, the down-counter, responsible for counting the number of data packets delivered to the destination memory, is set with an initial value from the burst register by SetCount = 1. Therefore, when clock cycle 6 starts, the second source memory address, AS2, is formed at the AddrS port along with the first destination address, AD1, at the AddrD port. The Status output displays the CONT code, specifying the ongoing data transfer. In this cycle, the first data, D1, is transferred from the address, AS1, to the address, AD1, with an active-high WriteD signal, and subsequently written to the destination memory. The CountOut output also shows the initial value from the burst register, defining the number of data packets to be written to the destination memory. This cycle sets the control signals, IncrAddrS, IncrAddrD and DecCount, to logic 1 in order to prepare the next source and destination addresses, and to decrement the CountOut by one in the next cycle.

Table 7.1 3-1 MUX port assignments in Fig. 7.7

INPUT	PORT	INPUT	PORT	INPUT	PORT
SetAddrS = 1	set	SetAddrD = 1	set	SetCount = 1	set
IAS [ReadyS = 1, ReadyD = 1, IncrAddrS = 1]	incr	IAD [ReadyS = 1, ReadyD = 1, IncrAddrD = 1]	incr	DC [ReadyS = 1, ReadyD = 1, DecCount = 1]	decr
ELSE	stall	ELSE	stall	ELSE	stall

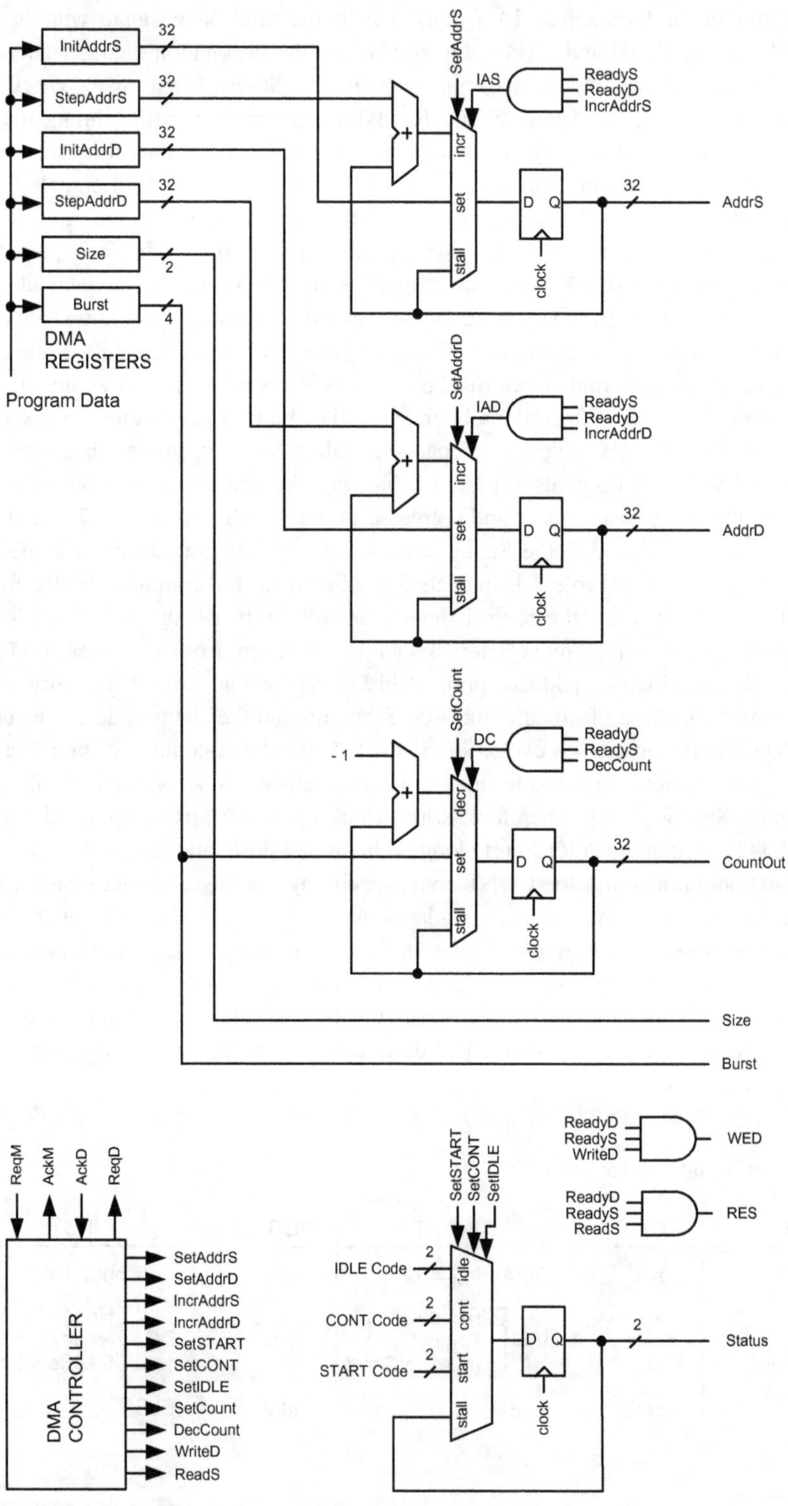

Fig. 7.7 Internal DMA architecture showing its data-path and controller

The routine data transfer continues until either the source or the destination Ready signal transitions to logic 0. When this happens, the entire data transfer stalls, disabling any write process taking place at the destination memory. The previous source and destination address values are repeated at the AddrS and AddrD ports as long as one of the Ready signals stays at logic 0. The down-counter is also forced to stall, and it displays the remaining number of data packets to be written to the destination memory.

The data transfer resumes when the ReadyS and ReadyD signals become logic 1. CountOut = 2 defines the end of the data transfer. In this cycle, the IncrAddrS signal also transitions to logic 0, indicating that there will be no more new source address generation at the AddS port. Similarly, the SetIDLE signal goes to logic 1, changing the bus master status code from CONT to IDLE in the next cycle. When CountOut = 1 in the following cycle, the last data is written to the destination address. From this point forward, the DMA handshakes with the CPU and the arbiter to terminate the data transfer. ReqD = 0 forces the arbiter to lower the acknowledge signal, AckD. AckM = 0 prompts the CPU to lower its DMA request signal, ReqM.

The controller design is a direct outcome of the timing diagram in Fig. 7.6. The first step in the controller design is to assign a name to each clock cycle in the timing diagram that produces a different set of outputs from the previous clock cycle. In other words, each clock cycle that produces a different set of outputs has to be labeled with a new state in the controller state diagram.

Cycle 1 is named as the IDLE state, producing ReqD = 0 and AckM = 0 in Fig. 7.8. When ReqM = 1 is received in cycle 1, ReqD switches from logic 0 to logic 1 in cycle 2, producing a new state, REQD. ReqD = 1, on the other hand, prompts AckD = 1 in cycle 3 which creates the ACKD state in this cycle. Cycle 4 also creates a new state, SET SRC, because a different set of control signals (AckM = 1, SetAddrS = 1 and SetSTART = 1) emerges in this cycle. The next cycle generates a new set of outputs (ReadS = SetAddrD = IncrAddrS = SetCONT = SetCount = 1) compared to the previous clock cycles, and therefore it is labeled as the SET DEST state. Between cycles 6 and 13, the controller outputs remain the same. Therefore, all these cycles can be grouped together under the same state name, INCR. In cycle 14, the IncrAddrS signal transitions to logic 0 and the SetIDLE signal transitions to logic 1. This new set causes the creation of a new state, END SRC. Cycle 15 changes the controller output values once more with respect to the previous cycle, and it is labeled as the END DEST state. In cycle 16, the down-counter output finally reaches zero, and all the controller outputs except the AckM and ReqD signals become logic 0. Therefore, this cycle is named as the END COUNT state. The controller lowers the AckM and ReqD signals to logic 0, which creates a new state, END REQD in cycle 17. In the next cycle, AckM = ReqD = 0 prompts the CPU and the arbiter to lower ReqM and AckD signals to logic 0, respectively, and the controller transitions to the IDLE state.

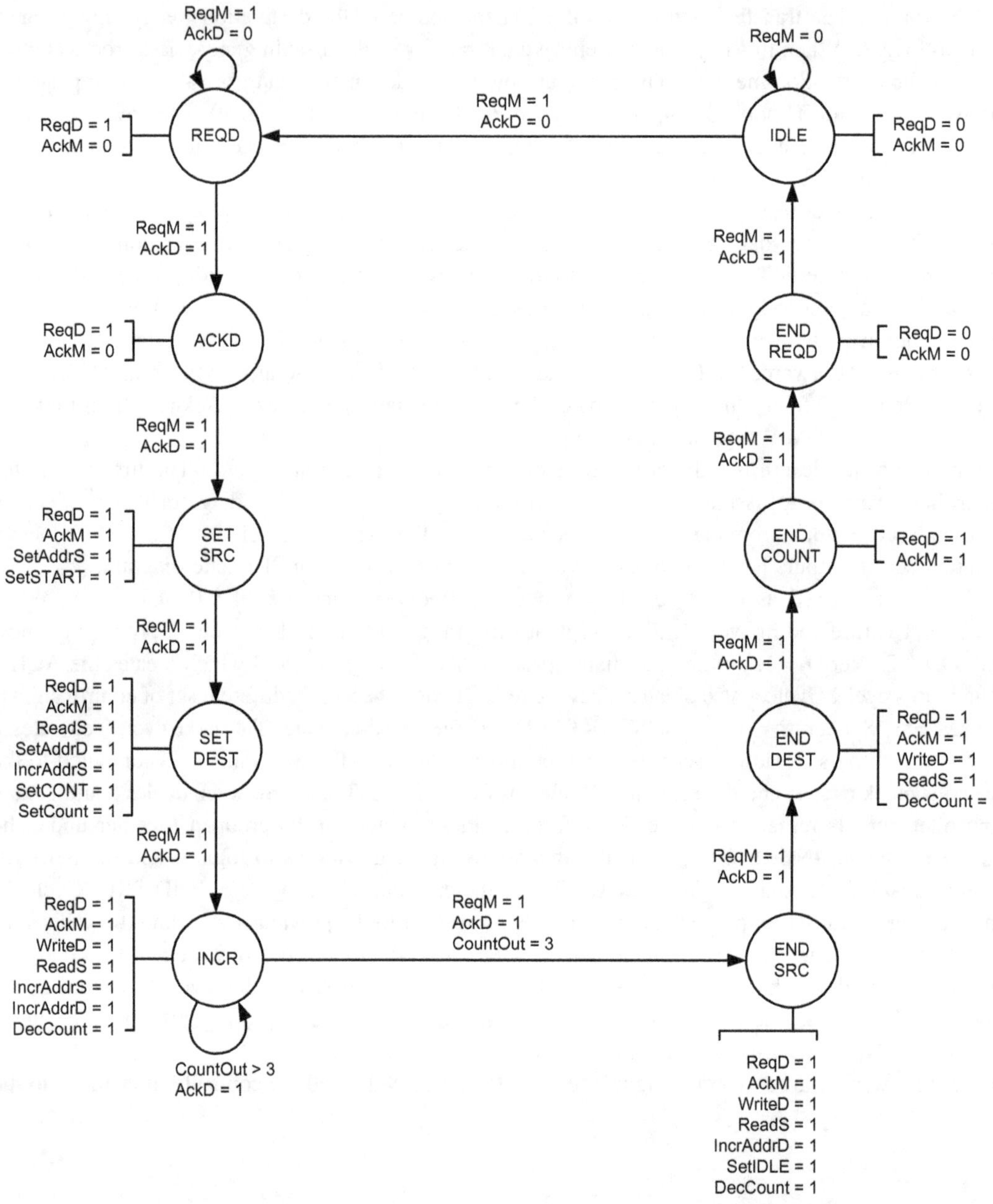

Fig. 7.8 DMA controller state diagram (control signals equal to logic 0 are omitted for simplicity)

7.3 Interrupt Controller

There are numerous events that may interrupt the normal flow of program execution. External events are created by I/O devices that operate under specific utility programs. These programs are configured such that an I/O device may deliver or require data from the CPU. There are also internal events within the CPU that result from occurances encountered when executing user programs, such as divide-by-zero or overflow conditions, which create exceptions. All these hardware-related external and software-related internal events are managed by the interrupt controller.

There are four types of interrupts according to their priority. The interrupt for resetting the CPU takes the precedence over all other interrupts because when reset occurs, the data in each CPU register needs to be preserved in a special memory to be restored later on. Internal interrupts take the second priority after the CPU reset. These interrupts generally originate from errors encountered in user programs or may result from breakpoints installed in a user program. Software interrupts take the third place in the priority list. These interrupts are actually vectored subroutine calls that stem from software emulation routines. Floating-point division produces one such example. Hardware interrupts are placed last in the priority list. Even though prioritizing hardware interrupts is completely user-programmable, the operating system may also manage the hardware priority list and communicate with a specific device though device drivers.

In this chapter, we will examine the sequence of events that take place to handle a hardware interrupt, and design a simple interrupt controller interface that serves up to 256 external I/O devices.

The interrupt process begins when one or more I/O devices submit interrupt requests to the interrupt interface in Fig. 7.9. In this figure, the interrupt interface is designed to handle up to 256 interrupt inputs, INTR0 through INTR255. Interrupt controller is a programmable state machine that prioritizes all pending interrupts, selects a highest priority device according to a priority list, and communicates with the CPU using the INTR output as shown in Fig. 7.9. When the acknowledgement signal, INTA, is received from the CPU, the interrupt interface transmits the device ID, INTRID, causing the interrupt on an eight-bit wide bus.

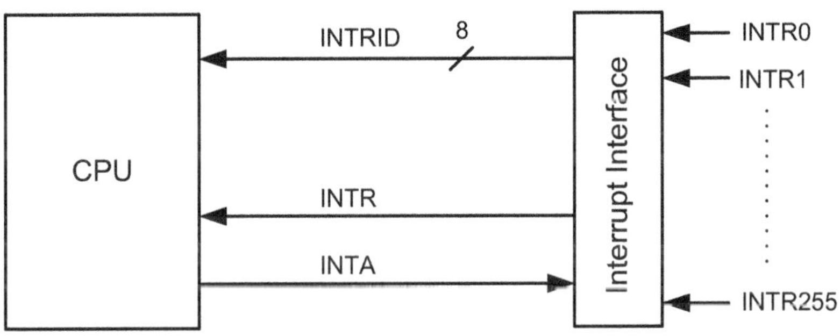

Fig. 7.9 Hardware interrupt interface input/output description

The interrupt ID in Fig. 7.9 matches the interrupt number at the input of the interface, and ranges between 0 and 255. Each interrupt ID correlates to a specific address in the Interrupt Address Table, IAT, in Fig. 7.10. Each address stored in IAT points the starting address of a particular Interrupt Service Routine (ISR) residing in the instruction memory. Therefore, when the interrupt interface generates an INTRID and accesses a specific memory address in the IAT, the contents at this address is immediately loaded to the Program Counter (PC). This prompts the CPU to pause executing the normal user program, and jump to the starting ISR address in the instruction memory to execute the corresponding ISR instructions. This basically translates to jumping from Instr3 of the user program to Intr1 of the ISR in the example in Fig. 7.10. While in the ISR, a return instruction, RET, indicates the end of interrupt service routine. At this point, the program returns to the address, ARET, to execute the rest of the user program.

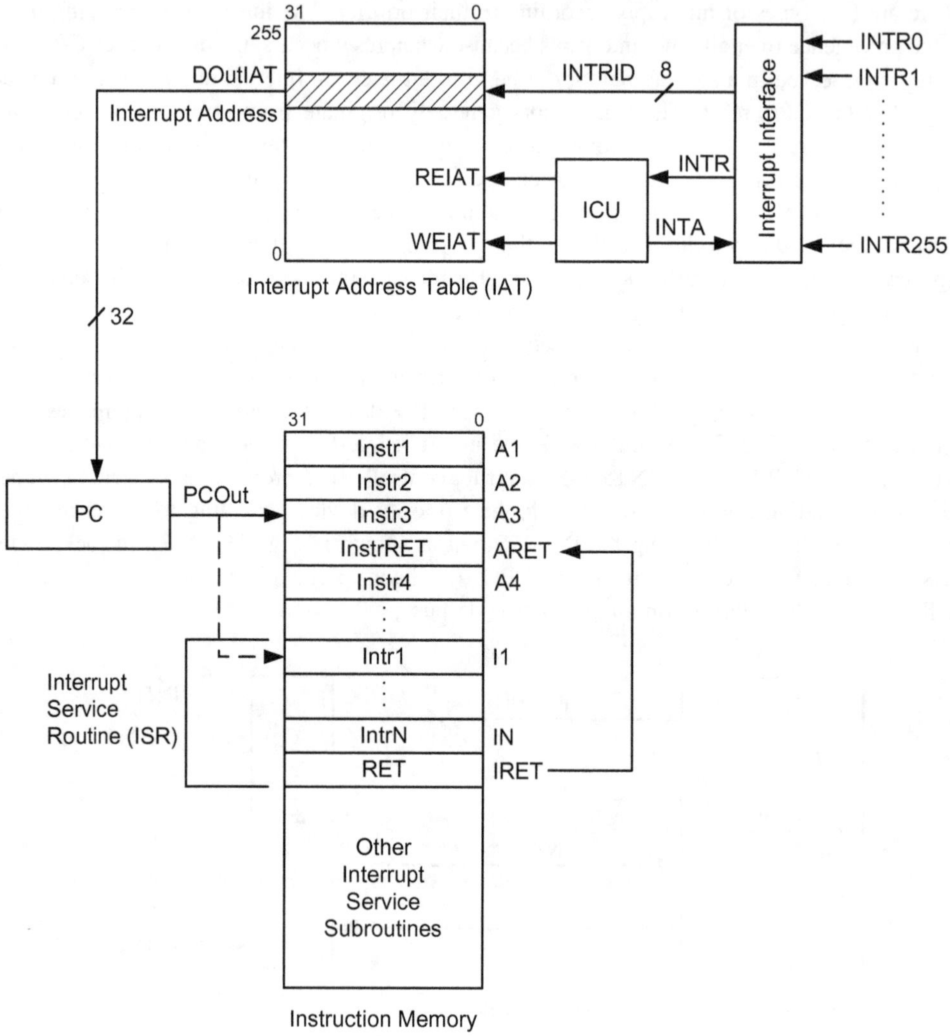

Fig. 7.10 A block diagram describing an interrupt

To convert the block diagram in Fig. 7.10 into a detailed data-path, each step of the interrupt service routine outlined above should be translated into a timing diagram. Creating a timing diagram, on the other hand, is usually accomplished in two steps. The first step includes all primary, bus-level signals such as handshake control signals, data and address in the timing diagram. The second step generates the necessary control signals to manage the data-flow.

To achieve the first step, let us consider the signals, INTRx (INTR0 to INTR255), INTR, INTA, INTRID, DOutIAT and PCOut, in Fig. 7.10 and the corresponding timing diagram in Fig. 7.11. The signals, INTRx, INTR and INTA, are not bus-level signals; but, they are considered as the primary I/O signals that indicate the start of an interrupt. Therefore, they will be grouped together with the other bus-level signals to show the complete interrupt sequence. In Fig. 7.11, an I/O device issues an interrupt, INTRx = 1, to the interrupt interface in clock cycle 1. In response, the interrupt interface generates INTR = 1 in cycle 2, for the Interrupt Control Unit (ICU) in charge of the handshake signals. In cycle 3, the ICU generates INTA = 1, and prompts the interrupt interface to transmit an eight-bit INTRID to the IAT in the following cycle. As mentioned earlier, the INTRID signal is also

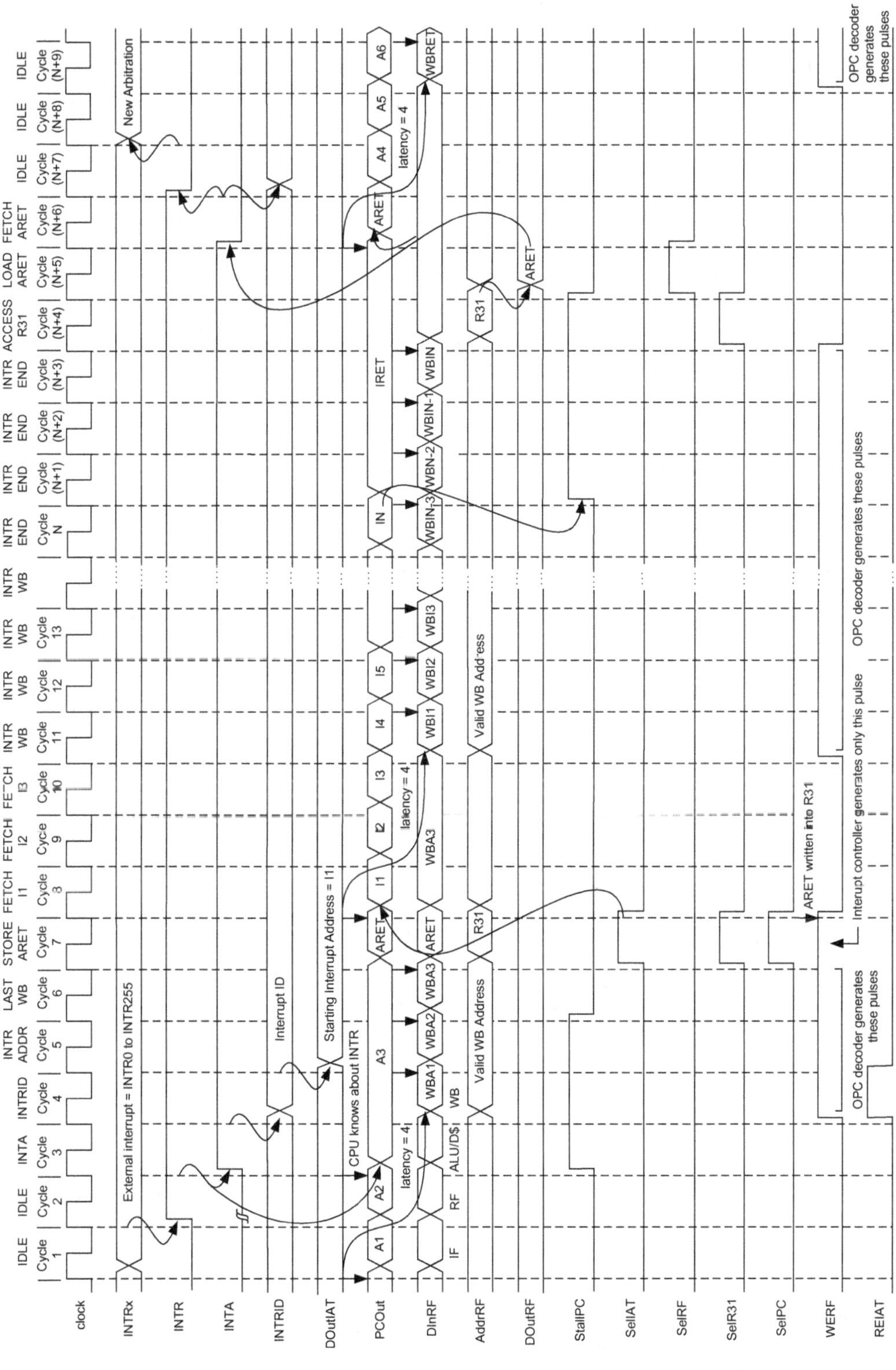

Fig. 7.11 Detailed timing diagram of an interrupt sequence

an address for the IAT. The data stored at this address is actually the starting address of an ISR in the instruction memory. Therefore, the interrupt controller simply reads the contents of the address, INTRID, from the DOutIAT port in cycle 5, and waits for the interrupt service routine to begin. In the mean time, at the beginning of cycle 3 when the ICU is aware of a pending interrupt by means of the INTR signal, it immediately stalls the CPU pipeline by stopping the PC from incrementing. However, the PC has already incremented to A3 at this point, and there are uncompleted instructions from the A1 and A2 addresses in the CPU pipeline. We know that from the onset of PC address generation to the end of the write-back cycle, a normal instruction takes four clock cycles to complete (latency = 4) according to the simplified CPU data-path in Fig. 7.12. Therefore, the PC output stays at the A3 address until the end of cycle 6 when the instructions, Instr1, Instr2 and Instr3, are completely flushed out of the CPU pipeline and written back to the CPU's register file (RF). However, this software interaction with the CPU adds the RF block and all the corresponding hardware to the interrupt controller's block diagram in Fig. 7.10, transforming this diagram into a more detailed architecture in Fig. 7.13.

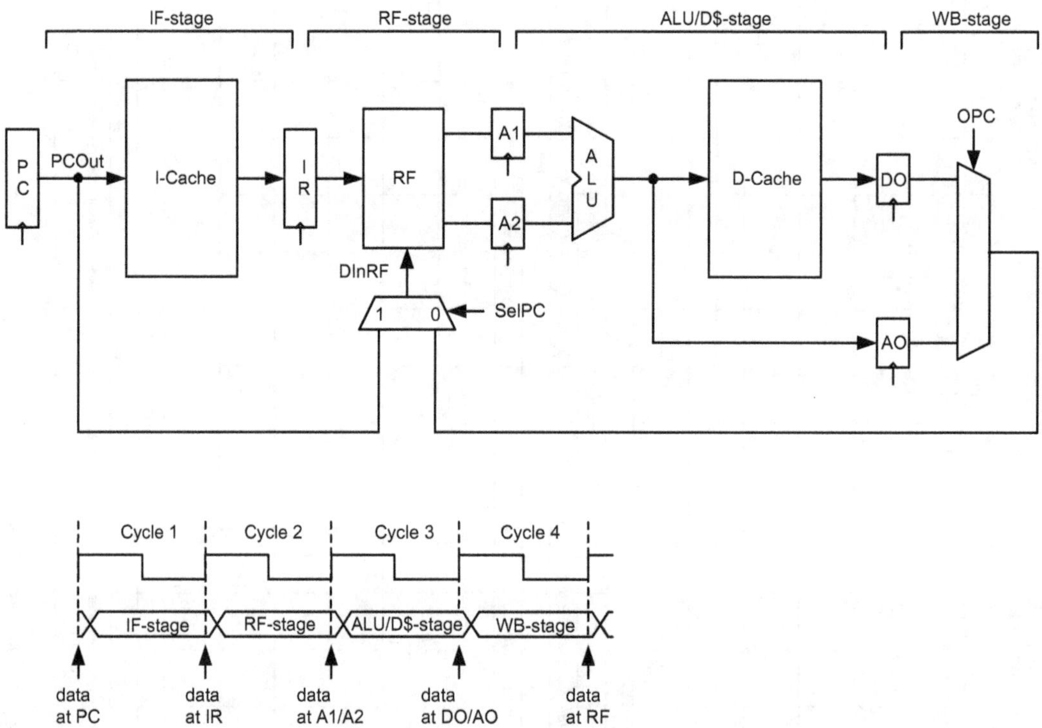

Fig. 7.12 A four-stage CPU employed with the interrupt sequence in Fig. 7.11

After the last write-back is completed in cycle 6, the PC is incremented by one, and the return address, ARET, is stored in a special register, R31, in the RF as shown in cycle 7. This step assumes that there must not be any Jump-and-Link (JAL) or Jump-and-Link-Register (JALR) instruction in the CPU instruction set because the return address following the ISR will simply be overwritten by one of these instructions. Before the CPU starts executing the interrupt service routine, it copies the contents of the entire register file into a temporary "shadow" memory identical to the register file. This step is called "context switching", and it is omitted from the timing diagram in Fig. 7.11 to maintain

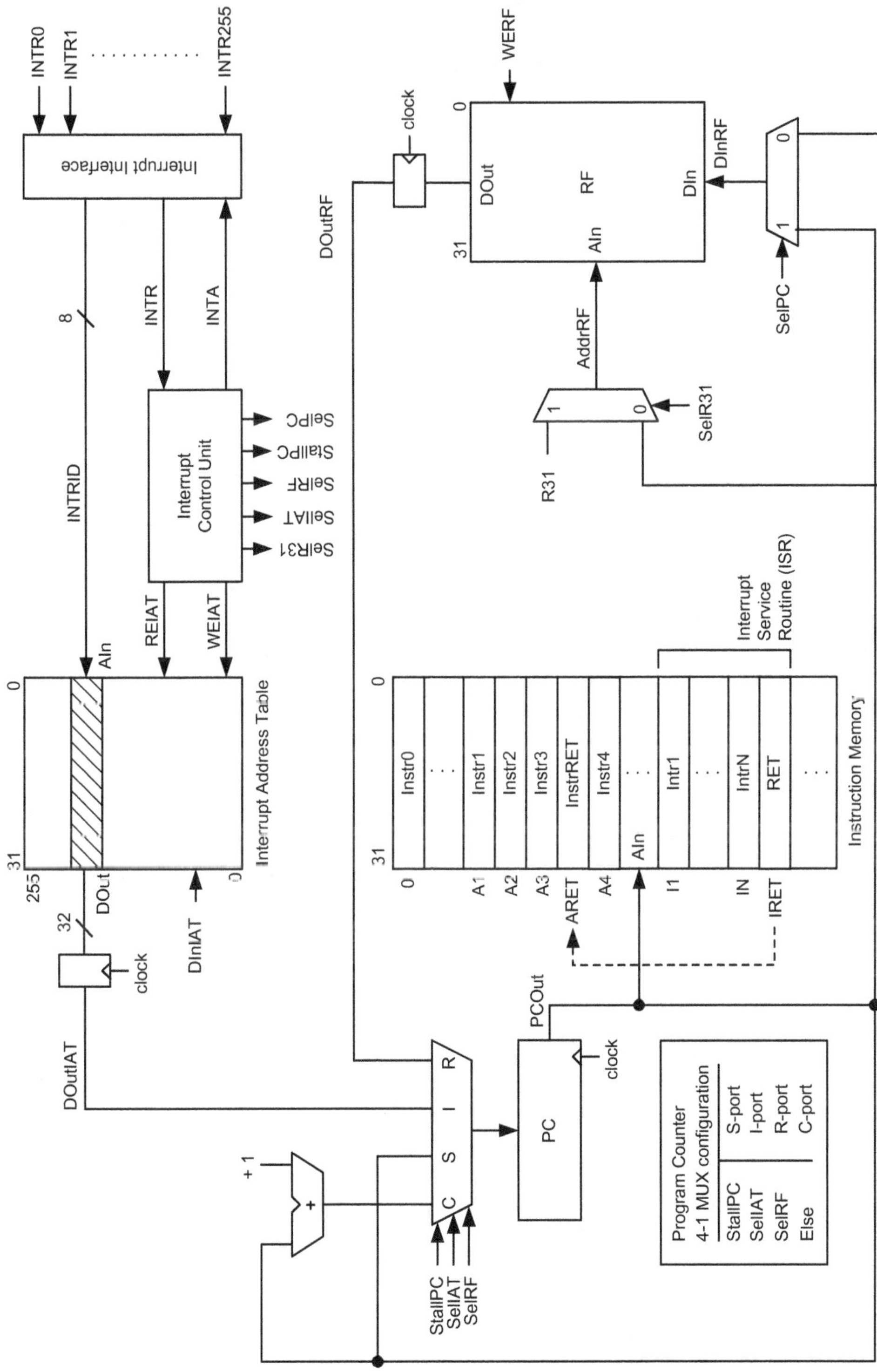

Fig. 7.13 Interrupt interface data-path producing the timing diagram in Fig. 7.11

simplicity. Once the steps of handling an ISR is fully explained, a more complex interrupt controller architecture, including the context switching process, will be introduced at the end of this section.

In cycle 8, the interrupt controller starts executing the instructions in the interrupt service subroutine. In this cycle, the first interrupt instruction address, I1, at the output of the IAT is loaded to the PC. Once loaded to the CPU data-path, it takes four clock cycles to execute the first interrupt instruction due to the CPU's write-back latency. The remaining interrupt instructions are similarly fetched from the instruction memory addresses I2 to IN, executed and written back to the RF until the end of cycle (N + 3). In cycle (N + 4), the register R31 is accessed. In the following cycle, the return address, ARET, is fetched from R31. This cycle also completes the interrupt service routine and prompts the ICU to transfer the control over to the CPU to execute the remaining user instructions. In cycle (N + 6), the ICU lowers the INTA signal to logic 0, and the interrupt controller loads the contents of the ARET to the PC. In response to the INTA, the interrupt interface also lowers the INTR signal to logic 0 and invalidates the INTRID in cycle (N + 7). New interrupt arbitration will take place in cycle (N + 8) to service the next interrupt.

Up to this point, we only explained how the address and data bus values changed once an interrupt signal is received from the interrupt interface. Now, we are ready to explain the second part of the timing diagram that includes the control signals to manage the data-flow. After examining the detailed data-path in Fig. 7.13, the control signals can be grouped into three categories. The first group supports the PC input control and contains the StallPC, SelIAT and SelRF signals. The StallPC signal simply routes the output of the PC, PCOut, to its input through the S-port of the 4-1 MUX to stall the PC. The SelIAT signal routes the output of the IAT, DOutIAT, to the input of the PC through the I-port of the 4-1 MUX so that an interrupt address can be loaded. The SelRF signal enables the R-port and connects the output of the RF, DOutRF, to the input of the PC to load the return address, ARET, once the ISR is over. If none of these control signals are generated, then the PC increments through the C-port. The second group controls the address and data inputs to the RF and consists of the SelR31 and SelPC inputs. The SelR31 input selects the register R31 to be the address for the RF at the AddrRF port. The SelPC input selects the contents of the PC to be the data for the RF at the DInRF port. The third group controls the read and the write enable signals, REIAT and WEIAT, for the IAT, respectively. Writing to the IAT does not take place during a routine interrupt service. However, the WEIAT signal is used to reprogram the IAT with a new set of ISR addresses.

All three groups of controls manage the proper data flow in Fig. 7.13. The StallPC signal transitions to logic 1 at the beginning of cycle 3 and stays there until cycle 6 to stop the PC from incrementing so that the CPU completes writing the instructions, Instr1, Instr2 and Instr3, back to the RF. Because of these write-backs, the write-enable signal for the RF, WERF, is also kept at logic 1 from cycle 4 to cycle 6. The read-enable signal for the IAT, REIAT, is kept high in cycle 4 because the first interrupt instruction address, I1, needs to be fetched from the IAT following a valid INTRID. Cycle 7 is a special cycle to load the register R31 with the program return address, ARET. Therefore, the signals, SelR31, SelPC and WERF, all become logic 1 during this cycle. The SelIAT signal is also kept at logic 1 during cycle 7 in order to load the first interrupt address, I1, to the PC in cycle 8. The WERF signal is kept at logic 1 from cycle 11 to cycle (N + 3) to be able to complete all interrupt-related write-backs to the RF. The StallPC signal is kept at logic 1 from cycle (N + 1) to cycle (N + 4) to stall the value of PCOut at IRET. Cycle (N + 5) is dedicated to retrieving the program return address, ARET, from the RF. Therefore, the SelRF signal is kept at logic 1 in this cycle to load the PC with the contents of ARET in the following cycle. The WERF signal transitions to logic 1 in cycle (N + 9) in order to write the result of the instruction, InstrRET, back to the RF.

Figure 7.14 shows the resultant state diagram for the interrupt controller. In this schematic, the control signals with zero values are omitted from the state machine to avoid congestion and improve readability. The name of each state in the state machine comes from the labels on top of the timing diagram in Fig. 7.11. The machine starts with the IDLE state where there is no INTR signal. Therefore, this state generates INTA = 0. When a valid INTR is received, the state machine transitions to the INTA state and produces two outputs, INTA = 1 and StallPC = 1. This state corresponds to cycle 3 of the timing diagram. As INTR = 1 continues, the machine goes though the INTRID, INTR ADDR, LAST WB and STORE ARET states, which correspond to cycles 4, 5, 6 and 7, respectively. These are the preparation states prior to an ISR. The FETCH I1 state indicates the first interrupt instruction fetch, which corresponds to cycle 8. The interrupt controller goes though the FETCH I2 and FETCH I3 states where it fetches the second and third interrupt instructions. These are indicated in cycles 9 and 10, respectively. A cycle later the machine enters the INTR WB state where it starts writing the results of interrupt instructions back to the RF. The interrupt controller stays at this state until the interrupt address reaches its last value, IN. When the last interrupt address is fetched, the machine transitions to the INTR END state where it performs three additional interrupt write-backs, and it stalls the PC at IRET until the last interrupt write-back, WBIN, completes. This state continues during cycles (N + 1), (N + 2) and (N + 3) in the timing diagram. Following the last interrupt write-back, the interrupt controller prepares the system to finish the current interrupt service before receiving another interrupt. The closing states are the ACCESS R31, LOAD ARET and

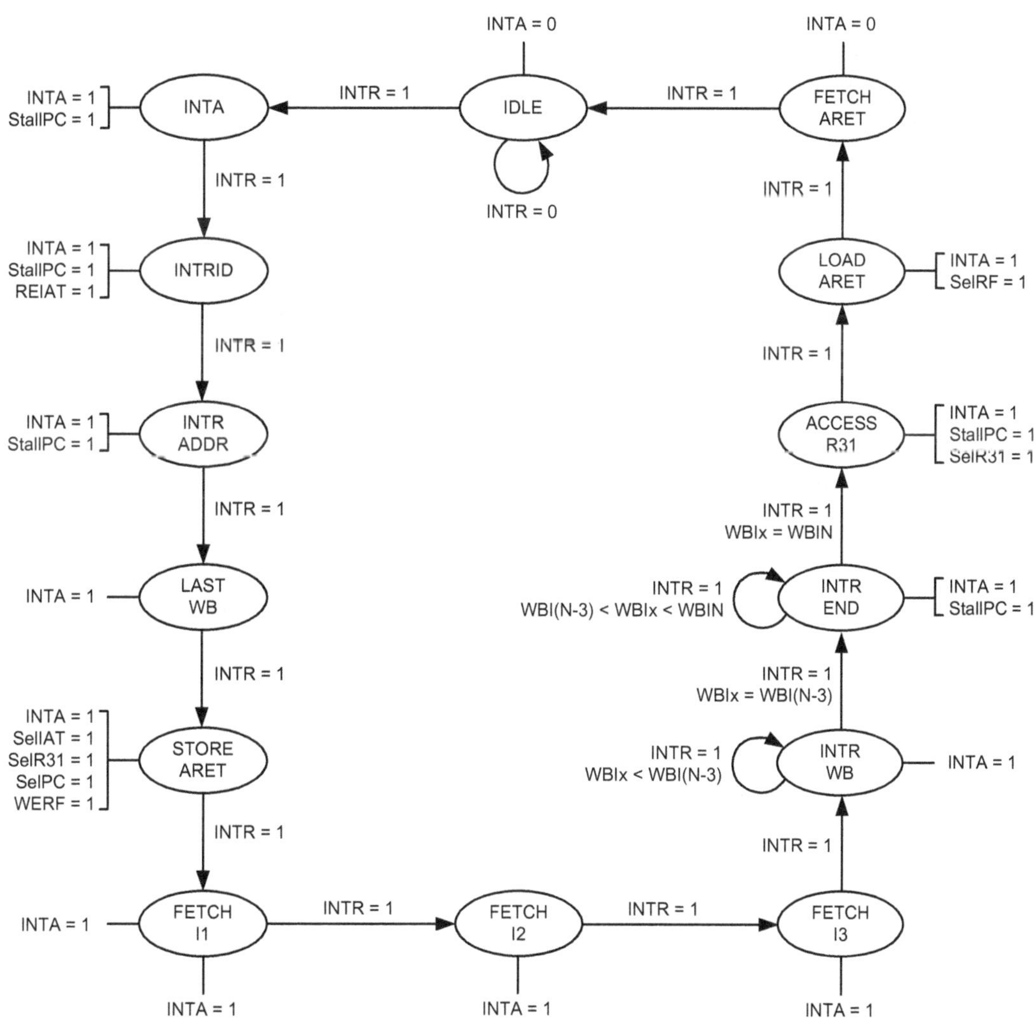

Fig. 7.14 Interrupt controller state diagram (control signals equal to logic 0 are omitted for simplicy)

FETCH ARET states, which correspond to cycles (N + 4), (N + 5) and (N + 6) in the timing diagram, respectively. The interrupt controller goes back to the IDLE state in cycle (N + 7) where the INTRID becomes no longer valid.

However, a crucial problem arises when implementing this state machine. The interrupt controller needs to know the end of an ISR. Somehow the number of instructions in the interrupt service routine must be determined in advance in order to continue the state transitions in the state machine. The states INTR WB and INTR END are the examples of this problem. The interrupt controller needs to stay in the INTR WB state from the first to the (N − 4)th interrupt write-backs, and similarly in the INTR END state from the (N − 3)th to the (N − 1)th interrupt write-backs during an ISR. Since the number of instructions varies in an ISR program, this state machine's implementation becomes impossible for the ICU design, necessitating a change in the original timing diagram which will affect the data-path in Fig. 7.13 and the state diagram in Fig. 7.14.

Figure 7.15 shows a slightly modified version of the interrupt controller data-path to circumvent this problem. In this figure, a decoder is added between the output of the instruction register, IROut, and the interrupt controller in order to detect the return opcode, RET, at the last interrupt address, IRET. This, however, creates an additional input, DetIRET, for the ICU. Therefore, when the return opcode is decoded in cycle (N + 2) in the new timing diagram in Fig. 7.16, the DetIRET signal becomes logic 1 and prompts the ICU to make preparations to end the current ISR. In this figure, the StallPC signal is also lowered to logic 0 between cycles (N + 1) and (N + 4) because not stalling the PC during this interval will simply generate invalid addresses at the PCOut port, which may be counterproductive for the operation of the interrupt controller prior to retrieving the return address, ARET, from R31 in cycle (N + 6).

This modification leads to a number of changes in the ICU's state diagram shown in Fig. 7.17. After transitioning to the INTR WB state in cycle 11, the state machine stays in this state until it detects DetIRET = 1. This input forces the ICU to move to the WBIN state to complete the last interrupt write-back. At this point, the machine goes through three more states, ACCESS R31, LOAD ARET and FETCH ARET, to load the user program return address back to the PC in order to resume the original program.

Besides the absence of context switching, the architecture in Fig. 7.15 also suffers from providing sufficient register space for both the regular program and the ISR. In other words, the compiler needs to allocate some register space for the main program and some for the ISR, and it should not let the instructions in the ISR overwrite the register values that belong to the main program if some of the registers in the main program and the ISR are named the same.

The inclusion of context switching, and therefore a shadow register file, brings forth a new architecture with a new timing diagram. These are shown in Figs. 7.18, 7.19 and 7.20. The initial steps of the timing diagram in Fig. 7.18 are basically the same as the timing diagram in Fig. 7.16.

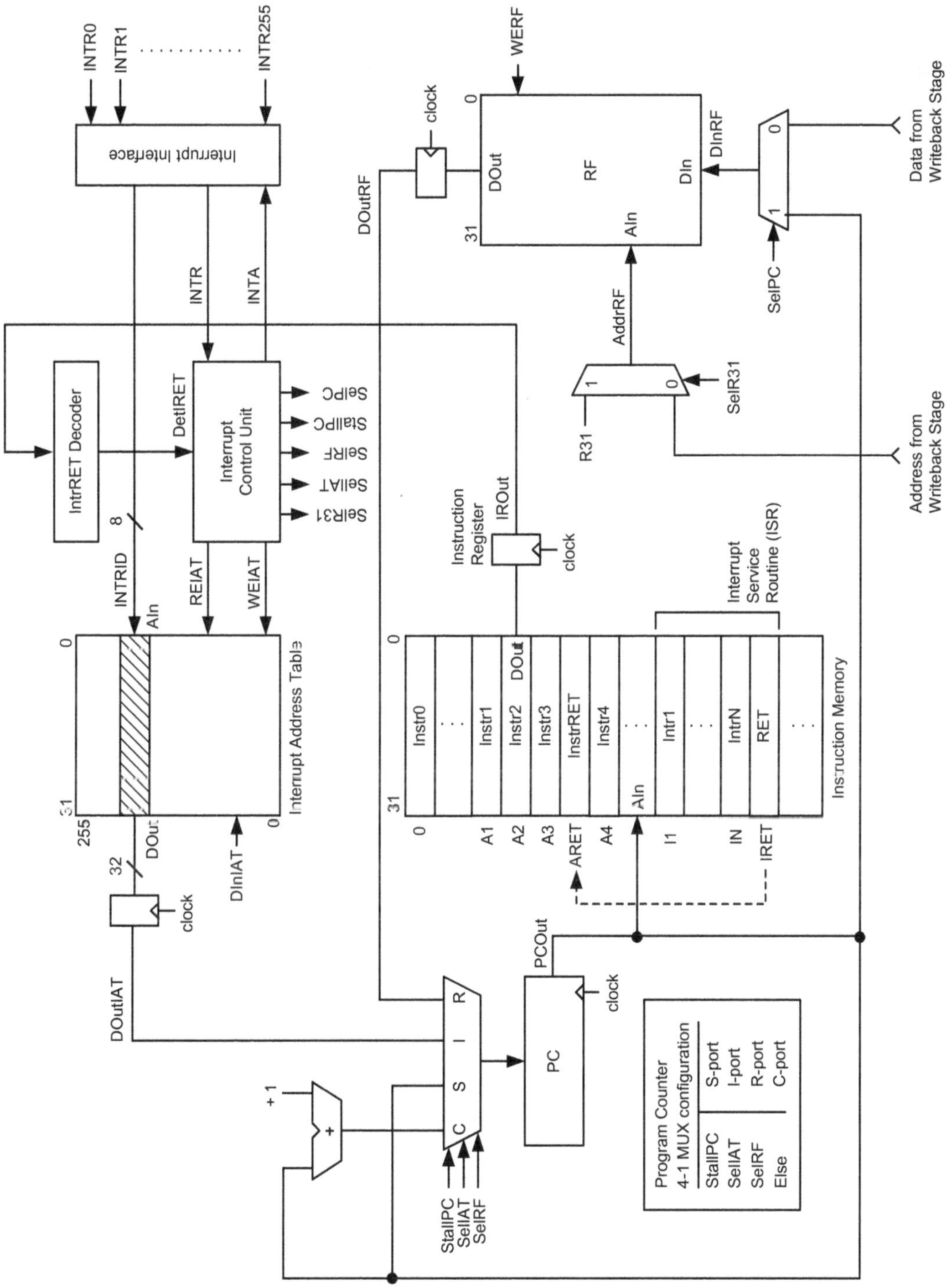

Fig. 7.15 Modified interrupt interface data-path

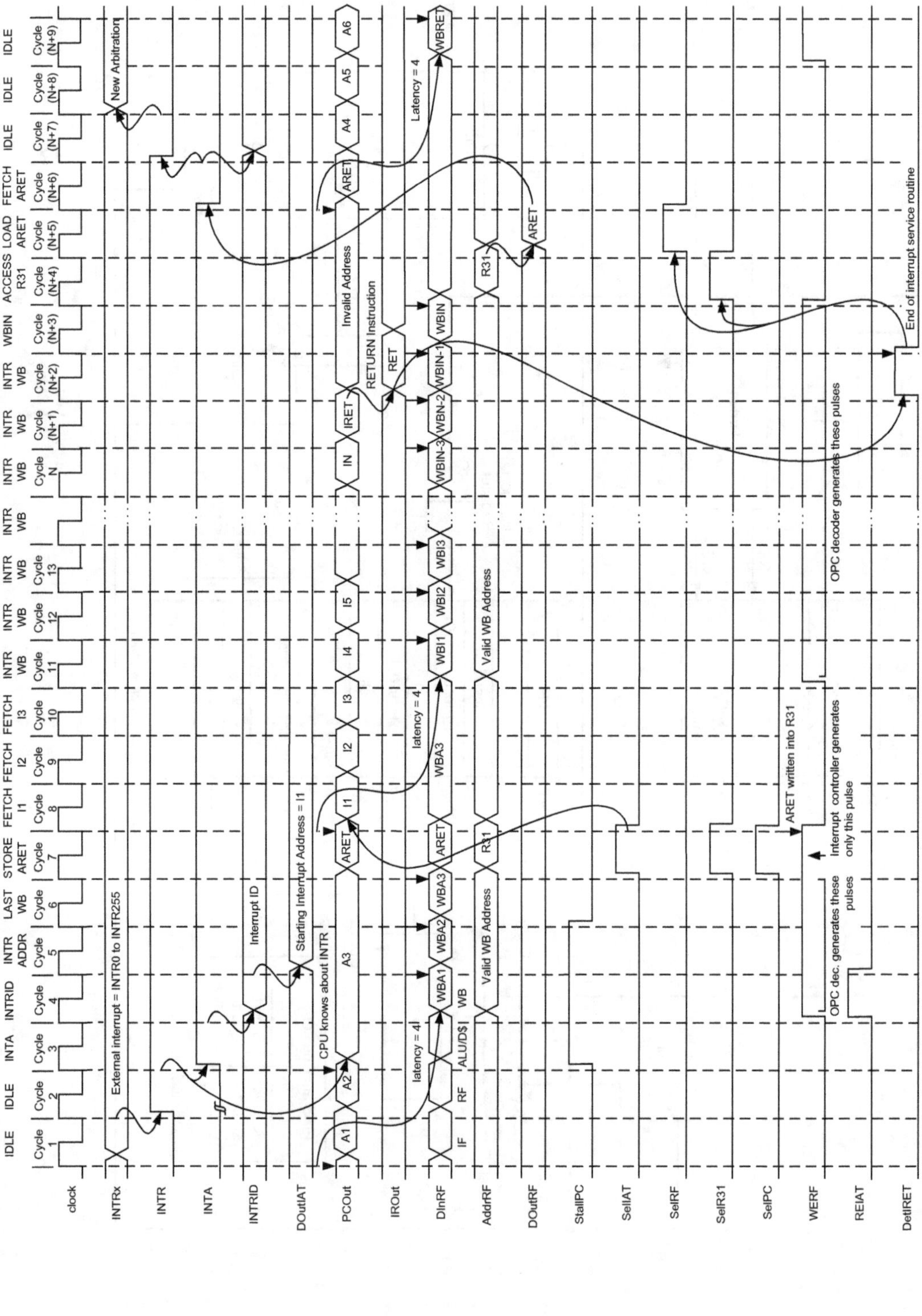

Fig. 7.16 Timing diagram of the modified interrupt interface in Fig. 7.15

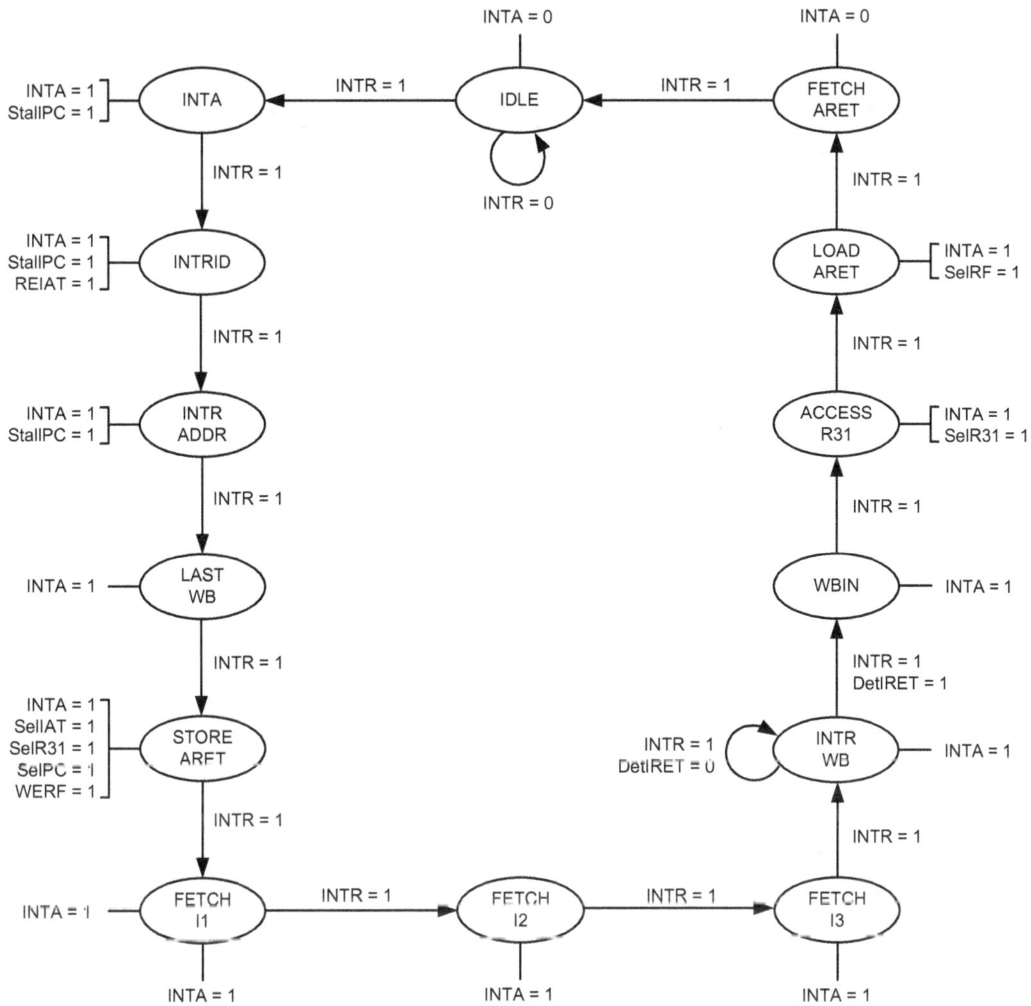

Fig. 7.17 Modified interrupt controller state diagram (control signals equal to logic 0 are omitted for simplicity)

As long as INTR = 0 there is no activity in the interrupt controller as before. The arrival of the interrupt (INTR = 1) starts the process and transitions INTA = 1. Generating INTA = 1 may take one cycle or many cycles after INTR = 1. However, when INTA = 1 (shown in cycle 3), the interrupt ID becomes available in cycle 4, and this input enables the IAT to generate a starting interrupt address at the DOutIAT node in cycle 5. But, before the ISR starts three steps must be completed. The first step is to write back the results of the instructions from the main user program, which ends at the end of cycle 6. The second step is to preserve the return address, ARET, in R31, which takes place in cycle 7. And, the third step is to finish the context switching, or transferring the contents of registers, R1 to R31, from the RF to the shadow register file, RFS, during cycles 8–37. In this process, no transfer is done on R0 since its contents are assumed zero. In this time interval, the address pointer for the RF, AddRF, is incremented by one while reading each register value. Every time the register data, DR1 to DR31(ARET), becomes available at the DOutRF node of the RF, it is promptly written to the corresponding shadow register from R1S to R31S. Cycle 38 signifies the end of the context switching where ARET is written to RFS31.

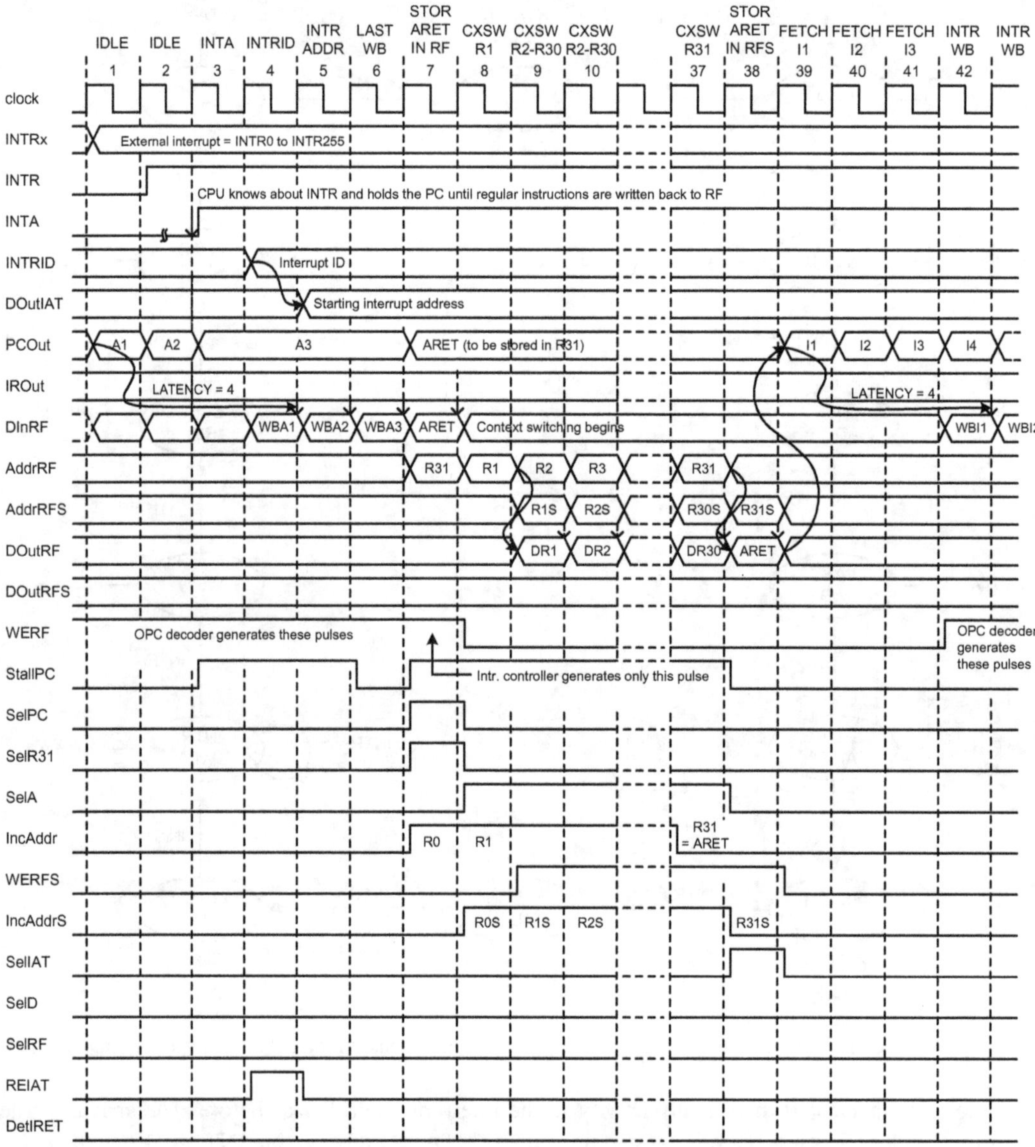

Fig. 7.18 Interrupt controller timing diagram with context switching

In cycle 39, the initial interrupt address routed through the I port of the 4-1 MUX shown in Fig. 7.20 becomes available at the output of the PC, and the first interrupt instruction is fetched from the address I1. Similarly, cycles 40 to 42 fetch interrupt instructions I2 to I4, respectively. The first interrupt write-back also takes place in cycle 42.

The process of fetching, processing and writing-back the results of ISR instructions to the RF continues for N number of cycles where N is an arbitrary number as shown in Fig. 7.19. In cycle (N + 39), the last interrupt instruction is fetched from the instruction memory. This is followed by fetching the return instruction, IRET, in cycle (N + 40). The contents of the return instruction, RET, is ultimately decoded in cycle (N + 41), and ends the active ISR. In cycle (N + 42), the contents of the last interrupt instruction, and in cycle (N + 43) the contents of the return instruction, RET, are written back to the RF (in actuality, RET instruction ends at the OPC decode stage, and nothing from

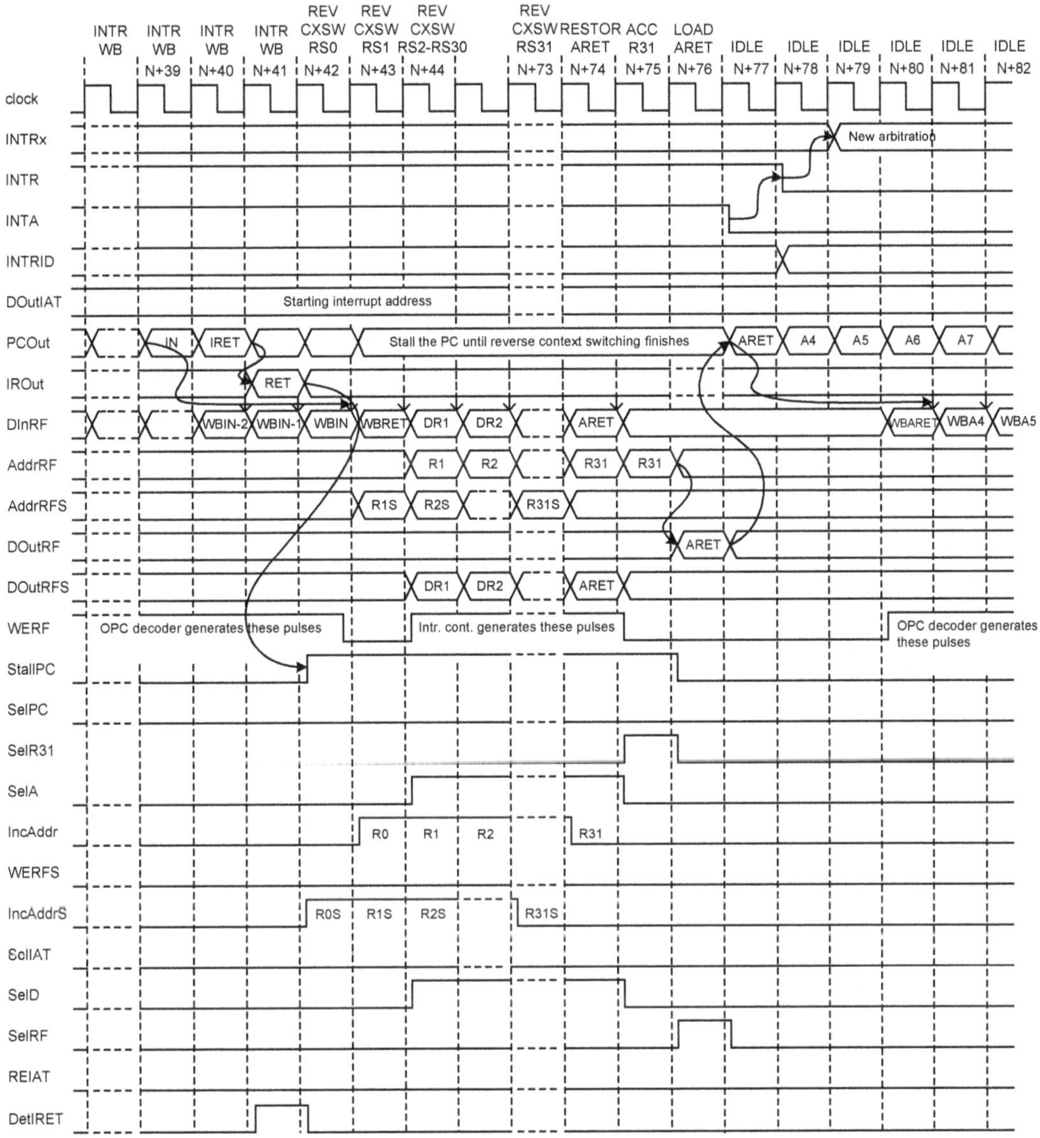

Fig. 7.19 Continuation of the interrupt controller timing diagram in Fig. 7.18

this particular instruction is written to the RF). In these last two cycles, the contents of instructions beyond IRET also become available at the IROut node and sent to the rest of the CPU pipeline. However, they are never written back to the RF because the RF is busy, retrieving the original register contents of the user program between cycles (N + 44) and (N + 74). In cycle (N + 44), the "reverse context switching" starts, and the contents of the shadow register file, RFS, are sequentially transferred back to the regular RF. For this process, the address pointer, AddrRFS, is incremented by one from R1S to R31S. Every time shadow register data, DR1 to DR31 (ARET), becomes available at the DOutRFS node, it is routed through the D port of the 3-1 MUX in Fig. 7.20 and written sequentially to the regular RF at addresses from R1 to R31, respectively. This process is shown between cycles (N + 44) and (N + 74) in Fig. 7.19. The retrieval of the return address, ARET, from R31 takes place in cycles (N + 75) and (N + 76). In cycle (N + 77), ARET is finally restored at the

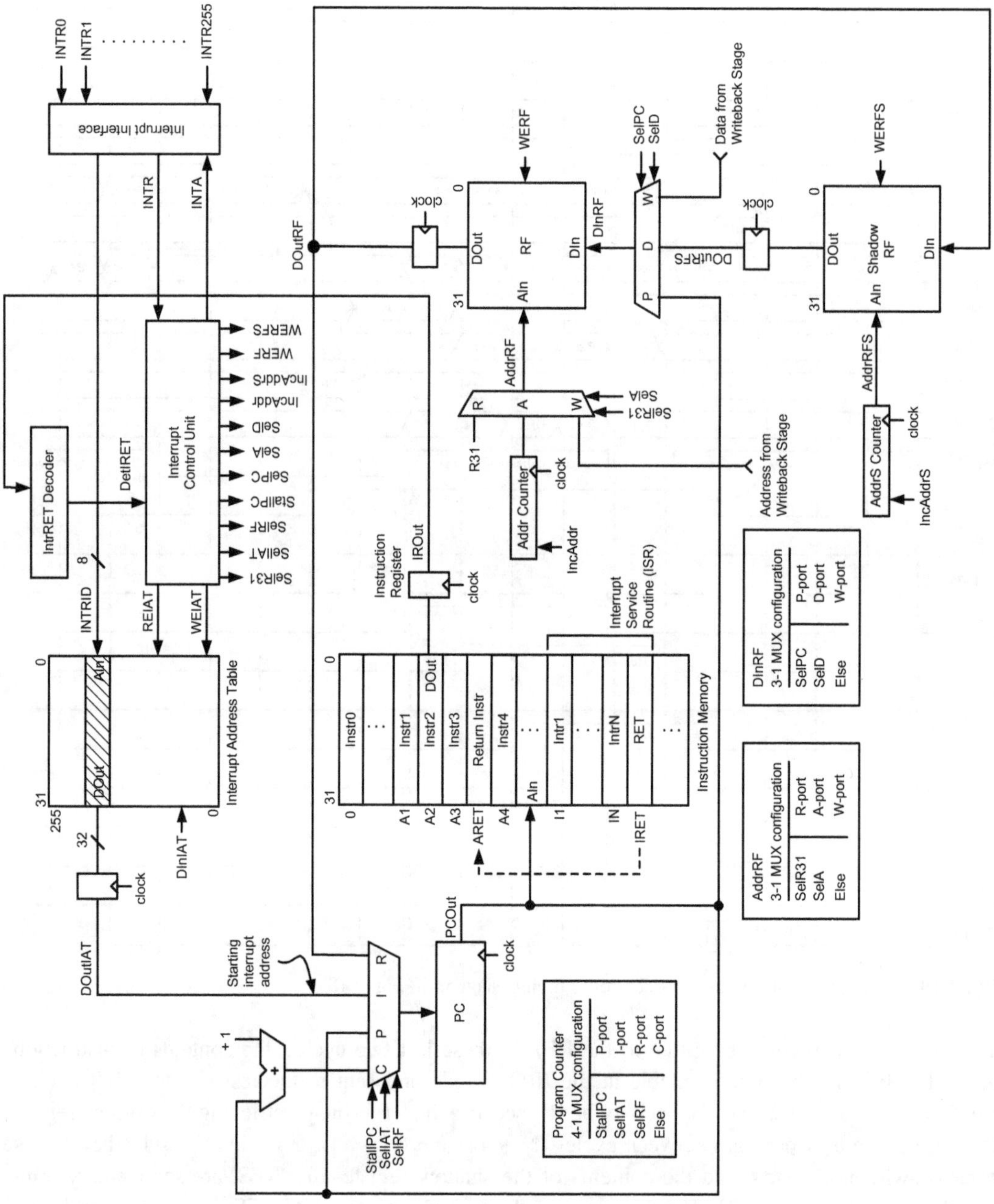

Fig. 7.20 Interrupt controller architecture with context switching

PCOut node. In this cycle, INTA also transitions to logic 0, informing the CPU to start servicing the next interrupt.

The Interrupt Control Unit (ICU) for the new architecture in Fig. 7.20 is shown in Fig. 7.21. In this figure, the ICU stays in the IDLE state while INTR = 0. When INTR = 1, the state machine transitions to the INTA state in cycle 3 where it generates INTA = 1 and StallPC = 1 as shown in Fig. 7.18. Note that all control outputs that stay at logic 0 are not shown in Fig. 7.21 for simplicity.

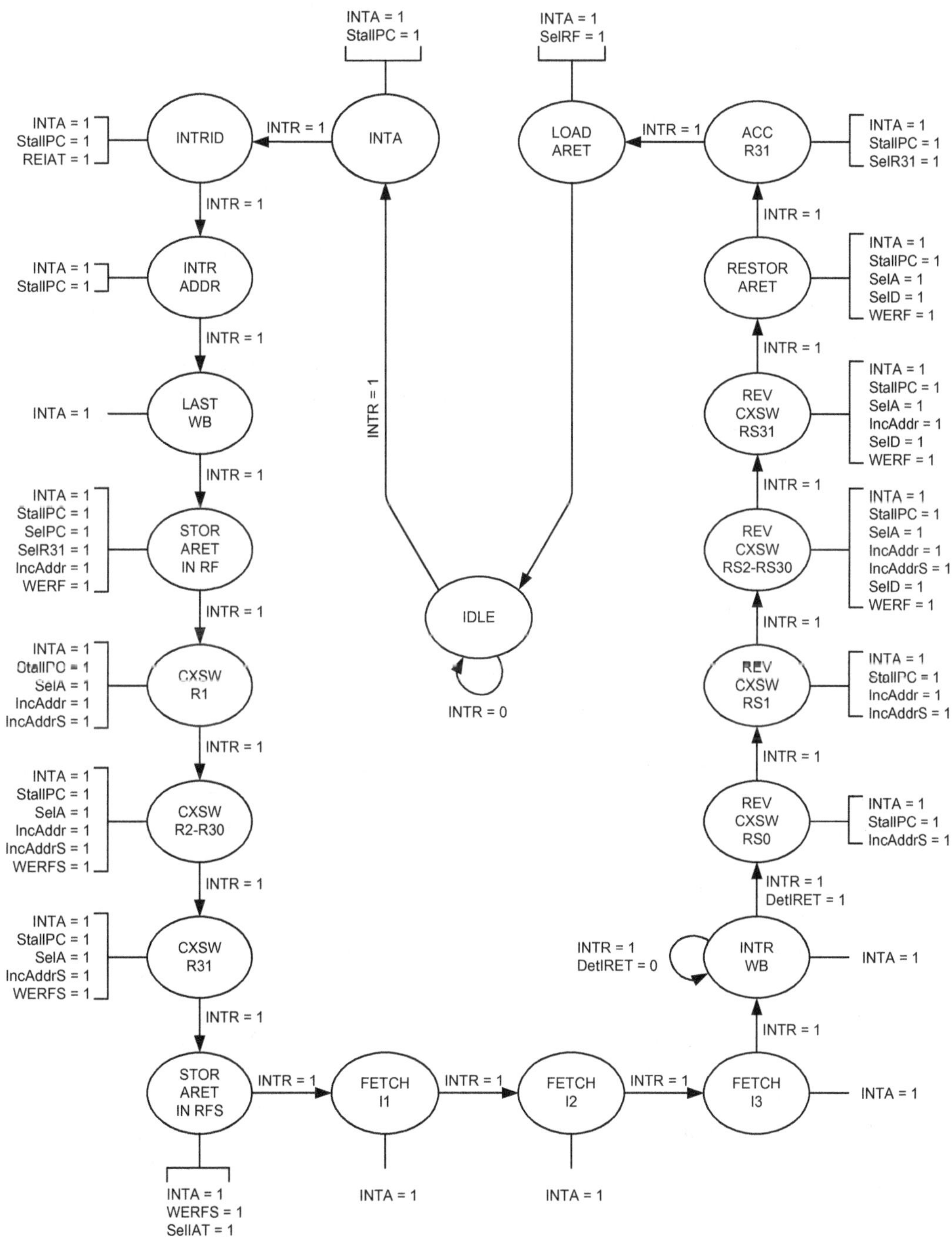

Fig. 7.21 Interrupt controller state machine with context switching (control signals equal to logic 0 are omitted for simplicy)

As long as INTR stays at logic 1 the state machine first transitions to the INTRID state in cycle 4, generating INTA = 1, StallPC = 1 and REIAT = 1, then to the INTR ADDR state with INTA = 1 and StallPC = 1 in cycle 5, and finally to the LASTWB state with only INTA = 1 in cycle 6. This completes all the write-backs from the main program. In cycle 7, the ICU transitions to the state STOR ARET IN RF where two separate events take place. The first event is to store the return address, ARET, in R31 by activating the P port of the 3-1 data MUX by SelPC = 1 and the R port of the 3-1 address MUX by SelR31 = 1. The second event is to start incrementing the address counter for the RF by IncAddr = 1. In this state, INTA and StallPC still stay at logic 1. In cycle 8, the state machine enters the CXSW R1 state where the context switching starts. In this state, the ICU activates the A port of the 3-1 address MUX by Sel A = 1 so that the output of the address counter connects to the address input of the RF, AddrRF. Also, both IncAddr and IncAddrS signals become logic 1 to generate the next address values for the RF and RFS while INTA = StallPC = 1. Cycles 9 to 36 represent the CXSW R2-R30 state where contents of each register in the RF are written to the corresponding register in RFS. In this state, Sel A is still at logic 1 to enable the A port of the 3-1 address MUX in order to receive address values generated by the address pointer, WERFS = 1 to write each register value from the RF to the corresponding register in the RFS, and finally IncAddr = IncAddrS = 1 to keep incrementing addresses for the RF and RFS while INTA = StallPC = 1. In cycle 37, the ICU transitions to the CXSW R31 state. In this state, Sel A = 1 sends the last address value, R31, to the AddrRF port, IncAddrS = 1 increments the RFS address to R31S, and WERFS = 1 stores DR30 in R30S in the RFS. Cycle 38 produces the STOR ARET IN RFS state and ends the context switching process. In this state, WERFS = 1 stores ARET in R31S, and SelIAT = 1 allows the starting interrupt address, I1, to be at the output of the PC in the next cycle. The states, FETCH I1, FETCH I2, FETCH I3, are generated to fetch the interrupt instructions, I1, I2 and I3, in cycles 39, 40 and 41, respectively. Interrupt instruction write-backs start in cycle 42 where the ICU enters the INTR WB state. The state machine stays in this state until it detects the return opcode, RET, in the form of DetIRET = 1 from the opcode decoder. The ICU leaves the INTR WB state in cycle (N + 41) and enters a new state, the REV CXSW RS0 state, in cycle (N + 42) to start the reverse context switching process in order to transfer the contents of each register from the RFS to the RF in sequential order. The REV CXSW RS0 state produces IncrAddrS = 1 to generate the first RFS address, RS1, in the next clock cycle. This state is followed by the REV CXSW RS1 state in cycle (N + 43) where IncAddr = IncrAddrS = 1 to increment both address pointers for the RF and RFS. This trend continues from cycle (N + 44) to cycle (N + 73) where the ICU stays in the REV CXSW RS2-RS30 state. During this interval, Sel A = 1 selects the A port of the 3-1 address MUX to propagate the addresses generated by the address pointer to the AddrRF node of the RF, Sel D = 1 activates the D port of the 3-1 data MUX to transfer each register data from the RFS to the DInRF port of the RF, and WERF = 1 writes each transferred register value to the RF while incrementing both address pointers by IncAddr = IncAddrS = 1. Cycle (N + 73) corresponds to the REV CXSW RS31 state where DR30 is written to R30 by SelA = SelD = WERF = 1 while the address pointer to the RF is incremented one last time by IncAddr = 1. The next cycle, (N + 74), corresponds to the RESTOR ARET state and restores the return address, ARET, in the RF. In this state, ARET is written to R31 by SelA = SelD = WERF = 1. In cycle (N + 75), the state machine transitions to the ACCR31 state where SelR31 = 1 selects the R port of the 3-1 address MUX to propagate R31 to the AddrRF node. In cycle (N + 76), the register value at R31, ARET, is read out from the RF to the DOutRF node, and the ICU activates the R port of the 4-1 MUX to allow this value to be loaded to the PC. This corresponds to the LOAD ARET state in Fig. 7.21. Finally, in cycle (N + 77), the state machine enters the IDLE state where the return address, ARET, is produced at the PC output, and the main program resumes.

7.4 Serial Transmitter Receiver Interface

There are times when the CPU needs to use its serial I^2C or SPI interface in order to communicate with an I/O device. The serial interface consists of a transmitter to send serial data to an I/O device, and a receiver to receive serial data from the same device on a one-bit bus. The following section describes the basic structure of a transceiver composed of a transmitter and a receiver to handle one-bit serial data.

Transmitter

Figure 7.22 shows the data-path of a transmitter where an incoming 32-bit data from the CPU is received at the TXIn[31:0] port, and stored in one of the two buffers before being serially sent out from the TXOut port. Each buffer is essentially a shift register. Once a 32-bit data packet is loaded into a shift register, the bits start shifting from the least significant bit position to the most significant bit position until all 32 bits are sent out. Transmitting serial data in the reverse order is also possible depending on the bus protocol the I/O device uses at the receiving end.

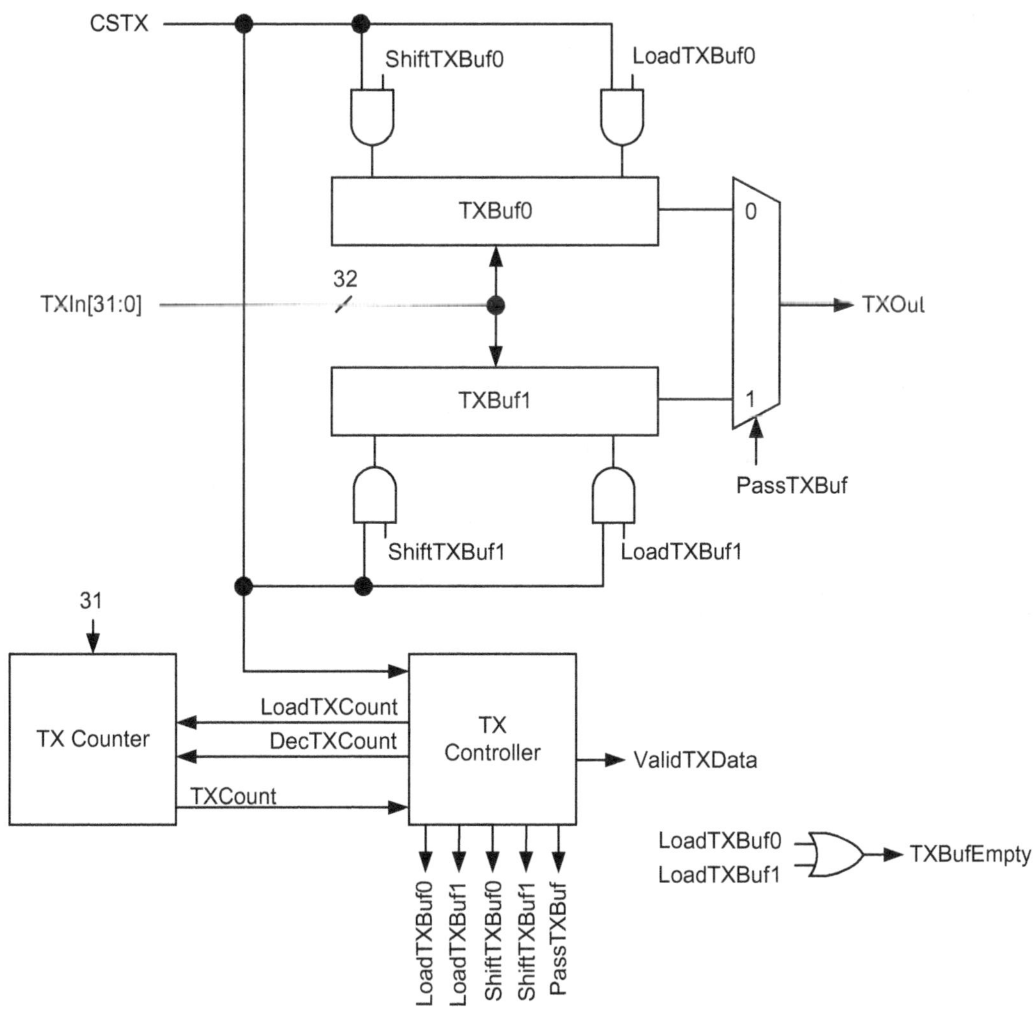

Fig. 7.22 Transmitter data-path

This design uses a dual buffer scheme to overcome the difficulties of bus traffic such as long waiting times when accessing system's main memory. A single buffer may also be sufficient for this particular architecture provided that every time data is needed for a specific buffer there will be an associated waiting period before data arrives from the main memory. Once the main memory is accessed, storing a 32-bit data in one of the transmitter buffers takes only a cycle. Streaming all 32 bits from a particular buffer, on the other hand, takes 32 consecutive clock cycles. The clock used to send serial data may also be a slower clock depending on the design constraints. Therefore, the transmitter uses all 32 clock periods to request, wait and receive data to its secondary buffer while it streams bits out of the first buffer. When the first buffer becomes empty, the transmitter immediately starts streaming data out of its secondary buffer while it fills the first buffer.

PassTxBuf input in Fig. 7.22 is a control signal for the 2-1 MUX that determines when to switch buffer outputs. LoadTxBuf0 and LoadTXBuf1 inputs determine when to load the first and the second buffers, respectively. ShiftTXBuf0 and ShiftTXBuf1 inputs control the beginning and the end of the serial data shift from buffer 0 and buffer 1, respectively. The CSTX is a Chip-Select input port for the transmitter, and it stays at logic 1 as long as the system uses the transmitter.

The timing diagram in Fig. 7.23 shows how data is stored and streamed out of the data buffers. It is constructed while the transmitter data-path in Fig. 7.22 is being developed. Once again, the top part of this diagram shows the bus-level data signals that describe the data-flow while the bottom part contains the control signals that govern this data-flow.

The transmitter wakes up when it receives an active-high CSTX signal from the CPU in cycle 1. In cycles 2 and 3, the 32-bit data packets, Buf0 and Buf1, fill the first and second transmitter buffers, TXBuf0 and TXBuf1, respectively. Once the first buffer is full in cycle 2, single bits start emerging from the least significant bit position of the buffer (the shifting mechanism can also be configured in the reverse order such that the buffer emits bits from the most significant bit position) in cycle 3. The first bit that comes out of TXBuf0 in cycle 3 is Bit0 which is the least significant bit of the data packet. This is followed by Bit1 through Bit31 between cycles 4 and 34, respectively. When the first buffer becomes empty, the transmitter immediately switches to its second buffer and starts streaming bits from TXBuf1. In the mean time, the transmitter fills TXBuf0 with a new 32-bit of data in cycle 35 as long as there is no bus traffic. Emptying the second buffer takes until cycle 66 when Bit31 is sent out from the TXOut terminal. When TXBuf1 is empty, the transmitter starts streaming out data from TXBuf0 in cycle 67 while filling TXBuf1. The process of filling one buffer while streaming bits out of the second continues as long as the data stored in the main memory is fully exhausted. Figure 7.23 shows how data packets from the main memory become available when the transmitter switches from its empty buffer to its full buffer, ignoring any delay associated with accessing the main memory. In reality, when the transmitter starts streaming data out of the full buffer, it immediately generates an interrupt for its empty buffer to fetch data from the main memory. The waiting period is 31 clock cycles. In the 32nd clock cycle, the empty buffer must be full. Otherwise, the transmitter stalls, and no new data can be transmitted.

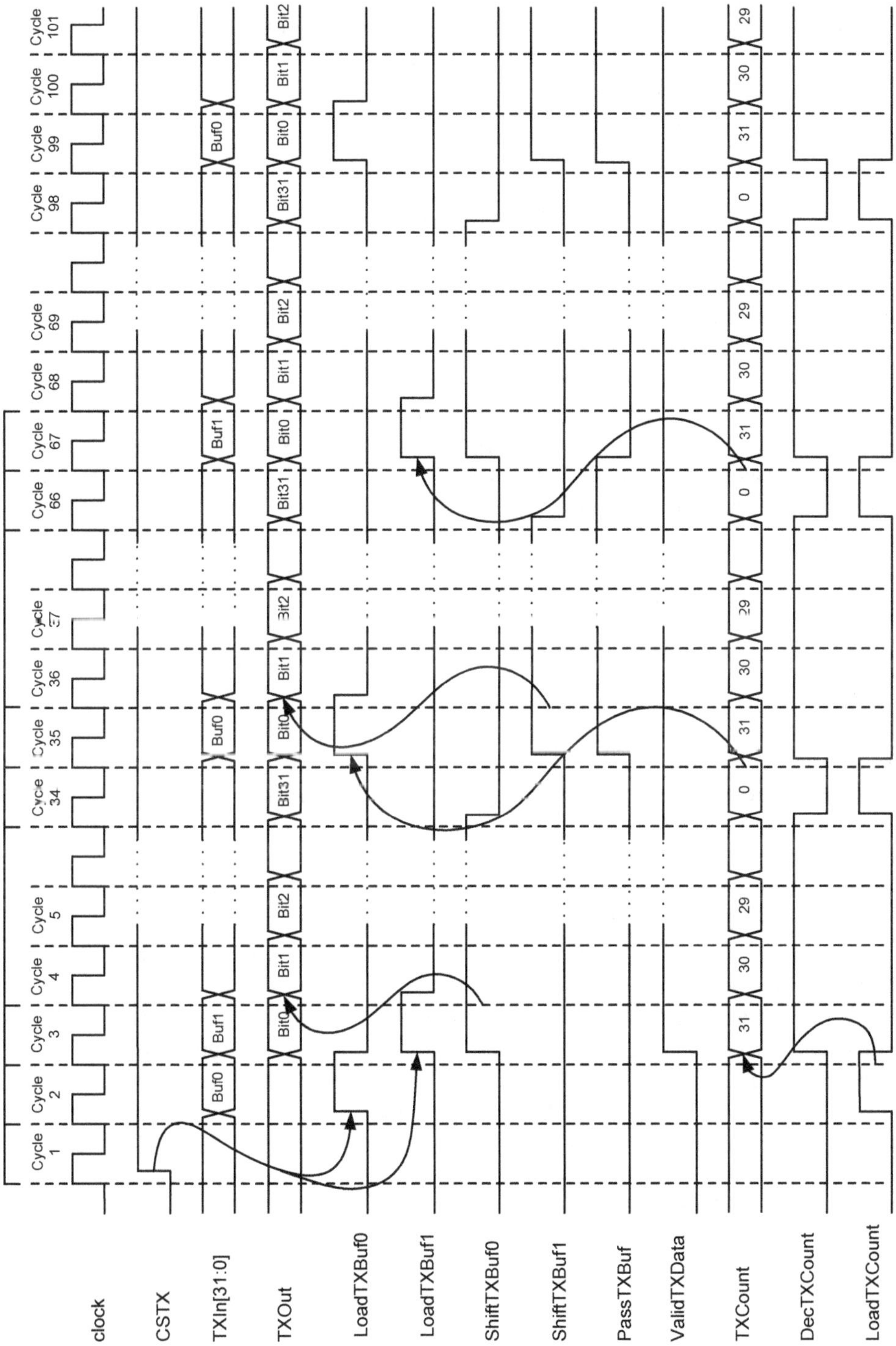

Fig. 7.23 Transmitter timing diagram

The control signals that govern the data-flow in Fig. 7.22 constitute the second part of the timing diagram. Once the active-high CSTX signal is received, the LoadTXBuf0 signal transitions to logic 1 to load the first 32-bit data packet, Buf0, to TXBuf0 in cycle 2. In cycle 3, the second 32-bit data packet, Buf1, is loaded to TXBuf1, which requires the LoadTXBuf1 signal to be at logic 1. From cycle 3 until cycle 33, TXBuf0 works as a shift register to stream bits from 0 to 31. Therefore, the ShiftTXBuf0 signal stays at logic 1 during this period. In cycle 34, the last bit is shifted out of TXBuf0, and therefore, all data-flow controls for this buffer transition to logic 0. In cycle 35, the LoadTXBuf0 signal transitions to logic 1 to fill the empty buffer, TXBuf0. The ShiftTXBuf1 signal also transitions to logic 1 in the same cycle to shift bits from 0 to 31 until cycle 66. Cycle 35 is also the time to switch the buffer outputs. Therefore, the PassTXBuf signal becomes logic 1 in this cycle until cycle 67 when all the data in TXBuf1 is streamed out. Cycle 66 is the cycle to deliver the last bit out of TXBuf1. Cycles 67 through 98 are exact replicas of cycles 3 to 34 when TXBuf1 is filled while bits are emitted from TXBuf0. The ValidTXData signal validates bits for external use as they are streamed out of the transmitter. Therefore, this signal stays at logic 1 from cycle 3 until the last transmitter bit.

Figure 7.24 shows the state diagram of the controller unit to load and shift data in each buffer, to switch buffer outputs, and to validate bits out of the transmitter. The state names in this figure follow the names depicted at the top of the timing diagram in Fig. 7.23. When there is no activity in CSTX signal, the state machine stays in the IDLE state. When CSTX = 1, the controller transitions to the PRELOAD BUF0 state where it produces LoadTXBuf0 = 1 to load TXBuf0, and LoadTXCount = 1 to load the TXcounter in Fig. 7.22 with the value of 31. This state is associated with cycle 2 in the timing diagram. Note that the TXCounter is a five-bit counter which is used to detect the end of the serial data stream, and it is essential for the controller to be able to make a transition to the next state. As long as CSTX = 1, the state machine transitions to the LOAD BUF1 state where it produces LoadTXBuf1 = ShiftTXBuf0 = ValidTXData = 1. This state corresponds to cycle 3 in the timing diagram. This state is also the beginning of the count-down stage where the TXCounter output starts decrementing from 31 towards 0 by DecTXCount = 1. Next, the state machine goes to the EMPTY BUF0 state, and stays in this state as long as the output of the TXCounter, TXCount, is greater than one. This state corresponds to cycles 4 through 33 in the timing diagram. When TXCount = 1, the controller goes to the EMPTY Last Bit BUF0 state where Bit31 is shifted out of TXBuf0. This state is equivalent to cycle 34 in the timing diagram when the TXCounter is reloaded with the value of 31 to start another count-down. As long as CSTX = 1, the state machine first transitions to the LOAD BUF0 state in cycle 35, and then to the EMPTY BUF1 state as the TXCounter decrements towards 1 from cycle 36 to cycle 65. When TXCount = 1, the state machine transitions to the EMPTY Last Bit BUF1 state in cycle 66 when the last bit of TXBuf1 is shifted out. From cycle 67 onwards, the state machine goes back to the LOAD BUF1 state, and traces through the previous five states since the control outputs generated in each state are identical to the ones in the timing diagram in each clock cycle. The state diagram in Fig. 7.24 does not show the transitions from an arbitrary state to the IDLE state when CSTX = 0 to improve readability. The case when the transmitter exhausts all of its valid data from both of its buffers and forced to stall is not shown in Fig. 7.24 either. If heavy bus traffic is expected, the reader should employ an additional STALL state in case new data is not yet loaded to TXBuf0 or TXBuf1 before the state machine transitions to the LOAD BUF0 or LOAD BUF1 states, respectively.

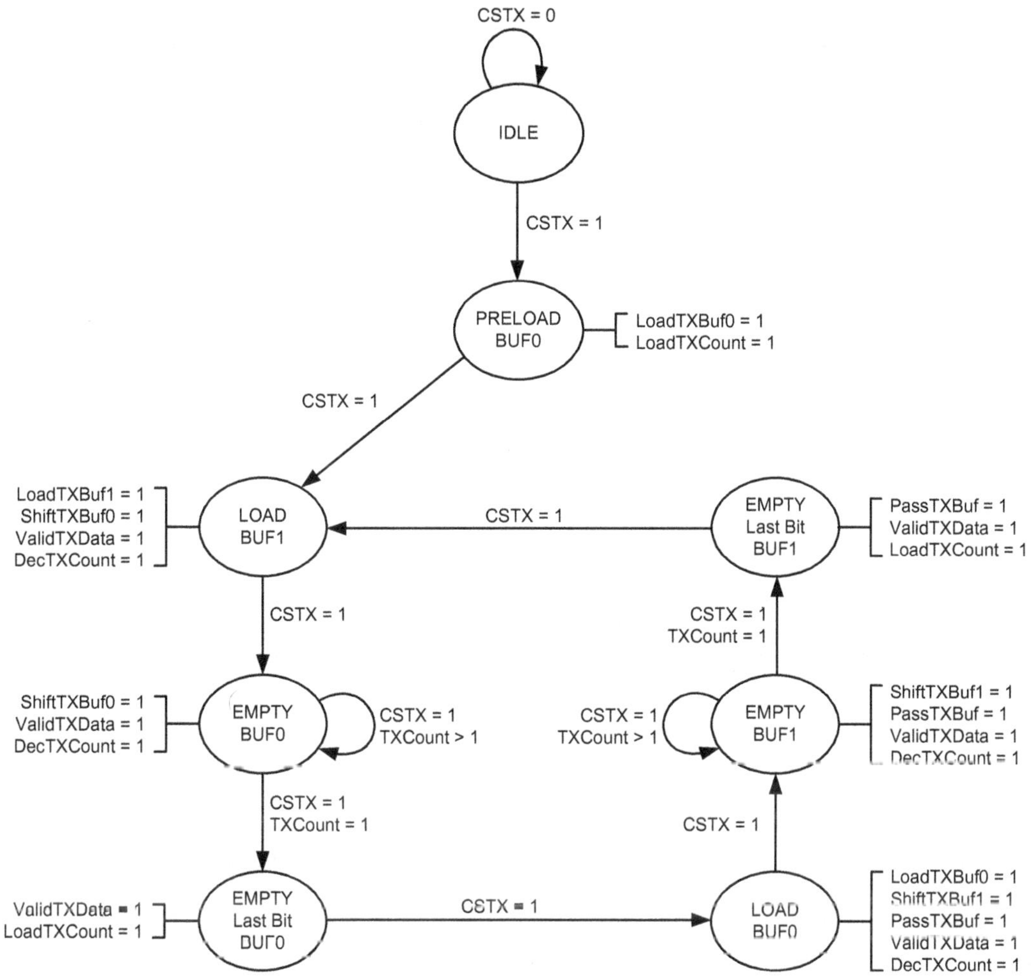

Fig. 7.24 Transmitter controller state diagram (control signals equal to logic 0 are omitted for simplicity)

Receiver

Figure 7.25 shows the data-path of a receiver where incoming data bits are serially received by the RX buffer, RXBuf, at the RXIn port before being packed as 32-bit data packets and sent to the CPU from the RXOut[31:0] port. This architecture can also be accomplished with multiple buffers in case the receive clock frequency is much higher than the processor clock frequency, or the CPU bus is often occupied by other data transactions.

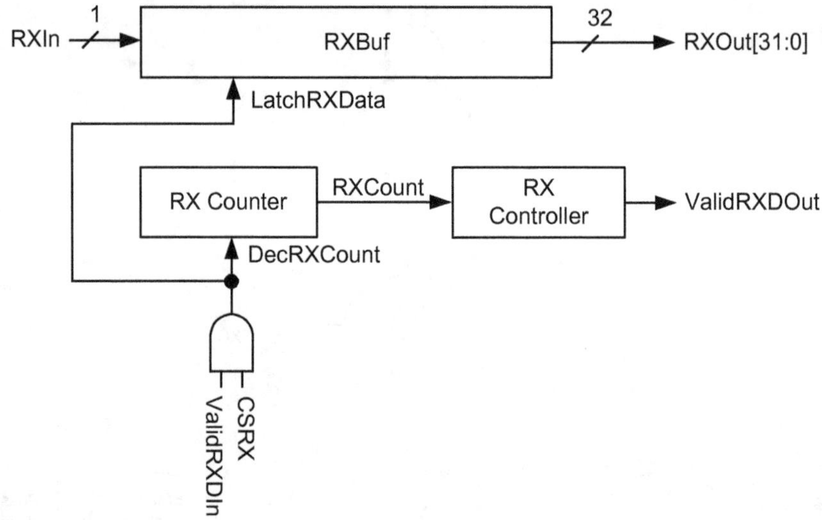

Fig. 7.25 Receiver data-path

Figure 7.26 summarizes the operation of the receiver in a timing diagram. Once the receiver is activated by CSRX = 1 in cycle 1, any incoming data bit is ignored until the ValidRXDIn signal transitions to logic 1. In other words, this external signal validates the data bit at the RXIn port, and indicates when to start latching data bits into the RX buffer. As a result, Bit0 is stored in RXBuf in cycle 2 and Bit31 in cycle 33 (the reverse bit order storage to this buffer is also possible provided that the I/O device sends the most significant bit, Bit31, first and the least significant bit, Bit0, last). In cycle 34, a new Bit0 is fetched by the receive buffer. This cycle is also the time period to pack all 32 bits, and send them out of the RXOut[31:0] port. The 32-bit data is accompanied by the ValidRXDOut signal for validation. In order to determine in which clock cycle the ValidRXDOut signal transitions to logic 1, a five-bit counter is used. This counter starts decrementing as soon as the DecRXCount or the LatchRXData signal goes to logic 1. When the counter reaches 0, the RX Controller produces ValidRXDOut = 1 in the following cycle to validate the 32-bit word at the RXOut[31:0] port. From cycle 34 to 67, the receiver keeps latching new valid bits into RXBuf. There may be a period where the serial bit stream may not be valid (ValidRXDIn = 0) such as the cycles 38 and 39. During this period, both the latching action at the RXIn port and the count-down mechanism at the RX Counter instantaneously stop by LatchRXData = 0 and DecRXCount = 0, respectively. The normal receiver operation resumes as soon as the ValidRXDIn signal transitions to logic 1 in cycle 40.

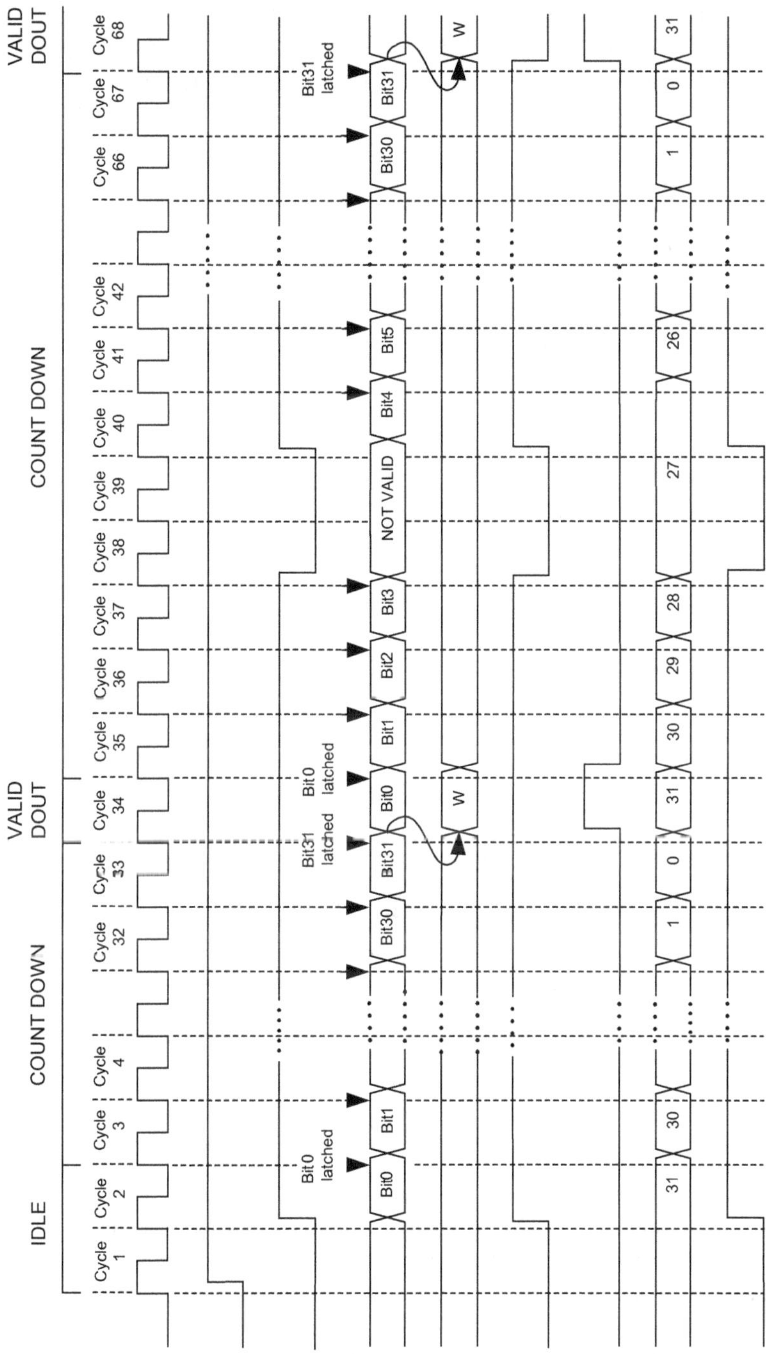

Fig. 7.26 Receiver timing diagram

The RX controller is a simple state machine with three distinct states as shown in Fig. 7.27. The IDLE state is the state when CSRX = 0 or ValidRXDIn = 0. Even when the CSRX signal goes to logic 1, the controller stays in this state as long as ValidRXDIn = 0. This translates to cycles 1 and 2 in Fig. 7.26. When the CSRX and ValidRXDIn signals both go to logic 1, the controller moves to the COUNT DOWN state to fill RXBuf. The controller stays in this state until the RXCount signal reaches 0. This state covers the cycles from 3 to 33 in the timing diagram. In the next cycle, the state machine goes to the VALID DOUT state where it stays for only one clock cycle and produces ValidRXDOut = 1 for the 32-bit data at the RXOut[31:0] port. Following the VALID DOUT state, the controller goes back to the COUNT DOWN state where it starts filling the receive buffer again. As long as valid bits arrive at the RXIn port, the state machine rotates between the COUNT DOWN and the VALID DOUT states in Fig. 7.27. When CSRX = 0 or ValidRXDIn = 0, the state machine transitions either from the COUNT DOWN or the VALID DOUT state to the IDLE state. These transitions are omitted in Fig. 7.27 to maintain simplicity.

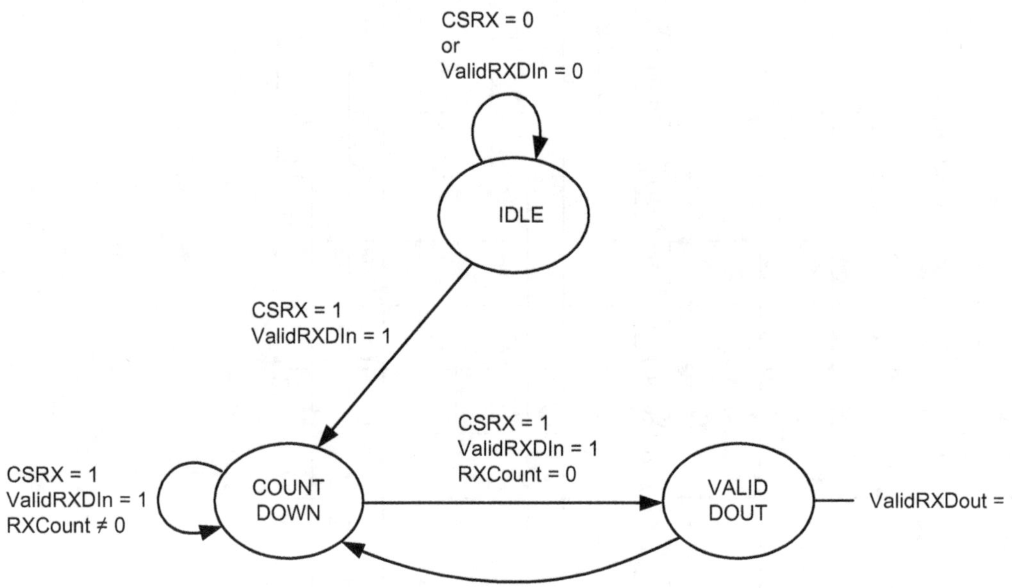

Fig. 7.27 Receiver controller state diagram (control signals equal to logic 0 are omitted for simplicity)

7.5 Timers

Every digital system contains programmable timers to handle a multitude of tasks. If an external event needs to be monitored, it calls for a timer. Periodic internal system tasks are also managed by timers. Timers can also be used to generate square waveforms or pulses with adjustable pulse widths such as Pulse Width Modulation (PWM) signals to control output devices or perform periodic tasks.

A basic system timer is shown in Fig. 7.28. This timer essentially consists of a counter, a compare unit and two registers. The first register stores the entire timer period after which the counter receives an automatic reset, and the other divides the clock frequency of the counter. The period and the divide by N registers are fully programmable. The compare unit is simply a subtractor which subtracts the counter output from the period value. As the counter starts incrementing from zero and ultimately reaches the value stored in the period register, the output of the subtactor and its sign bit become all zero. These bits are subsequently decoded by the compare unit which produces logic 1 at the timer output as shown in this figure.

The counter in Fig. 7.28 has an automatic reset. However, some counters do not have this feature, and they simply increment until all their output bits become logic 1. In the next clock cycle, the counter resets and starts counting up from zero again.

The basic timer configuration can be molded into various topologies. However, all modifications still contain a counter, a register that stores the cut-off period, and a comparator that compares the counter output against the value in the period register. The comparison can be achieved by a subtractor/decoder scheme as in Fig. 7.28.

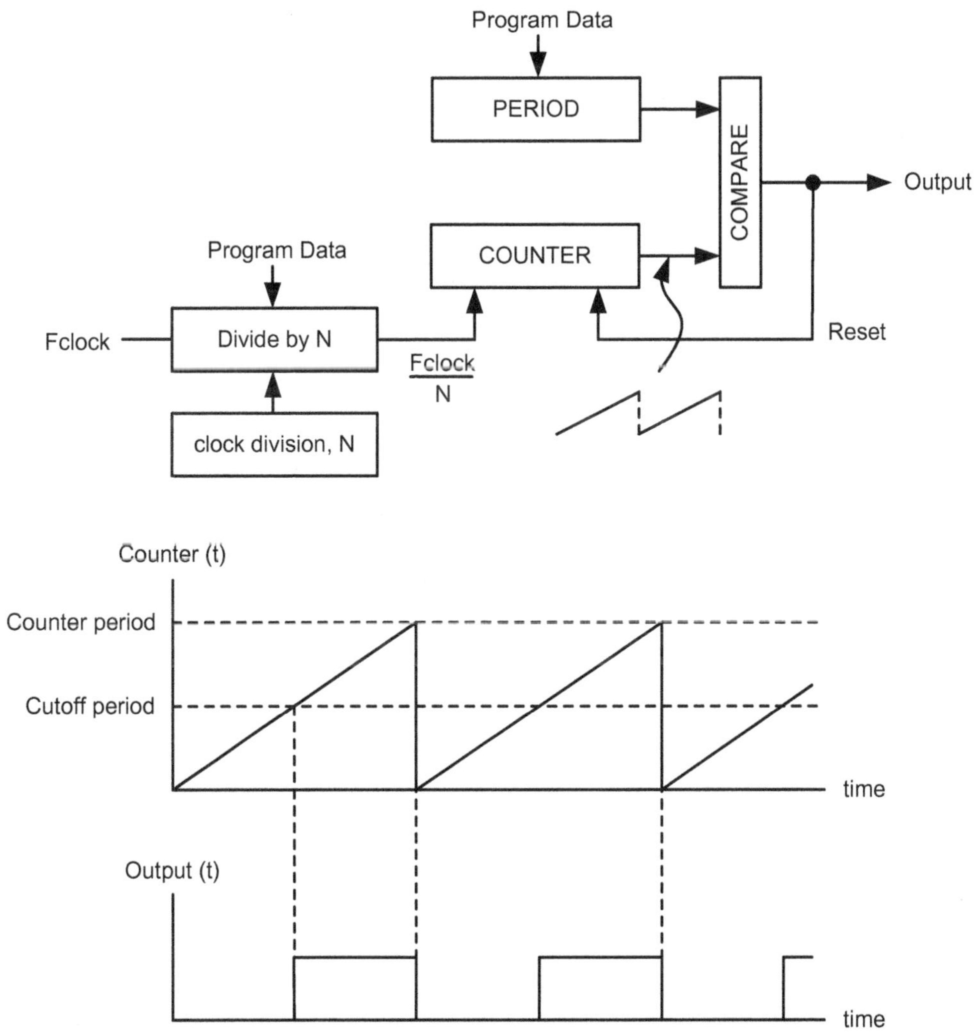

Fig. 7.28 Simplified timer block diagram

The following section presents many different forms of timers, each of which can still be modified and converted into various other forms that can produce additional features and functionality. The basic timers in this section are configured to produce a one-time pulse such as a one-shot timer, a periodic waveform with adjustable duty cycle such as a rate generator, a square waveform with fully programmable period, or a step function with an adjustable delay such as interrupt generator. There are subtle differences and incremental enhancements from one timer circuit to another, but in the end each timer uses a counter, a register and a comparator as pointed out before.

One-Shot Timer

The one-shot timer, as its name suggests, generates a single, non-repetitive pulse whose pulse width is programmed by the user. Figure 7.29 shows the micro-architecture of a typical one-shot timer.

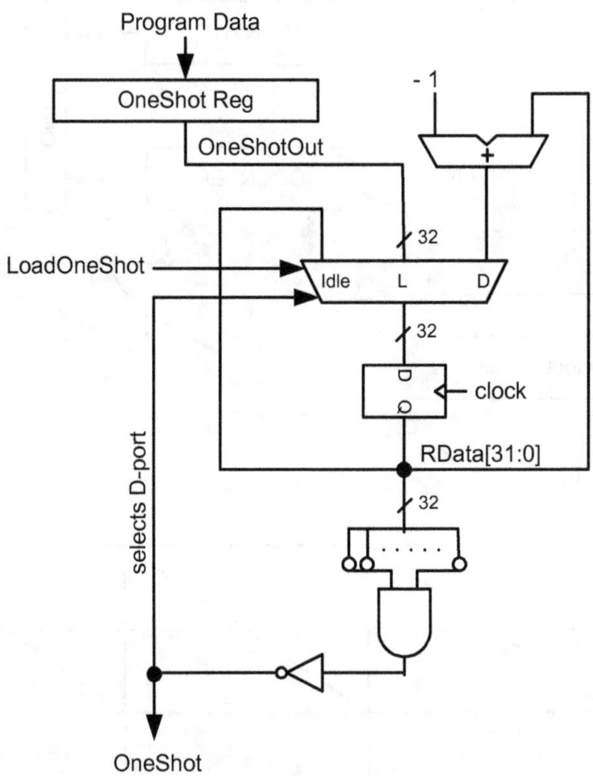

Fig. 7.29 Block diagram of the one-shot timer

The pulse width is stored in the OneShot register via a program bus. Once programmed, the data in this register is routed through the L-port of the 3-1 MUX by LoadOneShot = 1, and it provides an initial value for the down-counter. This is shown as OneShotOut = 4 in the timing diagram in Fig. 7.30 during cycle 1 as a numerical example. In cycle 2, the RData[31:0] node becomes 4. Since this value is different from 0, the decoder placed at the RData[31:0] node (a 32-input AND gate with an inverter) produces logic 1 at the OneShot output. This, in turn, activates the D-port of the 3-1 MUX and routes the decremented RData value, (RData – 1) = 3, to the input of the down-counter. In cycle 3, the RData

[31:0] node becomes 3, and the OneShot output stays at logic 1, keeping the D-port of the 3-1 MUX active. The decremented RData value, (RData – 1) = 2, is fed back to the timer input again. The D-port of the 3-1 MUX stays active until cycle 6 when the RData[31:0] node becomes zero. From this point forward, the Idle-port of the 3-1 MUX becomes active, and the timer output stays at zero. The timer stays in this state until it is reprogrammed with a new value.

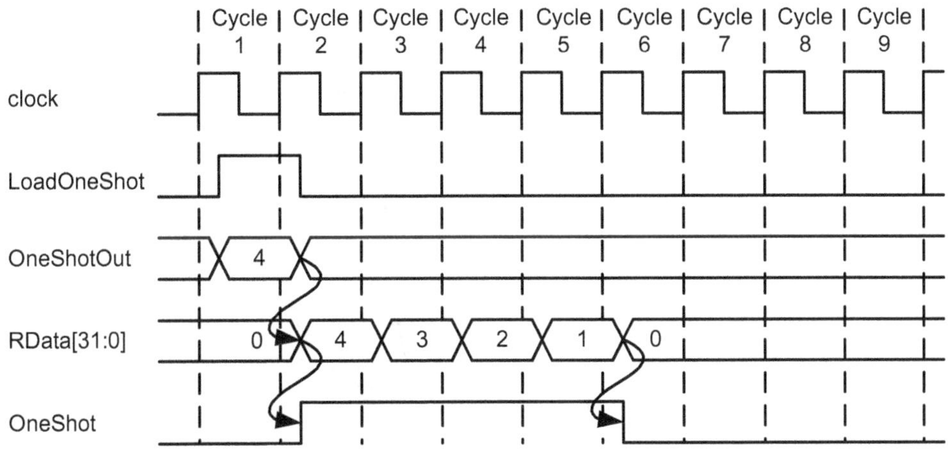

Fig. 7.30 Timing diagram of the one-shot timer (OneShot register in Fig. 7.29 is programmed with a value of 4 as an example)

Rate Generator

The rate generator is another type of timer which periodically generates single pulses separated by a programmable time duration. Once a desired rate is stored in the RateGen register via program bus, it is routed through the L-port of the 2-1 MUX to the input of the timer by LoadRateGen = 1 as shown in Fig. 7.31. This step is shown in cycle 1 of the timing diagram in Fig. 7.32 with RateGenOut = 4 as an example. In cycle 2, the RData[31:0] node becomes four, and the RateGen terminal becomes zero. Because neither the RateGen port nor the LoadRateGen input is at logic 1, the D-port of the 2-1 MUX becomes automatically active to allow the decremented RData, (RData – 1) = 3, to become the input of the down-counter. In cycle 3, RData[31:0] becomes equal to three, which keeps the D-port active because both RateGen and LoadRateGen are zero. The decremented RData, (RData – 1) = 2, is again routed to the input of the timer. This path stays active until RData[31:0] = 1. At this point, RateGen output becomes equal to one, and selects the L-port of the 2-1 MUX. The value in the RateGen register is reloaded to the input of the down-counter. This is shown in cycle 5 of the timing diagram. In cycle 6, RData[31:0] becomes four, and the RateGen output becomes zero. The cycles 7 through 9 are the exact replicas of the cycles 3 through 5. The RData node keeps decrementing until it reaches to one, at which point the RateGen output becomes one. Therefore, periodic single pulses are generated once in every four consecutive cycles at the RateGen output once the RateGen register is programmed with a value of four.

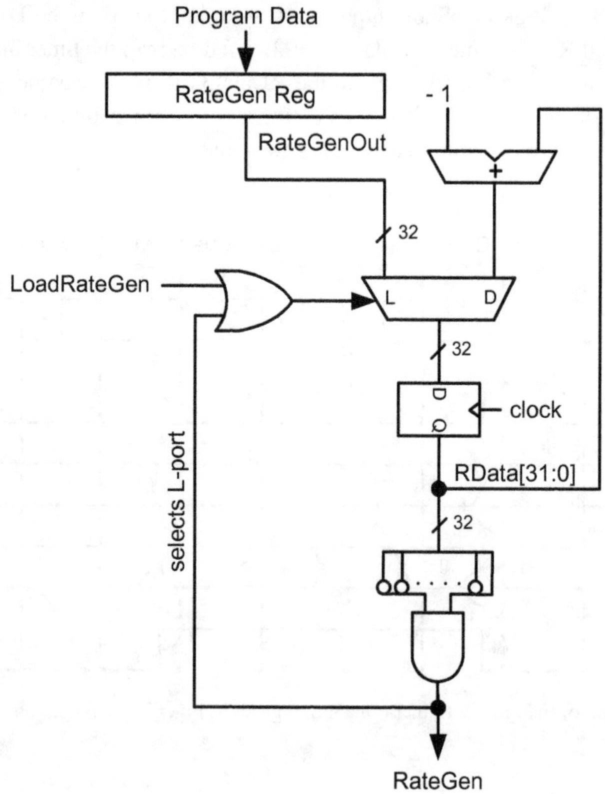

Fig. 7.31 Block diagram of the rate generator

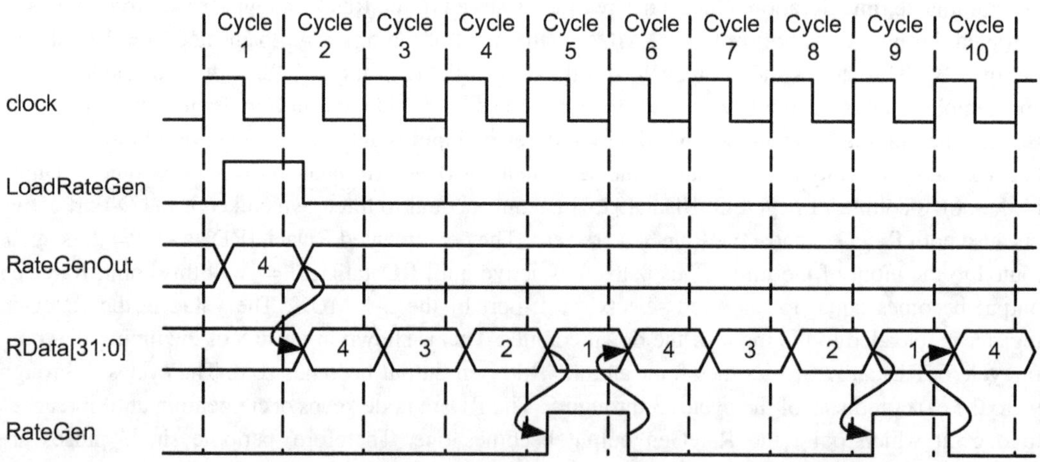

Fig. 7.32 Timing diagram of the rate generator (RateGen register in Fig. 7.31 is programmed with a value of 4 as an example)

Square Wave Generator

Square waveforms can also be generated by the timer as shown in Fig. 7.33. The pulse duration of the square wave is initially stored in the SqWave register through a program bus. Once the programming is finished, LoadSqWave = 1 loads the value of the Sqwave register through the L-port of the 2-1 MUX to the input of the down counter. This is shown in cycle 1 of the timing diagram in Fig. 7.34 with SqWaveOut = 3 as a numerical example. In cycle 2, RData[31:0] = 3, and RateOut = 0 since the 32-input AND gate can produce logic 1 only when RData[31:0] = 1. The decremented RData value, (RData – 1) = 2, is routed through the active D-port of the 2-1 MUX to the input of the down-counter. In cycle 3, the RData[31:0] node becomes two, but the RateOut node still stays at zero. When RData[31:0] = 1 in cycle 4, the 32-input AND gate produces RateOut = 1, and activates the L-port of the 2-1 MUX. As a result, the down-counter is reloaded with the value in the

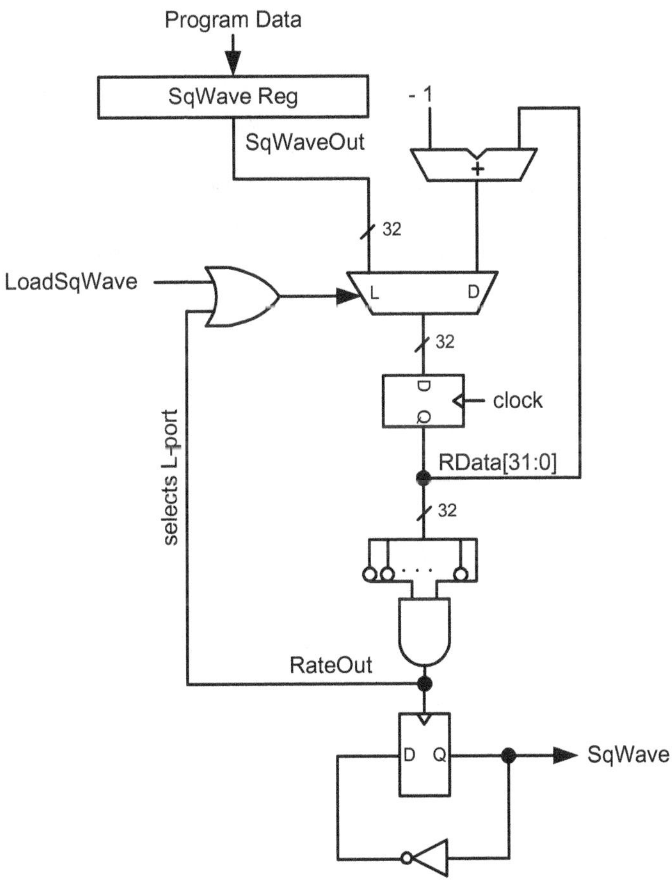

Fig. 7.33 Block diagram of the Square Wave Generator

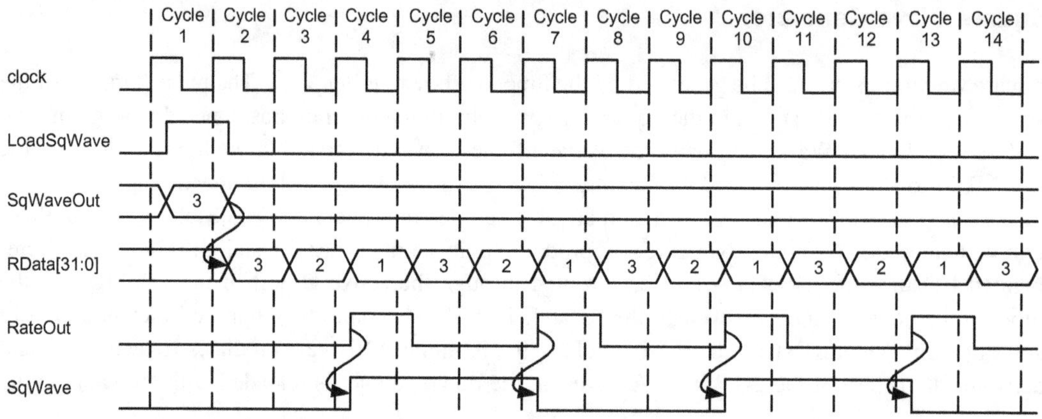

Fig. 7.34 Timing diagram of the square wave generator (SqWave register in Fig. 7.33 is programmed with a value of 3 as an example)

SqWave register. From this point forward, the circuit repeats the same pattern, producing a pulse in every three cycles at the RateOut node. Although the rate generator and the square wave circuits look identical, the square wave generator contains an additional state machine whose clock is controlled by the RateOut node. Therefore, at every positive edge of the RateOut signal, the value at the SqWave port alternates. If the SqWave output initially produces logic 0 between cycles 1 and 3, the positive edge of the RateOut signal in cycle 4 switches the value at the SqWave output from logic 0 to logic 1. Similarly, the RateOut pulse in cycle 7 changes the value of the SqWave output back to logic 0. Therefore, the circuit in Fig. 7.33 creates a square waveform whose frequency is fully programmable by the SqWave register.

Interrupt Generator

One of the most useful timer functions is to have a timer generate a predetermined interrupt signal for the system. If an external event needs to be observed at a specific time or if periodic sampling needs to be employed for an event, the system must have the means to generate an interrupt. It achieves this by using the circuit in Fig. 7.35. This circuit is composed of a basic down-counter whose output is configured to make a transition from logic 0 to logic 1 when the count-down reaches zero.

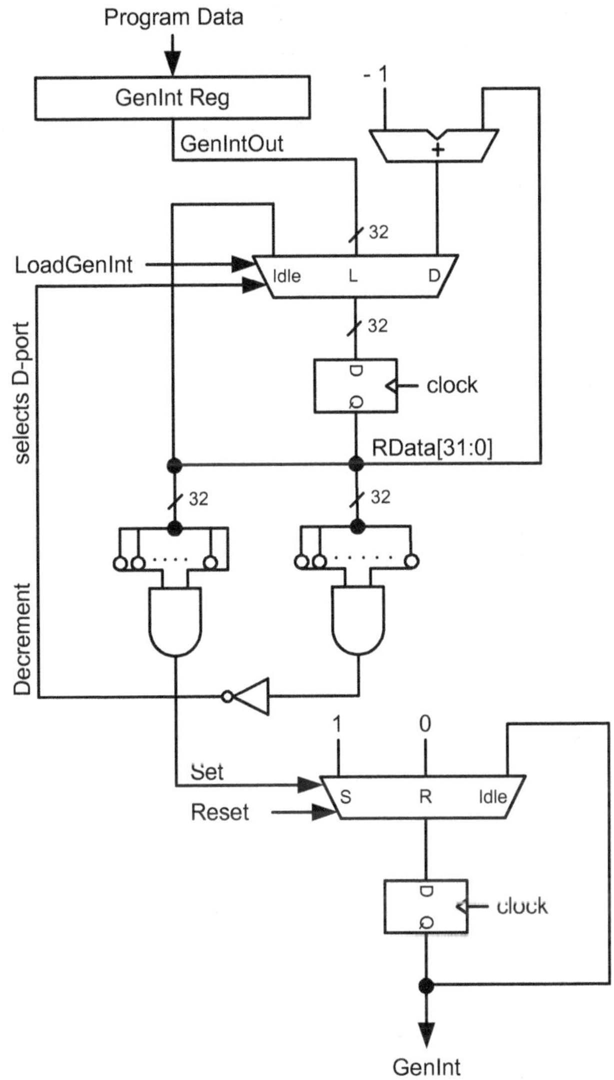

Fig. 7.35 Block diagram of the interrupt generator

As with the other timers, the count-down period is programmed in the GenInt register using a program bus as shown in Fig. 7.35. Once programming is finished, the L-port of the 3-1 MUX is activated by LoadGenInt = 1 to allow the value in the GenInt register to be loaded to the down-counter. This scenario is shown by GenIntOut = 4 in cycle 1 of the timing diagram in Fig. 7.36 as an example. In cycle 2, the contents of the down-counter input are transferred to the RData[31:0] node. Therefore, RData[31:0] = 4 produces logic 1 at the Decrement node, which activates the D-port of the 3-1 MUX, and allows the decremented value of RData, (RData – 1) = 3, to be loaded to the input of the down-counter. In cycle 3, RData[31:0] = 3, but the Decrement node stays at logic 1, keeping the D-port active for the rest of the count-down process. When the RData[31:0] node finally reaches one in cycle 5, the Set node in Fig. 7.35 transitions to logic 1, and turns on the S-port of the second 3-1 MUX. In cycle 6, the GenInt output transitions from logic 0 to logic 1 after four cycles of count-down. Because the LoadGenInt input and the Decrement node are both at logic 0, the Idle port of the 3-1 MUX automatically becomes active. As a result, RData[31:0] = 0 keeps circulating back to the timer input until another value is loaded to the GenInt register. The GenInt output needs to be reset by the active-high Reset signal in Fig. 7.35 in order to generate another interrupt signal.

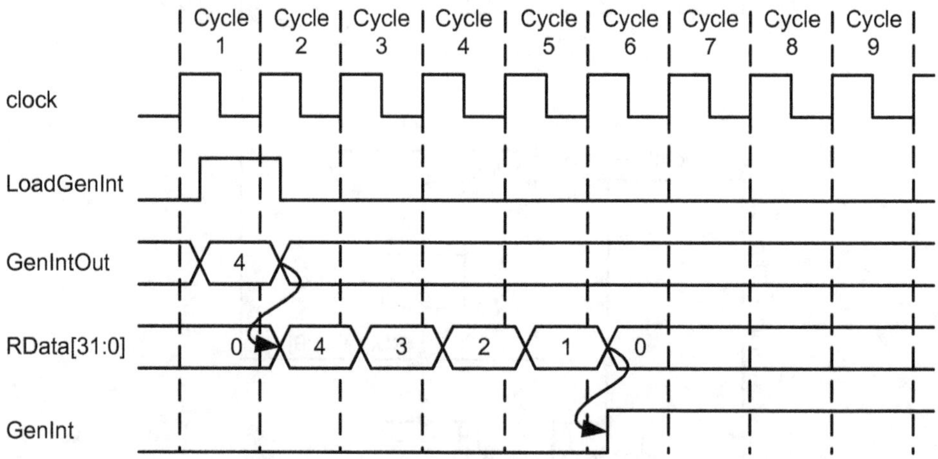

Fig. 7.36 Timing diagram of the interrupt generator (GenInt register in Fig. 7.35 is programmed with a value of 4 as an example)

7.6 Display Adaptor

The display is one of the most crucial peripherals in a system because it establishes a clear link between the user and the system. The format in all modern LCD or LED displays is composed of an active image area bounded by vertical (and horizontal if any) blanking region(s). The size of the blanking regions is adjusted according to the response time of a particular display. Slower displays require larger vertical (and horizontal if any) blanking regions to sync with active images that need to be displayed at a frame rate of 30 to 60 frames per second. Pixels of an active image are fetched from a system memory and displayed on a non-interlaced screen as shown in Fig. 7.37. In this section, we will assume there are vertical and horizontal blanking sections surrounding the active image area when constructing the architecture of the display unit.

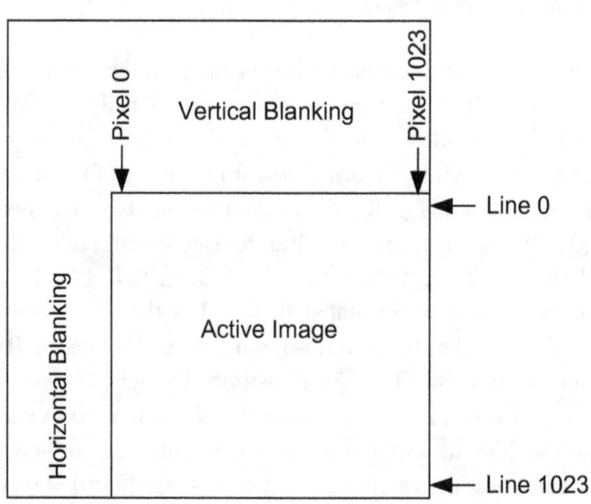

Fig. 7.37 Non-interlaced frame format of an active image consisting of 1024 pixels by 1024 lines

Vertical and horizontal blanking areas are made out of black pixels. On the other hand, each pixel in the active image is composed of eight-bit wide Red (R), Green (G) and Blue (B) components. The basic operation consists of fetching 24-bit pixels from the system memory and placing them in the active image area. Usually each display has a frame buffer to store pixels from the system memory. When the data in the frame buffer is exhausted, the display controller requests another block of data to be transferred from the system memory. As the system complexity increases, the bus activity between the main memory, the CPU and the system peripherals also increases. This produces a scenario where

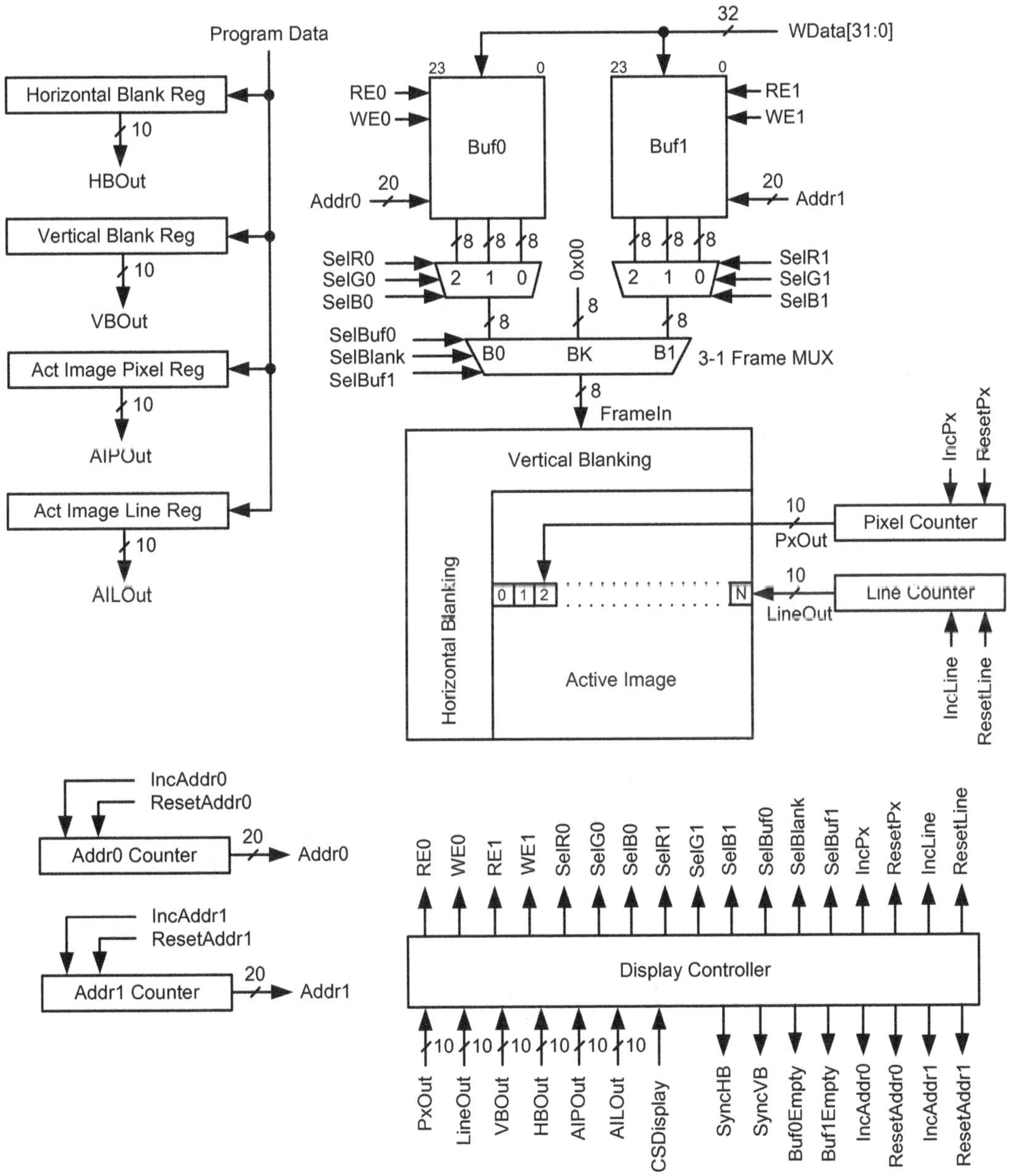

Fig. 7.38 Data-path of the display unit

the display unit may have to wait for many clock cycles before any data arrives at its buffer. However, this is not an acceptable solution since this situation also creates choppy images for the user. Therefore, dual, quadruple or even higher number of frame buffers need to be used in the display unit to maintain continuous stream of images on the monitor without any interruption. A dual frame buffer implementation is shown in Fig. 7.38 where image data is displayed from one frame buffer while the other buffer is being filled.

Prior to image processing, the Horizontal Blank register, Vertical Blank register, Active Image Pixel register and Active Image Line register are programmed using a separate program bus as shown in Fig. 7.38. This bus can even be in the form of a serial bus because register programming will take place during system booting when the image processing speed is not a critical factor. The reader should refer to the design examples and review questions in Chap. 5 to devise a serial interface to program these registers. The data stored in these four registers define the normal operational parameters of the display unit. The image data, on the other hand, is continuously fed to the display buffers, Buf0 and Buf1, by a 32-bit wide system bus, WData[31:0], as shown in Fig. 7.38. Even though the bus width is 32 bits, only the lower 24 bits are used in this architecture to transfer pixels to each buffer. The display controller first places the incoming pixels from the system bus to both Buf0 and Buf1, and then transfers pixels from one of these buffers to the image frame. The controller also generates two timing attributes, SycnHB and SyncVB, to indicate the start of the horizontal blanking and the start of the vertical blanking sections of the frame, respectively. This is done in order to synchronize the display adaptor with the monitor.

To illustrate the operation of the display unit, an active image composed of five pixels wide by nine lines tall (white area) is considered in Fig. 7.39 as an example. This image is surrounded by two lines of vertical blanking and three pixels of horizontal blanking sections (shaded area) in the same figure. The numbers in each box represents a pixel component, R, G or B, whether it belongs to the active image or the blanking section. Therefore, component numbers 0, 1 and 2 constitute the first blank pixel in the vertical blanking section whereas component numbers 57, 58 and 59 correspond to the first pixel of the active image. The blank pixels have no values and equal to 0x00 as shown in Fig. 7.38. The active image pixels, however, are fetched from one of the image buffers in Fig. 7.38 and placed in the active frame in ascending order. For example, the component 57 is fetched first and placed at the upper left corner of the image frame, and the component 263 is fetched last and placed at the lower right corner.

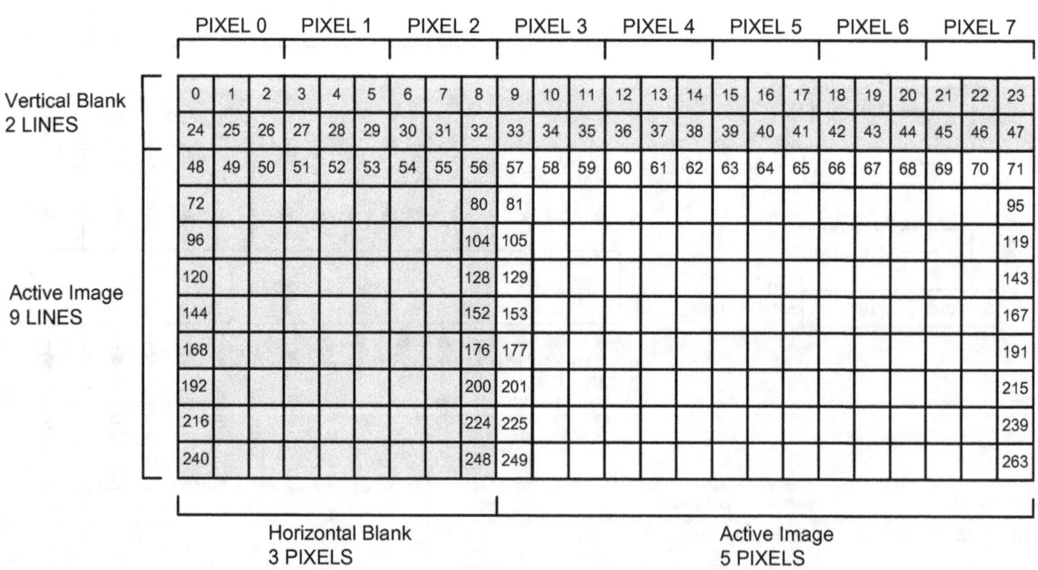

Fig. 7.39 An image frame composed of active image and blanking components

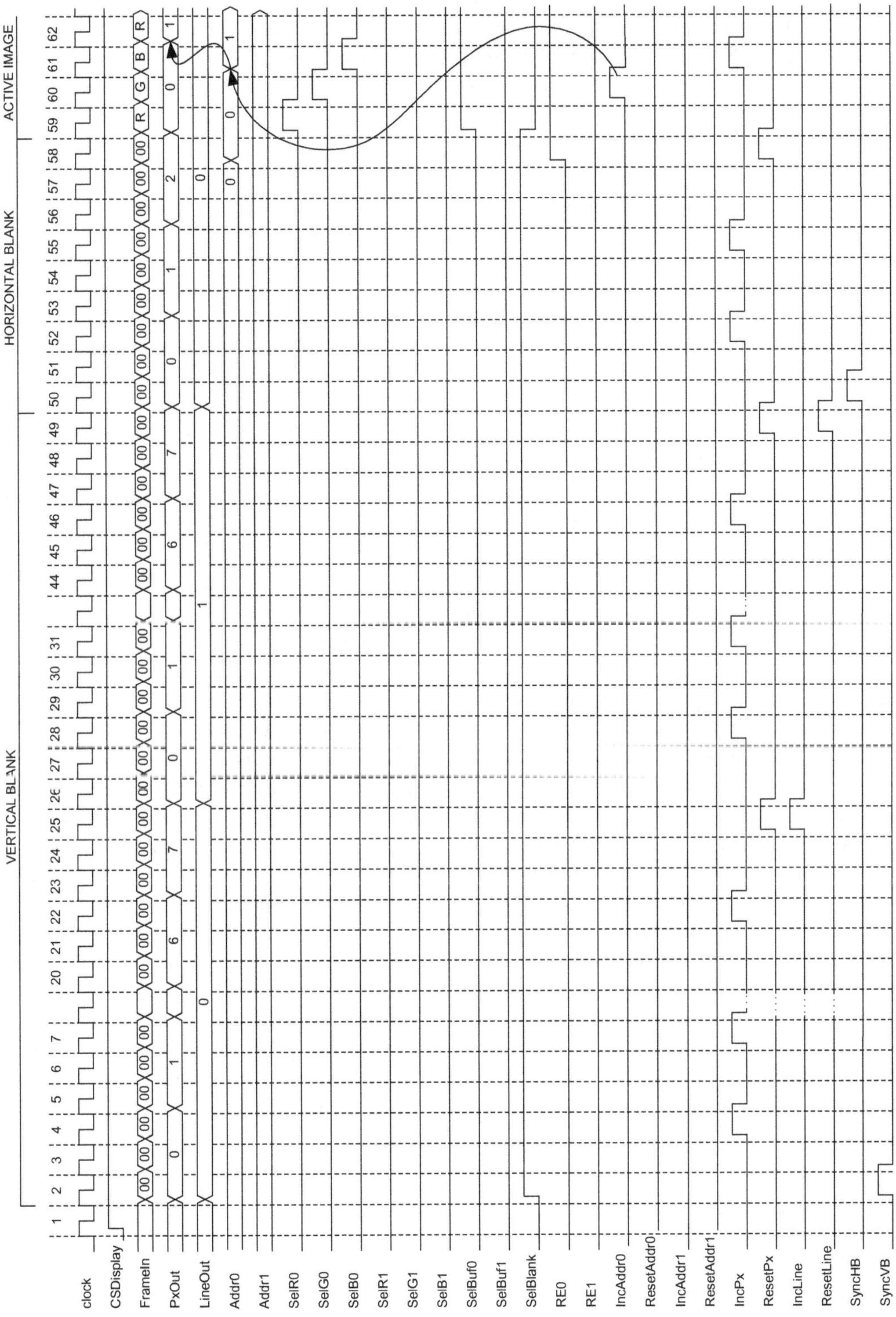

Fig. 7.40 Timing diagram of the display unit in Fig. 7.38

The display unit is activated by CSDisplay = 1 as shown in cycle 1 of the timing diagram in Fig. 7.40. This figure shows data-path signals in the upper rows and control signals in the lower rows. Once active, the first component of the blanking pixel 0, 0x00, arrives at the frame in cycle 2. Within the same cycle, the SycnVB signal becomes logic 1, indicating the start of the vertical blanking for the frame. In cycles 3 and 4, the other two components of the blanking pixel 0 arrive at the frame. The first line of the vertical blanking completes in cycle 25. The second blanking line follows the same pattern as the first one: the first component of the blanking pixel 0 arrives at the frame in cycle 26, and the last component of the blanking pixel 7 arrives in cycle 49. Cycle 49 is also the cycle that resets the line counter by ResetLine = 1 because the next line is the start of an active image. In cycle 50, SycnHB becomes logic 1, signifying the start of horizontal blanking. From cycle 50 to cycle 58, the first line of horizontal blanking is formed prior to the active image. Cycle 58 also defines the border between the horizontal blanking and the active image. In this cycle, the read enable signal for Buf0, RE0, becomes logic 1 in order to read the first image pixel from this buffer. Blanking pixels are supplied to the frame from the BK-port of the 3-1 MUX as shown in Fig. 7.38, and delivered to the frame through FrameIn input.

The timing diagram in Fig. 7.41 is the continuation of Fig. 7.40 and focuses on the active image pixel delivery. The R-component of the first image pixel comes to the image frame in cycle 59 followed by the G and B components in cycles 60 and 61, respectively. Since the image data comes from the first display buffer, Buf0, the port assignment in the 3-1 MUX must be changed from port BK to port B0 by SelBuf0 = 1 from cycle 59 onwards. Therefore, after the first image pixel is delivered to the frame, the address pointer for buffer 0, Addr0, is incremented by one in cycle 61 to be able to fetch the R-component of the next pixel from Buf0 in cycle 62 (one cycle memory read latency). From cycle 59 to cycle 73, the first line of the active image is delivered to the frame by incrementing Addr0 from 0 to 4. Cycle 73 also indicates the start of the second horizontal blanking line. In this cycle, RE0 transitions to logic 0, and Addr0 stops incrementing because pixel flow from Buf0 needs to be interrupted in order to start delivering blank pixels to the frame. To accommodate this, the port assignment in the 3-1 MUX is changed from port B0 to port BK. After completing the delivery of horizontal blanking pixels, the second line of the active image is delivered between cycles 83 and 97. Addr0 is incremented from 5 to 9 during this period to fetch pixels from Buf0 and form the second active image line. The rest of the image is delivered to the frame by the end of cycle 265. In cycle 266, a new frame is formed with SyncVB = 1. As in the previous frame, pixels that constitute the vertical blanking are followed by pixels that form the horizontal blanking and the active image. The new active image pixels are delivered from Buf1. In this example, both Buf0 and Buf1 are assumed to contain only 45 pixels, and therefore each buffer has 45 bytes of image data as opposed to the architecture in Fig. 7.38 that shows $1024 \times 1024 = 1{,}048{,}576$ active image pixels in each buffer.

Figure 7.42 shows the display controller design to manage the data-flow in Fig. 7.38. The state machine in Fig. 7.42 is a Moore-type composed of a string of states, each responsible for delivering blank or active image pixels to the frame. The state machine needs to keep track of the pixel and line numbers in the frame, and be able to define the boundaries between the blanking and the active image regions. Therefore, its functionality largely depends on the output values of the pixel and line counters in Fig. 7.38.

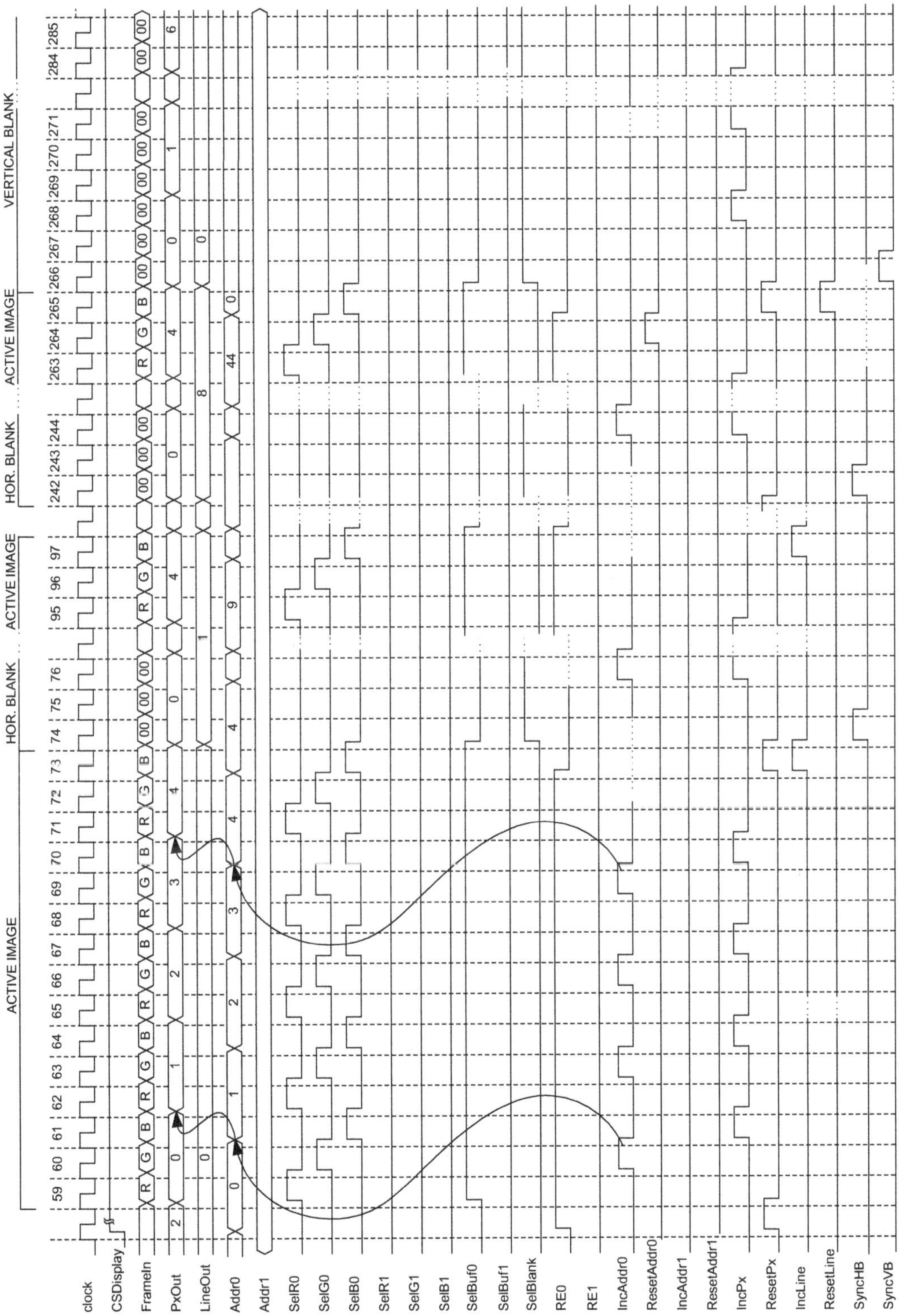

Fig. 7.41 Continuation of the display unit's timing diagram in Fig. 7.40

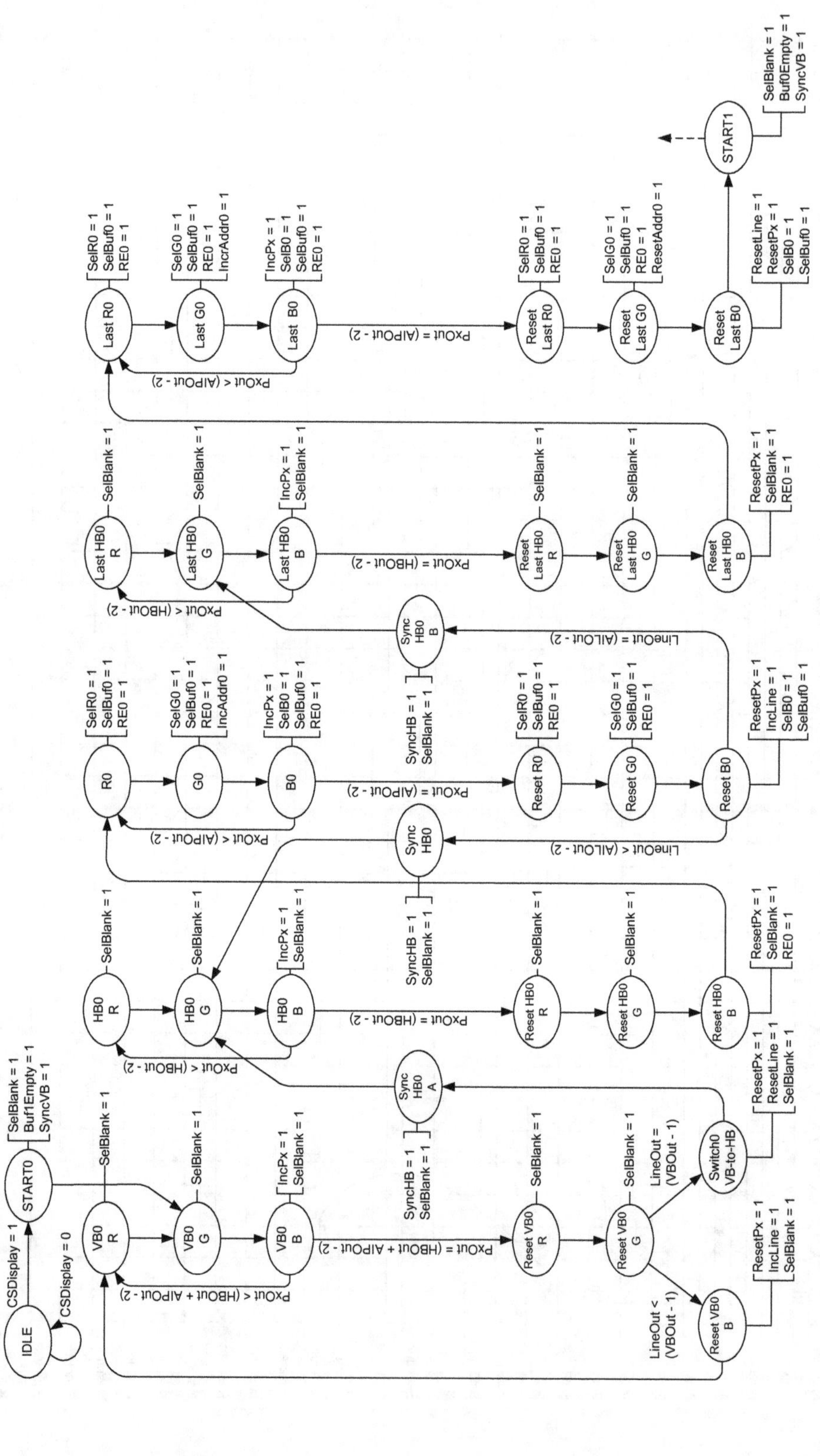

Fig. 7.42 The controller of the display unit for Buf0 (control signals equal to logic 0 are omitted for simplicity)

The state machine stays in the IDLE state until it is externally activated by CSDisplay = 1, which corresponds to cycle 1 of the timing diagram in Fig. 7.40. Once activated, the R-component of the first blank pixel enters the frame through the BK-port of the 3-1 MUX, and requires SelBlank = 1 in cycle 2. This refers to the START0 state in Fig. 7.42. In cycle 3, the G-component of the first blank pixel is delivered to the frame. This is shown as the VB0-G state in the state diagram. In this state, SelBlank = 1 in order to transmit the G-component of the first blank pixel to the frame. The B-component of the first blank pixel arrives in cycle 4, which corresponds to the VB0-B state. During this cycle, SelBlank = 1 and IncPx = 1 to increment the pixel counter by one. At this point, the controller checks if the end of the first vertical blanking line has been reached by forming PxOut = (HBOut + AIPOut – 2). Here, PxOut corresponds to the output of the pixel counter, HBOut corresponds to the number of horizontal blanking pixels at the output of the Horizontal Blanking register, and AIPOut corresponds to the number of active image pixels at the output of the Active Image Pixel register in Fig. 7.38. If the end has not been reached, the machine keeps circling around the VB0-R, VB0-G and VB0-B states until PxOut becomes equal to (HBOut + AIPOut – 2). This period translates from cycle 5 to cycle 22 in the timing diagram in Fig. 7.40. During this period, every time the B-component of a blank pixel is delivered to the frame, the pixel counter increments by one. When the end point is detected, the state machine goes to the Reset VB0-R state where it delivers the R-component of the last blanking pixel that belongs to the first blanking line. This state corresponds to cycle 23 in the timing diagram. The controller delivers the remaining G and B-components of the last blanking pixel in cycles 24 and 25, which translate to the Reset VB0-G and Reset VB0-B states, respectively. In cycle 25, the pixel counter is reset by ResetPx = 1, and the line counter is incremented by IncLine = 1 in order to produce the next vertical blanking line. However, the contents of the Vertical Blanking register, VBOut, needs to be checked prior to the start of the next vertical blanking line in case this register is programmed to have only one vertical blanking line. Therefore, while in the Reset VB0-G state, the line counter output, LineOut, is compared against (VBOut – 1). If the line counter output is less than (VBOut – 1), then the state machine first goes to the Reset VB0-B state and then back to the VB0-R state in order to generate another blanking line as described in cycles 26 to 49 in the timing diagram. If LineOut is equal to (VBOut – 1), the state machine goes to the Switch0 VB-to-HB state where it generates SelBlank = 1, ResetPx = 1 and ResetLine = 1 in order to terminate the vertical blanking and start the first line of the horizontal blanking.

Cycle 50 starts the beginning of horizontal blanking region, and delivers the R-component of the first horizontal blanking pixel to the frame. This cycle translates to the Sync HB0-A state because SyncHB = 1 is also generated in this state. The state machine moves through the HB0-G and HB0-B states in cycles 51 and 52, and checks if the end of the horizontal blanking region has been reached by comparing the PxOut with (HBOut – 2). If PxOut < (HBOut – 2), then more R, G and B blanking pixel components are brought to the frame through the BK-port of the 3-1 MUX. However, if PxOut = (HBOut – 2), then the state machine enters the Reset HB0-R state to deliver the R-component of the last horizontal blanking pixel in cycle 56 as this condition indicates the end of horizontal blanking. In cycles 57 and 58, the machine traverses through the Reset HB0-G and the Reset HB0-B states. The latter state resets the pixel counter by ResetPx = 1, and enables Buf0 by RE0 = 1 to start reading active image pixels.

In cycle 59, the state machine enters the R0 state to deliver the R-component of the first active image pixel from Buf0. In this state, port 0 of the 3-1 MUX at the output of Buf0 becomes active by SelR0 = 1, and port B0 of the 3-1 frame MUX becomes active by SelBuf0 = 1. The read enable input for Buf0 also stays at logic 1 by RE0 = 1. In cycle 60, the state machine goes to the G0 state where it delivers the G-component of the first active image pixel to the frame. This cycle requires SelG0 = 1 to turn on port 1 of the 3-1 MUX at the output of Buf0 while keeping SelBuf0 = 1 and RE0 = 1. Addr0 is also incremented in this state by IncAddr0 = 1. In cycle 61, the controller reaches the B0

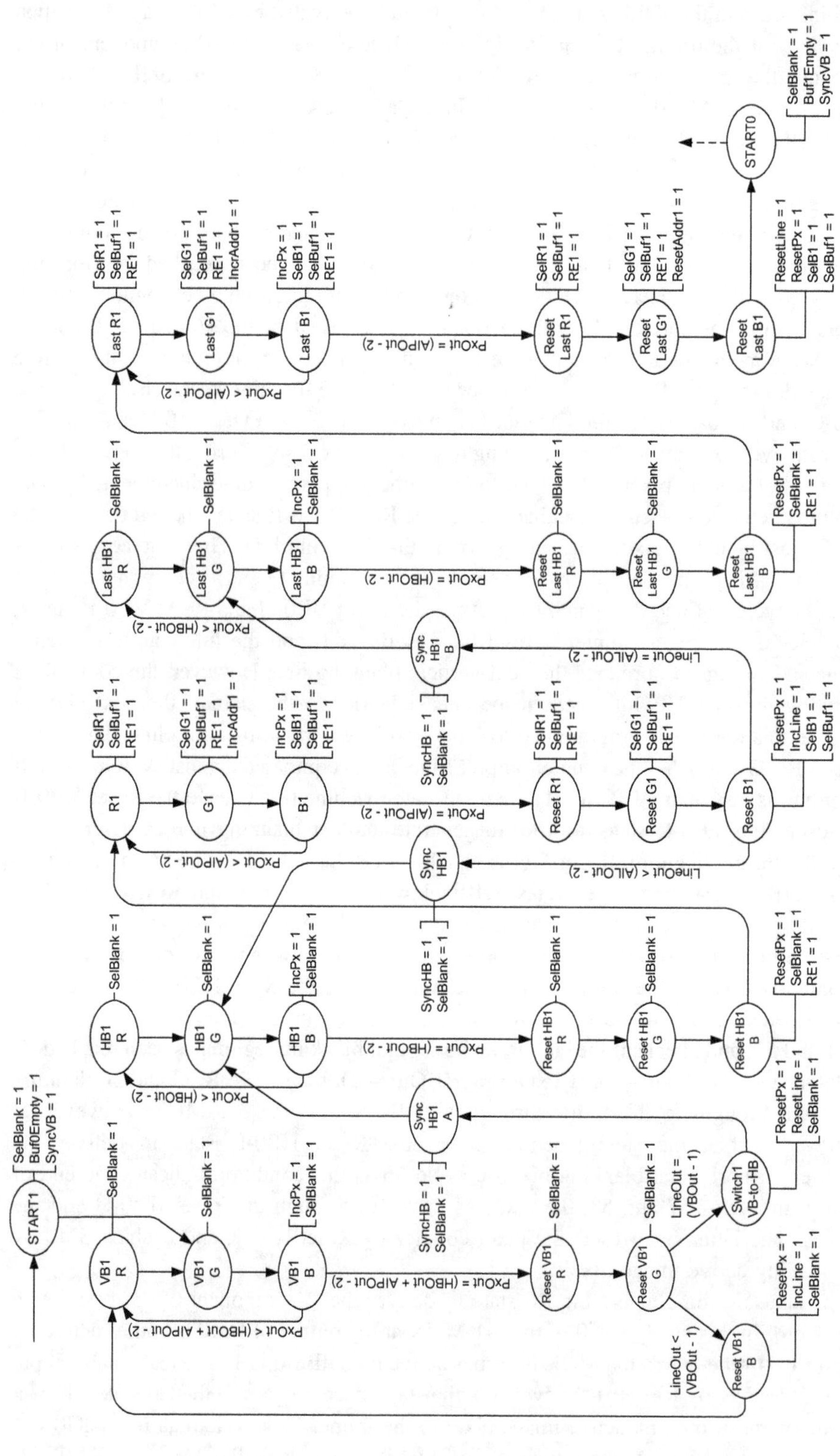

Fig. 7.43 The controller of the display unit for Buf1 (control signals equal to logic 0 are omitted for simplicity)

state where it increments the pixel counter by IncPx = 1, selects port 2 of the 3-1 MUX at the output of Buf0 by SelB0 = 1, and maintains both SelBuf0 = 1 and RE0 = 1. In this state, the controller checks if the end of active image has been reached by comparing PxOut against (AIPOut – 2). If the controller finds PxOut < (AIPOut – 2), it goes back to the R0 state to retrieve more image pixels from Buf0. This scenario corresponds to cycles 62 to 70 of the timing diagram in Fig. 7.41. If the controller finds PxOut = (AIPOut – 2), it moves to the Reset R0 state in cycle 71 to deliver the last R-component of the image pixel, and generates SelR0 = 1, SelBuf0 = 1 and RE0 = 1. The state machine then moves to the Reset G0 state in cycle 72 and the Reset B0 state in cycle 73. In the Reset B0 state, the controller resets the pixel counter by ResetPx = 1, increments the line counter by IncLine = 1, selects port 2 of the 3-1 Buf0 MUX by SelB0 = 1, and keeps SelBuf0 = 1. While in this state, the controller checks to see if the active image is more than a single line or not, and compares the output of the line counter, LineOut, against (AILOut – 2). If the controller finds that LineOut < (AILOut – 2), it first moves to the Sync HB0 state to generate SyncHB = 1, and then back to the HB0-G state to start fetching horizontal blanking pixels for the next line. However, if the controller finds that LineOut = (AILOut – 2), it realizes that it will be processing the last line of the current frame. First, it goes to the Sync HB0-B state and generates SyncHB = 1, and then to the Last HB0-G state to deliver the G-component of a horizontal blanking pixel.

The states from Last HB0-R to Reset Last-B0 are the exact replicas of the states from HB0-R to Reset B0 except that in the Reset Last-G0 state, Buf0 address pointer is reset by ResetAddr0 = 1, and in the Reset Last-B0 state, the line counter is reset by ResetLine = 1. Once the controller exhausts all the active image pixels in Buf0, it switches to Buf1 to construct the next image frame as shown in the state diagram of Fig. 7.43. In this figure, the states controlling the blanking and the image pixel delivery from Buf1 is exactly the same as in Buf0. Once all the pixels are delivered to the frame from Buf1, the state machine in Fig. 7.43 hands over the control of the display adaptor to the state machine in Fig. 7.42.

7.7 Data Converters

Analog-to-Digital Converter (ADC)

All analog domains interface with digital systems through Analog-to-Digital (ADC) or Digital-to-Analog Converters (DAC) as shown in Fig. 7.44. In this figure, analog signal from a sensor is amplified to a certain level before sampling takes place in the sample-and-hold circuit inside the ADC. The sampled analog signal is then converted into digital form and directed to the CPU for processing according to an embedded program.

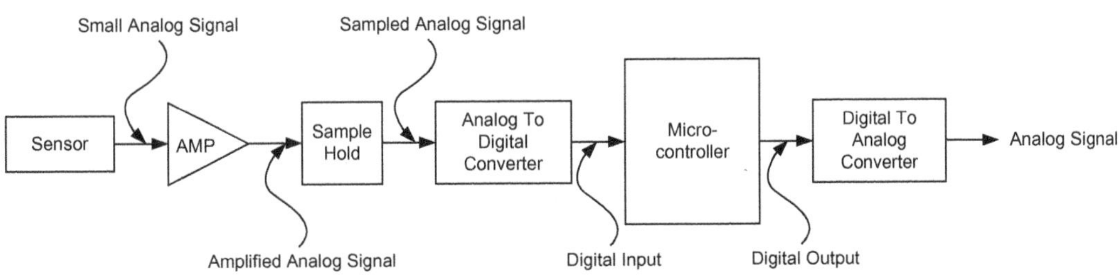

Fig. 7.44 Typical analog-digital and digital-analog converter data-paths

The signal resolution is an important factor to consider in an ADC design. It simply means dividing a sampled analog signal by 2^N number of voltage levels to represent its value where N is the number of bits in digital domain. The second important consideration is the range of analog values an ADC can capture and process.

Figure 7.45 describes the ADC resolution in a numerical example where an analog signal changes between 0 and 5 V. The bit resolution is only three bits. Therefore, the ADC divides the range of an analog signal into $2^3 = 8$ levels between its maximum and minimum value, and identifies each analog level with three output bits.

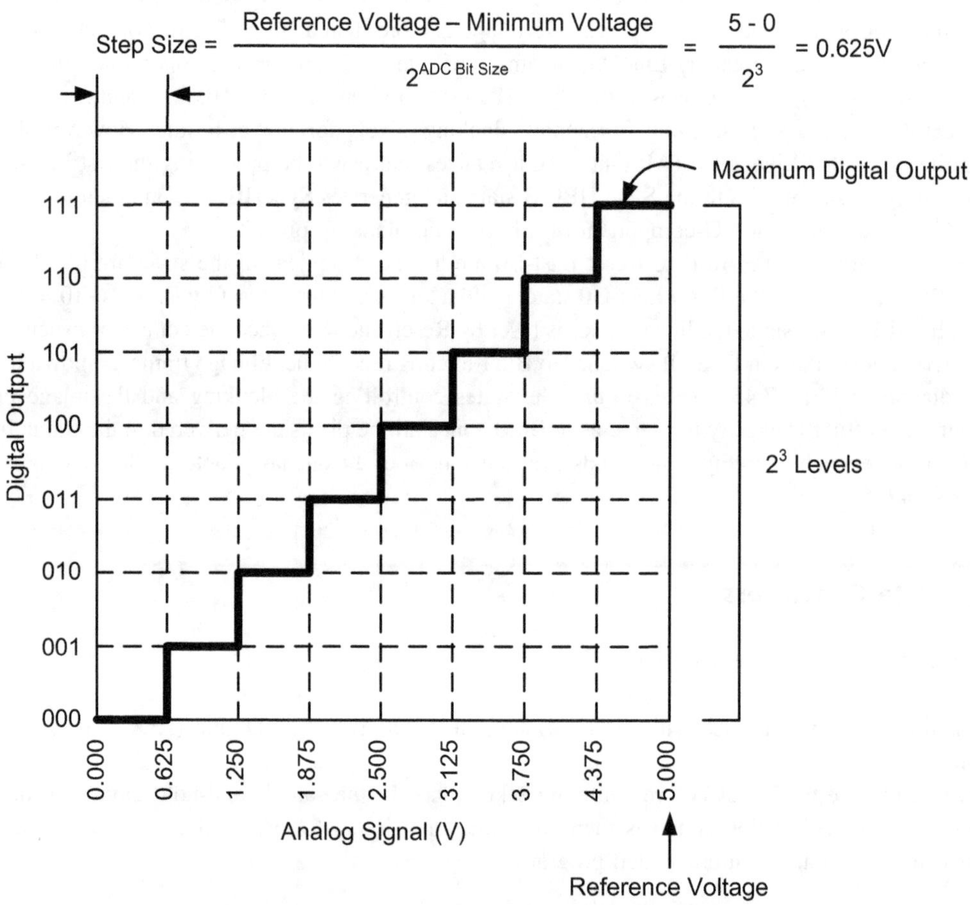

Fig. 7.45 Input-output description of a three-bit ADC (power supply voltage is 5 V)

For example, an analog signal of 2.501 V is identified by a digital output of 100. If the analog signal increases to 3.124 V, the digital output that represents this voltage value still stays at 100. In other words, in a three-bit ADC there is no difference between 2.501 and 3.124 V in terms of their digital representation. The 0.625 V step size is the natural occurring error in a three-bit ADC, but this error can be easily reduced by increasing the number of bits in the ADC. In general, increasing the number of ADC bits by one halves the error. Therefore, designing a four-bit ADC instead of a three-bit ADC reduces the quantization error by 0.3125 V.

Reference voltage in an ADC is generally determined by the maximum voltage level of the analog signal, and it is used to calculate the step size. In this three-bit ADC example in Fig. 7.45, the reference voltage is 5 V because the amplified analog voltage at the input of the ADC is limited not go beyond 5 V.

ADC samples non-periodic analog signals in regular time intervals as described in Fig. 7.46. The time interval between sampling points is called the sampling period. The sampling period is adjusted according to the processing speed of the ADC in order to generate accurate digital outputs.

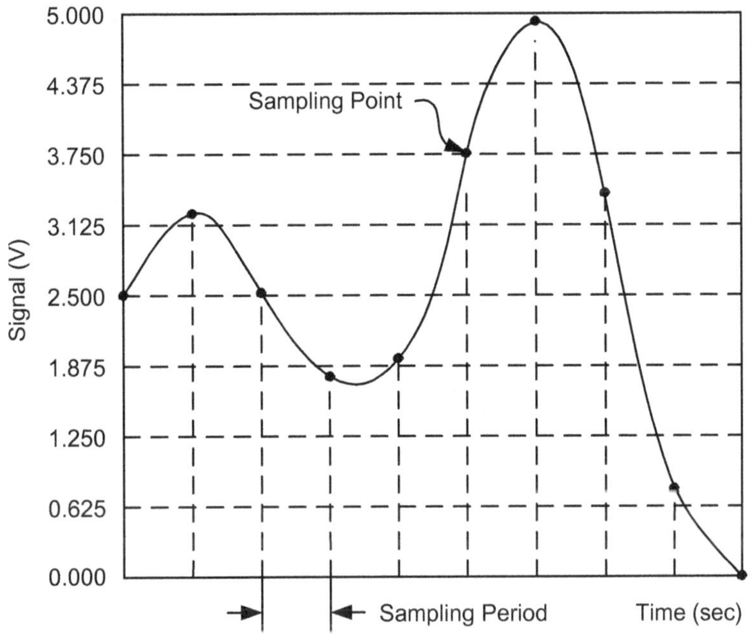

Fig. 7.46 Sampling a continuous analog signal

Once sampled, the analog voltage at the input of an ADC is held steady throughout the sampling period while the conversion takes place as shown in Fig. 7.47. The shape of the converted signal may be quite different from the original analog signal due to the ADC resolution and the time duration between samples. In a three-bit ADC, sampling takes place in 0.625 V increments. Therefore, each sampling point becomes subject to a dynamic quantization error which changes between 0 and 0.625 V. For example, a three-bit ADC samples 3.4 V according to its closest sampling level of 3.125 V, and produces a 0.275 V error. Arbitrary signals that change with a frequency faster than the sampling frequency are subject to much larger dynamic errors. When converted back to their analog form, these signals can exhibit large deviations from their original outlines.

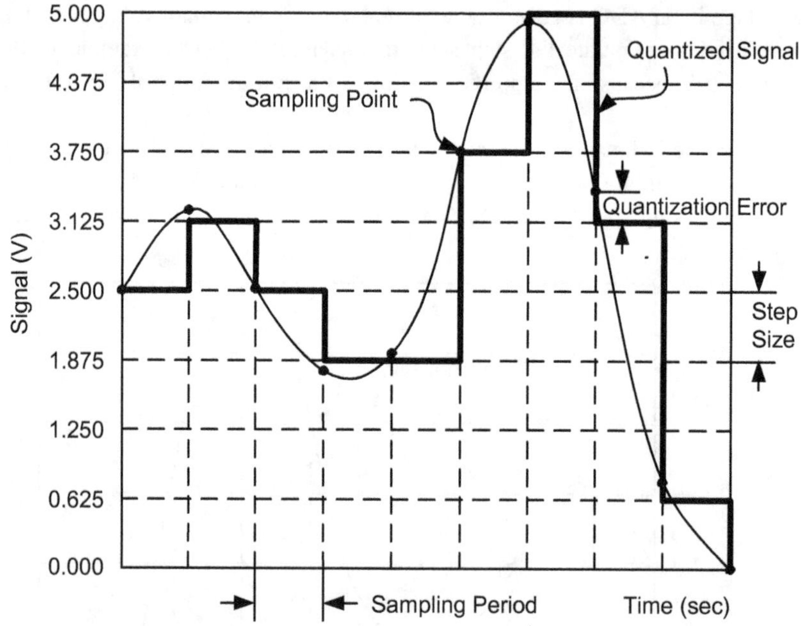

Fig. 7.47 Sampling period, hold concept and regeneration of an analog signal

A basic sample-and-hold circuit consists of an NMOS transistor and a capacitor as shown in Fig. 7.48. The control input simply turn on an N-channel MOSFET for a short period of time, called sampling width, during which the analog voltage level at the input is stored on the capacitor. When the transistor is turned off, this analog value is held constant until the next sampling point.

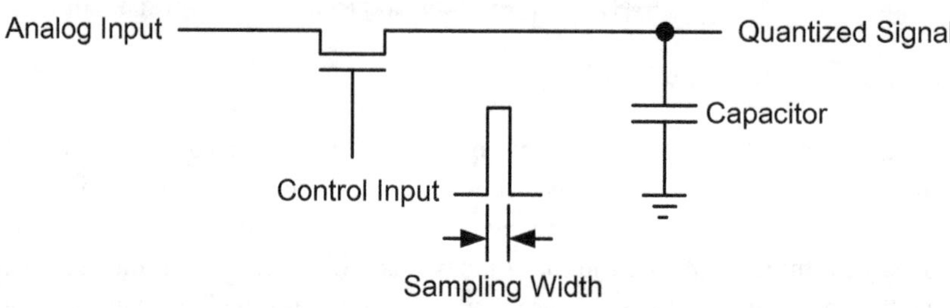

Fig. 7.48 A typical sample-and-hold circuit

Flash ADC

The simplest ADC is the flash-type as shown in Fig. 7.49. This three-bit ADC contains $2^3 = 8$ operational amplifiers. The analog signal is applied to all eight positive input terminals. The reference voltage is distributed to each negative input terminal via a voltage divider circuit. Each operational amplifier acts as a differential amplifier and amplifies the difference between a continuously changing analog signal and the portion of the reference voltage.

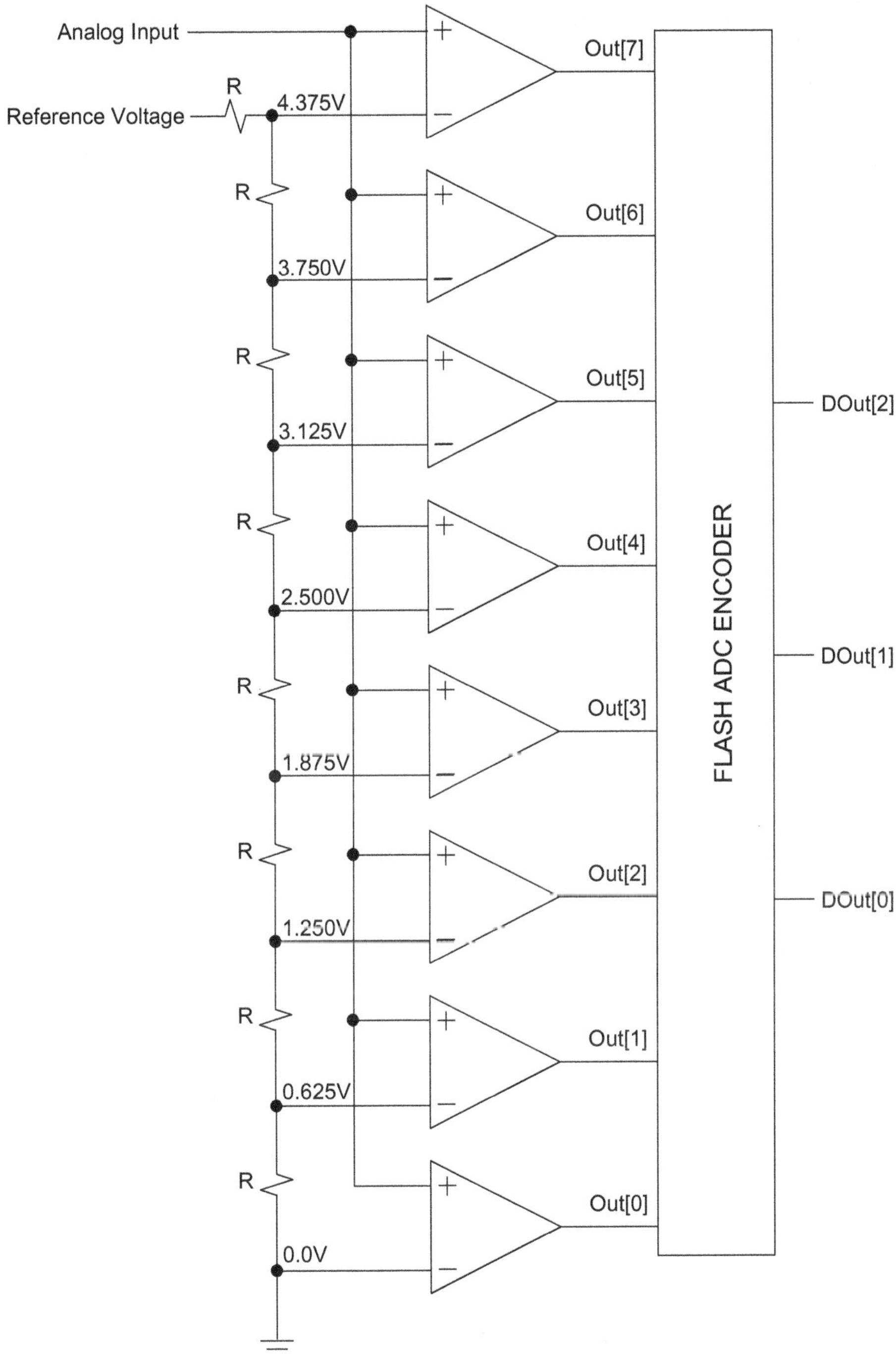

Fig. 7.49 Typical three-bit flash ADC schematic (power supply voltage is 5 V)

Figure 7.50 describes the operation of the three-bit flash ADC and its encoder in a truth table. When the analog voltage is less than or equal to 0.625 V, only Out[0] becomes logic 1, all other outputs from Out[1] to Out[7] become logic 0. When the analog signal exceeds 0.625 V but less than 1.25 V, only Out[0] and Out[1] become logic 1, and again all others become logic 0. Higher analog voltages at the input successively produce more logic 1 levels as shown in Fig. 7.50. An encoder is placed at the output stage of all operational amplifiers to transform the voltage levels at Out[7:0] into a three-bit digital output, DOut[2:0]. The digital output is subject to an error of 0.625 V because only three bits are used for conversion.

Analog Input	Out[7]	Out[6]	Out[5]	Out[4]	Out[3]	Out[2]	Out[1]	Out[0]	DOut[2]	DOut[1]	DOut[0]
$0.625 > V_{IN} > 0.000$	0	0	0	0	0	0	0	1	0	0	0
$1.250 > V_{IN} > 0.625$	0	0	0	0	0	0	1	1	0	0	1
$1.875 > V_{IN} > 1.250$	0	0	0	0	0	1	1	1	0	1	0
$2.500 > V_{IN} > 1.875$	0	0	0	0	1	1	1	1	0	1	1
$3.125 > V_{IN} > 2.500$	0	0	0	1	1	1	1	1	1	0	0
$3.750 > V_{IN} > 3.125$	0	0	1	1	1	1	1	1	1	0	1
$4.375 > V_{IN} > 3.750$	0	1	1	1	1	1	1	1	1	1	0
$5.000 > V_{IN} > 4.375$	1	1	1	1	1	1	1	1	1	1	1

Fig. 7.50 Three-bit flash ADC truth table describing its operation (power supply voltage is 5 V)

Ramp ADC

The ramp ADC uses only a single operation amplifier, but it employs an up-counter as well as a DAC in a loop structure shown in Fig. 7.51. The digital output is obtained from the C[3:0] terminals in this figure, and the result progressively forms within several clock cycles as opposed to being almost instantaneous as in the flash ADC.

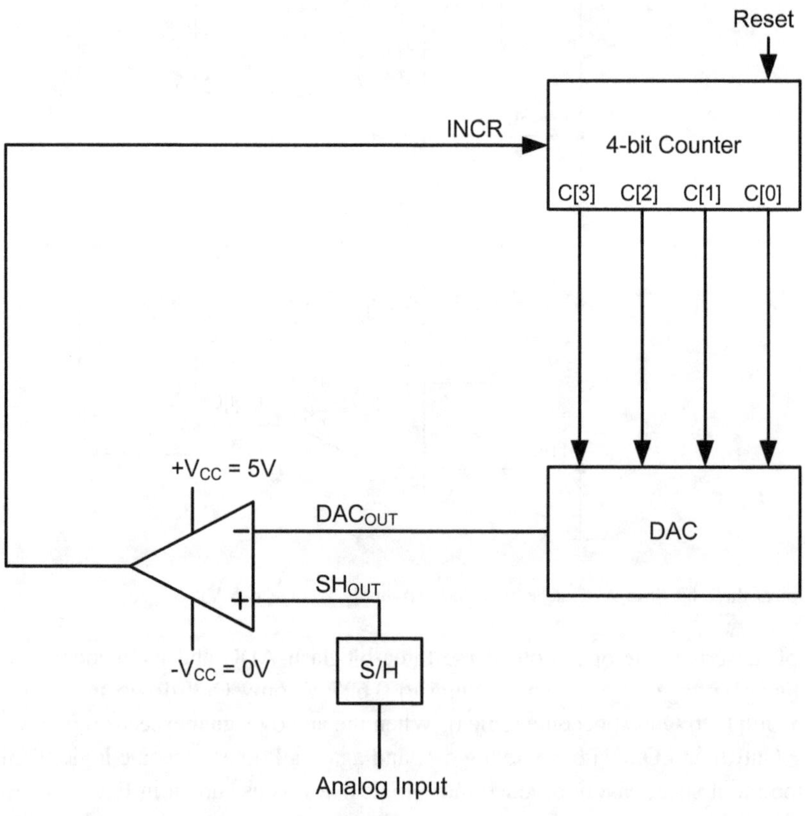

Fig. 7.51 Typical four-bit ramp ADC schematic

The top portion of Fig. 7.52 describes the output voltage assignments of a four-bit ramp ADC using two different types of number-rounding schemes. The down-rounding scheme assigns a lower analog value for each digital output compared to the up-rounding scheme. For example, the down-rounding scheme produces a digital output of C[3:0] = 0100 for analog voltages between 1.0937 V and 1.4062 V at the input. If the up-rounding scheme is used, the same digital output is produced by an analog voltage between 1.4062 V and 1.7187 V.

The middle table in Fig. 7.52 shows an example of the conversion process in this four-bit ADC if the down-rounding mechanism is used. Prior to its operation, the four-bit up-counter is reset and produces C[3:0] = 0000. Assuming an analog voltage of 2 V is applied to the input, which must be kept constant until the conversion is complete, C[3:0] = 0000 forces the DAC output, DAC_{OUT}, to be 0 V according to the down-rounding scheme. Since this value is less than 2 V at the sample/hold circuit output, SH_{OUT}, the output of the differential amplifier, INCR, transitions to the positive supply potential of the operational amplifier, $+V_{CC} = 5$ V, which prompts the four-bit counter to increment to C[3:0] = 0001. Consequently, the DAC generates $DAC_{OUT} = 0.3125$ V according to the truth table in Fig. 7.52. However, this value is still less than $SH_{OUT} = 2$ V. Therefore, the differential amplifier produces another INCR = 5 V which prompts the counter to increment again to C[3:0] = 0010. Up-counting continues until C[3:0] = 0111 or $DAC_{OUT} = 2.1875$ V. Since this last voltage is greater than $SH_{OUT} = 2$ V, the differential amplifier output switches back to its negative supply voltage, $-V_{CC} = 0$ V, and stops the up-counter from incrementing further. The digital output stays steady at C[3:0] = 0111 from this point forward, representing 2 V analog voltage with a dynamic error of 0.1875 V.

The table at the bottom part of Fig. 7.52 represents the conversion steps if the up-rounding mechanism is used in this ADC. External reset still produces C[3:0] = 0000 initially. However, the DAC output starts the conversion with an increased amount of 0.3125 V instead of 0 V. The counter increments until C[3:0] = 0110, and produces 2.1875 V at the DAC_{OUT} node. At this value INCR becomes 0 V, and the up-counter stops incrementing further. C[3:0] = 0110 becomes the ADC result for 2 V.

Successive Approximation ADC

The third type ADC is based on the successive approximation technique to estimate the value of the applied analog voltage. This type is a trade-off between the flash-type and the ramp-type ADCs in terms of speed and the number of components used in the circuit. As a numerical example, a typical four-bit successive approximation ADC is shown in Fig. 7.53. In this figure, the up-counter in the ramp ADC is replaced by a control logic which successively converts an analog input to a digital form by a trial-and-error method. The output of the ADC is obtained at the C[3:0] terminal.

Step Size = $5/2^4$ = 0.3125V

C[3]	C[2]	C[1]	C[0]	Down-Round(V)	Up-Round(V)
0	0	0	0	0.0000	0.3125
0	0	0	1	0.3125	0.6250
0	0	1	0	0.6250	0.9375
0	0	1	1	0.9375	1.2500
0	1	0	0	1.2500	1.5625
0	1	0	1	1.5625	1.8750
0	1	1	0	1.8750	2.1875
0	1	1	1	2.1875	2.5000
1	0	0	0	2.5000	2.8125
1	0	0	1	2.8125	3.1250
1	0	1	0	3.1250	3.4375
1	0	1	1	3.4375	3.7500
1	1	0	0	3.7500	4.0625
1	1	0	1	4.0625	4.3750
1	1	1	0	4.3750	4.6875
1	1	1	1	4.6875	5.0000

Analog Input = 2V with Down-Rounding Mechanism

Step	C[3]	C[2]	C[1]	C[0]	DAC_{OUT}(V)	INCR(V)
1	0	0	0	0	0.0000	5.0
2	0	0	0	1	0.3125	5.0
3	0	0	1	0	0.6250	5.0
4	0	0	1	1	0.9375	5.0
5	0	1	0	0	1.2500	5.0
6	0	1	0	1	1.5625	5.0
7	0	1	1	0	1.8750	5.0
8	0	1	1	1	2.1875	0.0

Final output with quantization error of 0.3125V

Analog Input = 2V with Up-Rounding Mechanism

Step	C[3]	C[2]	C[1]	C[0]	DAC_{OUT}(V)	INCR(V)
1	0	0	0	0	0.3125	5.0
2	0	0	0	1	0.6250	5.0
3	0	0	1	0	0.9375	5.0
4	0	0	1	1	1.2500	5.0
5	0	1	0	0	1.5625	5.0
6	0	1	0	1	1.8750	5.0
7	0	1	1	0	2.1875	0.0

Final output with quantization error of 0.3125V

Fig. 7.52 Four-bit ramp ADC truth table describing its operation (power supply voltage is 5 V)

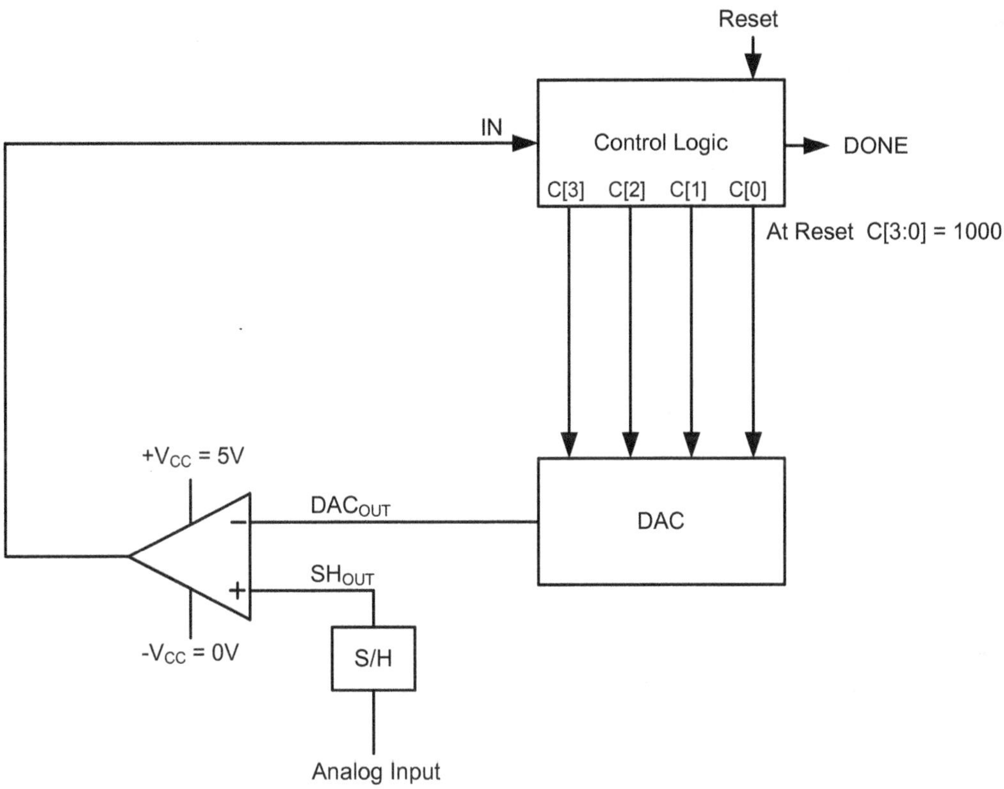

Fig. 7.53 Typical four-bit successive approximation ADC (power supply voltage is 5 V)

The top portion of Fig. 7.54 shows the truth table to operate a four-bit successive approximation ADC. Two numerical examples in this figure illustrate the down-rounding and the up-rounding schemes used during conversion.

The first example in Fig. 7.54 illustrates the down rounding mechanism in a four-bit successive approximation ADC. In this example, an analog voltage of 3.5 V is applied to the analog input of the ADC. An external reset starts the converter at C[3:0] = 1000, which is considered a mid point between C[3:0] = 0000, representing the minimum analog input of 0 V, and C[3:0] = 1111, representing the maximum analog input of 5 V for this ADC. For C[3:0] = 1000, the DAC generates an initial analog voltage of 2.5 V at the DAC_{OUT} node. Since this value is less than the sampled analog voltage of 3.5 V at the SH_{OUT} node, the operational amplifier produces IN = 5 V, prompting the control logic to try a slightly higher digital output. As a result, the control logic produces C[3:0] = 1100 as its first trial, which is equivalent to a midway point between C[3:0] = 1000 and 1111. DAC_{OUT} becomes 2.5 + (2.5/2) = 3.75 V. But, this new voltage is larger than SH_{OUT} = 3.5 V, and the operational amplifier produces IN = 0 V in return. The drop at IN node is an indication to the control logic that its initial attempt of C[3:0] = 1100 was too large, and it must lower its output. This time, the control logic tries C[3:0] = 1010, which is between C[3:0] = 1000 and 1100 and translates to DAC_{OUT} = 2.5 + (2.5/4) = 3.125 V. This value is smaller than the applied voltage at SH_{OUT}, therefore the operational amplifier generates IN = 5 V, and prompts the control logic to try a slightly higher output between C[3:0] = 1010 and 1100. In the third round, the control logic produces C[3:0] = 1011. The DAC_{OUT} node becomes 2.5 + (2.5/4) + (2.5/8) = 3.4375 V and generates IN = 5 V. This new input suggests the control logic to try even higher digital output, such as

Step Size = $5/2^4$ = 0.3125V

C[3]	C[2]	C[1]	C[0]	Down-Round (V)	Up-Round (V)
0	0	0	0	0.0000	0.3125
0	0	0	1	0.3125	0.6250
0	0	1	0	0.6250	0.9375
0	0	1	1	0.9375	1.2500
0	1	0	0	1.2500	1.5625
0	1	0	1	1.5625	1.8750
0	1	1	0	1.8750	2.1875
0	1	1	1	2.1875	2.5000
1	0	0	0	2.5000	2.8125
1	0	0	1	2.8125	3.1250
1	0	1	0	3.1250	3.4375
1	0	1	1	3.4375	3.7500
1	1	0	0	3.7500	4.0625
1	1	0	1	4.0625	4.3750
1	1	1	0	4.3750	4.6875
1	1	1	1	4.6875	5.0000

Analog Input = 3.5V with Down-Rounding Mechanism START with (5.0/2) = 2.5

Step	C[3]	C[2]	C[1]	C[0]	DAC_{OUT}(V)	Control Logic In
1	1	0	0	0	2.5000	3.5 > 2.5000 Thus try 2.5 + (2.5/2) = 3.75
2	1	1	0	0	3.7500	3.5 < 3.7500 Thus try 2.5 + (2.5/4) = 3.125
3	1	0	1	0	3.1250	3.5 > 3.1250 Thus try 2.5 + (2.5/4) + (2.5/8) = 3.4375
4	1	0	1	1	3.4375	3.5 > 3.4375 Stop at 3.4375 since the difference is below 0.15625V

Final output with quantization error is below 0.15625V

Analog Input = 3.5V with Up-Rounding Mechanism START with (5.0/2) = 2.8125

Step	C[3]	C[2]	C[1]	C[0]	DAC_{OUT}(V)	Control Logic In
1	1	0	0	0	2.8125	3.5 > 2.8125 Thus try 2.8125 + (2.5/2) = 4.0625
2	1	1	0	0	4.0625	3.5 < 4.0625 Thus try 2.8125 + (2.5/4) = 3.4375
3	1	0	1	0	3.4375	3.5 > 3.4375 Stop at 3.4375 since the difference is below 0.15625V

Final output with quantization error below 0.15625V

Fig. 7.54 Four-bit successive approximation ADC truth table describing down-rounding and up-rounding approximation techniques (power supply voltage is 5 V)

2.5 + (2.5/4) + (2.5/8) + (2.5/16) = 3.5937 V, in the fourth round. However, 2.5/8 = 0.3125 V is the resolution limit for this four-bit ADC, and as a result, the controller stalls at C[3:0] = 1011, revealing DAC_{OUT} = 3.4375 V. This voltage differs from SH_{OUT} = 3.5 V by only 0.0625 V.

The second example in Fig. 7.54 explains the successive approximation technique if the up-rounding scheme is employed. The conversion again starts at C[3:0] = 1000, but with an incremented value of 2.5 + 0.3125 = 2.8125 V at the DAC output. Since this voltage is below

SH_{OUT} = 3.5 V, IN node becomes 5 V and prompts the control logic to produce a larger digital output. The control logic responds to this with a digital output of C[3:0] = 1100, which corresponds to the analog voltage of 2.8125 + (2.5/2) = 4.0625 V at the DAC_{OUT} node. As a result, IN node becomes 0 V and forces the control logic to lower its digital output. This time, the control logic tries C[3:0] = 1010 which is equivalent to 2.8125 + (2.5/4) = 3.4375 at the DAC_{OUT} node. Due to the resolution limit of this four-bit ADC, this step also becomes the end of successive approximation.

The control circuit of the four-bit ADC with down-rounding scheme is shown in Fig. 7.55. In this figure, the approximation process starts at the midpoint, C[3:0] = 1000, corresponding to DAC_{OUT} = 2.5 V according to the table in Fig. 7.54. When the external reset is removed, and the difference between SH_{OUT} and DAC_{OUT}, Δv, is found to be greater than the 0.15625 V, the control logic either goes to the state 1.25 and produces C[3:0] = 0100 (equivalent to 1.25 V), or to the state 3.75 and produces C[3:0] = 1100 (equivalent to 3.75 V) at the first step of successive approximation. This decision depends on the value of the control logic input, IN. If IN = 0, which translates to the analog input to be less than 2.5 V, the next state becomes the state 1.25. However, if IN = 1, the analog input is considered to be greater than 2.5 V, and the next state becomes the state 3.75. If Δv is less than 0.15625 V, on the other hand, the control logic cannot proceed any further and moves to the DONE state. In the second step of successive approximation, the state 1.25 either transitions to the state 0.625 or to the state 1.875, depending on the value of the IN node. Similar transitions take place from the state 3.75 to either the state 3.125 or to the state 4.375, again depending on the value of IN. After this point, the state machine performs one last approximation to estimate the value of analog input voltage and reaches the DONE state with an output value as shown in Fig. 7.55.

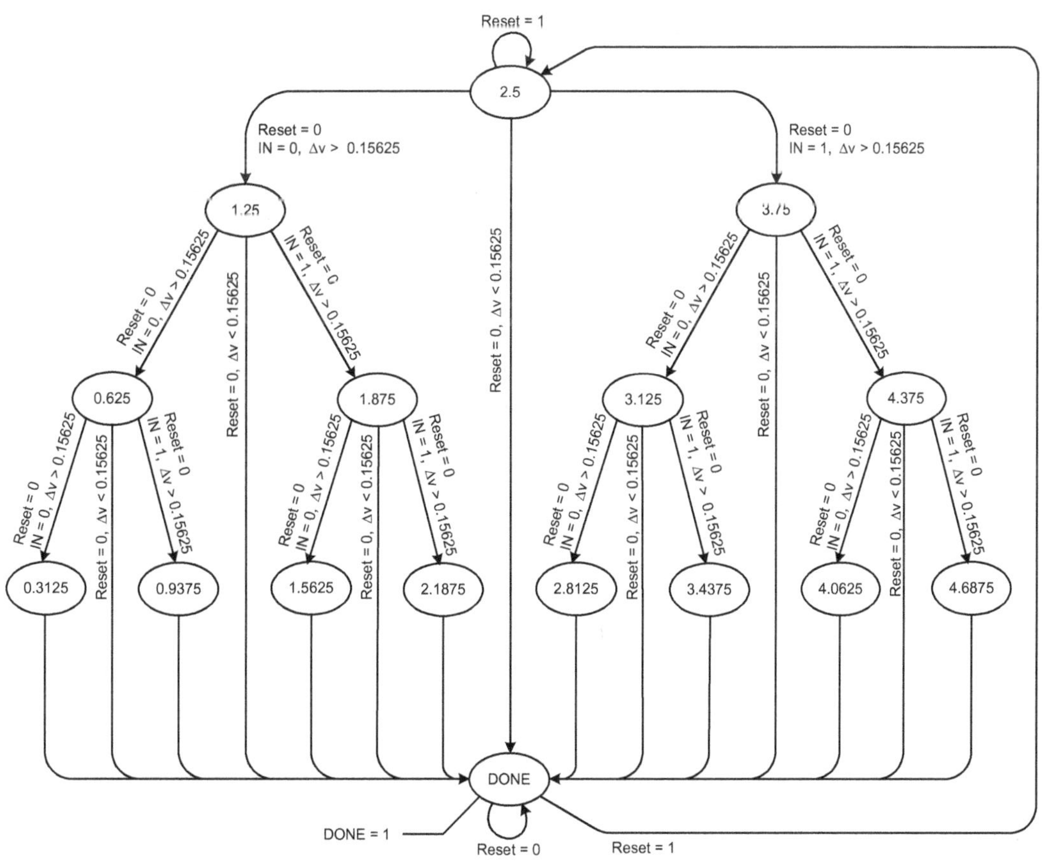

Fig. 7.55 Four-bit successive approximation control circuit

7.8 Digital-to-Analog Converter (DAC)

The most common DAC utilizes the weighted summation method of its digital inputs. A three-bit DAC with a weighted binary adder is shown in Fig. 7.56 as an example.

This circuit is composed of two parts. The first part adds all three binary input bits, IN[2] (the most significant bit), IN[1] and IN[0] (the least significant bit), and produces an output, $ADD_{OUT} = -(0.5$ IN[2] + 0.25 IN[1] + 0.125 IN[0]) according to the equation in Fig. 7.57. The second part is an analog inverter which forms OUT = $-ADD_{OUT}$.

Therefore, the circuit in Fig. 7.56 generates OUT = 0.5 IN[2] + 0.25 IN[1] + 0.125 IN[0], where each binary value of IN[2:0] input is multiplied by the coefficients, 2^{-1}, 2^{-2} and 2^{-3}, before they are added to produce an output. For example, the combination of IN[2] = 1, IN[1] = 0 and IN[0] = 1, with +5 V and 0 V logic levels generates OUT = 2.5 + 0.625 = 3.125 V. Similarly, all the other analog outputs in Fig. 7.57 can be generated using the equation at the top part of this figure with a maximum error of 0.625 V.

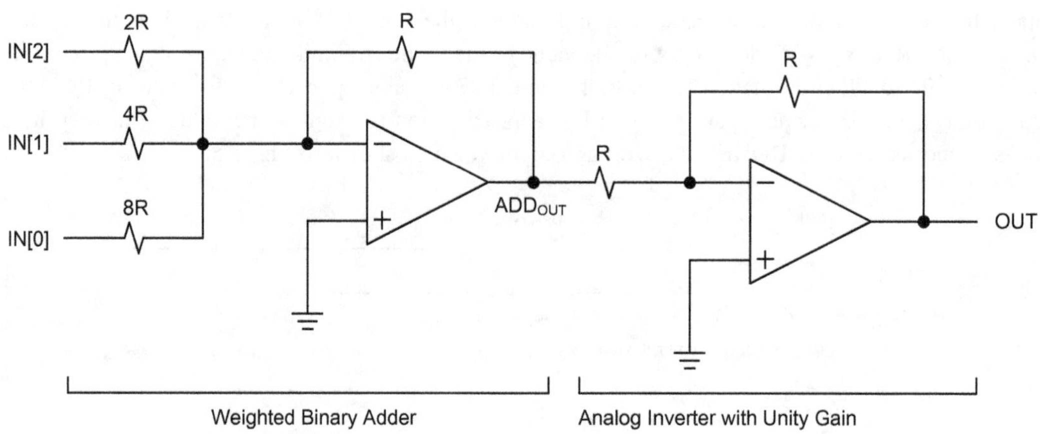

Fig. 7.56 Three-bit DAC schematic with weighted binary adder

$$ADD_{OUT} = -\frac{R}{2R} IN[2] - \frac{R}{4R} IN[1] - \frac{R}{8R} IN[0]$$

$$= -0.5\ IN[2] - 0.25\ IN[1] - 0.125\ IN[0]$$

$$OUT = -ADD_{OUT} = 0.5\ IN[2] + 0.25\ IN[1] + 0.125\ IN[0]$$

IN[2]	IN[1]	IN[0]	OUT(V)
0	0	0	0.000
0	0	1	0.625
0	1	0	1.250
0	1	1	1.875
1	0	0	2.500
1	0	1	3.125
1	1	0	3.750
1	1	1	4.375

Example:

IN[2] = 1, IN[1] = 0, IN[0] = 1 with +5V/0V logic levels

$ADD_{OUT} = -2.5 - 0 - 0.625 = -3.125V$

OUT = + 3.125V with 0.625V quantization error

Fig. 7.57 Three-bit DAC operation with weighted binary adder (power supply voltage is 5 V)

Review Questions

1. A DMA controller transfers four words of data, D1, D2, D3 and D4, from SRAM 1 (source memory) to SRAM 2 (destination memory) on a 32-bit wide bidirectional bus as shown below. D1, D2, D3 and D4 are fetched from addresses, AS1, AS2, AS3 and AS4, in the source memory and placed in the AD1, AD2, AD3 and AD4 addresses in the destination memory, respectively. Since each SRAM memory has a single I/O port, address and data cannot exist in the same clock cycle. Therefore, for the read operation, data becomes available at the SRAM I/O port a cycle after a valid address is presented. For the write operation, data has to be present at the SRAM port a cycle after the valid address. The active-high RE and WE enables the SRAM for read or write operations when a valid address is available.
   The DMA controller has two programming ports to program the initial address and the incremented address values. It uses ProgAddrS and ProgIncrS control inputs to initialize and increment the source address, and similarly ProgAddrD and ProgIncrD inputs to initialize and increment the destination memory address. The active-high StartS produces the first address at AddrOutS. IncrS input generates incremented addresses based on the initial address. Similarly, StartD produces the first address at AddrOutD followed by the incremented addresses produced by the active-high IncrD.
   Capture1 and Capture2 inputs capture and store data from the bus temporarily in the Buf1 and Buf2 data buffers respectively. All active-high enable inputs enable the tri-state buffers when they are at logic 1. Otherwise, there will be no connection between the address pointers and the bus or between the memories and the bus.

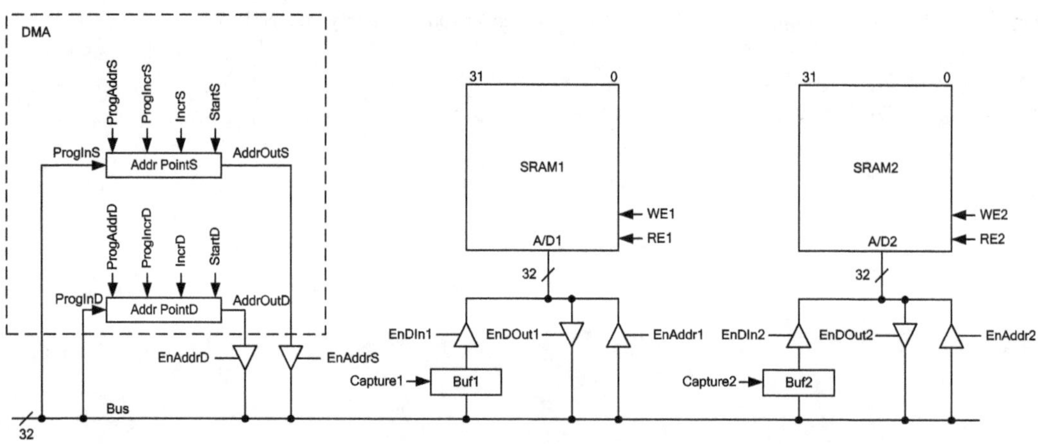

   (a) Draw the schematic for the source and destination address pointers.
   (b) Form a timing diagram to transfer four words of data from SRAM1 to SRAM2.

2. Below are the waveforms generated by two different timers.

   (a) A rate generating timer produces an active-high pulse in every 128 cycles as shown below. Design this timer and draw its schematic.

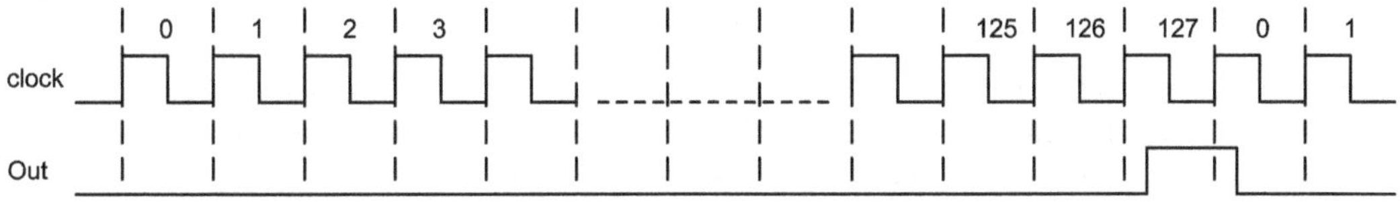

(b) A one-shot timer produces a continuous pulse 100 cycles after the start signal as shown below. Design this timer and draw its schematic.

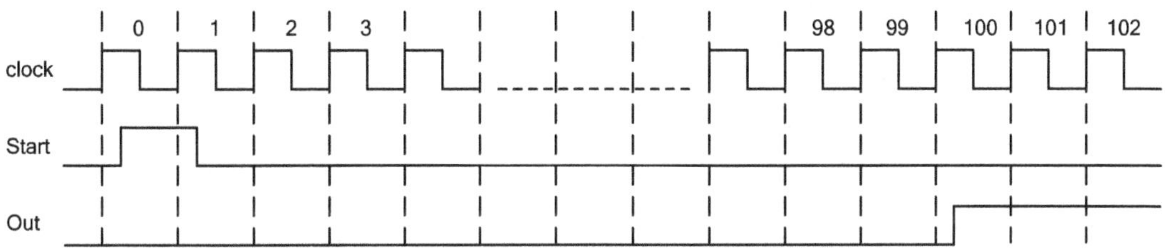

3. An interrupt controller serving four hardware interrupts is connected to a 16-bit CPU, which consists of 16-bit wide instruction memory, Imem, a data memory, Dmem, a program counter for the Imem, PC, a program counter for the data memory, DC, the data registers A and B, and a controller. This schematic is shown below.

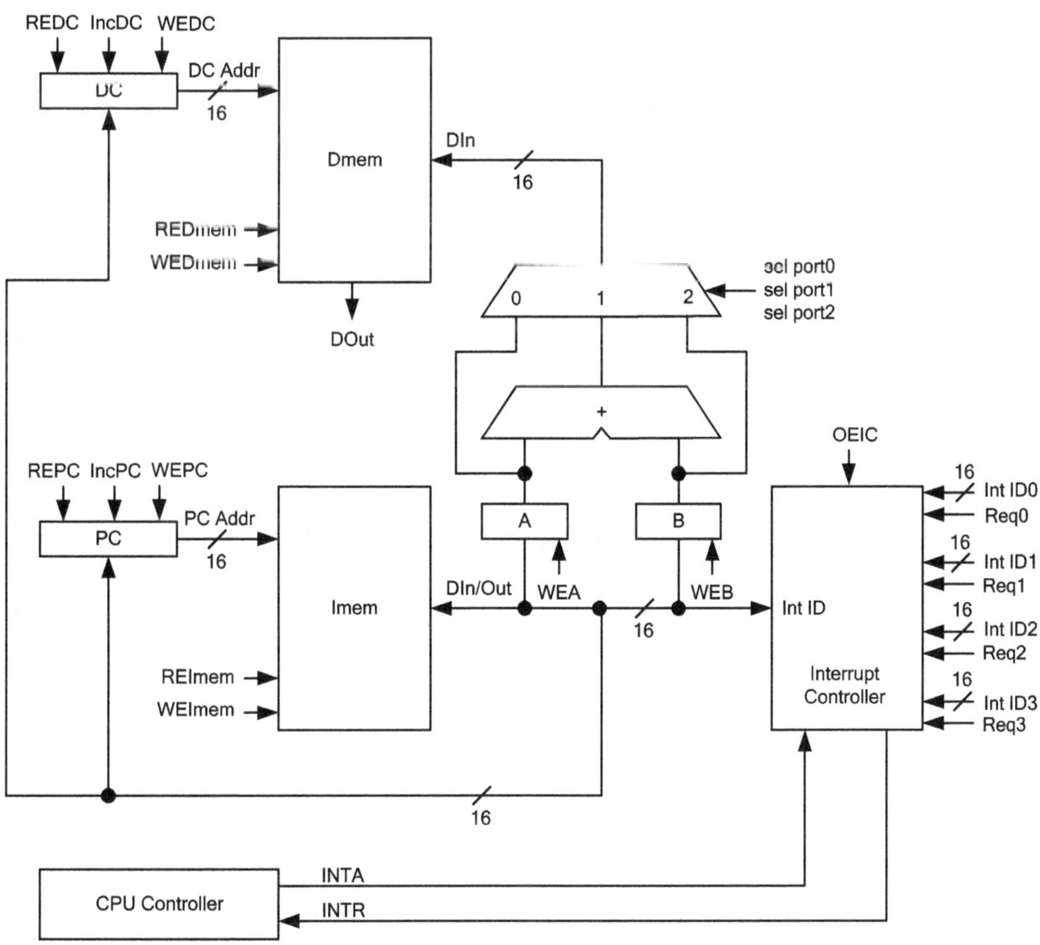

The interrupt protocol in this schematic is as follows:

Step 1: The active-high interrupt, INTR, is generated by the interrupt controller to inform the CPU about the presence of an interrupt. The interrupt controller must have a request from an external device with an interrupt ID before generating INTR.
Step 2: The CPU acknowledges INTR with the active-high interrupt acknowledge, INTA.
Step 3: The interrupt controller produces an interrupt ID on the 16-bit bi-directional data bus.
Step 4: The interrupt ID is loaded to the PC to access the interrupt service routine (ISR) address.
Step 5: The ISR address is loaded to the PC.
Step 6: One of the four interrupt service instructions is fetched from the instruction memory for a particular interrupt. The steps are as follows:

Step 6a: The first interrupt instruction specifies the data memory address to be loaded to the DC.
Step 6b: The second instruction contains the value of A to be loaded to the A register.
Step 6c: The third instruction contains the value of B to be loaded to the B register.
Step 6d: The fourth instruction provides either the contents of the A register or the contents of B register or the added results of both registers to be loaded to the data memory at a DC address. This value will later be used in the program (the related hardware is not shown in the schematic above).

Step 7: When the four-cycle interrupt service is complete, the CPU lowers INTA. In response, the interrupt controller lowers INTR a cycle after INTA transitions to logic 0, and waits for the next interrupt.

Note: Imem or Dmem have SRAM configurations. Storing data is achieved with WE = 1 within the same clock cycle as the valid address. Reading data from a memory is achieved with RE = 1 with a latency of two clock cycles (data is not read at the beginning of next clock cycle but the one after that).

(a) If the priority scheme in the interrupt controller is such that device 0 has the highest priority and device 3 has the lowest priority, design this controller with the I/O port description shown on the schematic. Note that this controller can support only hardware interrupts.
(b) Show a timing diagram outlining the complete interrupt service from Step 1 to Step 7 above.
(c) Show the state diagram of the CPU controller including all the control inputs to operate instruction and data memories, registers etc.

4. A display unit works on a unidirectional bus that transmits 24-bit video pixels. Each pixel is comprised of eight-bit R, G and B components which occupy the least significant three bytes of the write data bus (the most significant byte is always 0x0). The write bus fills a video buffer with a frequency of 2f. Similarly, the frequency of fetching data from a buffer to fill a video frame is f (buffer emptying rate). Assume the horizontal and the vertical blanking sections of the video frame are both zero.

The display unit is the highest priority peripheral on the bus because of the fact that it requires a minimum rate of 30 frames/s to process and display data. However, other peripherals in the system also use the same bus in order to send or receive data between video bursts.

Each timing diagram below contains three vital entries for a frame buffer. The top row indicates the ID number of a flag associated with an empty buffer. The middle row indicates the ID number of an empty buffer. The bottom row shows the ID number of a full buffer. The flag is a direct input to the CPU, and points out which buffer in the display unit needs to be filled. Note that all buffers are considered full before data transactions start. Each square in the timing diagram corresponds to filling or emptying an entire buffer. Since filling a buffer takes half the time to empty it, the number of squares doubles at the bottom row relative to the middle row.

Suppose you use a two buffer, a three buffer or a four buffer system in the video unit. The key consideration in this design is to be able to supply continuous data to the display unit from the frame buffers while other peripherals use the bus.

Once full, define the length of data burst from each buffer in each timing table below. Mark each entry with buffer numbers, and use the letters E or F to indicate whether each buffer is empty or full, respectively. The video unit empties buffers in the following order. In a two-buffer system, buffer 1 empties first, buffer 2 empties second. In a three-buffer system, buffer 1 empties first, buffer 2 s, and buffer 3 third. In a three-buffer system, buffer 1 empties first, buffer 2 s, buffer 3 third, and buffer 4 fourth. Indicate the flag number for each empty buffer inside the circle.

2 buffer system:

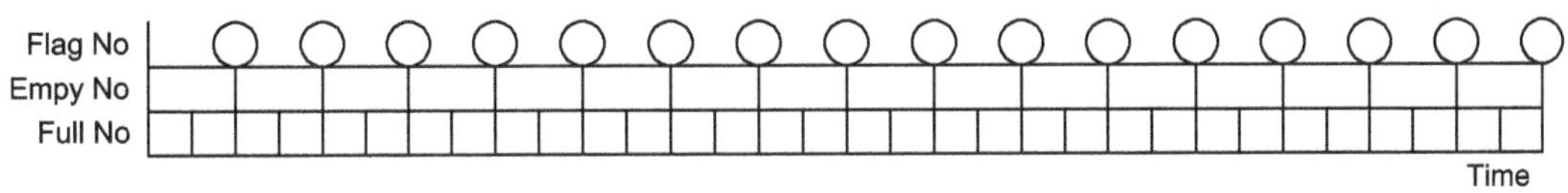

3 buffer system:

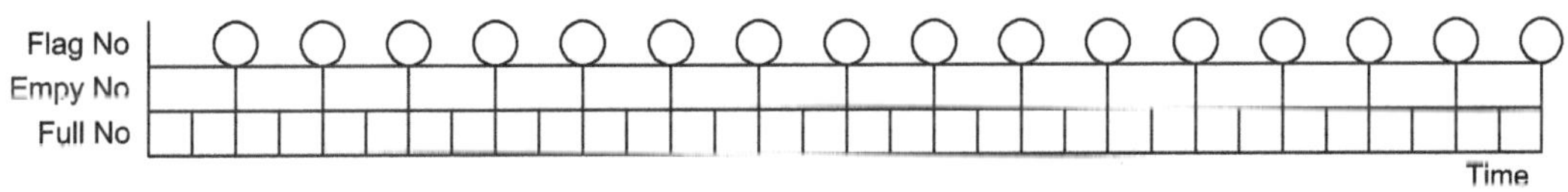

4 buffer system:

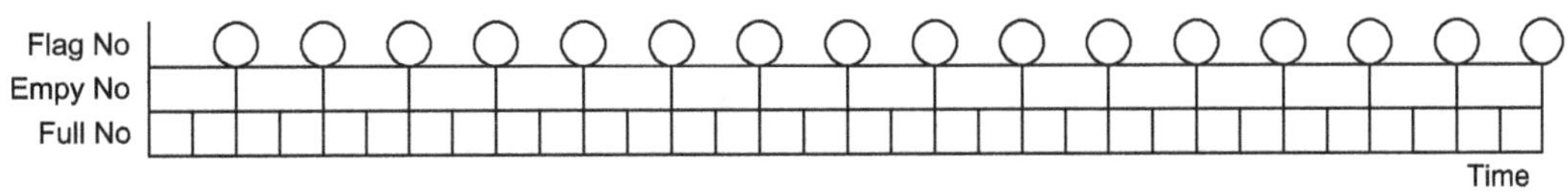

5. A three-bit successive approximation ADC is given below. A sample-Hold circuit (S/H) samples a varying voltage at the Analog Input port in periodic intervals and feeds the sampled voltage level to the operational amplifier.

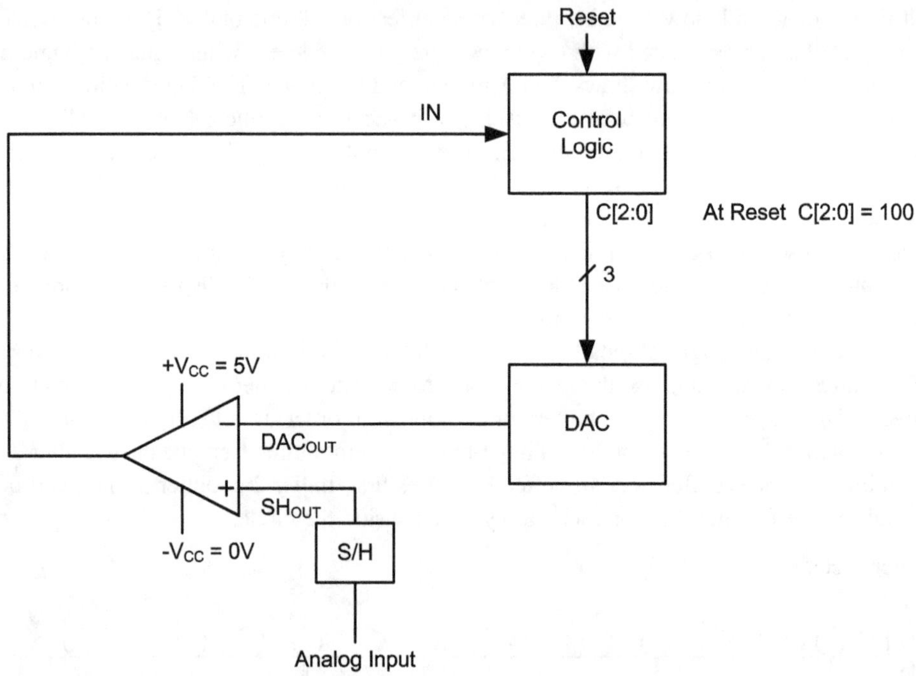

Use the empty timing diagram below, fill the blanks with analog or digital data for Analog Input = 0.5 V with down-rounding mechanism.

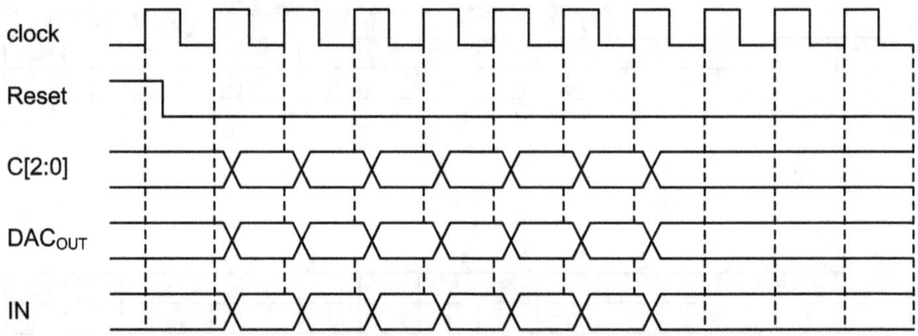

6. A three-bit ramp ADC below operates with $+V_{CC} = 3$ V and $-V_{CC} = 0$ V. Input to this ADC can take any value between 0 and 3 V.

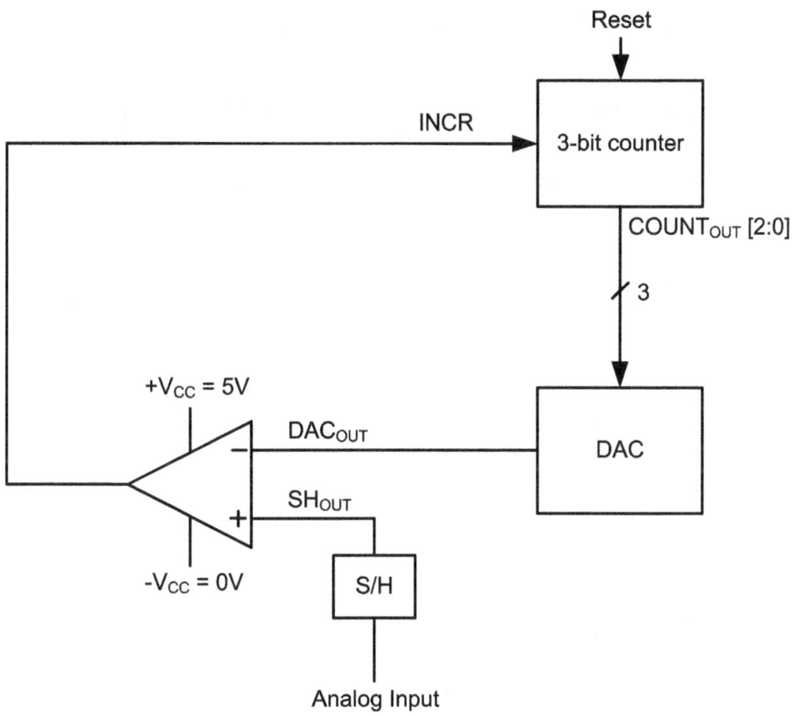

(a) Assume that DAC has an up-rounding scheme to generate analog outputs from digital inputs. Apply 1.2 V to the Analog Input port, and draw the timing diagram that contains the clock, Reset, counter output ($COUNT_{OUT}$), DAC output (DAC_{OUT}) and operational amplifier output (INCR). Show what happens to the timing diagram when the active-high Reset signal transitions to logic 1 after the ADC produces the desired digital output.

(b) Now apply 2.9 V to the Analog Input. Draw the timing diagram with the same inputs and outputs listed above. Show what happens to the timing diagram when the active-high Reset signal transitions to logic 1 after the ADC produces the desired digital output.

(c) Assume that the DAC rounding scheme is changed from up-rounding to down-rounding scheme. Apply 1.2 V to the Analog Input, and generate the timing diagram with the input and output signals listed above. Show what happens to the timing diagram when the active-high Reset signal becomes logic 1 after the ADC produces the desired digital output. Do you see any issues with the operation of this circuit?

(d) Now apply 2.9 V at the Analog Input port and generate the timing diagram with the input and output signals listed above. Do you see any issues in the operation of this circuit?

7. The following circuit shows the block diagram of a simple transmitter-receiver. The transmission protocol starts with the transmitter sending the request signal, Req, to the receiver. The receiver acknowledges the request by producing an acknowledgement signal, Ack, and starts reading data from the transmitter at the next positive edge of the clock following the Ack signal. Once the data transmission ends, the transmitter keeps sending the last data packet on the Data bus.

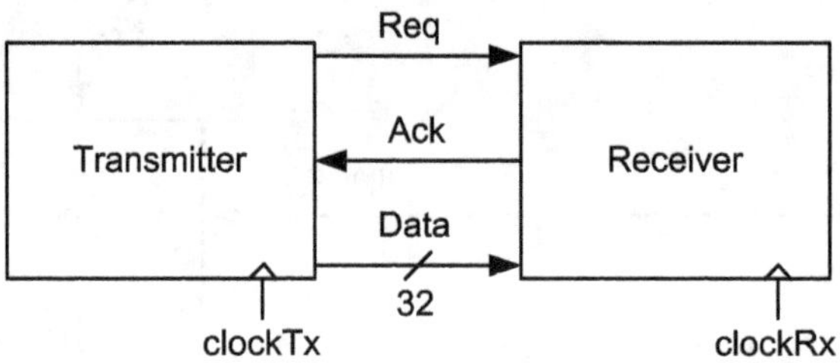

(a) Assume the transmitter sends out three consecutive data packets, D0, D1, D2. Complete the timing diagram below for the Ack and Data signals. Show which data the receiver actually receives.

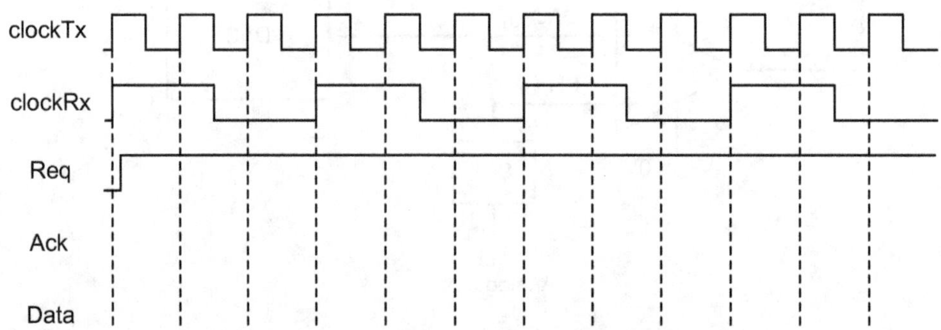

(b) Assume the transmitter sends out three consecutive data packets, D0, D1, D2. Complete the timing diagram below for the Ack and Data signals. Show which data the receiver actually receives.

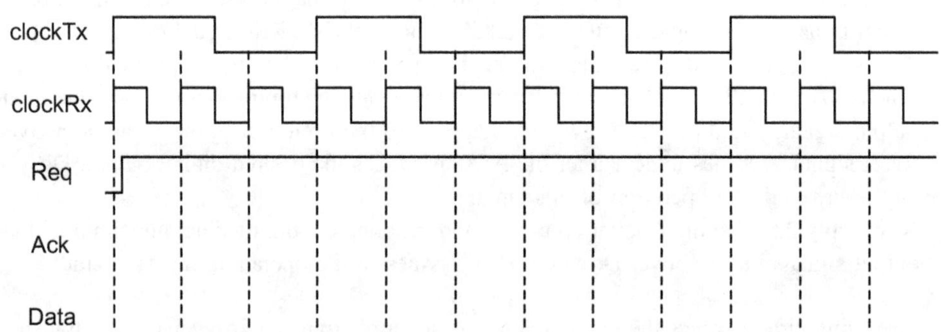

8. A black-and-white display supports 256 shades of gray, ranging from white to black. The physical display frame has 100 pixels in the x-direction and 100 lines in the y-direction, and it requires a frame rate of 100 frames/s. The display needs no horizontal or vertical blanking pixels to synch with the display adaptor. The clock frequency of the display and the adaptor are set at 1 MHz. The display adaptor is connected to an eight-bit wide high-speed bus operating at 10 MHz to receive data. Once the adaptor recognizes that one of its buffers is empty, it immediately sends a request signal to the bus arbiter to own the bus. The acknowledgement from the bus arbiter requires 18 ms delay due to heavy bus traffic. Once the acknowledgment is received, the display adaptor fills all of its buffers. Each buffer contains exact number of pixels to fill only one frame.

(a) With the timing specs defined above for the display unit and the high-speed bus, determine the number of buffers used in this system with a timing table that shows how these buffers are periodically emptied and filled following an 18 ms bus waiting period. Note that this is not a timing diagram that includes the frame or the bus clocks or the propagation of data.

(b) Draw an architectural data-path of the display adaptor including the buffers, the gray-scale frame and the related hardware (counters, multiplexers, controller etc.). Make sure to generate all the internal I/O signals of the controller to operate the arbiter and maintain the proper data flow in the display adaptor.

9. A display adaptor shown below has an overlay feature where an overlay image is mapped over the active image area as long as the overlay image is smaller than the active image in size. The system neither requires blanking space nor needs a dual buffering.

Assume pixels in the image and overlay buffers are not separated into RGB components, but fully integrated when they are taken out of these buffers to feed the frame buffer. There is a certain synchronization mechanism between the image and overlay address counters, and the pixel and line counters. As soon as a pixel is fetched from an image or overlay buffer, that pixel is placed in the frame buffer with the aid of the pixel and line counters.

There are no write-enable controls for data buffers as these buffers will not be replenished with new pixels once exhausted. The image is displayed only once. The read-enables for both buffers are also kept at logic 1 until the buffers are empty. Therefore, the only mechanism that transfers pixels out of these buffers into the frame buffer is incrementing the address counters and switching the selector inputs to the 2-1 buffer MUX.

After removing the external reset, the operation of the display unit starts as shown in the timing diagram below. The address counter for the image data buffer does not have an increment control input because this counter increments constantly as soon as the reset is removed from the circuit.

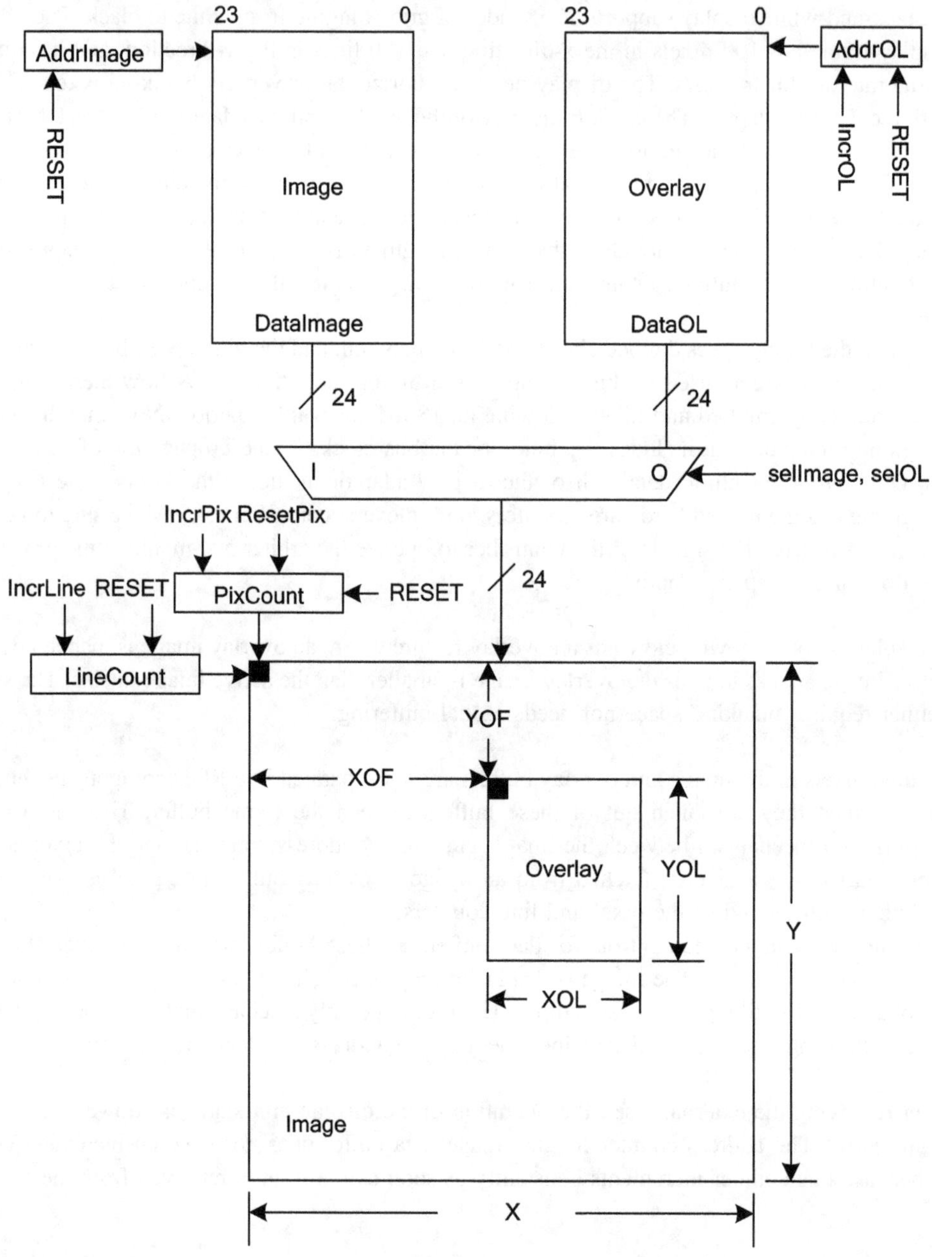
23
0
AddrImage
RESET
Image
DataImage
23
0
AddrOL
IncrOL
RESET
Overlay
DataOL
24
24
I
O
selImage, selOL
IncrPix ResetPix
PixCount
RESET
24
IncrLine RESET
LineCount
YOF
XOF
Overlay
YOL
XOL
Y
Image
X

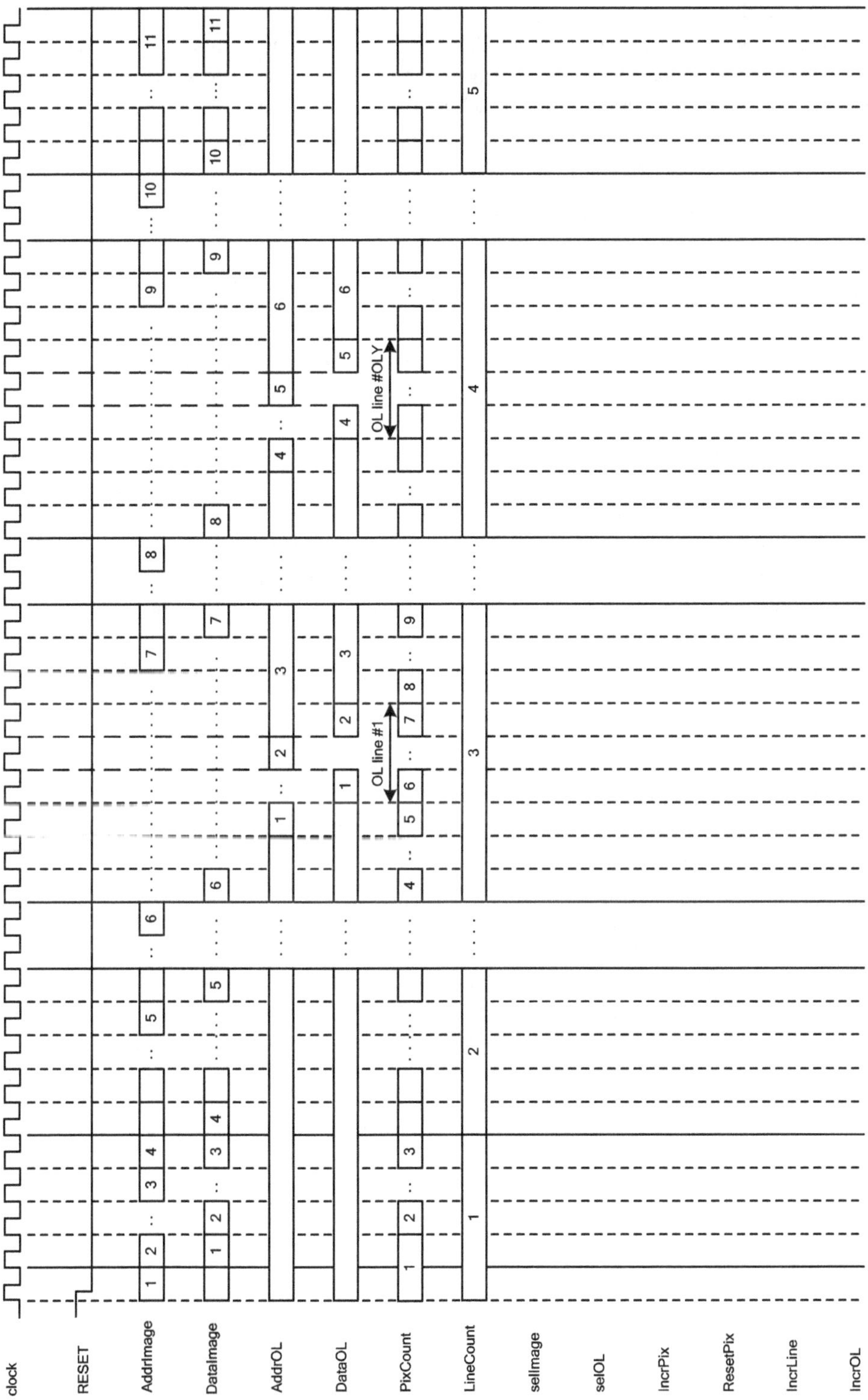
clock
RESET
AddrImage
DataImage
AddrOL
DataOL
PixCount
LineCount
selImage
selOL
IncrPix
ResetPix
IncrLine
IncrOL
OL line #1
OL line #OLY

(a) Build the register file. Indicate the programmable values in each register to support the operation of this unit.
(b) Fill in each numbered box in the timing diagram (if there are no numbers, ignore the box), and show how the control signals change. In this figure, there are six control signals: selImage (to select the image buffer), selOL (to select the overlay buffer), IncrPix (to increment the pixel counter), ResetPix (to reset the pixel counter to 0), IncrLine (to increment the line counter) and IncrOL (to increment the overlay counter).

Use the following notation for the data buffers:

For the Image buffer, Image Data = Im[Image Buffer Address]
For the Overlay buffer, Overlay Data = Olay[OL Buffer Address]

10. An interrupt controller interfaced with a three-stage RISC CPU is shown below. Once an external interrupt (INTR0 to INTR15) is generated, the interrupt interface selects the highest priority interrupt and generates a single INTR bit for the Interrupt Control Unit (ICU). The ICU acknowledges the interrupt by INTA, which prompts the interface to send a four-bit INTRID to the Interrupt Address Table (IAT). A 32-bit interrupt address is then produced from the IAT which causes the program counter (PC) to jump and execute an Interrupt Service Routine (ISR) program in the instruction memory. Before the ISR is executed, the remains of the original program in the CPU pipeline have to be to be executed and stored in the register file (RF). Also, the address of the next instruction in the user program is stored in R31 in the RF. Upon the completion of a particular ISR, the program returns to its original location by retrieving the address stored in the register R31. The rest of the user program executes promptly.

In this particular case, INTR0 as the highest priority interrupt comes in and prompts the interrupt interface to generate INTRID = 0. This is the beginning of a four instruction long ISR0 that starts at address 100 as shown in the instruction memory.

Assuming there are a total of seven instructions in the user program, each instruction written back to the RF is labeled as WUi. For example, Ins1 results are written back to the RF as WU1, Ins2 results as WU2 etc. as shown in the timing diagram. Similarly, each ISR instruction written back to the RF is labeled as WIi. For example, Intr1 results are written back to the RF as WI1, Intr2 results as WI2 etc.

Also, once the PC generates a value, each instruction produces an RF output as DUi. For example, Ins1 produces DU1, Ins2 produces DU2 etc. Similarly, each interrupt instruction produces an RF output as DIi. For example, Intr1 produces DI1, Intr2 produces DI2 etc.

Based on the inputs in the preceding paragraphs, fill the blanks in the rest of the timing diagram below. Indicate when each write takes place to the RF with a little arrow as shown in the timing diagram.

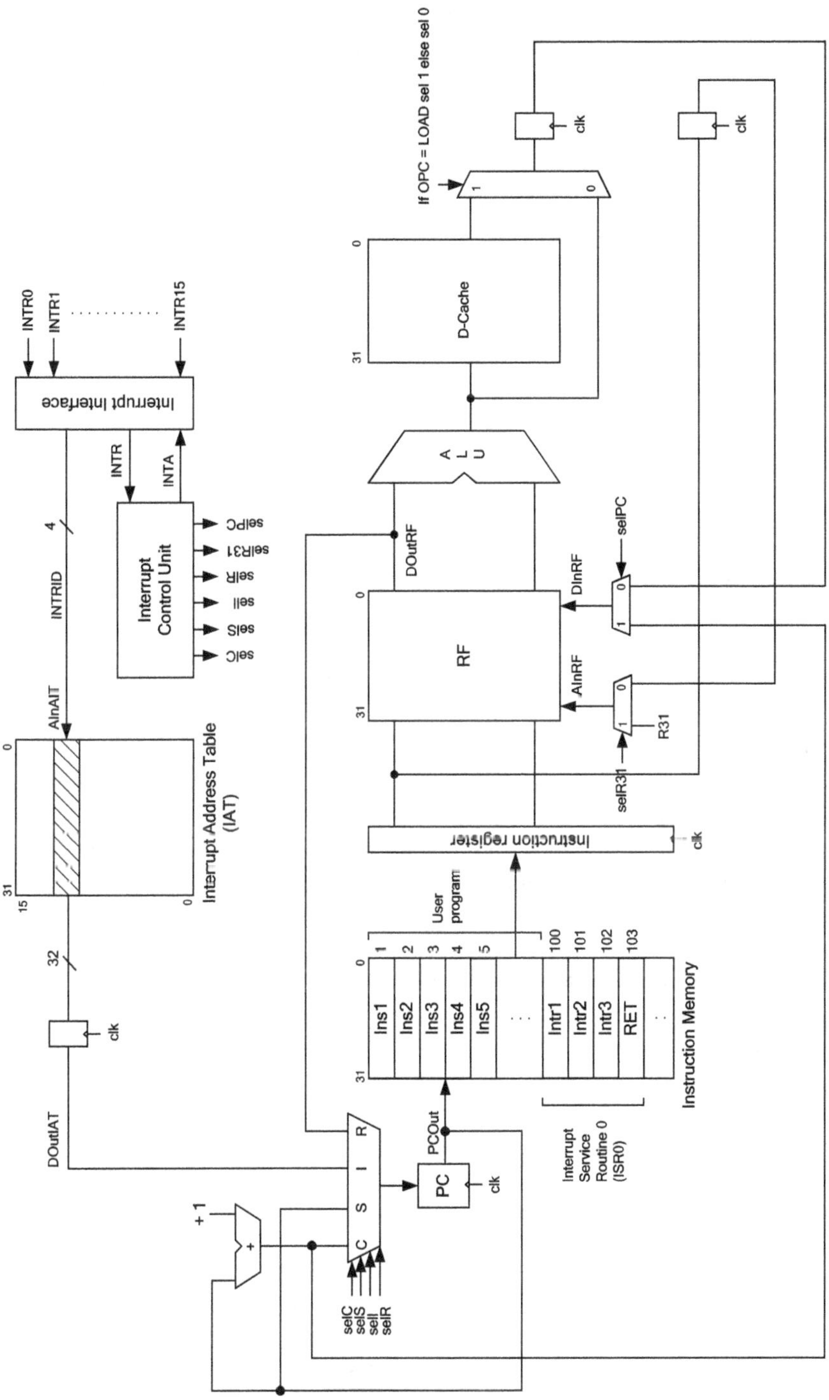
INTR0
INTR1
INTR15
Interrupt Interface
INTR
INTA
INTRID
4
Interrupt Control Unit
selPC
selR31
selR
selI
selS
selC
AInAIT
Interrupt Address Table (IAT)
32
clk
DOutIAT
+1
selC
selS
selI
selR
PC
PCOut
Ins1
Ins2
Ins3
Ins4
Ins5
Intr1
Intr2
Intr3
RET
Instruction Memory
User program
Interrupt Service Routine 0 (ISR0)
Instruction register
RF
DOutRF
AInRF
DInRF
R31
selR31
selPC
ALU
D-Cache
If OPC = LOAD sel 1 else sel 0

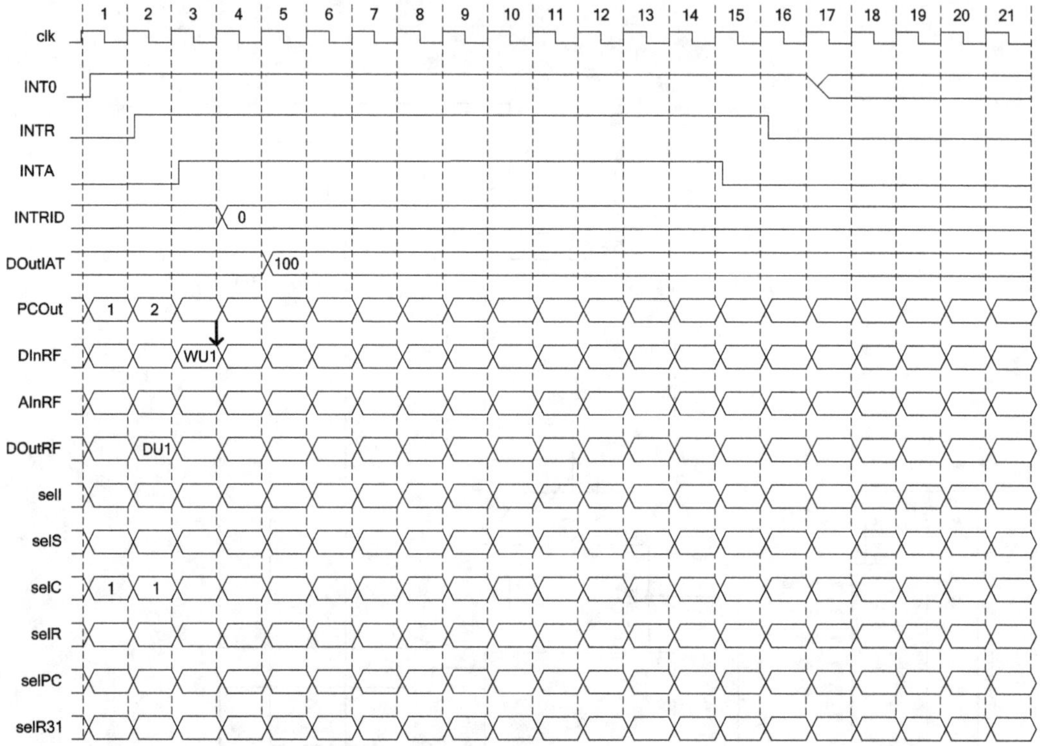

11. A parallel bus operating at 1 GHz clock frequency (1 ns clock period) employs a single peripheral, A, with two 100 byte buffers, A0 and A1. Each buffer requires 100 ns to fill. The process starts when the peripheral fills its first buffer, A0, before filling the second buffer, A1. The peripheral exhausts data from its buffer within 400 ns, and needs bus access to replenish it. To maintain data continuity the peripheral immediately starts fetching data from its secondary buffer.

(a) In the timing diagram below with 100 ns intervals, show when the peripheral needs bus access to fill either of its buffers with solid squares (in other words, fill in the square). Indicate when the peripheral starts fetching data from its buffer with crossed squares (in other words, fill in the square with an "X" sign).

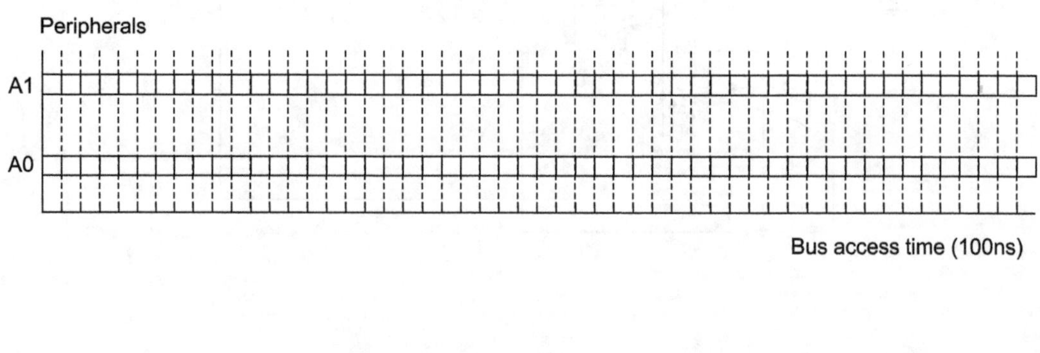

(b) Now, assume that there are two peripherals, A and B. Peripheral A still has two buffers and behaves exactly as described in part (a). Peripheral B, which has less priority than peripheral A, has four buffers, B0, B1, B2 and B3. This peripheral, when it finds the bus is free of activity, fills its B0 buffer first, followed B1, B2 and B3. Again, each buffer takes 100 ns to fill. However, unlike peripheral A, peripheral B exhausts data from any of its four buffers within 600 ns, and requires bus access to replenish the data in the exhausted buffer. In the timing diagram below with 100 ns intervals, show when each peripheral needs bus access to fill its buffers with solid squares, and when the peripheral starts fetching data from its buffers with crossed squares.

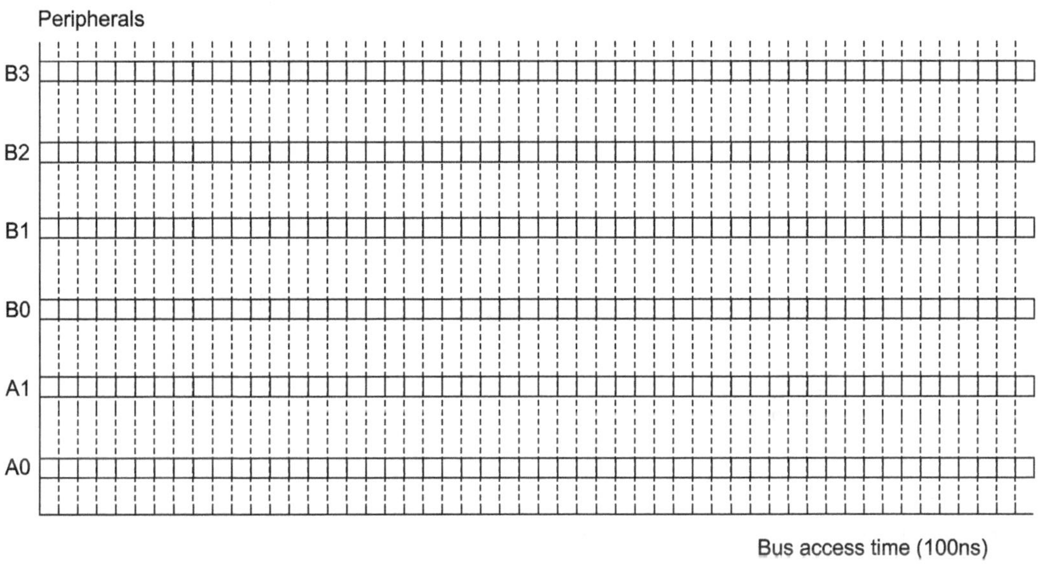

12. Design an SPI (see Chap. 4 on serial bus) interface using an integrated transmitter and receiver shown below. The eight-bit Shift Register at the interface transmits one-bit data from the SDO port at the positive edge of SCK, and simultaneously receives one-bit data from the SDI port at the following negative edge. The transmit data is first loaded to the SPI Register using an eight-bit system bus. Subsequently, the contents of the SPI Register are loaded to the Tx Buffer if the buffer is empty and then to the Shift Register when it requires new data. In a similar fashion, when the Shift Register acquires new eight new bits through its SDI port, it transfers its contents to Rx Buffer first and then to the SPI Register. When the transmit function is desired, the received data is considered junk data. When the receive function is desired, the transmit data is, in turn, assumed junk. The two flags, TxF and RxF, update the status register whether Tx Buffer and Rx Buffer are empty or not.

Design the SPI data-path and the controller using timing diagrams. SCK is assumed a slower clock, and it has a period eight times the clock period of the system clock. The designer should feel free to add additional hardware or signals to implement this design in different way.

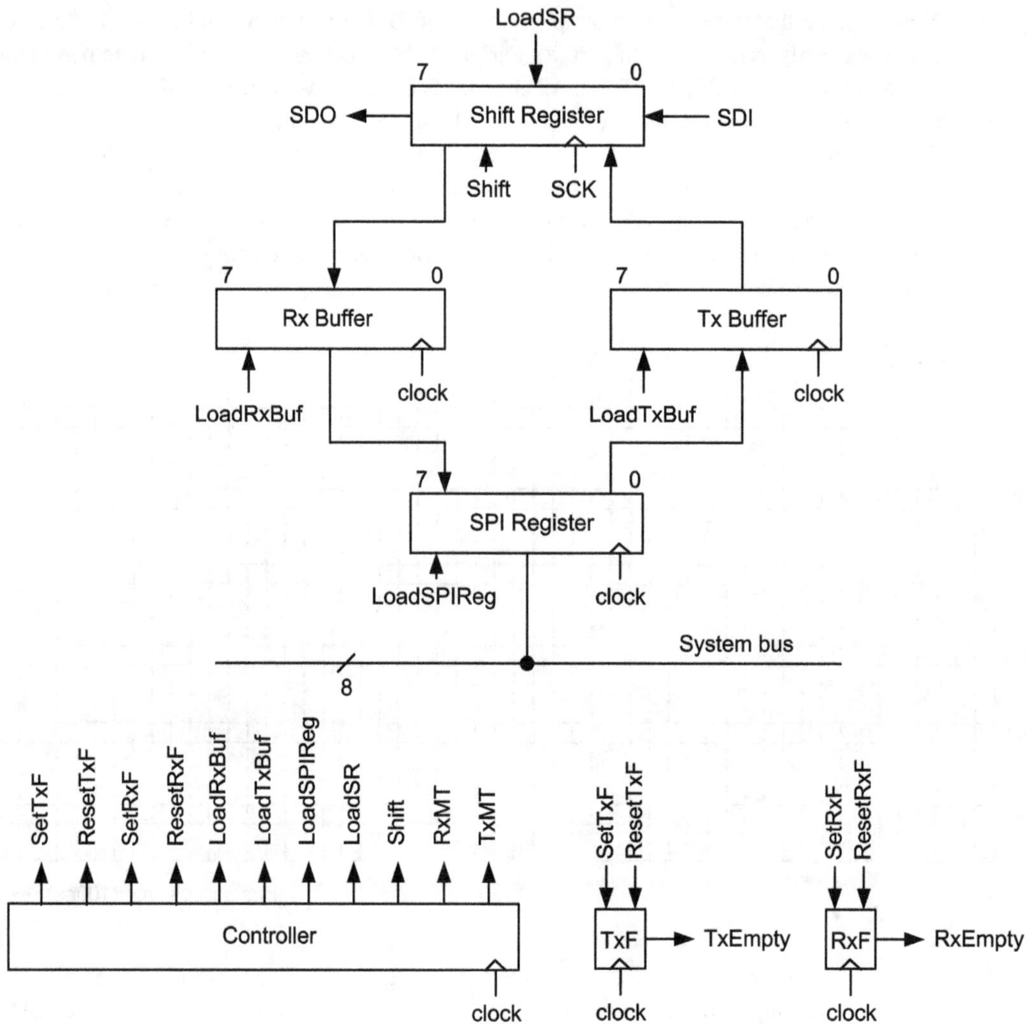

Projects

1. Implement and verify a DMA that supports two identical 32×64 SRAM memories using Verilog. Produce its timing diagram and the controller.
2. Implement and verify an interrupt controller that supports 256 hardware interrupts using Verilog. Include the hardware for context switching, i.e. transferring the contents of the entire register file to a temporary buffer prior to executing interrupt service routine instructions.
3. Implement the one-shot timer using Verilog. Produce its timing diagram.
4. Implement the rate generator using Verilog. Produce its timing diagram.
5. Implement and verify the display adaptor unit that supports a screen with eight pixels, two blanking lines and nine active image lines.

8 Special Topics

This chapter introduces two core topics that belong to a computing system. The first topic is a brief introduction to programmable logic. The second topic is the architectural description of a data-driven processor that operates with the arrival of new data.

When it comes to prototyping an application-specific digital block, the first thing that comes to mind is the Field-Programmable-Gate-Array (FPGA) platform. This platform is flexible enough to implement any combinatorial, sequential or asynchronous logic with ease. Using programmable logic, we can create mega cells such as ALU blocks or simple memories, logic blocks that perform specific functions, processors, and even an entire computing system using a Hardware Design Language (HDL).

The second topic in this chapter describes a data-driven architecture that works with a cluster of simple processors. Each processor in the cluster is designed to carry out specific task(s), and each becomes active when valid data arrives from a neighboring processor. In a data-driven system, either an individual processor carries out a specific task and transfers the result to the next processor or every processor in the cluster simultaneously execute many different tasks all at once to produce a single result.

8.1 Field-Programmable-Gate Array

The basic idea behind the Field-Programmable-Gate-Array (FPGA) architecture is the use of Look-Up-Tables (LUT). A typical three-input LUT in Fig. 8.1 contains eight registers to store bits, an 8-1 MUX to select one of the eight register outputs, and a flip-flop at the output of the 8-1 MUX to implement sequential logic. The programming phase consists of serially distributing the bit values of an input data, In[7:0], to all eight registers through the ProgIn port when the Prog input is set to logic 1. In order to achieve this, all eight LUT registers are connected in a shift register configuration, and In[7:0] is serially shifted in from Bit[0] to Bit[7]. The bottom register at the Bit[7] position has another output, ProgOut, connected to the ProgIn input of another LUT such that every LUT on the FPGA chip can be serially programmed using a single wire to save wiring space.

A. Bindal, *Fundamentals of Computer Architecture and Design*,
https://doi.org/10.1007/978-3-030-00223-7_8

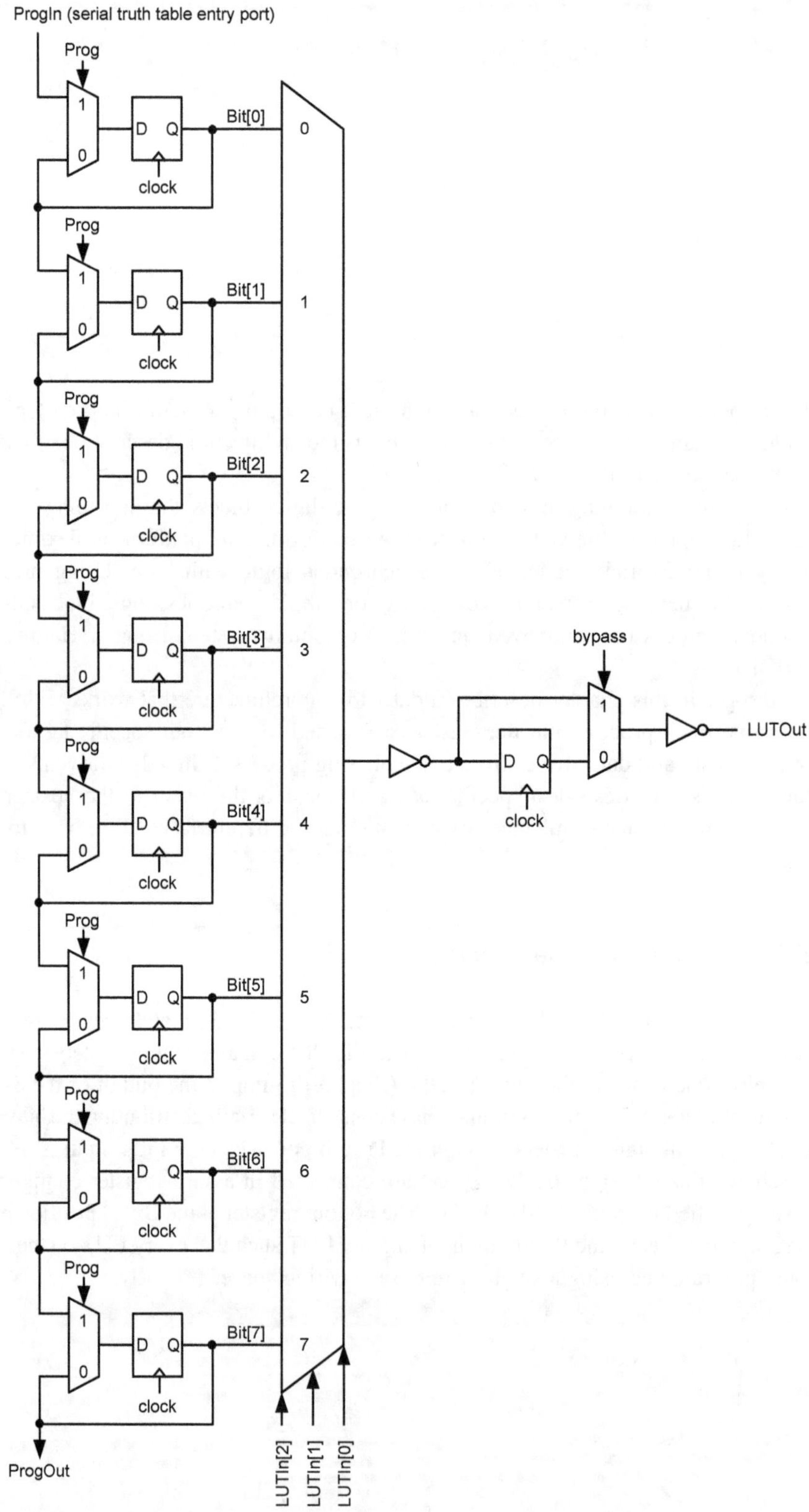

Fig. 8.1 Three-input look-up-table (LUT) block diagram

In normal operation, bits stored in the LUT registers constitute the output values of the truth table in Table 8.1. The inputs of the truth table, on the other hand, are the selectors of the 8-1 MUX, from LUTIn[0] to LUTIn[2], in Fig. 8.1. Therefore, any arbitrary truth table can be produced simply by programming the LUT registers, needing no other conventional logic gate. For example, when LUTIn [2] = LUTIn[1] = LUTIn[0] = 0 in Fig. 8.1, the 8-1 MUX routes the value stored in Bit[0] to its output. Similarly, LUTIn[2] = LUTIn[1] = LUTIn[0] = 1 combination routes Bit[7] to the output. The 2-1 MUX is used to bypass the flip-flop output if a combinational logic implementation is preferred.

The number of registers in a LUT is determined by the number of outputs in the truth table. For example, if there are three inputs in the truth table, this combination generates $2^3 = 8$ possible outputs. Therefore, the LUT must contain eight registers. In general, N inputs require 2^N registers in a LUT.

In summary, in order to implement a logic function using FPGA, the inputs of a logic function (truth table) must be applied to the MUX selectors, and the outputs must be stored in the LUT registers according to Table 8.1.

To demonstrate how a combinational logic block is implemented in an FPGA platform, we will design a four-bit Ripple-Carry-Adder (RCA) as shown in Fig. 8.2. The circuit consists of four full adders all connected serially to propagate the carry bit from right to left. The sum, SUM0 to SUM3, and the carry-out, Cout0 to Cout3, outputs have to be programmed in the LUT registers.

Table 8.1 Three-input LUT truth table (when bypass port is set to 1)

LUTIn[2]	LUTIn[1]	LUTIn[0]	LUTOut
0	0	0	Bit[0]
0	0	1	Bit[1]
0	1	0	Bit[2]
0	1	1	Bit[3]
1	0	0	Bit[4]
1	0	1	Bit[5]
1	1	0	Bit[6]
1	1	1	Bit[7]

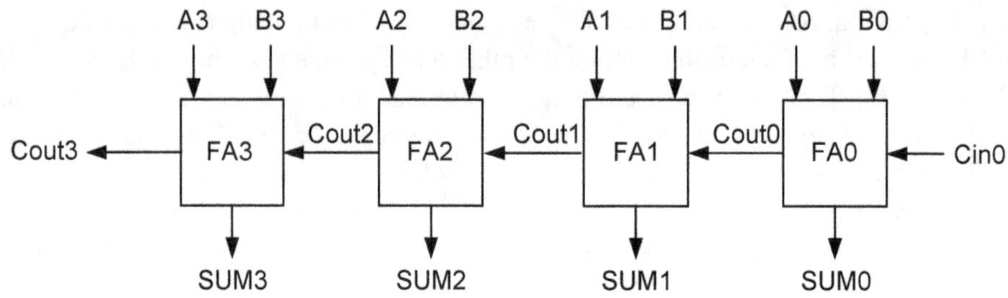

$$SUM0 = A0 \oplus B0 \oplus Cin0$$
$$SUM1 = A1 \oplus B1 \oplus Cin1$$
$$SUM2 = A2 \oplus B2 \oplus Cin2$$
$$SUM3 = A3 \oplus B3 \oplus Cin3$$

$$Cout0 = A0.B0 + Cin0.(A0 + B0)$$
$$Cout1 = A1.B1 + Cin1.(A1 + B1)$$
$$Cout2 = A2.B2 + Cin2.(A2 + B2)$$
$$Cout3 = A3.B3 + Cin3.(A3 + B3)$$

Fig. 8.2 Four-bit ripple-carry-adder

Figure 8.3 describes how a full adder sum output is stored in a three-input LUT. This process is the same for generating each sum output from SUM0 to SUM3. In this figure, the LUT output value at the first row is stored in the Bit[0] position, and the last output entry is stored in the Bit[7] position. This bit arrangement in the LUT registers implements the SUM function for any combination of Cin, A and B applied as the 8-1 MUX selector inputs. The bypass input at the output 2-1 MUX must also be set to logic 1 to bypass the flip-flop stage since this design is not a sequential circuit.

The Cout function of the full adder is implemented similarly as shown in Fig. 8.4. The Cout function in the last column of the truth table is programmed into the LUT registers while Cin, A and B are connected to the selector inputs. The bypass bit is also set to logic 1 to bypass the flip-flop since the implementation is purely combinational.

Figure 8.5 shows the FPGA implementation of the four-bit RCA in Fig. 8.2 after the programming phase is complete. In this design, each FPGA cell, called a cluster, is assumed to contain two LUTs. While A0, A1, A2, A3, B0, B1, B2, B3 and Cin0 are external input pins for the four-bit RCA, Cin1, Cin2 and Cin3 inputs are all internally generated from Cout0, Cout1 and Cout2 function blocks, and routed between clusters to maintain interconnectivity. All the bypass inputs, from bypass-Cout0 to bypass-SUM3, have to be at logic 1 and stored in a separate LUT during the programming phase.

LUTIn[2] = Cin	LUTIn[1] = A	LUTIn[0] = B	LUTOut[0] = SUM
0	0	0	0
0	0	1	1
0	1	0	1
0	1	1	0
1	0	0	1
1	0	1	0
1	1	0	0
1	1	1	1

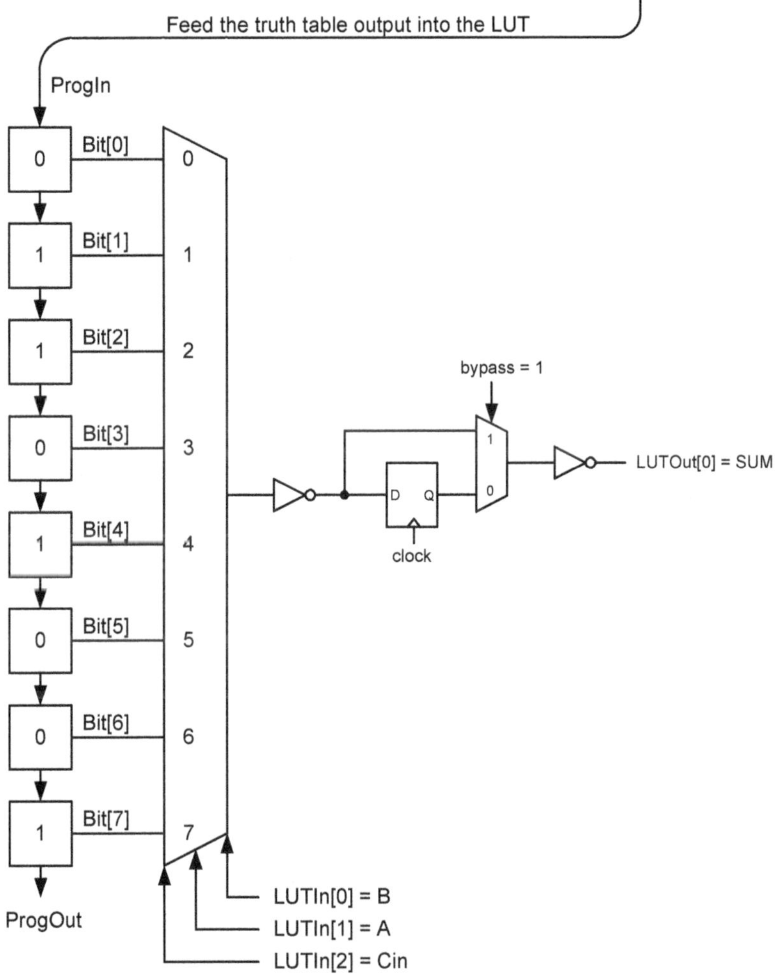

Fig. 8.3 Programming the full adder SUM output with a three-input LUT

LUTIn[2] = Cin	LUTIn[1] = A	LUTIn[0] = B	LUTOut[1] = Cout
0	0	0	0
0	0	1	0
0	1	0	0
0	1	1	1
1	0	0	0
1	0	1	1
1	1	0	1
1	1	1	1

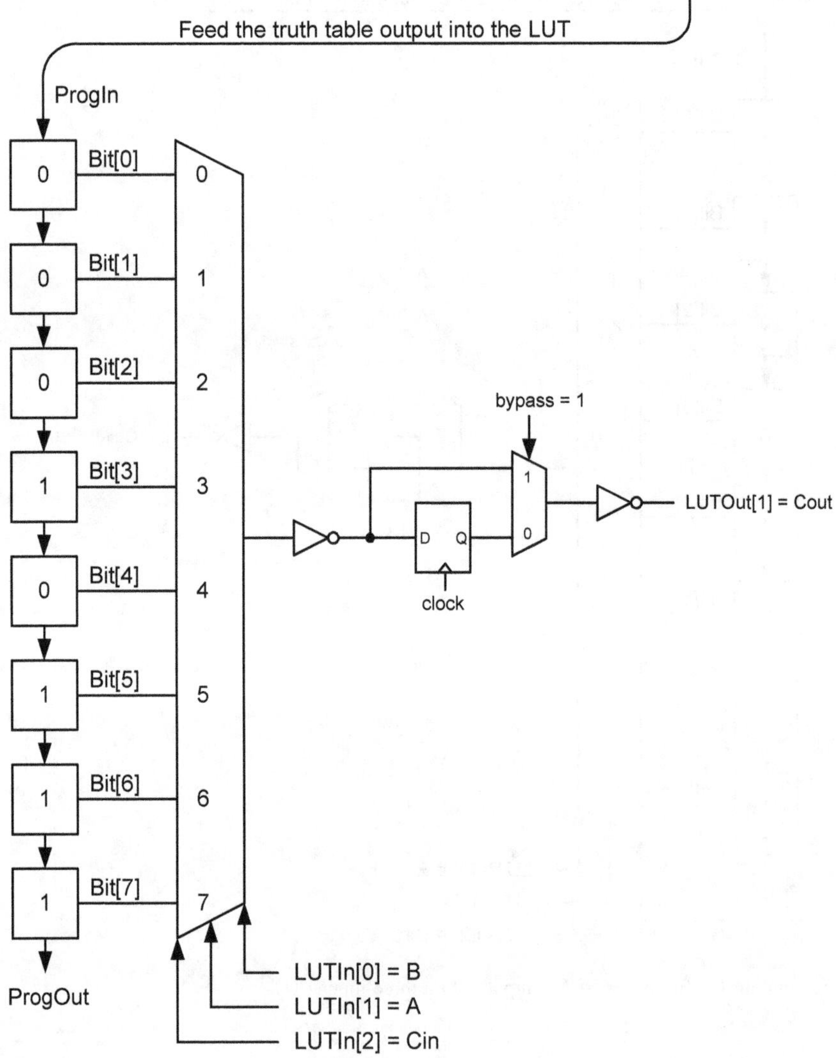

Fig. 8.4 Programming the full adder Cout output with a three-input LUT

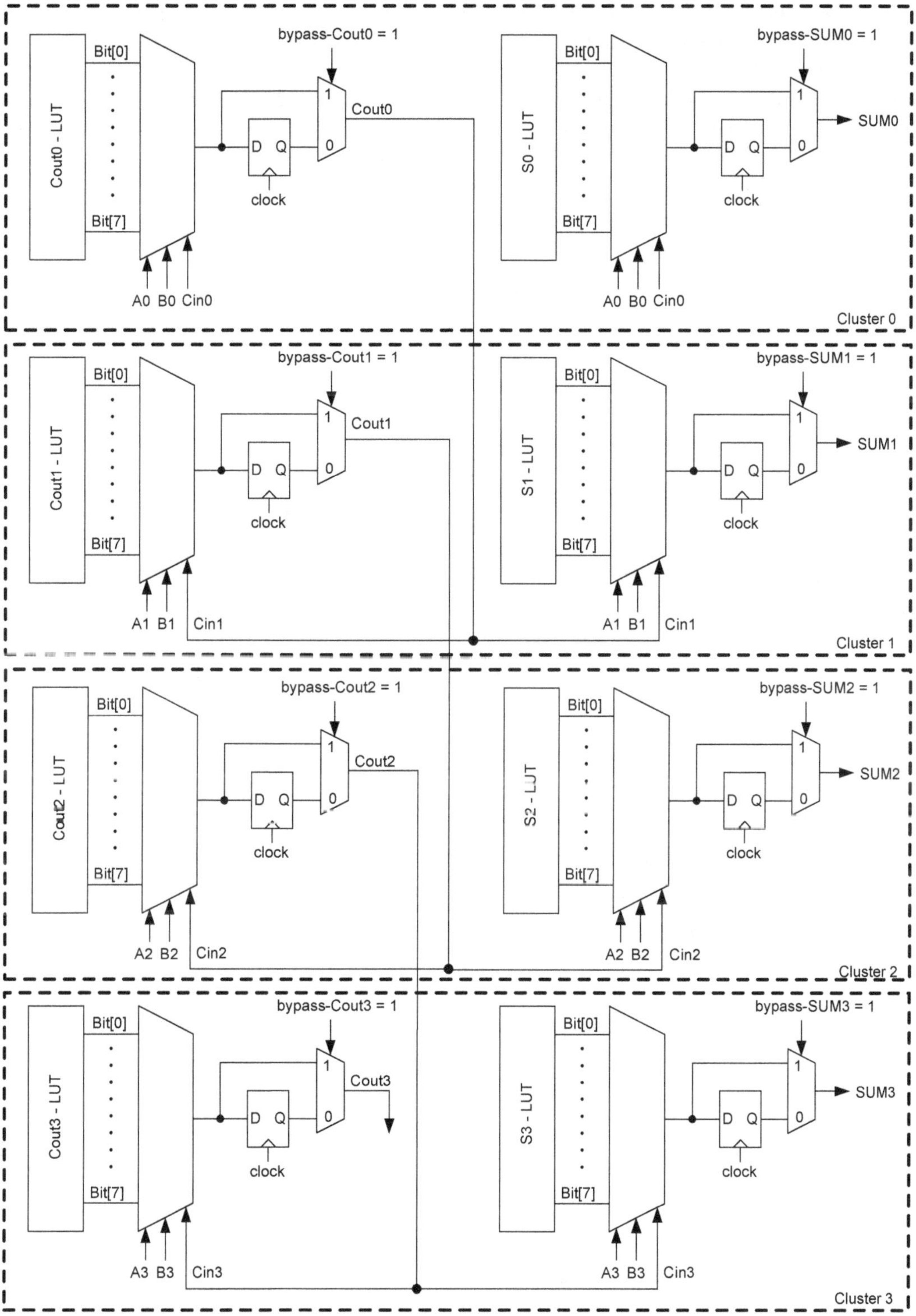

Fig. 8.5 Four-bit ripple-carry-adder data-path in FPGA

A commercial FPGA cluster contains a multitude of multiplexers connected to the LUT inputs to accomplish the maximum flexibility in logic configuration. Figure 8.6 shows one such cluster configuration that contains two LUTs, each with three inputs. A cluster configured this way is able to achieve maximum networking capability with other FPGA clusters in addition to implementing many types of combinational and sequential logic circuits. Programming these LUTs is achieved simply by connecting the ProgOut port of one LUT to the ProgIn port of the neighboring LUT, and distributing the serial data to two LUTs from the ProgIn port as shown in the figure.

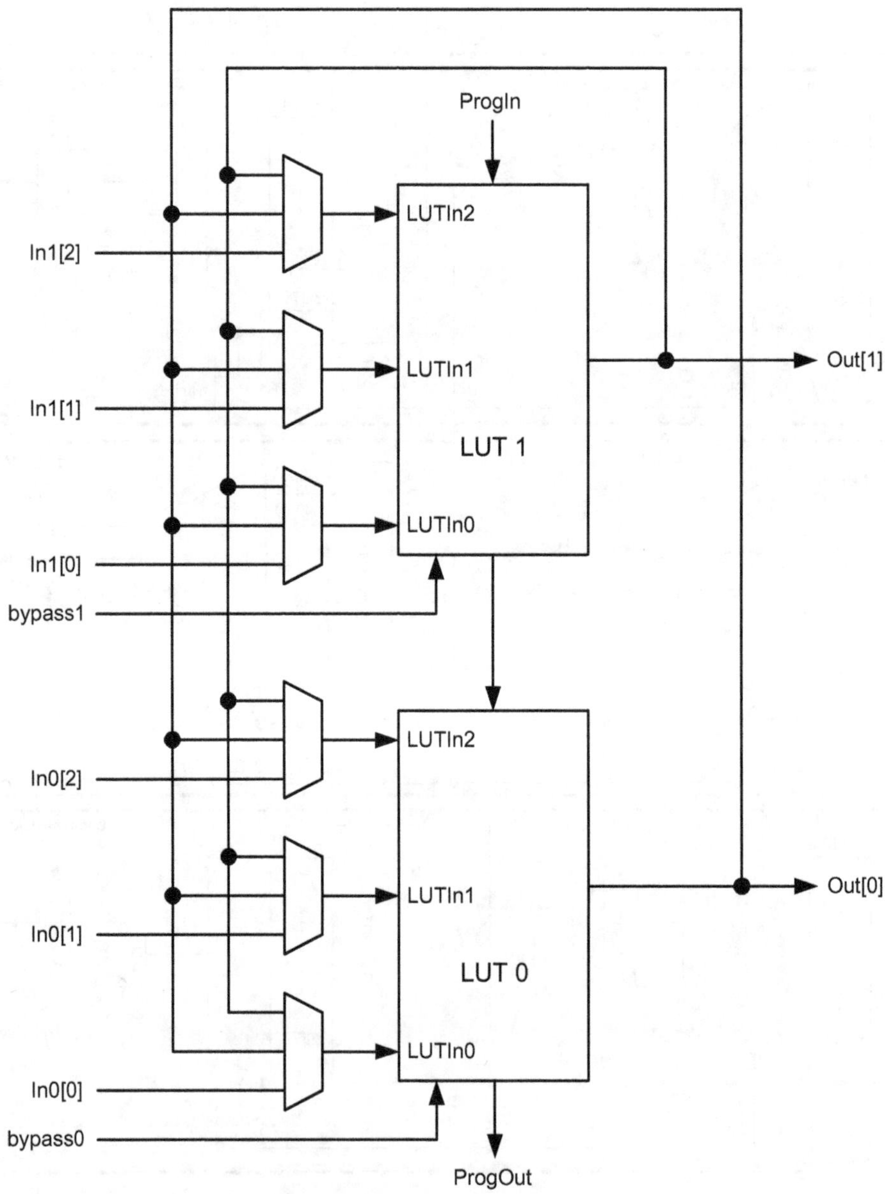

Fig. 8.6 A commercial FPGA cluster containing two LUTs per cluster

The detailed schematic in Fig. 8.7 shows the FPGA implementation of Cluster 0 in Fig. 8.5 but uses a commercial FPGA architecture in Fig. 8.6. Each 3-1 MUX selector input from SelOut0[0] to SelOut1[5] is stored in a 12-bit shift register (LUT2) in Fig. 8.7.

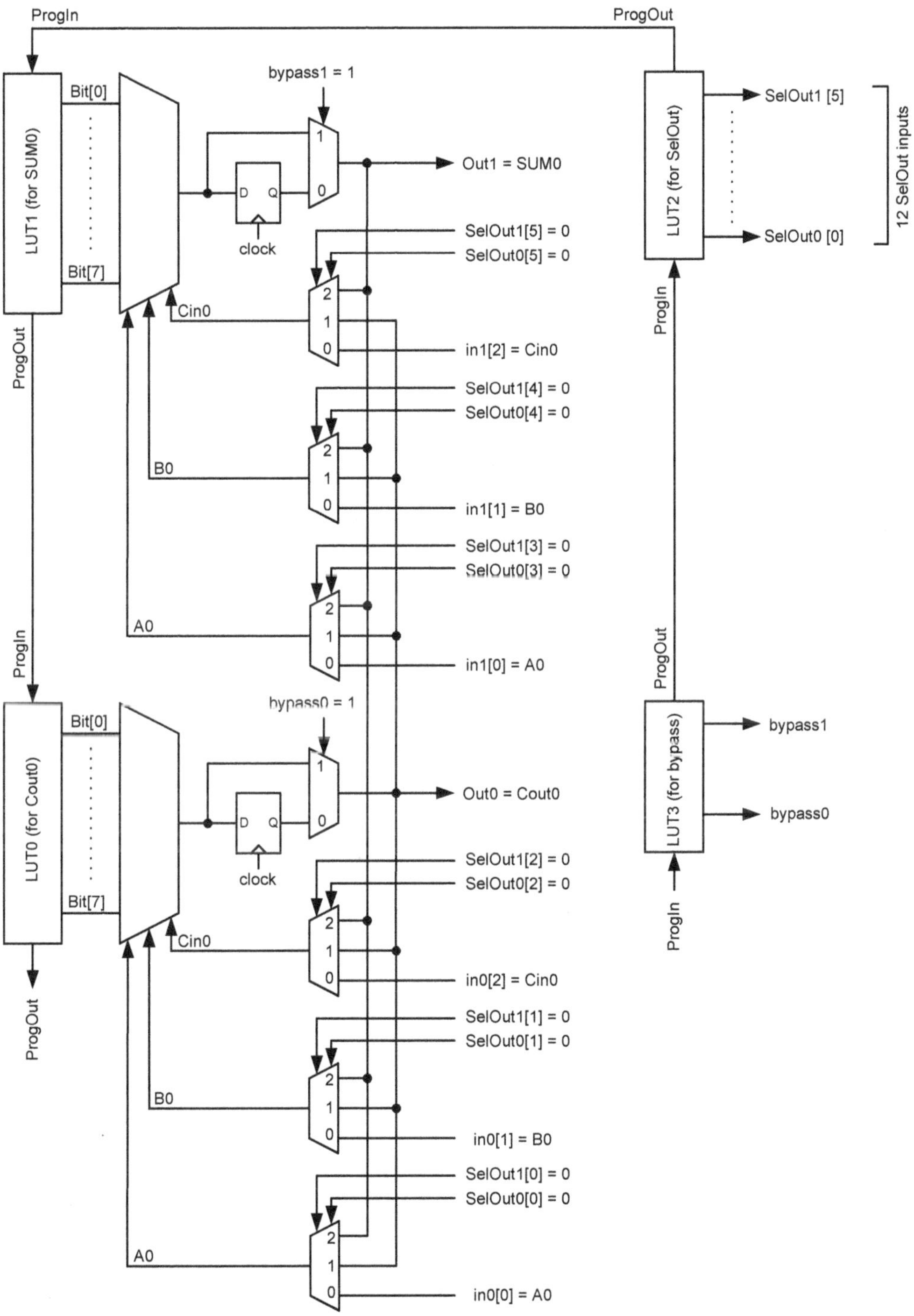

Fig. 8.7 Implementing SUM0 and Cout0 using a single cluster containing two LUTs

In this figure, SelOut1[i] = 0 and SelOut0[i] = 1 combination selects port 1 or Out0. Here, i changes from 0 to 5. SelOut1[i] = 1 and SelOut0[i] = 0 combination selects port 2 or Out1. When SelOut1 = SelOut0 = 0, the default port 0 is selected, and an external input becomes one of the selector inputs for the 8-1 LUT MUX. The bypass pins, bypass0 and bypass1, are also stored in a two-bit shift register (LUT3).

A Moore or Mealy type state machine can also be implemented using FPGA platforms. The state diagram in Fig. 8.8 produces the transition table in Table 8.2, which includes two next state outputs, NS0 and NS1, two present state inputs, PS0 and PS1, an external input, IN, and a present state output, OUT[2:0], to produce integer values between one and four. For state assignments, only one bit is allowed to change between neighboring states, i.e. S0 = 00, S1 = 01, S2 = 11 and S3 = 10. The resultant circuit in Fig. 8.9 shows the locations of the present and next states, the input and the output.

Programming the NS0 function in Table 8.2 in a three-input LUT configuration is explained in Fig. 8.10. In this figure, the NS0 column is distributed among eight LUT registers, storing the first bit of NS0 at the Bit[0] position, and the last bit at the Bit[7] position. The inputs, PS0, PS1 and IN, that generate NS0 are connected to the 8-1 MUX selector pins, LUTIn[0], LUTIn[1] and LUTIn[2],

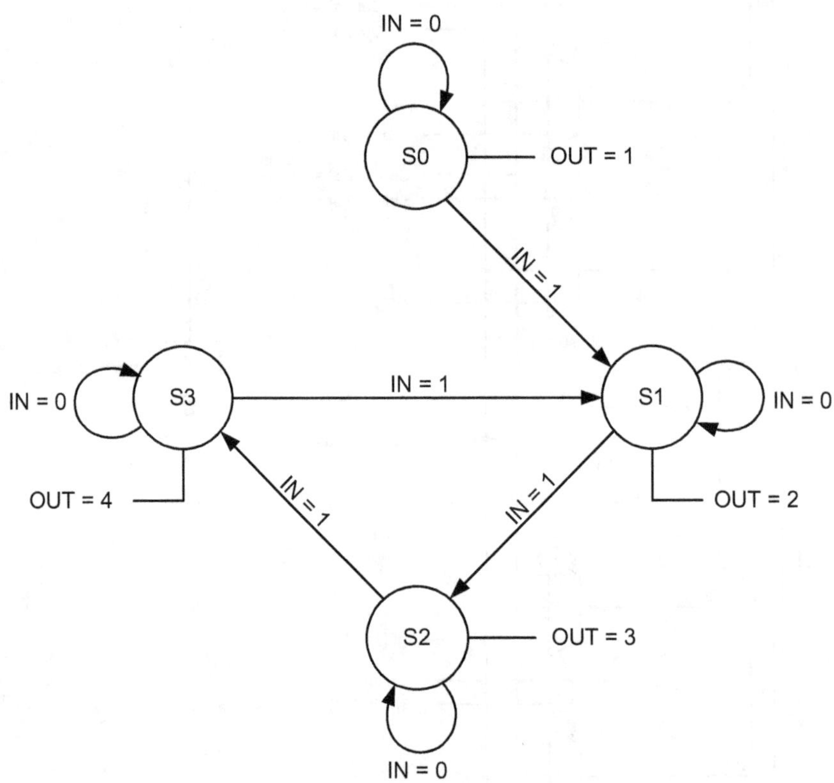

Fig. 8.8 A moore machine

Table 8.2 The transition table for the moore machine in Fig. 8.8

IN	PS1	PS0	NS1	NS0	OUT[2]	OUT[1]	OUT[0]
0	0	0	0	0	0	0	1
0	0	1	0	1	0	1	0
0	1	0	1	0	1	0	0
0	1	1	1	1	0	1	1
1	0	0	0	1	0	0	1
1	0	1	1	1	0	1	0
1	1	0	0	1	1	0	0
1	1	1	1	0	0	1	1

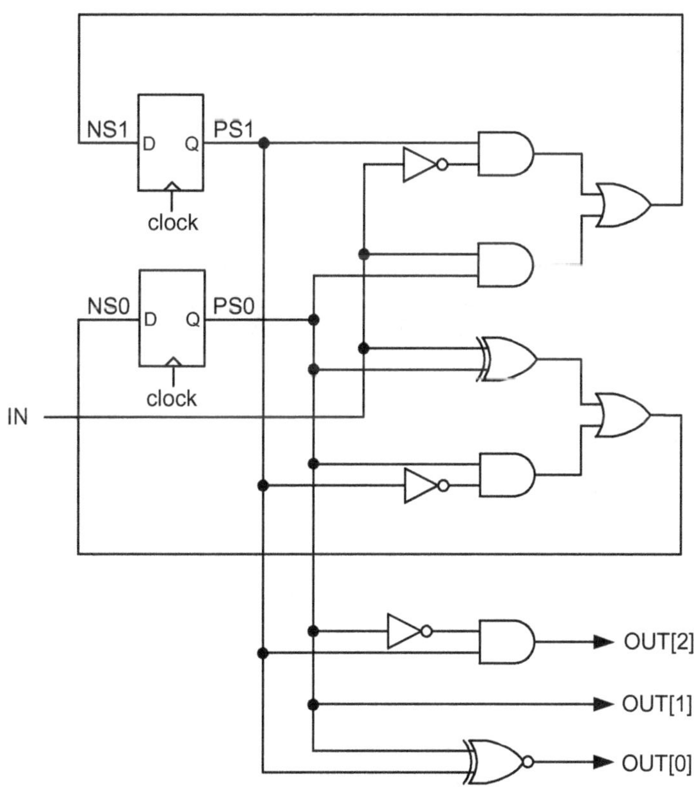

Fig. 8.9 The circuit diagram for the moore machine in Fig. 8.8

IN	PS1	PS0	NS0
0	0	0	0
0	0	1	1
0	1	0	0
0	1	1	1
1	0	0	1
1	0	1	1
1	1	0	1
1	1	1	0

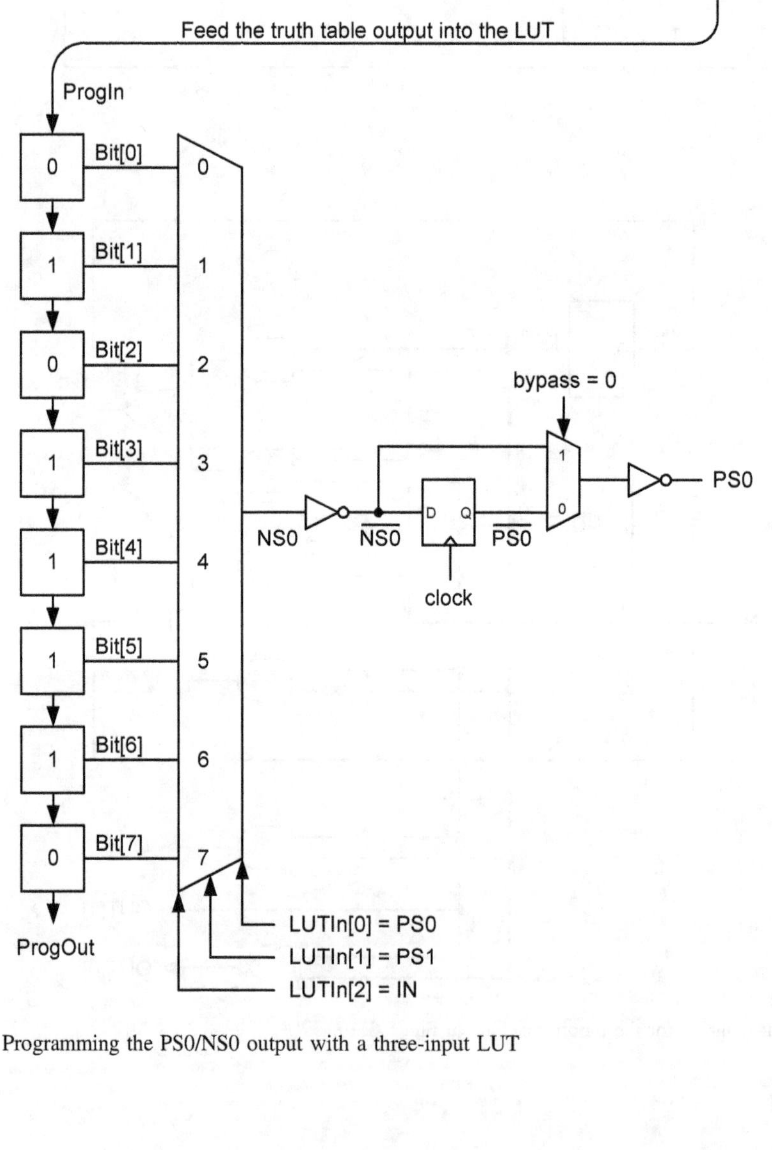

Fig. 8.10 Programming the PS0/NS0 output with a three-input LUT

respectively. Since the bypass input pin is set to logic 0, the NS0 node becomes the input of the flip-flop, and the PS0 node becomes the output.

The NS1 functionality is implemented in a similar fashion as shown in Fig. 8.11. The NS1 column in Table 8.2 is stored in the LUT registers. PS0, PS1 and IN inputs are connected to LUTIn[0], LUTIn[1] and LUTIn[2] MUX selector pins, respectively. The bypass pin is set to logic 0 in order to form NS1 node at the input of the flip-flop and PS1 node at the output.

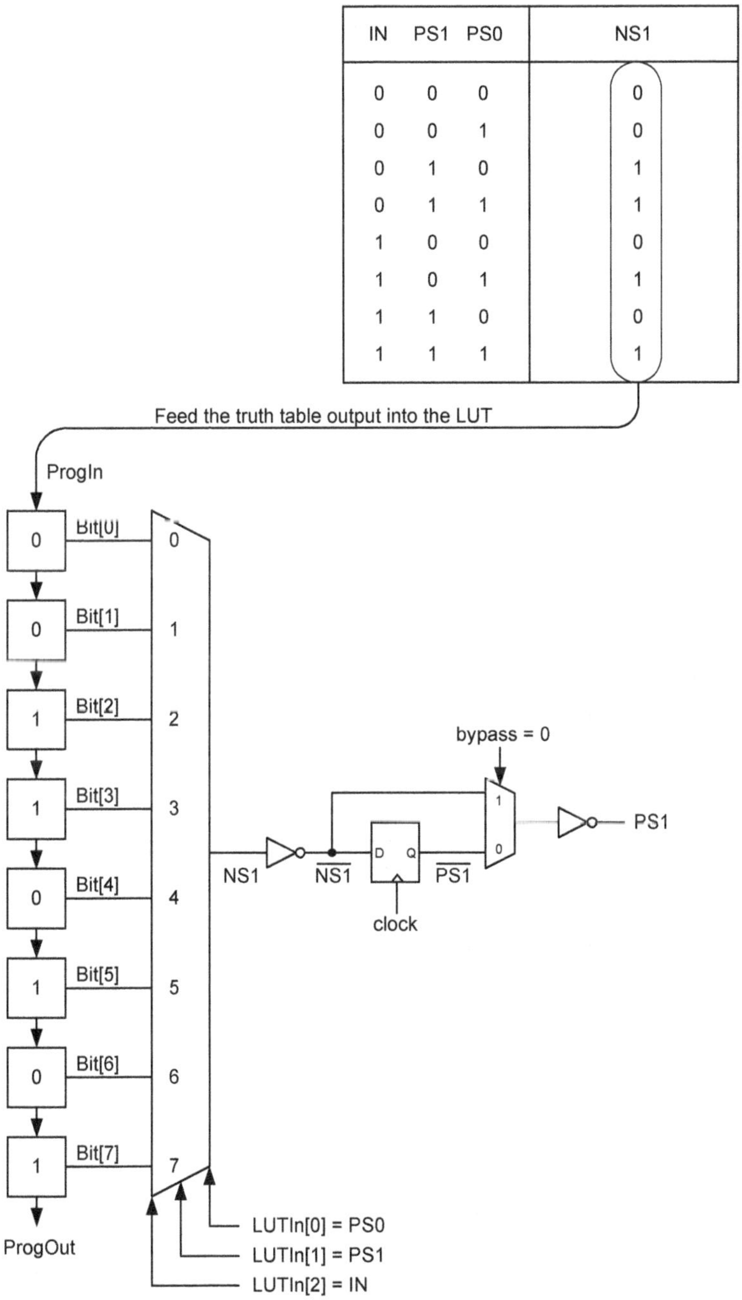

Fig. 8.11 Programming the PS1/NS1 output with a three-input LUT

The OUT[0], OUT[1] and OUT[2] outputs are also programmed in the LUT registers according to Table 8.2, and shown in Fig. 8.12, Fig. 8.13 and Fig. 8.14, respectively. However, the bypass input in each case must be set to logic 1 in order to bypass the flip-flop stage since these outputs are completely combinational and do not require any clock in their paths.

IN	PS1	PS0	OUT[0]
0	0	0	1
0	0	1	0
0	1	0	0
0	1	1	1
1	0	0	1
1	0	1	0
1	1	0	0
1	1	1	1

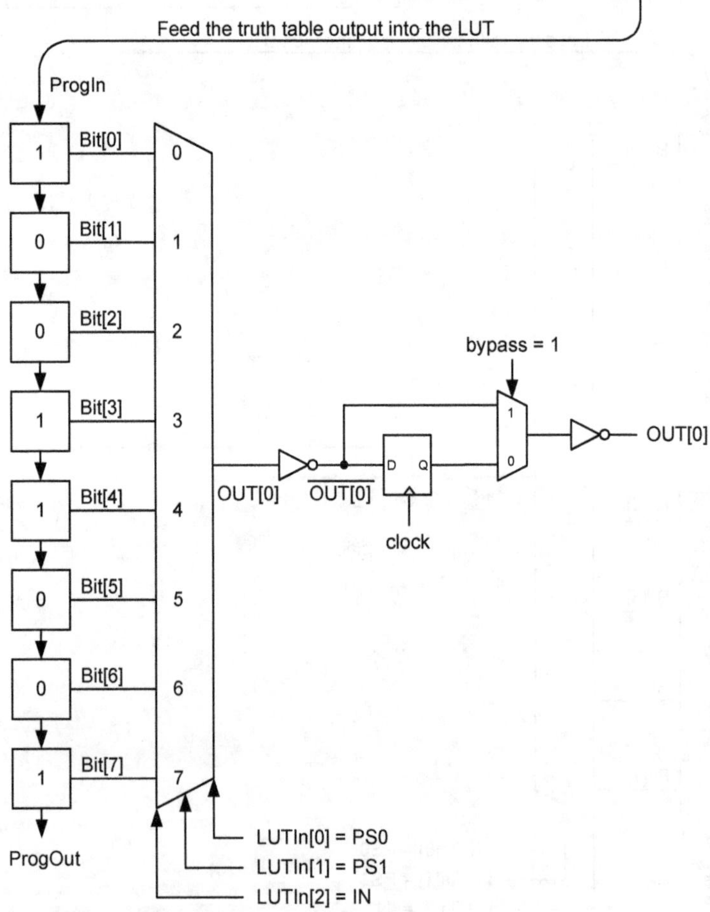

Fig. 8.12 Programming the OUT[0] output with a three-input LUT

IN	PS1	PS0	OUT[1]
0	0	0	0
0	0	1	1
0	1	0	0
0	1	1	1
1	0	0	0
1	0	1	1
1	1	0	0
1	1	1	1

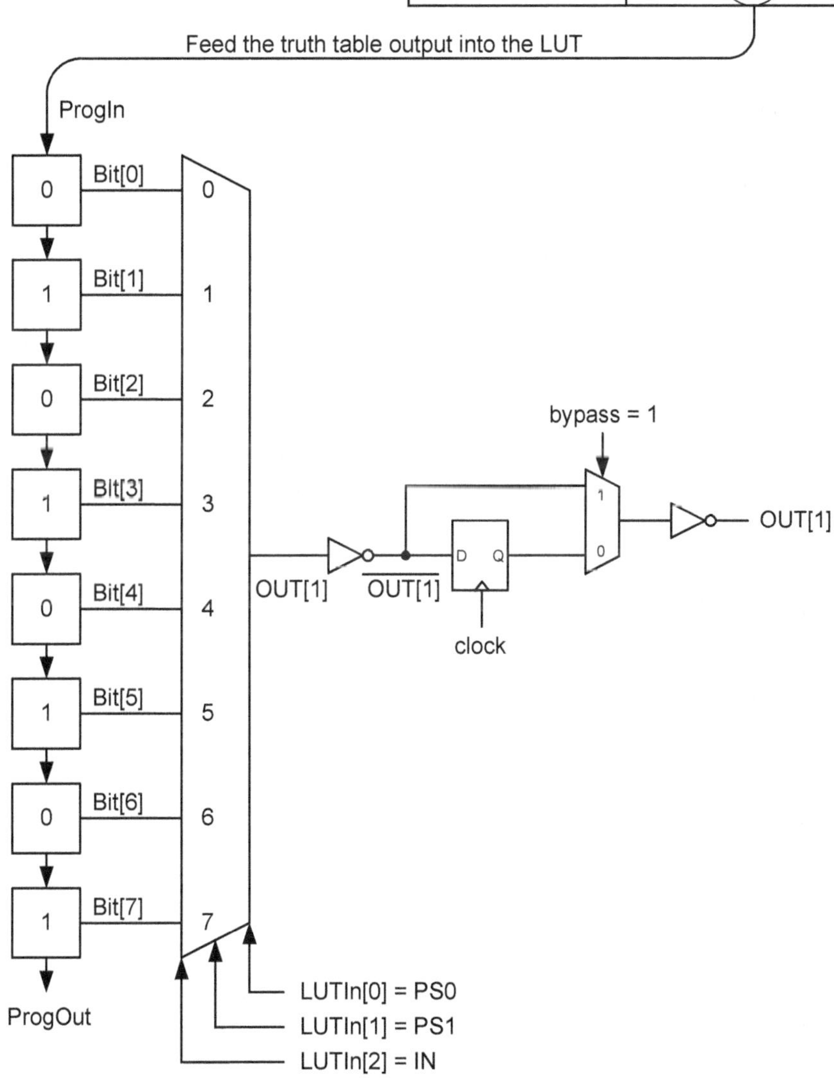

Fig. 8.13 Programming the OUT[1] output with a three-input LUT

IN	PS1	PS0	OUT[2]
0	0	0	0
0	0	1	0
0	1	0	1
0	1	1	0
1	0	0	0
1	0	1	0
1	1	0	1
1	1	1	0

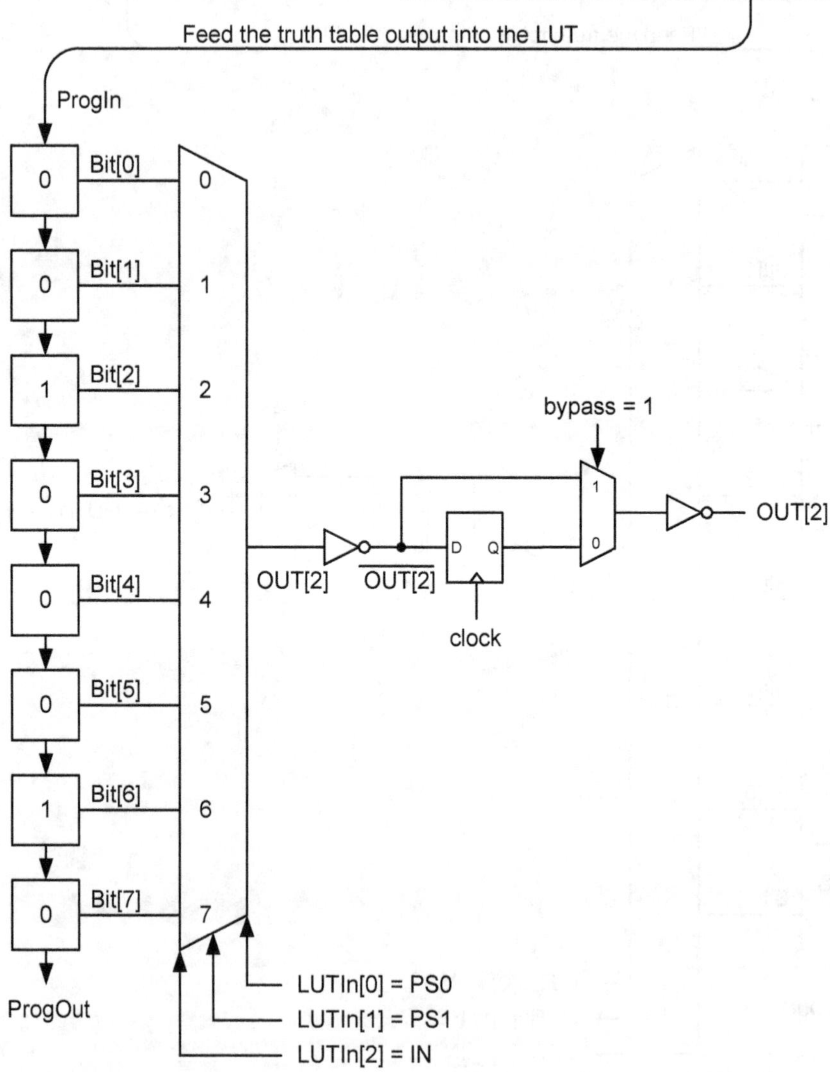

Fig. 8.14 Programming the OUT[2] output with a three-input LUT

Figure 8.15 describes the implementation of the Moore machine in Fig. 8.9 after programming each LUT in three different clusters. Cluster 0 generates NS0 and NS1 functions implicitly but also produces PS0 and PS1 outputs. Cluster 1 and Cluster 2 implement OUT[2:0]. The IN port is the only external input that goes to all three clusters to maintain the logic functionality. All the bypass inputs from bypass-PS0 to bypass-OUT2 are stored in a separate LUT.

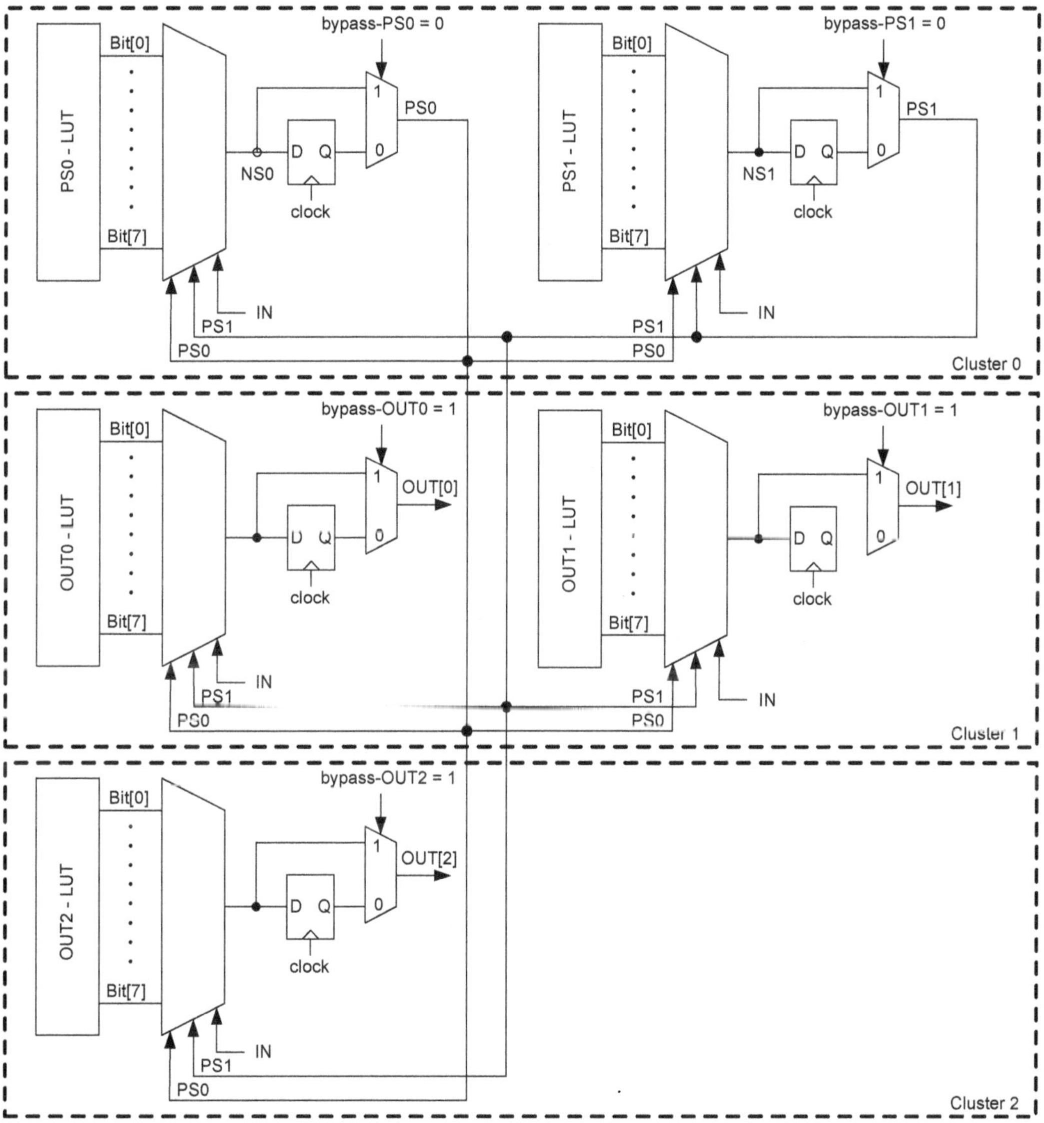

Fig. 8.15 Moore state machine data-path in FPGA

The schematic in Fig. 8.16 shows the implementation of Cluster 0 in Fig. 8.15 in a commercial FPGA platform in Fig. 8.6. In this schematic, all 3-1 MUX selector inputs from SelOut0[0] to SelOut1[5], are stored in LUT2. Similarly, the bypass inputs, bypass0 and bypass1, are stored in LUT3 to be used during normal operation.

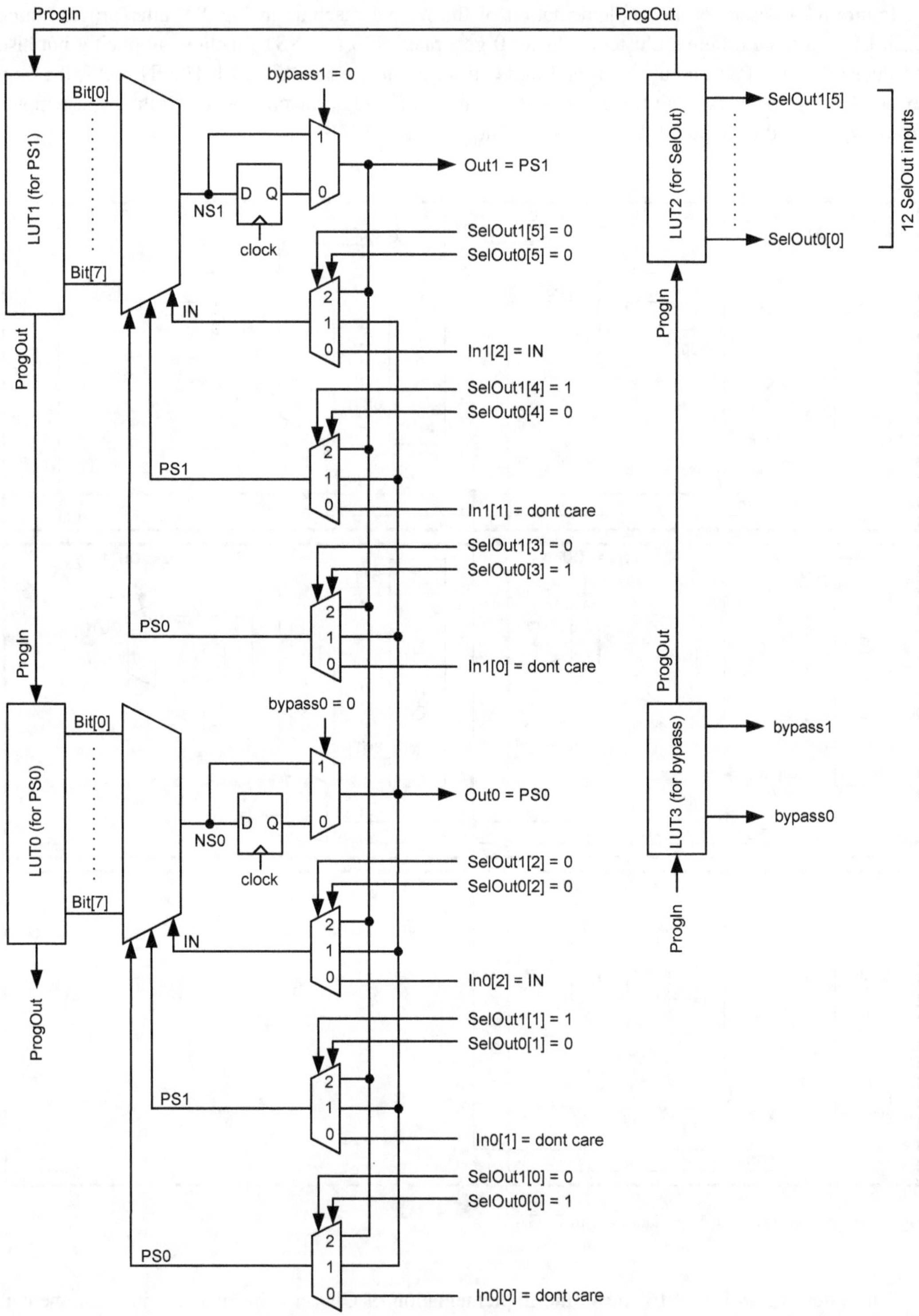

Fig. 8.16 Implementing PS0 and PS1 using a single cluster containing two LUTs

8.2 Data-Driven Processors

Data-Flow Graphs

Programming data-driven processors is achieved by data-flow graphs. Each graph consists of a group of functional nodes and communication paths, connecting the nodes. Each node in the flow-graph executes two incoming data tokens when they arrive, and produces an output operand according to the function defined in the node. The output operand is then forwarded to a neighboring functional node where it becomes a data token for that node. Therefore, a node function can be defined by an instruction for a processor. Each instruction representing a functional node contains an operation code (OPC), input operand(s) (OPER), and output operand node addresses. All input-output paths among functional nodes are connected with directed communication paths to guide the flow of data. With this picture in mind, operands that flow into a functional node are executed according to the node's operation code. Once executed in the node, new operands form and flow out of the node to other nodes. A simple example is given in Fig. 8.17. In this example, the operands, $OPER_{S1}$ and $OPER_{S2}$, are executed by the node's operation code, OPC_S, when they arrive at the functional node, N_S. After the execution, a new operand forms and flows out of the node to two new destination nodes, N_{D0} and N_{D1}, where it meets with two other operands, $OPER_{D0}$ and $OPER_{D1}$, respectively. The data-flow program stops when all operands are executed.

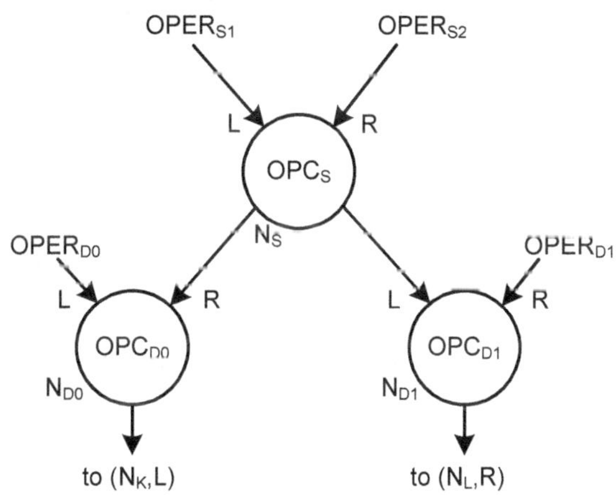

Fig. 8.17 Sample flow graph for a data-driven machine

The parallel nature of data-flow architecture makes parallel processing tasks quite achievable in data-driven machines. While conventional processors are set to be maximally serial to minimize hardware, data-driven architectures can be maximally parallel with up to one processor per operation to maximize performance. For the example in Fig. 8.17, two different data-driven processors can be used simultaneously to perform OPC_{D0} and OPC_{D1} following the operation at the node N_S. Multi-processor platforms formed by a group of conventional processors may have limitations to achieve certain parallel processing tasks. In contrast, data-driven processors can time-share the processing load of several functional nodes, and therefore reduce the hardware requirement to implement a data-flow graph.

Data-Flow Node Types

The types of data-flow nodes used in this architecture are classified according to the number of input operands fed into the functional nodes. Figure 8.18 illustrates this classification.

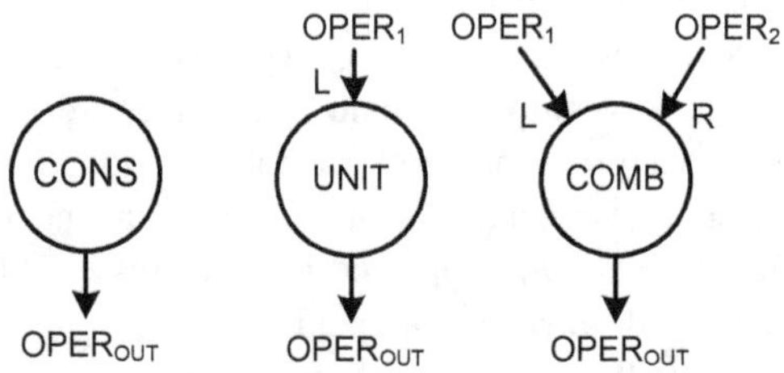

Fig. 8.18 Functional node types

Functional nodes that accept no inputs are nodes with constants (CONS). The contents of this node type do not change during programming. Functional nodes that accept one input are the unitary (UNIT) nodes. This node type transforms an operand as soon as it arrives at its input. Invert (negate), Set and Reset are the unitary nodes in this architecture that require a single input operand. Functional nodes that accept two input operands are the combinatorial (COMB) nodes. This node type executes incoming operands when they are both valid at the input. The operation of some combinatorial functional nodes, such as subtract and shift, depends on the relative placement of the input operands. This is called the operand polarity. It is reflected in the instruction format by defining input operands as left or right input operands. In this architecture, operands with changing data values are directed to the left side of a functional node, whereas constant operands or operands used as control signals are placed on the right side of the functional node in a data-flow diagram.

While operands emanating from functional nodes are forwarded only to one destination address in the earlier data-flow diagrams, this architecture offers the flexibility where the same operand can be forwarded to two different destination addresses. This unique feature saves the number of nodes used in the data-flow diagram as well as increases execution speed of the program.

Basic Data-Flow Program Structures

There are three basic programming structures in data-driven architectures when constructing data-flow diagrams: sequential, conditional and recursive. Each structure is illustrated in Fig. 8.19.

Sequential programming constructs imply that the data-flow is unidirectional, from one functional node to the next without any loops or paths related to a condition. The simple data-flow diagram in Fig. 8.17 is one such example of the sequential programming structure. Another example composed of multi-layer functional nodes is given at the top of Fig. 8.19.

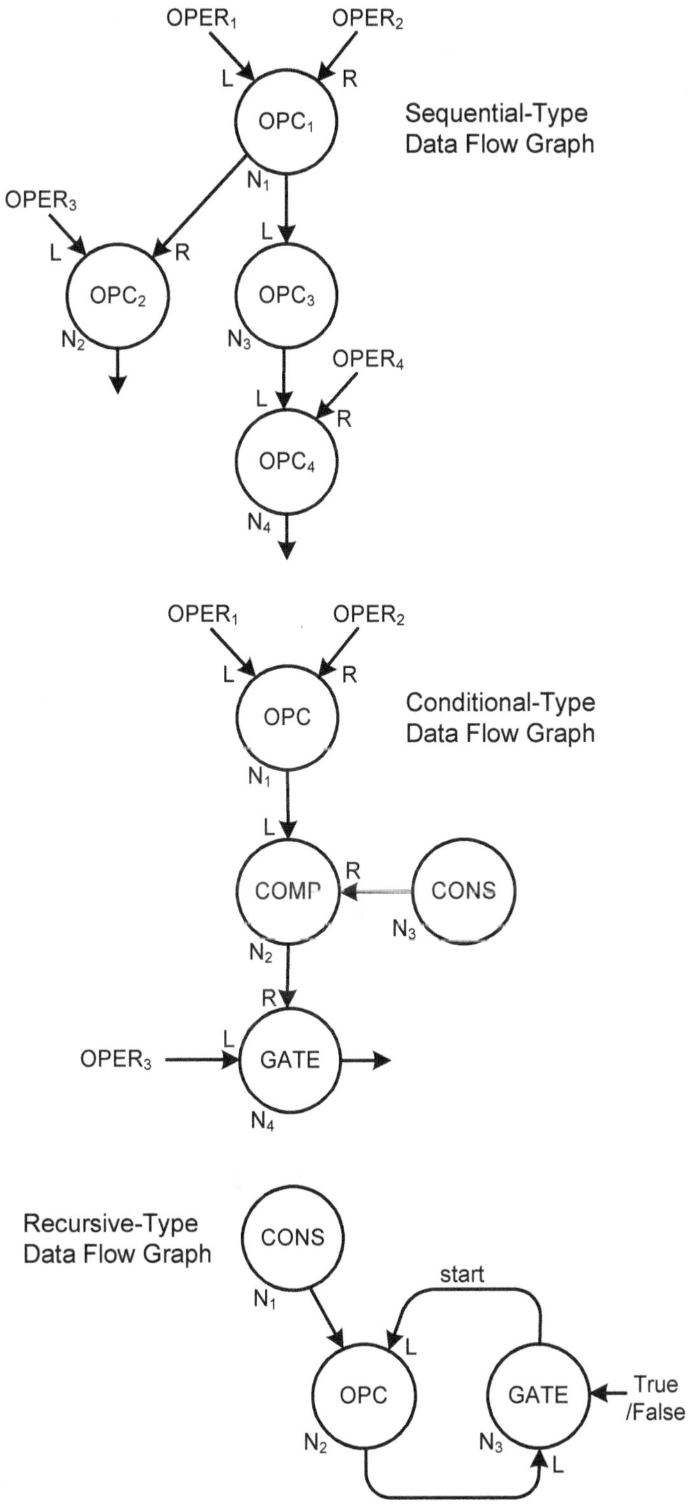

Fig. 8.19 Data-flow programming structures

The conditional programming structure consists of a functional node that accepts a conditional input besides a data input. If the condition is satisfied, a valid operand at the data input becomes a valid operand at the output. If the condition is not satisfied, the operand at the output retains its old value. The conditional input enters the node from the right side since it is considered to be a control input as mentioned earlier. Gate and Compare instructions are considered conditional since the data-flow produced at the output of the node depends on whether the condition is satisfied or not. The output operand does not change its value until the condition is satisfied. An example of this type is shown in Fig. 8.19.

The recursive programming structure contains looping constructs in the form of feedback around a functional node as shown at the bottom of Fig. 8.19. The number of iterations in a loop continues until all input operands are exhausted. The loop can be broken to allow a conditional node if needed, otherwise the loop is activated on the arrival of new operands either from the right or the left sides of the node.

A simple example in Fig. 8.20 combines all the programming structures mentioned above. This example calculates the area under a straight line, Y = (X − 1), from X = 2 to X = 3. The increment in the x-axis is defined to be DeltaX, which is equal to 0.1 in the flow chart.

The flow chart in Fig. 8.20 has been transformed into a data-flow graph shown in Fig. 8.21. All the nodes with constants in the data-flow graph in Fig. 8.21 are zero-input nodes, which accept no

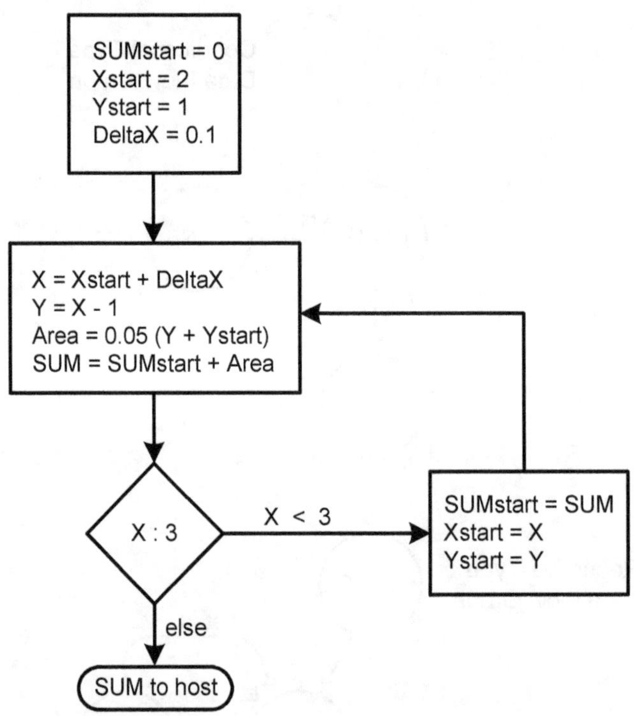

Fig. 8.20 Flow chart integrating the area under Y = (X − 1)

operands. The only unitary functional node is the one with the SET operation code. Upon the arrival of an operand to its single input, this node generates logic 1 at its output. Otherwise, its output stays at logic 0.

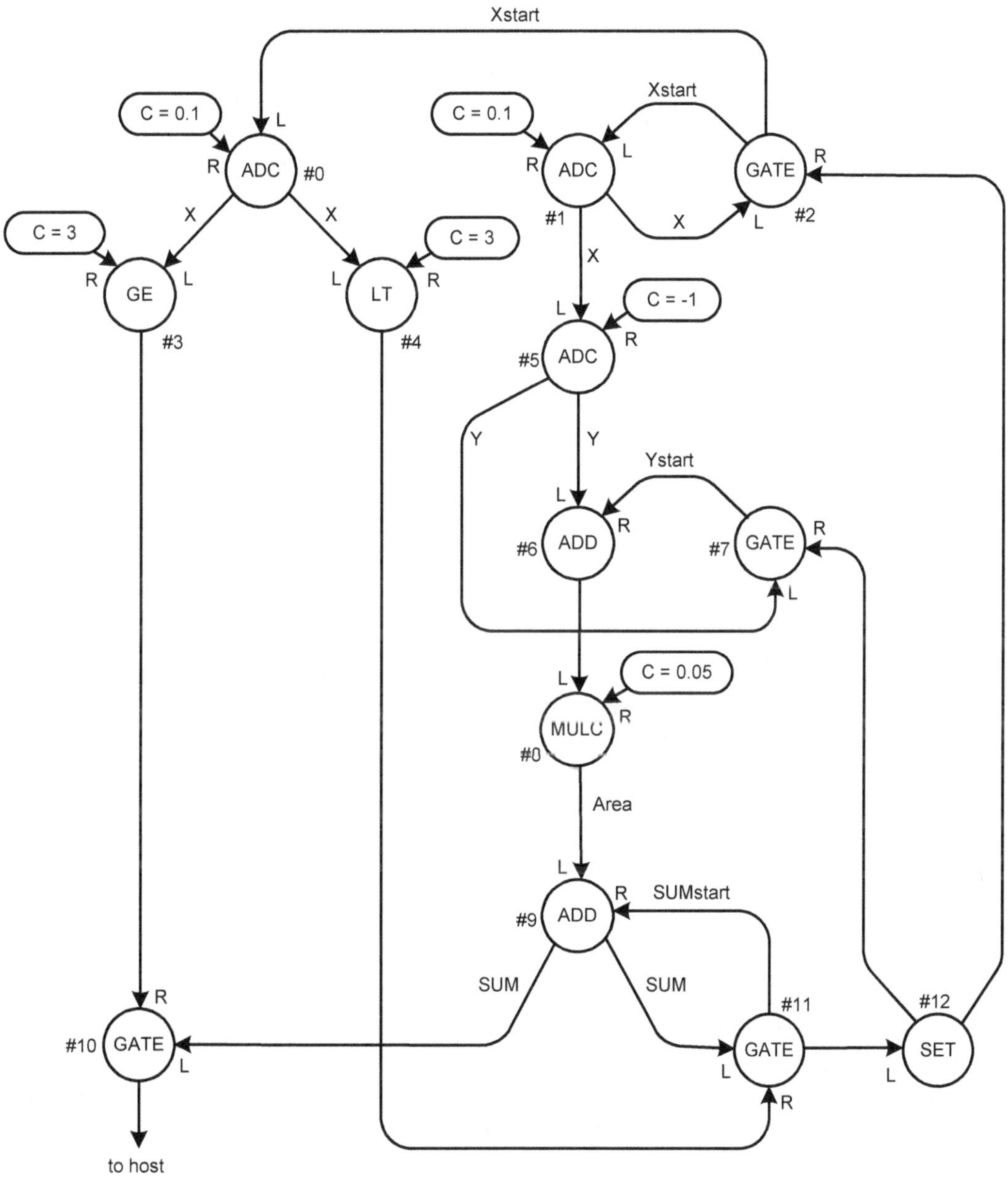

Fig. 8.21 Data-flow graph integrating the area under Y = (X − 1)

There are two types of two-input functional nodes in the same data-flow diagram. The majority of these nodes are conditional type: they either wait for a condition to arrive (GATE) or they have a permanent condition attached at their right port in terms of a constant. Greater-Than-Or-Equal-To

(GE), Less-Than (LT), Add-with-Constant (ADC), and Multiply-with-Constant (MULC) functional nodes belong to the latter category.

The rest of the two-input nodes are sequential such as Add node (ADD) where polarity information is not important for data execution.

During the programming phase, the initial values of Xstart, Ystart and SUMstart are stored at the proper arcs in Fig. 8.21. When program execution starts, Xstart = 2 is added to a floating-point constant, C = 0.1, at the nodes 0 and 1, generating the first value of X. While X is compared against C = 3 at the nodes 3 and 4, it is also directed to the node 2 for a Gate operation, and added to C = −1 at the node 5, producing Y. Subsequently, Y is directed to the node 6 to be added to Ystart = 1, and to the node 7 for another Gate operation. The output of the node 6 multiplies with C = 0.05 at the node 8, producing the first incremental area value, and it is directed to the node 9 to be added with SUMstart = 0. The output of the node 9, SUM, is then forwarded to the nodes 10 and 11 for two other Gate operations. Depending on comparisons at the nodes 3 and 4, the SUM output will either be forwarded outside of the processor or forwarded to the node 12 in order to set this node. If the node is set, Gate operations at the nodes 2, 7 and 11 take place, replacing the old values of Xstart, Ystart and SUMstart with X, Y and SUM, respectively. Iterations continue until the node 10 becomes active and the result is delivered to the user.

Input Flags

An input operand to a functional node contains an operand flag to indicate whether or not the data processing is complete. A high flag implies that the input operand is valid and ready to be processed. The input operand flag goes to logic 0 as soon as the functional node processes the input operand.

Nodal Networks

Data-flow diagrams in this architecture can be structured in three different ways. The first is a direct connection among functional nodes: data flows from one node to another freely in an unobstructed fashion as shown at the top left corner of Fig. 8.22. The only control mechanism for processing data at each functional node is that both input operand flags must be at logic 1. The second and third nodal networks use programmable routers to send operands from the source to the destination nodes. The simple router in Fig. 8.22 uses a local network to connect a group of functional nodes called a cluster. Note that the inputs to a functional node in a cluster can come from any functional node in this network. This decision is made by a simple arbitration scheme in the router, which dictates that any node in the process of generating a new input operand for itself has priority over the other nodes in a cluster. In other words, if a neighboring node produces an input operand for a particular node in a cluster while this particular node is in the process of generating an input operand for itself, the arbiter stalls any data processing in the neighboring node until the self-operand generation is complete. The bottom structure of Fig. 8.22 illustrates the hierarchical organization of clusters where an inter-cluster network manages many local cluster networks. While each cluster arbiter manages its own individual cluster, all cluster-to-cluster communication is maintained by a separate inter-cluster arbiter.

Processor Design Overview

The processor implements the node functionality by reading the node instruction from the memory, executing it, and writing the result back to the two destination node addresses specified in the

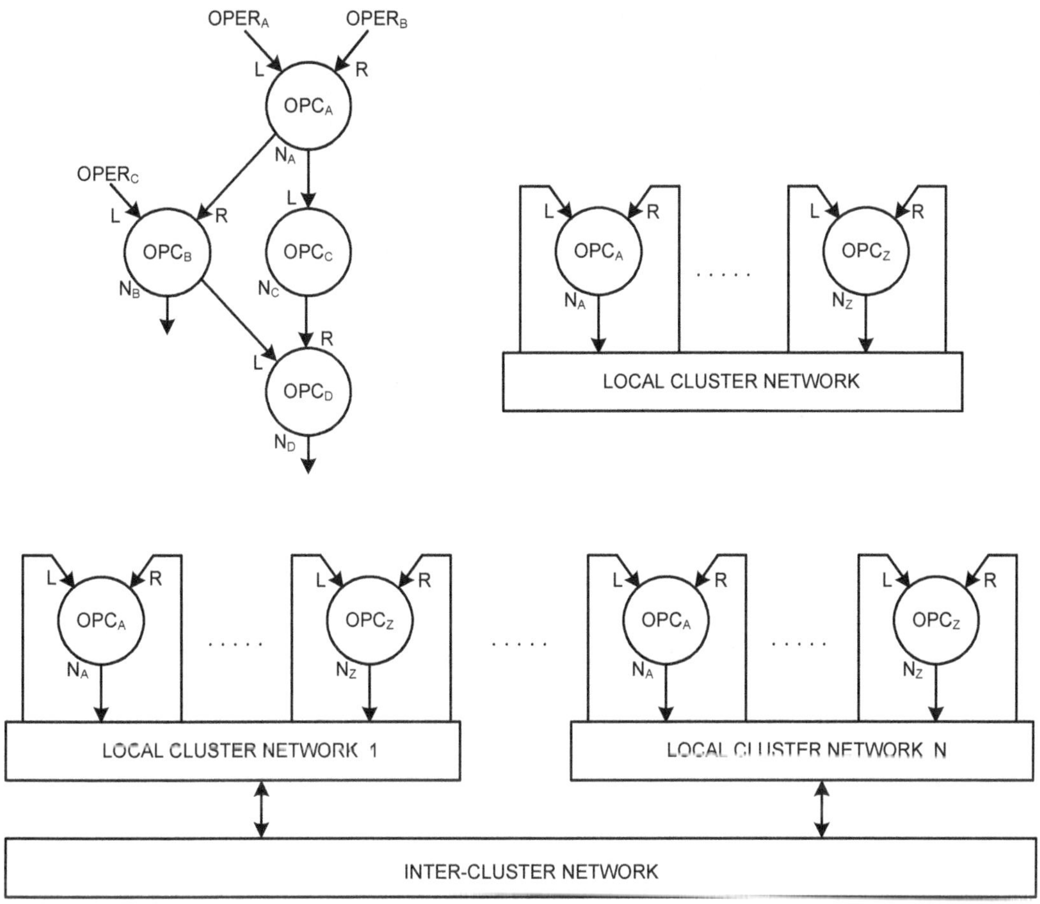

Fig. 8.22 Data-flow graphs without router, and with local and hierarchical cluster networks

instruction. In order to implement this sequence, each processor needs to have a memory, an ALU and a controller. The memory contains all nodal instructions. Each nodal instruction consists of two input operands with their valid flags, an operation code and the two destination node addresses where the results are sent as shown in Fig. 8.23.

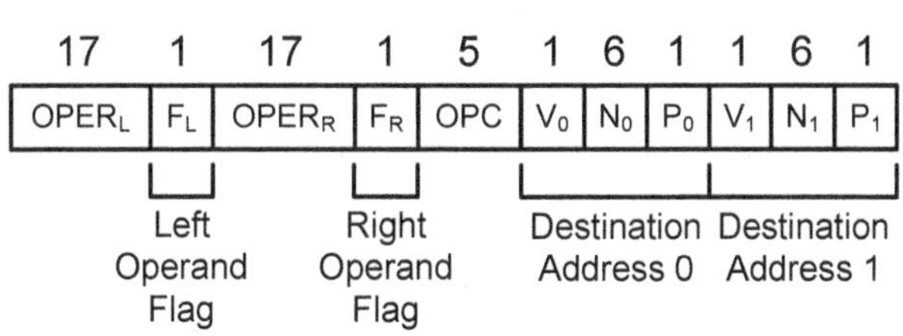

Fig. 8.23 Instruction format

Consider a processor implementing a single node. Initially, the processor is at idle. When both operand flags in the instruction become valid, the controller starts. In the first step, the controller generates a nodal address for the instruction that resides in the memory. In the second step, the controller fetches the input operands and operation code from this instruction and forwards them to the ALU. The ALU combines the input operands and generates an output operand. Subsequently, the controller directs this operand to the first destination address. In the third step, the controller writes the same output operand to the second destination address, sets the operand flag at the first destination address, and clears input operand flags at the source address. In the fourth and final step, the controller sets the operand flag at the second destination address.

If the processor needs to execute more than one node, then each nodal address in the flow graph must be mapped to a physical address in the memory. This approach automatically transfers the left and the right nodal operands, the operation code and the destination addresses of a particular node from the flow graph to an instruction in the memory. However, during this process each operand flag is stored in a separate tag memory to allow the controller to continuously search for valid operand flags. If the controller finds a node with valid left and right operand flags, it sends the corresponding operands to the ALU for execution. After the operands are processed and sent to the destination addresses in the instruction, the controller points the next node to be processed. If there is no other node with valid operand flags, the controller stalls the processor until an instruction with valid operand flags emerges in the instruction memory.

The processing efficiency and speed in the processor can be increased by pipelining. After an instruction is executed, sending the ALU result to a destination address can be overlapped with tasks such as generating an address or fetching a different instruction. This can be achieved using a dual port memory.

Instruction Format

In this architecture, the instruction format contains two 17-bit input operands, $OPER_L$ and $OPER_R$, two flags to validate the operands, F_L and F_R, one five-bit operation code, OPC, and two eight-bit destination address fields. Each destination address is composed of a valid bit, V, a six-bit node number, N, and a left-right polarity bit, P. The valid bits in the destination address fields, V_0 and V_1 indicate the validity of the corresponding nodal address. This instruction format is shown in Fig. 8.23.

Architecture and Operation

Figure 8.24 shows a simplified block diagram of a processor executing the simple program in Fig. 8.17. Each node number in the data-flow graph in Fig. 8.17 corresponds to an address in the instruction memory in Fig. 8.24.

The program execution starts when the controller detects an instruction with valid left and right operand flags, such as the one at the memory location N_S with $F_L = F_R = 1$. The controller reads out the instruction and sends the operands, $OPER_{S1}$ and $OPER_{S2}$, and the operation code, OPC_S, to the ALU. The ALU executes the operands according to the OPC_S, and the controller clears both operand flags of the instruction, underlining the completion of this instruction. The controller subsequently sends the ALU result to the first and second destination addresses, N_{D0} and N_{D1}, as data tokens as shown in Fig. 8.24. When the result is delivered to a valid destination address, the controller automatically sets the operand flag to specify the validity of data for further processing. For example, the right operand flag at N_{D0} is set when $OPER_A = OPER_{S1}\ (OPC_S)\ OPER_{S2}$ is delivered to this address.

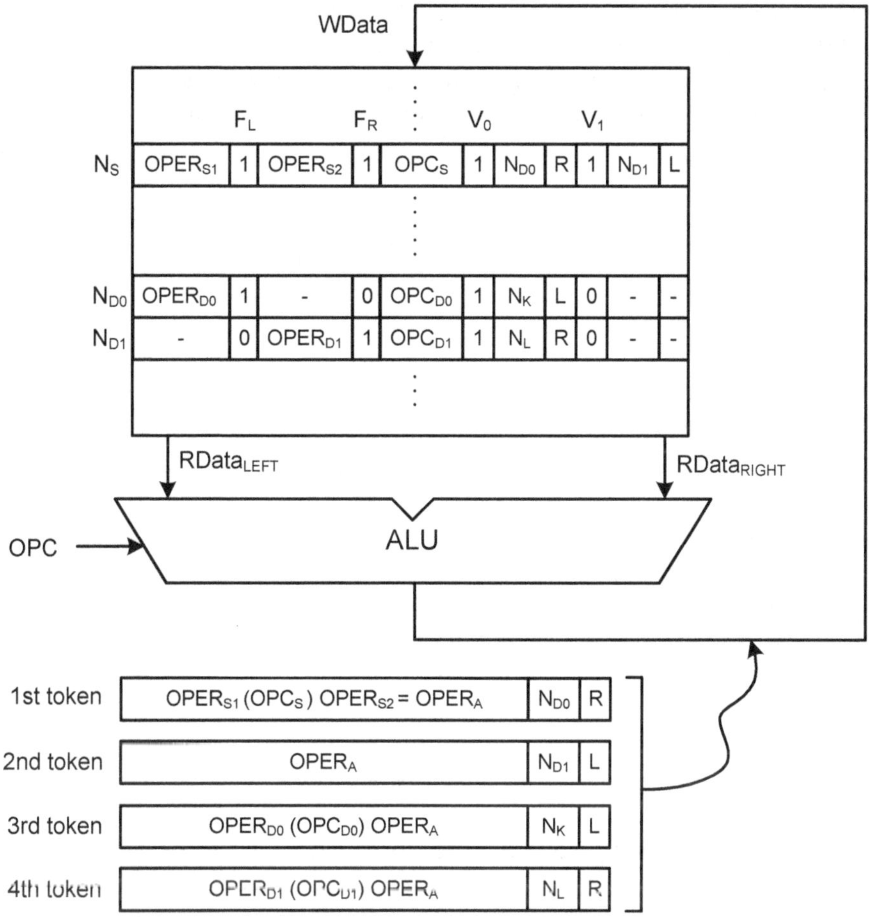

Fig. 8.24 Processor architecture executing the program in Fig. 8.17

Similarly, the controller sets the left operand flag at N_{D1} when it delivers the same ALU result to this address in the next cycle.

After executing the instruction at N_S, the controller detects the next instruction with valid operand flags at the memory location N_{D0}. Once again, the controller extracts the operation code, OPC_{D0}, and the operands, $OPER_{D0}$ and $OPER_A$, from the instruction and sends them to the ALU for execution. Subsequently, the controller clears the operand flags at N_{D0} and sends the ALU result as a left operand to the memory address N_K. This is shown as the third data token in Fig. 8.24. The controller performs the same set of tasks for the instruction located at the address N_{D1} and forms the fourth data token. The program execution stops when the controller can no longer find a pair of valid operand flags in the tag memory.

Implementation

Figure 8.25 shows the implemented instruction field format. Besides the operation code and the input operand fields, each destination address in Fig. 8.23 is now expanded to contain an additional seven-bit processor ID, ProcID, and cluster ID, ClusID, to allow multiple processor communication in a local network. In this architecture, we have the flexibility of choosing a network ranging from 128 processors in a single cluster to two processors per cluster for 64 different clusters. The presence of ClusID and ProcID enables independent but simultaneous networking activities to take place among clusters and processors. In other words, the source processor can write the same ALU result to two different destination processors within the same cluster or in different clusters.

17	1	17	1	5	7		1	6	1	7		1	6	1
$OPER_L$	F_L	$OPER_R$	F_R	OPC	$ClusID_0$	$ProcID_0$	V_0	N_0	P_0	$ClusID_1$	$ProcID_1$	V_1	N_1	P_1

Fig. 8.25 Instruction format in the implementation

Processor Micro-architecture

The simplified data-driven processor architecture shown in Fig. 8.24 is implemented in Fig. 8.26. One of the essential elements in Fig. 8.26 is the presence of a dual-port RAM. While the controller fetches an instruction from the first data port, it writes the ALU result of another instruction to the second port to increase processor performance and programming efficiency.

In this architecture, all operand flags are stored in a separate tag memory in the processor. The left and right operand flags at each tag address are AND-gated and connected to the node address generator as inputs. When the operand flags that belong to a specific tag memory address become

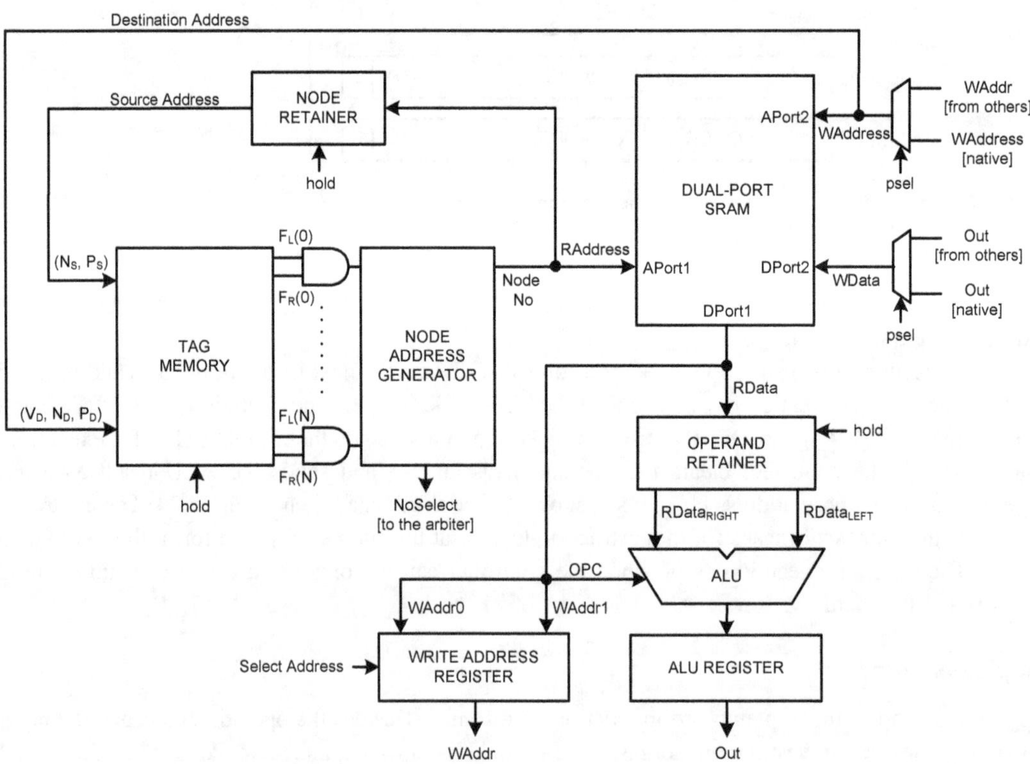

Fig. 8.26 Processor data-path and micro-architecture

valid, the address generator uses this tag address as a read address, RAddress, to read the corresponding instruction from the dual-port RAM. If there is more than one set of valid flags in the tag memory, the node address generator selects a pair of flags with the lowest nodal address value, and delivers this address to the memory. When there are no more valid flags, the address generator produces a No Select signal for the network arbiter which, in turn, stalls the processor. From this moment on, the processor suspends all its activity and waits for new operands to be delivered to the instruction memory.

When a processor interacts with other processors in a network, it is quite possible that it may receive a hold request from the network arbiter while attempting to write into a neighboring processor. This hold signal is generated because the neighboring processor may be busy processing data for itself or writing data to another processor as mentioned earlier. When the source processor receives a hold signal, it stalls all its processing functions and retains its internal status and output data values until the hold is removed. Therefore, RAddress is also stored in the node retainer to preserve the node address, N_S, and the source operand polarity, P_S, for the tag memory in case the network arbiter issues a hold.

Once the instruction at RAddress is read from the RData port of the dual-port memory, its right and left operands are routed to the ALU along with the operation code for execution. The source operands are also stored in the operand retainer in case the program execution is put on a momentary hold by the network arbiter. Both of the destination addresses, WAddr0 and WAddr1, are buffered in the write address register and used alternately to deliver the processed data at the Out terminal to destination processors. The write address register also keeps old write addresses for the duration of hold.

The processed data either produced locally or from other processors in a network is eventually written to the dual-port RAM through the WData port. The destination address at the WAddress port simply accompanies the newly arrived data. The destination address at this port is also directed to the tag memory to update the corresponding operand flags.

Processor Programming

Prior to the program execution, instructions are loaded to the dual-port RAM through the WData port in Fig. 8.26. While the program is loaded to the memory, operand flags of each instruction are also stored in the tag memory.

Inter-processor Arbiter and Router

In a data-driven architecture, the organization of processors in a network is hierarchical. A group of processors form a local cluster, in which each individual processor owns a processor ID, ProcID, to communicate with other processors using a simple arbitration protocol.

A local cluster has also an identification number, ClusID, in a network. Only one processor in a cluster can communicate with another processor in a different cluster at a given time.

The inter-processor router connects each processor's destination address and data output to all other processors with a massive multiplexing network as shown in Fig. 8.27. The arbiter is designed to give address and data transfer privileges to a single processor while issuing a hold to all lower-priority processors in a cluster. There are two general rules observed in the arbiter's priority scheme. The first rule states that if a processor issues a write to itself, it has the highest priority over the other processors in a cluster. The second rule is that if two or more processors issue write requests simultaneously to a processor at idle, the highest priority among these processors belongs to the one with the lowest ProcID.

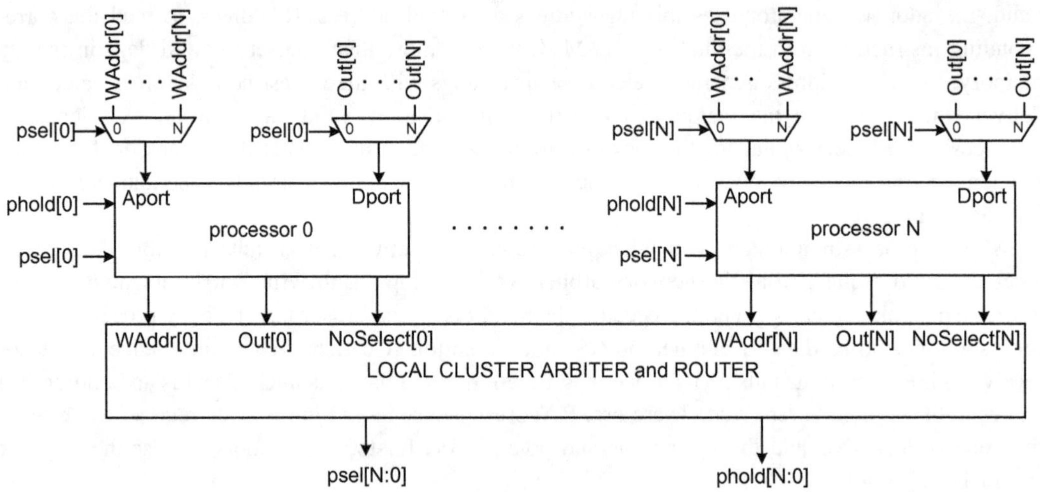

Fig. 8.27 Inter-processor arbiter and router

Review Questions

1. The following sum of products (SOP) function is given:

$$\text{out} = AC + AB\overline{C} + \overline{B}$$

Implement this function using three-input LUTs only.

2. The following product of sums (POS) function is given:

$$\text{out} = (A + B) \,.\, (\overline{B} + \overline{C}) \,.\, (\overline{A} + B + C)$$

Implement this function using two-input LUTs only.

3. The following truth table needs to be implemented in FPGA.

A	B	C	out
0	0	0	1
0	0	1	0
0	1	0	1
0	1	1	1
1	0	0	0
1	0	1	1
1	1	0	1
1	1	1	0

Assume the three-input LUT configuration is given below:

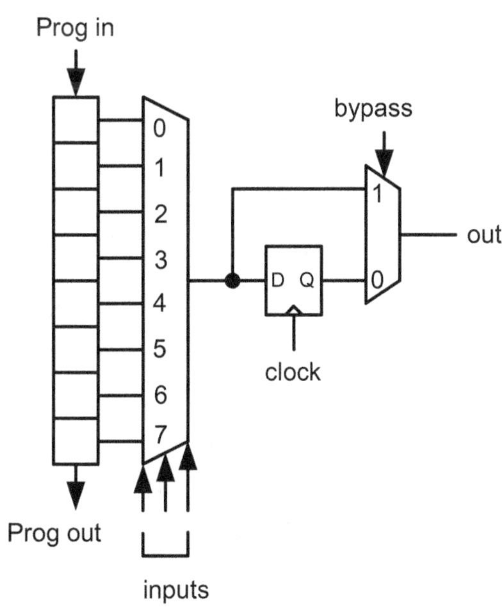

(a) Implement the truth table only with three-input LUTs.
(b) Implement the same truth table only with four-input LUTs.
(c) Implement the same truth table only with two-input LUTs.

4. Implement the 3-1 multiplexer below using three-input LUTs only.
The definition of the MUX is given below:
If sel[1:0] = 00 or 11 then out = a
If sel[1:0] = 01 then out = b
If sel[1:0] = 10 then out = c

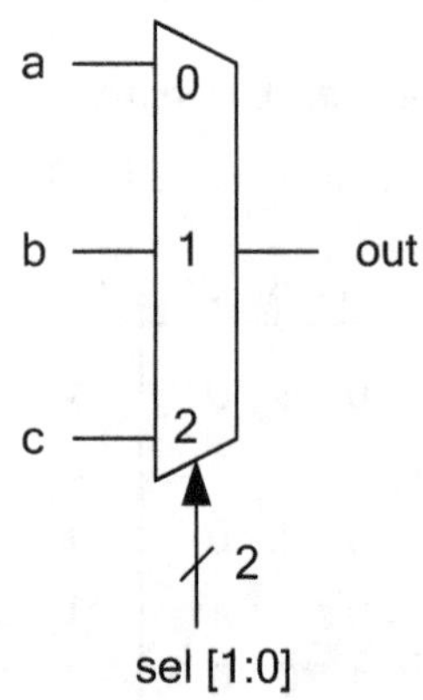

5. A simple Moore state machine that consists of two states is given below. After constructing the state and transitional tables, implement this machine using minimal number of primitive gates and flip-flops.
Use two-input LUTs. Program each two-input LUT to implement the Moore state machine in FPGA. Assume that there are three LUTs, each with two inputs in a cluster. Draw the complete architectural diagram of the FPGA, including interconnections.
Assign A = 0 and B = 1 in the state machine below.

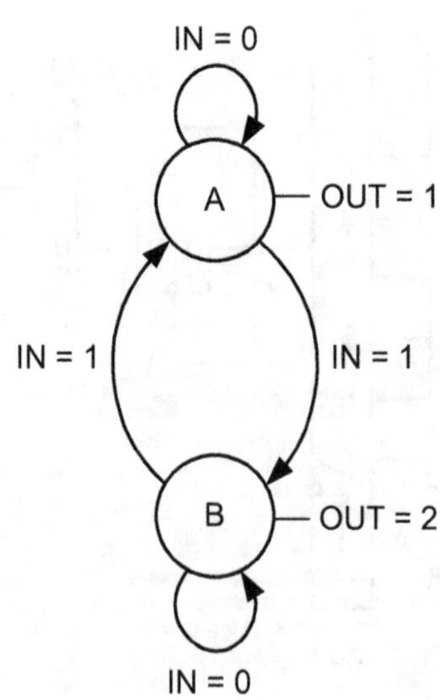

6. A three-bit counter is given below. The Hold input activates port 1 of the 2-1 MUX, thus retaining the output value. Otherwise, the counter keeps incrementing by one.

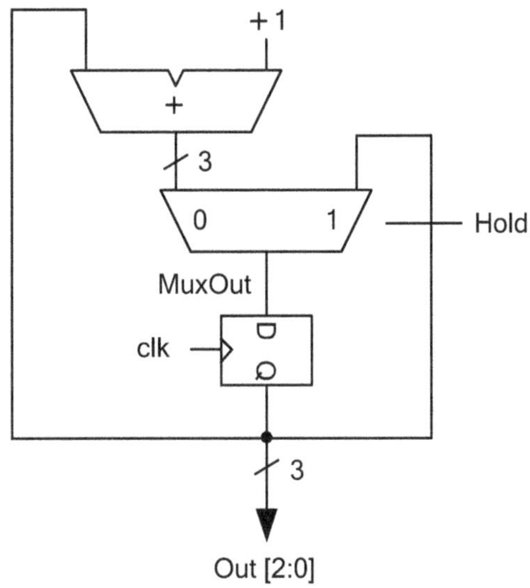

Assume that the three-bit adder is simply a ripple-carry adder composed of three full-adders (FA) with sum (Sum) and carry-out (Cout) outputs as shown below.

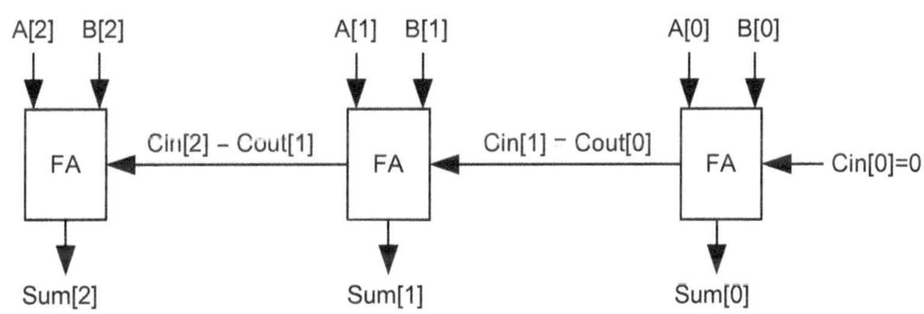

$$\mathrm{Sum}[i] = A[i] \oplus B[i] \oplus \mathrm{Cin}[i]$$

$$\mathrm{Cout}[i] = A[i]\,.\,B[i] + \mathrm{Cin}[i-1]\,.\,(A[i] + B[i])$$

To avoid the complexity of input MUXing, assume that this is a non-commercial FPGA platform. Implement this circuit using three-input LUTs in an FPGA.

Projects

1. Implement and verify the three-input LUT using Verilog.
2. Implement the four-bit ripple-carry adder given as an implementation example in this chapter using Verilog. Use the three-input LUTs from project 1 to create an FPGA implementation of the adder. Perform functional verification on the entire unit.
3. Implement the Moore state machine given as an implementation example in this chapter using Verilog. Use the three-input LUTs from project 1 to create an FPGA implementation. Perform functional verification on the entire unit.

Appendix: An Introduction to Verilog Hardware Design Language

A.1 Module Definition

A digital system in Verilog is defined in terms of modules as shown in Fig. A.1. Each module has inputs and outputs with different bit widths. Modules can be integrated to form bigger digital blocks as shown in Fig. A.2. In this figure, the top module contains four smaller modules, each of which has inputs and outputs. They are interconnected with each other in such a way to produce much larger system functionality and produce a new set of external input and output signals.

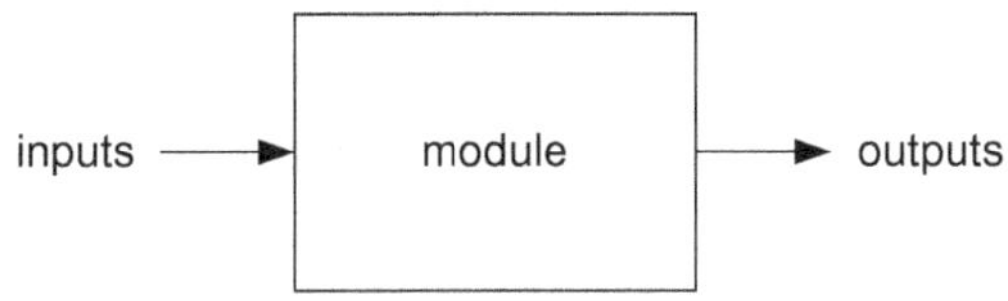

Fig. A.1 A typical module in Verilog

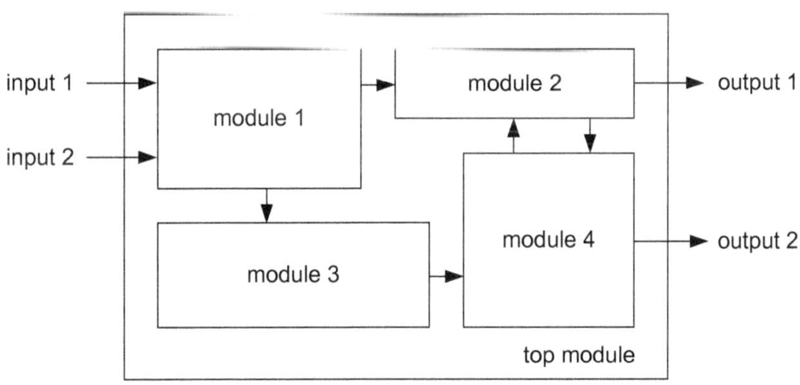

Fig. A.2 Module integration to form a much larger system

A. Bindal, *Fundamentals of Computer Architecture and Design*,
https://doi.org/10.1007/978-3-030-00223-7

When creating a Verilog code to represent a digital block such as in Fig. A.3, the module name is written first. The input and output names are written in parentheses next to the module name; however, their order is not important. This is followed by separate input-output (I/O) statements and the description of the module as shown below.

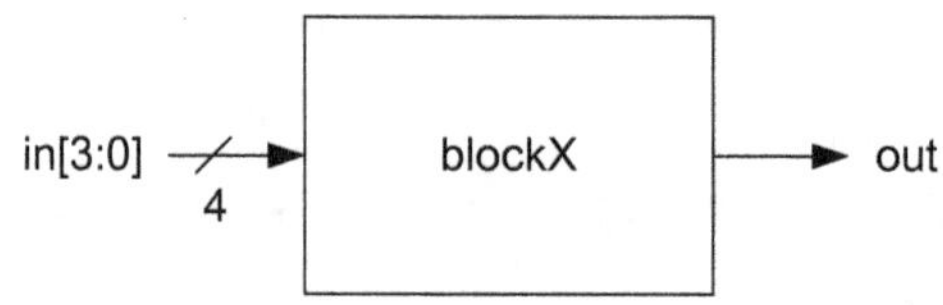

Fig. A.3 A Verilog module, blockX, with in[3:0] and out

```
module blockX (out, in);
output out;
input [3:0] in;

/* module description here */

endmodule
```

For a more specific example let us write a small Verilog code for the two-input AND gate in Fig. A.4.

```
module andgate (out, in1, in2);
output out;
input in1, in2;

/* module description here */

endmodule
```

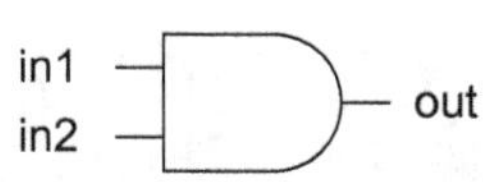

Fig. A.4 Two-input AND gate

Now, let us integrate blockX in Fig. A.3 with the two-input AND gate in Fig. A.4 to form a larger digital block as shown in Fig. A.5.

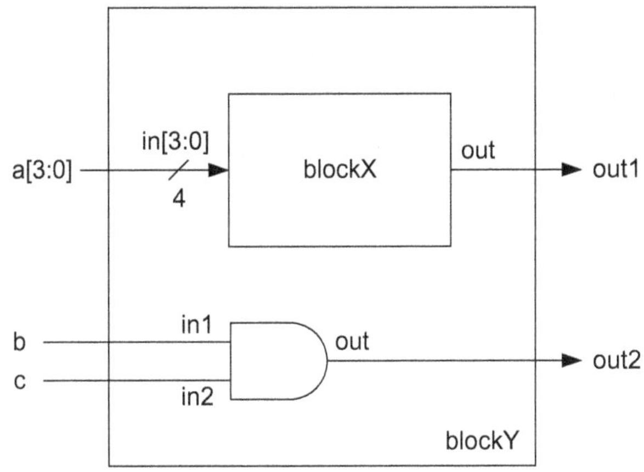

Fig. A.5 A Verilog module integrating blockX in Fig. A.3 and the two-input AND gate in Fig. A.4

This new module, called blockY, has different I/O names due to the naming convention used in the schematic in Fig. A.5. As a preference, the external port names in the module statement, and the internal port names in each sub-module are declared as outputs first, inputs second. However, this is not a requirement, but only a convention.

In addition, both the blockX and the two-input AND gate need to be instantiated inside the blockY. This is achieved by using the "dotted" convention as shown below. In this program, the sub-module port names are written outside the parentheses, and the I/O names used in the top module are written inside the parentheses. The names, i1 and i2, are associated with each sub-module instantiation, andgate and blockX, respectively.

```
module blockY (out1, out2, a, b, c);
output out1, out2;
input [3:0] a;
input b, c;

andgate  i1  (.out(out2),
              .in1(b),
              .in2(c));

blockX   i2  (.out(out1),
              .in[0](a[0]),
              .in[1](a[1]),
              .in[2](a[2]),
              .in[3](a[3]));

/* the rest of the code here */

endmodule
```

Basic logic gates (or Verilog primitives) do not need instantiations. The two-input AND gate in Fig. A.4 could have been written without any "dotted" instantiation in the top module above.

The Verilog primitives are the following gates:

```
AND, NAND, OR, NOR, XOR, XNOR, BUF, NOT.
```

Here, BUF corresponds to a buffer, and NOT corresponds to an inverter.

Example A.1: Let us implement one-bit full adder using Verilog.

In a full adder, the sum and carry-out functions are declared as:

sum = a ⊕ b ⊕ cin

cout = a . b + cin . (a + b)

In the functional equations above, cin corresponds to the carry-in input and cout corresponds to the carry-out output of the full adder. The terms, a and b, are the two inputs to the full-adder.

The schematic to produce sum and carry-out functions are shown in Fig. A.6. The intermediate nodes are named as node1, node2, node3 and node4 as interconnecting nodes.

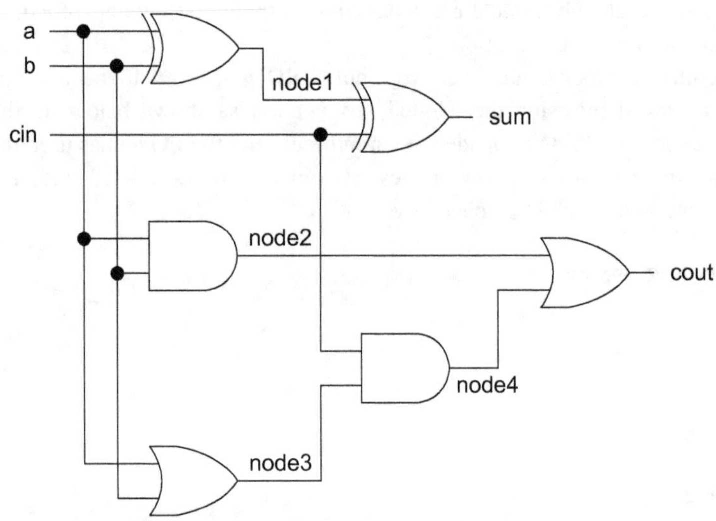

Fig. A.6 Hardware implementation of the full adder

The Verilog code below contains no "dotted" instantiations. This style of Verilog code is called structural where only primitive gates are used.

```
module fulladder (sum, cout, a, b, cin);
output sum, cout;
input a, b, cin;

/* structural style module description */
xor (node1, a, b);
xor (sum, node1, cin);
and (node2, a, b);
or (node3, a, b);
and (node4, cin, node3);
or (cout, node2, node4);

endmodule
```

After the Verilog code is written, a test fixture needs to be produced to verify the module. For our example of the one-bit full adder, Fig. A.7 illustrates the concept of building the test fixture.

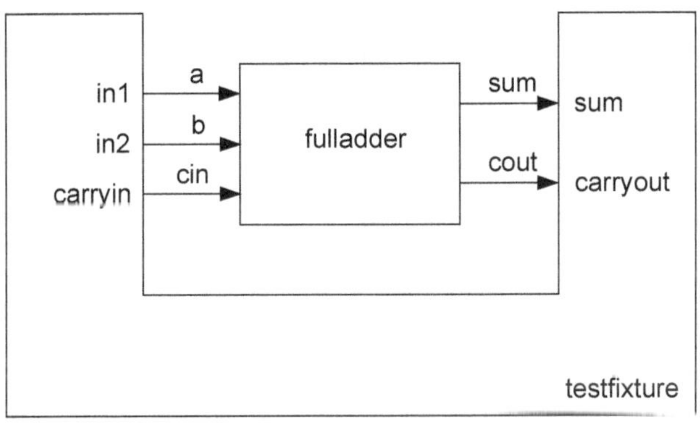

Fig. A.7 Test fixture formation to verify the functionality of the full adder

In Fig. A.7, the outputs of the full adder must be declared as the inputs of the test fixture using wire statement in Verilog. The outputs of the test fixture are the inputs for the full adder, and they should generate all the verification vectors needed to test the full adder thoroughly. For each test input to the full adder, the outputs of the test fixture must remain unchanged until a new set is introduced. Therefore, each output needs to be stored in memory, which calls for the reg (register) statement in Verilog. Also, the test fixture neither contains any I/O names in the module definition nor has any input and output statements. Thus, the general structure of the test fixture becomes as follows:

```
module testfixture;

/* all test fixture outputs are declared as reg statements */
/* all test fixture inputs are declared as wire statements */

/* module instantiation here */

/* verification vectors here */
/* display results here */

endmodule
```

The verification vectors are executed only once using the initial statement in Verilog. For a combinational logic circuit such as a full adder, these vectors are simply the inputs of the truth table. The outputs, on the other hand, need to be either displayed or stored in a file in order to be compared against the "expected" outputs. Therefore, the Verilog code of the test fixture for full adder becomes:

```
module testfixture;

/* all test fixture outputs are declared as reg statements */
reg in1, in2, carryin;

/* all test fixture inputs are declared as wire statements */
wire sum, carryout;

/* module instantiation here */
fulladder  i1   (.sum(sum),
                 .cout(carryout),
                 .a(in1),
                 .b(in2),
                 .cin(carryin));

/* verification vectors here are executed only once due to initial statement */
initial
begin
        carryin = 0; in1 = 0; in2 = 0;
#10                          in2 = 1;
#10                in1 = 1; in2 = 0;
#10                          in2 = 1;
#10     carryin = 1; in1 = 0; in2 = 0;
#10                          in2 = 1;
#10                in1 = 1; in2 = 0;
#10                          in2 = 1;
end

/* display results here */
endmodule
```

The "#" sign used in almost every statement inside the initial block indicates a delay function. For example, the code waits for 10 time units after the first three statements, cin = 0, in1 = 0, in2 = 0, are executed before the value of in2 is changed from logic 0 to logic 1 in the forth statement. Since in1, in2 and cin are all reg statements, their values are retained until changed.

Monitoring the simulation results are achieved by the $time and $monitor statements. $time displays the current simulation time. $monitor constantly observes the program variables and records any change in their value during simulation.

For example, $monitor ($time, output, input1, input2) statement displays simulation time and the values of output, input1 and input2. On the other hand, $monitor ($time, ,"output=%h input=%b", out, in \n) statement displays the simulation time, leaves a space between the simulation time and "output" because of the extra comma, displays "output" in hexadecimal format and "input" in binary format. After each set of simulation time, input and output, a new line starts for the second set due to carriage return, "\n", entry. One can also use "\t" to insert a tab between terms to achieve separation.

All Verilog files are stored with the ".v" extension after the file name, such as fulladder.v. When executing the Verilog command to simulate multiple files, including the test fixture, the command line should include the top module last and all the remaining modules inside the top module first.

In this full adder example, we need to write fulladder.v first and then testfixture.v because the test fixture contains the full adder module. Therefore, the command line becomes:

```
verilog fulladder.v testfixture.v
```

The notation for comments in Verilog is identical to the ones used in C-programming. For a single line, a pair of slashes, "//", is used for a single line comment. For multiple lines, the comment starts with a slash and a star, "/*", and ends with a star and a slash, "*/".

A.2 Numbers in Verilog

The numbers in Verilog are represented by three distinct terms:

```
SIZE `BASE VALUE
```

Note that the tick mark, "`", attached to the BASE entry is not apostrophe.

As an example, 32`h3C represents a 32-bit hexadecimal number, 0x0000003C. Similarly, 8`b1 represents an eight bit binary number, 00000001. Any time the size entry is omitted, the size defaults to 32 bits. For example, `h8A corresponds to 0x0000008A. The base entry to represent high impedance or floating wire is "`z". For example, 8`bz means an eight-bit bus with all its eight bits are floating or zzzzzzzz. Similarly, don't care bits are shown by "`x". For example, 4`bx means four wires are either at logic 0 or logic 1, and represents xxxx.

A.3 Time Directives for Compiler

To mimic propagation delays in Verilog simulation, a timing directive is used. The directive, `timescale, is preceded by a tick mark to enable the compiler to delay the execution of a Verilog statement by the amount following the pound mark, "#".

The `timescale directive is the first line in a Verilog code. The statement does not end with a semicolon and contains two entries separated by a "/" sign. The first entry corresponds to the actual delay. The second entry represents the simulation resolution.

For example, `timescale 10 ns / 100 ps means each time the compiler sees a "#" sign in a Verilog statement, it delays the execution of the statement by multiplies of 10 ns, depending on the number that follows the "#" sign. The resolution is 100 ps. Therefore, for a delay of 10 ns, the simulation accuracy is 1:100.

Let us consider the following structural Verilog code as an example.

```
`timescale 1ns / 100ps
module inverter (out, in);
output  out;
input   in;
not     #2 (out, in);      //2 ns delay between input and output
endmodule
```

This is a module that represents a single inverter. The propagation delay in the inverter is 2 ns because the number following the "#" sign in the "not" gate is 2, and this number is multiplied by 1 ns in the timescale directive. The simulation resolution is 100 ps. Therefore, there is 1:20 accuracy when generating a delay function for the inverter.

Other common compiler directives are the `define and `include statements, and neither ends with a semicolon. The `define statement defines a variable for the Verilog code. For example, in the following Verilog code a variable called inv_delay corresponds a delay of 30 ps with 1 ps simulation resolution.

```
`timescale 10ps / 1ps
`define inv_delay #3
module inverter (out, in);
output  out;
input   in;
not     inv_delay (out, in);       //30ps delay between input and output
endmodule
```

The `include statement fetches smaller modules from various directories and includes them into a bigger module for simulation. For example, the following Verilog code instantiates the full_adder.v module located in the verilog_modules directory to be used in bigger_module.v.

```
`include "verilog_modules/full_adder.v"
`timescale 10ps / 1ps
`define inv_delay #3
module bigger_module (out1, out2, in1, in2, in3);
        ...
        ...
endmodule
```

A.4 Parameters

Parameters are used to replace numbers for enhancing the readability of Verilog code. The parameter statement is not a compiler directive. Therefore, it is ended with a semicolon. The statement has only one entry, which attaches a name to a number.

```
parameter      name = value;
```

Assume the following example:

```
module         alu (out,in1,in2);
output         [31:0] out;
input          [31:0] in1,in2;
          ...
          ...
endmodule
```

The value, 31, can be replaced by the name, BUS, using the parameter statement to enhance readability of the Verilog code. Thus,

```
module         alu (out,in1,in2);
parameter      BUS=31;
output         [BUS:0] out;
input          [BUS:0] in1,in2;
          ...
          ...
endmodule
```

A.5 Basics of Structural Verilog Modeling

Structural modeling was introduced earlier in this chapter to describe how basic logic gates are used in a Verilog code. Structural Verilog eliminates the dotted convention when instantiating Verilog primitives. The logic gates supported by Verilog are:

```
AND    NAND
OR     NOR
XOR    XNOR
BUF    NOT
```

For example, a three-input NOR gate with inputs, in1, in2 and in3, is represented by:

```
nor (out, in1, in2, in3);
```

As another example, a buffer with a single input, in, is written as:

```
buf (out, in);
```

In each structural statement, the output of the logic gate is listed first followed by its inputs.

Tri-state buffers and inverters are represented by conditional structural statements. The statement bufif1 represents a tri-state buffer with an active-high enable as shown in Fig. A.8.

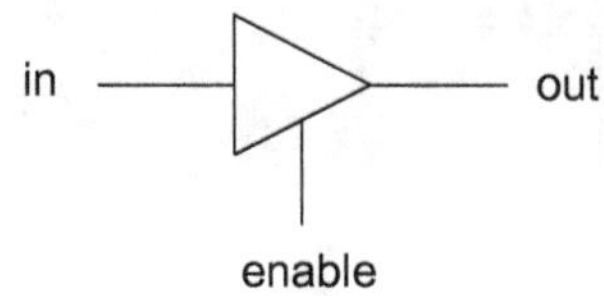

Fig. A.8 Tri-state buffer with active-high enable

This gate behaves like a buffer when enable = 1, and becomes an open circuit when enable = 0. The structural Verilog statement for the tri-state buffer becomes:

```
bufif1 (out, in, enable);
```

A tri-state buffer with an active-low enable signal shown in Fig. A.9 makes this logic gate behave like a buffer when enable = 0, and an open-circuit when enable = 1.

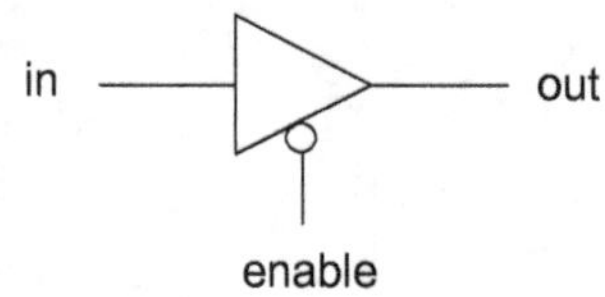

Fig. A.9 Tri-state buffer with active-low enable

The structural Verilog statement for this gate uses a bufif0 statement:

```
bufif0 (out, in, enable);
```

Tri-state inverters use the same active-high or active-low enable signals. To represent a tri-state inverter with an active-high signal in Fig. A.10, the notif1 statement is used.

```
notif1 (out, in, enable);
```

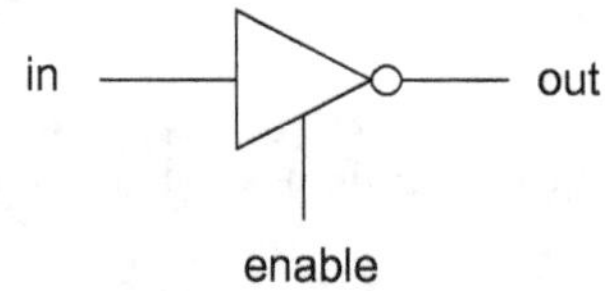

Fig. A.10 Tri-state inverter with active-high enable

The tri-state inverter with an active-low enable signal in Fig. A.11 uses the notif0 statement as shown below.

```
notif0 (out, in, enable);
```

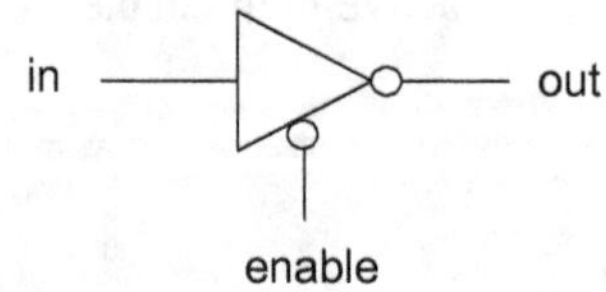

Fig. A.11 Tri-state inverter with active-low enable

A.6 Behavioral Modeling

There are two types of procedural blocks in behavioral Verilog coding. The first one is called the "initial" statement. Each Verilog statement included in the initial statement is executed only once. The procedural block below shows the general form of the initial statement. The statements in an initial block are enveloped with the "begin" and "end" clauses. The initial statement may or may not come with a condition(s) listed in parentheses after the symbol "@". If the initial statement comes with a condition, the statement is executed when the condition occurs. Otherwise, the program omits the initial statement.

```
initial @ (condition)
        begin
        statement 1;    // all statements within the initial statement are executed only once
        statement 2;
        ...
        ...
        end
```

As an example, let us consider a test fixture that verifies the functionality of a module with two inputs, in1 and in2, and three outputs, a, b and c. The test vectors applied to this module need to be executed only once. The form of the initial statement will be as follows:

```
module test;
reg input1, input2;
wire outa, outb, outc;
testmodule i1  (.in1(input1),
                .in2(input2),
                .a(outa),
                .b(outb),
                .c(outc))
initial
        begin
                input1 = 0; input2 = 0;
        #10     input2 = 1;
        #10     input1 = 1; input2 = 0;
        #10     input2 = 1;
        end
endmodule
```

The second type of procedural block is the "always" statement. This statement may also come with a condition in parentheses. Unlike the initial statement, the always statement is executed repeatedly. If the always statement comes with a condition, the execution of statements within the always statement takes place only when the condition is encountered. Otherwise, the program skips over the always statement. The general form of the always statement is shown below.

```
always @ (condition)
        begin
        statement 1;     // all statements within the always statement are executed repeatedly
        statement 2;
        ...
        ...
        end
```

Example A.2: Implement a flip-flop with two inputs, d and clock, and two outputs, q and qbar with the following Verilog code.

```
`timescale 10ps / 1ps
module flip_flop (q, qbar, d, clock);
output q, qbar;  // qbar is the inverted output, q
input clock, d;
reg q, qbar;
always @ (posedge clock)
        begin
        #2      q = d;
        #1      qbar = ~d;
        end
endmodule
```

In the Verilog code above, the logical value at the flip-flop output, q, becomes equal to the logical value at the input, d, 20 ps after the rising edge of the clock. The logical value at the qbar output waits for the completion of the first statement, and becomes equal to the inverted d input 10 ps after the first statement. This waiting period from one statement to the next arises because these two statements are the blocking type. In other words, when the statement uses a "=" sign for an assignment, the statement becomes a blocking statement, which blocks the execution of the next statement until it is fully executed.

The same Verilog code can be rewritten as:

```
`timescale 10ps / 1ps
module flip_flop (q, qbar, d, clock);
output q, qbar;
input clock, d;
reg q, qbar;
always @ (posedge clock)
        begin
        #2      q <= d;
        #3      qbar <= ~d;
        end
endmodule
```

In this program, the first and the second statements in the always block become non-blocking type due to the "<=" sign, and they are executed simultaneously. Therefore, the output, q, becomes equal to the input, d, 20 ns after the positive edge of the clock. Similarly, the output, qbar, becomes equal to the inverted input, ~d, 30 ps after the positive edge of the clock.

More than one condition can be included in an initial or always procedural block. For example, a flip-flop with an asynchronous active-low reset input can be modeled as follows:

```
`timescale      10ps/1ps
module flip_flop (q, qbar, d, clock, reset);
output          q, qbar;
input           d, clock, reset;
reg             q, qbar;
always @ (posedge clock or negedge reset)
        begin
        if (reset == 0)
                begin
                #2      q <= 0;
                #2      qbar <= 1;
                end
        else
                begin
                #4      q <= d;
                #5      qbar <= ~d;
                end
        end
endmodule
```

If the program encounters an active-low reset before the positive edge of clock, both q and qbar outputs become logic 0 20 ps after the negative edge of reset. Otherwise, without any reset, the flip-flop operates normally, and the output, q, becomes equal to the input, d, 40 ps after the positive edge of the clock. At the same time, the output, qbar, becomes equal to the inverted input, ~d, 50 ps after the positive edge of the clock due to the non blocking nature of these assignments.

A.7 Arithmetic and Logical Operators in Verilog

There are two types of operators used in Verilog: arithmetic and logical. The arithmetic operators simply add, subtract, multiply or divide the variables used in a program. The symbol for each operation is given below:

Add	+
Subtract	-
Multiply	*
Divide	/

The logical operators execute all the logic functions, comparisons, bit-shifting and concatenation. The symbol for each operation is given below:

Bitwise AND &
Bitwise NAND ~&
Bitwise OR |
Bitwise NOR ~|
Bitwise XOR ^
Bitwise XNOR ~^
Bitwise NOT ~
Less than <
Greater than >
Greater than or equal >=
Less than or equal <=
Equal ==
Logical left shift <<
Logical right shift >>
Conditional ? :
Concatenation { }

For example, if the variables, X and Y, are equal to 4'b0110 and 4'b1011, respectively, the logical operations on X and Y become as follows:

~Y = 0100
X & Y = 0010
X ~| Y = 0000

As another example, let us assume A = 4'b1101 and B = 8'b01110101. Shifting B one bit to the left becomes:

B << 1 = 11101010
A & (B << 1) = 00001101 & 11101010 = 00001000

A.8 Conditional Statement

Conditional statements are equivalent to the "if-then" statements in C-programming. They follow the same format in C-language but enveloped between the begin and end clauses.

The simplest form of a conditional statement is given below:

```
begin
if (condition1)
        if (condition2)
                if (condition3)
                statement 1;
                ...
                else
                statement 2;
                ...
        else
        statement 3;
        ...
else
statement 4;
...
end
```

Conditions may also be combined using logical operators. For example, the Verilog code below AND-gates all three conditions, condition 1, condition 2 and condition 3, and produces a single condition for the if-clause.

```
begin
if ((condition1) && (condition 2) && (condition 3))
        statement 1;
        ...
else
        statement 2;

end
```

The condition may also include several operators as shown in the example below.

```
begin
if (a > 0)
        if (x <= 0)
                y = 1;
        else            // x > 0 is implied
                y != 2;
else if (a == 0)
        if (x < = 0)
                y = 3;
        else            // x > 0 is implied
                y != 4;
else                    // no if statement, thus a < 0 is implied
        if (x <= 0)
                y = 5;
        else            // x > 0 is implied
                y != 6;
end
```

A.9 Case Statement

Conditional statements can be grouped in a compact form using the case statement.

Example A.3: Implement an 8-1 multiplexer in Fig. A.12 using a case statement.

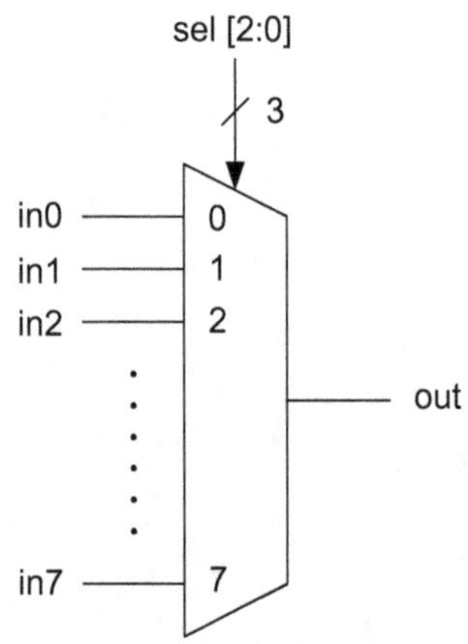

Fig. A.12 8-1 MUX

The case statement for this MUX can be written as follows:

```
`define sel_value0      3'b000
`define sel_value1      3'b001
`define sel_value2      3'b010
`define sel_value3      3'b011
`define sel_value4      3'b100
`define sel_value5      3'b101
`define sel_value6      3'b110
`define sel_value7      3'b111
module mux (out, sel, in0,in1, in2, in3, in4, in5, in6, in7);
output out;
input in0, in1, in2, in3, in4, in5, in6, in7;
input [2:0] sel;
reg out;
always @ (sel or in0 or in1 or in2 or in3 or in4 or in5 or in6 or in7)
begin
        case (sel)
        `sel_value0 : out = in0;
        `sel_value1 : out = in1;
        `sel_value2 : out = in2;
        ...
        `sel_value7 : out = in7;
        default:      out = in0;
        endcase
end
endmodule
```

In this code, the case statement is executed if any of the inputs, in0 to in7, or the select input, sel [2:0], changes. Once inside the case statement, the output of the MUX, out, becomes equal to one of the MUX inputs according to the input select signal, sel_value0 to sel_value7. The output of the MUX also needs to be declared with a reg statement because it needs to retain its current value until changed. The case statement is enclosed between the case and encase clauses. Since the case statement is in a procedural block it also needs the begin and end clauses enveloping the case statement.

Example A.4: Implement a 4-1 MUX in Fig. A.13 using case statement.

```
module mux (out, sel, in0, in1, in2, in3);
output out;
input in0, in1, in2, in3;
input [1:0] sel;
reg out;
always @ (sel or in0 or in1 or in2 or in3)
begin
        case (sel)
        2'b00 : out = in0;
        2'b01 : out = in1;
        2'b10 : out = in2;
        2'b11 : out = in3;
        default: out = in0;
        endcase
end
endmodule
```

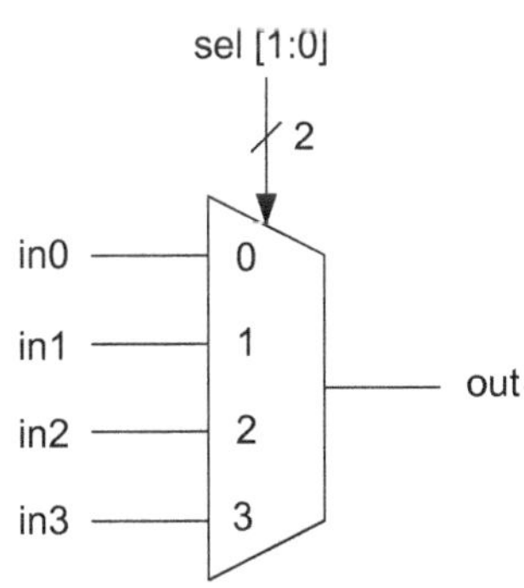

Fig. A.13 4-1 MUX

The case statement can also be used to implement an Arithmetic Logic Unit (ALU) as shown in Fig. A.14 because the ALU output is accompanied by a multiplexer.

```
`define XOR    2'b00
`define SHIFT  2'b01
`define ADD    2'b10
`define SUB    2'b11
module alu (out, opcode, a, b);
output  [7:0] out;
input   [7:0] a, b;
input   [1:0] opcode;
reg     [7:0] out;
always @ (opcode or a or b)
begin
        case (opcode)
        `XOR:   out = a ^ b;
        `SHIFT: out = a << b;
        `ADD:   out = a + b;
        `SUB:   out = a - b;
        default: out = a + b;
        endcase
end
endmodule
```

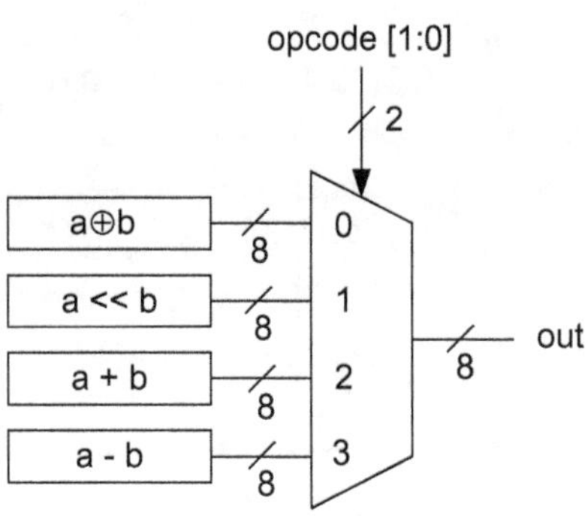

Fig. A.14 A simple ALU

A.10 Looping Statements

There are two useful looping statements in Verilog, the for statement and the while statement. Both statements have to be included in a procedural block.

Example A.5: Implement the for-loop given below.

In this example, the variable, i, starts from 0, and stops at 10, incrementing by 1. The variable, j, is defined in terms of the variable, i. The array simply determines a[j] in terms of a[i]. Thus,

```
for (i = 0; i <= 10; i ++)
        begin
        j = i + 1;
        a[j] = a[i] + 1;
        end
```

The while-loop waits for the occurrence of an event. When the event takes place, the statements in the while-loop are executed. In the for-loop below, the variable, x[i], is determined in terms of a[i] and b[i] as long as the variable, sum, is not equal to 0.

```
while (sum != 0)
        begin
                for (i = 0; i < 10; i++)
                begin
                x[i] = a[i] - b[i];
                end
        end
```

A.11 State Machine Implementations

There are two types of state machines: Mealy-type and Moore-type. Both types can easily be implemented in Verilog using case statements.

The present state of a state machine is defined by flip-flop outputs. The next state is defined by flip-flop inputs because at the positive edge of clock the next state becomes the present state.

A.12 Mealy Machine

The present state outputs of the Mealy machine stems from the present state and present state inputs. Therefore, if present state inputs change during the clock period, this change affects the present state outputs and the next state instantaneously as shown in Fig. A.15.

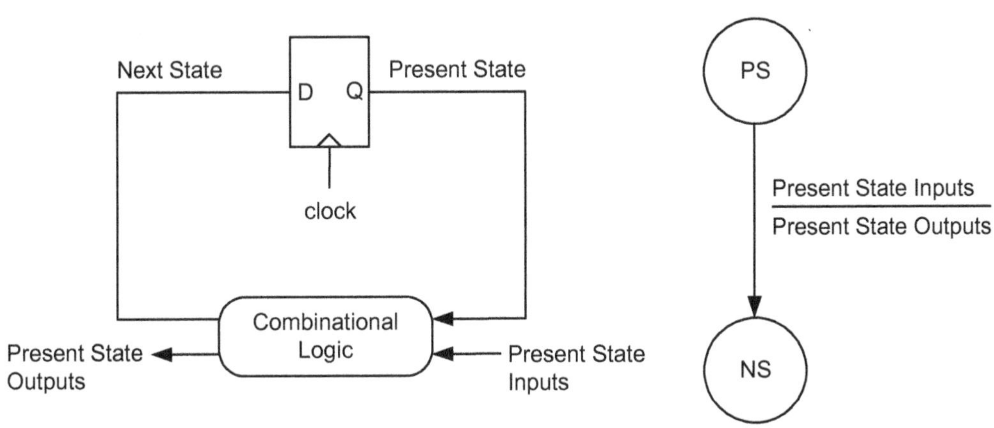

Fig. A.15 Block diagram and state representation of Mealy machine

Example A.6: Implement the Mealy-type state machine with four states in Fig. A.16.

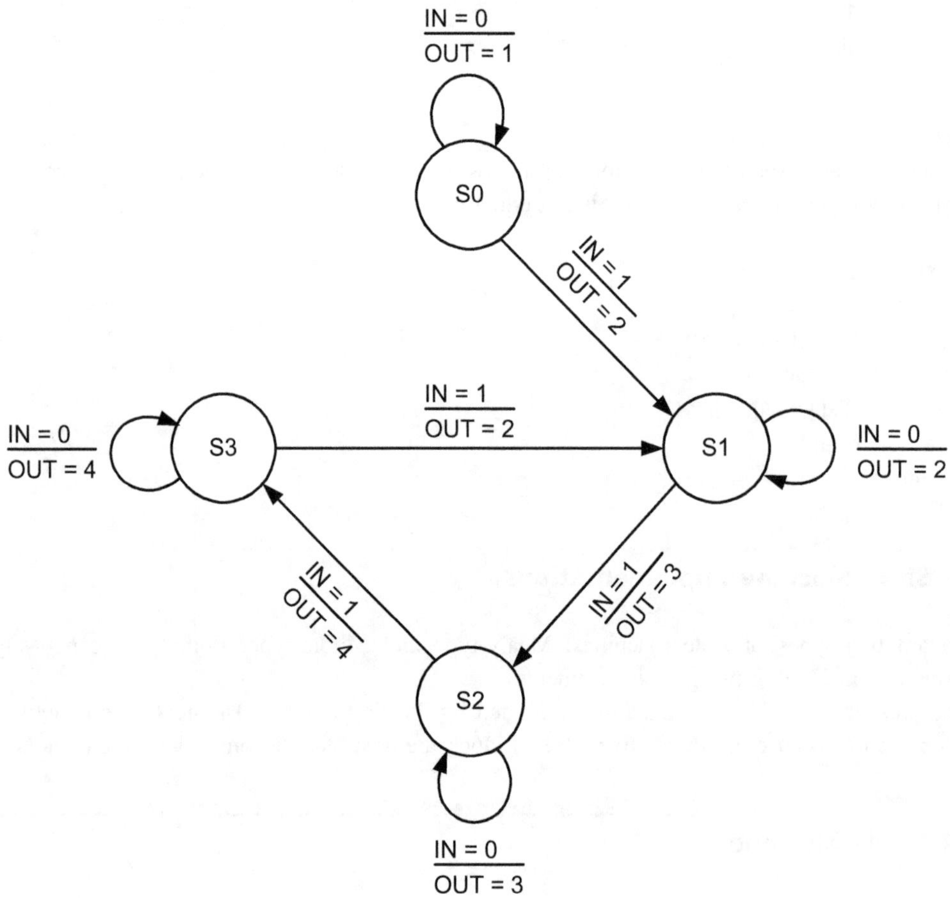

Fig. A.16 State diagram of a Mealy machine with four states (reset not shown for simplicity)

When implementing this state machine in Verilog, it is best to divide the overall circuit topology into two sections as shown in Fig. A.17.

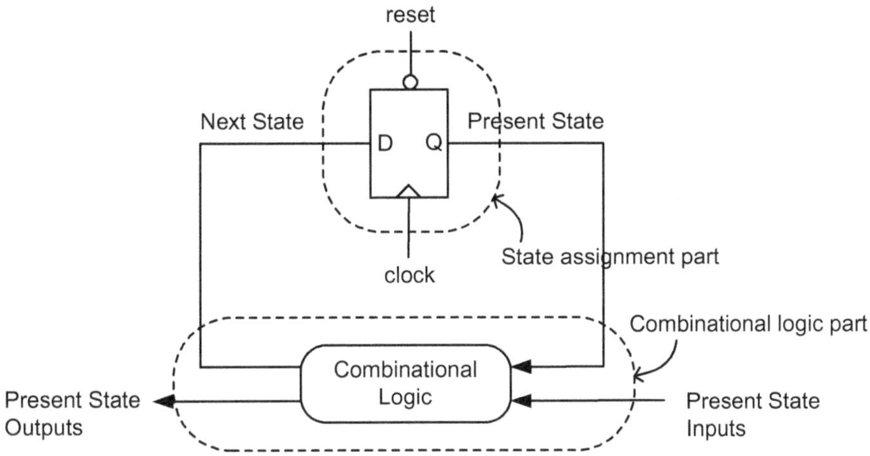

Fig. A.17 Segmentation of the Mealy machine (with asynchronous reset)

The first section is purely combinational: its inputs are the present state and the present state inputs; its outputs are the next state and the present state outputs. Since both the outputs and the next state are functions of the inputs and the present state, this section can conveniently be implemented with a multiplexer. The second section is sequential and consists of flip-flop inputs and outputs. This section, as we will see below, is implemented with an always statement.

The Verilog code below first implements the combinational section of the state machine then the sequential part. The numeric values corresponding to each of the four present states in Fig. A.16 are assigned to the parameters, s0, s1, s2 and s3, using parameter statement because this simplifies the observation of the input and output values at a particular state.

The combinational part of the state machine is implemented by a case statement inside an always procedural block, and executed if one of the multiplexer inputs, in and pstate, changes. Here, the input, in, corresponds the only input, IN, in Fig. A.17, and the input, pstate, corresponds to the present state. If the pstate input is assumed to be the selector input to a multiplexer, the case statement then lists the multiplexer output as a function of all possible combinations of pstate. The default in the case statement always corresponds to the initial state of the state machine. In each statement inside the case statement, the output assignments are always non-blocking type because in real hardware the outputs are produced concurrently and independent of each other.

The sequential part of the state machine is implemented by an always statement, which becomes active only at the positive edge of the clock and the negative edge of the reset.

```
// Mealy machine with asynchronous reset
module mealy (out, reset, in, clock);
output [2:0]      out;
input             reset, in, clock;
reg [2:0]         out;
reg [1:0]         nstate, pstate;

parameter  s0=2'b00, s1=2'b01, s2=2'b10, s3 =2'b11;
```

```
always @ (in or pstate)
begin
        case (pstate)

        s0:     begin
                if (in == 0)
                        begin
                        out <= 1;
                        nstate <= s0;
                        end
                else
                        begin
                        out <= 2;
                        nstate <= s1;
                        end
                end

s1:     begin
        if (in == 0 )
                begin
                out <= 2;
                nstate <= s1;
                end
        else
                begin
                out <= 3;
                nstate <= s2;
                end
        end
s2:     begin
        if (in ==0)
                begin
                out <= 3;
                nstate <=s2;
                end
        else
                begin
                out < = 4;
                nstate <= s3;
                end
        end
```

```
        s3:     begin
                if (in ==0)
                        begin
                        out <= 4;
                        nstate <= s3;
                        end
                else
                        begin
                        out < = 2;
                        nstate <= s1;
                        end
                end
        default: begin
                out <=1;
                nstate <= s0;
                end
        endcase
end

always @ (posedge clock or negedge reset)
begin
        if (reset == 0)
                pstate <= s0;
        else
                pstate <= nstate;
end
endmodule
```

A.13 Moore Machine

Implementing the Moore machine is not any different from the Mealy machine except the formation of present state outputs. Figure A.18 shows the present state outputs of the Moore machine to be a function of the present state, but totally independent of the present state inputs.

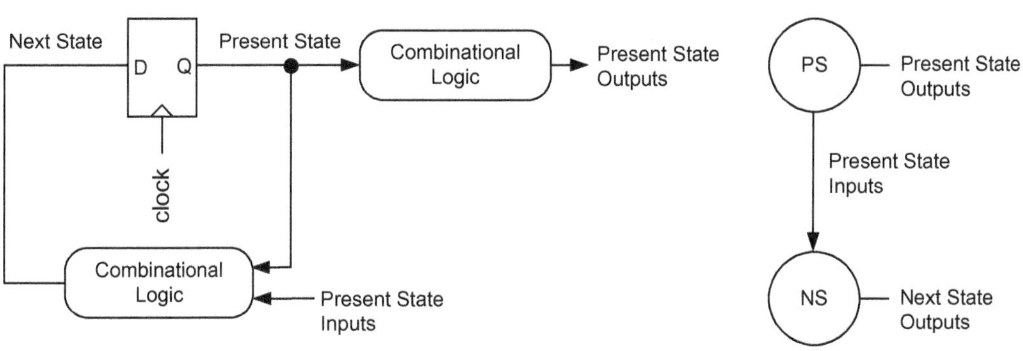

Fig. A.18 Block diagram and state representation of Moore machine

Example A.7: Implement a four-state Moore state machine in Fig. A.19.

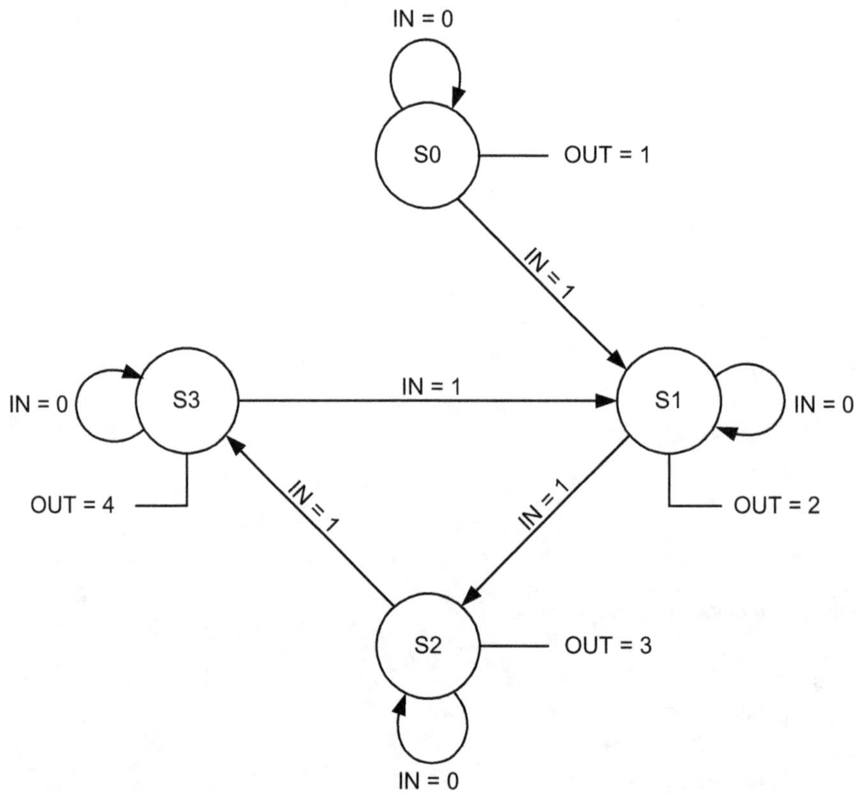

Fig. A.19 State diagram of a Moore machine with four states (reset not shown for simplicity)

The Verilog implementation of this state machine requires combining the two combinational sections of the circuit in Fig. A.20 with case statements, and separately implementing the sequential part using an always statement.

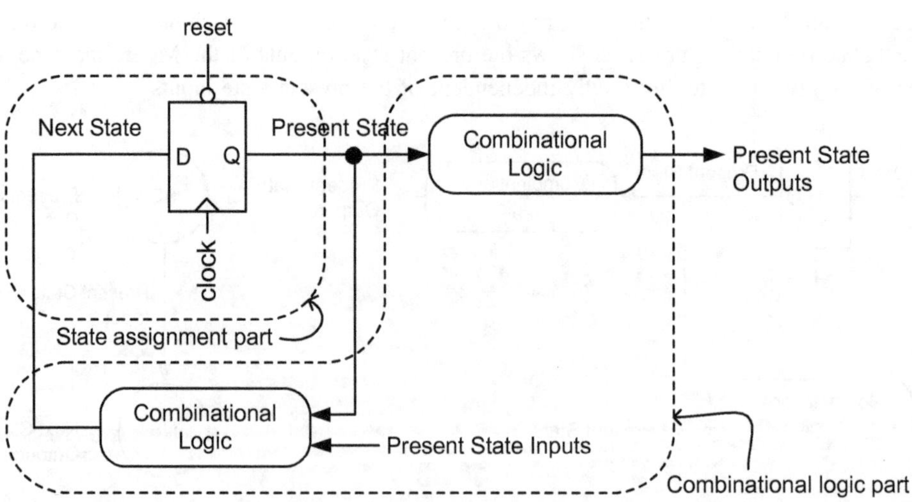

Fig. A.20 Segmentation of the Moore machine (with asynchronous reset)

Implementing the combinational parts is accomplished by a multiplexer: the first input of the state machine, in, is assigned as the input of the multiplexer, and the second input, pstate, is assigned as the selector. Implementing the sequential section, on the other hand, requires a complete picture of what happens to the flip-flop outputs if any of the flip-flop inputs change.

The code below assigns numeric values to all four states with a parameter statement. The first always statement includes a case statement to show what happens to the multiplexer outputs if one of the selector inputs changes. Again, all multiplexer outputs form concurrently. Therefore, all output assignments are defined to be non-blocking type. As opposed to the Mealy machine, the present state outputs of the Moore machine are solely generated from the present state. Therefore, for each state from s0 to s3 the output assignments are written first, independent of any present state input.

The sequential part is implemented by an always statement which includes the edge dependency of the clock and the reset. This statement is executed only if the indicated edges of these two inputs take place. Otherwise, it is ignored.

```
// Moore machine with asynchronous reset
module moore_async (out, reset, in, clock);
output [2:0]    out;
input           reset, in, clock;
reg [2:0]       out;
reg [1:0]       nstate, pstate;
parameter       s0=2'b00, s1= 2'b01, s2= 2'b10, s3= 2'b11;

always @ (in or pstate)
begin
        case (pstate)
        s0:     begin
                out <= 1;
                        if (in == 1)
                        nstate <= s1;
                        else
                        nstate <= s0;
                end
        s1:     begin
                out <= 2;
                        if (in == 1)
                        nstate <= s2;
                        else
                        nstate <= s1;
                end
```

```
        s2:     begin
                out <= 3;
                        if (in == 1)
                        nstate <= s3;
                        else
                        nstate <= s2;
                end
        s3:     begin
                out <= 4;
                        if (in == 1)
                        nstate <= s1;
                        else
                        nstate <= s3;
                end
        default: begin
                out <= 1;
                nstate <= s0;
                end
        endcase
end

always @ (posedge clock or negedge reset)
        begin
        if (reset == 0)
                pstate <= s0;
        else
                pstate <= nstate;
        end
endmodule
```

The Verilog code below implements the Moore machine with a synchronous reset. This time, reset is not an isolated input to the flip-flops as in Fig. A.20, but instead it is applied to the combinational logic block along with the other present state inputs.

```
//Moore machine with synchronous reset
module more_sync (out, reset, in, clock);

output [2:0]    out;
input           reset, in, clock;
reg [2:0]       out;
reg [1:0]       nstate, pstate;
parameter       s0= 2'b00, s1= 2'b01, s2 =2'b10, s3 = 2'b11;

always @ (in or reset or pstate)
begin
        case (pstate)
        s0:     begin
                out <= 1;
                if (reset == 0 )
                        nstate <=s0;
                else
                        begin
                        if (in == 1)
                        nstate <= s1;
                        else
                        nstate <= s0;
                        end
                end

s1:     begin
        out <= 2;
        if (reset == 0 )
                nstate <= s0;
        else
                begin
                if (in == 1)
                        nstate <= s2;
                else
                        nstate <= s1;
                end
        end
s2:     begin
        out <= 3;
        if (reset == 0)
                nstate <= s0;
        else
                begin
                if (in == 1)
                        nstate <= s3;
                else
                        nstate <= s2;
                end
        end
```

```
s3:     begin
        out <= 4;
        if (reset == 0)
                nstate <= s0;
        else
                begin
                if (in == 1)
                        nstate <= s1;
                else
                        nstate <= s3;
                end
        end
default: begin
        out <= 1;
        nstate <= s0;
        end
endcase

end

always @ (posedge clock)
        begin
        pstate <= nstate;
        end
endmodule
```

A.14 Principles of Register-Transfer-Logic Type Coding

The Register-Transfer-Logic (RTL) style of Verilog coding inherits many C-program constructs and implements the intended hardware with ease. Although structural or behavioral Verilog coding may be necessary for certain types of logic blocks, RTL is still the most common coding style to build hardware.

A.15 Wire Assignment

The most common RTL statement is the wire statement. This statement is either accompanied with an assign statement or declared by itself, and it resides outside of a procedural block. In the first example below, the inputs, a and b, form an XOR gate with an output, out. Separate wire and assign statements are used to implement the XOR gate.

```
wire out;
assign out = a ^ b;
```

However, the two statements can be combined to form a single wire statement.

```
wire out = a ^ b;
```

If the implementation requires multiple wires in the form of a bus, then the statements for the node, out, can be written as follows:

```
wire [7:0] out;
assign out = a ^ b;
```
or
```
wire [7:0] out = a ^ b;
```

A.16 Conditional Operator

Another useful RTL construct is the conditional operator. Assume a tri-state buffer in Fig. A.21.

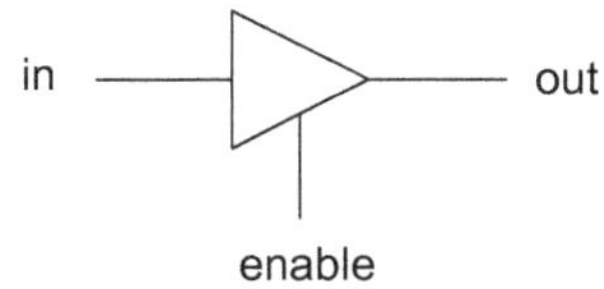

Fig. A.21 Tri-state buffer implemented by conditional operator

The wire statement that includes the conditional operator can be written as follows:

```
wire out;
assign out = enable ? in : 1'bz;
```

The "?" sign in the above statement signifies the condition for the input, enable, to be logic 1 or not. If enable is logic 0, then the output, out, becomes an open circuit. Since out is only one bit, the high impedance state is shown as 1'bz.

The wire and assign statements can also be combined to produce a single statement.

```
wire out = enable ? in : 1'bz;
```

Example A.8: Implement a 3-1 MUX in Fig. A.22 using conditional operator.

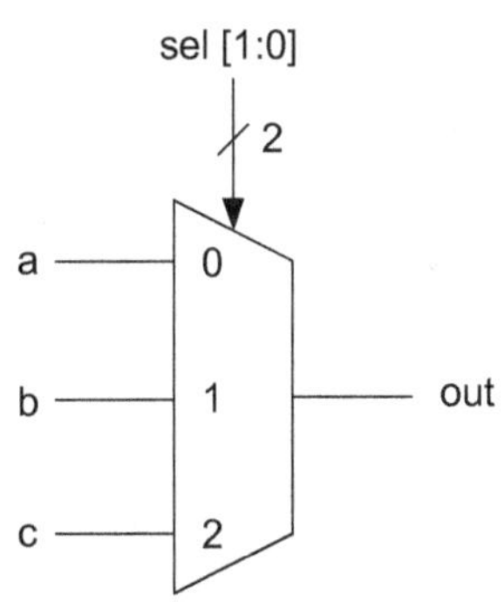

Fig. A.22 A 3-1 MUX implemented by conditional operator

```
wire out;
assign out = (sel == 2'b00) ? a : (sel == 2'b01) ? b : c;
```

In this statement, if sel [1:0] = 00 then the output, out, becomes a. If sel [1:0] = 01 then out becomes b. For all the other values of sel, out becomes c.

The same statement can also be written without the assign statement as:

```
wire out = (sel == 2'b00) ? a : (sel == 2'b01) ? b : c;
```

If this is a 3-1 MUX with multiple inputs and outputs as in Fig. A.23, then the wire statement needs to have a proper dimension.

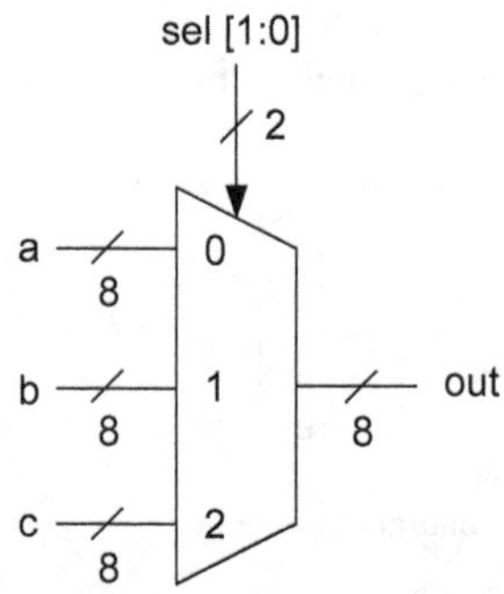

Fig. A.23 An eight-input 3-1 MUX implemented by conditional operator

```
wire [7:0] out = (sel == 2'b00) ? a : (sel == 2'b01) ? b : c;
```

The dimension of the inputs, a, b, c and sel, should be declared in the input statements prior to the wire statement as shown below.

```
...
input [7:0] a, b, c;
input [1:0] sel;
...
wire [7:0] out = (sel == 2'b00) ? a : (sel == 2'b01) ? b : c;
...
```

A.17 Memory Declaration

Memory is declared by a reg statement in Verilog. Assume an SRAM-like (x+1)-bit wide memory with (y+1) rows shown in Fig. A.24. This memory is declared as follows:

```
reg [x:0] fifo [y:0];
```

Here, "fifo" is the name of the memory with a dimension of [x:0] by [y:0].

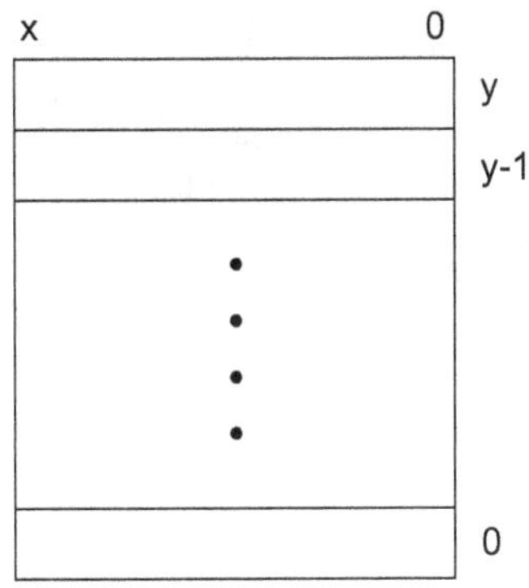

Fig. A.24 (x+1) wide (y+1) deep memory

A.18 Memory Addressing

Following the memory declaration, a memory address statement should be included in the program to read or to write to a specific location in the memory.

Example A.9: Implement an eight-bit wide memory with 16 rows in Fig. A.25 using Verilog.

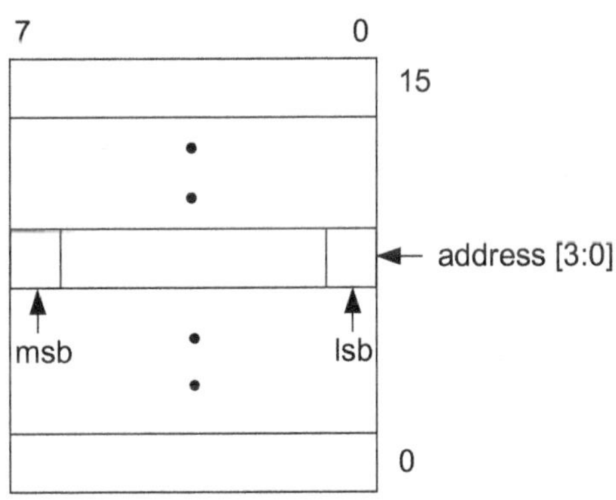

Fig. A.25 An 8 × 16 memory

In order to access the most significant bit (msb) and the least significant bit (lsb) of this memory at any row, an address needs to be formed in the Verilog code.

```
...
input [3:0] address;
reg [7:0] mem [15:0];
reg [7:0] row;
reg msb, lsb;
row = mem [address];
msb = row [7];
lsb = row [0];
...
```

In this example, the reg statement, reg [7:0] mem [15:0], declares a 8 × 16 memory with a name, mem.

If the memory address is externally supplied to the memory, this input needs to be declared in the input statement. Each bit in a row can be declared with a second reg statement, reg [7:0] row. The most and least significant bits are declared as the third reg statement, reg msb, lsb.

Therefore, each bit in an arbitrary row can be accessed by the statement, row = mem [address], in the Verilog code above. Accessing the most and the least significant bits are accomplished by msb = row [7] and lsb = row [0], respectively.

A.19 Memory Modeling

Different types of memory require different styles of memory modeling.

Example A.10: Implement the simple SRAM memory with a single bidirectional data port in Fig. A.26 with a clock input.

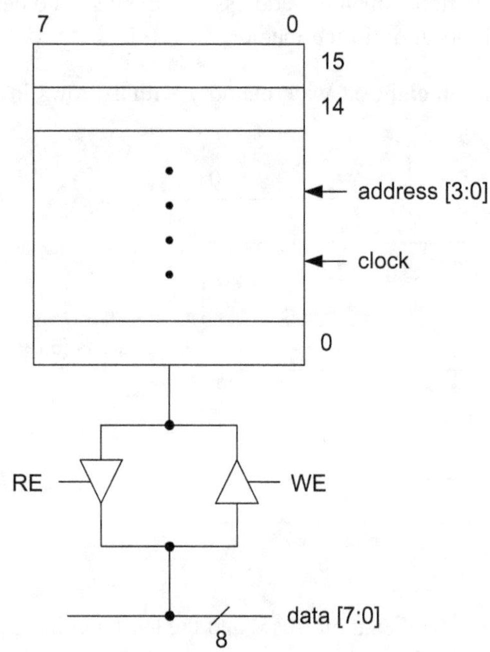

Fig. A.26 An 8 × 16 single port, bidirectional memory

In this memory, the data is written to an arbitrary row or read from an arbitrary row at the positive edge of the clock signal once the memory address is specified.

Since the data port is bidirectional, it needs to be declared as an inout statement. Therefore, the Verilog code can be written as follows:

```
module mem (data, address, WE, RE, clock);
inout [7:0]      data;
input [3:0]      address;
input            WE, RE, clock;
reg [7:0]        SRAM [15:0];
reg [7:0]        data;
reg [1:0]        read_write_state;

always @ (posedge clock)
begin
        read_write_state = {WE, RE};      // curly brackets are for concatenating WE and RE
        case (read_write_state)
        2'b00:    data = 8'bz;
        2'b01:    data = SRAM [address];
        2'b10:    SRAM [address] = data;
        2'b11:    $display ("error"); // WE and RE cannot be at logic 1 simultaneously
        default:  data = SRAM [address];    // SRAM needs to be in the read mode when idling
        encase
end
endmodule
```

In the example above, the case statement is formed in a procedural block because both read and write takes place at the positive edge of clock. When neither RE nor WE is at logic 1, the bidirectional data port must be at a high impedance state, data = 8'bz. When a read takes place, data is read out from a specified memory address and directed to the bidirectional bus, data [7:0]. When a write takes place, data from the bidirectional bus is written to a specified address, SRAM [address]. When both WE and RE are at logic 1, this should be indicated as an error.

If statements can also be used to replace the case statement except the case statement is more compact and includes all the possible cases to model a memory.

Example A.11: Implement a unidirectional, byte addressable memory shown in Fig. A.27 with two data ports.

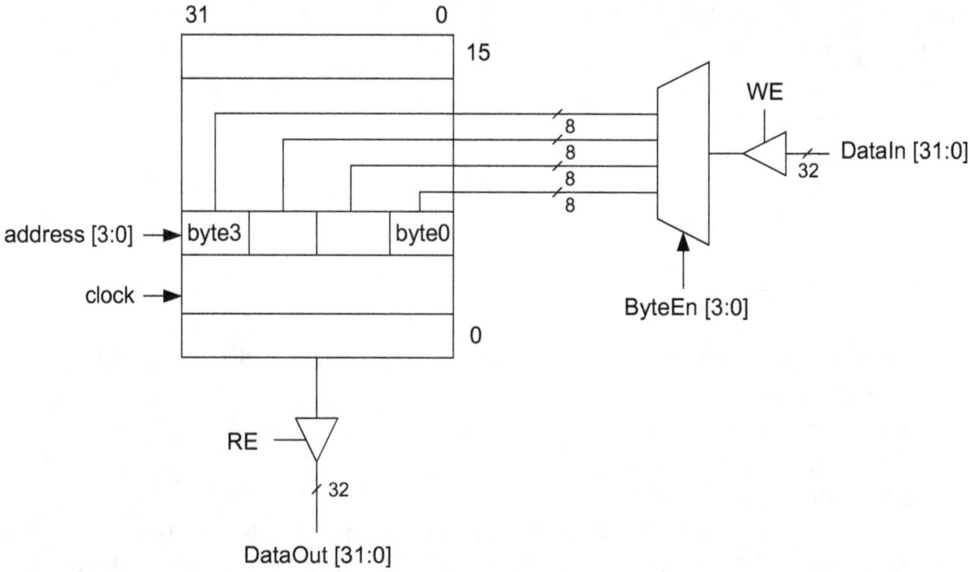

Fig. A.27 A 32 × 16 dual port, unidirectional memory

The input port, DataIn [31:0], is assumed byte-addressable and configured to write up to four bytes of data to a specified memory address when WE = 1. The output port, DataOut, is not byte-addressable, and it is used to read all 32 bits of data from the memory.

Thus,

```
module mem (DataOut, DataIn, address, clock, ByteEn, WE, RE);
output [31:0]    DataOut;
input [31:0]     DataIn;
input [3:0]      address, ByteEn;
input            clock, RE, WE;
reg [31:0]       SRAM [15:0];
reg [31:0]       temp;

always @ (posedge clock)
begin
if (WE == 1 && RE == 0)
        begin
        case (ByteEn)
        4'b0000:  temp [31:0] = 32'bz;
        4'b0001:  temp [7:0] = DataIn [7:0];
        4'b0010:  temp [15:8] = DataIn [15:8];
        4'b0011:  temp [15:0] = DataIn [15:0];
        4'b0100:  temp [23:16] = DataIn [23:16];
        4'b0101:  begin
                  temp [23:16] = DataIn [23:16];
                  temp [7:0] = DataIn [7:0];
                  end
        4'b0110:  temp [23:8] = DataIn [23:8];
```

```
        ...
        4'b1111:  temp [31:0] = DataIn [31:0];
        default:  begin
                  temp [31:0] = 32'bz;
                  $display ("no bytes are enabled");
                  end
        endcase
        SRAM [address] = temp;
        end
else if  (RE == 1 && WE == 0)
        DataOut = SRAM [address];
else if  (RE == 0 && WE == 0)
        DataOut = 32'bz;
else
        display ("Error - RE and WE are enabled");
end

endmodule
```

A.20 Few Words on Functional Verification

Functional verification is a very critical step in logic design and needs to cover every possible corner case and input combination to a Verilog module. However, when the circuit is not purely combinational but contains sequential components, the difficulty in functional verification increases. A proper process is to isolate the sequential components from the combinational sections of the circuit, and verify each section individually prior to overall system verification.

When a combinational circuit goes through a formal functional verification, the best method is to apply the inputs of the entire truth table as test vectors to the module, store the circuit's response in an output file, and then compare this output file with the one that contains the expected outputs (the outputs of the truth table) as shown in Fig. A.28.

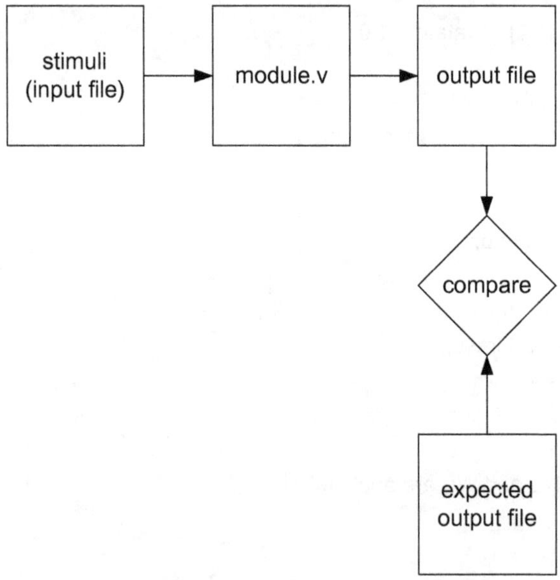

Fig. A.28 Formal functional verification process

However, if the circuit is sequential, each state-to-state transition needs to be examined in the state machine when the inputs to the state change. Furthermore, the outputs from each state need to match the expected output values.

It may be sufficient to do functional check using timing diagrams if the size of the circuit is small. However, for bigger circuits, including many combinational and sequential modules, the verification process is applied to each individual module, and then to the entire system. Both types of verification are essentially a file matching process as shown in Fig. A.28.

Index

A. Bindal, *Fundamentals of Computer Architecture and Design*,
https://doi.org/10.1007/978-3-030-00223-7

W

X

www.ingramcontent.com/pod-product-compliance
Lightning Source LLC
Chambersburg PA
CBHW080702190426
43356CB00056B/58
* 9 7 8 3 0 3 0 0 0 2 2 2 0 *